THE NUMERICAL ANALYSIS PROBLEM SOLVER®

REGISTERED TRADEMARK

Staff of Research and Education Association,
Dr. M. Fogiel, Director

Research and Education Association
505 Eighth Avenue
New York, N. Y. 10018

THE NUMERICAL ANALYSIS PROBLEM SOLVER®

Copyright © 1983 by Research and Education Association. All rights reserved. No part of this book may be reproduced in any form without permission of the publisher.

Printed in the United States of America

Library of Congress Catalog Card Number 83-62277

International Standard Book Number 0-87891-549-4

Revised Printing, 1986

PROBLEM SOLVER is a registered trademark of Research and Education Association, New York, N.Y. 10018

WHAT THIS BOOK IS FOR

Students have generally found numerical analysis a difficult subject to understand and learn. Despite the publication of hundreds of textbooks in this field, each one intended to provide an improvement over previous textbooks, students continue to remain perplexed as a result of the numerous conditions that must often be remembered and correlated in solving a problem. Various possible interpretations of terms used in numerical analysis have also contributed to much of the difficulties experienced by students.

In a study of the problem, REA found the following basic reasons underlying students' difficulties with numerical analysis taught in schools:

(a) No systematic rules of analysis have been developed which students may follow in a step-by-step manner to solve the usual problems encountered. This results from the fact that the numerous different conditions and principles which may be involved in a problem, lead to many possible different methods of solution. To prescribe a set of rules to be followed for each of the possible variations, would involve an enormous number of rules and steps to be searched through by students, and this task would perhaps be more burdensome than solving the problem directly with some accompanying trial and error to find the correct solution route.

(b) Textbooks currently available will usually explain a given principle in a few pages written by a professional who has an insight in the subject matter that is not shared by students. The explanations are often written in an abstract manner which leaves the students confused as to the application of the principle. The explanations given are not sufficiently detailed and extensive to make the student aware of the wide range of applications and different aspects of the principle being studied. The numerous possible variations of principles and their applications are usually not discussed, and it is left for the students to discover these for themselves while doing exercises. Accordingly, the average student is expected to rediscover that which has been long known and practiced, but not published or explained extensively.

(c) The examples usually following the explanation of a topic are too few in number and too simple to enable the student to obtain a thorough grasp of the principles involved. The explanations do not provide sufficient basis to enable a student to solve problems that may be subsequently assigned for homework or given on examinations.

The examples are presented in abbreviated form which leaves out much material between steps, and requires that students derive the omitted material themselves. As a result, students find the examples difficult to understand--contrary to the purpose of the examples.

Examples are, furthermore, often worded in a confusing manner. They do not state the problem and then present the solution. Instead, they pass through a general discussion, never revealing what is to be solved for.

Examples, also, do not always include diagrams/graphs, wherever appropriate, and students do not obtain the training to draw diagrams or graphs to simplify and organize their thinking.

(d) Students can learn the subject only by doing the exercises themselves and reviewing them in class, to obtain experience in applying the principles with their different ramifications.

In doing the exercises by themselves, students find that they are required to devote considerably more time to numerical analysis than to other subjects of comparable credits, because they are uncertain with regard to the selection and application of the theorems and principles involved. It is also often necessary for students to discover those "tricks" not revealed in their texts (or review books), that make it possible to solve problems easily. Students must usually resort to methods of trial-and-error to discover these "tricks", and as a result they find that they may sometimes spend several hours to solve a single problem.

(e) When reviewing the exercises in classrooms, instructors usually request students to take turns in writing solutions on the boards and explaining them to the class. Students often find it difficult to explain in a manner that holds the interest of the class, and enables the remaining students to follow the material written on the boards. The remaining students seated in the class are, furthermore, too occupied with copying the material from the boards, to listen to the oral explanations and concentrate on the methods of solution.

This book is intended to aid students in numerical analysis to overcome the difficulties described, by supplying detailed illustrations of the solution methods which are usually not apparent to students. The solution methods are illustrated by problems selected from those that are most often assigned for class work and given on examinations. The problems are arranged in order of complexity to enable students to learn and understand a particular topic by reviewing the problems in sequence. The problems are illustrated with detailed step-by-step explanations, to save the students the large amount of time that is often needed to fill in the gaps that are usually found between steps of illustrations in textbooks or review/outline books.

The staff of REA considers numerical analysis a subject that is best learned by allowing students to view the methods of analysis and solution techniques themselves. This approach to learning the subject matter is similar to that practiced in various scientific laboratories, particularly in the medical fields.

In using this book, students may review and study the illustrated problems at their own pace; they are not limited to the time allowed for explaining problems on the board in class.

When students want to look up a particular type of problem and solution, they can readily locate it in the book by referring to the index which has been extensively prepared. It is also possible to locate a particular type of problem by glancing at just the material within the boxed portions. To facilitate rapid scanning of the problems, each problem has a heavy border around it. Furthermore, each problem is identified with a number immediately above the problem at the right-hand margin.

To obtain maximum benefit from the book, students should familiarize themselves with the section, "How To Use This Book," located in the front pages.

To meet the objectives of this book, staff members of REA have selected problems usually encountered in assignments and examinations, and have solved each problem meticulously to illustrate the steps which are difficult for students to comprehend. Special gratitude is expressed to them for their efforts in this area, as well as to the numerous contributors who devoted brief periods of time to this work.

Gratitude is also expressed to the many persons involved in the difficult task of typing the manuscript with its endless changes, and to the REA art staff who prepared the numerous detailed illustrations together with the layout and physical features of the book.

The difficult task of coordinating the efforts of all persons was carried out by Carl Fuchs. His conscientious work deserves much appreciation. He also trained and supervised art and production personnel in the preparation of the book for printing.

Finally, special thanks are due to Helen Kaufmann for her unique talents to render those difficult border-line decisions and constructive suggestions related to the design and organization of the book.

Max Fogiel, Ph.D.
Program Director

HOW TO USE THIS BOOK

This book can be an invaluable aid to students in numerical analysis as a supplement to their textbooks. The book is subdivided into 19 chapters, each dealing with a separate topic. The subject matter is developed beginning with computer operations in numerical analysis, errors and approximation, series expansion, finite difference calculus, interpolation techniques, rootfinding, numerical integration, linear systems and programming, fourier transforms, and difference equations. Also included are heat equations, Monte Carlo methods and applications in Fortran. An extensive number of applications have been included, since these appear to be more troublesome to students.

TO LEARN AND UNDERSTAND A TOPIC THOROUGHLY

1. Refer to your class text and read the section pertaining to the topic. You should become acquainted with the principles discussed there. These principles, however, may not be clear to you at that time.

2. Then locate the topic you are looking for by referring to the "Table of Contents" in front of this book, "The Numerical Analysis Problem Solver."

3. Turn to the page where the topic begins and review the problems under each topic, in the order given. For each topic, the problems are arranged in order of complexity, from the simplest to the more difficult. Some problems may appear similar to others, but each problem has been selected to illustrate a different point or solution method.

To learn and understand a topic thoroughly and retain its contents, it will be generally necessary for students to review the problems several times. Repeated review is essential in order to gain experience in recognizing the principles that should be applied, and select the best solution technique

TO FIND A PARTICULAR PROBLEM

To locate one or more problems related to a particular subject matter, refer to the index. In using the index, be certain to note that the numbers given there refer to problem numbers, not to page numbers. This arrangement of the index is intended to facilitate finding a problem more rapidly, since two or more problems may appear on a page.

If a particular type of problem cannot be found readily, it is recommended that the student refer to the "Table of Contents" in the front pages, and then turn to the chapter which is applicable to the problem being sought. By scanning or glancing at the material that is boxed, it will generally be possible to find problems related to the one being sought, without consuming considerable time. After the problems have been located, the solutions can be reviewed and studied in detail. For this purpose of locating problems rapidly, students should acquaint themselves with the organization of the book as found in the "Table of Contents".

In preparing for an exam, it is useful to find the topics to be covered in the exam from the "Table of Contents," and then review the problems under those topics several times. This should equip the student with what might be needed for the exam.

CONTENTS

Chapter No. **Page No.**

1 INTRODUCTION TO NUMERICAL CALCULATION   1
    Fundamentals of Numerical Analysis   1
    Principles of Computer Operations   9
    Number Representations   12
    Fortran Rules and Vocabulary   19

2 ERRORS AND APPROXIMATIONS IN NUMERICAL ANALYSIS   26
    Significant Figures, Errors   26
    Absolute and Relative Errors   29
    Truncation and Round-off Errors   31
    Methods of Approximation   36

3 SERIES   51
    Infinite Series(Test For Convergence & Divergence)   51
    Taylor's Series Expansion   58
    MacLaurin's Series Expansion   74
    Power Series Expansion   80
    Laurent Series Expansion   97

4 FINITE DIFFERENCE CALCULUS   106
    Finite Differences   106

Difference Tables and Difference Formulas   115
Difference Operators   127

5 INTERPOLATION AND EXTRAPOLATION   132

Interpolation Techniques   132
Linear Interpolation   154
Inverse Interpolation Techniques   163
Extrapolation Techniques   169

6 SIMULTANEOUS LINEAR ALGEBRAIC EQUATIONS AND MATRICES   173

Gaussian Elimination Method   173
Jacobi and Gauss-Seidel Methods   181
Matrix Mathematics   194

7 ROOTFINDING FOR NONLINEAR EQUATIONS   210

The Iterative Method For Solving Nonlinear Equations   210
Newton's Method   220
Newton-Raphson Method   230
The Method of False Position(Regula Falsi)   240
Other Rootfinding Methods   244

8 LEAST SQUARES - CURVE FITTING   256

Least-Squares Approximation   256
Data Smoothing   285
Approximate Differentiation   287

9 NUMERICAL DIFFERENTIATION   297

General Applications of Numerical Differentiation   297
Truncation Error Using Taylor's Formula   300
Differentiation By Lagrange's Formula, Newton's Formula and Stirling's Formula   308
Approximate Differentiation   320

10 NUMERICAL INTEGRATION 328

    General Applications of Numerical Integration 328
    Numerical Integration: Simpson's Rule, Trapezoidal Rule, Weddle's Rule, and Romberg's Method 336
    Other Methods For Numerical Integration 363

11 LINEAR SYSTEMS 369

    Vector Spaces, Matrices 369
    Matrix Norms and Applications 399
    Matrix Factorization 408
    Iterative Methods 418

12 MATRIX EIGENVALUES AND EIGENVECTORS 431

    Eigenvalues and Eigenvectors of a matrix 431
    Eigenvalues and Eigenvectors of Linear Operators 469
    Applications of The Spectral Theorem 475

13 LINEAR PROGRAMMING 480

    Basic Problem 480
    Simplex Method 488
    Duality 503
    Other Optimization Problems 510

14 FOURIER TRANSFORMS 519

    Properties of Fourier Series 519
    Fourier Series Expansion 529
    Differentiation of Fourier Series 552
    Fourier Integrals 557

15 DIFFERENCE EQUATIONS 570

    General Applications of Difference Equations 570
    Boundary Value Problems 580
    Motivation: Particular Solutions 589

16 DIFFERENTIAL EQUATIONS 595

Applications of First Order and Second Order Differential Equations 595
Euler and Runge-Kutta Methods 613
Boundary Value Problems 634
Other Methods For Solving Intial-Value Problems in Ordinary Differential Equations 668

17 PARTIAL DIFFERENTIAL EQUATIONS 707

Solution of General Partial Differential Equations 707
Formulation of Partial Differential Equations 714
Heat Equations 718
Miscellaneous Problems and Applications 737

18 MONTE CARLO METHODS 766

Monte Carlo Random Number Distribution 766
Sampling Simulation 768
Queuing Models and Game Theory 772
Monte Carlo Applications to Integral Problems 783

19 APPLICATIONS OF FORTRAN LANGUAGE IN NUMERICAL ANALYSIS 786

Computer Manipulations with Matrices 786
Evaluating Functions 805
Roots of Equations 825
Numerical Integration Methods in Fortran Programs 837
Solutions of Differential Equations 848

INDEX 859

APPENDIX 873

CHAPTER 1

INTRODUCTION TO NUMERICAL CALCULATION

FUNDAMENTALS OF NUMERICAL METHODS

● **PROBLEM** 1-1

The digital computer is a basic tool used in arriving at the solution of numerical problems that would otherwise be extremely long and laborious. Give a brief explanation of how the digital computer operates.

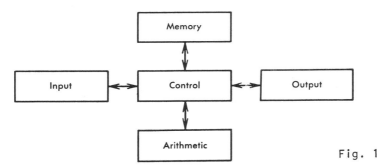

Fig. 1

Solution: The functions of a digital computer can be schematized as in Fig. 1.

Explanation of the functions:

Input:

Magnetic tape, paper tape, and punched cards are the common media for carrying information (data and program) from the outside world through the input mechanism to the internal functions of the computer. Figure 2 displays two punched cards and a piece of punched paper tape. The input mechanism of the computer is able to receive information from cards and tape by reading the punched holes. A particular computer may have only the mechanism for reading cards; another computer may be able to read from both cards and paper tape. Card (a) Fig. 2 is typical of a data card containing two pieces of data. On the other hand, card (b) contains one instruction or one statement of a program.

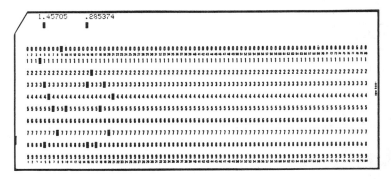

(a) Data card.

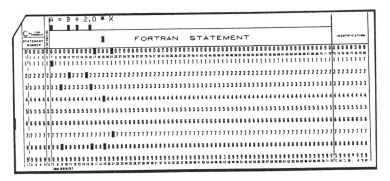

(b) Statement of a program on a card.

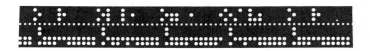

(c) Punched paper tape.

Two punched cards and a piece of punched paper tape. The computer reads the holes, black rectangles on cards and white circles on paper tape. Printed characters on cards aid humans to read punched information.

Fig. 2

Output:

Output information can be punched in cards and in paper tape. It is also common to output information through a line printer onto sheets of paper.

Memory:

Programs, input data, output data, and other data (needed temporarily as the computation proceeds) are stored in the memory. The memory is such that it is capable of storing both alphabetic and numerical information; however, the alphabetic information is stored in an equivalent numerical form. Of course, the sign and the decimal associated with

2

the number can be stored. Information must be stored in memory in an organized fashion so that the computer may refer to a particular memory cell for the particular information stored therein. In the computer vernacular, it can be said that each storage cell has an address.

Control Unit:

As the wording suggests, the control unit interprets the program in supervising, organizing, and sequencing the separate functions of input, output, arithmetic and memory.

In this very brief look into how the digital computer operates, many intricate details have been avoided.

● **PROBLEM 1-2**

In Algebra the expression N = N + 1 is meaningless, however, it is meaningful in Fortran. Explain.

Solution: It means "take the value stored under the name N, add 1 to it, and store the result under the name N again."

● **PROBLEM 1-3**

Given a sequence of n numbers, $x_1, x_2, \ldots, x_n$, find the largest and the smallest number of this sequence and write out the results.

Solution: The figure shown is a flow chart for the solution of this problem. The READ statement reads the number n into the location N and the numbers $x_1, x_2, \ldots, x_n$ into the

vector X. Locations XMAX and XMIN are next set equal to X(1). The iteration statement sets up control for doing the operations that follow up to α for I = 2, 3, . . . , N. The first time through the loop XMAX is compared with X(2) (I = 2). If X(2) is greater than XMAX, then X(2) is placed in XMAX; if it is equal to XMAX, control is sent directly to the end of the iteration loop. If X(2) is less than XMAX, it is compared with XMIN, which contains X(1). If X(2) is less than XMIN, it is placed in XMIN. If it is greater than or equal to XMIN, control is sent to the end of the iteration loop. The same sequence of operations is then carried out using X(3). This continues for each I up to and including I = N. Clearly when the iterations are completed, the contents of XMAX will have been compared with every number

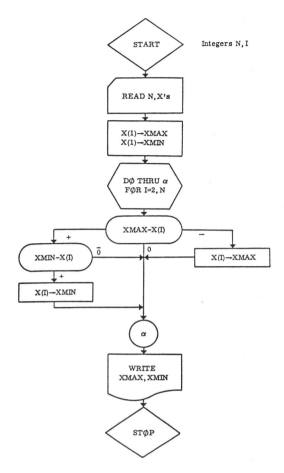

and the largest retained in XMAX. XMIN will likewise contain the smallest of the sequence. The contents of XMAX and XMIN are then written out.

● **PROBLEM 1-4**

Construct a flow chart that will find the sum of a sequence of numbers represented by $a_1, a_2, \ldots, a_n$.

Solution: This addition operation on the computer is done by finding first the partial sum of $a_1 + a_2$, then $a_1 + a_2 + a_3$, and so on, until each number has been added to the partial sum to obtain the total. Either of the two flow charts in the accompanying figure will accomplish the necessary operations, the two solutions being the same except for the control statements.

The first symbol in every flow chart is START. It indicates that appropriate control statements are to be placed ahead of the program for proper initiation of the machine. The READ statement means that the number n is read into a location labeled N and that the sequence of numbers $a_1, a_2, \ldots, a_n$ is read into the locations labeled

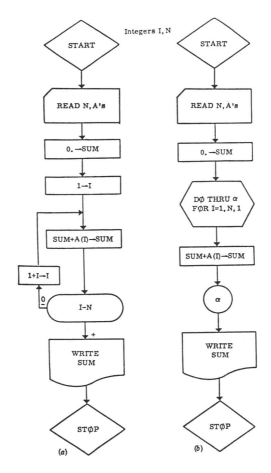

(a)   (b)

A(1), A(2), ..., A(N). The shorthand notation in this box will be used frequently. The number in location N indicates the number of items in the set. A(1), ..., A(N) contain the numbers $a_1, ..., a_n$, respectively. The next statement is the first arithmetic statement and places a floating-point zero in the contents of a location called SUM.

The iteration statement in Figure (b) says to do the statments between the box and the circle (in this case there is only one statement) N times, the first time for I = 1, the second time for I = 2, etc., and the last time for I = N. The box within the range of this iteration says to take the current value of SUM, add the number in the location A(I), using the current value of I, and store the result back into SUM. The first time through the iteration this adds zero (SUM was initially set to zero) and A(1) and stores the result, A(1), in SUM. The second time through the iteration, this statement adds A(1), the current value of SUM, and A(2), storing the result, A(1) + A(2), in SUM. The third time through the iteration, it adds SUM [currently A(1) + A(2)] to A(3) and stores the result, A(1) + A(2) + A(3), back into SUM. Each time through the iteration, the statement updates SUM by the next number in order from the set of numbers A(1), A(2), ..., A(N). At the conclusion of the iterations the location SUM contains the sum of all the numbers A(1) through A(N).

5

In Fig. (a) the iteration is controlled by a logic statement and illustrates the functions performed by the iteration statement. The integer 1 is placed into location I ahead of the loop. Then A(1) is added to the contents of SUM, which is zero prior to this operation. SUM now contains A(1). The logic statement tests the sign of I - N. If it is negative or equal to zero, I is updated by 1, and control is returned to the arithmetic statement. This sequence of operations is continued until I - N is greater than zero, at which point SUM is written out.

The last box in each chart says to display the number in SUM on the output. This is the conclusion to the problem. The flow chart is completed with a STØP.

● **PROBLEM 1-5**

Find the sum of the first n terms in the Taylor-series expansion for e^x.

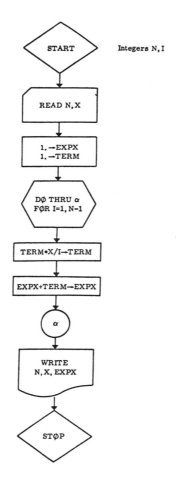

Solution: $e^x = 1 + x + \frac{1}{2!}x^2 + \frac{1}{3!}x^3 + \ldots + \frac{x^{n-1}}{(n-1)!}$

The arithmetic operations needed for evaluating this sum are simplified by noting that the (i + 1)st term in the series is obtained from the ith term by multiplying by x/i. The second term is equal to 1 · x/1, and the (i + 1)st term is equal to the product of

$$\frac{x^{i-1}}{(i-1)!} \quad \text{and} \quad \frac{x}{i}$$

The flow chart for the problem is shown in the figure. Numbers for N and X are read into the machine. Locations TERM and EXPX, set to contain floating point 1.0, will be used to store the results of the computations. The iteration loop sets up control to repeat the operations to α for I = 1, 2, . . . , N - 1. The first statement in the loop gives the next term in the series by updating the previous term. The first time through the loop TERM contains 1.0, which is multiplied by x/1, and then stored back in TERM, so that TERM now contains the second term of the series. Each time through the loop TERM is updated, so that it contains the (i + 1)st term in the series. The last statement in the loop adds the contents of TERM to the contents of EXPX to give the partial sum of the first i + 1 terms of the series. The first time through the loop EXPX contains 1.0, and TERM contains x/1.0 when this statement is executed. The updated contents of EXPX are 1.0 + x/1.0, which is the sum of the first two terms of the series. When the iteration has been completed n - 1 times, EXPX contains the sum of the first n terms of the series. Note that the integer I is used in a floating-point computation. Such statements may not be acceptable in computer programs but are permitted here for convenience.

• **PROBLEM 1-6**

Using

$$\sin(x) = x - \frac{x^3}{3!} + \frac{x^5}{5!} - \cdots + (-1)^{n-1} \frac{x^{2n-1}}{(2n-1)!}$$

$$+ (-1)^n \frac{x^{2n+1}}{(2n+1)!} \cos(\xi_x), \text{ solve } f(x) = \int_0^1 \sin(xt)\,dt.$$

Solution:

$$f(x) = \int_0^1 \left[xt - \frac{x^3 t^3}{3!} + \cdots + (-1)^{n-1} \frac{(xt)^{2n-1}}{(2n-1)!} \right.$$

$$\left. + (-1)^n \frac{(xt)^{2n+1}}{(2n+1)!} \cos(\xi_{xt}) \right] dt$$

$$= \sum_{j=1}^n (-1)^{j-1} \frac{x^{2j-1}}{(2j-1)!\,(2j)} + (-1)^n \frac{x^{2n+1}}{(2n+1)!} \int_0^1 t^{2n+1} \cos(\xi_{xt})\,dt$$

with ξ_{xt} between 0 and xt. The integral in the remainder is easily bounded by $1/(2n+2)$; but one can also convert it to a simpler form. Although it wasn't proven, it can be shown that $\cos(\xi_{xt})$ is a continuous function of t. Then applying the integral mean value theorem,

$$\int_0^1 \sin(xt)\,dt = \sum_{j=1}^{n} (-1)^{j-1} \frac{x^{2j-1}}{(2j-1!)(2j)}$$

$$+ (-1)^n \frac{x^{2n+1}}{(2n+1)!(2n+2)} \cos(\zeta_x)$$

for some ζ_x between 0 and x.

● **PROBLEM 1-7**

Consider a routine for evaluating $F(x)$; and, let $F^*(x)$ denote the computed result obtained by the program as an approximation to $F(x)$. Evaluate by the experimental method a probable error in the computed result.

Solution: It is more common to investigate by experimentation the accuracy of computed results obtained with a function evaluation routine. The simplest type of experimental testing consists of computing $F^*(x)$ for selected values of x and checking against known values of $F(x)$. Such testing, which involves personal inspection of results, cannot be very thorough. Fortunately the process of experimental testing can be made more automatic and more thorough by writing a test program in the following way.

With the aid of a random number generator compute a sequence of, say, n test arguments, $x_1, x_2, \ldots, x_n$. For each x_k compute (1) $F^*(x_k)$, the approximation to $F(x_k)$ produced by the function evaluation routine being tested, and (2) $F^{**}(x_k)$, another approximation to $F(x_k)$ that is sufficiently accurate to be used as a check value for $F^*(x_k)$. Compute statistics that will provide a ready indication of the magnitude of absolute error or relative error in the values of $F^*(x_k)$. Commonly used test statistics are the maximum relative error

$$\max_{1 \leq k \leq n} \left| \frac{F^*(x_k) - F^{**}(x_k)}{F^{**}(x_k)} \right|$$

and the root-mean-square relative error

$$\sqrt{\frac{1}{n} \sum_{k=1}^{n} \left[\frac{F^*(x_k) - F^{**}(x_k)}{F^{**}(x_k)} \right]^2}$$

The value of n, that is, the number of test arguments, can be very large. Some programs have been tested with as many as n = 100,000 random arguments; but smaller values of n are normally used, such as n = 5,000. A uniform random number generator can be used to generate test arguments, with the range of values adapted to the argument range of F(x). Thus, if the argument range for F(x) were a finite interval $\{a,b\}$, then the distribution of x_k could be uniform in $\{a,b\}$. An exponential random number generator is sometimes useful when the argument range of F(x) is infinite. For example, the distribution of x_k might be exponential with mean 1 for testing of a ln x routine for which the argument range was $(0,\infty)$. Sometimes it is desirable to make several tests and generate test arguments in different parts of the argument range of F(x) for different test runs. In this way, for example, regions where F(x) is unstable can be given special attention.

$F^{**}(x_k)$ must be computed in such a way that it is a sufficiently accurate approximation to $F(x_k)$ to be used as a check value for $F^*(x_k)$. Normally this means that $F^{**}(x_k)$ must be computed in higher precision than $F^*(x_k)$ is. Thus, if $F^*(x_k)$ is computed in single-precision arithmetic, then $F^{**}(x_k)$ might be computed in double-precision. A special-purpose, high-precision routine for evaluating F(x) might have to be written just to obtain $F^{**}(x_k)$ with sufficient accuracy. The higher the precision in which $F^*(x_k)$ is obtained, of course, the more difficult it is to compute a suitably accurate check value in still higher precision. If computing $F^{**}(x_k)$ does present difficulties on the computer in which tests are being made, it may be desirable to compute check values on another computer with which the necessary accuracy can be obtained more readily. This is obviously more trouble than using just one computer since test arguments and check values have to be transferred from one computer to another in some fashion. Variable-precision computers with which high-precision calculations can be performed with ease are especially useful for the computation of check values in high precision. (For example, Fortran programs can be run on the IBM 1620 with a precision of 28 decimal digits.)

PRINCIPLES OF COMPUTER OPERATIONS

• **PROBLEM 1-8**

Computers are often imagined as giant electronic "brains" which never make an error. This notion can be wrong. Interpret this simple program to demonstrate this concept.

```
      SUM = 0
      DO 10 I = 1,10000
   10 SUM = SUM + 0.0001
      WRITE (3, 20) SUM
   20 FORMAT (F15.8)
      END
```

Solution: Take a quantity SUM, make it equal to zero, and then add the quantity 0.0001 to it 10,000 times, and the answer is 1,00000000, exactly.

If one runs this program on a typical computer, the answer is likely to be

$$0.99935269$$

Granted that the error may be small--only 0.00065 or so, or about 0.065 percent--still, the answer is wrong. It certainly destroys the notion of a computer's infallibility, and creates concern about the results of other, more complicated calculations.

• **PROBLEM 1-9**

Solve the following two simultaneous equations for x_1 and x_2, and formulate the equations for the computer.

$$A_{11} X_1 + A_{12} X_2 = C_1$$

$$A_{21} X_1 + A_{22} X_2 = C_2$$

Solution: The novice might be tempted to write these two equations

$$A_{11} X_1 + A_{12} X_2 = C_1 \qquad (1)$$

$$A_{21} X_1 + A_{22} X_2 = C_2 \qquad (2)$$

as Fortran substitution statements into a program. But, the substitution statement is not an algebraic equation. One method of employing the computer in solving these two equations is to prepare algebraic solutions--using determinants, substitution method, or some other--to find solutions for X_1 and X_2 as shown in Eqs. (3) and (4)

$$X_1 = \frac{A_{22} C_1 - A_{12} C_2}{A_{11} A_{22} - A_{12} A_{21}} \qquad (3)$$

$$X_2 = \frac{A_{11} C_2 - A_{21} C_1}{A_{11} A_{22} - A_{12} A_{21}} \qquad (4)$$

By this method of solution, the computer is used only in performing the numerical calculations shown in Eqs. (3) and (4) in order to ultimately find the numerical value of X_1 and X_2. This might be a practical application of a digital computer if one wishes to solve this pair of equations five hundred times, each time for one set of values for the coefficients A_{11}, A_{12}, A_{21}, A_{22}, C_1, and C_2.

Now to proceed with the problem one recognizes that Eqs. (3) and (4) can be used as the basis for preparing the necessary Fortran program. Obviously, in order to find one pair of solutions—X_1 and X_2—the appropriate values of the six coefficients must be available in MEMORY. After X_1 and X_2 have been calculated, then, it will be necessary either to store these solutions in MEMORY or to output the appropriate values.

Because in Fortran alphabetic or numeric characters cannot be used in a sub or super position, it is necessary to give Fortran variables different names from those indicated in the above algebraic equations. See the table for one acceptable list of variable names.

TABLE
Equivalent Symbols

Algebraic	Fortran	Algebraic	Fortran
A_{11}	A11	C_1	C1
A_{12}	A12	C_2	C2
A_{21}	A21	X_1	X1
A_{22}	A22	X_2	X2

The Fortran program will be:

```
C   TWO EQUATIONS
    1 READ 2, A11, A12, A21, A22, C1, C2
    2 FORMAT (6F10.5)
      D = A11 * A22 - A12 * A21
      X1 = (A22 * C1 - A12 * C2)/D
      X2 = (A11 * C2 - A21 * C1)/D
      PUNCH 3, X1, X2
    3 FORMAT (2E15.7)
      GO TO 1
      END
```

• **PROBLEM 1-10**

> One of the most commonly used control statements in Fortran is the arithmetic IF statement; show how the programmer uses the arithmetic IF statement.

<u>Solution</u>: The IF statement allows the programmer to change the sequence of operations, depending upon whether the value of an expression is negative, zero, or positive. For example, the factorial of an integer N may be computed by means of the following program which contains an IF statement.

```
    . . .
    . . .
    I = 0
    NFACT = 1
```

```
10    I = I + 1
      NFACT = NFACT*I
      IF(I - N) 10, 20, 20
20    . . .
      . . .
```

The first two statements in the program set the initial values of I equal to 0 and of NFACT equal to 1. Statement 10 increases I by 1. The next statement computes NFACT for the new value of I. The IF statement checks whether I - N is less than, greater than, or equal to zero, i.e., whether I is less than, greater than, or equal to N.

• **PROBLEM 1-11**

Write the corresponding Fortran statements to the following mathematical expressions:

(a) $X = A + B \cdot C$
(b) $I = (J/2)^4$
(c) $Z = \sqrt{\sin x}$
(d) $X \cdot (B^D)$
(e) $\dfrac{D^E}{F} - G$
(f) $D^{E/(F-G)}$

Solution:

	Mathematics	FORTRAN
(a)	$X = A + B \cdot C$	X = A + B * C
(b)	$I = (J/2)^4$	I = (J/2) ** 4
(c)	$Z = \sqrt{\sin x}$	Z = SQRT(SIN(X))
(d)	$X \cdot (B^D)$	X*(B**D)
(e)	$\dfrac{D^E}{F} - G$	D**E/F-G
(f)	$D^{E/(F-G)}$	D**(E/(F-G))

NUMBER REPRESENTATIONS

• **PROBLEM 1-12**

In the computer the binary number system is used as opposed to the decimal number system used by humans to execute arithmetic operations. However, in both systems, there are some similarities and differences. Contrast and compare the two systems.

Solution: Humans execute arithmatic operations in the decimal number system, a system that has "ten" as its base; but when it comes to the computer, the story is quite different. In the computer, internal calculations are done in the binary system, a system that has "two" as its base.

The decimal number system has ten different digits, 0 through 9. The binary number system has only two different digits, 0 and 1. As an abbreviation, the two binary digits are usually called bits.

However, the binary number system is similar to the decimal system in that the position of a particular digit in a number has a great importance in both systems. Thus in both number systems the left digits of a number are more important than the right digits since they have a greater value.

In a decimal integer, each digit has 10 times the value of the digit to its right. Thus the last digit on the right of an integer is the "units" digit, the next is the "tens" digit, the next is the "hundreds" digit, and so on. Therefore, the decimal number 532 can be interpreted as

$$5 \text{ hundreds}$$
$$+ 3 \text{ tens}$$
$$+ 2 \text{ units.}$$

A more concise notation expressing the same thing is

$$532 = (5 \times 10^2) + (3 \times 10^1) + 2 \times 10^0).$$

The same principle applies to non-integer numbers such as 14.37 below

$$14.37 = (1 \times 10^1) + (4 \times 10^0) + (3 \times 10^{-1}) + (7 \times 10^{-2})$$
$$= 10 + 4 + 0.3 + 0.07.$$

Binary numbers can be expressed in exactly the same way, except that they only contain the digits 0 and 1, and that adjacent digits differ in value by a power of 2, rather than by a power of 10. For example, the binary integer 1011 can be expressed as

$$1011 = (1 \times 2^3) + (0 \times 2^2) + (1 \times 2^1) + (1 \times 2^0).$$

Changing into decimal numbers, we see that the binary 1011 is equal to $8 + 2 + 1$, or a decimal number 11. Because each digit of a binary number can only be a 0 or a 1 and therefore can carry less information than the 10 digits of the decimal number system, binary numbers are generally much longer than their decimal equivalents. For example, the decimal number 4094 is 111111111110 in binary.

Noninteger binary numbers can be expanded into powers of 2 and converted into decimal almost as easily as intergers. For example, the binary number 110.11 can be written as

$$110.11 = (1 \times 2^2) + (1 \times 2^1) + (0 \times 2^0) + (1 \times 2^{-1}) + (1 \times 2^{-2}).$$

where the exponent of 2 starts with 0 to the left of the decimal point and increases to the left, and starts with -1 to the right of the decimal point and becomes more negative to the right. In the case of the binary 110.11, the decimal equivalent is

$$4 + 2 + \frac{1}{2} + \frac{1}{4} = 6.75.$$

A decimal integer can always be exactly represented by a binary integer, for every integer can be expressed as a sum of powers of 2. But this is not true of fractional numbers.

In general, it can easily be shown that a rational fraction can be exactly expressed with a finite number of binary digits only if it can be expressed as the quotient of two integers p/q, where q is a power of 2; that is, $q = 2^n$ for some integer n. Obviously only a small proportion of rational fractions will satisfy this requirement.

Thus even some simple decimal fractions cannot be exactly expressed in the binary number system. For example, the simple decimal fraction 0.1 is an infinitely long binary number 0.000 110011 ... with two 1's and two 0's repeating themselves forever. Every repeating decimal (such as 0.33333333...) is also repeating in binary, but other numbers which are not repeating decimals in the decimal number system may become repeating in binary.

The decimal number 0.0001 is a repeating binary fraction which begins with 0.00000000000011010001101101110001 ... and lasts for 104 bits before starting to repeat with the part ...1101000110110111000l.... Every 104 bits from now on, this binary fraction starts to repeat itself.

Obviously an infinitely long binary fraction cannot be stored or used in a digital computer, and so some finite number of bits must be used and the rest discarded. This automatically leads to a small error which, by being repeated many times, can lead to a large error in the final answer.

● **PROBLEM** 1-13

Perform the following conversions:

a) $1 1 0 1 1 1 0 1 1_2$ into base 10

b) $4 5 7_8$ into base 2

c) $7 B 3_{16}$ into base 8

d) $1 2 4 2 5_{10}$ into base 16

<u>Solution</u>: $1 1 0 1 1 1 0 1 1_2$ may be converted into base 10 by considering this fact: Each digit in a base two number may be thought of as a switch, a zero indicating "off", and a one indicating "on". Also note that each digit

corresponds, in base 10, to a power of two. To clarify, look at the procedure:

$2^8 \quad 2^7 \quad 2^6 \quad 2^5 \quad 2^4 \quad 2^3 \quad 2^2 \quad 2^1 \quad 2^0$

$1 \quad\ \ 1 \quad\ \ 0 \quad\ \ 1 \quad\ \ 1 \quad\ \ 1 \quad\ \ 0 \quad\ \ 1 \quad\ \ 1_2$

$(1 \times 256) + (1 \times 128) + (0 \times 64) + (1 \times 32) + (1 \times 16) + (0 \times 4) + (1 \times 2) + (1 \times 1) = 443.$

If the switch is "on", then you add the corresponding power of two. If not, you add a zero. Follow the next conversion closely.

b) 457_8 may be converted into base two by using the notion of triads.

Triads are three-bit groups of zeros and ones which correspond to octal (base 8) and decimal (base 10) numbers. This table can, with some practice, be committed to memory:

Binary (2)	Octal (8)	Decimal (10)	Hexadecimal (16)
0 0 0	0	0	0
0 0 1	1	1	1
0 1 0	2	2	2
0 1 1	3	3	3
1 0 0	4	4	4
1 0 1	5	5	5
1 1 0	6	6	6
1 1 1	7	7	7

So, taking each digit of 457_8 separately, one gets:

$4_8 = 100_2$

$5_8 = 101_2$

$7_8 = 111_2$

which becomes $100101111_2 = 457_8$

c) $7B3_{16}$ is a hexadecimal number. Letters are needed to replace numbers, since the base here is 16. The following chart will help:

BINARY	OCTAL	DECIMAL	HEXADECIMAL
1 0 0 0	10	8	8
1 0 0 1	11	9	9
1 0 1 0	12	10	A
1 0 1 1	13	11	B
1 1 0 0	14	12	C
1 1 0 1	15	13	D
1 1 1 0	16	14	E
1 1 1 1	17	15	F
1 0 0 0 0	20	16	10

It may be easier to follow if $7B3_{16}$ is converted into decimal first.

$$16^2 \quad\quad 16^1 \quad\quad 16^0$$
$$7 \quad\quad\quad B \quad\quad\quad 3_{16}$$

$(7\times256) + (11\times16) + (3\times1) = 1971_{10}$

Now, convert to octal using this procedure: Divide 8, the base to be used, into 1971. Find the remainder of this division and save it as the first digit of the new octal number. Then divide 8 into the quotient of the previous division, and repeat the procedure. The following is an illustration:

```
8 | 1971
   -1968
      3    →  | 3       246 × 8 = 1968

     246
    -240
       6   →  | 6       30 × 8 = 240

      30
     -24
       6   →  | 6       3 × 8 = 24

       3   →  | 3       3 ÷ 8 ≠ an integer
```

$7B3_{16} = 1971_{10} = 3663_8$

d) $1 2 4 2 5_{10}$ may be converted to base 16 by this method: Take the highest power of 16 contained in 12425. Then, subtract that value and try the next lowest power. Continue the process until all the digits are accounted for.

$$
\begin{array}{rl}
 & 1\ 2\ 4\ 2\ 5 \\
16^3 \times 3 = & 1\ 2\ 2\ 8\ 8 \\
\hline
 & 1\ 3\ 7 \\
16^1 \times 8 = & 1\ 2\ 8 \\
\hline
 & 9 \\
16^0 \times 9 = & 9 \\
\hline
 & 0
\end{array}
$$

Answer: $12425_{10} = 3089_{16}$

• **PROBLEM 1-14**

Perform the following conversions from

a) 1 0 0 1 1 0 1 0

b) 1 0 1 . 1 0 1

c) 0 . 0 0 1 1

base 2 to base 8, 10, and 16.

Solution: a) To convert base 2 to base 8, use the triad method where three digits are taken together and converted to equivalent decimal value; this results in base 8 conversion

$$\underline{|0\ 1\ 0|}\ \underline{|0\ 1\ 1|}\ \underline{|0\ 1\ 0|}_2 = 232_8$$

Similarly for hexadecimal conversion 4 bits are taken together. Zeros are assumed for most significant grouping.

$$\underline{|0\ 0\ 0\ 0|}\ \underline{|1\ 0\ 0\ 1|}\ \underline{|1\ 0\ 1\ 0|} = 9A_{16}.$$

For binary to decimal conversion

$$\begin{array}{cccccccc} 1 & 0 & 0 & 1 & 1 & 0 & 1 & 0 \\ 1\times 2^7 & + 0\times 2^6 & + 0\times 2^5 & + 1\times 2^4 & + 1\times 2^3 & + 0\times 2^2 & + 1\times 2^1 & + 0\times 2^0 \end{array}$$

$$128 + 0 + 0 + 16 + 8 + 0 + 2 + 0 = 154_{10}$$

b) i) $101.101_2 = 5.5_8$

ii) $\underline{\underset{5}{0101}} . \underline{\underset{A}{1010}}_2 = 5.A_{16}$

iii) 101.101_2

$(1\times 2^2 + 0\times 2^1 + 1\times 2^0) . (1\times 2^{-1} + 0\times 2^{-2} + 1\times 2^{-3})$

$= 4 + 0 + 1 + 0.5 + 0 + 0.125 = 5.625_{10}$

c) i) $0.0011 = \underline{|000|} . \underline{|001|}\ \underline{|100|} = 0.14_8$

ii) $0.0011 = \underline{|0000|} . \underline{|0011|} = 0.3_{16}$

iii) $0.0011 = . (0\times 2^0 + 0\times 2^{-1} + 0\times 2^{-2} + 1\times 2^{-3} + 1\times 2^{-4})$

$= 0 + 0 + 0 + 0.125 + 0.0625 = 0.1875_{10}$

• **PROBLEM 1-15**

Describe the procedures involved in adding and subtracting hexadecimal numbers.

Solution: The process of adding and subtracting hexadecimal (Hex) digits is almost identical to that of adding and subtracting Arabic digits. Actually, there are a few ways in which these calculations may be performed. The most efficient one is the method used in the decimal calculations. The operations will be done from right to left. In using this method, each Hex symbol will be translated to a decimal digit before the calculations, then, upon completion of the calculations, each resulting decimal digit will be retranslated to a corresponding Hex symbol.

Addition: In decimal calculations, as one goes from one column to the next, one carries units of tens. For example:

```
    col. 2        col. 1        col. 0

      1             2             8

                    8             8
   ─────────────────────────────────────
      2             1             6
```

In Hex addition, instead of carrying units of ten, carry units of sixteen. Example 1

```
   col. 3      col. 2      col. 1      col. 0

     6           E           8           8

   +             1           D           3
   ──────────────────────────────────────────
```

For col. 0:

      Hex → Decimal → Decimal Sum → Hex Sum

       8   8   8

       3   3 $\frac{3}{11}$ → B

Note in this column the decimal sum is less than 16, and no units of sixteen are carried.

For col. 1:

      Hex → Decimal → Decimal Sum → Hex Sum

       8   8   8

       D   13 $\frac{13}{21}$ → 16 + 5

Here the decimal sum is greater than 16. After factoring out the units of sixteen (here only one), the remaining digit should be placed in this column.

For col. 2:

      Hex → Decimal → Decimal Sum → Hex Sum

       E   14   14

       1   1 $\frac{1}{15}$

                                 + $\frac{1}{16}$ (from col. 1)

                                    → 16 + 0

One unit of sixteen can be factored out, leaving zero as the remaining digit for col 2.

For col. 3:

    Hex → Decimal → Decimal Sum → Hex Sum

    6   6   6
                             ―
                             6

                          + 1   (from col.2)
                           ―
                            7 → 7

Thus, the final result is:

    6 E 8 8

    + 1 D 3
    ―――――
    7 0 5 B

Subtraction: In the subtraction procedure the units of sixteen are borrowed instead of being carried.

Example 2:

        col 2   col. 1   col. 0

          3   5   E

     -   2   B   8
       ――――――――――――――――――――

For col. 0

    Hex → Decimal → Decimal Difference - Hex Difference

    E   14   14

    8   8   - 8
                          ―――
                           6 → 6

For col. 1:

    Hex → Decimal → Decimal Difference → Hex Difference

    5   5   (5) +16 (Borrowed unit from col. 2)

    B   11   11
                       ―――
                        10 →   A

The difference in col. 2 is zero since one unit was borrowed from 3. The final result is:

        2   +16

        ~~3~~   5   E

    - 2   B   8
              ―――   ―
                A   6

• PROBLEM 1-16

a) Convert 764.301_8 to binary.
b) Convert 11011.10111_2 to octal.

Solution: a) The binary groups are used to replace individual digits, as indicated:

$$\begin{array}{cccccc} 7 & 6 & 4 & 3 & 0 & 1 \\ \downarrow & \downarrow & \downarrow & \downarrow & \downarrow & \downarrow \\ 111 & 110 & 100 & 011 & 000 & 001 \end{array}$$

TABLE OF CONVERSION

Decimal	Octal	Binary
1	1	1 (or 001)
2	2	10 (or 010)
3	3	11 (or 011)
4	4	100
5	5	101
6	6	110
7	7	111
8	10	1000
9	11	1001
10	12	1010
11	13	1011
12	14	1100
13	15	1101
14	16	1110
15	17	1111
16	20	10000

The resulting number is the required binary number. Some additional insight into the process can be obtained from the following diagram, which illustrates for the integral part the values of the numbers in decimal form:

$$\begin{array}{ccc}
7 & 6 & 4 \\
\downarrow & \downarrow & \downarrow \\
7 \times 8^2 & 6 \times 8^1 & 4 \times 8^0 \\
\downarrow & \downarrow & \downarrow \\
7 \times 2^6 & 6 \times 2^3 & 4 \times 2^0 \\
\downarrow & \downarrow & \downarrow \\
(1 \times 2^2 + 1 \times 2^1 + 1 \times 2^0) \times 2^6 & (1 \times 2^2 + 1 \times 2^1 + 0 \times 2^0) \times 2^3 & (1 \times 2^2 + 0 \times 2^1 + 0 \times 2^0) \times 2^0 \\
\downarrow & \downarrow & \downarrow \\
1 \times 2^8 + 1 \times 2^7 + 1 \times 2^6 & 1 \times 2^5 + 1 \times 2^4 + 0 \times 2^3 & 1 \times 2^2 + 0 \times 2^1 + 0 \times 2^0 \\
\downarrow & \downarrow & \downarrow \\
1\ 1\ 1 & 1\ 1\ 0 & 1\ 0\ 0
\end{array}$$

b) Grouping the digits by threes, starting from the binary point, we have

    11   011 . 101   11

It is necessary to introduce an additional zero in the first group and last group in order to have three binary bits in each group, thus:

    011   011 . 101   110

Each of these groups is now replaced by its octal equivalent, giving 33.56_8

• **PROBLEM 1-17**

Express 954.4 in the scale of three.

<u>Solution</u>: It is readily verified by division that

$954 = 1(3)^6 + 0(3)^5 + 2(3)^4 + 2(3)^3 + 1(3)^2 + 0(3) + 0;$

the successive divisions required are usually recorded in the following set-up.

$$
\begin{array}{r}
3\,\underline{|954} \\
3\,\underline{|318} + 0 \\
3\,\underline{|106} + 0 \\
3\,\underline{|35} + 1 \\
3\,\underline{|11} + 2 \\
3\,\underline{|3} + 2 \\
1 + 0
\end{array}
$$

Thus the integral part of the given number is 1022100, in the scale of three.

On the other hand, successive multiplications by 3 give

$$\frac{4}{10} = \frac{12}{30} = \frac{1}{3} + \frac{2}{30},$$

$$\frac{2}{30} = \frac{6}{90} = \frac{0}{9} + \frac{6}{90},$$

$$\frac{6}{90} = \frac{18}{270} = \frac{1}{27} + \frac{8}{270},$$

$$\frac{8}{270} = \frac{24}{810} = \frac{2}{81} + \frac{4}{810},$$

$$\frac{4}{810} = \frac{12}{2430} = \frac{1}{243} + \frac{2}{2430},$$

$$\frac{2}{2430} = \frac{6}{7290} = \frac{0}{729} + \frac{6}{7290},$$

. . . .

The process is now repeating and, collecting terms, one finds
954.4 (scale 10) = 1022100 . 101210121012 ... (scale 3).

● **PROBLEM** 1-18

If a = 111010, b = 1011 (scale 2), evaluate a + b, a - b, ab, a/b.

Solution:

```
            111010                       111010
             1011                         1011
   a + b = 1000101            a - b =   101111
```

```
            111010        a/b = 101.01000101 ...
             1011         1011 )111010
            111010              1011
            111010              1110
            000000              1011
            111010              1100
   ab = 1001111110               1011
                                10000
                                 1011
                                10100
                                 1011
                                 1001 etc.
```

FORTRAN RULES AND VOCABULARY

● **PROBLEM** 1-19

Using FORTRAN rules what would be the value of sum in each case?

a) SUM = 3 - 1 + 5 * 2/2

b) SUM = (4 - 2) ** 2

c) SUM = 9 ** 2 + 9/2 ** 2

d) SUM = (((1 + 2)/2 * 5) * 10/2 + 7)/3

Solution: The rules adopted by the creators of FORTRAN state that evaluation of arithmetic operations is conducted from left to right, except when the succeeding operation has a higher "binding power" than the one currently being considered.

The binding power of "+" and "-" is the lowest in strength; operators "*" and "/" are of middle strength, and raising to a power (**) has the highest priority.

The concept of parenthesizing subexpressions was added to the rules to achieve a sequence in operations different from the rules of binding power of operators.

Therefore, using this knowledge the solutions are found to be:

a) First division 2/2, then multiplication by 5, followed by adding to -1 and to 3. The result is 7.

b) First, solving the expression inside the brackets, then using exponent 2. The answer is 4. Note that the operation was done from left to right and exponent binding power wasn't taken into consideration because of the priority of parenthesis.

c) Using the binding power rule, 9**2 is 81 and 2**2 is 4. The third operation division: 9/4 - which is 2.25. The last operation is addition. The result is 83.25.

d) The result is 14.833.

● **PROBLEM** 1-20

Each of the following is not a correct FORTRAN statement. Give reasons in each case.

a)   READ (2,1) K, L, M + 1

b)   K + 5 = 37 * I - J * 13.

c)   WRITE (3,13) I, J, K, L.

d)   K = 2 × I

e)   Z ** Y ** W

Solution: a) The only acceptable characters in a variable name in any FORTRAN statement are the twenty-six letters of the alphabet and the ten digits 0 to 9 (some systems also accept the $ included in the variable name). Algebraic signs cannot be included, and therefore M + 1 is an invalid variable.

b) For FORTRAN statements involving arithmetic operations, use the following symbols:

-   for subtraction

+   for addition

** for exponentiation

*   for multiplication

/   for division

No arithmetic operations are allowed on the left side of the "=" sign.

c) In the WRITE statement the first number (3) indicates the device that is to be used to print the values obtained at the end of the program. (Typewriter, printer). The second number (13) is the number of the corresponding FORMAT statement. The listing of the variables should not be followed by a period. This is an error.

d) FORTRAN has no symbol "×" (presumably we mean multiplication, in which case the correct FORTRAN statement is K = 2 * I).

e) The expression is ambiguous. It could mean

$$(Z^Y)^W \quad \text{or} \quad (Z)^{(Y^W)}.$$

These are not always equal. For example, if Z = 2, Y = 3, W = 4:

$$(2^3)^4 \neq 2^{(3^4)}$$

● **PROBLEM 1-21**

Find the mistakes in the following FORTRAN expressions.

a)   (A - (B - (C + D(4.7))

b)   ((A/B)

c)   V = 1.63 * /D

<u>Solution</u>: a) In FORTRAN any constant or variable in an expression, unless it is the first constant or variable of that expression, must be preceded immediately by one of the following:

  a left parenthesis

  one of the operators +, -, *, /, or ** .

Any constant or variable in an expression, unless it is the last constant or variable of that expression, must be followed by one of the following:

  a right parenthesis

  one of the operators +, -, *, /, or ** .

The number of opened and closed parentheses must be equal. Therefore the correct expression is (A - (B - (C + D(4.7)))).

b) By the same argument that was used in part (a), the correct expressions are (A/B) or (A)/(B).

c) The statement is incorrectly formed, and has no meaning. Except for the unary minus, each operator +, *, /, or ** must have a term or factor both to its left and to its right in order for an expression to be correctly formed.

● **PROBLEM 1-22**

Pick out the errors in the following FORTRAN statements and explain briefly why they are incorrect:

a)   RESULT = SUM/FLOAT(NUM) + 2ERR

b)   IF (M*(N/M) = N) GO TO 20

c)   INTEGER CAPITAL, COST, INCOME

d)   AREA = LENGTH * WIDTH

e)   TAU = BETA/-3.0

<u>Solution</u>: a) FORTRAN variable names cannot begin with numbers, so 2ERR is an invalid variable.

b) In an IF statement, comparisons are made by using relational operators. An equality sign, which is used only in FORTRAN assignment statements, should be replaced by .EQ. TERM.

c) CAPITAL is a seven-letter word. FORTRAN allows a maximum of six characters in each variable name.

d) In FORTRAN, variables beginning with letters I through N are integer unless specified otherwise. Since WIDTH and AREA are real variable names, LENGTH cannot be used in order to avoid mixed mode multiplication.

e) Two arithmetic operators may not be juxtaposed in a FORTRAN assignment statement.

CHAPTER 2

ERRORS AND APPROXIMATIONS IN NUMERICAL ANALYSIS

SIGNIFICANT FIGURES, ERRORS

• **PROBLEM 2-1**

The number 31.546824 is known to have a relative error no worse than 1 part in 100,000. How many of the digits are known to be correct?

Solution: For this case the relative error is

$$\frac{\Delta Q}{|Q|} \leq \frac{1}{100,000}$$

or

$$\Delta Q \leq \frac{1}{100,000}|Q|$$

ΔQ is sought. In the below equation, Q is unknown and Q_1 is known (31.546824).

$$|Q| \leq |Q_1| + \Delta Q$$

Hence

$$\Delta Q \leq \frac{1}{100,000}(|Q_1| + \Delta Q)$$

or

$$\Delta Q \leq .00031546824 + \frac{1}{100,000}\Delta Q$$

or

$$\frac{99,999}{100,000}\Delta Q \leq .00031546824$$

or

$$\Delta Q \leq .00032$$

Since ΔQ is less than half a unit in the thousandths place, the significant digit in the thousandths place (the digit 6) is correct, so the number has at least five correct significant digits.

● **PROBLEM** 2-2

Find the error in the evaluation of the fraction $\cos 7° 10' \div \log_{10} 242.7$, assuming that the angle may be in error by 1' and that the number 242.7 may be in error by a unit in its last figure.

<u>Solution</u>: Since this is a quotient of two functions, it is better to compute the relative error from the formula $E_r \leq \Delta u_1/u_1 + \Delta u_2/u_2$ and then find the absolute error from the relation $E_a = NE_r$. Now write

$$N = \frac{\cos 7° 10'}{\log_{10} 242.7} = \frac{\cos x}{\log_{10} y} = u_1/u_2$$

and

$$\Delta u_1 = \Delta \cos x = -\sin x \Delta x,$$

$$\Delta u_2 = \Delta \log_{10} y = 0.43429 (\Delta y/y).$$

$$\therefore E_r \leq \frac{\sin x}{\cos x} \Delta x + \frac{0.43429}{y \log y} \Delta y,$$

or

$$E_r \leq \tan x \Delta x + \frac{0.435}{y \log y} \Delta y.$$

Now taking $x = 7° 10'$, $\Delta x = 1' = 0.000291$ radian, $y = 242$, $\Delta y = 0.1$, and using a slide rule for the computation,

$$E_r < 0.126 \times 0.000291 + \frac{0.435 \times 0.1}{242 \times 2.38} = 0.00011.$$

Since $N = \cos 7° 10'/\log 242.7 = 0.41599$,

$$E_a = 0.00011 \times 0.416 = 0.000046,$$

or $E_a < 0.00005$.

The value of the fraction is therefore between 0.41604 and 0.41594, and the mean of these numbers is taken to four figures as the best value of the fraction, or

$$N = 0.4160.$$

● **PROBLEM** 2-3

The following quadratic approximations to $f(x) = e^x$ were obtained by telescoping the Taylor series for e^x:

$$g_1(x) = 0.994571 + 1.130318x + 0.542990x^2, \quad -1 \leq x \leq 1$$

$$g_2(x) = 1.008129 + 0.860198x + 0.839882x^2, \quad 0 \leq x \leq 1$$

In each case a sufficient number of terms was employed in the original Taylor series to ensure that the coefficients of

27

these approximations have essentially "converged." That is, if additional terms were taken in the Taylor series, the coefficients of $g_1(x)$ and $g_2(x)$ would not change significantly. Note that $g_1(x)$ is a valid approximation over the interval $-1 \leq x \leq 1$ while $g_2(x)$ applies only over the interval $0 \leq x \leq 1$. Compare the accuracy of these two approximations on $0 \leq x \leq 1$ where both are valid.

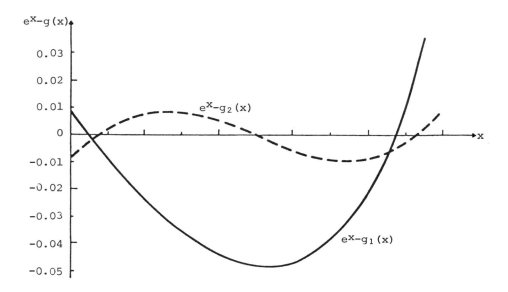

Solution: The errors in each of the approximations are plotted in the figure.

The errors in both approximations have nearly equal positive and negative peaks. However, the error in $g_2(x)$ is distributed much more uniformly over the interval. (The error in $g_1(x)$ is distributed quite uniformly over the interval $-1 \leq x \leq 1$ for which it was contructed, but $g_1(x)$ tends to overestimate e^x on much of $0 \leq x \leq 1$). The magnitude of the maximum error of $g_1(x)$ is approximately 0.054, while that of $g_2(x)$ is approximately 0.0101, or only about 1/5 that of $g_1(x)$. It is almost inevitable that the quadratic $g_1(x)$ which must provide a good approximation over the interval $-1 \leq x \leq 1$ will have a larger maximum error than will an approximation of the same type ($g_2(x)$) which is constructed to serve as a good approximation over an interval only half as large ($0 \leq x \leq 1$). It would be necessary to employ an economized approximation of higher degree than 2 in order to obtain the same accuracy on $-1 \leq x \leq 1$ that $g_2(x)$ provides on $0 \leq x \leq 1$. It should be apparent that the simplest, most effective approximations can be obtained by restricting the interval of approximation to the absolute minimum size required.

ABSOLUTE AND RELATIVE ERRORS

• **PROBLEM 2-4**

The following definitions apply to errors in a calculation:

The <u>absolute error</u> is defined as:

$$\text{Error}_{abs} = (\text{Calculated value}) - (\text{True value})$$

The <u>relative error</u> is defined as:

$$\text{Error}_{rel} = \frac{(\text{Calculated value}) - (\text{True value})}{(\text{True value})}$$

The percentage error is defined as

$$\text{Error}_{pct} = \text{Error}_{rel} \times 100$$

Assume the true value for a calculation should be 5.0, but the calculated value is 4.0: Calculate the absolute error, relative error and the percentage error.

<u>Solution</u>:

$$\text{Error}_{abs} = 4.0 - 5.0 = -1.0.$$

$$\text{Error}_{rel} = \frac{4.0 - 5.0}{5.0} = \frac{-1.0}{5.0} = -0.2,$$

$$\text{Error}_{pct} = -0.2 \times 100 = -20\%.$$

• **PROBLEM 2-5**

Show that the relative error in a quantity is approximately equal to the absolute error in its natural logarithm, since

$$\Delta(\ln y) \approx d(\ln y) = \frac{dy}{y} \approx \frac{\Delta y}{y}.$$

<u>Solution</u>: Using an exact formula, one has $\Delta(\ln y) = \ln(y + \Delta y) - \ln(y) = \ln(1 + \Delta y/y)$. A comparison follows:

$\frac{\Delta y}{y} =$	0.001	0.01	0.1	0.2	-0.2	-0.1
$\Delta(\ln y) =$	0.00100	0.00995	0.095	0.182	-0.223	-0.105

For logarithms to base 10 ("common logarithms") one has

$$\Delta(\log_{10} y) = \log_{10} e \cdot \Delta(\ln y) \approx 0.434 \frac{\Delta y}{y}.$$

● **PROBLEM** 2-16

Let x be a real number and F(x) it's floating-point k-digit representation. For example, if $x = \frac{2}{9}$ and $k = 4$, then $F\left(\frac{2}{9}\right) = .2222$. The operations of addition, subtraction, multiplication and division are defined as follows:

$$x \oplus y = F[F(x)+F(y)]$$
$$x \ominus y = F[F(x)-F(y)]$$
$$x \otimes y = F[F(x) \times F(y)] \quad (1)$$
$$x \oslash y = F\left[\frac{F(x)}{F(y)}\right]$$

Compute the absolute error and relative error for the operations defined in (1) for $x = \frac{1}{3}$, $y = \frac{4}{7}$. For arithmetic calculations apply 4-digit chopping. Suppose

$$a = .5711, \quad b = 4271, \quad c = .1001 \times 10^{-3},$$

compute

$$a \oplus b$$
$$(a \ominus y) \oslash c \quad (2)$$

and the corresponding absolute errors and relative errors.

Solution: With the help of a pocket calculator, we find

$$x = \frac{1}{3} \quad F(x) = .3333$$
$$y = \frac{4}{7} \quad F(y) = .5714$$

Then
$$x \oplus y = \frac{1}{3} \oplus \frac{4}{7} = F[F(x)+F(y)] = F[.3333 + .5714]$$
$$= F[.9047] = .9047$$

The results of the operations are listed in Table 1.

Table 1

Operation	Result	Actual Value
$x \oplus y$	.9047	$\frac{19}{21}$
$x \ominus y$	-.2381	$-\frac{5}{21}$
$x \otimes y$	.1904	$\frac{4}{21}$
$x \oslash y$	.5833	$\frac{7}{12}$

Let $\tilde{\omega}$ be an approximate value of ω; then absolute error is defined by

$$|\omega - \tilde{\omega}| \quad (3)$$

and the relative error is

$$\frac{|\omega - \tilde{\omega}|}{|\omega|} \qquad (4)$$

assuming $\omega \neq 0$.

From Table 1 and (3), (4) we compute

Table 2

Operation	Absolute Error	Relative Error
x ⊕ y	$.6190 \times 10^{-4}$	$.6842 \times 10^{-4}$
x ⊖ y	$.4762 \times 10^{-5}$	$.2 \times 10^{-4}$
x ⊗ y	$.7619 \times 10^{-4}$	$.4 \times 10^{-3}$
x ⊘ y	$.3333 \times 10^{-4}$	$.5714 \times 10^{-4}$

Note that the maximum relative error is $.4 \times 10^{-3}$.

If we apply, for example, 7-digit chopping then

$$F(x) = .3333333 \qquad F(y) = .5714286$$

and

$$x \otimes y = .1904762.$$

The absolute error is $.191 \times 10^{-7}$, and the relative error is $.131 \times 10^{-6}$. The accuracy has improved.

For a, b, c given in (2), we obtain

Table 3

Operation	Result	Actual Value	Absolute Error	Relative Error
a ⊕ b	4271	4271.5711	.5711	$.1123 \times 10^{-3}$
(a ⊖ y) ⊕ c	-.003	-.000328	.00267	8.1445

Note that the absolute error in a ⊕ b is large, while the relative error is small because we divide by a large actual value.

One should exercise caution in arranging the arithmetic operations and avoiding multiplication by large numbers (or division by small numbers) and subtraction of two numbers nearly equal.

TRUNCATION AND ROUND-OFF ERRORS

• **PROBLEM 2-7**

Give a brief explanation of mathematical truncation error.

Solution: Mathematical truncation error refers to the error of approximation in numerically solving a mathematical problem, and it is the error generally associated with the subject of numerical analysis. It involves the approximation of infinite process by finite ones, replacing noncomputable problems with computable ones. Following are some examples to make the idea more precise.

(a) Using the first two terms of the Taylor series from

$$(1 + x)^\alpha = 1 + \binom{\alpha}{1}x + \binom{\alpha}{2}x^2 + \ldots + \binom{\alpha}{n}x^n$$

$$+ \binom{\alpha}{n+1}\frac{x^{n+1}}{(1+\xi_x)^{n+1-\alpha}}$$

with $\binom{\alpha}{k} = \frac{\alpha(\alpha-1)\ldots(\alpha-k+1)}{k!}$ $k = 1,2,3\ldots$

for any real number α (for all cases the unknown ξ_x is located between x and 0), one can write

$$\sqrt{1+x} \approx 1 + \frac{1}{2}x$$

which is a good approximation when x is small.

(b) For the differential equation problem

$$Y'(t) = f(t,Y(t)) \qquad Y(t_0) = Y_0$$

use the approximation of the derivative

$$Y'(t) \approx \frac{Y(t+h) - Y(t)}{h}$$

for some small h. Let $t_j = t_0 + jh$ for $j \geq 0$, and define an approximate solution function $y(t_j)$ by

$$\frac{y(t_{j+1}) - y(t_j)}{h} = f(t_j, y(t_j))$$

so one gets

$$y(t_{j+1}) = y(t_j) + hf(t_j, y(t_j)) \qquad j \geq 0$$

This is Euler's method of solving a differential equation initial value problem.

• **PROBLEM 2-8**

Use exact, chopping and rounding to evaluate the following:

$$f(x) = x^3 - 4x^2 + 2x - 2.2 \quad \text{at } x = 2.41.$$

For chopping and rounding, apply three-digit arithmetic. Compute the relative errors. To decrease their values, try to carry out the same calculations for the same polynomial, but written in a different form.

Solution: The numerical data necessary for the solution of the problem are summarized in the following table.

	x	x^2	x^3	$4x^2$	$2x$
exact	2.41	5.8081	13.997521	23.2324	4.82
3-digit chopping	2.41	5.80	13.9	23.2	4.82
3-digit rounding	2.41	5.81	14.0	23.2	4.82

Note that the three-digit chopped numbers retain the leading three digits and may differ from the three-digit rounded numbers. For example, if $x = 14.7124$, then $X_{chopped} = 14.7$, and $X_{rounded} = 14.7$. On the other hand, if $x = 14.7824$, then $X_{chopped} = 14.7$, while $X_{rounded} = 14.8$.

We obtain

$$f(2.41)_{exact} = 13.997521 - 23.2324 + 4.82 - 2.2$$
$$= -6.614879$$

$$f(2.41)_{chopped} = 13.9 - 23.2 + 4.82 - 2.2$$
$$= -6.68$$

$$f(2.41)_{rounded} = 14.0 - 23.2 + 4.82 - 2.2 = -6.58$$

The corresponding relative errors are:

relative error, 3-digit chopping = $\left|\frac{-6.614879 + 6.68}{-6.614879}\right| = .0098$

relative error, 3-digit rounding = $\left|\frac{-6.614879 + 6.58}{-6.614879}\right| = .0052$

In order to obtain smaller values for the relative errors we can write $f(x)$ in an equivalent form

$$f(x) = x[x(x-4)+2] - 2.2$$

Then, the exact value of $f(x)$ at $x = 2.41$ is, of course, the same, but

$$f(2.41)_{chopped} = 2.41[2.41(2.41-4)+2] - 2.2$$
$$= 2.41[-3.83+2] - 2.2 = -4.41 - 2.2$$
$$= -6.61$$

$$f(2.41)_{rounded} = 2.41[2.41(2.41-4)+2] - 2.2$$
$$= -6.61$$

The relative error for both values is now

$$\left|\frac{-6.614879 + 6.61}{-6.614879}\right| = .00073$$

The relative error is now much smaller. The reason for this is that there is a decrease in the number of error-producing computations. Note that in $x^3 - 4x^2 + 2x - 2.2$, we had five multiplications and three additions, while in $x[x(x-4)+2] - 2.2$, we had two multiplications and three additions.

● **PROBLEM 2-9**

When one gives the number of digits in a numerical value one should not include zeros in the beginning of the number, as these zeros only help to denote where the decimal point should be. If one is counting the number of decimals, one should of course include leading zeros to the right of the decimal point.

For example, 0.001234 ± 0.000004 has five correct decimals and three significant digits, while 0.001234 ± 0.000006 has four correct decimals and two significant digits.

The number of correct decimals gives one an idea of the magnitude of the absolute error, while the number of significant digits gives a rough idea of the magnitude of the relative error.

Consider the following decimal numbers: 0.2397, -0.2397, 0.23750, 0.23650, 0.23652. Shorten to three decimals by the Round-off and Chopping methods.

Solution:

    0.2397 rounds to 0.240 (is chopped to 0.239),

   -0.2397 rounds to -0.240 (is chopped to -0.239),

    0.23750 rounds to 0.238 (is chopped to 0.237),

    0.23650 rounds to 0.236 (is chopped to 0.236),

    0.23652 rounds to 0.237 (is chopped to 0.236).

Observe that when one rounds off a numerical value one produces an error; thus it is occasionally wise to give more decimals than those which are correct.

One consequence of these rounding conventions is that numerical results which are not followed by any error estimations should often, though not always, be considered as having an uncertainty of ½ unit in the last decimal place.

● **PROBLEM 2-10**

Assume that floating-point arithmetic with three significant digits is used on the system

    $x + 400y = 801,$

    $200x + 200y = 600.$

Solve it by rounding off the errors.

Solution: Multiplying the first equation by -200 and adding the result to the second equation, one gets

    $-7.98 \times 10^4 y = -15.9 \times 10^4.$

This gives the approximate value of 1.99 for y. Substitution of this value in the first equation yields the approximate value x = 5. The relative error in the value for x is 400% since the correct solution is (1,2). The problem here lies in the equation x = 801 - 400y. Any rounding error in y (in this case 0.01) gets magnified by a factor of 400, which can be quite significant because x is small.

• **PROBLEM 2-11**

Scaling is one way to get around errors that can be caused by improportionate computations. Scale the system

$$x + 230y + 3460z = 20,000$$
$$30x + 5y + 0.1z = 300 \qquad (1)$$
$$0.001x + 0.002y + 0.003z = 7,$$

by columns.

Hint: Column scaling is similar to row scaling except that it alters the solution. For example, if the first column is divided by 10, then the new solution will be in terms of $10x_1$ instead of x_1.

Solution: In order to scale the system (1) by columns, let $x = x^*/10^2$, $y = y^*/10^3$, and $z = z^*/10^4$. This results in the system

$$0.01x^* + 0.230y^* + 0.3460z^* = 20,000,$$
$$0.3x^* + 0.005y^* + 0.00001z^* = 300, \qquad (2)$$
$$0.00001x^* + 0.000002y^* + 0.0000003z^* = 7.$$

Of course, one must remember to replace x^*, y^*, and z^* by their values in terms of x, y, and z after solving the system (2).

• **PROBLEM 2-12**

On the system

$$x + 400y = 801$$
$$200x + 200y = 600,$$

use the technique of partial pivoting to round off the solution's error to three significant digits.

Solution: Scaling the system by rows, one gets

$$0.001x + 0.4y = 0.801,$$
$$0.2x + 0.2y = 0.6.$$

The technique of partial pivoting requires that the second equation be used as the pivot since 0.2 > 0.001. Carrying three significant digits, one obtains the equation

$0.399y = 0.798$.

Thus $y = 2.00$ and $x = 1.00$. Note that a small round-off error in y would have produced only a small error in x, since the second equation is less sensitive than the first to an error in y.

METHODS OF APPROXIMATION

● **PROBLEM 2-13**

Approximate $y = \log_e x = \ln x$ by polynomials of respective max-degrees 1,2,3, in the neighborhood of the point (1,0).

<u>Solution</u>: Here, $x_0 = 1$; and $y' = x^{-1}$, $y'' = -x^{-2}$, $y''' = 2x^{-3}$. Hence

$$f(1) = 0, \; f'(1) = 1, \; f''(1) = -1, \; f'''(1) = 2,$$

and

$$y = x - 1,$$

$$y = (x-1) - \frac{1}{2}(x-1)^2,$$

$$y = (x-1) - \frac{1}{2}(x-1)^2 + \frac{1}{3}(x-1)^3,$$

are the required approximating polynomials.

● **PROBLEM 2-14**

Let $f(x) = 1/(1 + x)$, then $y_k = 1/(k + 1)$, $k = 0,1,2,\ldots, n - 1$, and

$$\int_0^{n-1} f(x)\,dx = \ln n.$$

(It is convenient to stop at the index $n - 1$ rather than the index n.) Given the above, use the Euler Summation formula

$$\sum_{i=0}^{n} y_i = \int_0^n f(x)\,dx + \frac{1}{2}(y_0 + y_n) + \int_0^n P_1(x) f'(x)\,dx$$

where

$$P_n(x) = a + a_1(x - x_0) + a_2(x - x_0)^2 + \ldots + a_n(x - x_0)^n$$

to derive the formula for the Euler's Constant.

<u>Solution</u>:

$$\sum_{k=0}^{n-1} \frac{1}{k+1} = \ln n + \frac{1}{2}\left(1 + \frac{1}{n}\right) - \int_0^{n-1} \frac{P_1(x)}{(x+1)^2}\,dx$$

Here $P_1(x)$ is a polynomial used to approximate the function of x.

Since
$$|P_1(x)| \le \frac{1}{2}$$
for every x and
$$\int_k^{k+1} \frac{P_1(x)}{(x+1)^2} dx < 0,$$
then
$$0 > \int_0^{n-1} \frac{P_1(x)}{(x+1)^2} dx \ge -\frac{1}{2} \int_0^{n-1} \frac{1}{(x+1)^2} dx = -\frac{1}{2}\left(1 - \frac{1}{n}\right).$$

It follows that
$$\int_0^\infty \frac{P_1(x)}{(x+1)^2} dx = \lim_{n\to\infty} \int_0^{n-1} \frac{P_1(x)}{(x+1)^2} dx$$

exists and is between 0 and $-\frac{1}{2}$. Therefore the number C defined by

$$C = \lim_{n\to\infty} \left[\sum_{k=0}^{n-1} \frac{1}{k+1} - \ln n\right] = \frac{1}{2} - \int_0^\infty \frac{P_1(x)}{(x+1)^2} dx$$

exists. The number C is called Euler's constant; it can be shown that

$$C = 0.577216-.$$

● **PROBLEM** 2-15

If $f(x) = \frac{1}{2}(x + |x|)$, find the parabola

$$y = a_0 + a_1 x + a_2 x^2$$

which minimizes the integral I_s given by

$$I_s = \int_a^b (f(x) - P_n(x))^2 dx$$

where

$$P_n(x) = y$$

in the interval $[-1, 1]$.

Solution: Thus, (the subscript s is omitted);

$$I = \int_{-1}^1 \left[\frac{1}{2}(x + |x|) - (a_0 + a_1 x + a_2 x^2)\right]^2 dx$$

$$= \int_{-1}^{0} (a_0 + a_1 x + a_2 x^2)^2 \, dx + \int_{0}^{1} (a_0 + (a_1 - 1) x + a_2 x^2)^2 \, dx$$

$$= \frac{1}{3} + 2a_0^2 - a_0 + \frac{4}{3} a_0 a_2 + \frac{2}{3} a_1^2 - \frac{2}{3} a_1 + \frac{2}{5} a_2^2 - \frac{1}{2} a_2 .$$

Now

$$I_{a_0} = 4a_0 - 1 + \frac{4}{3} a_2 ,$$

$$I_{a_1} = \frac{4}{3} a_1 - \frac{2}{3} ,$$

$$I_{a_2} = \frac{4}{3} a_0 + \frac{4}{5} a_2 - \frac{1}{2} ,$$

where

$$I_{a_i} = \partial I / \partial a_i ,$$

and

$$I_{a_0 a_0} = 4, \quad I_{a_0 a_1} = I_{a_1 a_0} = 0 ,$$

$$I_{a_1 a_1} = \frac{4}{3} , \quad I_{a_0 a_2} = I_{a_2 a_0} = \frac{4}{3} ,$$

$$I_{a_2 a_2} = \frac{4}{3} , \quad I_{a_1 a_2} = I_{a_2 a_1} = 0 ,$$

where

$$I_{a_i a_j} = \partial^2 I / \partial a_i \partial a_j .$$

Sufficient conditions for a minimum are

$$I_{a_0} = I_{a_1} = I_{a_2} = 0,$$

$$\begin{vmatrix} I_{a_0 a_0} & I_{a_0 a_1} & I_{a_0 a_2} \\ I_{a_1 a_0} & I_{a_1 a_1} & I_{a_1 a_2} \\ I_{a_2 a_0} & I_{a_2 a_1} & I_{a_2 a_2} \end{vmatrix} > 0 .$$

These conditions are satisfied if

$$a_0 = \frac{3}{32} ,$$

$$a_1 = \frac{16}{32} ,$$

$$a_2 = \frac{15}{32} ,$$

and therefore

$$y = \frac{1}{32}(3 + 16x + 15x^2)$$

is the required polynomial. It is suggested to graph this parabola and

$$y = \frac{1}{2}(x + |x|)$$

on the same set of axes.

• **PROBLEM 2-16**

Approximate

$$y = e^{\sin x}$$

at $x = 0$ by a function of type

$$C(x) = a_0 + a_1 \sin \frac{2\pi}{p}(x - x_0) + \ldots + a_n \sin n \frac{2\pi}{p}(x - x_0)$$
$$+ b_1 \cos \frac{2\pi}{p}(x - x_0) + \ldots + b_n \cos m \frac{2\pi}{p}(x - x_0),$$

where $a_0, \ldots, a_n, b_0, \ldots, b_m$ and x_0

are constants; $C(x)$ is everywhere continuous and periodic with period p and hence is suitable for approximating the function $y = f(x)$ at a point (x_0, y_0) if the function $f(x)$ is itself continuous and periodic with period p. As a rule, m is chosen to be n, n - 1, or n + 1. The function $C(x)$ has n + m + 1 arbitrary constants and hence can be determined by the imposition of an equivalent number of conditions. Solve for n = m = 1 and for n = m = 3.

Solution: For n = m = 1, take $x_0 = 0$ and put

$$C_1(x) = a_0 + a_1 \sin x + b_1 \cos x.$$

(The function to be approximated has the period $p = 2\pi$.) Determine the three constants a_0, a_1, and b_1 by equating the values of $e^{\sin x}$ and $C_1(x)$ and the values of their first and second derivatives at $x = x_0$. Obtain

$$C_1(x) = 2 + \sin x - \cos x.$$

If the graphs of $y = C_1(x)$ and the given function were drawn, they would show the periodicity of the two curves and indicate that $C_1(x)$ is a good approximation to $e^{\sin x}$ from about $-2\pi/9$ to about $7\pi/18$ (roughly from -40° to 70°) in the period interval from $-\pi$ to π.

For n = m = 3, start with

$$C_3(x) = a_0 + a_1 \sin x + a_2 \sin 2x + a_3 \sin 3x$$
$$+ b_1 \cos x + b_2 \cos 2x + b_3 \cos 3x.$$

Since six derivatives are needed it is convenient to write $e^{\sin x}$ as a power series in x, namely,

$$y = e^{\sin x} = 1 + x + \frac{x^2}{2!} - \frac{3x^4}{4!} - \frac{8x^5}{5!} - \frac{3x^6}{6!} + \frac{56x^7}{7!} + \ldots$$

Obtain on substituting 0 for x, equating the values of corresponding derivatives, and solving,

$$C_3(x) = \frac{1}{180} (200 + 210 \sin x - 6 \sin 2x - 6 \sin 3x + 45 \cos x - 72 \cos 2x + 7 \cos 3x).$$

The graphs of $y = C_3(x)$ and the given function would show that this time the approximation is good from about $-4\pi/9$ to about $11\pi/18$ (-80° to 110°) in the period interval from $-\pi$ to π.

● **PROBLEM** 2-17

Approximate $y = e^{\sin x}$ by a function of type

$$C(x) = a_0 + a_1 \sin \frac{2\pi}{p} (x - x_0) + \ldots + a_n \sin n \frac{2\pi}{p} (x - x_0)$$
$$+ b_1 \cos \frac{2\pi}{p} (x - x_0) + \ldots + b_n \cos m \frac{2\pi}{p} (x - x_0),$$

where

$a_0, \ldots, a_n, b_0, \ldots, b_m$ and x_0 are constants.

$C(x)$ coincides with $y = e^{\sin x}$ at $x = 0, \pi/3, \pi/2, \pi, 3\pi/2$.

Solution: First determine the constants in

$$C_2(x) = a_0 + a_1 \sin x + a_2 \sin 2x + b_1 \cos x + b_2 \cos 2x$$

so that the graph of this equation passes through the five points

$$(0,1), (\pi/3, e^{\frac{1}{2}\sqrt{3}}), (\pi/2, e), (\pi, 1), (3\pi/2, e^{-1}).$$

Substituting the coordinates of the points into $C_2(x)$, equating corresponding values, and solving,

$$C_2(x) = 1.272 + 1.175 \sin x - 0.057 \sin 2x - 0.272 \cos 2x.$$

Its graph and the graph of $y = e^{\sin x}$ indicate that $C_2(x)$ is a good approximation to $y = e^{\sin x}$ for all values of x.

Note that because of the periodicity of the trigonometric

functions, one is more likely to run into a system of inconsistent linear equations (in the a's and b's) than in polynomial approximation.

● **PROBLEM 2-18**

Let $P_n^*(x)$ denote the polynomial of degree $\leq n$ that approximates $F(x) = 16^{-x}$ with minimax absolute error in $[0,a]$, and let $Q_n^*(x)$ denote the polynomial of degree $\leq n$ that approximates $F(x) = 16^{-x}$ with minimax relative error in $[0,a]$. Use the theorem:

Let $P_n^*(x)$ denote the polynomial of degree $\leq n$ that approximates $F(x)$ with minimax absolute error in $[a,b]$. Suppose that $F(x)$ possesses an $(n+1)$st derivative $F^{(n+1)}(x)$ for x in $[a,b]$. If for two nonnegative numbers m and M and for $a \leq x \leq b$ either

$$m \leq F^{(n+1)}(x) \leq M$$

or

$$m \leq -F^{(n+1)}(x) \leq M,$$

then

$$\frac{m(b-a)^{n+1}}{2^{2n+1}(n+1)!} \leq \max_{[a,b]} |P_n^*(x) - F(x)| \leq \frac{M(b-a)^{n+1}}{2^{2n+1}(n+1)!};$$

to find bounds for both

$$\max_{[0,a]} |P_n^*(x) - 16^{-x}|$$

and

$$\max_{[0,a]} \left|\frac{Q_n^*(x) - 16^{-x}}{16^{-x}}\right|.$$

Solution: Since

$$F^{(n+1)}(x) = (-1)^{n+1}(\ln 16)^{n+1} 16^{-x},$$

one gets $m \leq \pm F^{(n+1)}(x) \leq M$ for all x in $[0,a]$, where

$$m = 16^{-a}(\ln 16)^{n+1}$$

and
$$M = (\ln 16)^{n+1},$$

thus satisfying the condition of the theorem. It follows that

$$\frac{16^{-a}(\ln 16)^{n+1}a^{n+1}}{2^{2n+1}(n+1)!} \leq \max_{[0,a]} \left| P_n^*(x) - 16^{-x} \right| \leq \frac{(\ln 16)^{n+1}a^{n+1}}{2^{2n+1}(n+1)!}$$

Applied directly, the theorem cannot give bounds for

$$\max_{[0,a]} \left| \frac{Q_n^*(x) - 16^{-x}}{16^{-x}} \right|.$$

With the additional analysis shown below, however, one can also obtain such bounds.

$$\max_{[0,a]} \left| \frac{Q_n^*(x) - 16^{-x}}{16^{-x}} \right| \geq \max_{[0,a]} \left| Q_n^*(x) - 16^{-x} \right| \geq \max_{[0,a]} \left| P_n^*(x) - 16^{-x} \right|$$

$$\geq \frac{16^{-a}(\ln 16)^{n+1}a^{n+1}}{2^{2n+1}(n+1)!}.$$

Likewise,

$$\max_{[0,a]} \left| \frac{Q_n^*(x) - 16^{-x}}{16^{-x}} \right| \leq \max_{[0,a]} \left| \frac{P_n^*(x) - 16^{-x}}{16^{-x}} \right|$$

$$\leq 16^a \max_{[0,a]} \left| P_n^*(x) - 16^{-x} \right|$$

$$\leq \frac{16^a (\ln 16)^{n+1}a^{n+1}}{2^{2n+1}(n+1)!}.$$

Therefore,

$$\frac{16^{-a}(\ln 16)^{n+1}a^{n+1}}{2^{2n+1}(n+1)!} \leq \max_{[0,a]} \left| \frac{Q_n^*(x) - 16^{-x}}{16^{-x}} \right|$$

$$\leq \frac{16^a (\ln 16)^{n+1}a^{n+1}}{2^{2n+1}(n+1)!}.$$

● **PROBLEM 2-19**

Find the polynomial $P_2^*(x)$ of degree ≤ 2 that aproximates $F(x) = \sqrt{x}$ with minimax relative error in $\left[\frac{1}{16}, 1\right]$. Use Remez' method.

As initial estimates for the critical points of

$$\frac{P_2^*(x) - \sqrt{x}}{\sqrt{x}},$$

use

$$x_1 = \frac{1}{16}, \quad x_2 = .4, \quad x_3 = .8, \text{ and } x_4 = 1.$$

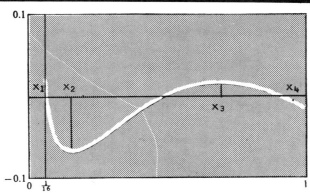

Fig. 1

Graph of relative-error function $\frac{P_2(x) - \sqrt{x}}{\sqrt{x}}$ after one iteration with Remez' method.

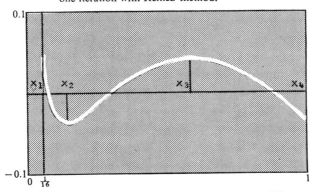

Fig. 2

Graph of relative-error function $\frac{P_2(x) - \sqrt{x}}{\sqrt{x}}$ after two iterations with Remez' method.

<u>Solution</u>: Applying Remez' method:

Step (1) Choose the initial estimates for the critical points.

Step (2) The system of linear equations

$$a_0 + a_1 x_k + \ldots + a_n x_k^n - (-1)^k \mu F(x_k) = F(x_k),$$

$$k = 1, 2, \ldots, n + 2,$$

becomes

$$a_0 + a_1 x_1 + a_2 x_1^2 + \mu\sqrt{x_1} = \sqrt{x_1}$$

$$a_0 + a_1 x_2 + a_2 x_2^2 - \mu\sqrt{x_2} = \sqrt{x_2}$$

$$a_0 + a_1 x_3 + a_2 x_3^2 + \mu\sqrt{x_3} = \sqrt{x_3}$$

$$a_0 + a_1 x_4 + a_2 x_4^2 - \mu\sqrt{x_4} = \sqrt{x_4}$$

On each iteration this system of equations is solved numerically by means of Gaussian elimination.

Step (3) On each iteration it is necessary to locate the extreme points in

$$\left[\frac{1}{16}, 1\right]$$

of the relative-error function

$$\frac{P_2(x) - \sqrt{x}}{\sqrt{x}},$$

where the coefficients of

$$P_2(x) = a_0 + a_1 x + a_2 x^2$$

are obtained from step (2). Two of the extreme points are

$$y_1 = \frac{1}{16}$$

and

$$y_4 = 1$$

on every iteration. The other two, y_2 and y_3, are the two roots of the quadratic equation

$$3a_2 x^2 + a_1 x - a_0 = 0$$

and are computed with the quadratic formula on every interation.

The quadratic equation in step (3) was derived as follows:

Let

$$E(x) = \frac{P_2(x) - \sqrt{x}}{\sqrt{x}}.$$

If one replaces $P_2(x)$ by $a_1 + a_1 x + a_2 x^2$, differentiates $E(x)$, and set the derivative equal to zero, then one obtains after some simplifications the condition

$$3a_2 x^2 + a_1 x - a_0 = 0.$$

Since $E(x)$ must have at least four extreme points in

$$\left[\frac{1}{16}, 1\right],$$

whereas the quadratic equation has only two roots, it follows that y_1 and y_4 must be the endpoints of the interval and that the two interior extreme points, y_2 and y_3, must be the roots of the quadratic equation. Note, incidentally, that this argument also establishes that

$$\frac{P_2^*(x) - \sqrt{x}}{\sqrt{x}}$$

must be a standard error function.

TABLE 1

After iteration	x_2	x_3	a_0	a_1	a_2	μ
0	.400000	.800000	—	—	—	—
1	.157582	.692428	.171896	1.339080	−.525123	−.014147
2	.146345	.588965	.170415	1.453820	−.659052	−.034816
3	.148921	.590636	.171509	1.442061	−.649966	−.036396
4	.148917	.590599	.171509	1.442106	−.650022	−.036407

Table 1 shows the results of four iterations with Remez' method. The numbers shown in this table have been rounded to six decimal places. Values of the iterates converged to six decimals in these four iterations.

• **PROBLEM** 2-20

Find the polynomial $P_7^*(x)$ of degree ≤ 7 that approximates

$$\cos \frac{1}{4} \pi x$$

with minimax absolute error in $[-1, 1]$. Use the Remez' method. The initial values are prescribed by

$$x_k = \frac{1}{2}(b - a)\cos\frac{(n - k + 2)\pi}{n + 1} + \frac{1}{2}(b + a),$$

$$k = 1, 2, \ldots, n + 2. \quad (1)$$

Solution: Since

$$\cos \frac{1}{4} \pi x$$

is an even function and since the approximation interval is symmetric about the origin, $P_7^*(x)$ contains only even-degree terms.

Let

$$P_7^*(x) = a_0^* + a_2^* x^2 + a_4^* x^4 + a_6^* x^6.$$

The absolute-error function $P_7^*(x) - \cos\frac{1}{4}\pi x$ is a standard error function. It is also an even function, and its nine critical points are therefore symmetrically located about the origin. That is,

$$-x_1^* = x_9^*, \quad -x_2^* = x_8^*, \quad -x_3^* = x_7^*, \quad -x_4^* = x_6^*$$

and

$$x_5^* = 0.$$

If the starting values in step 1 of Remez' method are chosen so that analogous conditions hold, then this symmetry will be preserved throughout subsequent iterations, and the odd-degree coefficients of $P_7(x)$ in step 2 will be zero on every iteration. Taking advantage of these facts, apply Remez' method in the following way:

Step 1) Because of the symmetries mentioned above, it is

necessary to use only x_5, x_6, x_7, x_8, and x_9. For initial values use (1), that is,

$$x_k = \cos \frac{(9-k)\pi}{8},$$

$$k = 5,6,7,8,9.$$

Step 2) The system of linear equations

$$a_0 + a_1 x_k + \ldots + a_n x_k^n - (-1)^k \mu = F(x_k),$$

$$k = 1, 2, \ldots, n+2,$$

for the $n+2$ unknowns $a_0, a_1, \ldots, a_n$, and μ, simplifies to the following:

$$a_0 + a_2 x_5^2 + a_4 x_5^4 + a_6 x_5^6 + \mu = \cos \tfrac{1}{4}\pi x_5$$

$$a_0 + a_2 x_6^2 + a_4 x_6^4 + a_6 x_6^6 - \mu = \cos \tfrac{1}{4}\pi x_6$$

$$a_0 + a_2 x_7^2 + a_4 x_7^4 + a_6 x_7^6 + \mu = \cos \tfrac{1}{4}\pi x_7$$

$$a_0 + a_2 x_8^2 + a_4 x_8^4 + a_6 x_8^6 - \mu = \cos \tfrac{1}{4}\pi x_8$$

$$a_0 + a_2 x_9^2 + a_4 x_9^4 + a_6 x_9^6 + \mu = \cos \tfrac{1}{4}\pi x_9.$$

On each iteration this system of equations is solved numerically by means of Gaussian elimination.

Step 3) Since $P_7(x) - \cos \tfrac{1}{4}\pi x$ is an even function, its extreme points are distributed symmetrically about the origin. Therefore, as was said earlier, it is necessary to locate only y_5, y_6, y_7, y_8, and y_9. On every iteration $y_5 = 0$ and $y_9 = 1$. To obtain y_6, y_7, and y_8, differentiate the absolute-error function $P_7(x) - \cos \tfrac{1}{4}\pi x$ and compute the zeros of the derivative. This can be done by isolating each y_k by searching and then computing it accurately by bisection. Since the starting values chosen for step 1 are very good, y_k is always close to x_k and can be isolated easily by searching near x_k.

Because the initial choices for the x_k's happen to be very accurate, after only one iteration a_0, a_2, a_4, a_6, and μ become stabilized to ten decimal places. The x_k's do not converge quite so rapidly. The results after one iteration are shown below, rounded to ten decimal places.

$a_0 = .9999999724$ $a_6 = -.0003188805$

$a_2 = -.3084242536$ $\mu = .0000000276$

$a_4 = .0158499153$

● **PROBLEM 2-21**

Develop the power series expansion of $\cos \frac{1}{4} \pi x$.

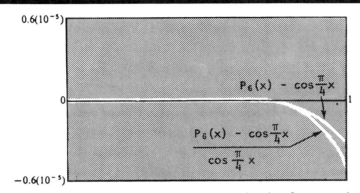

Graphs of absolute-error and relative-error functions for truncated power series approximation to $\cos \frac{1}{4}\pi x$.

Solution:

$$\cos \frac{1}{4} \pi x = \sum_{k=0}^{\infty} \frac{(-1)^k \left(\frac{1}{4}\pi\right)^{2k}}{(2k)!} x^{2k}$$

$$= 1 - .3084251375 x^2 + .0158543442 x^4 - .0003259919 x^6$$

$$+ .0000035909 x^8 - .0000000246 x^{10} + \ldots,$$

where the coefficients have been rounded to ten decimal places. Let $P_6(x)$ denote the polynomial of degree six obtained by truncating this series after the fourth term, and consider this polynomial as an approximation to $\cos \frac{1}{4} \pi x$ for $-1 \leq x \leq 1$. Graphs of the absolute-error function $P_6(x) - \cos \frac{1}{4} \pi x$ and the relative-error function

$$\frac{P_6(x) - \cos \frac{1}{4} \pi x}{\cos \frac{1}{4} \pi x}$$

for $0 \leq x \leq 1$ are shown in the figure. For both absolute error and relative error, the magnitude of the error is greatest when $x = 1$. Simple calculations show that the maximum absolute error is approximately $.357(10^{-5})$ and that the maximum relative error is approximately $.504(10^{-5})$.

● **PROBLEM 2-22**

Using Remez' method, determine the rational function $R_{5,2}(x)$ in $V_{5,2}[-1, 1]$ that approximates $\cos \frac{1}{4}\pi x$ with minimax absolute error in $[-1, 1]$. Hint: Since $\cos \frac{1}{4}\pi x$ is an even function and the approximation interval is symmetric about the origin, the odd-degree coefficients in

$$R^*_{m,n}(x) = \frac{p^*_0 + p^*_1 x + p^*_2 x^2 + \ldots + p^*_m x^m}{q^*_0 + q^*_1 x + q^*_2 x^2 + \ldots + q^*_n x^n}, \qquad (1)$$

vanish so that it becomes just

$$R^*_{5,2}(x) = \frac{p^*_0 + p^*_2 x^2 + p^*_4 x^4}{1 + q^*_2 x^2}$$

<u>Solution</u>: Approximate numerical values for p^*_0, p^*_2, p^*_4, and q^*_2 need to be computed. Since the absolute-error function for this approximation is an even function, too, the nine critical points are symmetrically located about the origin. That is, $-x^*_1 = x^*_9$, $-x^*_2 = x^*_8$, $-x^*_3 = x^*_7$, $-x^*_4 = x^*_6$, and $x^*_5 = 0$. Choose the starting values x_k, $k = 1, 2, \ldots, 9$, so that similar conditions hold for them; then this will continue to be true on every iteration of Remez' method, and the odd-degree coefficients of $R_{5,2}(x)$ in step 2 will vanish on every iteration. Now apply the steps in Remez' method in the following way:

Step 1) Because of the symmetries mentioned above, one needs to use only x_5, x_6, x_7, x_8, and x_9. Using

$$x_k = \frac{1}{2}(b - a) \cos \frac{(m + n + 2 - k)\pi}{m + n + 1} + \frac{1}{2}(b + a),$$
$$k = 1, 2, \ldots, m + n + 2,$$

to determine starting values, one gets

$$x_k = \cos \frac{(9 - k)\pi}{8}, \quad k = 5, 6, 7, 8, 9.$$

Step 2) The system of nonlinear equations

$$p_0 + p_1 x_k + \ldots + p_m x_k^m + q_1 x_k [- F(x_k) - (-1)^k \mu]$$
$$+ q_2 x_k^2 [- F(x_k) - (-1)^k \mu]$$
$$+ \ldots + q_n x_k^n [- F(x_k) - (-1)^k \mu]$$
$$- (-1)^k \mu = F(x^k), \quad k = 1, 2, 3, \ldots, m + n + 2,$$

for the $m + n + 2$ unknowns $p_0, p_1, \ldots, p_m, q_1, q_2, \ldots, q_n$, and μ; becomes

$$p_0 + p_2 x_5^2 + p_4 x_5^4 + q_2 x_5^2 (- \cos \tfrac{1}{4}\pi x_5 + \mu) + \mu = \cos \tfrac{1}{4}\pi x_5$$

$$p_0 + p_2 x_6^2 + p_4 x_6^4 + q_2 x_6^2 (- \cos \tfrac{1}{4}\pi x_6 - \mu) - \mu = \cos \tfrac{1}{4}\pi x_6$$

48

$$p_0 + p_2 x_7^2 + p_4 x_7^4 + q_2 x_7^2 (-\cos \tfrac{1}{4}\pi x_7 + \mu) + \mu = \cos \tfrac{1}{4}\pi x_7$$

$$p_0 + p_2 x_8^2 + p_4 x_8^4 + q_2 x_8^2 (-\cos \tfrac{1}{4}\pi x_8 - \mu) - \mu = \cos \tfrac{1}{4}\pi x_8$$

$$p_0 + p_2 x_9^2 + p_4 x_9^4 + q_2 x_9^2 (-\cos \tfrac{1}{4}\pi x_9 + \mu) + \mu = \cos \tfrac{1}{4}\pi x_9.$$

On each iteration with Remez' method, this system is solved for p_0, p_2, p_4, q_2, and μ by the iterative method.

Step 3) Since the absolute-error function $R_{5,2}(x) - \cos \tfrac{1}{4}\pi x$ is an even function, it is sufficient to locate its extreme points in [0, 1], instead of [-1, 1]. The coefficients of $R_{5,2}(x)$ are the numbers obtained in step 2. Always have $y_5 = 0$ and $y_9 = 1$. To obtain y_6, y_7, and y_8, differentiate the error function and compute the zeros of the derivative. This can be done by isolating each y_k by searching and then computing it accurately by bisection. Since the starting values for the x_k's chosen in step 1 happen to be very good, y_k is close to x_k on every iteration and can easily be isolated by searching in the neighborhood of x_k.

Results after two iterations with Remez' method are shown below, with constants rounded to ten decimal places. After two iterations the numbers have stabilized to ten decimals.

$p_0 = 1.0000000241$ $q_2 = .0209610796$

$p_2 = - .2874648358$ $\mu = -.0000000241$

$p_4 = .0093933390$

● **PROBLEM 2-23**

Using a rational approximation method, approximate $\tanh \mu x$ for

$$-1 \leq x \leq 1,$$

where

$$\mu = \tfrac{1}{2} \ln 3.$$

Solution: First, find a rational approximation

$$R_{2,4}(x) \text{ to}$$

$$\frac{\tan \mu x}{x}$$

by Maehly's method and use $xR_{2,4}(x)$ as an approximation to $\tanh \mu x$. The object of using this procedure (instead of applying Maehly's method to $\tanh \mu x$ itself) is to obtain an approximation to tanh μx with low relative error around the origin, at which tanh μx is zero. Consider the equations

$$p_0 = q_0 c_0 + \frac{1}{2} \sum_{r=1}^{n} q_r c_r ,$$

$$p_k = q_0 c_k + \frac{1}{2} q_k c_0 + \frac{1}{2} \sum_{r=1}^{n} q_r (c_{r+k} + c_{|r-k|}) ,$$

$$k = 1, 2, \ldots, m + n. \qquad (1)$$

In this case use $m = 2$ and $n = 4$. The c_k's are the coefficients of the Chebyshev series for

$$\frac{\tanh \mu x}{x} ;$$

all the odd-numbered c_k's are zero because

$$\frac{\tanh \mu x}{x}$$

is an even function. It is not difficult to show that the unknowns p_1, q_1, and q_3 equal zero. If all the zero terms are omitted and one simplifies the equations, then (1) becomes

$$p_0 = q_0 c_0 + \tfrac{1}{2} q_2 c_2 + \tfrac{1}{2} q_4 c_4$$

$$p_2 = q_0 c_2 + q_2 (c_0 + \tfrac{1}{2} c_4) + q_4 (\tfrac{1}{2} c_2 + \tfrac{1}{2} c_6)$$

$$0 = q_0 c_4 + q_2 (\tfrac{1}{2} c_2 + \tfrac{1}{2} c_6) + q_4 (c_0 + \tfrac{1}{2} c_8)$$

$$0 = q_0 c_6 + q_2 (\tfrac{1}{2} c_4 + \tfrac{1}{2} c_8) + q_4 (\tfrac{1}{2} c_2 + \tfrac{1}{2} c_{10}).$$

Approximate numerical values for the c_k's appearing in the equations above can be computed by the method of Chebyshev series expansion. In order to obtain a particular solution of the equations, set $q_0 = 1$ and solve for p_0, p_2, q_2, and q_4, computing approximate numerical values for these unknowns. Then

$$P_2(x) = .52320016427\, T_0(x) + .00739002644\, T_2(x)$$

$$= .51581013784 + .01478005287 x^2$$

and

$$Q_4(x) = T_0(x) + .06107956977\, T_2(x) + .00010081226\, T_4(x)$$

$$= .93902124249 + .12135264149 x^2 + .00080649804 x^4,$$

where the coefficients are rounded to 11 decimal places. Then $R_{2,4}(x) = P_2(x)/Q_4(x)$ can be expressed in the form

$$R_{2,4}(x) = \frac{.54930614399 + .01573984933 x^2}{1 + .12923311636 x^2 + .00085887093 x^4},$$

where the numerator and denominator have been scaled so that the constant term in the denominator equals one and the other coefficients are rounded to 11 decimal places. The approximation to tanh μx for $-1 \leq x \leq 1$ is $x R_{2,4}(x)$.

CHAPTER 3

SERIES

INFINITE SERIES (TEST FOR CONVERGENCE AND DIVERGENCE)

• **PROBLEM 3-1**

Define the term series with computational considerations.

Solution: A series is a group of terms, often infinite, where the terms are formed according to some general rule. Series are very useful for numerical computations of such constants as e, π, etc.; and for the computations of terms such as log x, sin x, etc.

Note that a series can be used for computational purposes only if it converges, that is the only kind of series that can be used for computation is a convergent series. The reason for this is that if a series were not convergent but divergent, as one would add more and more terms, a continually different result would be obtained. With a convergent series, however, the series approaches a limit, and gives an ever more accurate result, the more terms added. Adding more terms does not change the result substantially after the first several terms, it only makes the result more accurate.

Assuming, then, that a series converges, it can be differentiated, integrated, added to other series, subtracted from other series, multiplied by a constant, etc.

For example, if the series for sin x is known, the series for cos x can be found by differentiating that series term by term, since it is known that

$$\frac{d}{dx} (\sin x) = \cos x$$

In a computation, once convergence is established, the number of terms to be used in the computation depends only on the accuracy desired.

• **PROBLEM 3-2**

Show that:

a) $\sum_{n=2}^{\infty} \frac{(\log n)^2}{n^3}$ is a convergent series.

b) $\sum_{n=2}^{\infty} \frac{1}{n(\log n)^p}$ is convergent if $p > 1$, divergent for $p \leq 1$.

<u>Solution</u>: a) To show

$$\sum_{n=2}^{\infty} \frac{(\log n)^2}{n^3}$$

is a convergent series apply the integral test. Hence, it must be shown that

$$\lim_{b \to \infty} \int_2^b \frac{(\log x)^2}{x^3} dx < \infty.$$

To do this, use the method of integration by parts. Therefore, take

$$dv = x^{-3} dx;$$

$$v = \frac{-1}{2x^2}, \quad u = (\log x)^2, \quad du = \frac{2(\log x)}{x} dx.$$

Then,

$$\int_2^b \frac{(\log x)^2}{x^3} dx = \frac{-(\log x)^2}{2x^2} \Big|_2^b + \int_2^b \frac{2 \log x}{2x^3} dx. \tag{1}$$

Now, applying integration by parts again, let

$$dv = x^{-3} dx;$$

$$v = \frac{-1}{2x^2}, \quad u = \log x, \quad du = \frac{dx}{x}.$$

Then (1) becomes

52

$$\int_2^b \frac{(\log x)^2}{x^3} dx = \left.\frac{-(\log x)^2}{2x^2}\right|_2^b - \left.\frac{\log x}{2x^2}\right|_2^b + \int_2^b \frac{dx}{2x^3}.$$

Hence

$$\lim_{b\to\infty} \int_2^b \frac{(\log x)^2}{x^3} dx$$

equals

$$\lim_{b\to\infty} \left(\frac{-(\log b)^2}{2b^2} - \frac{(\log b)}{2b^2} - \frac{1}{4b^2} + \frac{(\log 2)^2}{8} + \frac{(\log 2)}{8} + \frac{1}{16}\right)$$

Then, by L'Hospital's Rule the first three terms tend to 0 as $b\to\infty$.

Therefore

$$\lim_{b\to\infty} \int_2^b \frac{(\log x)^2}{x^3} dx = \frac{(\log 2)^2}{8} + \frac{(\log 2)}{8} = \frac{1}{16} < \infty$$

Thus, the series is convergent.

b) To show

$$\sum_{n=2}^{\infty} \frac{1}{n(\log n)^p}$$

is convergent if $p > 1$, divergent for $p \leq 1$, again apply the integral test. Therefore set up the integral

$$\int_2^b \frac{dx}{x(\log x)^p}.$$

This yields

$$\int_{\log 2}^{\log b} \frac{du}{u^p} = \left.\frac{u^{-p+1}}{-p+1}\right|_{\log 2}^{\log b}$$

Therefore,

$$\lim_{b\to\infty} \int_2^b \frac{dx}{x(\log x)^p} = \lim_{b\to\infty} \left(\frac{(\log b)^{-p+1}}{-p+1} - \frac{(\log 2)^{-p+1}}{-p+1}\right) \quad (p \neq 1).$$

If $p > 1$,

the first term goes to 0 as $b\to\infty$ and the integral equals

$$\frac{(\log 2)^{-p+1}}{p-1} < \infty.$$

Thus, the series is convergent.

If $p < 1$, the first term grows without bound as $b \to \infty$ and the integral is unbounded. Thus, the series is divergent. For $p = 1$,

$$\lim_{b \to \infty} \int_2^b \frac{dx}{x(\log x)^p} = \lim_{b \to \infty} \int_2^b \frac{dx}{x(\log x)} =$$

$$\lim_{b \to \infty} \log(\log b) - (\log(\log 2))$$

Thus, the series is divergent. This shows that the series is convergent for $p > 1$; and divergent for $p \leq 1$.

• **PROBLEM 3-3**

Determine if the series

$$\frac{1}{2} + \frac{1}{3} + \frac{1}{2^2} + \frac{1}{3^2} + \frac{1}{2^3} + \frac{1}{3^3} + \ldots$$

is convergent or divergent, using the ratio test and the root test.

Solution: The series can be rewritten as the sum of the sequence of numbers given by

$$a_n = \begin{cases} \dfrac{1}{2^{(n+1)/2}} & \text{if } n \text{ is odd } (n > 0) \\ \\ \dfrac{1}{3^{n/2}} & \text{if } n \text{ is even } (n > 0) \end{cases}$$

Now the ratio test states:

If $a_k > 0$ and

$$\lim_{k \to \infty} \frac{a_{k+1}}{a_k} = \ell < 1,$$

then

$$\sum_{k=1}^{\infty} a_k \quad \text{converges.}$$

Similarly, if
$$\lim_{k \to \infty} \frac{a_{k+1}}{a_k} = \ell \quad (1 < \ell \leq \infty)$$
then
$$\sum_{k=1}^{\infty} a_k \text{ diverges.}$$

If $\ell = 1$, the test fails. Therefore, applying this test gives:

If n is odd,

$$\lim_{n \to \infty} \frac{a_{n+1}}{a_n} = \lim_{n \to \infty} \frac{\frac{1}{3^{n/2}}}{\frac{1}{2^{(n+1)/2}}} = \lim_{n \to \infty} \frac{2^{(n+1)/2}}{3^{n/2}} = \lim_{n \to \infty} \left(\frac{2}{3}\right)^{n/2} 2^{\frac{1}{2}} = 0.$$

If n is even,

$$\lim_{n \to \infty} \frac{a_{n+1}}{a_n} = \lim_{n \to \infty} \frac{\frac{1}{2^{(n+1/2)}}}{\frac{1}{3^{n/2}}} = \lim_{n \to \infty} \left(\frac{3}{2}\right)^{n/2} 2^{\frac{1}{2}}$$

and no limit exists.

Hence, the ratio test gives two different values, one < 1 and the other > 1, therefore the test fails to determine if the series is convergent. Thus, another test, known as the root test is now applied. This test states:

Let

$$\sum_{k=1}^{\infty} a_k$$

be a series of nonnegative terms, and let

$$\lim_{n \to \infty} \left(\frac{n}{\sqrt{a_n}}\right) = S,$$

where

$$0 \leq S \leq \infty.$$

If:

1) $0 \leq S < 1$, the series converges
2) $1 < S \leq \infty$, the series diverges
3) $S = 1$, the series may converge or diverge.

Applying this test yields, if n is odd

$$\lim_{n\to\infty} \sqrt[n]{a_n} = \lim_{n\to\infty} \sqrt[n]{\frac{1}{2^{(n+1)/2}}} = \lim_{n\to\infty} \sqrt[n]{\frac{1}{2^{n/2}}} \sqrt[n]{\frac{1}{2^{1/2}}}$$

$$= \lim_{n\to\infty} \frac{1}{\sqrt{2}} \frac{1}{2^{1/2n}} = \frac{1}{\sqrt{2}} < 1 .$$

If n is even

$$\lim_{n\to\infty} \sqrt[n]{a_n} = \lim_{n\to\infty} \sqrt[n]{\frac{1}{3^{n/2}}} = \frac{1}{\sqrt{3}} < 1 .$$

Thus, since for both cases, the

$$\lim_{n\to\infty} \sqrt[n]{a_n} < 1 ,$$

the series converges.

• **PROBLEM 3-4**

Determine if the series:

a)

$$\sum_{k=1}^{\infty} \frac{(k+1)^{1/2}}{(k^5 + k^3 - 1)^{1/3}}$$

converges or diverges. Use the limit test for convergence.

b)

$$\sum_{k=1}^{\infty} \frac{k \log k}{7 + 11k - k^2}$$

converges or diverges. Use the limit test for divergence.

Solution: a) To determine if the given series converges or diverges, the following test called the limit test for convergence is used. This test states:

If

$$\lim_{k\to\infty} k^p U_k = A \text{ for } p > 1,$$

then

$$\sum_{k=1}^{\infty} U_k$$

converges absolutely.

To apply this test let $p = \frac{7}{6} > 1$. Then since

$$U_k = \frac{(k+1)^{1/2}}{(k^5 + k^3 - 1)^{1/3}},$$

$$\lim_{k\to\infty} k^p U_k = \lim_{k\to\infty} \frac{k^{7/6}(k+1)^{1/2}}{(k^5 + k^3 - 1)^{1/3}} = \lim_{k\to\infty} \frac{k^{7/6} k^{1/2}(1 + 1/k)^{1/2}}{(k^5 + k^3 - 1)^{1/3}}$$

$$= \lim_{k\to\infty} \frac{k^{5/3}(1 + 1/k)^{1/2}}{(k^5 + k^3 - 1)^{1/3}} = \lim_{k\to\infty} \frac{(1 + 1/k)^{1/2}}{\left[k^{-5}(k^5 + k^3 - 1)\right]^{1/3}}$$

$$= \lim_{k\to\infty} \frac{(1 + 1/k)^{1/2}}{(1 + 1/k^2 - 1/k^5)^{1/3}} = \frac{(1)^{1/2}}{(1)^{1/3}} = 1$$

Therefore, the series converges absolutely.

b) For the series

$$\sum_{k=1}^{\infty} \frac{k \log k}{7 + 11k - k^2},$$

the following test called the limit test for divergence is used. This test states, if

$$\lim_{k\to\infty} k U_k = A \neq 0 \text{ (or } \pm \infty) \qquad \text{then}$$

$$\sum_{k=1}^{\infty} U_k$$

diverges. If $A = 0$, the test fails.

Since

$$U_k = \frac{k \log k}{7 + 11k - k^2}$$

$$\lim_{k\to\infty} k U_k = \lim_{k\to\infty} \frac{k^2 \log k}{7 + 11k - k^2} = \lim_{k\to\infty} \frac{k^2 \log k}{k^2\left(\frac{7}{k^2} + \frac{11}{k} - 1\right)}$$

$$= \lim_{k\to\infty} \frac{\log k}{\frac{7}{k^2} + \frac{11}{k} - 1}.$$

As $k \to \infty$, $\log k \to \infty$

while

$$\frac{7}{k^2} + \frac{11}{k} - 1 \to -1.$$

Thus

$$\sum_{k=1}^{\infty} \frac{k \log k}{7 + 11k - k^2} \qquad \text{diverges.}$$

TAYLOR'S SERIES EXPANSION

• **PROBLEM** 3-5

> Find the Taylor series expansion of $f(x)$ about $x = x_0$ and show that a complete knowledge of the behavior of $f(x)$ at a point x_0 determines the value of the function at any point x at which the series converges.

<u>Solution</u>: Let $f(x)$ be a function with derivatives of all orders at $x = x_0$, and assume that in a neighborhood of x_0, $f(x)$ may be represented by the power series in $x - x_0$:

$$f(x) \equiv c_0 + c_1(x - x_0) + c_2(x - x_0)^2 + \ldots =$$

$$\sum_{n=0}^{\infty} c_n(x - x_0)^n. \tag{a}$$

The derivatives of $f(x)$ may be obtained by differentiating the series (a) term by term:

$$f'(x) = \sum_{n=1}^{\infty} n c_n (x - x_0)^{n-1} = c_1 + 2c_2(x - x_0) + 3c_3(x - x_0)^2 + \ldots, \tag{b}$$

$$f''(x) = \sum_{n=2}^{\infty} n(n-1) c_n (x - x_0)^{n-2} = 2 \cdot 1 c_2 + 3 \cdot 2 c_3 (x - x_0) + \ldots, \tag{c}$$

. . . .

$$f^{(m)}(x) = \sum_{n=m}^{\infty} n(n-1)(n-2) \ldots (n-m+1) c_n (x-x_0)^{n-m}$$

$$= \sum_{n=m}^{\infty} \frac{n!}{(n-m)!} c_n (x - x_0)^{n-m} \tag{d}$$

$$= m! c_m + (m+1)! c_{m+1} (x - x_0) + \ldots$$

Setting $x = x_0$ on both sides of (a), (b), (c), and (d), one finds

$$f(x_0) = c_0 \quad \text{or} \quad c_0 = f(x_0),$$
$$f'(x_0) = c_1 \quad \text{or} \quad c_1 = f'(x_0),$$
$$f''(x_0) = c_2 \quad \text{or} \quad c_2 = \frac{f''(x_0)}{2},$$

. . . .

$$f^{(m)}(x_0) = m! c_m \quad \text{or} \quad c_m = \frac{f^{(m)}(x_0)}{m!},$$

by means of which (a) becomes

$$f(x) = \sum_{n=0}^{\infty} \frac{f^{(n)}(x_0)}{n!} (x - x_0)^n, \tag{e}$$

where

$$f^{(n)}(x_0) \equiv \left. \frac{d^n f}{dx^n} \right|_{x=x_0},$$

$$f^{(0)}(x_0) \equiv f(x_0)$$

and $0! = 1$.

Equation (e) is the Taylor series expansion. Note also that by setting $x_0 = 0$ in (e), one gets the Maclaurin series expansion:

$$f(x) = \sum_{n=0}^{\infty} \frac{f^{(n)}(0)}{n!} x^n.$$

• **PROBLEM 3-6**

Evaluate $\sqrt{1+x}$ to two decimal places at $x = .4$ by Taylor series expansion. Consider the error expression:

$$|R_m| \leq \left| \frac{f^{(m)}(x_0)}{m!} (x - x_0)^m \right|,$$

the mth derivative is computed at x_0. $|R_m|$ is the truncation error after the mth term.

<u>Solution</u>: $\sqrt{1+x} = (1+x)^{1/2} = 1 + \frac{1}{2}x - \frac{1}{8}x^2 + \frac{1}{16}x^3 - \frac{5}{128}x^4 + \ldots$.

To evaluate $\sqrt{1 + .4}$ to two decimal places, at $x = .4$

$$1 + \frac{1}{2}(.4) - \frac{1}{8}(.4)^2 = 1.18 \quad [1.183];$$

$$|R_3| < \frac{(.4)^3}{16} = .004 \quad [.003].$$

• **PROBLEM 3-7**

Using the Taylor series expansion for e^x about $x = 0$, find $e^{0.5}$ to $\mathcal{O}(0.5)^3$. Bound the error using the error expression

$$\left| \frac{d^{n+1} f}{dx^{n+1}} \right|_{max} \cdot \left(\frac{(|x-a|)^{n+1}}{(n+1)!} \right) \quad (1)$$

(The subscript "max" denotes the maximum magnitude of the derivative on the interval from a to x and $\mathcal{O}$ means error order), and compare the result with the actual error.

59

Solution: From equation

$$f(x) = f(a) + (x-a)f'(a) + \frac{(x-a)^2}{2!}f''(a)$$
$$+ \frac{(x-a)^3}{3!}f'''(a) + \ldots + \frac{(x-a)^n}{n!}f^{(n)}(a) + \ldots$$

$$e^x = e^{(0)} + xe^{(0)} + \frac{x^2}{2!}e^{(0)} + \frac{x^3}{3!}e^{(0)} + \frac{x^4}{4!}e^{(0)} + \ldots$$

or

$$e^x = 1 + x + \frac{x^2}{2!} + \frac{x^3}{3!} + \frac{x^4}{4!} + \ldots$$

and if $x = 0.5$,

$$e^{0.5} = 1 + 0.5 + \frac{(0.5)^2}{2!} + \frac{(0.5)^3}{3!} + \frac{(0.5)^4}{4!} + \ldots$$

or, to $\mathcal{O}(0.5)^3$,

$$e^{0.5} = 1 + 0.5 + \frac{(0.5)^2}{2!} = 1.625$$

Now according to (1), the error in this quantity should not be greater than

$$\left|\frac{d^3(e^x)}{dt^3}\right|_{max} \frac{(0.5)^3}{3!} = |e^x|_{max} (0.0208333)$$

where max denotes the maximum magnitude on $0 \leq x \leq 0.5$.

$$|e^x|_{max} = e^{0.5} = 1.6487213,$$

so the error is no greater in magnitude than

$(1.6487213)(0.0208333) = 0.0343831$

The actual error is

$$e^{0.5} - 1.625 = 1.6487213 - 1.6250000 = 0.0237213$$

which lies within the error bound. Notice that in this case the error which was $\mathcal{O}(0.5)^3$ was actually 0.0237213 or about $0.19(0.5)^3$.

• **PROBLEM 3-8**

Show that $f(x) = (x - 1)^{\frac{1}{2}}$ cannot be expanded in a Taylor series about $x = 0$ or $x = 1$, but can be expanded about $x = 2$. Carry out the expansion about $x = 2$.

Solution:
$$f(x) = (x - 1)^{\frac{1}{2}}$$
$$f'(x) = \frac{1}{2}(x - 1)^{-\frac{1}{2}}$$
$$f''(x) = -\frac{1}{4}(x - 1)^{-3/2}$$
$$f'''(x) = \frac{3}{8}(x - 1)^{-5/2}$$

For an expansion about x = 0, the quantities f(0), f'(0), f"(0), etc. are needed. These involve noninteger powers of (-1), e.g. $(-1)^{1/2}$. These cannot be evaluated to give real values and thus this expansion is impossible.

For an expansion about x = 1, one needs quantities such as

$$f(1) = (0)^{1/2}$$

$$f'(1) = \frac{1}{2}(0)^{-1/2}$$

$$f''(1) = -\frac{1}{4}(0)^{-3/2}$$

While f(1) is bounded, all of the derivatives f'(1), f"(1), etc. are unbounded. Thus the expansion about x = 1 is impossible.

The Taylor series expansion about x = 2 is

$$f(x) = (2-1)^{1/2} + (x-2)\left[\frac{1}{2}(2-1)^{-1/2}\right]$$
$$+ \frac{(x-2)^2}{2}\left[-\frac{1}{4}(2-1)^{-3/2}\right] + \frac{(x-2)^3}{3!}\left[\frac{3}{8}(2-1)^{-5/2}\right] + \ldots$$

All derivatives are bounded and finite and the series is

$$f(x) = 1 + \frac{(x-2)}{2} - \frac{(x-2)^2}{8} + \frac{(x-2)^3}{16} - \ldots$$

This series is convergent for $|x-2| < 1$.

● **PROBLEM 3-9**

Write the Taylor expansion for ln x about the point x = 1. Use n terms.

Solution:

$f(x) = \ln x$ $f(1) = 0$

$f'(x) = 1/x$ $f'(1) = 1$

$f''(x) = -1/x^2$ $f''(1) = -1$

$f'''(x) = +2/x^3$ $f'''(1) = 2$

$f^{iv}(x) = -2 \times \frac{3}{x^4}$ $f^{iv}(1) = -2 \times 3$

$\vdots$

$f^{(n-1)}(x) = (-1)^n (n-2)!/x^{n-1}$ $f^{(n-1)}(1) = (-1)^n (n-2)!$

$$f^{(n)}(x) = (-1)^{n+1}(n-1)!/x^n \qquad f^{(n)}[1 + \theta(x-1)]$$
$$= (-1)^{n+1}(n-1)!/[1 + \theta(x-1)]$$

Substituting these values into the expression

$$f(x) = f(a) + (x-a)f'(a) + \frac{(x-a)^2}{2!}f''(a) + \ldots$$
$$+ \frac{(x-a)^{n-1}}{(n-1)!}f^{(n-1)}(x) + \frac{(x-a)^n}{n!}f^{(n)}[a + \theta(x-a)],$$
$$0 < \theta < 1,$$

and remembering that $a = 1$ in this case, one obtains

$$\ln x = (x-1) - \frac{1}{2!}(x-1)^2 + \frac{2}{3!}(x-1)^3 - \frac{2 \times 3}{4!}(x-1)^4 + \ldots$$
$$+ \frac{(-1)^n(n-2)!(x-1)^{n-1}}{(n-1)!} + \frac{(-1)^{n+1}(n-1)!(x-1)^n}{n![1 + \theta(x-1)]^n},$$
$$0 < \theta < 1$$

or, simplifying the expression,

$$\ln x = (x-1) - \frac{(x-1)^2}{2} + \frac{(x-1)^3}{3} - \frac{(x-1)^4}{4} + \ldots$$
$$+ \frac{(-1)^n(x-1)^{n-1}}{n-1} + \frac{(-1)^{n+1}(x-1)^n}{n(1-\theta+\theta x)^n},$$
$$0 < \theta < 1$$

• **PROBLEM 3-10**

(a) Derive the first three (non-zero) terms of Taylor's series expansion for the function $f(x) = \sin x$, about the origin. Since it is about the origin, this series is also called a Maclaurin series.

(b) Use the above result to compute the value of $\sin x$ for $x = 0.2$ rad, correctly to five decimal places.

Solution: (a) With the expansion about the origin the Taylor's series is:

$$f(x) = f(0) + xf'(0) + (x^2/2!)f''(0) + (x^3/3!)f^{(3)}(0) +$$
$$(x^4/4!)f^{(4)}(0) + (x^5/5!)f^{(5)}(0) + \ldots$$

For

$f(x) = \sin x$	$f(0) = 0$
$f'(x) = \cos x$	$f'(0) = 1$
$f''(x) = -\sin x$	$f''(0) = 0$
$f^{(3)}(x) = -\cos x$	$f^{(3)}(0) = -1$

$$f^{(4)}(x) = \sin x \qquad f^{(4)}(0) = 0$$
$$f^{(5)}(x) = \cos x \qquad f^{(5)}(0) = 1$$

Therefore,
$$\sin x = x - x^3/3! + x^5/5! + \ldots$$

(b) For $x = 0.2$ rad,
$$\sin 0.2 = 0.2 - (0.2)^3/6 + (0.2)^5/120 + \ldots$$
$$\sin 0.2 = 0.2 - 0.0013333 + 0.0000027 + \ldots$$

To five decimal places,
$$\sin 0.2 = 0.19867$$

It is noted that this is the correct answer for sin 0.2, to five decimal places. Actually, one can tell from the above computations that this answer is correct to five places by using the theorem which says that in a converging alternating series the error committed in stopping with any term is always less than the first term neglected. The first term omitted in the calculation above for sin 0.2 is the third term, and it contributes only 3 in the sixth place.

● **PROBLEM 3-11**

Obtain a first-degree polynomial approximation to
$$f(x) = (1 + x)^{1/2}$$
by means of the Taylor expansion about $x_0 = 0$.

Solution:
$$f'(x) = \frac{1}{2(1 + x)^{1/2}}$$
$$f''(x) = \frac{-1}{4[(1 + x)^{1/2}]^3}$$
$$f(0) = 1$$
$$f'(0) = \frac{1}{2}$$

Thus the Taylor expansion with remainder term is
$$(1 + x)^{1/2} = 1 + \frac{x}{2} - \frac{x^2}{8[(1 + \xi)^{1/2}]^3}$$

(Here ξ lies between 0 and x.) Thus the first-degree polynomial approximation to $(1 + x)^{1/2}$ is $1 + x/2$.

The accuracy of this approximation depends on what set of values x is allowed to assume. If $x \in [0, 1]$, for example,

63

one gets

$$\left|(1+x)^{1/2} - \left(1 + \frac{x}{2}\right)\right| \leq \frac{.1^2}{8[(1+\xi)^{1/2}]^3} \leq \frac{.1^2}{8} = .00125$$

● **PROBLEM 3-12**

Obtain a second-degree polynomial approximation to $f(x) = e^{-x^2}$ over $[0,.1]$ by means of the Taylor expansion about $x_0 = 0$. Use the expansion to approximate $f(.05)$, and bound the error.

Solution:

$$f(x) = e^{-x^2} \qquad\qquad f'(x) = -2xe^{-x^2}$$

$$f''(x) = (-2 + 4x^2)e^{-x^2} \qquad f'''(x) = (12x - 8x^3)e^{-x^2}$$

$$f(0) = 1 \qquad f'(0) = 0 \qquad f''(0) = -2$$

Thus
$$f(x) = 1 - x^2 + \frac{f'''(t)x^3}{6}$$

where
$$0 < t < .1$$

Using $f(x) \cong 1 - x^2$, one gets $f(.05) \cong .9975$. The truncation error is bounded by

$$|f(.05) - .9975| \leq \frac{(.05)^3}{6} \max_{t\epsilon I} |f'''(t)|$$

where $I = [0,.05]$. To obtain a bound on $|f'''(t)|$, use the result that, for $t \epsilon I$,

$$|f'''(t)| = |12 - 8t^2| \cdot |t| \cdot \left|e^{-t^2}\right|$$

$$< \max_{t\epsilon I} |12 - 8t^2| \cdot \max_{t\epsilon I} |t| \cdot \max_{t\epsilon I} \left|e^{-t^2}\right|$$

Hence

$$|f(0.5) - .9975| < \frac{(.05)^3}{6}(12)(.05)(1.0) = 1.25 \times 10^{-5}$$

● **PROBLEM 3-13**

Obtain an approximate value of

$$\frac{1}{2}\int_{-1}^{1} \frac{\sin x}{x}\,dx$$

by means of the Taylor expansion.

Solution: The expansion of $\sin x$ in a Taylor series about $x_0 = 0$ through terms of degree 6 is

$$\sin x = x - \frac{x^3}{3!} + \frac{x^5}{5!} - \frac{x^7}{7!} \cos \xi$$

$$\frac{\sin x}{x} = 1 - \frac{x^2}{6} + \frac{x^4}{120} - \frac{x^6}{7!} \cos \xi$$

Thus

$$\frac{1}{2} \int_{-1}^{1} \frac{\sin x}{x} dx = \frac{1}{2} \left(x - \frac{x^3}{18} + \frac{x^5}{600} \right) \Big|_{-1}^{1} - \frac{1}{2} \int_{-1}^{1} \frac{x^6}{7!} \cos \xi \, dx$$

This simplifies to

$$\frac{1}{2} \int_{-1}^{1} \frac{\sin x}{x} dx = \frac{1,703}{1,800} - \frac{1}{2} \int_{-1}^{1} \frac{x^6}{7!} \cos \xi \, dx$$

To bound the error, compute

$$\left| \frac{1}{2} \int_{-1}^{1} \frac{x^6}{7!} \cos \xi \, dx \right| \leq \frac{1}{2} \int_{-1}^{1} \frac{x^6}{7!} |\cos \xi| \, dx \leq$$

$$\frac{1}{2} \int_{-1}^{1} \frac{x^6}{7!} dx = \frac{1}{(7)(7!)}$$

Hence

$$\left| \frac{1}{2} \int_{-1}^{1} \frac{\sin x}{x} dx - \frac{1,703}{1,800} \right| \leq \frac{1}{(7)(7!)}$$

$$< 3.0 \times 10^{-5}$$

● **PROBLEM 3-14**

Find an approximation to the solution to $x = \cos x$ by use of Taylor's series.

Solution: The general idea here is that the equation

$$x = \cos x$$

is to be replaced by an equation of the form

65

$$x = 1 - \frac{x^2}{2!} + \frac{x^4}{4!} - \frac{x^6}{6!} \cdots + (-1)^n \frac{x^{2n}}{(2n)!} \qquad (1)$$

Taking $n = 1$, for example, solve

$$x = 1 - \frac{x^2}{2}$$

to obtain $x = -1 \pm 3^{1/2}$. Since the only solution to $x = \cos x$ lies in the interval $[0,1]$, one obtains

$$-1 + 3^{1/2}$$

as an approximate solution. This value can be used as an initial guess in a fixed-point iteration to solve the equation. The iteration might be performed on the original equation $x = \cos x$ or on the approximate equation (1) for some $n > 1$.

● **PROBLEM 3-15**

Find $\sin .1$, using Taylor's series and $t = 2$ floating-point arithmetic.

Solution:

$$\sin x = x - \frac{x^3}{3!} + \frac{x^5}{5!} - \frac{x^7}{7!} \cdots$$

Hence

$$\sin .1 = .1 - \frac{.1^3}{6} + \frac{.1^5}{120} - \frac{.1^7}{5{,}040} \cdots$$

Now in floating-point arithmetic

$$\frac{.1^3}{6} = \frac{(.10 \times 10^0)(.10 \times 10^0)(.10 \times 10^0)}{.60 \times 10^1}$$

$$= \frac{.10 \times 10^{-2}}{.60 \times 10^1} = .17 \times 10^{-3}$$

The sum of the first two terms of the series is

$$.10 \times 10^0 - .17 \times 10^{-3} = .10 \times 10^0$$

Thus, the second term and all subsequent terms of the series do not affect the result.

● **PROBLEM** 3-16

Expand the function cos x, in powers of x - a, where

$$a = -\frac{\pi}{4},$$

and determine the interval of convergence.

Solution: Expanding the given function in powers of x-a is equivalent to finding the Taylor Series for the function. To find the Taylor Series determine f(x), f(a), f'(x), f'(a), f"(x), f"(a), etc. One finds:

$$f(x) = \cos x; \qquad f(a) = f\left(-\frac{\pi}{4}\right) = \frac{\sqrt{2}}{2}.$$

$$f'(x) = -\sin x; \qquad f'(a) = f'\left(-\frac{\pi}{4}\right) = \frac{\sqrt{2}}{2}.$$

This is a positive value because the value of sin x in the 2nd quadrant is negative, therefore -sin x is positive.

$$f''(x) = -\cos x; \qquad f''(a) = f''\left(-\frac{\pi}{4}\right) = -\frac{\sqrt{2}}{2}.$$

$$f'''(x) = \sin x; \qquad f'''(a) = f'''\left(-\frac{\pi}{4}\right) = -\frac{\sqrt{2}}{2}.$$

Develop the series as follows:

$$f(x) = f(a) + f'(a)[x-a] + \frac{f''(a)}{2!}[x-a]^2$$

$$+ \frac{f'''(a)}{3!}[x-a]^3 + \ldots .$$

By substitution:

$$\cos x = \frac{\sqrt{2}}{2} + \frac{\sqrt{2}}{2}\left(x + \frac{\pi}{4}\right) - \frac{\frac{\sqrt{2}}{2}}{2!}\left(x + \frac{\pi}{4}\right)^2$$

$$- \frac{\frac{\sqrt{2}}{2}}{3!}\left(x + \frac{\pi}{4}\right)^3 + \ldots .$$

To determine the law of formation, examine the terms of this series. The nth term of the series is found to be:

$$\frac{\frac{\sqrt{2}}{2}}{(n-1)!}\left(x+\frac{\pi}{4}\right)^{n-1} ;$$

Then, the (n+1)th term is:

$$\frac{\frac{\sqrt{2}}{2}}{n!}\left(x+\frac{\pi}{4}\right)^{n} .$$

Therefore the Taylor series is:

$$\cos x = \frac{\sqrt{2}}{2} + \frac{\sqrt{2}}{2}\left(x+\frac{\pi}{4}\right) - \frac{\frac{\sqrt{2}}{2}}{2!}\left(x+\frac{\pi}{4}\right)^2 - \frac{\frac{\sqrt{2}}{2}}{3!}\left(x+\frac{\pi}{4}\right)^3 + \ldots$$

$$+ \frac{\frac{\sqrt{2}}{2}}{(n-1)!}\left(x+\frac{\pi}{4}\right)^{n-1} + \frac{\frac{\sqrt{2}}{2}}{n!}\left(x+\frac{\pi}{4}\right)^n + \ldots$$

To find the interval of convergence, use the Ratio Test. Set up the ratio

$$\frac{U_{n+1}}{U_n} ,$$

obtaining:

$$\frac{\sqrt{2}\left(x+\frac{\pi}{4}\right)^n}{2(n!)} \times \frac{2(n-1)!}{\sqrt{2}\left(x+\frac{\pi}{4}\right)^{n-1}} =$$

$$\frac{\left(x+\frac{\pi}{4}\right)^n}{n(n-1)!} \times \frac{(n-1)!}{\left(x+\frac{\pi}{4}\right)^{n-1}} = \frac{x+\frac{\pi}{4}}{n} .$$

Now, one finds

$$\lim_{n\to\infty}\left|\frac{x+\frac{\pi}{4}}{n}\right| = |0| = 0 .$$

By the ratio test it is known that if

$$\lim_{n\to\infty}\left|\frac{U_{n+1}}{U_n}\right| < 1$$

the series converges. Since $0 < 1$, the series converges for all values of x.

● **PROBLEM 3-17**

Give a Taylor expansion of $f(x,y) = e^x \cos y$ on some compact convex domain E containing $(0,0)$.

Solution: Assume $f(x,y) = \phi(x)\psi(y)$ where $\phi(x) = e^x$ and $\psi(y) = \cos y$. Also assume $E = E_x \times E_y$ where E_x and E_y are compact convex subsets of R, i.e., closed and bounded intervals. Note that

$$\sup_{x \in E_x} \left\| \frac{d^k \phi(x)}{dx^k} \right\| = e^{\max E_x}$$

and

$$\sup_{y \in E_y} \left\| \frac{d^k \psi(y)}{dy^k} \right\| \leq 1.$$

Hence $\phi(x)$ and $\psi(y)$ are real analytic on E. Furthermore, note that

$$e^x = \sum_{i=0}^{\infty} \frac{x^i}{i!} \left. \frac{d^i e^x}{dx^i} \right|_{x=0} = \sum_{i=0}^{\infty} \frac{x^i}{i!} \quad (1)$$

and

$$\cos y = \sum_{i=0}^{\infty} \frac{y^i}{i!} \left. \frac{d^i \cos y}{dy^i} \right|_{y=0} = \sum_{i=0}^{\infty} (-1)^i \frac{y^{2i}}{(2i)!} \quad (2)$$

Since $f = \phi\psi$ and ϕ and ψ are real analytic on E, f is real analytic on E:

$$\sup_{x,y \in E} \left\| \frac{\partial^k f(x,y)}{\partial x^i \partial y^{k-1}} \right\| \leq e^{\max E_x}.$$

hence, from (1) and (2)

$$e^x \cos y = \left(\sum_{i=0}^{\infty} \frac{x^i}{i!} \right) \left(\sum_{j=0}^{\infty} (-1)^j \frac{y^{2j}}{(2j)!} \right)$$

$$= \sum_{k=0}^{\infty} \left(\sum_{\substack{i+j=k \\ i,j \geq 0}} (-1)^j \frac{x^i y^{2j}}{i!(2j)!} \right).$$

The first three terms corresponding to k=0, k=1, and k=2 give the approximation:

$$e^x \cos y \approx 1 + x - \frac{1}{2} y^2 + \frac{1}{2} x^2 - \frac{1}{2} xy^2 + \frac{1}{24} y^4$$

which is known to be accurate near (0,0).

● **PROBLEM 3-18**

> Rewrite the polynomial
> $$\sum_{i=0}^{n} \alpha_i t^i$$
> as a polynomial in $x = t-1$. Verify this for the polynomial $1 + t + 3t^4$.

Solution: To rewrite the polynomial
$$f(t) = \sum_{i=0}^{n} \alpha_i t^i$$
as a polynomial in $x = t-1$, i.e., as
$$g(x) = \sum_{i=1}^{m} b_i x^i ,$$
use Taylor's Theorem. Note that f is C^∞ and that
$$f'(t) = \sum_{i=0}^{n} i\alpha_i t^{i-1} ,$$
$$f''(t) = \sum_{i=0}^{n} i(i-1) \alpha_i t^{i-2} , \ldots ,$$
$$f^{(n)}(t) = \sum_{i=0}^{n} i(i-1) \ldots (i-n+1)\alpha_i t^{i-n} ,$$
and
$$0 = f^{(n+1)}(t) = f^{(n+2)}(t) = \ldots .$$

To get a polynomial in $t-1$, expand about 1. Hence,
$$f(1) = \sum_{i=0}^{n} \alpha_i, \quad f'(1) = \sum_{i=1}^{n} i\alpha_i , \quad f''(1) = \sum_{i=2}^{n} i(i-1)\alpha_i, \ldots ,$$
$$f^{(k)}(1) = \sum_{i=k}^{n} i(i-1) \ldots (i-k+1)\alpha_i , \ldots ,$$
$$f^{(n)}(1) = n!\alpha_n .$$

Therefore, the Taylor expansion

$$f(t) = \sum_{j=0}^{n} \frac{f^{(j)}(1)}{j!} (t-1)^j$$

$$= \sum_{j=0}^{n} \frac{1}{j!} \sum_{i=j}^{n} i(i-1) \cdots (i-j+1) \alpha_i (t-1)^j$$

$$= \sum_{j=0}^{n} \frac{1}{j!} \sum_{i=j}^{n} \frac{i!}{(i-j)!} \alpha_i (t-1)^j$$

$$= \sum_{j=0}^{n} \left(\sum_{i=j}^{n} \frac{i!}{(i-j)!j!} \alpha_i \right) (t-1)^j$$

$$= \sum_{j=0}^{n} \left(\sum_{i=j}^{n} \binom{i}{j} \alpha_i \right) (t-1)^j$$

$$= \sum_{j=0}^{m} b_j x_j = g(x) .$$

Note that the degree of the new polynomial (1) is also n and the j-th coefficient is in terms of the last n-j+1 of the α_i.

For $f(t) = 1 + t + 3t^4$, according to this method,

$$\alpha_0 = \alpha_1 = 1, \alpha_2 = \alpha_3 = 0$$

and

$$\alpha_4 = 3.$$

Hence

$$f(t) = \sum_{j=0}^{4} \left[\sum_{i=j}^{4} \frac{i!}{(i-j)!j!} \alpha_i \right] (t-1)^j$$

$$= \left[\frac{0!}{0!0!} \alpha_0 + \frac{1!}{1!0!} \alpha_1 + \frac{2!}{2!0!} \alpha_2 + \frac{3!}{3!0!} \alpha_3 + \frac{4!}{4!0!} \alpha_4 \right] (t-1)^0$$

$$(t-1)^0 +$$

$$+ \left[\frac{1!}{0!1!} \alpha_1 + \frac{2!}{1!1!} \alpha_2 + \frac{3!}{2!1!} \alpha_3 + \frac{4!}{3!1!} \alpha_4 \right] (t-1)^1$$

$$+ \left[\frac{2!}{0!2!} \alpha_2 + \frac{3!}{1!2!} \alpha_3 + \frac{4!}{2!2!} \alpha_4 \right] (t-1)^2 + \left[\frac{3!}{0!3!} \alpha_3 + \right.$$

$$\left. \frac{4!}{1!3!} \alpha_4 \right] (t-1)^3 + \frac{4!}{0!4!} \alpha_4 (t-1)^4$$

$$= (1+1+0+0+3) + (1+0+0+12)(t-1) + (0+0+18)(t-1)^2$$

$$+ (0+12)(t-1)^3 + 3(t-1)^4$$

$$= 5 + 13(t-1) + 18(t-1)^2 + 12(t-1)^3 + 3(t-1)^4. \qquad (2)$$

To show that (2) is actually equal to f(t), expand:

$$f(t) = 5 + 13(t-1) + 18(t^2-2t+1) + 12(t^3-3t^2+3t-1)$$

$$+ 3(t^4-4t^3+6t^2-4t+1)$$

$$= (5-13+18-12+3) + t(13-36+36-12) + t^2(18-36+18)$$

$$+ t^3(12-12) + t^4(3)$$

$$= 1 + t + 3t^4.$$

So, indeed,

$$1 + t + 3t^4 = 5 + 13x + 18x^2 + 12x^3 + 3x^4$$

where

$$x = t - 1.$$

● **PROBLEM 3-19**

A. Find the Taylor series expansion of $f(x,y)$ about x_0, y_0.

B. Expand $f(x,y) = \cos xy$ about $x_0 = \pi/2$, $y_0 = \frac{1}{2}$ up to second-order terms.

Solution: A. The Taylor series of a function $z = f(x,y)$ of two independent variables x, y, about the point (x_0, y_0) is obtained from

$$f(x_0 + x) = \sum_{n=0}^{\infty} \frac{f^{(n)}(x_0)}{n!} x^n \qquad (a)$$

by incrementing one of the variables at a time. Indicating the partial derivatives of $z(x,y)$ by

$$f_x = \frac{\partial z}{\partial x}, \quad f_y = \frac{\partial z}{\partial y}, \quad f_{xx} = \frac{\partial^2 z}{\partial x^2}, \quad f_{xy} = \frac{\partial^2 z}{\partial x \, \partial y}$$

$$f_{yy} = \frac{\partial^2 z}{\partial y^2}, \quad \ldots, \qquad (b)$$

(a) applied to $f(x_0 + x, y_0)$, i.e., to the function f considered as a function of x only while y_0 is kept constant, gives

$$f(x_0 + x, y_0) = f(x_0,y_0) + f_x(x_0,y_0)\frac{x}{1!} + f_{xx}(x_0,y_0)\frac{x^2}{2!} + \ldots,$$

from which, changing y_0 into $y_0 + y$ on both sides of the equation,

$$f(x_0 + x, y_0 + y) = f(x_0, y_0 + y) + f_x(x_0, y_0 + y)\frac{x}{1!}$$

$$+ f_{xx}(x_0, y_0 + y)\frac{x^2}{2!} + \ldots . \qquad (c)$$

Expanding by means of (a) the coefficient of each power of x in (c) into a power series in y about y_0, while x_0 is kept constant, yields

$$f(x_0, y_0 + y) = f(x_0,y_0) + f_y(x_0,y_0)\frac{y}{1!} + f_{yy}(x_0,y_0)\frac{y^2}{2!} + \ldots,$$

$$f_x(x_0, y_0 + y) = f_x(x_0,y_0) + f_{xy}(x_0,y_0)\frac{y}{1!} + f_{xyy}(x_0,y_0)\frac{y^2}{2!}$$

$$+ \ldots, \qquad (d)$$

$$f_{xx}(x_0, y_0 + y) = f_{xx}(x_0,y_0) + f_{xxy}(x_0,y_0)\frac{y}{1!}$$

$$+ f_{xxyy}(x_0,y_0)\frac{y^2}{2!} + \ldots$$

. . . .

Substituting (d) in (c), the Taylor series expansion of $f(x,y)$ about x_0, y_0 becomes

$$f(x_0 + x, y_0 + y)$$

$$= f(x_0,y_0) + [f_x(x_0,y_0)x + f_y(x_0,y_0)y]$$

$$+ \frac{1}{2!}[f_{xx}(x_0,y_0)x^2 + 2f_{xy}(x_0,y_0)xy + f_{yy}(x_0,y_0)y^2]$$

$$+ \frac{1}{3!}[f_{xxx}(x_0,y_0)x^3 + 3f_{xxy}(x_0,y_0)x^2y$$

$$+ 3f_{xyy}(x_0,y_0)xy^2 + f_{yyy}(x_0,y_0)y^3] + \ldots .$$

B. Here

$$f_x = -y \sin xy; \quad f_y = -x \sin xy; \quad f_{xx} = -y^2 \cos xy;$$

$$f_{xy} = -\sin xy - xy \cos xy; \quad f_{yy} = -x^2 \cos xy;$$

$$f\left(\frac{\pi}{2}, \frac{1}{2}\right) = \frac{\sqrt{2}}{2};$$

$$f_x\left(\frac{\pi}{2}, \frac{1}{2}\right) = -\frac{1}{2}\frac{\sqrt{2}}{2}; \quad f_y\left(\frac{\pi}{2}, \frac{1}{2}\right) = -\frac{\pi}{2}\frac{\sqrt{2}}{2};$$

$$f_{xx}\left(\frac{\pi}{2}, \frac{1}{2}\right) = -\frac{1}{4}\frac{\sqrt{2}}{2}; \quad f_{xy}\left(\frac{\pi}{2}, \frac{1}{2}\right) = -\frac{\sqrt{2}}{2}\left(1 + \frac{\pi}{4}\right);$$

$$f_{yy}\left(\frac{\pi}{2}, \frac{1}{2}\right) = -\frac{\pi^2}{4}\frac{\sqrt{2}}{2};$$

$$f\left(\frac{\pi}{2} + x, \frac{1}{2} + y\right)$$

$$= \frac{\sqrt{2}}{2}\left\{1 - \left(\frac{1}{2}x + \frac{\pi}{2}y\right) - \left[\frac{1}{8}x^2 + \left(1 + \frac{\pi}{4}\right)xy \right.\right.$$

$$\left.\left. + \frac{\pi^2}{8}y^2\right]\right\}.$$

MACLAURIN'S SERIES EXPANSION

• **PROBLEM 3-20**

Write the Maclaurin expansion for e^x.

Solution:

$$f(x) = e^x \qquad f(0) = 1$$
$$f'(x) = e^x \qquad f'(0) = 1$$
$$f''(x) = e^x \qquad f''(0) = 1$$
$$f'''(x) = e^x \qquad f'''(0) = 1$$
$$f^{iv}(x) = e^x \qquad f^{iv}(0) = 1$$
$$\vdots$$
$$f^{(n-1)}(x) = e^x \qquad f^{(n-1)}(0) = 1$$

$$f^{(n)}(x) = e^x \qquad\qquad f^{(n)}(\theta x) = e^{\theta x}$$

Substituting these quantities in the expression for the Maclaurin series:

$$f(x) = f(0) + xf'(0) + \frac{x^2}{2!} f''(0) + \ldots$$

$$+ \frac{x^{n-1}}{(n-1)!} f^{(n-1)}(0) + \frac{x^n}{n!} f^{(n)}(\theta x),$$

$$0 < \theta < 1$$

obtain

$$e^x = 1 + x + \frac{x^2}{2!} + \frac{x^3}{3!} + \ldots + \frac{x^{n-1}}{(n-1)!} + \frac{x^n}{n!} e^{\theta x}, \quad 0 < \theta < 1$$

● **PROBLEM 3-21**

Evaluate $1/2 e^{0.25}$ correct to five decimal places by use of the Maclaurin series for xe^{-x^2}.

Solution: The successive derivatives of this function soon become very unwieldy; hence it is desirable to compute the coefficients of the power series and to estimate the magnitude of the error committed by stopping at a certain term by other means. If in series

$$e^x = 1 + \frac{x}{1!} + \frac{x^2}{2!} + \ldots + \frac{x^n}{n!} + \ldots ,$$

$$R = \infty.$$

$-x^2$ replaces x and then multiplying through by x, yields

$$xe^{-x^2} = x - \frac{x^3}{1!} + \frac{x^5}{2!} - \frac{x^7}{3!} \pm \ldots$$

$$+ (-1)^n \frac{x^{2n+1}}{n!} + \ldots .$$

This is an alternating series and it can be shown that six terms are sufficient to yeild, for

$$x = \tfrac{1}{2}, \quad 1/2 e^{0.25} = 0.38940,$$

a value which is correct to five decimal places.

75

• **PROBLEM 3-22**

Find the Maclaurin series for the function:

$$\frac{1}{2}(e^x + e^{-x}),$$

and the interval of convergence.

Solution: Let

$$f(x) = \frac{1}{2}(e^x + e^{-x}).$$

To find the Maclaurin series, determine $f(0)$, $f'(x)$, $f'(0)$, $f''(x)$, $f''(0)$, etc. One finds:

$f(x) = \frac{1}{2}(e^x + e^{-x})$ $f(0) = 1$

$f'(x) = \frac{1}{2}(e^x - e^{-x})$ $f'(0) = 0$

$f''(x) = \frac{1}{2}(e^x + e^{-x})$ $f''(0) = 1$

$f'''(x) = \frac{1}{2}(e^x - e^{-x})$ $f'''(0) = 0$

$f^4(x) = \frac{1}{2}(e^x + e^{-x})$ $f^4(0) = 1$, etc.

Now, the series is developed as follows:

$$f(x) = f(0) + f'(0)x + \frac{f''(0)}{2!}x^2 + \frac{f'''(0)}{3!}x^3 + \ldots .$$

By substitution:

$$\frac{1}{2}(e^x + e^{-x}) = 1 + 0 + \frac{x^2}{2!} + 0 + \frac{x^4}{4!} + \ldots$$

$$= 1 + \frac{x^2}{2!} + \frac{x^4}{4!} + \ldots .$$

To determine the law of formation examine the terms of this series. The nth term of the series is found to be:

$$\frac{x^{2n-2}}{(2n-2)!};$$

Then, the (n+1)th term is:

$$\frac{x^{2n}}{(2n)!}.$$

Therefore the Maclaurin series is:

$$\frac{1}{2}(e^x + e^{-x}) = 1 + \frac{x^2}{2!} + \frac{x^4}{4!} + \ldots + \frac{x^{2n-2}}{(2n-2)!} + \frac{x^{2n}}{(2n)!}.$$

To find the interval of convergence use the ratio test. Set up the ratio

$$\frac{u_{n+1}}{u_n},$$

obtaining:

$$\frac{x^{2n}}{(2n)!} \cdot \frac{(2n-2)!}{x^{2n-2}} = \frac{x^{2n}}{(2n)(2n-2)!} \cdot \frac{(2n-2)!}{x^{2n-2}} = \frac{x^2}{2n}.$$

One finds:

$$\lim_{n \to \infty} \left| \frac{x^2}{2n} \right| = |0| = 0.$$

By the ratio test it is known that, if

$$\lim_{n \to \infty} \left| \frac{u_{n+1}}{u_n} \right| < 1,$$

the series converges. Since 0 is always less than 1, the series converges for all values of x.

● **PROBLEM 3-23**

Find the Maclaurin series and the interval of convergence for the function $f(x) = \cos x$.

Solution: To find the Maclaurin series for the given function, determine $f(0)$, $f'(x)$, $f'(0)$, $f''(x)$, $f''(0)$, etc.
One finds:

$f(x) = \cos x$ $f(0) = 1$

$f'(x) = -\sin x$ $f'(0) = 0$

$f''(x) = -\cos x$ $f''(0) = -1$

$f'''(x) = \sin x$ $f'''(0) = 0$

$f^4(x) = \cos x$ $f^4(0) = 1$

$$f^5(x) = -\sin x \qquad f^5(0) = 0$$

$$f^6(x) = -\cos x \qquad f^6(0) = -1.$$

Develop the series as follows:

$$f(x) = f(0) + f'(0)x + \frac{f''(0)}{2!}x^2 + \frac{f'''(0)}{3!}x^3 + \frac{f^4(0)}{4!}x^4$$

$$+ \frac{f^5(0)}{5!}x^5 + \frac{f^6(0)}{6!}x^6 + \ldots$$

By substitution:

$$\cos x = 1 + 0 - \frac{x^2}{2!} + 0 + \frac{x^4}{4!} + 0 - \frac{x^6}{6!} + \ldots$$

$$= 1 - \frac{x^2}{2!} + \frac{x^4}{4!} - \frac{x^6}{6!} + \ldots .$$

To determine the law of formation, examine the terms of this series. The nth term of the series is found to be

$$\frac{x^{2n-2}}{(2n-2)!} .$$

Then the (n+1)th term is

$$\frac{x^{2n}}{(2n)!} .$$

Therefore, the Maclaurin series is:

$$\cos x = 1 - \frac{x^2}{2!} + \frac{x^4}{4!} - \frac{x^6}{6!} + \ldots \pm \frac{x^{2n-2}}{(2n-2)!}$$

$$\pm \frac{x^{2n}}{(2n)!} \ldots .$$

To find the interval of convergence use the ratio test. Set up the ratio

$$\frac{u_{n+1}}{u_n} ,$$

obtaining:

$$\frac{x^{2n}}{(2n)!} \times \frac{(2n-2)!}{x^{2n-2}} = \frac{x^{2n}}{(2n)(2n-2)!} \times \frac{(2n-2)!}{x^{2n-2}} = \frac{x^2}{2n} .$$

Now, one finds

$$\lim_{n \to \infty} \left| \frac{x^2}{2n} \right| = |0| = 0.$$

By the ratio test it is known that if

$$\lim_{n \to \infty} \left| \frac{u_{n+1}}{u_n} \right| < 1$$

the series converges.

Since 0 is always less than 1, the series converges for all values of x.

● **PROBLEM 3-24**

Using the Maclaurin expansion, find the value of the sine function to eight significant figures.

Solution: The remainder term in the Maclaurin expansion of the sine function

$$\sin x = x - \frac{x^3}{3!} + \frac{x^5}{5!} + \ldots, \qquad 0 < \theta < 1$$

is

$$(-1)^p x^{2p+1} \cos \theta x / (2p + 1)!,$$

where p is the number of terms required. If one desires to compute the sine function to eight significant figures, the relative error would be less than $1/(2 \times 10^8)$. Then, note that $\cos \theta x < 1$, so it is required that

$$\frac{1}{\sin x} \frac{x^{2p+1}}{(2p + 1)!} < \frac{1}{2 \times 10^8}$$

The value of p, the number of terms required, depends on how large a value of x must be accommodated. Because of the periodicity of the sine function, it is certainly sufficient to consider only $x < 2\pi$. Also, since $\sin(\pi + x) = -\sin x$, values of x between π and 2π can be replaced by values between 0 and π. Further, since $\sin(\pi - x) = \sin x$, angles between $\pi/2$ and π can be replaced by angles between 0 and $\pi/2$. For values of x between 0 and $\pi/2$, the sin x in the denominator is no problem, since $1 \leq x/(\sin x) \leq \pi/2$ for x in this range. The largest value of the left-hand side of the inequality occurs when $x = \pi/2$. For this value of x, the inequality is satisfied by $2p + 1 = 15$, or $p = 7$. Thus the value of the sine function will be given in eight significant figures by the expression

$$\sin x = x - \frac{x^3}{3!} + \frac{x^5}{5!} - \frac{x^7}{7!} + \frac{x^9}{9!} - \frac{x^{11}}{11!} + \frac{x^{13}}{13!}$$

or, in a form which can be evaluated with a minimum of multiplication.

$$\sin x = x\left(1 - x^2\left(\frac{1}{3!} - x^2\left(\frac{1}{5!} - x^2\left(\frac{1}{7!} - x^2\left(\frac{1}{9!} - x^2\left(\frac{1}{11!} - \frac{x^2}{13!}\right)\right)\right)\right)\right)\right)$$

A pair of FORTRAN statements which will perform this calculation are

```
U = X*Y
Y = -((((((U*A(13) - A(11))*U - A(9))*U - A(7))*U -
    A(5))*U -A(3))*U-A(1))*X
```

POWER SERIES EXPANSION

• **PROBLEM** 3-25

Compute ln 2 correct to five decimal places by putting $x = \frac{1}{2}$ in the power series

$$\ln \frac{1}{1-x} = x + \frac{x^2}{2} + \frac{x^3}{3} + \frac{x^4}{4} + \ldots + \frac{x^n}{n} + \ldots,$$

$$R = 1.$$

Solution: A simple calculation shows that

$$f^{(n+1)}(x) = n!/(1-x)^{n+1}.$$

Hence, the expression

$$E(x) = \frac{(x - x_0)^{n+1}}{(n + 1)!} f^{(n+1)}(X),$$

for the error term becomes (x_0 is again 0),

$$\frac{x^{n+1}}{(n+1)!} \frac{n!}{(1-X)^{n+1}} = \frac{x^{n+1}}{(n+1)(1-X)^{n+1}}.$$

We have $x = \frac{1}{2}$; further, X must be chosen between 0 and $\frac{1}{2}$ so that the magnitude of the error is as large as possible, hence X is also $\frac{1}{2}$. The error term consequently reduces to

$1/(n+1)$ and the first n for which $1/(n+1) < 0.000005$ must be found. The first n is n = 200,000.

On the other hand,

$$E_n = \ln \frac{1}{1-x} - \sum_{i=1}^{n} \frac{x^i}{i} = \frac{x^{n+1}}{n+1} + \frac{x^{n+2}}{n+2} + \cdots$$

$$< \frac{x^{n+1}}{n+1} + \frac{x^{n+2}}{n+1} + \cdots$$

$$= \frac{x^{n+1}}{n+1} (1 + x + x^2 + \cdots)$$

$$= \frac{x^{n+1}}{n+1} \frac{1}{1-x}.$$

If $x = \frac{1}{2}$, the last expression becomes $1/2^n(n+1)$ and the first n for which this is less than 0.000005 must now be found. This time, n = 14. This estimate of the error gives a far better result in the shape of a much smaller n than the previous one.

Taking the first fourteen terms of the series, one finds ln 2 = 0.693143^+.

If the error committed in neglecting terms from $x^{15}/15$ onward is 0.000002^-, or less, the value of ln 2 is 0.69314; but if the error is between 0.000002 and 0.000005, then ln 2 is 0.69315. Then one or two terms more of the series must be computed. Since $1/15 \cdot 2^{15} = 0.00002^+$, ln 2 = 0.69315, correct to five decimal places.

● **PROBLEM 3-26**

Compute sin 40° correct to five significant figures from

$$\sin x = x - \frac{x^3}{3!} + \frac{x^5}{5!} \pm \cdots + (-1)^n \frac{x^{2n+1}}{(2n+1)!} + \cdots ;$$

$$R = \infty. \tag{1}$$

Solution: Equation (1) yields 40° = $2\pi/9$ = 0.6981317 rad. Thus, using

$$E(x) = \frac{(x-x_0)^{n+1}}{(n+1)!} f^{(n+1)}(x)$$

for the error, since x = 0.7 (approximately), $x_0 = 0$,

$f^{(n+1)}(X) = 1$ (at most),

$$\frac{(0.7)^{n+1}}{(n+1)!} < 0.000005.$$

The first positive integral n for which this inequality holds (obtained by trial and error) is n = 7. Hence, a polynomial of max-degree 7 is necessary to attain the desired precision. On the other hand, using

$$|E_n| = \left| A - \sum_{i=0}^{n} (-1)^i a_i \right| \le |a_{n+1}|$$

to estimate the error,

$$\frac{(0.7)^{2n+1}}{(2n+1)!} < 0.000005,$$

whence n = 4. That is, the first four terms of (1) are sufficient to yield sin 40° correct to five significant figures; a result equivalent to the preceding one. We have

$$\sin 40° \quad 0.6981317 - \frac{(0.6981317)^3}{3!} + \frac{(0.6981317)^5}{5!} - \frac{(0.6981317)^7}{7!}$$

$$= 0.6427875,$$

or, to five significant figures,

$$\sin 40° = 0.64279.$$

The symbol $\approx$ is read "is approximately equal to."

• **PROBLEM 3-27**

Compute, with at most 1 percent error, $f'''(10)$, where $f(x) = (x^3 + 1)^{-1/2}$.

Solution: x is large, but x^{-1} is fairly small. Expand in powers of x^{-1}:

$$f(x) = (x^3 + 1)^{-1/2} = x^{-3/2}(1 + x^{-3})^{-1/2}$$

$$= x^{-1.5}\left(1 - 0.5 \cdot x^{-3} + \frac{0.5 \cdot 1.5}{2} x^{-6} - \ldots\right)$$

$$= x^{-1.5} - \frac{1}{2} x^{-4.5} + \frac{3}{8} x^{-7.5} - \ldots$$

Differentiate three times:

$$f'''(x) = - x^{-4.5} \left(\frac{105}{8} - \frac{1,287}{16} x^{-3} + \ldots \right).$$

For $x = 10$ the second term is less than 1 percent of the first; the terms after the second decrease quickly and are negligible. One can show that the magnitude of each term is less than $8 \cdot x^{-3}$ of the previous term. Hence one gets $f'''(10) = -4.14 \cdot 10^{-4}$ to the desired accuracy.

● **PROBLEM 3-28**

Using power series representation, find the solution of $y'' = x + y^2$ which passes through the point $(0,1)$.

Solution: Assume the solution can be expressed as a power series

$$y = a_0 + a_1 x + a_2 x^2 + \ldots + a_n x^n + \ldots ,$$

which converges in some interval about $x_0 = 0$. Then

$$y' = a_1 + 2a_2 x + \ldots + n a_n x^{n-1} + (n+1) a_{n+1} x^n + \ldots ,$$

and

$$y^2 = a_0^2 + (a_0 a_1 + a_1 a_0) x + (a_0 a_2 + a_1 a_1 + a_2 a_0) x^2$$

$$+ \ldots + (a_0 a_n + a_1 a_{n-1} + \ldots + a_n a_0) x^n + \ldots .$$

Therefore,

$$a_1 + 2a_2 x + \ldots + (n+1) a_{n+1} x^n + \ldots = a_0^2 + (a_0 a_1 + a_1 a_0 + 1) x$$

$$+ (a_0 a_2 + a_1 a_1 + a_2 a_0) x^2$$

$$+ \ldots + (a_0 a_n + \ldots + a_n a_0) x^n$$

$$+ \ldots .$$

Equate the coefficients of like powers of x and obtain the following system of simultaneous equations:

$$a_1 = a_0^2$$
$$2a_2 = 2a_0 a_1 + 1$$
$$3a_3 = 2a_0 a_2 + a_1^2$$

$$\cdots\cdots\cdots\cdots\cdots\cdots\cdots\cdots\cdots\cdots\cdots\cdots$$

$$(n+1)a_{n+1} = a_0 a_n + a_1 a_{n-1} + \ldots + a_n a_0 .$$

Consequently,

$$a_1 = a_0^2$$
$$a_2 = a_0^3 + \frac{1}{2}$$
$$a_3 = a_0^4 + \frac{1}{3} a_0$$
$$a_4 = a_0^5 + \frac{5}{12} a_0^2$$
$$a_5 = a_0^6 + \frac{1}{2} a_0^3 + \frac{1}{20} ,$$

$\cdots\cdots\cdots\cdots\cdots$

Substituting these values for a_1, a_2, a_3, ..., into the power series for y gives a general solution of the differential equation. This might have been anticipated because no use has been made as yet of the point (0,1) through which the graph of the solution must pass. If this point is used, one finds, since $x = 0$, and $y = a_0 = 1$, that $a_1 = 1$, $a_2 = 3/2$, $a_3 = 4/3$, $a_4 = 17/12$, $a_5 = 31/20$, ... Consequently,

$$y = 1 + x + \frac{3}{2} x^2 + \frac{4}{3} x^3 + \frac{17}{12} x^4 + \frac{31}{20} x^5 + \ldots .$$

One can obtain the same result somewhat differently, and perhaps more briefly, recalling that

$$n! a_n = y_0^{(n)} = y^{(n)}(x_0) = \frac{d^n f(x_0)}{dx^n} ,$$

where, to repeat, $d^n f(x_0)/dx^n$ is the nth derivative of f(x) evaluated at x_0. By successive differentiation (with respect to x) of the given differential equation, $y' = x + y^2$,

$$y'' = 1 + 2yy',$$

$$y^{(3)} = 2(yy'' + Y'^2),$$

$$y^{(4)} = 2(yy^{(3)} + 3y'y''),$$

$$y^{(5)} = 2(yy^{(4)} + 4y'y^{(3)} + 3y''^2),$$

$$\ldots\ldots\ldots\ldots\ldots\ldots\ldots\ldots\ldots\ldots;$$

whence, at

$(0,1)$, $y' = 1$

and

$y'' = 3$, $y^{(3)} = 8$, $y^{(4)} = 34$, $y^{(5)} = 186$,

... .

Hence,

$a_0 = 1$, $a_1 = 1$, $a_2 = 3/2$, $a_3 = 4/3$, $a_4 = 17/12$, $a_5 = 31/20$

..., as before.

● **PROBLEM 3-29**

Find the radius of convergence of the power series

$$\sum_{n=1}^{\infty} \frac{x^n}{n^2}.$$

Then determine if the convergence is uniform for

$-R \leq x \leq R$.

Solution: Here the ratio gives

$$\lim_{n \to \infty} \left| \frac{U_{n+1}}{U_n} \right| = \lim_{n \to \infty} |x| \frac{n^2}{(n+1)^2} = |x|,$$

so that $R = 1$. That is, the series converges absolutely for $-1 < x < 1$ and diverges for $|x| > 1$. Note that this result could have also been found by the relation,

$$R = \frac{1}{\alpha}$$

where

$$\alpha = \lim_{n \to \infty} \sup \sqrt[n]{|a_n|}$$

where

$$a_n = \frac{1}{n^2}$$

which yields

$$\alpha = \lim_{n \to \infty} \sup \sqrt[n]{|1/n^2|} = \lim_{n \to \infty} \sup \sqrt[n]{\frac{1}{n^2}}$$

$$= \lim_{n \to \infty} \sup \frac{1}{n^{2/n}} = \lim_{n \to \infty} \sup \frac{1}{e^{(2/n) \log n}}$$

$$= \frac{1}{e^{\lim_{n \to \infty} \sup (2/n) \log n}} = \frac{1}{e^0} = 1,$$

therefore,

$$\frac{1}{R} = 1,$$

so that as before $R = 1$. For $x = \pm 1$ the series converges by comparison with the harmonic series of order 2, that is

$$\left| \frac{(\pm 1)^n}{n^2} \right| \leq \frac{1}{n^2}$$

Hence the series converges for $-1 \leq x \leq 1$. To determine if the convergence is uniform in this interval, the Weierstrass M-test for uniform convergence is needed. That is, one must find a convergent series of constants

$$\sum_{n=1}^{\infty} M_n$$

such that

$$\left| \frac{x^n}{n^2} \right| \leq M_n$$

for all x in $-1 \leq x \leq 1$ if one is to determine that

$$\sum_{n=1}^{\infty} \frac{x_n}{n^2}$$

is uniformly convergent in this interval.

Since $-1 \leq x \leq 1$,

$$\left|\frac{x^n}{n^2}\right| \leq \frac{1}{n^2}$$

for all x in the range.

Therefore since

$$\sum_{n=1}^{\infty} M_n$$

converges, this shows the given power series converges uniformly on $-1 \leq x \leq 1$.

● **PROBLEM 3-30**

Find a power series for small values of x, for:

a) $\tan x$

b) $\dfrac{\sin x}{\sin 2x}$ $(x \neq 0)$

Solution: To find the power series of the given functions, one needs the following theorem:

Given the two power series

$$\sum_{n=0}^{\infty} a_n x^n = a_0 + a_1 x + a_2 x^2 + \ldots + a_n x^n + \ldots$$

and

$$\sum_{n=0}^{\infty} b_n x^n = b_0 + b_1 x + b_2 x^2 + \ldots + b_n x^n + \ldots ,$$

where $b_0 \neq 0$, and where both of the series are convergent in some interval $|x| < R$, let f be a function defined by

$$f(x) = \frac{a_0 + a_1 x + a_2 x^2 + \ldots + a_n x^n + \ldots}{b_0 + b_1 x + b_2 x^2 + \ldots + b_n x^n + \ldots}.$$

Then for sufficiently small values of x the function f can be represented by the power series

$$f(x) = c_0 + c_1 x + c_2 x^2 + \ldots + c_n x^n + \ldots,$$

where the coefficients $c_0, c_1, c_2, \ldots, c_n, \ldots$ are found by long division or equivalently by solving the following relations successively for each c_i ($i = 0$ to ∞):

$$b_0 c_0 = a_0$$

$$b_0 c_1 + b_1 c_0 = a_1$$

$$\vdots$$

$$b_0 c_n + b_1 c_{n-1} + \ldots + b_n c_0 = a_n$$

$$\vdots$$

a) To find the power series expansion of tan x, Taylor's series for sin x and cos x are needed. That is

$$\sin x = x - \frac{x^3}{3!} + \frac{x^5}{5!} - \ldots$$

and

$$\cos x = 1 - \frac{x^2}{2!} + \frac{x^4}{4!} - \ldots$$

Then,

$$\tan x = \frac{\sin x}{\cos x} = \frac{x - \frac{x^3}{3!} + \frac{x^5}{5!} - \ldots}{1 - \frac{x^2}{2!} + \frac{x^4}{4!} - \ldots} \qquad (1)$$

Therefore, by the theorem the power series expansion of tan x can be found by dividing the numerator by the denominator on the right side of (1). Hence, using long division one gets

$$\begin{array}{r} x + \frac{1}{3} x^3 + \frac{2}{15} x^5 + \ldots \\ 1 - \frac{1}{2} x^2 + \frac{1}{24} x^4 - \ldots \overline{\smash{\big)}\, x - \frac{1}{6} x^3 + \frac{1}{120} x^5 - \ldots} \\ x - \frac{1}{2} x^3 + \frac{1}{24} x^5 - \ldots \\ \hline \frac{1}{3} x^3 - \frac{1}{30} x^5 + \ldots \\ \frac{1}{3} x^3 - \frac{1}{6} x^5 + \ldots \\ \hline \end{array}$$

$$\frac{2}{15}x^5 - \ldots$$
$$\frac{2}{15}x^5 - \ldots$$

Thus
$$\tan x = x + \frac{1}{3}x^3 + \frac{2}{15}x^5 + \ldots$$

b) Since
$$\sin x = x - \frac{x^3}{3!} + \frac{x^5}{5!} - \ldots$$

one gets
$$\sin(2x) = 2x - \frac{(2x)^3}{3!} + \frac{(2x)^5}{5!} - \ldots,$$

so that
$$\frac{\sin x}{\sin 2x} = \frac{x - \frac{x^3}{3!} + \frac{x^5}{5!} - \ldots}{2x - \frac{(2x)^3}{3!} + \frac{(2x)^5}{5!} - \ldots} \quad . \tag{2}$$

Now multiplying the numerator and denominator on the right side of (2) by $1/x$ yields

$$\frac{\sin x}{\sin 2x} = \frac{1 - \frac{x^2}{6} + \frac{x^4}{120} - \ldots}{2 - \frac{4}{3}x^2 + \frac{4}{15}x^4 - \ldots} \quad .$$

Now by long division

$$\begin{array}{r}
\frac{1}{2} + \frac{1}{4}x^2 + \frac{5}{48}x^4 + \ldots \\
2 - \frac{4}{3}x^2 + \frac{4}{15}x^4 \overline{\smash{\big)}\, 1 - \frac{1}{6}x^2 + \frac{1}{120}x^4 - \ldots} \\
\underline{\frac{1}{2} - \frac{4}{6}x^2 + \frac{4}{30}x^4 - \ldots} \\
\frac{1}{2}x^2 - \frac{15}{120}x^4 + \ldots \\
\underline{\frac{1}{2}x^2 - \frac{1}{3}x^4 + \ldots} \\
\frac{25}{120}x^4 - \ldots \\
\frac{25}{120}x^4 - \ldots
\end{array}$$

Thus, for $x \neq 0$, $\quad \dfrac{\sin x}{\sin(2x)} = \dfrac{1}{2} + \dfrac{1}{4}x^2 + \dfrac{5}{48}x^4 + \ldots \quad .$

• **PROBLEM 3-31**

Show that the series representation

$$e^x = 1 + x + \frac{x^2}{2!} + \frac{x^3}{2!} + \ldots + \frac{x^n}{n!} + \ldots$$

is valid for all values of x.

Solution: To do this problem make use of Taylor's formula. That is if

$$f'(x), f''(x), \ldots, f^{(n)}(x)$$

exist and are continuous in the interval $a \leq x \leq b$ and if $f^{(n+1)}(x)$ exists in the interval $a < x < b$, then

$$f(x) = f(a) + f'(a)(x-a) + \frac{f''(a)(x-a)^2}{2!} + \ldots$$

$$+ \frac{f^{(n)}(a)(x-a)^n}{n!} + R_n \qquad (1)$$

where R_n is the remainder and which is written in the Lagrange form as

$$R_n = \frac{f^{(n+1)}(\xi)}{(n+1)!}(x-a)^{n+1}$$

where $a < \xi < x$. Note that as n changes, ξ also changes in general. In addition, if for all x and ξ in [a,b],

$$\lim_{n \to \infty} R_n = 0,$$

then (1) can be written in the form

$$f(x) = f(a) + f'(a)(x-a) + \frac{f''(a)}{2!}(x-a)^2$$

$$+ \frac{f'''(a)(x-a)^3}{3!} + \ldots . \qquad (2)$$

Note that (2) is the Taylor series or expansion of $f(x)$.
For the given problem $f(x) = e^x$, so that $f^{(n)}(x) = e^x$ for all orders n. Now taking $a = 0$ in (1),

$$f(x) = f(0) + f'(0)x + \frac{f''(0)x^2}{2!} + \ldots + \frac{f^{(n)}(0)x^n}{n!} + R_n$$

which yields upon substitution of $f(x) = \cdot e^x$,

$$e^x = 1 + x + \frac{x^2}{2!} + \ldots + \frac{x^n}{n!} + R_n$$

where

$$R_n = \frac{f^{(n+1)}(\xi)}{(n+1)!} x^{n+1} = \frac{e^\xi}{(n+1)!} x^{n+1}$$

where

$$0 < \xi < x.$$

Now we must prove that

$$\lim_{n \to \infty} R_n = 0,$$

so that we will have the Taylor expansion for e^x. However,

$$-|x| < \xi < |x|$$

and

$$0 < e^\xi < e^{|x|};$$

hence

$$\left| R_n \right| = \left| \frac{e^\xi}{(n+1)!} x^{n+1} \right| \leq \frac{e^{|x|} |x|^{n+1}}{(n+1)!} \qquad (3)$$

Thus by (3) to prove

$$\lim_{n \to \infty} R_n = 0,$$

it is sufficient to prove that

$$\lim_{n \to \infty} \frac{|x|^n}{n!} = 0.$$

To do this choose an integer N such that $N \geq 2|x|$. Then if $n > N$,

$$\frac{|x|^n}{n!} = \frac{|x|^N}{N!} \cdot \frac{|x|^{n-N}}{(N+1)(N+2)\ldots n}$$

$$= \frac{|x|^N}{N!} \left(\frac{|x|}{N+1}\right)\left(\frac{|x|}{N+2}\right) \cdots \left(\frac{|x|}{n}\right)$$

$$\leq \frac{|x|^N}{N!} \left(\frac{N}{N+1}\right)\left(\frac{N}{N+2}\right) \cdots \left(\frac{N}{n}\right) \frac{1}{2^{n-N}} \leq \frac{|x|^N}{N!} \left(\frac{1}{2}\right)^{n-N}$$

Therefore

$$\frac{|x|^n}{n!} \leq \frac{|x|^N}{N!} \left(\frac{1}{2}\right)^{n-N}. \qquad (4)$$

91

Now keeping N fixed we have

$$\lim_{n\to\infty}\left(\frac{1}{2}\right)^{n-N} = 0.$$

Therefore by (4)

$$\lim_{n\to\infty} \frac{|x|^n}{n!} = 0.$$

This means that the series representation

$$e^x = 1 + x + \frac{x^2}{2!} + \ldots + \frac{x^n}{n!} + \ldots$$

is valid for all values of x.

• **PROBLEM 3-32**

a) Using power series, show that

$$\frac{d(\sin x)}{dx} = \cos x$$

and

$$\frac{d(\cos x)}{dx} = -\sin x.$$

b) Then show that $\sin a \cos b + \cos a \sin b = \sin(a+b)$

and

$$\cos a \cos b - \sin a \sin b = \cos(a+b).$$

Solution: The power series expansion for sin x and cos x are for all values x

$$\sin x = x - \frac{x^3}{3!} + \frac{x^5}{5!} - \frac{x^7}{7!} + \ldots \qquad (1)$$

$$\cos x = 1 - \frac{x^2}{2!} + \frac{x^4}{4!} - \frac{x^6}{6!} + \ldots . \qquad (2)$$

Since, by theorem, a power series can be differentiated term-by-term in any interval lying entirely within its radius of convergence, one gets by (1) and (2), for all values x

$$\frac{d(\sin x)}{dx} = 1 - \frac{3x^2}{3!} + \frac{5x^4}{5!} - \frac{7x^6}{7!} + \ldots$$

92

$$= 1 - \frac{x^2}{2!} + \frac{x^4}{4!} - \frac{x^6}{6!} + \ldots = \cos x.$$

and

$$\frac{d(\cos x)}{dx} = \frac{-2x}{2!} + \frac{4x^3}{4!} - \frac{6x^5}{5!} + \ldots$$

$$= -x + \frac{x^3}{3!} - \frac{x^5}{5!} + \ldots = -\sin x$$

b) Using the result from part (a)

$$\{\sin x \cos(h-x) + \cos x \sin(h-x)\}'$$

$$= \cos x \cos(h-x) + \sin x \sin(h-x)$$

$$- \sin x \sin(h-x) - \cos x \cos(h-x)$$

$$= 0$$

Thus, $\sin x \cos(h-x) + \cos x \sin(h-x)$ is constant and this constant equals the value $\sin h$ (this was found by letting $x=0$).

Hence

$$\sin x \cos(h-x) + \cos x \sin(h-x) = \sin h.$$

Now replacing x by a and h by a+b yields

$$\sin a \cos b + \cos a \sin b = \sin(a+b).$$

From which differentiation with respect to a yields

$$\cos a \cos b - \sin a \sin b = \cos(a+b).$$

Note that from this result one obtains

$$\cos^2 x - \sin^2 x = \cos 2x.$$

• **PROBLEM 3-33**

Write the function

$$f(z) = \frac{z}{e^z - 1}$$

in terms of a power series.

Solution: The function

$$f(z) = \frac{z}{e^z - 1} \qquad (1)$$

is analytic everywhere except at those points where $e^z - 1$ vanishes and z does not, i.e., except at the points $\pm 2\pi i$, $\pm 4\pi i$, ... Substituting the power series

$$e^x - 1 = \frac{z}{1!} + \frac{z^2}{2!} + \ldots + \frac{z^n}{n!} + \ldots.$$

into (1), and dividing both numerator and denominator by z, obtain

$$f(z) = \frac{1}{1 + \frac{z}{2!} + \ldots + \frac{z^n}{(n+1)!} + \ldots}$$

(note that this expression is meaningful for $z = 0$). The series in the denominator converges for all z and does not vanish for $z = 0$, but otherwise has the same zeros as $e^z - 1$. The two zeros closest to the origin are at $\pm 2\pi i$, and therefore $f(z)$ has a power series representation of the form

$$\sum_{n=0}^{\infty} c_n (z - z_0)^n$$

for $z_0 = 0$

on the disk $K: |z| < 2\pi$. To find this representation, use the technique of division of power series. That is, given an expression of the form

$$f(z) = \frac{g(z)}{h(z)} = \frac{\sum_{n=0}^{\infty} a_n z^n}{\sum_{n=0}^{\infty} b_n z^n} = \sum_{n=0}^{\infty} c_n z^n,$$

the coefficient c_n may be found by using the coefficients $c_0, c_1, c_2, \ldots, c_{n-1}$ in the formula

$$c_n = \frac{a_n - c_0 b_n - c_1 b_{n-1} - \ldots - c_{n-1} b_1}{b_0} \qquad (2)$$

In the present case, the coefficients a_n, b_n appearing in this problem are just

$$a_0 = 1, \quad a_n = 0 \quad (n = 1, 2, \ldots),$$

$$b_n = \frac{1}{(n+1)!} \quad (n = 0, 1, 2, \ldots).$$

Therefore the first of the equations (2) gives

$$c_0 = 1,$$

and the rest reduce to the recurrence relation

$$c_0 \frac{1}{(n+1)!} + c_1 \frac{1}{n!} + \ldots + c_n = 0 \quad (n = 1, 2, \ldots), \quad (3)$$

relating c_n to the values $c_0, c_1, \ldots, c_{n-1}$.

The numbers $c_n n!$ are called the Bernoulli numbers and are denoted by B_n. To calculate B_n, use the recurrence relation (3), which now takes the form

$$B_0 \frac{1}{0!(n+1)!} + B_1 \frac{1}{1!n!} + \ldots + B_n \frac{1}{n!1!} = 0$$

$$(n = 1, 2, \ldots) \quad (4)$$

(obviously $B_0 = c_0 0! = 1$). Multiplying (4) by $(n+1)!$ and introducing the notation

$$\frac{(n+1)!}{k!(n+1-k)!} = \binom{n+1}{k}$$

(the familiar binomial coefficient), find that

$$B_0 \binom{n+1}{0} + B_1 \binom{n+1}{1} + \ldots + B_n \binom{n+1}{n} = 0$$

$$(n = 1, 2, \ldots). \quad (5)$$

Equation (5) can be written symbolically as

$$(1 + B)^{n+1} - B^{n+1} = 0, \quad (6)$$

where after raising $1 + B$ to the $(n+1)$th power, every B^k is changed to B_k ($k = 1, 2, \ldots, n+1$). Using (6) and the fact that $B_0 = 1$, deduce step by step that

$$B_0 + 2B_1 = 0,$$

$$B_0 + 3B_1 + 3B_2 = 0,$$

$$B_0 + 4B_1 + 6B_2 + 4B_3 = 0,$$

$$B_0 + 5B_1 + 10B_2 + 10B_3 + 5B_4 = 0,$$

$$B_1 = -\frac{1}{2} B_0 = -\frac{1}{2},$$

$$B_2 = -\frac{1}{3} B_0 - B_1 = \frac{1}{6},$$

$$B_3 = -\frac{1}{4} B_0 - B_1 - \frac{3}{2} B_2 = 0,$$

$$B_4 = -\frac{1}{5} B_0 - B_1 - 2B_2 - 2B_3 = -\frac{1}{30},$$

$$B_0 + 6B_1 + 15B_2 + 20B_3 + 15B_4 + 6B_5 = 0,$$

$$B_5 = -\frac{1}{6} B_0 - B_1 - \frac{5}{2} B_2 - \frac{10}{3} B_3 - \frac{5}{2} B_4 = 0,$$

$$B_0 + 7B_1 + 21B_2 + 35B_3 + 35B_4 + 21B_5 + 7B_6 = 0,$$

$$B_6 = -\frac{1}{7} B_0 - B_1 - 3B_2 - 5B_3 - 5B_4 - 3B_5 = \frac{1}{42},$$

Collecting these results,

$$B_0 = 1, \; B_1 = -\frac{1}{2}, \; B_2 = \frac{1}{6}, \; B_3 = 0, \; B_4 = -\frac{1}{30}, \; B_5 = 0,$$

$$B_6 = \frac{1}{42}, \; \ldots \tag{7}$$

As suggested by (7), the Bernoulli numbers with odd indices greater than 1 vanish. To see this, write

$$f(z) = \frac{z}{e^z - 1} = c_0 + c_1 z + c_2 z^2 + c_3 z^3 + \ldots + c_n z^n + \ldots \tag{8}$$

$$= B_0 + \frac{B_1}{1!} z + \frac{B_2}{2!} z^2 + \frac{B_3}{3!} z^3 + \ldots + \frac{B_n}{n!} z^n + \ldots$$

and then replace z by $-z$, obtaining

$$f(-z) = \frac{-z}{e^{-z} - 1} = \frac{-ze^z}{(e^{-z} - 1)e^z} = \frac{ze^z}{e^z - 1} \tag{9}$$

$$= B_0 - \frac{B_1}{1!} z + \frac{B_2}{2!} z^2 - \frac{B_3}{3!} z^3 + \ldots + \frac{B_n}{n!} (-1)^n z^n$$

$$+ \ldots .$$

Subtraction of (9) from (8) gives

$$f(z) - f(-z) = \frac{z}{e^z - 1} - \frac{ze^z}{e^z - 1} = -z$$

$$= 2 \frac{B_1}{1!} z + 2 \frac{B_3}{3!} z^3 + \ldots + 2 \frac{B_{2m+1}}{(2m+1)!} z^{2m+1}$$

$$+ \ldots ,$$

and hence, by the uniqueness of power series expansions,

$$2B_1 = -1, \quad B_3 = B_5 = \ldots = B_{2m+1} = \ldots = 0,$$

as asserted. Using this fact, write the expansion (8) in the form

$$f(z) = \frac{z}{e^z - 1} = 1 - \frac{z}{2} + \sum_{m=1}^{\infty} \frac{B_{2m}}{(2m)!} z^{2m}, \qquad (10)$$

where the series converges on the disk K: $|z| < 2\pi$. Moreover, the fact that $f(z)$ becomes infinite for $z = \pm 2\pi i$ implies that K is the largest disk on which (10) converges.

LAURENT SERIES EXPANSION

● **PROBLEM 3-34**

Obtain series expansions of the function

$$f(z) = \frac{-1}{(z-1)(z-2)}$$

about the point $z_0 = 0$ in the regions

$$|z| < 1, \quad 1 < |z| < 2, \quad |z| > 2.$$

Solution: The function f has singular points, or points at which it is not analytic, at $z_1 = 1$, $z_2 = 2$. Taylor's theorem tells that f has a valid Taylor series expansion about $z_0 = 0$ within the circle $|z| = 1$ since it is analytic there. Hence

$$f(z) = \sum_{n=0}^{\infty} \frac{f^{(n)}(z_0)}{n!}(z-z_0)^n \qquad |z| < 1.$$

It is convenient, however, to use partial fractions to write

$$f(z) = \frac{-1}{(z-1)(z-2)} = \frac{1}{z-1} - \frac{1}{z-2} = \frac{1}{2}\frac{1}{1-z/2} - \frac{1}{1-z}. \qquad (1)$$

For $|z| < 1$, we have $|z/2| < 1$ and note that these are the conditions under which each of the terms in (1) can be represented by a geometric series. Hence, since

$$\frac{1}{1-w} = \sum_{n=0}^{\infty} w^n, \quad (|w| < 1),$$

$$f(z) = \frac{1}{2}\sum_{n=0}^{\infty}\left(\frac{z}{2}\right)^n - \sum_{n=0}^{\infty} z^n = \sum_{n=0}^{\infty}\left[\frac{1}{2}\left(\frac{z}{2}\right)^n - z^n\right].$$

or

$$f(z) = \sum_{n=0}^{\infty} (2^{-n-1} - 1) z^n \qquad (|z| < 1). \qquad (2)$$

Since the Taylor coefficients are unique this must be the Taylor expansion for $f(z)$ in the region $|z| < 1$ and as a by-product of the calculations, it has been deduced that

$$\frac{f^{(n)}(0)}{n!}$$

must be equal to the coefficients in (2). Thus

$$f^{(n)}(0) = n!(2^{-n-1} - 1).$$

In the region $1 < |z| < 2$ one must use Laurent's theorem which states that if f is analytic in the region bounded by two concentric circles C_1 and C_2 centered at z_0, then at each point z in that region f(z) is represented by its Laurent series expansion

$$f(z) = \sum_{n=0}^{\infty} a_n (z-z_0)^n + \sum_{n=1}^{\infty} \frac{b_n}{(z-z_0)^n} \qquad (3)$$

where the radius of C_1 is assumed to be less than that of C_2, and

$$a_n = \frac{1}{2\pi i} \int_{C_1} \frac{f(s)\,ds}{(s-z_0)^{n+1}} \qquad n = 0,1,2,\ldots \qquad (4)$$

$$b_n = \frac{1}{2\pi i} \int_{C_2} \frac{f(s)\,ds}{(s-z_0)^{-n+1}} \qquad n = 1,2,\ldots \qquad (5)$$

Formulas (4) and (5) are not very useful since the integrals are usually difficult to do. Therefore, note that in the region $1 < |z| < 2$, $|1/z| < 1$ and $|z/2| < 1$ so that one may use the geometric series expansion to write

$$f(z) = \frac{1}{z}\frac{1}{1-1/z} + \frac{1}{2}\frac{1}{1-z/2} = \sum_{n=0}^{\infty} \frac{1}{z^{n+1}} + \frac{1}{2}\sum_{n=0}^{\infty} \frac{z^n}{2^{n+1}}, \qquad (6)$$

which is valid for $1 < |z| < 2$. Now the Laurent coefficients in (4) and (5) are unique, i.e., any representation such as that in (6) must be the Laurent expansion regardless of how it is arrived at. Therefore, the coefficients in the expansion must be equal to those of (4) and (5) so that as a by-product of the calculations we have been able to find the values of the integrals there. E.g., for $n = 1$ in (5) see from (6) that $b_1 = 1$ so

$$\int_{|z|=2} f(z)\,dz = 2\pi i$$

Finally, for $|z| > 2$ one may write

$$f(z) = \frac{1}{z}\left(\frac{1}{1-1/z} - \frac{1}{1-2/z}\right) \qquad (7)$$

and $|1/z| < 1$ as well as $|2/z| < 1$ in the region $|z| > 2$. Therefore, one may use the geometric series in this region to find from (7) that

$$f(z) = \sum_{n=0}^{\infty} \frac{1}{z^{n+1}} - \sum_{n=0}^{\infty} \frac{2^n}{z^{n+1}}$$

or

$$f(z) = \sum_{n=0}^{\infty} \frac{1-2^n}{z^{n+1}} \tag{8}$$

Again, since the Laurent coefficients are unique it is deduced that the coefficients in (8) are the Laurent coefficients. In particular since $b_1 = 0$ in (8) one finds from (4) that

$$\int_C f(s)\,ds = 0,$$

where in this case, C can be any circle centered at $z_0 = 0$ with radius $r > 2$.

● **PROBLEM 3-35**

Find the first three nonzero terms in the Laurent series expansion of csc z about $z_0 = 0$.

Solution: The Laurent series expansion of a function $f(z)$ which is analytic in an annulus bounded by the concentric circles C_1 and C_2 with centers at z_0 is given by

$$f(z) = \sum_{n=0}^{\infty} a_n(z-z_0)^n + \sum_{n=1}^{\infty} \frac{b_n}{(z-z_0)^n} \tag{1}$$

where

$$a_n = \frac{1}{2\pi i} \int_{C_1} \frac{f(s)\,ds}{(s-z_0)^{n+1}} \qquad (n = 0,1,2,\ldots) \tag{2}$$

$$b_n = \frac{1}{2\pi i} \int_{C_2} \frac{f(s)\,ds}{(s-z_0)^{-n+1}} \qquad (n = 1,2,\ldots). \tag{3}$$

Here the radius of C_2 is assumed to be larger than the radius of C_1. The integrals in (2) and (3) are usually difficult and sometimes impossible to do by elementary methods so that a simpler way to proceed is to find the Taylor series expansion for sin z and use long division to obtain a series representing csc z = 1/sin z. Then observing that the Laurent expansion is unique, it can be stated that the resulting series is indeed the Laurent series.

Now since sin z is analytic $\forall z$ with $|z| < \infty$ its Taylor expansion can be written as

$$\sin z = \sum_{n=0}^{\infty} \frac{f^{(n)}(0) z^n}{n!} = \sum_{n=1}^{\infty} \frac{(-1)^{n-1}}{(2n-1)!} z^{2n-1} \qquad (4)$$

Now sin z = 0 only for z = nπ, n = 0, ±1, ±2, ...

Hence,

$$\csc z = \frac{1}{\sin z}$$

is analytic in the annulus $0 < |z| < \pi$ can be represented by a series expansion there. Thus

$$\csc z = \frac{1}{\sin z} = \left[\frac{1}{z - \frac{z^3}{3!} + \frac{z^5}{5!} - \frac{z^7}{7!} + \cdots} \right]. \qquad (5)$$

Thus, by long division

$$
\begin{array}{r}
1/z + z/3! + z^3/(3!\cdot 3! - 5!) + \cdots \\
z - \frac{z^3}{3!} + \frac{z^5}{5!} - \frac{z^7}{7!} + \cdots \overline{\smash{\big)}\, 1 + 0 + 0 + 0 + 0} \\
\underline{-\left(1 - z^2/3! + z^4/5! - z^6/7! + \cdots\right)} \\
z^2/3! - z^4/5! + z^6/7! - \cdots \\
\underline{-\left(z^2/3! - z^4/3!\cdot 3! + z^6/3!\cdot 5! - \cdots\right)} \\
\frac{z^4}{(3!\cdot 3! - 5!)} - \cdots
\end{array}
$$

Therefore, from equation (5),

$$\csc z = \frac{1}{z} + \frac{1}{3!} z + \left(\frac{1}{3!\cdot 3!} - \frac{1}{5!}\right) z^3 + \cdots$$

or

$$\csc z = \frac{1}{z} + \frac{1}{6} z + \frac{7}{360} z^3 + \cdots \quad (0 < |z| < \pi). \tag{6}$$

As noted before, this must be the Laurent series for csc z since that series is unique. As a by-product of this calculation, one may therefore equate the coefficients in (6) with the corresponding Laurent coefficients given by (2) and (3) with C_2 being the circle $|z| = \pi$, and C_1 the degenerate circle $|z| = 0$. This provides a useful way to calculate the integrals of equations (2) and (3).

• **PROBLEM 3-36**

Find the principal part of the function

$$f(z) = \frac{e^z \cos z}{z^3}$$

at its singular point and determine the type of singular point it is.

Solution: The given function has an isolated singular point at $z = 0$. Such a function may be represented by a Laurent series

$$f(z) = \sum_{n=0}^{\infty} a_n (z-z_0)^n + \sum_{n=1}^{\infty} \frac{b_n}{(z-z_0)^n} \tag{1}$$

where a_n and b_n are the Laurent coefficients. Since a Laurent expansion is unique, one needn't calculate these coefficients directly but may proceed as follows, noting that any series of the form (1) that are obtained must be the Laurent series. First, expanding the numerator of f in a Taylor series about $z_0 = 0$ yields

$$e^z \cos z = \left(1 + z + \frac{z^2}{2!} + \cdots\right)\left(1 - \frac{z^2}{2!} + \cdots\right)$$

$$= 1 + z - \frac{z^3}{3} + \cdots . \qquad |z| < \infty \tag{2}$$

This is valid by Taylor's theorem since $e^z \cos z$ is analytic for all z. Hence

$$\frac{e^z \cos z}{z^3} = \frac{1}{z^3} + \frac{1}{z^2} - \frac{1}{3} + \ldots \qquad 0 < |z| < \infty .$$

As noted above, this must be the Laurent series for $f(z)$.

The principal part of a function f at a point z_0 is defined as the portion of its Laurent series involving negative powers of $z - z_0$. Hence, the principal part of

$$f(z) = \frac{e^z \cos z}{z^3}$$

at 0, its only singular point, can be seen from (3) to be

$$\frac{1}{z^3} + \frac{1}{z^2} .$$

If the principal part of f at z_0 contains at least one non-zero term but the number of such terms is finite, the isolated singular point z_0 is then called a pole of order m, where m is the largest of the powers

$$\frac{b_j}{(z - z_0)^j}$$

in the principal part of f. Hence, in this case, it is said that f has a pole of order 3 at $z_0 = 0$.

● **PROBLEM 3-37**

Find the principal part of the function

$$f(z) = \frac{z}{(z + 1)^2 (z^3 + 2)}$$

at $z_0 = -1$. What type of singular point is z_0?

Solution: Since the singular point $z_0 = -1$ is isolated, the function has a Laurent series expansion about -1 which is valid at every point except -1 in the circular domain centered at -1 with radius r equal to the distance between -1 and the next closest singularity, i.e.,

$$r = \left| 2^{1/3} \right| .$$

To find this expansion first expand

$$f_1(z) = \frac{z}{z^3 + 2}$$

in a Taylor series about $z_0 = -1$ to obtain

$$f_1(z) = \sum_{n=0}^{\infty} \frac{f^{(0)}(z_0)}{n!}(z - z_0)^n. \qquad (1)$$

Now

$$f_1(z) = \frac{z}{z^3 + 2}$$

so

$$f_1^{(1)}(z) = \frac{2 - 2z^3}{(z^3 + 2)^2}$$

and

$$f_1^{(2)} = \frac{6(z^5 - 4z^2)}{(z^3 + 2)^3}$$

etc. Using these results in (1) one finds

$$\frac{z}{z^3 + 2} = -1 + 4(z + 1) - 15(z + 1)^2 + \ldots . \qquad (2)$$

Now divide equation (2) by $(z + 1)^2$ to obtain

$$f(z) \qquad \frac{z}{(z + 1)^2(z^3 + 2)} = \frac{-1}{(z + 1)^2} + \frac{4}{z + 1} - 15 + \ldots . \qquad (3)$$

The principal part of a function at a point z_0 is defined as that part of its Laurent series involving negative powers of $(z - z_0)$. Since (3) is the Laurent series representing

$$f(z) = \frac{z}{(z + 1)^2(z^3 + 2)}$$

for $0 < |z + 1| < |2^{1/3}|$, it is concluded that the principal part of f at $z_0 = -1$ is

$$\frac{-1}{(z + 1)^2} + \frac{4}{z + 1} .$$

If the principal part of f at z_0 contains at least one

nonzero term but the number of such terms is finite, then the isolated singular point z_0 is called a pole of order m where m is the largest of

$$\frac{b_j}{(z-z_0)^j}$$

in the principal part of f. Hence, in this case,

$$f(z) = \frac{z}{(z+1)^2(z^3+2)}$$

is said to have a pole of order 2 at $z_0 = -1$.

CHAPTER 4

FINITE DIFFERENCE CALCULUS

FINITE DIFFERENCES

• **PROBLEM 4-1**

A third-degree polynomial P(x) is passed through the points (0, -1), (1, 1), (2, 1), and (3, -2). Find its value at x = 1.2.

Solution: First Method: Let the third-degree polynomial be

$$P(x) = a_3 x^3 + a_2 x^2 + a_1 x + a_0$$

If it is to pass through the above points, this equation must be satisfied for the above four pairs of values for x and y; therefore

$$a_0 = -1$$

$$a_3 + a_2 + a_1 + a_0 = 1$$

$$8a_3 + 4a_2 + 2a_1 + a_0 = 1$$

$$27a_3 + 9a_2 + 3a_1 + a_0 = -2$$

Solving these four equations in four unknowns, one gets $a_0 = -1$, $a_1 = \frac{8}{3}$, $a_2 = -\frac{1}{2}$, $a_3 = -\frac{1}{6}$ or

$$P(x) = -\frac{1}{6}(x^3 + 3x^2 - 16x + 6)$$

Thus, at x = 1.2, by synthetic division

```
    1     3      -16        6      | 1.2
          1.2    5.04    -13.152
    1     4.2   -10.96    -7.152
```

$$P(1.2) = \frac{7.152}{6} = 1.192$$

This method is not recommended, as the solution of a set of linear equations usually involves a considerable amount of computation.

Second Method: Tabulate the values of P(x) given, and form all possible differences, as in the table

TABLE

x	$y = P(x)$	Δy	$\Delta^2 y$	$\Delta^3 y$
0	−1			
		2		
1	1		−2	
		0		−1
2	1		−3	
		−3		
3	−2			

If one sets $x_0 = 0$, the differences required in

$$P_n(x) = P_n(x_0) + u\,\Delta P_n(x_0) + \frac{u^{[2]}}{2!}\Delta^2 P_n(x_0) + \cdots$$

$$+ \frac{u^{[n]}}{n!}\Delta^n P_n(x_0)$$

$$= \sum_{r=0}^{n} \frac{u^{[r]}}{r!}\Delta^r P_n(x_0) \qquad (1)$$

are seen to be just the leading differences −1, 2, −2, −1 in this table. With these differences known, Eq. (1) permits the polynomial to be written down. Since $h = 1$ and $x_0 = 0$, $u = x$, and this polynomial is

$$P(x) = -1 + 2x^{[1]} - x^{[2]} - \frac{1}{6}x^{[3]}$$

therefore the required value of P(x) at $x = 1.2$ is

$$P(1.2) = -1 + (2)(1.2) - (1.2)(0.2)$$

$$- \frac{1}{6}(1.2)(0.2)(-0.8) = 1.192$$

• **PROBLEM 4-2**

Express the polynomial

$$P(x) = 3x^5 - 7x^4 + 87x^3 + 28x^2 + 176x - 77$$

in terms of factorials.

Solution: Evidently the factorial of highest degree will be $x^{(5)}$ so subtract $3x^{(5)}$ from the given polynomial leaving a polynomial of degree 4 whose leading term is $23x^4$.

From this subtract $23x^{(4)}$ leaving a polynomial of degree 3 with leading term $120x^3$. Proceeding step-by-step in this manner, a zero remainder is finally arrived at and the original polynomial must therefore be equal to the sum of the terms subtracted. The work, systematically arranged, is shown:

$$
\begin{array}{rl}
P(x) = & 3 - 7 + 87 + 28 + 176 - 77 \\
3x^{(5)} = & 3 - 30 + 105 - 150 - 72 \\
\hline
\text{Diff.} = & 23 - 18 + 178 + 104 - 77 \\
23x^{(4)} = & 23 - 138 + 253 - 138 \\
\hline
\text{Diff.} = & 120 - 75 + 242 - 77 \\
120x^{(3)} = & 120 - 360 + 240 \\
\hline
\text{Diff.} = & 285 + 2 - 77 \\
285x^{(2)} = & 285 - 285 \\
\hline
\text{Diff.} = & 287 - 77 \\
287x^{(1)} = & 287 \\
\hline
\text{Diff.} = & -77 \\
-77x^{(0)} = & -77 \\
\end{array}
$$

Hence, the desired form of the polynomial is

$$P(x) = 3x^{(5)} + 23x^{(4)} + 120x^{(3)} + 285x^{(2)} + 287x^{(1)} - 77x^{(0)}.$$

• **PROBLEM 4-3**

Construct a table of forward differences for

$$y = 2x^4 - 3x^3 + 5x^2 - 2x + 3$$

starting with $x = 1$ and proceeding at intervals of 0.1. Use the alternative method of repeated additions to calculate values for $x > 1.4$. Calculate to $x = 2.0$.

TABLE 1

x	y	Δy	$\Delta^2 y$	$\Delta^3 y$	$\Delta^4 y$
1	5.0000				
		7852			
1.1	5.7852		1928		
		9780		372	
1.2	6.7632		2300		48
		12080		420	
1.3	7.9712		2720		
		14800			
1.4	9.4512				

Solution: Since the polynomial is of degree 4 the 4th differences are constant. Hence, first calculate five values by direct substitution and form the differences (Table 1).

Since the 4th difference will be constant the next line of differences can be formed from right to left as follows:

    1.5   11.2500   17988   3188   468   48

Then the next line is calculated, etc. An alternative method is to complete the column of 4th differences, then by repeated additions complete the column of 3rd differences, etc. This has some advantages where the differences are not all positive. Whichever method is used, the result should be checked from time to time with values calculated directly from the original polynomial. The computation to x = 2 appears in Table 2.

TABLE 2

					48
				420	
			2720		48
		14800		468	
1.4	9.4512		3188		48
		17988		516	
1.5	11.2500		3704		48
		21692		564	
1.6	13.4192		4268		48
		25960		612	
1.7	16.0152		4880		48
		30840		660	
1.8	19.0992		5540		48
		36380		708	
1.9	22.7372		6248		
		42628			
2.0	27.0000				

The value for x = 2.0 checks with that given by direct substitution.

In the foregoing problem the functional values were <u>exact</u>, no significant figures having been dropped. In practice, however, the values are ordinarily not exact, being rounded off after a specified number of decimal places, and in consequence the <u>tabulated</u> n-th differences of a polynomial of n-th degree will not be constant. In such cases the cumulative effect of small errors in the differences may seriously affect the accuracy of the y itself.

● **PROBLEM 4-4**

Obtain the divided differences for the values

x	-3	-1	2	4	6	7
y	-1584	216	-144	96	-288	-504

Solution: Write the x's and y's in parallel columns and calculate the divided differences in succeeding columns. The completed calculations are given in the table.

TABLE

x	y						
-3	-1584						
		900					
-1	216		-204				
		-120		36			
2	-144		48		-6		
		120		-18		1	
4	96		-78		4		
		-192		14			
6	-288		-8				
		-216					
7	-504						

For instance to obtain the first entry in the first column of differences one gets

$$\frac{(216) - (-1584)}{(-1) - (-3)} = 900.$$

To obtain the second entry in the third column of differences one gets

$$\frac{(-78) - (48)}{6 - (-1)} = -18.$$

Divided differences may be expressed in a symmetric form that has important applications. It is evident that

$$[x_1 x_2] = \frac{y_1}{x_1 - x_2} + \frac{y_2}{x_2 - x_1}$$

$$[x_0 x_1] = \frac{y_0}{x_0 - x_1} + \frac{y_1}{x_1 - x_0}$$

and from these one gets

$$[x_0 x_1 x_2] = \frac{[x_0 x_1] - [x_1 x_2]}{x_0 - x_2}$$

$$= \frac{1}{x_0 - x_2} \left[\frac{y_0}{x_0 - x_1} + \frac{[x_0 - x_2] y_1}{[x_1 - x_0][x_1 - x_2]} - \frac{y_2}{x_2 - x_1} \right]$$

$$= \frac{y_0}{(x_0 - x_1)(x_0 - x_2)} + \frac{y_1}{(x_1 - x_0)(x_1 - x_2)} + \frac{y_2}{(x_2 - x_0)(x_2 - x_1)}.$$

In the same way one finds

$$[x_0 x_1 x_2 x_3] = \frac{y_0}{(x_0 - x_1)(x_0 - x_2)(x_0 - x_3)}$$

$$+ \frac{y_1}{(x_1 - x_0)(x_1 - x_2)(x_1 - x_3)}$$

$$+ \frac{y_2}{(x_2 - x_0)(x_2 - x_1)(x_2 - x_3)}$$

$$+ \frac{y_3}{(x_3 - x_0)(x_3 - x_1)(x_3 - x_2)}$$

and in general

$$[x_0 x_1 \ldots x_n] = \frac{y_0}{(x_0 - x_1) \ldots (x_0 - x_n)} \tag{1}$$

$$+ \frac{y_1}{(x_1 - x_0)(x_1 - x_2) \ldots (x_1 - x_n)} + \ldots$$

$$+ \frac{y_n}{(x_n - x_0)(x_n - x_1) \ldots (x_n - x_{n-1})}.$$

From the expression (1), it can be seen at once that the value of any divided difference is entirely independent of the x's involved in the difference.

• **PROBLEM 4-5**

Find the forward differences (at unit interval) of the polynomial $p(r) = 2r^4 - 5r^3 - 8r^2 + 17r + 2$, using the coefficients in Table 1.

Table 1
Value of s_m^j

j \ m	1	2	3	4	5	6	7	8	9	10	11	12
1	1	1	1	1	1	1	1	1	1	1	1	1
2		1	3	7	15	31	63	127	255	511	1023	2047
3			1	6	25	90	301	966	3025	9330	28501	86526
4				1	10	65	350	1701	7770	34105	145750	611501
5					1	15	140	1050	6951	42525	246730	1379400
6						1	21	266	2646	22827	179487	1323652
7							1	28	462	5880	63987	627396
8								1	36	750	11880	159027
9									1	45	1155	22275
10										1	55	1705
11											1	66
12												1

Solution: Using the coefficients in Table 1:

$$2r^4 = 2r^{(4)} + 12r^{(3)} + 14r^{(2)} + 2r$$

$$-5r^3 = \quad\quad\quad - 5r^{(3)} - 15r^{(2)} - 5r$$

$$-8r^2 = \quad\quad\quad\quad\quad\quad - 8r^{(2)} - 8r$$

$$17r = \quad\quad\quad\quad\quad\quad\quad\quad 17r$$

and hence

$$p(r) = 2r^{(4)} + 7r^{(3)} - 9r^{(2)} + 6r + 2$$

$$\Delta p(r) = 8r^{(3)} + 21r^{(2)} - 18r + 6$$

$$\Delta^2 p(r) = 24r^{(2)} + 42r - 18$$

$$\Delta^3 p(r) = 48r + 42$$

$$\Delta^4 p(r) = 48$$

Also $\sum_{r=0}^{n-1} p(r) = (2/5)n^{(5)} + (7/4)n^{(4)} - 3n^{(3)} + 3n^{(2)} + 2n.$

To express these differences or sums of p(r) in powers of r, the inverse expansion of factorials in powers is needed, i.e., the coefficients $s_m^{\ j}$ such that

$$r^{(m)} = s_m^{\ 1} r + s_m^{\ 2} r^2 + \ldots + s_m^{\ m} r^m. \tag{1}$$

Table 2
Value of $s_m^{\ j}$

j	m							
	1	2	3	4	5	6	7	8
1	1	−1	2	−6	24	−120	720	−5040
2		1	−3	11	−50	274	−1764	13068
3			1	−6	35	−225	1624	−13132
4				1	−10	85	−735	6769
5					1	−15	175	−1960
6						1	−21	322
7							1	−28
8								1
9								
10								
11								
12								

j	m			
	9	10	11	12
1	40320	−362880	3628800	−39916800
2	−109584	1026576	−10628640	120543840
3	118124	−1172700	12753576	−150917976
4	−67284	723680	−8409500	105258076
5	22449	−269325	3416930	−45995730
6	−4536	63273	−902055	13339535
7	546	−9450	157773	−2637558
8	−36	870	−18150	357423
9	1	−45	1320	−32670
10		1	−55	1925
11			1	−66
12				1

These, the Stirling numbers of the first kind, are given in Table 2, for m up to 12. The same coefficients are useful if derivatives or integrals of factorials are wanted, as the first step is then conversion to a power series.

One finds easily, by means of (1) that

$$\sum_{r=0}^{n-1} p(r) = (2/5)n^5 - (9/4)n^4 + (1/2)n^3 + (45/4)n^2 - (79/10)n.$$

• **PROBLEM 4-6**

Given the following pairs of values of x and y = f(x)

x	1	2	4	8	10
y	0	1	5	21	27

determine numerically the first and second derivatives of f(x) at x = 4.

Solution: Since the order of the terms is immaterial in a divided-difference table, one may write the x = 4 entry last and construct a divided-difference table from these five pairs of values. The result is that shown above the line in the accompanying table. One may now approximate f(x) by a fourth-degree polynomial P(x) passing through these five pairs of values. The derivatives of f(x) are then given approximately by the derivatives of P(x), which are given in terms of the divided differences by

$$\frac{d^k f(x)}{dx^k} = k! f(x, x, \ldots, x) \quad \text{(k+1 arguments)} \tag{1}$$

(kth derivative of f(x) at x expressed in terms of divided differences); in particular,

$$f'(x) \cong P'(x) = P(x,x)$$

$$f''(x) \cong P''(x) = 2P(x,x,x). \tag{2}$$

DETERMINATION OF DERIVATIVES DIRECTLY
FROM A DIVIDED-DEFFERENCE TABLE

```
x    y
1    0
              1
2    1                1/3
         10/3              -1/24
8    21              -1/24            -1/144
         3                 -1/16
10   27              -1/16         P(2,8,10,4,4)
         11/3         P(8,10,4,4)
4    5         P(10,4,4)        P(8,10,4,4,4)
         P(4,4)        P(10,4,4,4)
4    P(4)   P(4,4,4)
         P(4,4)
4    P(4)
```

If one repeats the argument 4, as in the table shown, one should obtain P(4,4) and P(4,4,4) at the locations shown and hence, by Eq. (2), the first and second derivatives of f(x). One can, however, fill in the numerical values of the divided differences of P(x) below the line only by making use of the fact that P(x) is a fourth-degree polynomial and has therefore a constant fourth divided difference, i.e.,

$$P(2,8,10,4,4) = P(8,10,4,4,4) = -\frac{1}{144}.$$

This permits one to work from the right to complete the table; thus by the fundamental property of divided differences

$$P(8,10,4,4) = -\frac{1}{16} + (4-2)P(2,8,10,4,4)$$

$$= -\frac{11}{144}$$

$$P(10,4,4,4) = P(8,10,4,4) + (4-8)P(8,10,4,4,4)$$

$$= -\frac{7}{144}.$$

Having in this manner obtained the third divided differences from the fourth differences, one proceeds to obtain the second differences from the third, and then the first differences from the second in a completely similar way; thus

$$P(10,4,4) = -\frac{1}{6} + (4-8)(-\frac{11}{144}) = \frac{5}{36}$$

$$P(4,4,4) = \frac{5}{36} + (4-10)(-\frac{7}{144}) = \frac{31}{72}$$

$$P(4,4) = \frac{11}{3} + (4-10)\frac{5}{36} = \frac{17}{6}$$

Thus, by Eqs. (2)

$$f'(4) \cong \frac{17}{6} = 2.833$$

$$f''(4) \cong \frac{31}{36} = 0.861.$$

● **PROBLEM 4-7**

Find a polynomial which coincides with sin x at x = 21°, 22°, 24°, 25°. From it calculate sin 23°, to five decimal places.

Solution: Here the table of divided differences is

TABLE

x	sin x°			
21	.35837			
		1624		
22	.37461		-6	
		1606		0,
24	.40674		-6	
		1588		
25	.42262			

and the required polynomial is

y = 0.35837 + 0.01624(x − 21) − 0.00006(x − 21)(x − 22).

Substituting x = 23 into this expression gives y = 0.39073, which checks with sin 23° to five places.

● **PROBLEM 4-8**

Find approximately the real root of the equation

$$y^3 - 2y - 5 = 0.$$

Calculate to five decimal places.

Solution: Let
$$x = y^3 - 2y - 5.$$
This relation defines a function $y = f(x)$. The value of $f(0)$ is wanted. Attributing suitable values to y the following table of divided differences is obtained.

x	f(x)			
-1.941	1.9			
		+.10627		
-1.000	2.0		-.0060	
		+.09425		+.0005
+0.061	2.1		-.0044	
		+.08425		+.0003
+1.248	2.2		-.0034	
		+.07582		
+2.567	2.3			

Thus approximately, using

$$f(x) = f(x_1) + \sum_{s=1}^{n-1} (x - x_1) \cdots (x - x_s)[x_1 x_2 \cdots x_{s+1}]$$

one gets

y = 2.0 + 1 x .09425 + 1 x .061 x .0044 + 1 x .061 x 1.248

   x .0003 = 2.09454.

The corresponding value of x is -0.00013 and the above value of y is in error by about one unit in the last digit.

DIFFERENCE TABLES AND DIFFERENCE FORMULAS

● **PROBLEM 4-9**

Using the difference table given, find sin 86°, correct to five decimal places.

Solution: Since 86° occurs near the end of the table, use Newton's backward formula. We have h = 10, $x_0 = 90$, x - 90 = 10s, so that for x = 86°, s = -0.4. Then Newton's backward formula, using the bottom diagonal of the array, is

$$y = 1.00000 - (.01519)(.4) + (.02993)(.12000)$$
$$+ (.00139)(.06400) - (.00082)(.04160)$$
$$+ (.00004)(.02995) = 0.99757$$

The correct value to 5 places is 0.99756.

TABLE

x	$y = \sin x$	Δy	$\Delta^2 y$	$\Delta^3 y$	$\Delta^4 y$	$\Delta^5 y$
$0°$	.00000					
		17365				
$10°$	.17365		-528			
		16837		-511		
$20°$	.34202		-1039		31	
		15798		-480		14
$30°$	.50000		-1519		45	
		14279		-435		18
$40°$	.64279		-1954		63	
		12325		-372		2
$50°$	.76604		-2326		65	
		9999		-307		21
$60°$	.86603		-2633		86	
		7366		-221		-4
$70°$	.93969		-2854		82	
		4512		-139		
$80°$	.98481		-2993			
		1519				
$90°$	1.00000					

● **PROBLEM 4-10**

Find an approximate value of $\log_{10} 4.01$ from the given divided difference table.

TABLE

y	$\log_{10} y$			
4.0002	0.6020 817			
		+.108431		
4.0104	.6031 877		-.0136	
		+.108116		
4.0233	.6045 824		-.0130	
		+.107869		
4.0294	.6052 404			

Solution: The divided differences are as shown. Thus approximately log 4.01 = .6020817 + .0098 x .108431 + .0098 x .0004 x .0136 = .6031444, which is correct to seven places.

The error due to the remainder term is of order

$$\frac{.0098 \times .0004 \times .0133 \times .4343 \times 2}{y^3 \times 3!}$$

where y varies between 4.0002 and 4.0294, which is less than 2 in the 10th decimal place. The above value could therefore be affected only by errors of rounding in the seventh place.

● **PROBLEM 4-11**

Given that the following data is for a third-degree polynomial f(x), use an ordinary difference table to find the single error made in recording the data:

x	4	6	8	10	12	14	16	18	20	22	24
f(x)	123	137	154	179	217	273	353	459	599	777	998

TABLE 1

x	f(x)	Δf	$\Delta^2 f$	$\Delta^3 f$
4	123	14		
6	137	17	3	5
8	154	25	8	5
10	179	38	13	5
12	217	56	18	6
14	273	80	24	2
16	353	106	26	8
18	459	140	34	4
20	599	178	38	5
22	777	221	43	
24	998			
		$\Sigma = \overline{875}$	$\overline{207}$	$\overline{40}$
		= 998 −123	= 221 − 14	= 43 − 3

Solution: First, form an ordinary difference table, the result of which is shown in Table 1. Each of the entries in the three difference columns (Δf, $\Delta^2 f$, $\Delta^3 f$) are obtained by taking the difference between the two closest elements of the previous column. A partial check is shown for the construction of the table as it can be shown that the sum of the entries in each difference column must be equal to the difference between the last and first elements of the previous column.

The basis for finding the stated error in the data is the fact that the nth difference of an nth-degree polynomial is known to be a constant; i.e., $\Delta^n f(x)$ = constant if f(x) is a polynomial of degree n (n is a positive integer). Table 1 suggests that the correct value of $\Delta^3 f$ might be 5 and that the error is perhaps in f(16), the middle of the values of $\Delta^3 f$ which did not come out to be equal to 5. Therefore, construct a new difference table based on the assumption that all of the values of $\Delta^3 f$ = 5. Table 2 is constructed by working backwards from the 5's and using the top previous values for $\Delta^2 f$ (=3) and Δf (=14).

TABLE 2

x	f(x)	Δf	Δ²f	Δ³f
4	123	14		
6	137	17	3	5
8	154	25	8	5
10	179	38	13	5
12	217	56	18	5
14	273	79	23	5
16	352	107	28	5
18	459	140	33	5
20	599	178	38	5
22	777	221	43	
24	998			
		875	207	40

It can now be seen that the correct value of f(16) is 352, not 353 as recorded. That is, with f(16) changed to 352, and the other values of f(x) left unchanged, all the data points will exactly fit a certain third-degree polynomial (which could be found). This is known to be true because every value of Δ³f is equal to the same constant; the value of this constant in this particular problem happens to be 5.

● **PROBLEM 4-12**

For the function tabulated below, use an ordinary difference table to determine an estimate for its third derivative in the given interval.

x	1.00	1.05	1.10	1.15	1.20
f(x)	367.879	349.938	332.871	316.637	301.194

Solution: From the definition of ordinary differences, namely Δf(x) = f(x + h) − f(x) with the constant h being the difference between consecutive values of the variable x, one can write

$$\frac{\Delta f(x)}{h} = \frac{f(x+h) - f(x)}{h}$$

It is noted that the right-hand side of this equation becomes the derivative of f(x), with respect to x, as the value of h becomes zero, in the limit. Thus,

$$f'(x) \doteq \Delta f(x)/h$$

and this approximation will usually give good results for a relatively small h and with appropriate values for f(x).

Also, this result for f'(x) can be extended for the nth derivative, giving

$$f^{(n)}(x) \doteq \Delta^n f(x)/h^n$$

To use this result for the problem above, calculate the difference table for the given data.

TABLE

x	f(x)	Δf	Δ²f	Δ³f
1.00	367.879	-17.941	0.874	-0.041
1.05	349.938	-17.067	0.833	-0.042
1.10	332.871	-16.234	0.791	
1.15	316.637	-15.443		
1.20	301.194			

Using the average value of $\Delta^3 f$, one gets

$$f^{(3)}(x) \doteq \frac{\Delta^3 f}{h^3} = \frac{-0.0415}{(0.05)^3} = -332$$

That is, the approximation for the third derivative of the tabulated function, with respect to x, is -332.

The values of f(x) in the above given table are for $f(x) = 1000\,e^{-x}$. For this function, the value of the third derivative at the midpoint of the interval is $f^{(3)}(1.10) = -333$, whereas the average of the third derivatives at the ends (1.00 and 1.20) is -335. It is seen that the approximate value computed above is quite good, for an approximate value in the interval. The third derivative actually varies from -368 (at x = 1.00) to -301 (at x = 1.20).

• **PROBLEM 4-13**

From the data

x	2	3	5	8	9	12
f(x)	48	100	294	1080	1630	5824

find values of f(x) when x = 0,4,6,14 using the divided difference table.

TABLE OF DIVIDED DIFFERENCES

x_k	f_k					
2	48					
		52				
3	100		15			
		97		3.0		
5	294		33		0.5	
		262		6.5		0.1
8	1080		72		1.5	
		550		20.0		
9	1630		212			
		1398				
12	5824					

Solution: To calculate f(6) take the arguments in order of increasing $|x - x_k|$. Arrange the calculation as follows-

$6 - x_k$		$f(5) = 294$			294
	1	1 x	262		262
	-2	-2 x	33		-66
	3	-6 x	6.5		-39
	-3	18 x	0.5		9.0
	4	72 x	0.1		7.2
					$467.2 = f(6)$

The first column contains the X_k factors taking $x_k = 5$, 8, 3, 9, 2, in order; the second contains successive products of these factors with the appropriate d.d.'s, and the third resultant values of the terms in Newton's series, which are then summed.

Similarly for $f(4)$, introduce the f_k's in the order of $x_k = 3, 5, 2, 8, 9, 12$, which involves the d.d's in heavy type; for $f(0)$, take f_k's in order of increasing x_k, and therefore differences along the forward diagonal; for $f(14)$, the f_k's in order of decreasing x_k, and differences along the backward diagonal. The computations are as follows-

$f(4)$	$4 - x_k$	$f(3) = 100$			100
	1	1 x	97		97
	-1	-1 x	15		-15
	2	-2 x	3.0		-6
	-4	8 x	0.5		4
	-5	-40 x	0.1		-4
				$f(4) =$	176
$f(0)$	$-x_k$	$f(2) = 48$			48
	-2	-2 x	52		-104
	-3	6 x	15		90
	-5	-30 x	3.0		-90
	-8	240 x	0.5		120
	-9	-2160 x	0.1		-216
				$f(0) =$	-152

f(14)	14 - x_k	f(12) = 5824		5824
	2	2 x 1398		2796
	5	10 x 212		2120
	6	60 x 20		1200
	9	540 x 1.5		810
	11	5940 x 0.1		594
		f(14) =		13344

• **PROBLEM 4-14**

From the tabulated values of csc x for x = 1°, 2°, 3°, 4° calculate csc 1°30'.

TABLE OF RECIPROCAL DIFFERENCES

x	CSC x	ρ_1	ρ_2	ρ_3
1°	57.298677			
2°	28.653706	-0.034910142		
3°	19.107321	-0.10475169	0.017457	
4°	14.335588	-0.20956747	0.026225	342.05

Solution: The continued fraction formed for the top diagonal and with x = 1.5 is

$$y = 57.298677 + \cfrac{0.5}{-0.034910142 + \cfrac{-0.5}{-57.281220 + \cfrac{-1.5}{342.08}}}$$

which reduces to
$$y = 38.201548$$

The value given in the tables for csc 1°30' is 38.201547. Instead of calculating the interpolated value of y directly from the continued fraction as was done above, another method is to employ the successive convergents of the continued fraction

$$y = a_0 + \cfrac{x - x_0}{a_1 + \cfrac{x - x_1}{a_2 + \cfrac{x - x_2}{a_3 + \ldots}}}$$

where for brevity let

$$a_n = \rho_n(x_n x_{n-1} \ldots x_0) - \rho_{n-2}(x_{n-2} \ldots x_0).$$

If $z_n = p_n/q_n$ denotes the nth convergent of the foregoing fraction, one gets by the theory of continued fractions

$$p_0 = 1 \quad , \quad q_0 = 0 \quad ,$$
$$p_1 = a_0 p_0 \quad , \quad q_1 = 1 \quad , \quad z_1 = p_1/q_1$$
$$p_2 = a_1 p_1 + (x - x_0)p_0 \quad , \quad q_2 = a_1 q_1 \quad , \quad z_2 = p_2/q_2$$
$$p_3 = a_2 p_2 + (x - x_1)p_1 \quad , \quad q_3 = a_2 q_2 + (x - x_1)q_1, z_3 = p_3/q_3$$
$$p_{n+1} = a_n p_n + (x - x_{n-1})p_{n-1}, \quad q_{n+1} = a_n q_n + (x - x_{n-1})q_{n-1},$$
$$z_{n+1} = p_{n+1}/q_{n+1}$$

Accordingly, the computation may be arranged in the tabular form

n	a_n	$x - x_n$	p_n	q_n	z_n
0	a_0	$x - x_0$	1	0	
1	a_1	$x - x_1$	p_1	1	z_1
2	a_2	$x - x_2$	p_2	q_2	z_2
3	a_3	$x - x_3$	p_3	q_3	z_3
--	---	---	---	---	--

The values of the a's are taken from the table of reciprocal differences, the values of $x - x_i$ for the desired x are entered, the p's and q's can then be obtained by the recurrence relations above, and finally the successive convergents are found by $z_n = p_n/q_n$. For this problem the computation is shown below.

a_n	$x - x_n$	p_n	q_n	z_n
57.298677	0.5	1	0	
-0.034910142	-0.5	57.298677	1	57.298677
-57.281220	-1.5	-1.50030495	-0.034910142	42.976192
342.08		57.289957	1.4996955	38.201059
		19599.9989	513.06820	38.201547

While this procedure effects no saving of labor, it has the virtue of indicating whether or not the process is converging toward a fixed value.

● **PROBLEM 4-15**

Consider the reciprocal difference problem. Find the rational fraction for the values

x: 0 1 2 3
---
y: 1 1 2 3

Solution: In reciprocal difference and approximation problems by rational fractions, difficulties may arise in arriving at solutions. For instance assume there are five given points and that the determinant is

$$\begin{vmatrix} 1 & y & x & xy & x^2 & x^2y \end{vmatrix} = 0. \qquad (1)$$

Assume further that all five points satisfy a relationship of the type

$$a + by + cx + dxy = 0 \qquad d \neq 0$$

Then they also satisfy

$$ax + bxy + cx^2 + dx^2y = 0.$$

Using these two relations one can combine columns in (1) so as to have two columns with only zeros in the first five rows. It is now apparent that all the cofactors of the last row vanish and the determinant vanishes identically. Conversely if the determinant vanishes identically, all cofactors of the last row vanish. In particular

$$\begin{vmatrix} 1 & y_i & x_i & x_iy_i & x_i^2 \end{vmatrix} = 0 \qquad (2)$$

and

$$\begin{vmatrix} 1 & y_i & x_i & x_iy_i & x_i^2y_i \end{vmatrix} = 0 \qquad (3)$$

where $i = 0, 1, 2, 3$ in the first, second, ..., fourth rows respectively. Let the cofactors of the elements of some row, say the first row, of (2) be A, B, C, D, E, and for (3) let them be A', B', C', D', E. Note that E is the same in both. Then

$$A + By + Cx + Dxy + Ex^2 = 0,$$

$$A' + B'y + C'x + D'xy + Ex^2y = 0,$$

will be satisfied by the 5 pairs (x_i, y_i). The equation in x obtained by eliminating y is of the form

$$E^2x^4 + \text{lower powers} = 0.$$

Since this equation of fourth degree is satisfied by five distinct values of x it vanishes identically and in particular $E = 0$.

The determinant set up to obtain the rational fraction may factor into two rational factors. Since the determinant is linear in y the factorization must be of the form

$$P(x) \; Q(x, y) = 0 \qquad (4)$$

where without loss of generality assume that $Q(x, y)$ is irreducible. Let the degree of $P(x)$ be p and let $Q(x,y) = 0$ define a rational fraction $y = R(x)$ of order q. If $k + 1$ points were used in forming the original determinant one must have $q + 2p \leq k$. Also since all these points satisfy (4), and at most p of them can satisfy $P(x) = 0$ the remainder, $k + 1 - p$, must satisfy $y = R(x)$. Now $q + 1$ points are required to determine $R(x)$, so that the equation $y = R(x)$ is satisfied by $k + 1 - p - (q + 1) = k - p - q$ points in excess of the number required to determine $R(x)$. This excess is designated as S "surplus"

points. Hence
$$S = k - p - q$$
and by the inequality above
$$S \geq p .$$
Here the determinant is
$$\begin{vmatrix} 1 & 1 & 0 & 0 & 0 \\ 1 & 1 & 1 & 1 & 1 \\ 1 & 2 & 2 & 4 & 4 \\ 1 & 3 & 3 & 9 & 9 \\ 1 & y & x & xy & x^2 \end{vmatrix} = 0$$
whence
$$xy = x^2$$
This equation is satisfied by all four values but the equation $y = x$ obtained from it is not. For this problem $k = 3$, $q = 1$, $p = 1$, $S = 1$.

The process of reducing the determinant or of constructing a table of divided differences may be halted by the presence of zero divisors.

When this occurs two or more divided differences of the same order must be equal, say
$$\rho_3(x_1 x_2 x_3 x_4) = \rho_3(x_0 x_1 x_2 x_3) .$$

This implies the existence of a rational fraction $y = R_3(x)$ of order three satisfied by all five pairs (x_0, y_0), ..., (x_4, y_4).

The continued fraction by which the desired representation of y is expressed may fail to give back the original values for which it was constructed.

It can, however, be shown that such a situation will not occur for a fraction terminating with ρ_k if no two reciprocal differences of the same order less than k are equal.

The foregoing discussion of exceptional cases is by no means exhaustive but only points out the most common difficulties. If the term "degenerate set" is used to describe a set of n + 1 points which satisfy a rational fraction of order less than n one may summarize the results as follows:

Theorem: If a set of n + 1 points is not degenerate and contains no degenerate subset there exists an irreducible rational fraction $y = R_n(x)$ of order n which is satisfied

by the given points. Aside from a common constant factor in numerator and denominator the fraction is unique.

● **PROBLEM 4-16**

Given the following equally spaced data:

x	0	1	2	3	4
$f(x)$	30	33	28	12	-22

Find f'(0), f'(2), f'(4), and f"(0) using difference representations which are of $\mathcal{O}(h)^2$ with the aid of the following formulas:

Central difference representation:

$$f'(x) = \frac{f_{j+1} - f_{j-1}}{2h} + \mathcal{O}(h^2) \tag{1}$$

Equation (1) states that the first derivative of f with respect to x is accurate to within an error of the order of $\mathcal{O}(h^2)$.

Higher order forward and backward difference representation:

$$f'(x) = \frac{-f_{j+2} + 4f_{j+1} - 3f_j}{2h} + \mathcal{O}(h^2) \tag{2}$$

Solution: At x = 0, a forward difference representation must be used since no points are available in the backward direction:

$$f'(0) = \frac{-f(2) + 4f(1) - 3f(0)}{2(1)} + \mathcal{O}(1)^2$$

$$f'(0) = \frac{-28 + 4(33) - 3(30)}{2} = 7 \text{ to } \mathcal{O}(1)^2$$

At x = 2 there are a choice of several representations. Arbitrarily select a central difference representation of $\mathcal{O}(h)^2$:

$$f'(2) = \frac{f(3) - f(1)}{2} + \mathcal{O}(1)^2$$

$$f'(2) = \frac{12 - 33}{2} = -10.5 \text{ to } \mathcal{O}(1)^2$$

At x = 4, a backward difference representation must be employed:

$$f'(4) = \frac{3f(4) - 4f(3) + f(2)}{2(1)} + \mathcal{O}(1)^2$$

$$f'(4) = \frac{3(-22) - 4(12) + 28}{2} = -43 \text{ to } \mathcal{O}(1)^2.$$

• **PROBLEM 4-17**

Given $f(x) = \sin x$. Find $f'(1)$ by using a central difference representation of $\mathcal{O}(h)^2$ with $h = 0.2$. Is this a sufficiently small mesh size for this problem?

Solution: It should immediately be apparent that the problem statement is ambiguous. Assume that the result should be accurate to two decimal places. The central difference with $h = 0.2$ is

$$f'(1) = \frac{\sin(1.2) - \sin(0.8)}{2(0.2)} + \mathcal{O}(0.2)^2$$

$$f'(1) = \frac{0.932039 - 0.717356}{0.4} = 0.53671 + \mathcal{O}(0.2)^2$$

Decide if the value of h chosen is sufficiently small. There is no way of knowing how good this result is since there is nothing to compare it with (presuming for the moment that one does not know how to differentiate sin x analytically). Thus take another difference representation using $h = 0.1$, one-half of its previous value:

$$f'(1) = \frac{\sin(1.1) - \sin(0.9)}{2(0.1)} + \mathcal{O}(0.1)^2$$

$$f'(1) = \frac{0.891207 - 0.783327}{0.2} = 0.53940 \text{ to } \mathcal{O}(0.1)^2.$$

The change in the answer was approximately 3 in the third decimal place. To be sure of the answer to two digits, it would be safest to cut the mesh in half once more and examine the change. Using $h = 0.05$.

$$f'(1) = \frac{\sin(1.05) - \sin(0.95)}{2(0.05)} + \mathcal{O}(0.05)^2$$

$$f'(1) = \frac{0.867423 - 0.813416}{0.1} = 0.54007 + \mathcal{O}(0.05)^2.$$

The change this time was only about 6 in the fourth decimal place and it would appear one can have confidence in the first two significant digits. (The exact answer is $\cos(1) = 0.54030$.) So while the original mesh size gave the answer correct to two places when rounded off, one could not be sure of this without also obtaining results for two smaller mesh sizes.

• **PROBLEM 4-18**

Given the function $f(x) = \tan 40x$. Find $f'(0.175)$ using a backward difference representation of $\mathcal{O}(h)$ with $h = 0.075$.

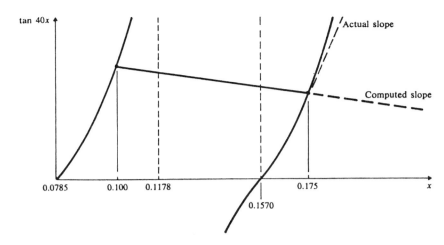

Solution:

$$f'(0.175) = \frac{f(0.175) - f(0.100)}{0.075} + \mathcal{O}(0.075)$$

$$= \frac{\tan(7.000) - \tan(4.000)}{0.075} + \mathcal{O}(0.075)$$

$$f'(0.175) = \frac{0.87145 - 1.15782}{0.075} = -3.81831 \text{ to } \mathcal{O}(0.075).$$

The only problem with this result is that it is absolute nonsense.

Examining a plot of tan 40x near x = 0.175 in the accompanying figure, find that the differencing has spanned a discontinuity in f(x), and the answer even has the wrong sign. Note, the moral is to be aware of the character of the function before blindly using difference representations.

DIFFERENCE OPERATORS

● **PROBLEM 4-19**

Sum to n terms the series whose nth term is

$(n - c)\sin(2an + b)$.

Solution: Here $\phi(x) = x - c$, $v_x = \sin(2ax + b)$ and the required sum is

$$P_{(n)}^{-1}(x + 1 - c)\sin(2ax + 2a + b)$$

$$= (n - c) P_{(n)}^{-1} \sin(2ax + 2a + b)$$

$$- P_{(n)}^{-2} \sin(2ax + 2a + b).$$

Now

$$\Delta \sin(2ax + b) = \sin(2ax + 2a + b) - \sin(2ax + b)$$
$$= 2\sin a \, \sin(2ax + b + a + \frac{\pi}{2}).$$

• **PROBLEM 4-20**

Find the sum of n terms of the series

$$1^2 + 2^2 + 3^2 + \ldots .$$

Solution: Here $u_x = x^2$, $\Delta u_x = 2x + 1$, $\Delta^2 u_x = 2$ and the required sum is

$$n + \frac{3n(n-1)}{2} + \frac{n(n-1)(n-2)}{3}$$

$$= \frac{n(n+1)(2n+1)}{6} .$$

• **PROBLEM 4-21**

Find the sum of n terms of the series

$$\frac{2}{1 \cdot 3 \cdot 4} + \frac{3}{2 \cdot 4 \cdot 5} + \frac{4}{3 \cdot 5 \cdot 6} + \ldots .$$

Solution: Here

$$u_x = \frac{x+1}{x(x+2)(x+3)}$$

$$= \frac{x(x+1) + x + 1}{x(x+1)(x+2)(x+3)} ,$$

$$u_{x+1} = (x+2)^{(-2)} + (x+1)^{(-3)} + x^{(-4)},$$

$$P_{(n)}^{-1} u_{x+1} = 2^{(-1)} + \frac{1}{2} 1^{(-2)} + \frac{1}{3} \cdot 0^{(-3)} - (n+2)^{(-1)}$$

$$- \frac{1}{2}(n+1)^{(-2)} - \frac{1}{3} n^{(-3)}$$

$$= \frac{1}{3} + \frac{1}{12} + \frac{1}{18} - \frac{1}{n+3} - \frac{1}{2(n+2)(n+3)}$$

$$- \frac{1}{3(n+1)(n+2)(n+3)} .$$

• **PROBLEM 4-22**

Evaluate $\sum_{s=1}^{n} (-1)^s s^v$.

Solution: One gets by
$$\nabla E_v(x) = x^v, \quad DE_v(x) = vE_{v-1}(x),$$

$$\sum_{s=1}^{n}(-1)^s s^v = \sum_{s=1}^{n}(-1)^s \nabla E_v(s)$$

$$= \frac{1}{2}\sum_{s=1}^{n}(-1)^s \left[E_v(s+1) + E_v(s)\right]$$

$$= \frac{1}{2}(-1)^n E_v(n+1) - \frac{1}{2}E_v(1).$$

• **PROBLEM 4-23**

Assume $u(t) = 2t/(t^2 + t - 1)$. From this show by expansion functions that $r_1 = (1/2)(-1 + \sqrt{5})$ and $r_2 = (1/2)(-1 - \sqrt{5})$.

Solution: One gets
$$\phi(r_1) = -1 + \sqrt{5} \quad \text{and} \quad \phi(r_2) = -1 - \sqrt{5}.$$

Since $D\psi(t) = 2t + 1$ it follows that
$$D\psi(r_1) = \sqrt{5} \quad \text{and} \quad D\psi(r_2) = -\sqrt{5},$$

consequently

$$f(x) = \frac{1-\sqrt{5}}{\sqrt{5}}\left[\frac{2}{-1+\sqrt{5}}\right]^{x+1}$$

$$- \frac{1+\sqrt{5}}{\sqrt{5}}\left[\frac{2}{-1-\sqrt{5}}\right]^{x+1};$$

this gives
$$f(x) = \frac{1}{2^{x-1}\sqrt{5}}\left[(1-\sqrt{5})^x - (1+\sqrt{5})^x\right],$$

and finally

(1) $\quad f(x) = \frac{-1}{2^{x-2}}\sum_{m=0}\binom{x}{2m+1}5^m.$

• **PROBLEM 4-24**

Given that
$$\Delta^k y_n = y_{n+k} - \binom{k}{1}y_{n+k-1} + \binom{k}{2}y_{n+k-2} + \ldots + (-1)^k y_n,$$

and

$$y_{n+k} = y_n + \binom{k}{1}\Delta y_n + \binom{k}{2}\Delta^2 y_n + \ldots + \binom{k}{k}\Delta^k y_n.$$

Symbolically, $\Delta^k = (E - 1)^k$, and $E^k = (1 + \Delta)^k$, where E is the shifting operator; $E_y = \{y_{n+1}\}$, and Δ is the forward difference operator; $\Delta y = \{y_{n+1} - y_n\}$.

Then the Taylor series
$$f(x + h) = f(x) + hf'(x) + \frac{h^2}{2!}f''(x) + \frac{h^3}{3!}f'''(x) + \ldots$$
can be written symbolically as
$$Ef(x) = (1 + hD + \frac{(hD)^2}{2!} + \frac{(hD)^3}{3!} + \ldots)f(x).$$

Thus
$$E = e^{hD}.$$

It follows that
$$e^{hD} = 1 + \Delta.$$

Now take the logarithm of both sides:
$$hD = \ln(1 + \Delta) = \Delta - \frac{1}{2}\Delta^2 + \frac{1}{3}\Delta^3 - \ldots, \tag{1}$$

$$hf'(x) = (\Delta - \frac{1}{2}\Delta^2 + \frac{1}{3}\Delta^3 - \ldots)f(x).$$

(a) Check the correctness of the result, and

(b) Show that the relation of Eq. 1 holds for all polynomials.

Solution: (a) Consider the special case $f(x) = e^{\alpha x}$, where α is an arbitrary (complex) constant. Then
$$Df(x) = \alpha e^{\alpha x} = \alpha f(x),$$
$$\Delta f(x) = e^{\alpha(x+h)} - e^{\alpha x} = (e^{\alpha h} - 1)f(x),$$
$$\Delta^p f(x) = (e^{\alpha h} - 1)^p f(x),$$
$$\sum_{p=1}^{n} (-1)^p \frac{\Delta^p f(x)}{p} = \sum_{p=1}^{n} (-1)^p \frac{(e^{\alpha h} - 1)^p}{p} f(x).$$

If $|e^{\alpha h} - 1| < 1$, then as $n \to \infty$ the right-hand side converges to
$$\ln(1 + e^{\alpha h} - 1)f(x) = \alpha h e^{\alpha x} = hDf(x).$$

Thus, Eq. 1 holds for $f(x) = e^{\alpha x}$, assuming that $|e^{\alpha h} - 1| < 1$, i.e.,
$$hDe^{\alpha x} = (\Delta - \frac{1}{2}\Delta^2 + \frac{1}{3}\Delta^3 \ldots)e^{\alpha x}. \tag{2}$$

(b) Considering part (a) of the question, in this case the expansion on the right-hand side has only a finite number of terms. The first n terms give, then, a formula for numerical differentiation which is exact for all polynomials of degree n. This is characteristic of formulas which are derived with the operator method.

Now expand the two sides of Eq. 2 into power of α, and compare coefficients of $\alpha^p/p!$.

$$hDx^p = (\Delta - \frac{1}{2}\Delta^2 + \frac{1}{3}\Delta^3 - \ldots)x^p, \quad p = 0,1,2,\ldots .$$

By linear combination of these equations for various values of p, then, it follows that Eq. 1 is valid for all polynomials, which was to be shown.

In fact, Eq. 2 demonstrates that Eq. 1 holds for all functions of the form

$$f(x) = \sum_{j=0}^{n} c_j \exp(\alpha_j x), \text{ if } |\exp(\alpha_j h) - 1| < 1.$$

CHAPTER 5

INTERPOLATION AND EXTRAPOLATION

INTERPOLATION TECHNIQUES

● **PROBLEM** 5-1

Using polynomial interpolation, with the aid of Newton's Forward Difference Formula, determine (a) $u(0.5)$, (b) $u(1.1)$, and (c) $u(x)$, consistent with the following data:

x	0	1	2	3	4
u(x)	5	8	17	44	101

Newton's Forward Difference Formula is

$$u(x) = (1 + \Delta)^x = u_0 + x\Delta u_0 + \frac{x(x-1)}{2!}\Delta^2 u_0 + \frac{x(x-1)(x-2)}{3!}\Delta^3 u_0 + \frac{x(x-1)(x-2)(x-3)}{4!}\Delta^4 u_0 + \ldots$$

<u>Solution</u>: First, form a forward difference table using the given data.

	x	u(x)	Δu	$\Delta^2 u$	$\Delta^3 u$	$\Delta^4 u$
x_0	0	5				
x_1	1	8	3			
x_2	2	17	9	6	12	
x_3	3	44	27	18	12	0
x_4	4	101	57	30		

132

Since $\Delta^4 u = 0$, the given data can be fit exactly with a third-degree polynomial. Using Newton's Forward Difference Formula, one gets

(a) $u(0.5) = 5 + (1/2)(3) + (-1/8)(6) + (1/16)(12) + 0$

$= 5 + 1.5 - 0.75 + 0.75 = 6.5$.

(b) $u(1.1) = 5 + 1.1(3) + 0.055(6) + (-0.0165)(12)$

$= 5 + 3.3 + 0.33 - 0.198 = 8.432$.

(c) $u(x) = 5 + x(3) + \frac{x(x-1)}{2}(6) + \frac{x(x-1)(x-2)}{6}(12)$

$= 5 + 3x + 3(x^2 - x) + 2(x^3 - 3x^2 + 2x)$

$u(x) = 2x^3 - 3x^2 + 4x + 5$.

It can be seen that this equation and the original data are in agreement by substituting the various values of x and computing the corresponding values of u.

It is noted that u(x) for specific values of x, such as u(0.5) and u(1.1) above, do not require the actual determination of the polynomial u(x), as later found in part (c). But, of course, after u(x) has been determined it may be used to find such values.

It is very important to understand that the independent variable, x, in Newton's formula above is only for x = 0, 1, 2, 3, ... That is, the first value of x is zero, and the interval of differencing h ($h = x_{i+1} - x_i$) must be equal to 1.

• **PROBLEM 5-2**

Using polynomial interpolation, with the aid of Newton's Forward Difference Formula, determine (a) u(12) and (b) u(t) consistent with the following data:

t	10	14	18	22
u(t)	5	1	-3	89

Newton's Forward Difference Formula is

$$u(x) = (1 + \Delta)^x = u_0 + x\Delta u_0 + \frac{x(x-1)}{2!}\Delta^2 u_0 +$$

$$\frac{x(x-1)(x-2)}{3!}\Delta^3 u_0 + \ldots$$

Solution: First form a forward difference table using the given data. Also, introduce a new independent variable x, defined so that its values are x = 0, 1, 2, 3. This can be done with the transformation equation

$$x = \frac{t - t_0}{h} = \frac{t - 10}{4}.$$

x	t	u(t)	Δu	$\Delta^2 u$	$\Delta^3 u$
0	10	5			
			-4		
1	14	1		0	
			-4		96
2	18	-3		96	
			92		
3	22	89			

Using Newton's Forward Difference Formula, one gets

(a) $u(t = 12) = u(x = 0.5)$

$$= 5 + (1/2)(-4) + (-1/8)(0) + (1/16)(96)$$

$$= 5 - 2 + 0 + 6 = 9.$$

Note that for this computation it was not really necessary to introduce the variabe x explicitly, but only to find the one value

$$x = (12 - 10)/4 = 0.5.$$

This value is merely the proportionate distance from $t = 10$ to $t = 12$, compared to $h = 14 - 10 = 4$.

(b) $u(x) = 5 + x(-4) + \frac{x(x-1)}{2}(0) + \frac{x(x-1)(x-2)}{6}(96)$

$$= 5 - 4x + 0 + 16(x^3 - 3x^2 + 2x)$$

$$u(x) = 16x^3 - 48x^2 + 28x + 5.$$

Note that this answer is in terms of x, not t as desired. The answer u(t) may now be found, though, by substituting

$$x = (t - 10)/4$$

into the polynomial u(x). This will give

$$u(t) = 0.25t^3 - 10.5t^2 + 142t - 615.$$

The above results may be checked by using this u(t) to calculate its values for $t = 10, 12, 14, 18,$ and 22.

• **PROBLEM 5-3**

Using the difference table for f, compute the polynomial $p \in P_H$ which interpolates f at x_j.

DIFFERENCE TABLE FOR f

j	x_j	$f(x_j)$	$\Delta f(x_j)$	$\Delta^2 f(x_j)$	$\Delta^3 f(x_j)$	$\Delta^4 f(x_j)$
0	5.0	1	4	-2	-3	5
1	5.5	5	2	-5	2	
2	6.0	7	-3	-3		
3	6.5	4	-6			
4	7.0	-2				

Solution: Given f has the values 1, 5, 7, 4, -2 at 5.0, 5.5, 6.0, 6.5, 7.0, respectively. One likes to compute the polynomial $p \in P_4$ which interpolates f at these points.

Introduce a new independent variable z. Then the transformation equation becomes

$$z = \frac{x - x_0}{h} = \frac{x - 5}{0.5}.$$

Consider $z = (x - 5)/0.5 = 2x - 10$.

Then

$$p(x) = 1 + 4z - 2\frac{z(z-1)}{2} - \frac{3z(z-1)(z-2)}{6} \qquad (1)$$

$$+ \frac{5z(z-1)(z-2)(z-3)}{24}.$$

Replacing z by 2x - 10, obtain

$$p(x) = 1 + 4(2x - 10) - (2x - 10)(2x - 11)$$

$$- \frac{(2x - 10)(2x - 11)(2x - 12)}{2}$$

$$+ \frac{5(2x - 10)(2x - 11)(2x - 12)(2x - 13)}{24}.$$

For many computational purposes it is more convenient to work with the expression (1) which gives p(x) in terms of z.

● **PROBLEM 5-4**

The following data and difference table are given:

x	t	y	Δy	$\Delta^2 y$	$\Delta^3 y$
x_{-1}	0.0	0			
			1987		
x_0	0.2	1987		-80	
			1907		-75
x_1	0.4	3894		-155	
			1752		
x_2	0.6	5646			

(a) Using only the first three points (x = 0.0, 0.2, 0.4), determine a value for y(0.24) by polynomial interpolation, with the aid of Stirling's Interpolation Formula.

(b) Using all four of the points, determine a value for y(0.3) by polynomial interpolation, with the aid of Bessel's Interpolation Formula.

Stirling's Interpolation Formula:

$$y = y_0 + x \frac{\Delta y_{-1} + \Delta y_0}{2} + \frac{1}{2}x^2 \Delta^2 y_{-1} + \ldots$$

Bessel:

$$y = \frac{y_0 + y_1}{2} + (x - \tfrac{1}{2})\Delta y_0 + \frac{x(x-1)}{2!}\left(\frac{\Delta^2 y_{-1} + \Delta^2 y_0}{2}\right)$$

$$+ \frac{x(x-\tfrac{1}{2})(x-1)}{3!}\Delta^3 y_{-1} + \ldots$$

HINT: In each of these formulas, x is the proportionate distance from t_0 to the value of t for which y is desired, compared to the interval of differencing h. The base point, t_0, may be taken anywhere. It is taken at t = 0.2 as indicated in the table. Note that the subscripts on the differences are constant going down a straight line to the right. For example, the first values given in the table are

$$\Delta y_{-1} = 1987, \quad \Delta^2 y_{-1} = -80, \quad \Delta^3 y_{-1} = -75.$$

The subscript comes from the level of the corresponding value of y; $y_{-1} = 0$.

Solution: (a) Since the interval of differencing is h = 0.2, for t = 0.24 one gets

$$x = \frac{0.24 - 0.2}{0.2} = 0.2.$$

Using Stirling's formula, one gets (using only the first three points)

$$y(0.24) = 1987 + 0.2 \frac{1987 + 1907}{2} + \frac{(0.2)^2}{2}(-80)$$

$$= 1987 + 389.4 - 1.6 = 2375.$$

(b) Using Bessel's formula and all four points, and with

$$x = \frac{0.3 - 0.2}{0.2} = 0.5,$$

$$y(0.3) = \frac{1987 + 3894}{2} + 0 + \frac{0.5(0.5 - 1)}{2} \left(\frac{-80 - 155}{2}\right) + 0$$

$$= 2940.5 + 14.7 = 2955.$$

It is noted that Bessel's formula is particularly convenient for x = 1/2 since, in that case, every other term is then zero.

The data in the table actually came from $y = 10000 \sin t$, with t in radians. For this function $y(0.24) = 2377$ and $y(0.3) = 2955$. It is seen that the answer in (a) is off by 2 in the last digit while the answer in (b) is correct. Remember that one more point was used in (b) than (a); actually part (a) is done with a second-degree polynomial and (b) with a third-degree polynomial.

• **PROBLEM** 5-5

Using polynomial interpolation, with the aid of Newton's Divided Difference Formula, find a value for y(-1) and the polynomial y(x) from the following data:

x	-2	0	1	2
y	4	10	10	16

Solution: First, construct a divided difference table for the given data. It is noted that only a divided difference table, not an ordinary (forward) table, can be formed when the interval of differencing is not a constant, as in the above data.

	x	y	δy	$\delta^2 y$	$\delta^3 y$
x_0	-2	4			
			3		
x_1	0	10		-1	
			0		1
x_2	1	10		3	
			6		
x_3	2	16			

Note that the first divided differences are

$$\delta(x_0, x_1) = \frac{y_1 - y_0}{x_1 - x_0}, \text{ etc.}$$

the second divided differences are

$$\delta^2(x_0, x_1, x_2) = \frac{\delta(x_1, x_2) - \delta(x_0, x_1)}{x_2 - x_0}, \text{ etc.}$$

etc.

Newton's Divided Difference Formula is

$$y = y_0 + (x - x_0)\dot{\delta}(x_0, x_1) +$$
$$(x - x_0)(x - x_1)\delta^2(x_0, x_1, x_2) +$$
$$(x - x_0)(x - x_1)(x - x_2)\delta^3(x_0, x_1, x_2, x_3) + \ldots$$

Thus, for $x = -1$, $y(-1) = 4 + (-1 + 2)3 + (-1 + 2)(-1 - 0)(-1)$
$$+ (-1 + 2)(-1 - 0)(-1 - 1)1$$
$$y(-1) = 4 + (1)(3) + (1)(-1)(-1) +$$
$$(1)(-1)(-2)(1)$$
$$= 4 + 3 + 1 + 2 = 10.$$

Similarly, $y(x) = 4 + (x + 2)3 + (x + 2)(x - 0)(-1)$
$$+ (x + 2)(x - 0)(x - 1)(1)$$
$$y(x) = 4 + (3x + 6) - (x^2 + 2x) + (x^3 + x^2 - 2x)$$
$$= x^3 - x + 10.$$

It is noted that values of $y(x)$ can be found for specific values of x without actually finding the interpolating polynomial. The above results can be checked by using the $y(x)$ determined to compute $y(-2)$, $y(-1)$, $y(0)$, $y(1)$, and $y(2)$.

It is also noted that Sheppard's Zigzag Rule can be used to obtain different paths for an interpolation using a divided difference table. The path used above starts with y_0 and goes down a straight line to $\delta^3 y = 1$. As an example, the following path is another possibility:

$$y = y_1 + (x - x_1)\dot{\delta}(x_1, x_2) +$$
$$(x - x_1)(x - x_2)\delta^2(x_0, x_1, x_2) +$$
$$(x - x_1)(x - x_2)(x - x_d)\delta^3(x_0, x_1, x_2, x_3)$$

and

$$y(-1) = 10 + (-1 - 0)(0) + (-1 - 0)(-1 - 1)(-1) +$$
$$(-1 - 0)(-1 - 1)(-1 + 2)(1) = 10.$$

In using Sheppard's Rule, in moving to the right one may only move up, or down, a half-line in selecting the next difference element to use at the end of each term. In larger-scale work, and with the numbers being rounded off, etc., it is usually more accurate to select a path which is fairly near a horizontal line through the value of x_i nearest the value of x for which $y(x)$ is desired.

• **PROBLEM 5-6**

Consider Table 1 of the function

$$f(x;y) = \frac{e^{xy} - 1}{x}$$

and interpolate for f(x;y) by the formula Bessel-Bessel with all the accuracy obtainable from the table.

TABLE 1

x	$y = 20$	22	24	26
0.035	28.9644	33.1362	37.6105	42.4092
0.040	30.6385	35.2725	40.2924	45.7304
0.045	32.4356	37.5830	43.2151	49.3776
0.050	34.3656	40.0833	46.4023	53.3859

Solution: First form the difference-table in x (Table 2).

TABLE 2

31.5370	36.4278	41.7538	47.5540
1.7971	2.3105	2.9227	3.6472
0.1280	0.1820	0.2526	0.3436
0.0099	0.0156	0.0237	0.0351

The first column contains the central differences, or means, obtainable from the first column of the function-table, that is

$$31.5370 = \frac{1}{2}(30.6385 + 32.4356),$$

$$1.7971 = 32.4356 - 30.6385,$$

etc. The second column of the difference-table in x contains the corresponding values, formed by means of the second column of the function-table, and so on.

From the difference-table in x now form, line by line, but otherwise by the same method, the difference-table in x and y (Table 3).

TABLE 3

39.0908	5.3260	0.4547	0.0390
2.6166	0.6122	0.1056	0.0135
0.2173	0.0706	0.0185	0.0038
0.0197	0.0081	0.0029	0.0009

Here one has, then,

$$39.0908 = \frac{1}{2}(36.4278 + 41.7538),$$

$$5.3260 = 41.7538 - 36.4278,$$

and so on.

Now write down the result, which is

$$f\left(x + \tfrac{1}{2};\ y + \tfrac{1}{2}\right) =$$

$$39.0908 + 2.6166x + 0.2173\frac{x^2 - \tfrac{1}{4}}{2} + 0.0197\frac{x(x^2 - \tfrac{1}{4})}{6}$$

$$+ y\left(5.3260 + 0.6122x + 0.0706\frac{x^2 - \tfrac{1}{4}}{2} + 0.0081\frac{x(x^2 - \tfrac{1}{4})}{6}\right)$$

$$+ \frac{y^2 - \tfrac{1}{4}}{2}\left(0.4547 + 0.1056x + 0.0185\frac{x^2 - \tfrac{1}{4}}{2} + 0.0029\frac{x(x^2 - \tfrac{1}{4})}{6}\right)$$

$$+ \frac{y(y^2 - \tfrac{1}{4})}{6}\left(0.0390 + 0.0135x + 0.0038\frac{x^2 - \tfrac{1}{4}}{2} + 0.0009\frac{x(x^2 - \tfrac{1}{4})}{6}\right)$$

$$+ R,$$

where R is given by

$$R = \frac{(x^2 - \tfrac{1}{4})(x^2 - \tfrac{9}{4})}{24} D_\xi^4 + \frac{(y^2 - \tfrac{1}{4})(y^2 - \tfrac{9}{4})}{24} D_\eta^4$$

$$- \frac{(x^2 - \tfrac{1}{4})(x^2 - \tfrac{9}{4})}{24} \frac{(y^2 - \tfrac{1}{4})(y^2 - \tfrac{9}{4})}{24} D_\xi^4 D_\eta^4.$$

D_ξ and D_η denote partial differentiation.

The formula assumes that the table interval is unity; it should therefore be noted that x and y have different meanings in the formula and in the function-table.

● **PROBLEM 5-7**

Given Table 1, find $\sin^{-1} 0.4$ in degrees and decimals using Bessel's Formula

$$p = [f - f_0 - (\delta_m^2 f_0 + \delta_m^2 f_1)(B_2(p))$$

$$- (\delta^3 f_{\tfrac{1}{2}})(B_3(p))]/\delta f_{\tfrac{1}{2}}. \qquad (1)$$

TABLE 1

x	$f(x) = \sin x$		$\delta^2 f$		$\delta^4 f$	$-0.184\delta^4 f$	$\delta_m^2 f$
0	0		0		0	0	0
		17365		−528			
10°	0·17365		− 528		17	−3	− 531
		16837		−511			
20°	·34202		−1039		31	−6	−1045
		15798		−480			
30°	·50000		−1519		45	−8	−1527
		14279		−435			
40°	·64279		−1954				
		12325					
50°	0·76604						

Solution: For interpolation in the interval $x = 20°$ to $30°$, one gets

$$f_0 = 34202$$

$$\delta f_{\frac{1}{2}} = 15798$$

$$\delta_m^2 f_0 + \delta_m^2 f_1 = -2572$$

$$\delta^3 f_{\frac{1}{2}} = -480$$

in terms of the fifth decimal as unit. One wants p for $f = 0.4$, $f - f_0 = 5798$ in terms of the fifth decimal. Hence, substituting in (1),

$$p = [5798 + 2572 B_2(p) + 480 B_3(p)]/15798$$

$$= 0.3670_1 + 0.1628 B_2(p) + 0.0304 B_3(p). \quad (2)$$

A nominal fifth decimal is kept here in $(f - f_0)/\delta f_{\frac{1}{2}}$, but not in the other terms since the quantities $B_2(p)$, $B_3(p)$ by which they are multiplied are less than $\frac{1}{10}$. The first term (2) is the value of p which would be obtained by linear interpolation. Taking it as a first approximation to p, the iterative process is as follows:

p	r.h.s. of (2)
0.367	$0.3670_1 + (0.1628)(-0.05808) + (0.0304)(+0.0051)$
	$= 0.3670_1 - 0.0094_6 + 0.0001_5 = 0.3577_0$
0.3577	$0.3670_1 + (0.1628)(-0.05744) + (0.0304)(+0.0054)$
	$= 0.3670_1 - 0.0093_5 + 0.0001_6 = 0.3578_2.$

The change in the value of the right-hand side of (2) is only about $\frac{1}{80}$ of the change in the value of p, so the value 0.3578₂ would not be changed by more than 1 in the fifth decimal (due to rounding errors) if the right-hand side were evaluated for the better approximation p = 0.3578₂. The number of figures in $\delta f_{\frac{1}{2}}$ is not enough to determine the fifth decimal in p to several units.

According to the purpose for which the value of $\sin^{-1} 0.4$ was wanted, it could be rounded off to four decimals, or the fifth retained as a guarding figure; if the latter course is taken it would be advisable to write it as a suffix, as a reminder that it is subject to an uncertainty of several units. Thus the result would be written $\sin^{-1} 0.4 = 23.5782°$.

● **PROBLEM 5-8**

Given the values in the table, find $f(\frac{2}{3})$ using Bessel's Formula and Everett's Formula.

x	$f(x)$	δf	$\delta^2 f$	$\delta^3 f$	$\delta^4 f$	$-0.184\delta^4 f$	$\delta_m^2 f$	$\delta_m^2 f_0 + \delta_m^2 f_1$
0.50	2.00000							
		−18182						
.55	1.81818		3031					
		−15151		−701				
.60	.66667		2330		203	−37	2293	
		−12821		−498				
.65	.53846		1832		131	−24	1808	3256
		−10989		−367				
.70	.42857		1465		93	−17	1448	
		−9524		−274				
.75	.33333		1191		63	−12	1179	
		−8333		−211				
.80	.25000		980					
		−7353						
0.85	1.17647							

Solution: Here $(\delta x) = 0.05$, $x = \frac{2}{3} = 0.65 + \frac{1}{3}(\delta x)$, so $p = \frac{1}{3}$, $q = \frac{2}{3}$.

(i) By Bessel's formula

$$B_2(p) = \frac{1}{4}p(p-1) = -\frac{1}{18}$$

$$B_3(p) = \frac{1}{6}p(p-1)(p-\frac{1}{2}) = +\frac{1}{162}$$

$$f_0 = 1.53846$$

$$p\delta f_{\frac{1}{2}} = -0.03663_0$$

$$B_2(p)(\delta_m^2 f_0 + \delta_m^2 f_1) = -0.00180_9$$

$$B_2(p)\delta^3 f_{\frac{1}{2}} = -0.00002_3$$

$$\overline{1.49999_8}$$

Rounded off to five decimals

$$= 1.50000$$

(ii) By Everett's formula

$$E_2(p) = -\frac{1}{6}q(1-q^2) = -\frac{5}{81}$$

$$F_2(p) = -\frac{1}{6}p(1-p^2) = -\frac{4}{81}$$

$$f_0 = 1.53846$$

$$p\delta f_{\frac{1}{2}} = -0.03663_0$$

$$E_2(p)\delta_m^2 f_0 = -0.00111_6$$

$$F_2(p)\delta_m^2 f_1 = -0.00071_5$$

$$\overline{1.49999_9}$$

Rounded off to five decimals

$$= 1.50000$$

In this problem modified second differences have been used, and the residual contributions from fourth differences are negligible. If modified second differences had not been used, it would have been necessary to include the fourth difference terms in each case.

• **PROBLEM 5-9**

Using polynomial interpolation, with the aid of Lagrange's Interpolation Formula, find a value for $y(-1)$ and the polynomial $y(x)$ from the data given:

x	-2	0	1	2
y	4	10	10	16

Solution: Lagrange's Interpolation Formula for 4 data points is:

$$y(x) = y_0 \frac{(x-x_1)(x-x_2)(x-x_3)}{(x_0-x_1)(x_0-x_2)(x_0-x_3)} + y_1 \frac{(x-x_0)(x-x_2)(x-x_3)}{(x_1-x_0)(x_1-x_2)(x_1-x_3)}$$

$$+ y_2 \frac{(x-x_0)(x-x_1)(x-x_3)}{(x_2-x_0)(x_2-x_1)(x_2-x_3)} + y_3 \frac{(x-x_0)(x-x_1)(x-x_2)}{(x_3-x_0)(x_3-x_1)(x_3-x_2)}.$$

$$y(-1) = 4\frac{(-1-0)(-1-1)(-1-2)}{(-2-0)(-2-1)(-2-2)} + 10\frac{(-1+2)(-1-1)(-1-2)}{(0+2)(0-1)(0-2)}$$

$$+ 10\frac{(-1+2)(-1-0)(-1-2)}{(1+2)(1-0)(1-2)} + 16\frac{(-1+2)(-1-0)(-1-1)}{(2+2)(2-0)(2-1)}$$

$$= 1 + 15 - 10 + 4 = 10,$$

$$y(x) = 4\frac{(x-0)(x-1)(x-2)}{(-2-0)(-2-1)(-2-2)} + 10\frac{(x+2)(x-1)(x-2)}{(0+2)(0-1)(0-2)}$$

$$+ 10\frac{(x+2)(x-0)(x-2)}{(1+2)(1-0)(1-2)} + 16\frac{(x+2)(x-0)(x-1)}{(2+2)(2-0)(2-1)}$$

$$= (-1/6)(x^3-3x^2+2x) + (5/2)(x^3-x^2-4x+4)$$

$$+ (10/3)(x^3-4x) + 2(x^3+x^2-2x)$$

$$y(x) = x^3 - x + 10.$$

All points of the data are used, and there is no round-off error in this problem. To put it another way, the basis of the formula is putting $\delta^4 y = 0$, with there being 4 points used.

• **PROBLEM 5-10**

Construct a polynomial having the values given by the table

x :	1	3	4	6
y :	-7	5	8	14

with the aid of the Lagrange's Interpolation Formula.

Solution: By substituting the given values of x and y into Lagrange's Interpolation Formula

$$I(x) = y_0 L_0^{(n)}(x) + y_1 L_1^{(n)}(x) + \ldots + y_n L_n^{(n)}(x),$$

where

$$L_j^{(n)}(x) = \frac{(x-x_0) \ldots (x-x_{j-1})(x-x_{j+1}) \ldots (x-x_n)}{(x_j-x_0) \ldots (x_j-x_{j-1})(x_j-x_{j+1}) \ldots (x_j-x_n)},$$

gives the interpolating polynomial for the n + 1 points (x_0, y_0), (x_1, y_1), . . . , (x_n, y_n), one gets

$$y = (-7)\frac{(x-3)(x-4)(x-6)}{(-2)(-3)(-5)} + (5)\frac{(x-1)(x-4)(x-6)}{(2)(-1)(-3)}$$

$$+ (8)\frac{(x-1)(x-3)(x-6)}{(3)(1)(-2)} + (14)\frac{(x-1)(x-3)(x-4)}{(5)(3)(2)}$$

which upon a simplification reduces to

$$y = \frac{1}{5}(x^3 - 13x^2 + 69x - 92).$$

It is readily verified by substitution that this polynomial satisfies the required conditions.

If the interpolation formula is to be used simply for calculating values of the function corresponding to assigned values of x it is unnecessary to reduce the polynomial to the simplified form as the substitution may be made directly into the original expression in product form.

● **PROBLEM 5-11**

Find $e^{0.9}$, given:

x	0.6	0.7	0.8	1.0
$y = e^x$	1.82212	2.01375	2.22554	2.71828

x	1.3	1.4
$y = e^x$	3.66930	4.05520

using the Lagrange, Aitken and Neville processes.

Solution: The Lagrange form yields, for the interpolation,

$$\frac{(2)(1)(1)(4)(5)}{(1)(2)(4)(7)(8)} 1.82212 - \frac{(3)(1)(1)(4)(5)}{(1)(1)(3)(6)(7)} 2.01375$$

$$+ \frac{(3)(2)(1)(4)(5)}{(2)(1)(2)(5)(6)} 2.22554 + \frac{(3)(2)(1)(4)(5)}{(4)(3)(2)(3)(4)} 2.71828$$

$$- \frac{(3)(2)(1)(1)(5)}{(7)(6)(5)(3)(1)} 3.66930 + \frac{(3)(2)(1)(1)(4)}{(8)(7)(6)(4)(1)} 4.05520$$

$$= 2.45960.$$

Some 150 steps are required (each addition, subtraction, multiplication, and division is counted as one step; the tabulation or recording of a number is not counted), but many of the steps here can be done mentally. If this form were to be used for the evaluation of

e^x for several values of x, the fractions $- 1.82212/(0.1)(0.2)(0.4)(0.7)(0.8)$, ..., which remain constant, would be calculated first, and then the multipliers $(x - 0.6)(x - 0.7) \ldots (x - 1.4)$, ..., for each value of x. After the fractions are evaluated, each y can be found with the expenditure of about 65 steps, most of which can be done mentally.

To evaluate $e^{0.9}$ by means of divided differences, first form the array as in Table 1.

TABLE 1

0.6	1.82212							
		1.91630						
0.7	2.01375		1.00800					
		2.11790		0.36167				
0.8	2.22554		1.15267		0.10254			
		2.46370		0.43345			0.01919	
1.0	2.71828		1.41274		0.11789			
		3.17007		0.51597				
1.3	3.66930		1.72232					
		3.85900						
1.4	4.05520							

Then evaluate

$$\frac{1}{1} \quad \frac{x - x_0}{x - x_0} \quad \frac{x - x_1}{(x - x_0)(x - x_1)} \quad \left| \ldots \right| \quad \frac{x - x_{n-1}}{(x - x_0)(x - x_1) \ldots (x - x_{n-1})}$$

Here, for x = 0.9,

$$\frac{1}{1} \quad \frac{0.3}{0.3} \quad \frac{0.2}{0.06} \quad \frac{0.1}{0.006} \quad \frac{-0.1}{-0.0006} \quad \frac{-0.4}{0.00024}$$

The entries in the second row are multiplied into the corresponding entries of the uppermost descending diagonal of the divided difference table, and the results added to yield 2.45960. It takes 45 steps to compute the difference table and about 20 more steps to reach the final result. Since the same divided difference table can be used to evaluate additional values of

e^x (in the range $0.5 \leq x \leq 1.5$), only 20 steps are necessary to calculate each additional value.

The tables 2 and 3 for the Aitken and Neville processes respectively, take about 65 steps each to reach the answer 2.45960, and, because of the nature of the process,

TABLE 2

0.6	0.3	1.82212					
0.7	0.2	2.01375	2.39701				
0.8	0.1	2.22554	2.42725	2.45759			
1.0	−0.1	2.71828	2.49424	2.46183	2.45966		
1.3	−0.4	3.66930	2.61377	2.46926	2.45984	2.45960	
1.4	−0.5	4.05520	2.65952	2.47201	2.45991	2.45960	2.45960

TABLE 3

0.6	0.3	1.82212					
			2.39701				
0.7	0.2	2.01375		2.45749			
			2.43733		2.45966		
0.8	0.1	2.22554		2.46038		2.45960	
			2.47191		2.45951		2.45960
1.0	−0.1	2.71828		2.45778		2.45960	
			2.40127		2.45984		
1.3	−0.4	3.66930		2.47016			
			2.12570				
1.4	−0.5	4.05520					

will take exactly the same number for the calculation of additional values. Note that a column of differences $0.9 - 0.6 = 0.3$, $0.9 - 0.7 = 0.2$, ..., $0.9 - 1.4 = -0.5$, has been inserted between the column for x and that for y.

• **PROBLEM 5-12**

(a) Given four discrete data points, (x_i, y_i), $i = 1, 2, 3, 4$, write down the equations of the Lagrange quadratics $p_2(x)$ using first the values at x_1, x_2, and x_3 and then the values at x_2, x_3, and x_4.

(b) Demonstrate that the Neville formula for the quadratic polynomial $p_{13}(x)$ through the points (x_1, y_1), (x_2, y_2), and (x_3, y_3) is equivalent to the Lagrange quadratic polynomial found in (a) for the same points.

(c) Interpolate the function value at $x = 2$ from the following data using a cubic Lagrange polynomial and a hand calculator:

x	0	1	3	4
y	8	9	35	72

(d) Repeat (c) using the Neville method. Rearrange the data to prevent extrapolations.

(e) Interpolate the function value at $x = 3.5$ from the following data using the (natural) cubic splines method and a hand calculator:

x	1	2	3	4	5
y	1	2	1	1.5	1

(f) Show that when the data (x_i, y_i), $i = 1, 2, \ldots, n$, are approximated by a least-squares polynomial

$$F(x) = \sum_{j=0}^{m} a_j x^j,$$

then the solution of the normal equations can be written as

$$a = (X^T X)^{-1} X^T y$$

Solution: (a) Notice how the formulae for the coefficients $L_i(x)$, $i = 1, 2, 3$ depend upon the choice of points. For the points x_1, x_2, and x_3

$p_2(x) = [(x - x_2)(x - x_3)] / [(x_1 - x_2)(x_1 - x_3)]$

$\quad + [(x - x_1)(x - x_3)] / [(x_2 - x_1)(x_2 - x_3)]$

$\quad + [(x - x_1)(x - x_2)] / [(x_3 - x_1)(x_3 - x_2)]$.

For the points x_2, x_3, and x_4,

$p_2(x) = [(x - x_3)(x - x_4)] / [(x_2 - x_3)(x_2 - x_4)]$

$\quad + [(x - x_2)(x - x_4)] / [(x_3 - x_2)(x_3 - x_4)]$

$\quad + [(x - x_2)(x - x_3)] / [(x_4 - x_2)(x_4 - x_3)]$

(b) The result follows directly by substituting the equations for the Neville polynomials $p_{12}(x)$ and $p_{23}(x)$, through the paired data points, $(x_1, y_1)(x_2, y_2)$ and (x_2, y_2), (x_3, y_3) respectively, into the expression for the polynomial $p_{13}(x)$.

(c) The Lagrange coefficients are $L_1(2) = -1/6$, $L_2(2) = 2/3$, $L_3(2) = 2/3$, and $L_4(2) = -1/6$ giving $p_3(2) = 16$. The discrete data are predicted from the functional relationship $y = 8 + x^3$.

(d) For the Neville method, extrapolations are avoided by the following tableau:

Tableau

			Neville polynomial functions		
x	x̄ - x	y	1st-degree	2nd-degree	3rd-degree
1	1	9			
			$P_{12}(2) = 22$		
3	-1	35		$P_{13}(2) = 18$	
			$P_{23}(2) = 26$		$P_{14}(2) = 16$
0	2	8		$P_{24}(2) = 12$	
			$P_{34}(2) = 40$		
4	-2	72			

(e) Notice that for the natural cubic spline, the second-order derivatives $M_1 = M_5 = 0$ and the matrix equations can be reduced to a 3 x 3 system. Solve this system by hand in two steps using Gauss elimination and determine the equation for the piecewise polynomial $p_{3,3}(x)$ over the third subinterval. Then insert x = 3.5 into this equation to find the required function value.

The interval widths have a constant value $h_i = 1$, i = 1,2,3,4. The matrix coefficients are

$a_2 = a_3 = a_4 = \frac{1}{2}$; $c_2 = 1$; $c_3 = -1$; $c_4 = \frac{1}{2}$; $c_5 = -\frac{1}{2}$;

$b_2 = b_3 = b_4 = \frac{1}{2}$; $d_2 = -6$, $d_3 = 9/2$; $d_4 = -3$.

Thus, we have the following matrix system:

$$\begin{bmatrix} 2 & 0 & 0 & 0 & 0 \\ \frac{1}{2} & 2 & \frac{1}{2} & 0 & 0 \\ 0 & \frac{1}{2} & 2 & \frac{1}{2} & 0 \\ 0 & 0 & \frac{1}{2} & 2 & \frac{1}{2} \\ 0 & 0 & 0 & 0 & 2 \end{bmatrix} \begin{bmatrix} 0 \\ M_2 \\ M_3 \\ M_4 \\ 0 \end{bmatrix} = \begin{bmatrix} 0 \\ -6 \\ 9/2 \\ -3 \\ 0 \end{bmatrix}$$

which reduces to

$$\begin{bmatrix} 2 & \frac{1}{2} & 0 \\ \frac{1}{2} & 2 & \frac{1}{2} \\ 0 & \frac{1}{2} & 2 \end{bmatrix} \begin{bmatrix} M_2 \\ M_3 \\ M_4 \end{bmatrix} = \begin{bmatrix} -6 \\ 9/2 \\ -3 \end{bmatrix}$$

Two-stage Gauss elimination gives the second-order derivatives $M_2 = -3.96429$, $M_3 = 3.85714$, $M_4 = -2.46429$.

Over the interval [3,4] the interpolating polynomial becomes

$$p_{3,3}(x) = y_4(x - x_3) - y_3(x - x_4) - (M_3/6)[(x - x_4)^3 - (x - x_4)] + (M_4/6)[(x - x_3)^3 - (x - x_3)].$$

Substitution of the numerical values into this last expression yields the solution $p_{3,3}(3.5) = 1.16294$.

Notice that M_2 is found by solving the matrix equation but not used in the polynomial evaluations over the interval $[3,4]$.

(f) In the equation $a = (X^T X)^{-1} X^T y$
the symbol T indicates a transposed matrix and

$$X = \begin{bmatrix} 1 & x_1 & x_1^2 & \cdots & x_1^m \\ 1 & x_2 & x_2^2 & & x_2^m \\ \cdot & \cdot & \cdot & & \cdot \\ \cdot & \cdot & \cdot & & \cdot \\ \cdot & \cdot & \cdot & & \cdot \\ 1 & x_n & x_n^2 & \cdots & x_n^m \end{bmatrix} \qquad y = \begin{bmatrix} y_1 \\ y_2 \\ \cdot \\ \cdot \\ \cdot \\ y_n \end{bmatrix}$$

$$n \times m + 1 \qquad\qquad n \times 1$$

The required matrix equations are obtained directly by matrix multiplication and transposition. Note that matrix formulation is of considerable value in designing efficient algorithms for accessing and storing data and carrying out row operations.

To demonstrate that the normal equations derived in (f) can be modified by introducing a matrix of weight factors,

$$W = \begin{bmatrix} w_1 & & & & \\ & w_2 & & & \\ & & \cdot & & \\ & & & \cdot & \\ & & & & \cdot \\ & & & & & w_n \end{bmatrix}$$

$$n \times n$$

to give the weighted, least-squares solution as

$$a = (X^T W X)^{-1} X^T W y,$$

it may be useful to give m a numerical value, say two or three.

• **PROBLEM 5-13**

Calculate exp (.0075) from the following table:

x	e^x	Δ	Δ^2
0.006	1.00601 80361		
		10065212	
.007	1.00702 45573		10070
		10075282	
.008	1.00803 20855		

Use Stirling's formula.

Solution: Here $p = .5$.

The coefficient of the second difference is .125, and since $\omega = .001$ the remainder term is $-.625 \times (.001)^3 e^x = -6 \times 10^{-11}$ approximately.

Thus, using Stirling's formula, one gets

$$\exp(.0075) = 10070245573$$
$$+ \tfrac{1}{4}(10065212 + 10075282) + \tfrac{1}{8}(10070) - .6$$
$$= 1.0075281955.$$

• **PROBLEM 5-14**

Calculate the rate of interest if annuity, certain for 30 years, is 20, the annual payment being 1.

Solution: The equation to be solved in order to calculate the rate of interest is

$$\phi(y) \equiv \frac{1 - (1+y)^N}{y} - A \qquad (1)$$

where N = number of years

A = annuity-certain.

Then

$$\phi(y) \equiv \frac{1 - (1+y)^{-30}}{y} - 20 = 0$$

Now find in an interest-table

Rate of interest	30-years annuity
$2\frac{1}{2}\%$	20.9303
3%	19.6004
$3\frac{1}{2}\%$	18.3920

If, therefore, $\phi(y)$ is denoted by x, and y by $f(x)$, the following table may be formed.

x	f(x)	f_1	f_2
0.9303	0.025		
		-0.0037597	
-0.3996	0.030		0.00014892
		-0.0041377	
-1.6080	0.035		

It is not difficult to ascertain that, within the interval considered, f(x) satisfies the conditions mentioned above, so that (1) may be applied. Thus obtain, leaving aside the remainder-term,

$$f(0) = 0.025 + 0.9303 \times 0.0037597 - 0.9303 \times 0.3996$$
$$\times 0.00014892 = 0.028442,$$

being the required rate of interest.

If this value is inserted in the equation, one sees that it is a trifle too small, while it is found that 0.028450 is a trifle too large. The correct value to 6 places of decimals is, in fact, 0.028446.

● **PROBLEM 5-15**

Given the table of x versus y as follows:

x	2	4	6	8	10	12	14
y	23	93	259	569	1071	1873	2843

Using interpolation techniques, construct a forward difference table to find y when x = 4.2, using the Newton forward formula to find a cubic polynomial of the form $y = P_n(x)$.

Solution: Construct a forward difference table as shown. Then to find y when x = 4.2, pick x_0 anyplace in the table, so pick $x_0 = 4$. Now the differences necessary for hand calculation of y fall on a diagonal line down from x_0 as shown. (Of course h = 2.) Now, using the usual form for the Newton Forward Formula:

Table

x	y	Δy	Δ²y	Δ³y
2	23			
		70		
4	93		96	
		166		48
6	259		144	
		310		48
8	569		192	
		502		48
10	1071		240	
		742		48
12	1813		288	
		1030		
14	2843			

$$y = P_n(x) = P_n(x_0 + hu) = y_0 + \frac{u\Delta y_0}{1!} + \frac{u(u-1)}{2!}\Delta^2 y_0 + \ldots$$

$$+ \frac{u(u-1)(u-2)\ldots(u-n+1)}{n!}\Delta^n y_0 \qquad (2)$$

and recall:

$$u = \frac{x - x_0}{h} = \frac{y - y_0}{\Delta y_0}$$

hence

$$u = \frac{4.2 - 4}{2} = \frac{.2}{2} = .1$$

Substituting into equation (2), one gets

$$y = 93 + \frac{.1(166)}{1!} + \frac{.1(.1 - 1)(144)}{2!} + \frac{.1(.1 - 1)(.1 - 2)(48)}{3!}$$

$$y = 104.488$$

Since the accompanying table has constant third differences, one may conclude that it is derived from a cubic polynomial

$$y = x^3 + 7x + 1$$

If one substitutes x = 4.2 into the above cubic, one gets y = 104.488.

The construction of a table of forward differences permits us to see that interpolation to a high degree toward the bottom of the table with Newton's forward formula will not be as accurate, since the higher differences disappear.

In digital computer interpolation it is necessary to store only the table of x and y. From this one can compute the necessary differences with simple expressions for the various differences. This saves much time and space over storing the differences also.

LINEAR INTERPOLATION

• PROBLEM 5-16

Given $\sin \theta = 0.56432$ find $\cos \theta$.

Solution: From the trigonometric tables one gets

i	$x_i = \sin \theta$	$y_i = \cos \theta$	$x_i - x$
0	0.56425	0.82561	-7
1	0.56449	0.82544	17

Evaluating the determinant and dividing by 24, $\cos \theta = 0.82556$ is obtained. Note that any common factor may be removed from $(x_0 - x)$, $(x_1 - x)$, and $(x_1 - x_0)$.

Hence, one may ignore the decimal point and treat those quantities as integers. Note also that common digits on the left in y_0 and y_1 will also occur in y. Thus in this problem one could have ignored the digits 825 and written only

$$\frac{1}{24} \begin{vmatrix} 61 & -7 \\ 44 & 17 \end{vmatrix} = 56$$

so that again $\cos \theta = 0.82556$. In the case of repeated interpolations with numbers carried to many figures this may effect a worth-while saving of labor.

• PROBLEM 5-17

From a table of sin x for x given in radians at intervals of 0.1 find the value of sin 4.238.

Solution: The calculation is shown

x	sin x	1st degree	2nd degree	3rd degree	4th degree	5th degree	$x_i - x$
4.0	-0.75680250						-238
4.1	.81827711	-.90311207					-138
4.2	.87157577	-.89338269	-.88968553				-38
4.3	.91616594	-.88323083	939401	-0.88957475			62
4.4	.95160207	-.87270824	912631	7928	-0.88957194		162
4.5	.97753012	-.86186885	888326	8390	1	-0.88957199	262

The value should be -0.88957200.

• **PROBLEM 5-18**

Assume $f(x) = x^2 + x - 12$. Prepare a table of $f(x)$ and $f^{-1}(y), (y = f(x))$ for $x = 2.7, 2.9, 3.2, 3.5$. Find some approximations to $f^{-1}(0)$.

Solution:

x	f(x)	y	$f^{-1}(y)$
2.7	-3.01	-3.01	2.7
2.9	-.69	-.69	2.9
3.2	1.44	1.44	3.2
3.5	3.75	3.75	3.5

In this problem it can be seen that the equation $f(x) = 0$ has a solution in [2.7, 3.5]. To find the solution is equivalent to finding the value of $f^{-1}(0)$. Thus a reasonable way to approximate the solution to $f(x) = 0$ is to find the approximate value of $f^{-1}(0)$ by interpolation in the $f^{-1}(y)$ table. Using iterated interpolation, one gets

-3.01	3.01	2.7			
-.69	.69	2.9	2.9595		
1.44	-1.44	3.2	3.0382	2.9850	
3.75	-3.75	3.5	3.0562	2.9745	2.9915

Thus three successive approximations to the desired solution of $f(x) = 0$ are

$$P_{1,2} = 2.9595$$
$$P_{1,2,3} = 2.9850$$
$$P_{1,2,3,4} = 2.9915$$

The exact solution is $x = 3.000$.

● **PROBLEM 5-19**

Calculate the positive root of the equation

$$e^x - \frac{1}{x} = 0.$$

Solution: There is evidently only one such root. In order to find it, one may leave the equation as it stands; but it is slightly more convenient to write the equation in the form

$$x \log e + \log x = 0.$$

Now put

$$y = x \log e + \log x$$

and calculate, step by step, the accompanying table.

TABLE

x	y
0.5	−0.0839
0.6	0.0387
0.5684	0.0015
0.5671	−0.000052
0.5672	0.000068

It is first found, by trial, that the root is situated between 0.5 and 0.6. In thus locating the root, compute with the smallest possible number of figures; two figures will do. The two first values of y are now calculated to four places. Thereafter, the next argument x is calculated by linear interpolation, that is

$$x = 0.5 + 0.1 \frac{0.0839}{0.1226} = 0.5684,$$

and the corresponding value of y can be calculated to four places.

By means of the last value of x found and the one that is nearest to it, or x = 0.6, interpolate a new x

$$x = 0.5684 - 0.0015 \frac{0.0316}{0.0372} = 0.5671,$$

and the corresponding value of y is calculated to six places.

It is seen that we are now very close to the desired root, and that this must be either 0.5671 or 0.5672, if we are content with four figures in the root. Therefore there is no need to interpolate any more, but in order to decide between the two values, still calculate the value of y corresponding to x = 0.5672, and find that x = 0.5671 is preferable.

In this way, the approximation may evidently be carried as far, as one wishes. At the same time one has, in the direct calculation of y, a most efficient check on the correctness of the calculation. It is evidently of importance, not to make the calculation with more figures than necessary and, therefore, to begin with few figures and only to introduce more figures when they are actually required.

• **PROBLEM** 5-20

From the given values of the elliptic function $sn(x|0.2)$, find by interpolation the value of $sn(0.3|0.2)$ with the aid of the Aitken's process.

x	$sn(x\|0.2)$	(1)	(2)	(3)	(4)	(5)	Parts
0.0	0.00000						-3
.1	.09980	29940					-2
.2	.19841	29761.5	29583				-1
.4	.38752	29064	29356	...469.5			+1
.5	.47595	28557	29248.5	...471.5	...467.5		+2
.6	.55912	27956	29146.4	...473.85	...467.3	..7.9	+3

Solution: Aitken's Linear Process of Interpolation by Iteration:

Let u_a, u_b, u_c,..., dentoe the values of a function corresponding to the arguments a, b, c, ...

Denote the divided differences by [ab], [abc],... .
Let f(x; a, b, c), for example, denote the interpolation polynomial which coincides in value with u_x at the points a, b, c. Then

$$f(x; a, b) = u_a + (x - a)[ab],$$

$$f(x; a, b, c) = u_a + (x - a)[ab] + (x - a)(x - b)[abc],$$

$$f(x; a, b, c, d) = u_a + (x - a)[ab] + (x - a)(x - b)[abc] + (x - a)(x - b)(x - c)[abcd],$$

and so on.

Eliminating [abcd] yields

$$f(x; a, b, c, d) = \frac{(d - x)f(x; a, b, c) - (c - x)f(x; a, b, d)}{(d - x)-(c - x)}$$

$$= \begin{vmatrix} f(x; a, b, c) & c - x \\ f(x; a, b, d) & d - x \end{vmatrix} \div (d - c).$$

(1)

Thus f(x; a, b, c, d) is obtained by the ordinary rule of proportional parts from the values of f(x; a, b, y) for y = c, y = d.

Applying this rule, write down the following scheme:

```
Argu-   func-
ment    tion      (1)            (2)              (3)         ...   Parts

a       u_a                                                         a - x
b       u_b     f(x; a,b)                                           b - x
c       u_c     f(x; a,c)     f(x; a,b,c)                           c - x
d       u_d     f(x; a,d)     f(x; a,b,d)     f(x; a,b,c,d)         d - x
```

Each entry is formed by cross-multiplication and division, with the numbers in their actual positions, thus

$$f(x; a,b) = \begin{vmatrix} u_a & a-x \\ u_b & b-x \end{vmatrix} \div (b-a),$$

$$f(x; a,c) = \begin{vmatrix} u_a & a-x \\ u_c & c-x \end{vmatrix} \div (c-a),$$

$$f(x; a,d) = \begin{vmatrix} u_a & a-x \\ u_d & d-x \end{vmatrix} \div (d-a),$$

$$f(x; a,b,c) = \begin{vmatrix} f(x; a,b) & b-x \\ f(x; a,c) & c-x \end{vmatrix} \div (c-b),$$

and so on.

The above scheme constitutes Aitken's process.

The members of column (1) are linear interpolation polynomials, those of column (2) quadratic interpolation polynomials, those of column (3) cubic interpolation polynomials, and so on.

With regard to the column headed "parts," one may replace the entries by any number proportional to them, as is obvious from (1). In particular, if the arguments are equidistant, divide each entry in this column by the argument interval ω. Moreover, when the arguments are equidistant, this division by ω will make them differ by integers.

Considering the above problem, the "parts" are $-.3$, $-.2$, $-.1$, $+.1$, $+.2$, $+.3$, which one replaces by integers. Also treat the tabular numbers as integers and carry extra figures as a guard. After column (2) one can drop the figures 29. One could likewise treat the entries of column (3) as 9.5, 11.5, 13.85. The following are examples showing how the numbers are obtained.

$$29940 = \begin{vmatrix} 0 & -3 \\ 9980 & -2 \end{vmatrix} \div 1,$$

$$29064 = \begin{vmatrix} 0 & -3 \\ 38752 & +1 \end{vmatrix} \div 4,$$

$$469.5 = \begin{vmatrix} 583 & -1 \\ 356 & +1 \end{vmatrix} \div 2,$$

$$7.9 = \begin{vmatrix} 7.5 & 2 \\ 7.5 & 3 \end{vmatrix} \div 1.$$

The result is 0.29468, which is correct to five places.

● **PROBLEM** 5-21

Find sin 0.25 from the values given below with the aid of the Neville's Process of Iteration.

0.1	1.5	0.0998			
			2481.5		
0.2	.5	.1987		...73.6	
			2471		...4.1
0.3	-.5	.2955		...74.6	
			2485.5		
0.4	-1.5	.3894			

Solution: The Neville's Process of Iteration is indicated by the following scheme:

Argument | Parts | Function

a   x - a u_a
                    f(x; a,b)
b   x - b u_b f(x; a,b,c)
                    f(x; b,c)   f(x; a,b,c,d)
c   x - c u_c f(x; b,c,d)   f(x; a,b,c,d,e)
                    f(x; c,d)   f(x; b,c,d,e)
d   x - d u_d f(x; c,d,e)
                    f(x; d,e)
e   x - e u_e

Here

$$f(x; a,b,c) = \begin{vmatrix} x - a & f(x; a,b) \\ x - c & f(x; b,c) \end{vmatrix} \div (c - a) \qquad (1)$$

in order that the "parts" may be identified as lying at the base of a triangle of which the interpolate is the vertex.

The process of course leads to the same interpolation polynomial as Aitken's process when founded on the same arguments.

In the case of equal intervals the parts may be most conveniently treated by division with the tabular interval ω.

Considering the above problem, $x = 0.25$, $\omega = 0.1$; the parts are given in the second column.

$$2471 = \begin{vmatrix} .5 & 1987 \\ -.5 & 2955 \end{vmatrix},$$

$$74.6 = \begin{vmatrix} .5 & 71 \\ -1.5 & 85.5 \end{vmatrix} \div 2,$$

$$\sin 0.25 = 0.2474.$$

● **PROBLEM 5-22**

Determine the spacing h in a table of equally spaced values of the function $f(x) = \sqrt{x}$ between 1 and 2, so that interpolation with a second-degree polynomial in this table will yield a desired accuracy.

Solution: By assumption, the table will contain $f(x_i)$, with $x_i = 1 + ih$, $i = 0,\ldots,N$, where $N = (2-1)/h$. If $\bar{x} \in [x_{i-1}, x_{i+1}]$, then approximate $f(\bar{x})$ by $p_2(\bar{x})$, where $p_2(x)$ is the quadratic polynomial which interpolates $f(x)$ at x_{i-1}, x_i, x_{i+1}. The error is then

$$e_n(\bar{x}) = f(\bar{x}) - p_2(\bar{x})$$

$$= \frac{f^{(n+1)}(\xi)}{(n+1)!} \prod_{j=0}^{n} (\bar{x} - x_j),$$

$$e_n = f(\bar{x}) - p_2(\bar{x})$$

$$= (\bar{x} - x_{i-1})(\bar{x} - x_i)(\bar{x} - x_{i+1}) \frac{f'''(\xi)}{3!}$$

for some ξ in (x_{i-1}, x_{i+1}). ξ is unknown, hence $f'''(\xi)$ can merely be estimated.

$$|f'''(\xi)| \leq \max_{1 \leq x \leq 2} |f'''(x)|.$$

One calculates $f'''(x) = (3/8)x^{-5/2}$; hence $|f'''(\xi)| \leq 3/8$. Further,

$$\max_{x \in [x_{i-1}, x_{i+1}]} |(x - x_{i-1})(x - x_i)(x - x_{i+1})|$$

$$= \max_{y \in [-h, h]} |(y + h)y(y - h)|$$

$$= \max_{y \in [-h, h]} |y(y^2 - h^2)|$$

using the linear change of variables $y = x - x_i$. Since the function $\psi(y) = y(y^2 - h^2)$ vanishes at $y = -h$ and $y = h$, the maximum of $|\psi(y)|$ on $[-h, h]$ must occur at one of the extrema of $\psi(y)$. These extrema are found by solving the equation $\psi'(y) = 3y^2 - h^2 = 0$, giving $y = \pm h/\sqrt{3}$. Hence

$$\max_{x \in [x_{i-1}, x_{i+1}]} |(x - x_{i-1})(x - x_i)(x - x_{i+1})|$$

$$= \frac{2h^3}{3\sqrt{3}}.$$

So, for any $\bar{x} \in [1, 2]$,

$$|f(\bar{x}) - p_2(\bar{x})| \leq \frac{(2h^3/[3\sqrt{3}])(3/8)}{6} = \frac{h^3}{24\sqrt{3}}$$

if $p_2(x)$ is chosen as the quadratic polynomial which interpolates $f(x) = \sqrt{x}$ at the three tabular points nearest $\bar{x}$. If one wishes to obtain seven-place accuracy this way, one would have to choose h so that

$$\frac{h^3}{24\sqrt{3}} < 5 \cdot 10^{-8}$$

giving $h \cong 0.0128$, or $N \cong 79$.

INVERSE INTERPOLATION TECHNIQUES

• **PROBLEM** 5-23

Use iterative inverse interpolation to find x when y = 6.25 from the table.

x	2	3	4	5	6	7	8	9	10
y	4	9	16	25	36	49	64	81	100

Solution: The finite difference table is

x	y		
2	4		
		5	
3	9		2
		7	
4	16		2
		9	
5	25		2
		11	
6	36		2
		13	
7	49		2
		15	
8	64		2
		17	
9	81		2
		19	
10	100		

The formula is

$$u = \frac{y - y_0}{\Delta y_0} - \frac{1}{\Delta y_0} \frac{u(u-1)}{2} \Delta^2 y_0$$

since higher differences are zero. Now

$$u_0 = \frac{6.25 - 4}{5} = \frac{2.25}{5} = .45$$

$$u_1 = .45 - \frac{1}{5}\left[\frac{.45(.45-1)(2)}{2}\right] = .4995$$

$$u_2 = .45 - \frac{1}{5}\left[\frac{(.4995)(.4995-1)(2)}{2}\right] = .49999995.$$

Since u_2 is close to u_1, now accept from the equation that

$$x = 2 + (1)(.49999995) = 2.49999995.$$

Since the table is a table of squares, the result should be

$$x = 2.5.$$

● **PROBLEM 5-24**

From the following tabulated values of the function $y = Si(x)$ determine x corresponding to $y = 1.8519336917$.

x	y
3.12	1.8518625083
3.13	9156107
3.14	9366481
3.15	9258225
3.16	8833367

Solution: Using the values for $x = 3.13, 3.14, 3.15$ the following equation is set up:

$$318630 s^2 - 102118 s - 59128 = 0.$$

The two roots, taken to two decimal places, are

$$s = -0.30$$

and

$$s = 0.62,$$

to which correspond $x' = 3.1370$ and $x' = 3.1462$. Now treat each one of these separately by Aitken's Method in order to obtain closer approximations.

Direct interpolation with $x' = 3.1370$ gives the desired value of y immediately and no further refinement is required.

For the second value the successive approximations are obtained as follows

$$x' = 3.1462, \ y' = 1.8519336768,$$

$$x'' = 3.14619, \ y'' = 1.8519336915,$$

$$x''' = 3.1461899, \ y''' = 1.8519336917.$$

Note that the curve is so flat in the vicinity of the desired points that x cannot be determined accurately beyond seven decimal places. Accordingly, take $x = 3.1370000$ and $x = 3.1461899$ as the answers to the problem.

If the roots of the quadratic equation are imaginary one may infer that either the given problem has no solution or has two or more solutions so close together that the quadratic does not fit closely enough to locate them.

It is important to observe that this process of inverse interpolation by successive approximations is in no way dependent on the particular method used for performing the direct interpolations. Thus it may be used equally well with Aitken's, Neville's, or Lagrange's methods, or with any of the various methods based on differences.

● **PROBLEM 5-25**

Calculate

$$\int_{.37}^{1} \frac{dx}{\sqrt{(1 - x^2)(3/5 + (2/5)x^2)}}$$

from the following table of $cn(u | 2/5)$, with the aid of inverse interpolation by divided differences.

$cn(u\|2/5)$	u			
0.44122	1.2			
		−1.34048		
.36662	1.3		−.132	
		−1.32066		−.18
.29090	1.4		−.091	
		−1.30685		−.16
.21438	1.5		−.055	
		−1.29836		
.13736	1.6			

Solution: The problem of interpolation briefly stated consists of finding, from a table of the function, the value of the function which corresponds to a given argument. The problem of inverse interpolation is that of finding from the same table the argument corresponding to a given value of the function. Thus if y be a function of the argument x, given the table

Argument	Function
x_1	y_1
x_2	y_2

the argument x corresponding to a given functional value
y is required. A numerical table by its nature determines
a single-valued function of the argument but the inverse
function may very well be many-valued.

The given table, by interchanging the roles of the argument x and the function y, becomes

Argument	Function		
y_1	x_1		
		$[y_1 y_2]$	
y_2	x_2		$[y_1 y_2 y_3]$
		$[y_2 y_3]$	.
y_3	x_3	.	
.	.		

where the divided differences

$$[y_1 y_2] = (x_1 - x_2) \div (y_1 - y_2), \text{ etc, have been formed.}$$

Then obtain

$$x = x_1 + (y - y_1)[y_1 y_2] + (y - y_1)(y - y_2)[y_1 y_2 y_3] + \ldots,$$

where, if one stops at the divided difference $[y_1 y_2 \ldots y_n]$, the remainder term is

$$(y - y_1)(y - y_2) \ldots (y - y_n)[y y_1 y_2 \ldots y_n].$$

Hence, it follows that the required integral is the inverse function $cn^{-1}(.37|2/5)$. The divided differences regarding the left-hand column as the argument are shown. Therefore, the value

$$1.2 + .07122 \times 1.34048 + .07122 \times .00338 \times .132$$

$$+ .07122 \times .00338 \times .07910 \times .18$$

$$= 1.29550.$$

● **PROBLEM 5-26**

Find the value of m corresponding to q = 0.01 from the following table, which gives values of the nome q as a function of the squared modulus $k^2 = m$; with the aid of inverse interpolation by successive approximation (that is, using these functional values find more figures different from those already obtained).

TABLE

m	q	Δ	Δ^2	Δ^3	Δ^4
0·12	0·00798 89058				
		71 40944			
·13	·00870 30002		82195		
		72 23139		1887	
·14	·00942 53141		84082		68
		73 07221		1955	
·15	·01015 60362		86037		67
		73 93258		2022	
·16	·01089 53620		88509		
		74 81317			
·17	·01164 34937				

Solution: As a first approximation

$$m = .14 + \frac{575}{7307} = .14787.$$

Using Gauss' formula, one finds

m	q
.14787	.00999 96780
.14788	.01000 04112

The interval is now 1/1000 of the original interval, so that by 4.0 the second difference is negligible and one gets, dividing 3220 by the new first difference 7332,

$$m = .14787\ 4392.$$

● **PROBLEM 5-27**

Find an approximate value of coth 0.6 from the following table with the aid of inverse interpolation by reversal of series.

x	$\coth^{-1} x$	Δ	Δ^2	Δ^3
1·85	0·6049 190			
		−40968		
1·86	·6008 222		+616	
		−40352		−16
1·87	·5967 870		+600	
		−39752		
1·88	·5928 118			

Solution: Inverse Interpolation by Reversal of Series:

The relation between the function y and the argument x, which is obtained from an interpolation formula by neglect of the remainder term, can be written in the form

$$y - y_1 = a_1 p + a_2 p^2 + a_3 p^3 + \ldots + a_n p^n,$$

where $p = (x - x_1)/\omega$ is the phase.

This (finite) power series can be reversed in the form

$$p = b_1(y - y_1) + b_2(y - y_1)^2 + b_3(y - y_1)^3 + \ldots ,$$

where

$$b_1 = \frac{1}{a_1},$$

$$b_2 = -\frac{a_2}{a_1^3},$$

$$b_3 = \frac{-a_1 a_3 + 2a_2^2}{a_1^5},$$

$$b_4 = \frac{-a_1^2 a_4 + 5 a_1 a_2 a_3 - 5 a_2^3}{a_1^7}.$$

Thus

$$p = \frac{y - y_1}{a_1} - \frac{a_2 (y - y_1)^2}{a_1^3} + \frac{(2a_2^2 - a_1 a_3)(y - y_1)^3}{a_1^5} + \ldots . \quad (1)$$

where

$$a_1 = (\delta - \tfrac{1}{2}\mu\delta^2 + \tfrac{1}{12}\delta^3) y_{3/2},$$

$$a_2 = (\tfrac{1}{2}\mu\delta^2 - \tfrac{1}{4}\delta^3) y_{3/2},$$

$$a_3 = \tfrac{1}{6}\delta^3 y_{3/2}.$$

Then by the procedure just described, take $y_1 = .6008\,222$, one gets

$$y - y_1 \doteq -8222, \quad a_1 = -40657, \quad a_2 = 308, \quad a_3 = -2.7.$$

Substituting in (1), one gets $p = .20254$.

Since $\omega = .01$, one gets therefore the approximation

$$\coth 0.6 = 1.862025,$$

EXTRAPOLATION TECHNIQUES

• **PROBLEM** 5-28

Given the following function:

x	1	2	3	4	5
f(x)	100.000	25.000	11.111	6.250	4.000

Extrapolate to find f(5.7). Using the actual function

$$f(x) = 100/x^2,$$

compare the value obtained. What type of polynomial is better suited for extrapolation in this case?

TABLE 1

x	f(x)	∇f	$\nabla^2 f$	$\nabla^3 f$	$\nabla^4 f$
1	100.000				
2	25.000	−75.000			
3	11.111	−13.889	61.111		
4	6.250	−4.861	9.028	−52.083	
5	4.000	−2.250	2.611	−6.417	45.666

Solution: Initially, one needs a difference table. Since x = 5 will be used as a base, a backward difference table will provide the most entries in that line. The result is Table 1.

It is clear from Table 1 that the differences all increase in magnitude toward the right of the table. The function is thus very poorly suited to polynomial interpolation (or extrapolation). Polynomial extrapolation is too dangerous in this case, but it is instructive to see just how bad such extrapolation can be, and also whether it is possible to salvage a reasonably accurate answer. Apply the Gregory-Newton backward formula

$$f(x) = f(0) + x(\nabla f_0) + \frac{x(x+1)}{2!} \nabla^2 f_0$$

$$+ \frac{x(x+1)(x+2)}{3!} \nabla^3 f_0 + \cdots$$

to the line x = 5 as a base, and use all entries in the base line. The resulting fourth degree polynomial is the one which fits exactly all five points in the table. Since x = 5.7 is 0.7 units below the base line, use x = 0.7 in the interpolation formula. This yields

$$f(5.7) = 4.000 + 0.7(-2.250) + \frac{(0.7)(1.7)}{2}(2.611)$$

$$+ \frac{(0.7)(1.7)(2.7)}{6}(-6.417) +$$

$$\frac{(0.7)(1.7)(2.7)(3.7)}{24}(45.666)$$

$$= 4.000 - 1.575 + 1.554 - 3.436 + 22.620 = 23.163$$

This value for f(5.7) certainly appears absurd in terms of the other tabulated values of f(x), but in general it is impossible to estimate by how much an extrapolated value is in error. However, in this case one has the advantage of being able to disclose the function which was used to make up the original table. Using this function,

f(5.7) = 3.078.

The extrapolation based on the fourth degree polynomial is thus completely worthless. In deciding whether polynomial interpolation is of any value in this case, consider the following table:

		$f(5.7)$
Original function		3.078
	linear	2.425
Type of	second degree	3.979
polynomial	third degree	0.543
	fourth degree	23.163

The linear extrapolation was obtained by taking two terms of the interpolation formula, the second degree by taking three terms, etc. While none of the extrapolated values could be called accurate by any means, it is clear that the linear and second degree extrapolations are the "best."

If polynomial extrapolation must be done with poorly behaved functions, then very low degree extrapolation is usually the safest, but even this should be carried out only for values of x very close to the tabulated region.

• **PROBLEM** 5-29

If a function to be extrapolated cannot be well approximated by a polynomial, a useful device can be to plot f(x) vs. x on log-log graph paper. This reduces an amazingly large variety of functions to essentially straight lines or to smooth curves which are easy to extrapolate. The numerical equivalent of this graphical procedure is to tabulate $\log_e f(x)$ vs. $\log_e x$, and then carry out polynomial extrapolation. Describe this procedure and evaluate its accuracy for the following function:

x	1	2	3	4	5
$f(x)$	100.000	25.000	11.111	6.250	4.000

Extrapolate to find f(5.7)

Solution: The tabulation of the function becomes

$x^* = \log_e x$	0	0.693	1.099	1.386	1.609
$f^*(x^*) = \log_e f(x)$	4.605	3.219	2.408	1.832	1.387

and one wishes to find

$$f^*(\log_e 5.7) = f^*(1.740).$$

This extrapolation could be accomplished by using the Lagrange interpolation formula (note that x* is not equally spaced). However, first check to see if f*(x*) is nearly linear. An estimate of the first derivative at

$$x^* = 1.609$$

is the simple backward difference

$$\frac{1.387 - 1.832}{1.609 - 1.386} = \frac{-0.445}{0.223} = -1.996$$

A similar difference at x* = 1.386 yields

$$\frac{1.832 - 2.408}{1.386 - 1.099} = \frac{-0.576}{0.287} = -2.007$$

These values are close enough to use linear extrapolation:

$$f^*(1.740) = 1.387 + (-1.996)(1.740 - 1.609)$$

$$= 1.126 = \log_e f(5.7)$$

$$f(5.7) = 3.081$$

This extrapolated value is virtually identical to the value of the original analytical function at x = 5.7, which is 3.078. In fact, the error is due only to roundoff. If the original function is considered,

$$f = 100/x^2,$$

and the natural log is taken,

$$\log_e f = \log_e 100 - 2 \log_e x$$

$$f^*(x^*) = \log_e 100 - 2x^*$$

so

$$f^*(x^*)$$

is linear in x^*, with a slope of -2.

CHAPTER 6

SIMULTANEOUS LINEAR ALGEBRAIC EQUATIONS AND MATRICES

GAUSSIAN ELIMINATION METHOD

• **PROBLEM** 6-1

Solve the following system by the Gaussian Elimination Method.

$$2x_1 + 4x_2 + 2x_3 = 16 \tag{1}$$

$$2x_1 - x_2 - 2x_3 = -6 \tag{2}$$

$$4x_1 + x_2 - 2x_3 = 0 \tag{3}$$

<u>Solution</u>: By subtracting Eq. (1) from Eq. (2) one obtains

$$-5x_2 - 4x_3 = -22 \tag{4}$$

If Eq. (1) is multiplied by 2 and then subtracted from (3) one obtains

$$-7x_2 - 6x_3 = -32 \tag{5}$$

Equations (4) and (5) are a set of two equations in two unknowns from which another of the unknowns can be eliminated to give one equation in one unknown. Multiply Eq. (4) by 7/5 and subtract the result from Eq. (5). The result is

$$-(2/5)x_3 = -6/5$$

from which $x_3 = 3$. By successively substituting back, first into (4) then into (1), one finds

$$x_2 = 2$$

$$x_1 = 1$$

In the elimination calculations only the coefficients a_{ij} and the constants b_i have been used. The x's serve only to identify the particular unknown to which the coefficient applies. If the coefficients and constant terms are arranged so that there is no necessity for carrying the unknowns for identification, the latter can be dropped. Such is the case when the problem is represented by the following array of numbers:

$$\begin{bmatrix} 2 & 4 & 2 & 16 \\ 2 & -1 & -2 & -6 \\ 4 & 1 & -2 & 0 \end{bmatrix} \qquad (6)$$

In this array, called a matrix, the first row represents the first equation, the second row the second equation, and so on. The first number in each row is the coefficient on the first unknown in the equation, the second number the coefficient on the second unknown, and so on. The last number in each row is the constant from that equation. This matrix can be visualized as the equations with the plus signs, unknowns, and equal signs removed. Thus, the second row of the matrix

$$2 \quad -1 \quad -2 \quad -6$$

represents the equation

$$2x_1 - 1x_2 - 2x_3 = -6$$

The elimination procedure used in the problem is modified somewhat to make the computations more routine for machine computations. The first objective is to eliminate x_1 from the last two equations, i.e., to place zeros in the first-column position of rows 2 and 3. This is done in such a way that the value of the coefficients has no influence on the routine except in cases where they lead to division by zero.

The first operation on the matrix is with the first row, which is

$$2 \quad 4 \quad 2 \quad 16$$

Each element (number) in the row is divided by the first number to give

$$1 \quad 2 \quad 1 \quad 8$$

This sequence of numbers represents the first equation after it has been multiplied by 1/2. The equality of the equation is unchanged.

The second step in the elimination is to multiply this modified first row by the first coefficient in the second row and subtract the result from the second row. The second row, which is initially

$$2 \quad -1 \quad -2 \quad -6$$

now becomes

$$(2 - 2) \quad (-1 - 4) \quad (-2 - 2) \quad (-6 - 16)$$

or

```
         0    -5    -4    -22
```

A zero is placed in the first column of the second row by this operation. This is equivalent to eliminating the x_1 term between the first and second equations and is the same result as found previously in Eq. (4). The next operation is to place a zero in the first-column position of row 3, which is done by multiplying the modified row 1 by the leading coefficient of row 3 and then subtracting the resulting terms from row 3, which then becomes

```
    (4 - 4)      (1 - 8)      (-2 - 4)      (0 - 32)
```

or

```
         0    -7    -6    -32
```

This set of numbers represents the equation obtained when x_1 is eliminated from the first and third equations and is exactly equivalent to Eq. (5) obtained earlier. The new matrix is now

$$\begin{bmatrix} 1 & 2 & 1 & 8 \\ 0 & -5 & -4 & -22 \\ 0 & -7 & -6 & -32 \end{bmatrix} \quad (7)$$

and represents the set of equations:

$$x_1 + 2x_2 + x_3 = 8$$
$$-5x_2 - 4x_3 = -22 \quad (8)$$
$$-7x_2 - 6x_3 = -32$$

The unknown x_1 has been eliminated from the last two equations, which now represent a system of two equations in two unknowns. The last two rows of the matrix represent these equations and can now be handled independently of the first row. The following operations are made on the last two rows:

1. Divide row 2 by its first element (here -5).

2. Multiply this new row 2 by the first element of row 3 (here -7) and subtract the result from row 3. The matrix then becomes

$$\begin{bmatrix} 1 & 2 & 1 & 8 \\ 0 & 1 & 4/5 & 22/5 \\ 0 & 0 & -2/5 & -6/5 \end{bmatrix} \quad (9)$$

The array (9) has the number 1 on the diagonal in all but the last row and has zeros in each position below the diagnoal. It represents the following set of equations:

$$x_1 + 2x_2 + x_3 = 8$$
$$x_2 + (4/5)x_3 = 22/5 \quad (10)$$
$$-(2/5)x_3 = -6/5$$

The value of x_3 is found from the third row of (9) or the third equation of (10).

$$x_3 = \frac{-6/5}{-2/5} = 3$$

When x_3 is known, the second row of (9) can be used to solve for x_2:

$$x_2 = 22/5 - (4/5)x_3 = 22/5 - (4/5)(3) = 2$$

and when x_3 and x_2 are known, the first row of (9) can be used to solve the for x_1:

$$x_1 = 8 - 2x_2 - x_3 = 8 - (2)(2) - 3 = 1$$

The set of numbers

$$x_1 = 1 \qquad x_2 = 2 \qquad \text{and} \qquad x_3 = 3$$

is the solution to the equations represented by the array (6).

● PROBLEM 6-2

Solve the system of equations
$$\begin{aligned} 3x + 2y - z + w &= 8 \\ x + 4y - 3z - 2w &= 1 \\ 6x - y - z + 4w &= 20 \\ 5x - 3y + 2z + 2w &= 7. \end{aligned} \qquad (1)$$

Use Gauss elimination.

Solution: Multiply the first equation by $-3, -1, 2$, in turn, and add to the second, third, and fourth equations to obtain

$$\begin{aligned} 3x + 2y - z + w &= 8 \\ -8x - 2y - 5w &= -23 \\ 3x - 3y + 3w &= 12 \\ 11x + y + 4w &= 23. \end{aligned} \qquad (2)$$

Next, multiply the second equation by $-\frac{3}{2}$ and $\frac{1}{2}$, in turn, and add to the third and fourth equations. Obtain

$$\begin{aligned} 3x + 2y - z + w &= 8 \\ -8x - 2y - 5w &= -23 \\ 15x + \tfrac{21}{2}w &= \tfrac{93}{2} \\ 7x + \tfrac{3}{2}w &= \tfrac{23}{2}. \end{aligned} \qquad (3)$$

Multiply the third equation by $-\frac{1}{7}$ and add to the fourth equation; obtain

$$3x + 2y - z + w = 8$$
$$-8x - 2y - 5w = -23$$
$$15x + \frac{21}{2}w = \frac{93}{2} \qquad (4)$$
$$\frac{34}{7}x = \frac{34}{7}.$$

The last equation yields $x = 1$; substituting into the third equation, find

$$w = 3;$$

substituting into the second, find

$$y = 0;$$

and finally, from the first equation,

$$z = -2.$$

After the first stage, the second equation of (2) could have been divided by -2 before eliminating y from the third and fourth equations; this would have given

$$3x + 2y - z + w = 8$$
$$4x + y + \frac{5}{2}w = \frac{23}{2}$$
$$15x + \frac{21}{2}w = \frac{93}{2} \qquad (5)$$
$$7x + \frac{3}{2}w = \frac{23}{2}.$$

Divide the third equation by $21/2$, multiply the result by $3/2$ and subtract from the last equation to obtain

$$3x + 2y - z + w = 8$$
$$4x + y + \frac{5}{2}w = \frac{23}{2}$$
$$\frac{10}{7}x + w = \frac{31}{7}$$
$$\frac{34}{7}x = \frac{34}{7}.$$

Then solve backward as before.

After obtaining the system (5), one could have eliminated the y term in the first equation by multiplying the second equation by 2 and subtracting the result from the first equation, giving

$$-5x \quad -z - 4w = -15$$

$$4x + y \quad + \frac{5}{2}w = \frac{23}{2}$$

$$15x \quad + \frac{21}{2}w = \frac{93}{2}$$

$$7x \quad + \frac{3}{2}w = \frac{23}{2}$$

As before, divide the third equation by 21/2 and then multiply the result by 3/2 and subtract from the fourth equation; but in addition, to eliminate the w terms from the first two equations, multiply the resulting third equation by 4 and add to the first equation, and by 5/2 and subtract from the second equation. This gives

$$\frac{5}{7}x \quad -z \quad = \frac{19}{7}$$

$$\frac{3}{7}x + y \quad = \frac{3}{7}$$

$$\frac{10}{7}x \quad + w = \frac{31}{7}$$

$$\frac{34}{7}x \quad = \frac{34}{7}.$$

Divide the last equation by 34/7; eliminate the x terms in the first three equations by multiplying the resulting fourth equation by 5/7 and subtracting from the first equation, multiplying by 3/7 and subtracting from the second equation, and by 10/7 and subtracting from the third equation. One ends up with the solution.

• **PROBLEM 6-3**

Solve the system of equations

$$10^{-2} x + y = 1$$

$$x + y = 2$$

(a) exactly, (b) by Gaussian elimination simulating two-decimal floating arithmetic, and (c) as in (b) but interchanging the equations.

<u>Solution:</u>

(a) $$10^{-2}x + y = 1$$
$$x + y = 2$$

Subtracting

$$\frac{90}{100} x = 1, \quad x = 1 + \frac{1}{99}$$

$$y = 1 - \frac{1}{99}.$$

(b)

$$.10 \times 10^{-1} x + .10 \times 10^1 y = .10 \times 10^1 \qquad (1)$$

$$.10 \times 10^1 x + .10 \times 10^1 y = .20 \times 10^1. \qquad (2)$$

Multiply (1) by 10^2 to get

$$.10 \times 10^1 x + .10 \times 10^3 y = .10 \times 10^3. \qquad (3)$$

Subtract (3) from (2) to get

$$.10 \times 10^3 y = .10 \times 10^3$$

so that

$$y = .10 \times 10^1$$

and substituting this in (1) gives

$$x = 0.$$

(c)

Multiply (2) by 10^{-2} to get

$$.10 \times 10^{-1} x + .10 \times 10^{-1} y = .20 \times 10^{-1}. \qquad (4)$$

Subtract (4) from (1) to get

$$y = .10 \times 10^1$$

and substituting this in (2) gives

$$x = .10 \times 10^1.$$

• **PROBLEM 6-4**

Show that the total number of multiplications required by the Gaussian scheme for inverting an n x n matrix A is approximately n^3.

Solution: Approach the problem in the following way. To invert A the systems of equations

$$Ac_i = e_i$$

where c_i is the i-th column of A^{-1} and e_i the i-th unit vector must be solved.

Assume, for simplicity, that the triangularization can be carried out without rearrangements. This requires, in the first stage, the determination of the factors $-\frac{a_{21}}{a_{11}}, -\frac{a_{31}}{a_{11}}, \ldots, -\frac{a_{n1}}{a_{11}}$ and then the multiplication of the first row by $-a_{r_1}/a_{11}$ and its addition to the r-th row, so as to kill the (r,1) term.

About n multiplications are needed and then a further (n-1) × (n-1), in all about n^2.

For the whole process about $\sum r^2 = n^3/3$ are needed.

Consider these operations carried out on the right hand sides. Take the case of e_i. No action is needed until the i-th stage when multiples of 1 to the zeros in the i + 1, ..., n-th position are added. At the next stage it is necessary to add multiples of the (i + 1)-st component to those in the i + 2, ..., n-th position. In all about

$$\sum_{r=1}^{n} (n - r) \doteq \frac{(n - i)^2}{2}$$

multiplications are needed. To deal with all the right hand sides therefore requires about

$$\sum_{i=1}^{n} \frac{(n - i)^2}{2} \doteq \sum_{j=1}^{n} \frac{j^2}{2} \doteq \frac{n^3}{6}$$

multiplications.

Finally n triangular systems must be solved; each involves about $n^2/2$ multiplications and so, in all about $n^3/2$ multiplications.

The grand total is $\frac{n^3}{3} + \frac{n^3}{6} + \frac{n^3}{2} = n^3$.

• **PROBLEM 6-5**

Using the Gauss elimination solve the following set of four simultaneous equations:

$$-x_1 + 3x_2 + 5x_3 + 2x_4 = 10$$
$$x_1 + 9x_2 + 8x_3 + 4x_4 = 15$$
$$x_2 \qquad\qquad + x_4 = 2$$
$$2x_1 + x_2 + x_3 - x_4 = -3.$$

<u>Solution</u>: The characteristic determinant of this system is

$$\begin{vmatrix} -1 & 3 & 5 & 2 \\ 1 & 9 & 8 & 4 \\ 0 & 1 & 0 & 1 \\ 2 & 1 & 1 & -1 \end{vmatrix} = 14,$$

so that a solution exists for this nonhomogeneous set, and in fact, the solution is

$$x_1 = -1, \quad x_2 = 0, \quad x_3 = 1, \quad \text{and} \quad x_4 = 2.$$

The resulting augmented matrix is

$$\begin{bmatrix} -1 & 3 & 5 & 2 & 10 \\ 1 & 9 & 8 & 4 & 15 \\ 0 & 1 & 0 & 1 & 2 \\ 2 & 1 & 1 & -1 & -3 \end{bmatrix}$$

JACOBI AND GAUSS-SEIDEL METHODS

• **PROBLEM** 6-6

How can scaling before Gaussian elimination be applied on the equilibrated $Ax = b$, where

$$A = \begin{pmatrix} \varepsilon & -1 & 1 \\ -1 & 1 & 1 \\ 1 & 1 & 1 \end{pmatrix}, \quad A^{-1} = \frac{1}{4}\begin{pmatrix} 0 & -2 & 2 \\ -2 & 1-\varepsilon & 1+\varepsilon \\ 2 & 1+\varepsilon & 1-\varepsilon \end{pmatrix}, \quad |\varepsilon| \ll 1?$$

Solution: This is a well-conditioned matrix, $\kappa(A) = 3$ in maximum norm, and therefore Gaussian elimination with partial pivoting will give an accurate solution. However, the choice of $a_{11} = \varepsilon$ as pivot will have a disastrous effect on the accuracy of the computed solution.

Now consider the scaling $x'_2 = x_2/\varepsilon$, $x'_3 = x_3/\varepsilon$. If the resulting system is again equilibrated, one gets $A'x' = b'$, where

$$A' = \begin{pmatrix} 1 & -1 & 1 \\ -1 & \varepsilon & \varepsilon \\ 1 & \varepsilon & \varepsilon \end{pmatrix}.$$

Here partial, and even complete, pivoting will select $a'_{11} = 1$ as the first pivot. However, this will have the same unfortunate effects on accuracy as did the same choice of pivots in the system $Ax = b$, since all the results will differ only in the exponents.

It is difficult to give a general rule as to how a linear system should be scaled before Gaussian elimination. It can be verified that the condition number in maximum norm for the matrices in the problem satisfies

$$\kappa(A') \approx \frac{3}{\varepsilon} \gg \kappa(A).$$

It seems, then that a possible approach to the scaling is to determine D_1 and D_2 so that $\kappa(D_2AD_1)$ is minimized. However, it turns out that these optimal D_1 and D_2 essentially depend on A^{-1}, which in practice is unknown. Another objection to this approach is that the scaling of the unknowns will change the norm in which the error is measured. Thus a sensible approach in most cases is to choose D_1 in a way which reflects the importance of the unknowns and to use D_2 to equilibrate the system.

● **PROBLEM 6-7**

Solve the following set by Gauss-Seidel iteration:

$$\begin{bmatrix} 3 & -5 & 47 & 20 \\ 11 & 16 & 17 & 10 \\ 56 & 22 & 11 & -18 \\ 17 & 66 & -12 & 7 \end{bmatrix} \begin{bmatrix} x_1 \\ x_2 \\ x_3 \\ x_4 \end{bmatrix} = \begin{bmatrix} 18 \\ 26 \\ 34 \\ 82 \end{bmatrix},$$

with an initial guess of

$$x^{(0)} = \begin{bmatrix} 1 \\ 1 \\ 1 \\ 1 \end{bmatrix},$$

and an absolute convergence criterion of $\varepsilon = 0.0001$.

Solution: Gauss-Seidel iteration is one of the most powerful iteration techniques for the solution of sets of linear equations. Consider the following set of three linear equations:

$$\begin{aligned} c_{11}x_1 + c_{12}x_2 + c_{13}x_3 &= r_1 \\ c_{21}x_1 + c_{22}x_2 + c_{23}x_3 &= r_2 \\ c_{31}x_1 + c_{32}x_2 + c_{33}x_3 &= r_3 \end{aligned} \quad (1)$$

Now, the first equation for x_1, the second for x_2, etc. are solved to yield

$$x_1 = \frac{r_1 - c_{12}x_2 - c_{13}x_3}{c_{11}}$$

$$x_2 = \frac{r_2 - c_{21}x_1 - c_{23}x_3}{c_{22}} \qquad (2)$$

$$x_3 = \frac{r_3 - c_{31}x_1 - c_{32}x_2}{c_{33}}$$

Initial guesses are needed for x_1, x_2, and x_3. Call these $x_1^{(0)}$, $x_2^{(0)}$, and $x_3^{(0)}$. From the first equation in (2) now find the value of x_1 on the first iteration as

$$x_1^{(1)} = \frac{r_1 - c_{12}x_2^{(0)} - c_{13}x_3^{(0)}}{c_{11}} \qquad (3)$$

The second equation in (2) gives $x_2^{(1)}$ as

$$x_2^{(1)} = \frac{r_2 - c_{21}x_1^{(1)} - c_{23}x_3^{(0)}}{c_{22}} \qquad (4)$$

Notice that $x_1^{(1)}$, which is the new value obtained from (3) on the current iteration, has been used instead of $x_1^{(0)}$. This use of the most recently obtained value of each of the unknowns is the distinguishing feature of Gauss-Siedel iteration.* The solution for x_3 from (2) is thus

$$x_3^{(1)} = \frac{r_3 - c_{31}x_1^{(1)} - c_{32}x_2^{(1)}}{c_{33}} \qquad (5)$$

*A different iterative process, called Jacobi iteration, employs all of the old values of the unknowns until the sweep through all equations has been completed, and then replaces the old values with the newly-computed values in a block.

The iterative process consists of repeatedly cycling through the solutions for the unknowns. As each new value of an unknown is computed, it replaces the old value. Only one computer storage location is thus required for each unknown. Programming is also greatly simplified, since whenever an unknown is used, it is automatically the most recently computed value. If the equations have the proper characteristics, then the iterative process will eventually converge to the solution vector.

The iteration is terminated when a convergence criterion is satisfied. The two commonly-used types of convergence criteria are absolute criteria and relative criteria. An absolute convergence criterion is of the form

$$|x_i^{(L+1)} - x_i^{(L)}| \leq \varepsilon \qquad (6)$$

If (6) is satisfied for all x_i, then the change in each unknown from the previous iteration (L) to the current iteration (L+1) is no more than ε. This type of criterion is most useful when the approximate magnitudes of the x_i are known beforehand. With such a criterion it is possible to choose ε such that the solution is considered converged when the change in each x_i is less than 1 unit in the fourth decimal place on two successive iterations. This does not mean that the fourth decimal place is accurate to 1 unit for each of the x_i; the actual accuracy of the x_i is dependent on the convergence rate of the process, which can vary widely for different sets of equations. Some knowledge of the accuracy of the converged values in any given problem can usually be obtained by observing the results of several iterations near convergence.

A relative convergence criterion is of the form

$$\left| \frac{x_i^{(L+1)} - x_i^{(L)}}{x_i^{(L+1)}} \right| \leq \varepsilon \qquad (7)$$

This type of criterion is the safest choice if the magnitudes of the x_i are not known beforehand, and corresponds to specifying the maximum allowable percentage change in each unknown on two successive iterations.

For very large sets of equations, it may be impractical to test every unknown for convergence since excessive amounts of computer time may be involved in the testing procedure. Convergence testing in such cases is usually individually tailored to the problem at hand and may consist of testing only certain critical unknowns or of using

$$\sum_{i=1}^{n} |x_i^{(L+1)} - x_i^{(L)}|$$

or $\sum_{i=1}^{n} |x_i^{(L+1)} - x_i^{(L)}|^2$ as quantities to be compared

with some predetermined convergence criterion.

Whether the iterative process is convergent or divergent does not depend on the initial guess supplied for the unknowns, but depends only on the character of the equations themselves. However, if the process is convergent, then a good first estimate of the unknowns will make it possible for the convergence criterion to be satisfied in a relatively small number of iterations. A poor first guess can prolong the iterative process considerably (but will not cause divergence). Now proceeding to the problem the set does not appear to be suitable for an iterative solution, since the main diagonal elements are not the largest elements in each row. However, by simply reordering the equations this can be partially remedied:

$$\begin{bmatrix} 56 & 22 & 11 & -18 \\ 17 & 66 & -12 & 7 \\ 3 & -5 & 47 & 20 \\ 11 & 16 & 17 & 10 \end{bmatrix} \begin{bmatrix} x_1 \\ x_2 \\ x_3 \\ x_4 \end{bmatrix} = \begin{bmatrix} 34 \\ 82 \\ 18 \\ 26 \end{bmatrix}$$

The main diagonal elements are now the largest elements in magnitude in each row except for the last row. The diagonal dominance in the first three equations is sufficiently strong that the small diagonal element in the fourth equation may not cause divergence. The first iteration gives

$$X^{(1)} = \begin{bmatrix} 0.339286 \\ 1.230789 \\ 0.066725 \\ 0.144090 \end{bmatrix}$$

After 10 iterations,

$$X^{(10)} = \begin{bmatrix} -0.930569 \\ 1.901519 \\ 1.359500 \\ -1.729954 \end{bmatrix}$$

The process satisfies the convergence criterion after 35 iterations and gives

$$X = \begin{bmatrix} -1.076888 \\ 1.990028 \\ 1.474477 \\ -1.906078 \end{bmatrix}$$

The Gauss-Seidel procedure clearly converges with no problems for this set with $C_{44} = 10$. However, it is interesting to note that if C_{44} is 9 or smaller then the procedure is divergent. Clearly, the presence of any small main diagonal elements can pose a significant threat to the convergence of Gauss-Seidel iteration. However, if iterative techniques are indicated for other reasons, they are definitely worth trying even in the presence of a few small main diagonal elements.

● **PROBLEM 6-8**

Solve the system $Ax = b$, where

$$A = \begin{pmatrix} 4 & -1 & -1 & 0 \\ -1 & 4 & 0 & -1 \\ -1 & 0 & 4 & -1 \\ 0 & -1 & -1 & 4 \end{pmatrix}, \quad b = \begin{pmatrix} 1 \\ 2 \\ 0 \\ 1 \end{pmatrix}.$$

Use Jacobi and Gauss-Seidel methods.

Solution: Jacobi's method yields the following sequence of approximations:

k	$x_1^{(k)}$	$x_2^{(k)}$	$x_3^{(k)}$	$x_4^{(k)}$
1	0.25	0.5	0	0.25
2	0.375	0.625	0.125	0.375
3	0.4375	0.6875	0.1875	0.4375
4	0.46875	0.71875	0.21875	0.46875
5	0.48344	0.73438	0.23438	0.48344
6	0.49219	0.74172	0.24172	0.49219
7	0.49586	0.74609	0.24609	0.49586
8	0.49805	0.74793	0.24793	0.49805
.	...	...	...	...
∞	0.5	0.75	0.25	0.5

The iteration converges, but rather slowly.

Instead, using Gauss-Seidel's method, one obtains:

k	$x_1^{(k)}$	$x_2^{(k)}$	$x_3^{(k)}$	$x_4^{(k)}$
1	0.25	0.5625	0.0625	0.40625
2	0.40625	0.70312	0.20312	0.47656
3	0.47656	0.73828	0.23828	0.49414
4	0.49414	0.74707	0.24707	0.49854
5	0.49854	0.74927	0.24927	0.49963
.	...	...	...	...

The convergence with Gauss-Seidel's method is, in this problem, about twice as fast as with Jacobi's method. This is often, but not always, true. Indeed, there are problems for which Gauss-Seidel's method diverges and Jacobi's method converges.

● **PROBLEM 6-9**

Show that the Gauss-Seidel method is convergent when the matrix is positive definite.

Solution: Begin by proving the following:

If F and G are matrices such that F is non-singular and F + G and F - G* are positive definite hermitian then the characteristic values of $F^{-1}G$ are all inside the unit circle.

Let λ, x be a characteristic pair for $F^{-1}G$. Then one can find that $x^*Gx = \lambda x^*Fx$ and from this, by adding x^*Fx to each side

$$x^*(F+G)x = (1+\lambda)x^*Fx. \qquad (1)$$

Since F+G is positive definite hermitian, it follows that $\lambda \neq -1$.

Since F + G is hermitian the right-hand side of (1) is equal to its conjugate transposed which gives

$$(1 + \bar{\lambda}) x^*F^*x = (1 + \lambda)x^*Fx$$
$$= (1 + \lambda)[x^*(F - G^*)x + x^*G^*x]$$
$$= (1 + \lambda)[x^*(F - G^*)x + \bar{\lambda}x^*F^*x],$$

i.e. $(1 - |\lambda|^2)x^*F^*x = (1 + \lambda)x^*(F - G^*)x.$

Multiply across by $1 + \bar{\lambda}$ and get

$$(1 - |\lambda|^2)\{(1 + \bar{\lambda})x^*Fx\} = |1 + \lambda|^2 x^*(F - G^*)x.$$

The bracketed factors on the left can be replaced by $x^*(F + G)x$ because F + G is positive definite hermitian and (1) can be starred. Hence

$$(1 - |\lambda|^2)x^*(F + G)x = |1 + \lambda|^2 x^*(F - G^*)x. \qquad (2)$$

The hermitian forms in (2) are positive if $x \neq 0$. Hence $1 - |\lambda|^2 > 0$ which is the result required.

The problem is now easily solved. Since A is hermitian A = L + D + L* with D real. If F = L + D, G = L* then A = L + G is positive definite hermitian and so clearly is its diagnoal D = F - G*. The result just established gives the conclusion wanted.

● **PROBLEM 6-10**

Given the following set of equations:

$$\begin{bmatrix} 1.1348 & 3.8326 & 1.1651 & 3.4017 \\ 0.5301 & 1.7875 & 2.5330 & 1.5435 \\ 3.4129 & 4.9317 & 8.7643 & 1.3142 \\ 1.2371 & 4.9998 & 10.6721 & 0.0147 \end{bmatrix} \begin{bmatrix} x_1 \\ x_2 \\ x_3 \\ x_4 \end{bmatrix} = \begin{bmatrix} 9.5342 \\ 6.3941 \\ 18.4231 \\ 16.9237 \end{bmatrix}$$

Solve this set by using Gauss-Jordan elimination with and without maximization of pivot elements and compare the result.

Solution: The exact answers to this set are

$x_1 = 1 \quad x_3 = 1$

$x_2 = 1 \quad x_4 = 1$

The answers obtained by using Gauss-Jordan elimination in single precision on the IBM 360/67 are:

Without Maximization of Pivot Elements	With Maximization of Pivot Elements
$x_1 = 0.9991369$	$x_1 = 1.000006$
$x_2 = 1.000077$	$x_2 = 1.000003$
$x_3 = 1.000001$	$x_3 = 1.000000$
$x_4 = 1.000076$	$x_4 = 1.000001$

The gain in accuracy when the pivot elements are maximized (by column shifting) should be apparent.

This small set of equations has been deliberately formulated to produce rather poor results unless maximization of pivot elements is used. It should be clear that maximization of pivot elements may be necessary for small sets of equations as well as large sets, and that it is virtually impossible to identify the need for maximization by simply examining the coefficient matrix. The best practice is simply to employ maximization of pivot elements routinely for all equation solving unless there is some reason for not doing so (as is the case with sets having banded coefficient matrices). A banded coefficient matrix has all zero elements except for a band centered on the main diagnoal. Thus, for example

$$C = \begin{bmatrix} c_{11} & c_{12} & & \\ c_{21} & c_{22} & c_{23} & \\ & c_{32} & c_{33} & c_{34} \\ & & c_{43} & c_{44} \end{bmatrix}$$

is a banded matrix of bandwidth three, also called a tridiagonal matrix.

• **PROBLEM 6-11**

Show that both the Jacobi and Gauss--Seidel methods are convergent when the matrix has a strictly dominant diagonal.

Solution: Jacobi: Assume the matrix A normalized to have units on the diagonal. The condition for convergence of the Jacobi process is that $\rho(I - A) < 1$. Since A is strictly diagonally dominant this yields $\Lambda_i = \Sigma'|a_{ij}| < 1$ for each i. This means that the Gerschgorin circles of $I - A$, which are all centered at the origin, have radii < 1. Hence $\rho(I - A) < 1$.
Gauss-Seidel: It is necessary now to show that $\rho((I - L)^{-1}U) < 1$, again assuming normalization. If λ, x is a characteristic pair for $(I - L)^{-1}U$ then $(I - L)^{-1}Ux = \lambda x$ which gives

$$(U + \lambda L)x = \lambda x. \tag{1}$$

Let $x_M = \max |x_i|$. The M-th equation in (1) gives

$$\sum_{j<M} a_{Mj} x_j + \lambda \sum_{j>M} a_{Mj} x_j = \lambda x_M,$$

i.e.

$$\sum_{j<M} a_{Mj}(x_j/x_M) + \lambda \sum_{j>M} a_{Mj}(x_j/x_M) = \lambda. \tag{2}$$

If $|\lambda| \geq 1$ the relation (2) is impossible since the absolute value of the left hand side

$$\leq \sum_{j<M} |a_{Mj}| + \lambda \sum_{j>M} |a_{Mj}|, \quad \text{by choice of M,}$$

$$\leq \lambda \sum' |a_{Mj}|, \quad \text{since } \lambda \geq 1,$$

$$< \lambda, \quad \text{since A is strictly diagoally dominant.}$$

Hence $|\lambda| < 1$, as required.

● **PROBLEM 6-12**

Show that trigonometrical tables are not essential for the carrying out of Jacobi rotations, square root tables sufficing.

Solution: The basic equation is

$$\begin{bmatrix} c & s \\ -s & c \end{bmatrix} \begin{bmatrix} A & H \\ H & B \end{bmatrix} \begin{bmatrix} c & -s \\ s & c \end{bmatrix} = \begin{bmatrix} a & 0 \\ 0 & b \end{bmatrix}$$

where c, s are the cosine and sine of the angle θ defined by

$$\tan 2\theta = 2H/(A - B),$$

and the values of a, b are given by

$$a = Ac^2 + Bs^2 + 2Hsc, \quad b = As^2 + Bc^2 - 2Hsc.$$

(Observe that $a + b = A + B$ and that $a^2 + b^2 = A^2 + B^2 + 2H^2$.)

Writing $n = 2H$, $d = A - B$ it follows from elementary trigonometry that

$$2c^2 = \{(n^2 + d^2)^{\frac{1}{2}} + d\}(n^2 + d^2)^{-\frac{1}{2}}.$$

$$2s^2 = \{(n^2 + d^2)^{\frac{1}{2}} - d\}(n^2 + d^2)^{-\frac{1}{2}} \tag{1}$$
$$2sc = n(n^2 + d^2)^{-\frac{1}{2}}$$

The case $A = 2$, $B = 5$, $H = -3$ yields

$$\tan 2\theta = 2/1, \quad n = 2, \quad d = 1, \quad \text{say.}$$

Hence $2c^2 = \{\sqrt{5} + 1\}\sqrt{5}$, $2s = \{\sqrt{5} - 1\}/\sqrt{5}$, $2sc = 2/\sqrt{5}$, which gives, in particular,

$$a = (7 - 3\sqrt{5})/2, \quad b = (7 + 3\sqrt{5})/2,$$

results which check with the fact that a, b, are the characteristic values of the matrix

$$\begin{bmatrix} A & H \\ H & B \end{bmatrix}.$$

Care must be taken with the ambiguities in c, s, by taking

$$\frac{-\pi}{2} \leq 2\theta \leq \frac{\pi}{2}$$

and then taking c positive and s to have the sign of $\tan 2\theta$. Further, the formulas (1) are numerically unsatisfactory if $n \ll d$ and special tricks must be used.

Since $\arctan 2 = 1.1071$ the required angle of rotation of the axes is $\theta = .5536 \sim 31.72°$.

● **PROBLEM 6-13**

Find the inverse of the matrix

$$\begin{bmatrix} 1 & 0 & 0 \\ -a & 1 & 0 \\ -b & -c & 1 \end{bmatrix}.$$

Dicuss the convergence of the Jacobi and Gauss-Seidel processes for the solution of $Ax = b$ when

$$A = \begin{bmatrix} 1 & -2 & 2 \\ -1 & 1 & -1 \\ -2 & -2 & 1 \end{bmatrix} \quad \text{and when} \quad A = \begin{bmatrix} 1 & \frac{1}{2} & -\frac{1}{2} \\ -1 & 1 & -1 \\ \frac{1}{2} & \frac{1}{2} & 1 \end{bmatrix}.$$

<u>Solution</u>: The inverse is

$$\begin{bmatrix} 1 & 0 & 0 \\ a & 1 & 0 \\ ac + b & c & 1 \end{bmatrix}.$$

In the first case it is necessary to find the spectral radii of

$$\begin{bmatrix} 0 & 2 & -2 \\ 1 & 0 & 1 \\ 2 & 2 & 0 \end{bmatrix}$$

and

$$\begin{bmatrix} 1 & 0 & 0 \\ -1 & 1 & 0 \\ -2 & -2 & 1 \end{bmatrix}^{-1} \begin{bmatrix} 0 & 2 & -2 \\ 0 & 0 & 1 \\ 0 & 0 & 0 \end{bmatrix}$$

$$= \begin{bmatrix} 1 & 0 & 0 \\ 1 & 1 & 0 \\ 4 & 2 & 1 \end{bmatrix} \begin{bmatrix} 0 & 2 & -2 \\ 0 & 0 & 1 \\ 0 & 0 & 0 \end{bmatrix}$$

$$= \begin{bmatrix} 0 & 2 & -2 \\ 0 & 2 & -1 \\ 0 & 8 & -6 \end{bmatrix}.$$

The first matrix has characteristic polynomial $-\lambda^3$ and the second $-\lambda\{\lambda^2 + 4\lambda - 4\}$ so that the spectral radii are $0, 2(1 + \sqrt{2})$ respectively. Thus, the Jacobi process converges and the Gauss-Seidel process does not.

In the second case it is necessary to find the spectral radii of

$$\begin{bmatrix} 0 & -\tfrac{1}{2} & \tfrac{1}{2} \\ 1 & 0 & 1 \\ -\tfrac{1}{2} & \tfrac{1}{2} & 0 \end{bmatrix}$$

and

$$\begin{bmatrix} 1 & 0 & 0 \\ -1 & 1 & 0 \\ \tfrac{1}{2} & \tfrac{1}{2} & 1 \end{bmatrix}^{-1} \begin{bmatrix} 0 & -\tfrac{1}{2} & \tfrac{1}{2} \\ 0 & 0 & 1 \\ 0 & 0 & 0 \end{bmatrix}$$

$$= \begin{bmatrix} 1 & 0 & 0 \\ 1 & 1 & 0 \\ 0 & -\tfrac{1}{2} & 1 \end{bmatrix} \begin{bmatrix} 0 & -\tfrac{1}{2} & \tfrac{1}{2} \\ 0 & 0 & 1 \\ 0 & 0 & 0 \end{bmatrix}$$

$$= \begin{bmatrix} 0 & -\tfrac{1}{2} & \tfrac{1}{2} \\ 0 & -\tfrac{1}{2} & \tfrac{3}{2} \\ 0 & 0 & -\tfrac{1}{2} \end{bmatrix}.$$

The first matrix has characteristic polynomial $-\lambda(\lambda^2 + 5/4)$ and the second $-\lambda(\lambda + \tfrac{1}{2})^2$ so that the spectral radii are $\sqrt{5}/2$ and $\tfrac{1}{2}$ respectively. Thus the Gauss-Seidel process converges and the Jacobi process does not.

• **PROBLEM 6-14**

Solve the system
$$x + y + z = 0$$
$$x^2 + y^2 + z^2 + 2xz - 1 = 0 \tag{1}$$

using Jacobian Determinants. Show whether x and y can be considered as functions of z.

Solution: To investigate whether x and y can be considered as functions of z, denote the left-hand members by f and g, respectively, and calculate the Jacobian

$$\frac{\partial(f,g)}{\partial(x,y)} = \begin{vmatrix} 1 & 1 \\ 2x + 2z & 2y \end{vmatrix} = -2(x + z - y). \tag{2}$$

Thus, except on the surface $x + z - y = 0$, x and y can be considered as functions of z. That is, z can be taken as the independent variable. When $y = x + z$, the equations become $2(x + z) = 0$ and $2(x + z)^2 = 1$ and are hence incompatible. To investigate whether x and z can be taken as the dependent variables, calculate the Jacobian

$$\frac{\partial(f,g)}{\partial(x,z)} = \begin{vmatrix} 1 & 1 \\ 2x + 2z & 2x + 2z \end{vmatrix} = 0 \tag{3}$$

Since this determinant is identically zero, one sees that x and z cannot be taken as the dependent variables. It is readily verified directly that the system (1) cannot be solved for x and z in terms of y. This situation follows from the fact that both equations involve only y and the combination $x + z$, and hence cannot be solved for x and z separately.

• **PROBLEM 6-15**

Apply the Givens and the Householder method to reduce the matrix

$$\begin{bmatrix} 5 & 7 & 6 & 5 \\ 7 & 10 & 8 & 7 \\ 6 & 8 & 10 & 9 \\ 5 & 7 & 9 & 10 \end{bmatrix}$$

to triple diagonal form. The result obtained by one program is

$$W_1 = \begin{bmatrix} 5 & 10.488089 & 0 & 0 \\ 10.488089 & 25.472729 & 3.521903 & 0 \\ 0 & 3.521898 & 3.680571 & -.185813 \\ 0 & 0 & -.185813 & .846701 \end{bmatrix}.$$

Theoretically this should be symmetric, and it is unnecessary to calculate the elements below the diagonal. However, if these are calculated, the differences in symmetric elements gives some idea of the errors occurring.

Find the characteristic values of W_1, e.g., by drawing a rough graph of the characteristic polynomial of W_1, and then using Newton's method to estimate the characteristic values more accurately.

Solution: The successive reductions are

$$\begin{bmatrix} 5 & 9.2195446 & 0 & 5 \\ 9.2195446 & 17.9058830 & 1.2235291 & 11.1719188 \\ 0 & 1.2235291 & 2.0941177 & 2.2777696 \\ 5 & 11.1719188 & 2.2777696 & 10 \end{bmatrix}$$

$$\begin{bmatrix} 5 & 10.4880889 & 0 & 0 \\ 10.4880889 & 25.4727293 & 2.1614262 & 2.7806542 \\ 0 & 2.1614262 & 2.0941177 & 1.4189767 \\ 5 & 2.7806542 & 1.4189767 & 2.4331552 \end{bmatrix}$$

and that given. The rotations used are given by
$c = .7592566 \quad s = .6507914$
$c = .8790491 \quad s = .4767313$
$c = .6137097 \quad s = .7895317$.

As a check compute the determinant of W_1 using the recurrence method and get
$\det W_1 = 1.0000053$.

The characteristic roots of W are approximately

.0105, .8431, 3.858, 30.29.

The Householder vectors are

[0, .9131, .3133, .2611]'

and
[0,0, .8533, .5215]'.

MATRIX MATHEMATICS

● **PROBLEM 6-16**

If
$$A = \begin{pmatrix} 1 & 3 & 1 \\ -2 & 1 & -1 \end{pmatrix}, \quad B = \begin{pmatrix} 1 \\ 2 \\ 3 \end{pmatrix}$$
find AB.

Solution: Since A has 3 columns and B has 3 rows, they are comformable in the order AB. One can expedite the process of finding the product by writing the matrices side by side, and then going across a row of A and down a column of B forming products by pairs, thus:

$$\begin{pmatrix} 1 & 3 & 1 \\ -2 & 1 & -1 \end{pmatrix} \begin{pmatrix} 1 \\ 2 \\ 3 \end{pmatrix} = \begin{pmatrix} 1 \times 1 + 3 \times 2 + 1 \times 3 \\ -2 \times 1 + 1 \times 2 - 1 \times 3 \end{pmatrix} = \begin{pmatrix} 10 \\ -3 \end{pmatrix}$$

● **PROBLEM 6-17**

For the matrices A and B
$$A = \begin{pmatrix} 1 & 3 & 1 \\ -2 & 1 & -1 \end{pmatrix}, \quad B = \begin{pmatrix} 1 \\ 2 \\ 3 \end{pmatrix}$$
find BA.

Solution: Since B has 1 column and A has 2 rows, they are not comformable in the order BA. The product BA is not defined.

● **PROBLEM 6-18**

Find A + B and A − B, where
$$A = \begin{pmatrix} 3 & 0 & -2 \\ 1 & 3 & 2 \end{pmatrix}, \quad B = \begin{pmatrix} 2 & -1 \\ 1 & 3 \\ 2 & -2 \end{pmatrix}$$

Solution: Since there are not the same number of rows or columns in A and B they cannot be added or subtracted.

● **PROBLEM 6-19**

Find A + B and A − B, where
$$A = \begin{pmatrix} 3 & 0 & -2 \\ 1 & 3 & 1 \end{pmatrix}, \quad B = \begin{pmatrix} 2 & 1 & 2 \\ -1 & 3 & -2 \end{pmatrix}.$$

Solution:

$$A + B = \begin{pmatrix} 5 & 1 & 0 \\ 0 & 6 & -1 \end{pmatrix},$$

$$A - B = \begin{pmatrix} 1 & -1 & -4 \\ 2 & 0 & 3 \end{pmatrix}$$

• **PROBLEM 6-20**

If

$$A = \begin{pmatrix} 1 & 2 \\ 3 & -1 \end{pmatrix}, \quad B = \begin{pmatrix} 3 & -2 \\ 2 & 1 \end{pmatrix}$$

find AB.

Solution: Since A has 2 columns and B has 2 rows, A and B are conformable in the order AB, so the product is indeed defined. To find the element in the first row, first column of the product matrix, take the first row of A, which is

1 2

and the first column of B, which is

3

2

and form the sum of the products by pairs:

1 x 3 + 2 x 2 = 7

Hence 7 is the element in the first row, first column of the product.

In like manner, the element in the first row and second column of the product is obtained from combining the first row of A with the second column of B, thus:

1 x (-2) + 2 x 1 = 0

and for the second row, first column,

3 x 3 + (-1) x 2 = 7

and the second row, second column,

3 x (-2) + (-1) x 1 = -7

Hence the product is

$$\begin{pmatrix} 1 & 2 \\ 3 & -1 \end{pmatrix} \begin{pmatrix} 3 & -2 \\ 2 & 1 \end{pmatrix} = \begin{pmatrix} 7 & 0 \\ 7 & -7 \end{pmatrix}$$

• **PROBLEM 6-21**

Find the rank of the matrix

$$\begin{pmatrix} -1 & 1 & 2 \\ -3 & 3 & 1 \end{pmatrix}.$$

Solution: The largest-order determinant one can construct is second order, whose rank is 2 or less. To see if it is 2, all second-order determinants must be checked. If the third column is crossed out, one can construct the determinant

$$\begin{vmatrix} -1 & 1 \\ -3 & 3 \end{vmatrix}$$

which has the value zero. Since this one vanishes, the second-order determinant must be checked. Crossing out the second column in the matrix obtain the determinant

$$\begin{vmatrix} -1 & 2 \\ -3 & 1 \end{vmatrix}$$

which has the value 5. Since there is a nonvanishing second-order determinant, the rank is 2.

• **PROBLEM 6-22**

Find the rank of the matrix

$$\begin{pmatrix} 1 & 2 & 3 \\ -1 & -2 & -3 \\ 2 & 4 & 6 \end{pmatrix}$$

Solution: The largest-order determinant one can construct is third order so the rank is 3 or less. The only third-order determinant is

$$\begin{vmatrix} 1 & 2 & 3 \\ -1 & -2 & -3 \\ 2 & 4 & 6 \end{vmatrix} = 0$$

so the rank is not 3. If one crosses out the third row and third column, one gets the determinant

$$\begin{vmatrix} 1 & 2 \\ -1 & -2 \end{vmatrix} = 0$$

Similarly, if all other second-order determinants are checked, one finds they all vanish.

Hence the rank is less than 2. If one crosses out the second and third rows and the second and third columns, the determinant of the matrix can be formed. Since the highest-order nonvanishing determinant is first order, the rank of the matrix is 1.

• **PROBLEM 6-23**

Solve the following set of equations by using the inverse of the coefficient matrix.

$$1x_1 + 2x_2 + 0x_3 = 7$$

$$2x_1 - 1x_2 + 1x_3 = 4$$

$$0x_1 + 4x_2 - 2x_3 = 2$$

Solution: Writing the equations as $AX = C$, the solution is known to be $X = A^{-1}C$ where the inverse can be determined from the formula $A^{-1} = \text{adj} A/|A|$. The numerator of this formula is the adjoint of the coefficient matrix A and can be found by forming the transpose of the cofactor matrix (A_{ij}). The elements of the cofactor matrix are the cofactors of the corresponding elements a_{ij} of A. Each cofactor A_{ij} is equal to $(-1)^{i+j}$ times the determinant of the matrix obtained from A by striking out the ith row and jth column of A. Thus, one gets

$$A_{11} = (-1)^2 \begin{vmatrix} -1 & 1 \\ 4 & -2 \end{vmatrix} = -2, \quad A_{12} = (-1)^3 \begin{vmatrix} 2 & 1 \\ 0 & -2 \end{vmatrix} = 4,$$

$$A_{13} = (-1)^4 \begin{vmatrix} 2 & 1 \\ 0 & 4 \end{vmatrix} = 8$$

$$A_{21} = (-1)^3 \begin{vmatrix} 2 & 0 \\ 4 & -2 \end{vmatrix} = 4, \quad A_{22} = (-1)^4 \begin{vmatrix} 1 & 0 \\ 0 & -2 \end{vmatrix} = -2,$$

$$A_{23} = (-1)^5 \begin{vmatrix} 1 & 2 \\ 0 & 4 \end{vmatrix} = -4$$

$$A_{31} = (-1)^4 \begin{vmatrix} 2 & 0 \\ -1 & 1 \end{vmatrix} = 2, \quad A_{32} = (-1)^5 \begin{vmatrix} 1 & 0 \\ 2 & 1 \end{vmatrix} = -1,$$

$$A_{33} = (-1)^6 \begin{vmatrix} 1 & 2 \\ 2 & -1 \end{vmatrix} = -5$$

Expanding along the first row,

$$|A| = 1 \begin{vmatrix} -1 & 1 \\ 4 & -2 \end{vmatrix} + (-1)(2) \begin{vmatrix} 2 & 1 \\ 0 & -2 \end{vmatrix} + 0 = 6.$$

Remembering to take the transpose of the cofactor matrix (A_{ij}), now one gets

$$A^{-1} = (1/6) \begin{pmatrix} -2 & 4 & 2 \\ 4 & -2 & -1 \\ 8 & -4 & -5 \end{pmatrix}$$

and

$$X = (1/6) \begin{pmatrix} -2 & 4 & 2 \\ 4 & -2 & -1 \\ 8 & -4 & -5 \end{pmatrix} \begin{pmatrix} 7 \\ 4 \\ 2 \end{pmatrix} = \begin{pmatrix} 1 \\ 3 \\ 5 \end{pmatrix} = \begin{pmatrix} x_1 \\ x_2 \\ x_3 \end{pmatrix}.$$

The solution, (x_i = 1,3,5), can easily be checked by substituting these three values back into the original three equations.

In practice, it is suggested that $|A|$ be determined before finding (A_{ij}) since there is a solution iff $|A| \neq 0$, for a nonhomogeneous set of equations. One would not want to do the work of determining the cofactor matrix and then later find out there is no inverse in a particular problem.

● **PROBLEM 6-24**

Compute A^{-1}.

$$A = \begin{pmatrix} 1 & 2 & -1 \\ 2 & 1 & 0 \\ -1 & 1 & 2 \end{pmatrix}.$$

Use matrix multiplication techniques.

Solution: Assuming A^{-1} exists, let B_j be the jth column of the n x n matrix B,

Form the product

$$B_j = \begin{bmatrix} b_{1j} \\ b_{2j} \\ \vdots \\ b_{nj} \end{bmatrix}$$

$$AB_j = \begin{bmatrix} a_{11} & a_{12} & \cdots & a_{1n} \\ a_{21} & a_{22} & \cdots & a_{2n} \\ \vdots & \vdots & & \vdots \\ a_{n1} & a_{n2} & \cdots & a_{nn} \end{bmatrix} \begin{bmatrix} b_{1j} \\ b_{2j} \\ \vdots \\ b_{nj} \end{bmatrix} = \begin{bmatrix} \sum_{k=1}^{n} a_{1k}b_{kj} \\ \sum_{k=1}^{n} a_{2k}b_{kj} \\ \vdots \\ \sum_{k=1}^{n} a_{nk}b_{kj} \end{bmatrix}$$

If AB = C, then the jth column of C is given by:

$$C_j = \begin{bmatrix} c_{1j} \\ c_{2j} \\ \vdots \\ c_{nj} \end{bmatrix} = \begin{bmatrix} \sum_{k=1}^{n} a_{1k}b_{kj} \\ \sum_{k=1}^{n} a_{2k}b_{kj} \\ \vdots \\ \sum_{k=1}^{n} a_{nk}b_{kj} \end{bmatrix}$$

Hence, the jth column of the product AB is the product of A and the jth column of B.

Suppose that A^{-1} exists and that $A^{-1} = B = (b_{ij})$; then AB = I and

$$AB_j = \begin{bmatrix} 0 \\ \vdots \\ 0 \\ 1 \\ 0 \\ \vdots \\ 0 \end{bmatrix} \quad \text{where the value 1 appears in the jth row.}$$

To actually find B, solve n linear systems in which the jth column of the inverse is the solution of the linear system with righthand side the jth column of I.

To compute A^{-1} for this problem, three linear systems have to be solved:

$$x_1 + 2x_2 - x_3 = 1, \qquad x_1 + 2x_2 - x_3 = 0,$$

$$2x_1 + x_2 \qquad\quad = 0, \qquad 2x_1 + x_2 \qquad\quad = 1,$$

$$-x_1 + x_2 + 2x_3 = 0; \qquad -x_1 + x_2 + 2x_3 = 0;$$

$$x_1 + 2x_2 - x_3 = 0,$$

$$2x_1 + x_2 \qquad\quad = 0,$$

$$-x_1 + x_2 + 2x_3 = 1.$$

Using Gaussian elimination, the computations are conveninetly performed upon the larger augmented matrix, formed by combining the matrices

$$\begin{bmatrix} 1 & 2 & -1 & | & 1 & 0 & 0 \\ 2 & 1 & 0 & | & 0 & 1 & 0 \\ -1 & 1 & 2 & | & 0 & 0 & 1 \end{bmatrix}$$

To elaborate, since the actual coefficient matrix does not change, perform the same sequence of row operations for each linear system. First, performing $(E_2 - 2E_1) \to (E_2)$ and $(E_3 + E_1) \to (E_3)$,

$$\begin{bmatrix} 1 & 2 & -1 & | & 1 & 0 & 0 \\ 0 & -3 & 2 & | & -2 & 1 & 0 \\ 0 & 3 & 1 & | & 1 & 0 & 1 \end{bmatrix}.$$

Next, performing $(E_3 + E_2) \to (E_3)$, one gets:

$$\begin{bmatrix} 1 & 2 & -1 & | & 1 & 0 & 0 \\ 0 & -3 & 2 & | & -2 & 1 & 0 \\ 0 & 0 & 3 & | & -1 & 1 & 1 \end{bmatrix}.$$

Backward substitution could be performed on each of the three augmented matrices,

$$\begin{bmatrix} 1 & 2 & -1 & | & 1 \\ 0 & -3 & 2 & | & -2 \\ 0 & 0 & 3 & | & -1 \end{bmatrix}, \begin{bmatrix} 1 & 2 & -1 & | & 0 \\ 0 & -3 & 2 & | & 1 \\ 0 & 0 & 3 & | & 1 \end{bmatrix},$$

$$\begin{bmatrix} 1 & 2 & -1 & | & 0 \\ 0 & -3 & 2 & | & 0 \\ 0 & 0 & 3 & | & 1 \end{bmatrix},$$

to find all the entires of A^{-1}, but it is often more convenient to use further row reduction. In particular, the operation $(1/3\, E_3) \to (E_3)$ yields:

$$\begin{bmatrix} 1 & 2 & -1 & | & 1 & 0 & 0 \\ 0 & -3 & 2 & | & -2 & 1 & 0 \\ 0 & 0 & 1 & | & -\frac{1}{3} & \frac{1}{3} & \frac{1}{3} \end{bmatrix}$$

and $(E_2 - 2E_3) \to (E_2)$ and $(E_1 + E_3) \to (E_1)$ produce:

$$\begin{bmatrix} 1 & 2 & 0 & | & \frac{2}{3} & \frac{1}{3} & \frac{1}{3} \\ 0 & -3 & 0 & | & -\frac{4}{3} & \frac{1}{3} & -\frac{2}{3} \\ 0 & 0 & 1 & | & -\frac{1}{3} & \frac{1}{3} & \frac{1}{3} \end{bmatrix}.$$

Performing $(-\frac{1}{3}E_2) \to (E_2)$, obtain:

$$\begin{bmatrix} 1 & 2 & 0 & | & \frac{2}{3} & \frac{1}{3} & \frac{1}{3} \\ 0 & 1 & 0 & | & \frac{4}{9} & -\frac{1}{9} & \frac{2}{9} \\ 0 & 0 & 1 & | & -\frac{1}{3} & \frac{1}{3} & \frac{1}{3} \end{bmatrix};$$

and finally, $(E_1 - 2E_2) \to (E_1)$ gives:

$$\begin{bmatrix} 1 & 0 & 0 & | & -\frac{2}{9} & \frac{5}{9} & -\frac{1}{9} \\ 0 & 1 & 0 & | & \frac{4}{9} & -\frac{1}{9} & \frac{2}{9} \\ 0 & 0 & 1 & | & -\frac{3}{9} & \frac{3}{9} & \frac{3}{9} \end{bmatrix}.$$

The final augmented matrix represents the solutions to the three linear systems

$$x_1 = -\frac{2}{9}, \quad x_1 = \frac{5}{9}, \quad x_1 = -\frac{1}{9},$$
$$x_2 = \frac{4}{9}, \quad x_2 = -\frac{1}{9}, \quad x_2 = \frac{2}{9},$$
$$x_3 = -\frac{3}{9}, \quad x_3 = \frac{3}{9}, \quad x_3 = \frac{3}{9},$$

so

$$A^{-1} = \begin{bmatrix} -\frac{2}{9} & \frac{5}{9} & -\frac{1}{9} \\ \frac{4}{9} & -\frac{1}{9} & \frac{2}{9} \\ -\frac{3}{9} & \frac{3}{9} & \frac{3}{9} \end{bmatrix}$$

$$= \frac{1}{9}\begin{bmatrix} -2 & 5 & -1 \\ 4 & -1 & 2 \\ -3 & 3 & 3 \end{bmatrix}$$

● **PROBLEM 6-25**

Consider the matrix

$$A = \begin{bmatrix} 1 & 0 & 1 \\ 2 & 2 & 3 \\ 1 & 2 & 4 \end{bmatrix}$$

Find the inverse matrix A^{-1}.

Solution: In actual numerical work one should use a pivoting strategy in order to help control round-off error. To simplify the manipulations, reduce the augmented matrix to triangular form without pivoting on the largest element. This yields the matrix

$$\begin{bmatrix} 1 & 0 & 1 & 1 & 0 & 0 \\ 0 & 2 & 1 & -2 & 1 & 0 \\ 0 & 0 & 2 & 1 & -1 & 1 \end{bmatrix}.$$

Thus $x_{31} = 1/2$, $x_{32} = -1/2$, $x_{33} = 1/2$. These values lead to the equations

$$2x_{21} + 1/2 = -2, \quad 2x_{22} - 1/2 = 1, \quad 2x_{23} + 1/2 = 0,$$

$$x_{21} = -5/4, \quad x_{22} = 3/4, \quad x_{23} = -1/4.$$

Substituting in the first row,

$$x_{11} + 1/2 = 1, \quad x_{12} - 1/2 = 0, \quad x_{13} + 1/2 = 0,$$

$$x_{11} = 1/2, \quad x_{12} = 1/2, \quad x_{13} = -1/2.$$

Finally, $$A^{-1} = \begin{bmatrix} 1/2 & 1/2 & -1/2 \\ -5/4 & 3/4 & -1/4 \\ 1/2 & -1/2 & 1/2 \end{bmatrix}.$$

• **PROBLEM 6-26**

For what value(s) of the parameter λ will the equations

$$4x + 2y + z = \lambda x$$

$$2x + 4y + 2z = \lambda y$$

$$x + 2y + 4z = \lambda z$$

have non-zero solutions?

Solution: Collecting the coefficients of x, y, and z, obtain as the determinant D of the homogeneous equations

$$D = \begin{vmatrix} (4 - \lambda) & 2 & 1 \\ 2 & (4 - \lambda) & 2 \\ 1 & 2 & (4 - \lambda) \end{vmatrix}.$$

In order that the homogeneous equations have a solution not zero the determinant D must vanish. Upon expanding the determinant the following cubic equation which the parameter λ must satisfy is obtained

$$(4 - \lambda)^3 - 9(4 - \lambda) + 8 = 0.$$

Solving this equation for λ, one finds three values $\lambda_1 = 1.62772$, $\lambda_2 = 3$, $\lambda_3 = 7.37228$, for each of which the given equations have a set of non-zero solutions.

● **PROBLEM 6-27**

Find latent roots and vectors for the matrix

$$A = \begin{bmatrix} 2 & 1 & -1 & 0 \\ 1 & -2 & 3 & 2 \\ -1 & 3 & 4 & -1 \\ 0 & 2 & -1 & 3 \end{bmatrix}$$

Solution: The solution can be divided into five stages.

Preliminary Estimation of λ_1 and u_1: A^2 is useful and is therefore evaluated first.

$$A^2 = \begin{bmatrix} 6 & -3 & -3 & 3 \\ -3 & 18 & 3 & -1 \\ -3 & 3 & 27 & -1 \\ 3 & -1 & -1 & 14 \end{bmatrix}$$
$$\;\;\, 3 \quad\;\; 17 \quad\;\; 26 \quad\;\; 15$$

Observe that

$$\Sigma \lambda_i = 7, \quad \Sigma \lambda_i^2 = 65$$

Assuming $b_0 = \{0, 0, 1, 0\}$, successive multiplications by A^2 give the following sequence—

27.00 $\{-0.1111, 0.1111, 1, -0.0370\}$

27.7027$\{-0.1484, 0.1939, 1, -0.0708\}$

28.0977$\{-0.1667, 0.2494, 1, -0.0936\}$

28.3419$\{-0.1774, 0.2852, 1, -0.1080\}$

28.4958$\{-0.1840, 0.3079, 1, -0.1168\}$

28.5925$\{-0.1881, 0.3221, 1, -0.1222\}$

The numbers on the left are approximations to λ_1^2. Convergence is distinctly slow, and indicates strongly that λ_2 is comparable with λ_1. This can be verified from the sequence for λ_1^2 by deriving the sequence giving $\lambda_1^2 \lambda_2^2$. It follows that if $M_k \sim \lambda_1^2$ and $M_{2,k} \sim \lambda_1^2 \lambda_2^2$ then

203

$$M_{2,k}/M_k = M_{k-1}(M_{k+1} - M_k)/(M_k - M_{k-1}) \sim \lambda_2^2$$

M_k	ΔM_k	$M_{2,k}/M_k$
28.0977		
	0.2442	
28.3419		17.7
	0.1539	
28.4958		17.8
	0.0967	
28.5925		

This shows that $\lambda_2^2 \sim 17.8$, $\lambda_2 \sim \pm 4.21$.

Approximate Values of Latent Roots: It is easily found, by multiplying any of the above column matrices by A, that $\lambda_1 > 0$, i.e. $\lambda_1 \sim 5.35$. The sign of λ_2 needs investigation. Now
$$\lambda_2 + \lambda_3 + \lambda_4 \sim 1.65$$

so if $\lambda_2 = -4.2$, both λ_3 and λ_4 must be positive, remembering that neither can exceed 4.2 in magnitude. For the matrix A - 5I, λ_2 then becomes the dominant root by far. Therefore examine A - 5I.

$$A_1 = A - 5I = \begin{bmatrix} -3 & 1 & -1 & 0 \\ 1 & -7 & 3 & 2 \\ -1 & 3 & -1 & -1 \\ 0 & 2 & -1 & -2 \end{bmatrix}$$

....

$$\begin{array}{cccc} -3 & -1 & 0 & -1 \end{array}$$

Both the second and third columns suggest trying
$$b_0 = \{0, 1, 0, 0\}$$
Multiplying by A_1 gives the sequence

$$-7.0\{-0.14, 1, -0.43, -0.29\}$$

$$-9.0\{-0.21, 1, -0.43, -0.33\}$$

$$-9.2\{-0.22, 1, -0.43, -0.34\}$$

Thus $\lambda_2 - 5 \sim -9.2$, which clearly establishes that $\lambda_2 \sim -4.2$. Now using the values for $\Sigma\lambda_i$, $\Sigma\lambda_i^2$ it follows that
$$\lambda_3 + \lambda_4 = 5.85, \quad \lambda_3^2 + \lambda_4^2 = 18.45$$
and finally that the complete set of roots is roughly 5.35, -4.2, 3.7, 2.1.

Improvement of λ_1 and u_1: This could be done by means of the polynomial
$$P_1(A) = (A + 4.2)(A - 3.7)(A - 2.1) = A^3 - 1.6A^2 - 16.5A - 32.5I$$

or more simply $P_1(A) = A^3 - 2A^2 - 16A - 32I$

A simpler expedient, avoiding the calculation of A^3, is to use
$$P_1(A) = A^2 - 11I$$
This has approximate roots 17, 7, 3, -7, which are fairly well separated.

$$A^2 - 11I = \begin{bmatrix} -5 & -3 & -3 & 3 \\ -3 & 7 & 3 & -1 \\ -3 & 3 & 16 & -1 \\ 3 & -1 & -1 & 3 \end{bmatrix}$$

....

$$\begin{array}{cccc} -8 & 6 & 15 & 4 \end{array}$$

Starting with $b_0 = \{-0.17, 0.25, 1, -0.10\}$

obtain

17.36 {-0.1843, 0.3088, 1, -0.1187}1.0058

17.5980 {-0.1910, 0.3315, 1, -0.1260}1.0045

17.6935 {-0.1932, 0.3402, 1, -0.1290}1.0181

17.7292 {-0.1941, 0.3435, 1, -0.1301}1.0193

By a rough extrapolation, using $X_1 = X_2 - (X_2 - X_1)^2/(X_2 - 2X_1 + X_0)$ for each element, obtain the vector

{-0.1947, 0.3455, 1, -0.1305}

and by one more multiplication

17.7511{-0.1946, 0.3455, 1, -0.1306}

Hence $\lambda_1^2 = 17.7511 + 11$, $\lambda_1 = 5.36200$, and the last estimate of u_1 is probably nearly accurate to 4D.

Calculation of λ_2, u_2; λ_3, u_3: Precise results for λ_2 are obtained most easily by continuing the sequence with $A - 5I$, as follows-

{-0.22, 1, -0.43, -0.34 }

-9.19 {-0.2274, 1, -0.4342, -0.3384}

-9.2068 {-0.2299, 1, -0.4345, -0.3379}

-9.2092 {-0.2307, 1, -0.4346, -0.3377}

-9.2099 {-0.2309, 1, -0.4346, -0.3377}

-9.2101 {-0.2310, 1, -0.4346, -0.3377}

-9.2102 {-0.2310, 1, -0.4346, -0.3377}

Thus $\lambda_2 = 5 - 9.2102 = -4.2102$, and u_2 is well established. To obtain λ_3 it is advisable to use a matrix polynomial,

$$Q_3(A) = (A + 4.2)(A - 2.1) = A^2 + 2.1A - 8.82I$$

or more simply

$$Q_3(A) = A^2 + 2A - 9I$$

This assumes that the sequence leading to u_3 is kept orthogonal to u_1', which may be done as follows. Assume $u_1' \cdot b_k = 0$, where

$$b_k = \{b_{1k}, b_{2k}, \ldots, b_{nk}\}$$

$$u_1 = \{u_{11}, u_{21}, \ldots, u_{n1}\}$$

All the elements of b_k can be left arbitrary except one, say b_{ik}; this is set equal to

$$-(u_{11}b_{1k} + \ldots + u_{i-1,1}b_{i-1,k} + u_{i+1,1}b_{i+1,k}$$

$$+ \ldots + u_{n1}b_{nk})/u_{i1}$$

and the condition is satisfied. b_{ik} is best chosen to correspond to the largest element u_{i1} of u_1, which may conveniently be made unity.

Now $u_1 = \{-0.1946, 0.3455, 1, -0.1306\}$

Let $b_0 = \{0, 0, 0.1306, 1\}$

Then by repeatedly multiplying by the matrix

$$A^2 + 2A - 9I = \begin{bmatrix} 1 & -1 & -5 & 3 \\ -1 & 5 & 9 & 3 \\ -5 & 9 & 26 & -3 \\ 3 & 3 & -3 & 11 \end{bmatrix}$$

$$\cdots\cdots\cdots\cdots\cdots\cdots\cdots\cdots\cdots\cdots$$

$$-2 \quad 16 \quad 27 \quad 14$$

obtain the sequence

$10.6082\{0.2213, 0.3936, 0.0377, 1\}$

$12.7316\{0.2073, 0.3994, 0.0329, 1\}$

$12.7217\{0.2078, 0.3998, 0.0329, 1\}$

$12.7241\{0.2078, 0.3998, 0.0329, 1\}$

At each multiplication, only the 4th, 1st and 2nd elements are found in the usual way; the 3rd is chosen to maintain the orthogonality. The vector u_3 is thus obtained, and λ_3 is obtained from the equation

$$Q_3(\lambda_3) = \lambda_3^2 + 2\lambda_3 - 9 = 12.7241$$

The correct root, easily distinguished, is $\lambda_3 = 3.7670$.

Derivation of u : Using $\Sigma \lambda_i$ one finds $\lambda_4 = 2.0812$. The final latent vector may be best obtained by making it orthogonal to u_1', u_2', u_3'. Thus assuming

$$u_4 = \{1, a, b, c\}$$

there are three orthogonality conditions which give linear simultaneous equations to determine a, b, and c. By this procedure, one finds

$$u_4 = \{1, 0.1777, 0.0964, -0.2820\}$$

● **PROBLEM 6-28**

Solve the equations by elimination:

$$-23x_1 + 11x_2 + x_3 = b_1,$$

$$11x_1 - 3x_2 - 2x_3 = b_2,$$

$$x_1 - 2x_2 + x_3 = b_3$$

for general values of b_1, b_2, and b_3. Hint: The elimination process should be carried out by avoiding divisions, and the rounding errors associated with them.

Solution:

Equation no. and operation	Coefficients of x_1	x_2	x_3	Coefficients of b_1	b_2	b_3	Check sum	Notes
(1)	-23	11	1	1	0	0	-10	
(2)	11	-3	-2	0	1	0	7	
(3)	1	-2	1	0	0	1	1	
(4) = (1) − (3)	-24	13	0	1	0	-1	-11	Cross sum checks
(5) = (2) + 2 × (3)	13	-7	0	0	1	2	9	Cross sum checks
(6) = (4) + 2 × (5)	2	-1	0	1	2	3	7	Cross sum checks
(7) = (4) + 12 × (6) = x_2	0	1	0	13	24	35	73	Cross sum checks
(8) = (6) + (7)	2	0	0	14	26	38	80	
(9) = ½(8) = x_1	1	0	0	7	13	19	40	Cross sum checks
(10) = 2 × (7) + (3)	1	0	1	26	48	71	147	
(11) = (10) − (9) = x_3	0	0	1	19	35	52	107	Cross sum checks

Hence

$$x_1 = 7b_1 + 13b_2 + 19b_3,$$

$$x_2 = 13b_1 + 24b_2 + 35b_3,$$

$$x_3 = 19b_1 + 35b_2 + 52b_3,$$

and

$$\begin{bmatrix} -23 & 11 & 1 \\ 11 & -3 & -2 \\ 1 & -2 & 1 \end{bmatrix}^{-1} = \begin{bmatrix} 7 & 13 & 19 \\ 13 & 24 & 35 \\ 19 & 35 & 52 \end{bmatrix}.$$

Note: (i) Advantage has been taken of the simple numerical values of the coefficients to lighten the numerical work of the elimination process. The particularly simple values of the coefficients of x_3 suggest that this is the unknown to eliminate first.

(ii) In line (6) no elimination is carried out, but a linear combination of the equations is made so as to keep down the magnitudes of the numbers occurring in the calculation.

(iii) By avoiding division and so keeping the work free from rounding errors, the exact solution is obtained without any attention having to be given to the number of figures kept at the various stages of the work. Further, the ill-conditioned nature of the equations (see note (iv) below) gives no difficulty in obtaining a solution. Also the numbers occurring are simple enough in this case for the whole calculation to be done without the aid of a desk machine.

(iv) The large values of the elements of the inverse matrix show why such a poor approximation to the solution gives such small residuals.

If (ξ_1, ξ_2, ξ_3) is an approximation to the solution, and R_1, R_2, R_3 are the residuals obtained on substituting $x_1 = \xi_1, x_2 = \xi_2, x_3 = \xi_3$ into the equations, then the corrections to the approximate solution are

$$(x_1 - \xi_1) = 7R_1 + 13R_2 + 19R_3,$$

$$(x_2 - \xi_2) = 13R_1 + 24R_2 + 35R_3,$$

$$(x_3 - \xi_3) = 19R_1 + 35R_2 + 52R_3,$$

so that if $R_1 = R_2 = R_3 = 0.01$, then $x_3 - \xi_3 = 1.06$; that is, the error in an approximate value of x_3 may be over 100 times the residuals in the equations, although in the equations this unknown only occurs with coefficients 1 and 2.

● **PROBLEM 6-29**

Find the equation of the parabola which coincides with the sine curve $y = \sin x$ at $x = 0, \pi/2, \pi$.

Solution: The values of y corresponding to the given values of x are y = 0, 1, 0; hence the equation of the parabola is

$$\begin{vmatrix} y & 1 & x & x^2 \\ 0 & 1 & 0 & 0 \\ 1 & 1 & \frac{\pi}{2} & \frac{\pi^2}{4} \\ 0 & 1 & \pi & \pi^2 \end{vmatrix} = 0$$

or $y = 4x/\pi - 4x^2/\pi^2$.

CHAPTER 7

ROOTFINDING FOR NONLINEAR EQUATIONS

THE ITERATIVE METHOD FOR SOLVING NONLINEAR EQUATIONS

● **PROBLEM 7-1**

Generally, iteration means the repeated application of a numerical process or pattern of action.

To illustrate a specific use of the idea of iteration, consider solving an equation of the form $x = F(x)$. Here F is a differentiable function whose value can be computed for any given value of the real variable x (within a certain interval). Using the method of iteration, one starts with an initial approximation x_0, and computes the sequence $x_1 = F(x_0)$, $x_2 = F(x_1)$, $x_3 = F(x_2)$, ...

Each computation of the type $x_{n+1} = F(x_n)$ is called an iteration. If the sequence $\{x_n\}$ converges to a limiting value α, then one gets $\lim F(x_n) = F(\alpha)$, so $x = \alpha$ satisfies the equation $x = F(x)$. As n grows, one would like the numbers x_n to be better and better estimates of the desired root. One stops the iteration when sufficient accuracy has been attained.

Consider the problem shown in the accompanying figure. Two curves, $y_1 = e^x$ and $y_2 = 3x$ intersect in the two places shown. The problem is to find the x coordinates of the two intersections. Find the root at $x = 0.62$ with an initial guess $x_0 = 0$ or $x_0 = 1$ and also the root at $x = 1.51$ with an initial guess $x_0 = 2$. Use the method of iteration discussed above.

Solution: At the intersections

$$y_1 = y_2,$$
$$e^x = 3x,$$
$$f(x) = e^x - 3x = 0,$$

and the two roots of $f(x) = 0$ are sought.

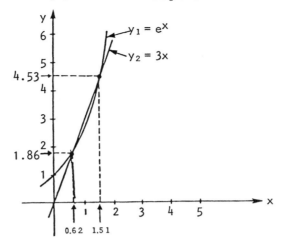

To separate into the form $x = g(x)$, write

$$0 = e^x - 3x$$

$$3x = e^x$$

$$x = \frac{e^x}{3},$$

or

$$x_{i+1} = \frac{e^{x_i}}{3}.$$

Note that this is only one of the many ways of writing $x = g(x)$, and not necessarily the best--only the most obvious.

To find the root at $x = 0.62$, with an initial guess $x_0 = 0$ or perhaps $x_0 = 1$, use the value $x_0 = 0$,

$$x_1 = \frac{e^0}{3} = 0.333 \ldots,$$

$$x_2 = \frac{e^{0.333}}{3} = 0.465,$$

$$x_3 = \frac{e^{0.465}}{3} = 0.530,$$

$$x_4 = \frac{e^{0.530}}{3} = 0.567,$$

which seems to converge quite well.

To get the root at $x = 1.51$, with an initial guess $x_0 = 2$:

$$x_1 = \frac{e^2}{3} = 2.46,$$

$$x_2 = \frac{e^{2.46}}{3} = 3.91,$$

$$x_3 = \frac{e^{3.91}}{3} = 16.7,$$

which diverges very quickly since x_4 would be almost 6 million.

The reason for the divergence is quite obvious if the derivative $g'(x)$ is examined:

$$g(x) = \frac{e^x}{3}$$

$$g'(x) = \frac{e^x}{3}.$$

To find the range of x for which the absolute value of the derivative is less than 1, let

$$\left|\frac{e^x}{3}\right| < 1$$

$$e^x < 3$$

$$x < \ln 3 \approx 1.1.$$

Whenever $x < 1.1$ the process will converge. Thus the initial guess $x_0 = 2$ will probably cause divergence, but moreover the solution $x = 1.51$ itself is out of the range of guaranteed convergence, and so most likely cannot be found by this method.

• **PROBLEM 7-2**

The Millikan oil-drop experiement for the determination of the ratio of the charge to the mass of an electron leads to the equation (Stokes' Law)

$$v = \frac{2}{9}g\frac{r^2}{n}(\mu - \mu_1)\left(1 + \frac{.000617}{pr}\right).$$

Find r, the radius of the oil-drop, for an experiment in which $g = 980$, $n = 1.832 \times 10^{-4}$, $\mu_1 = .0012$, $\mu = .9052$, $p = 72$, $v = .00480$. Use the method of iteration.

<u>Solution</u>: Substituting the given values, obtain

$$r^2 = \frac{9nv}{2g(\mu - \mu_1)}\left[1 + \frac{.000617}{pr}\right]^{-1}$$

$$= 4.4667 \times 10^{-9}\left[1 + \frac{.000617}{72r}\right]^{-1}.$$

$$r = 6.6833 \times 10^{-5}\left[1 + \frac{8.5694}{10^6 r}\right]^{-\frac{1}{2}}.$$

Clearly

$$r_1 = 6.6833 \times 10^{-5},$$

$$r_2 = 6.6833 \times 10^{-5} \left[1 + \frac{8.5694}{66.833}\right]^{-\frac{1}{2}} = \frac{6.6833 \times 10^{-5}}{1.06236}$$

$$= 6.2910 \times 10^{-5},$$

$$r_3 = 6.6833 \times 10^{-5} \left[1 + \frac{8.5694}{62.910}\right]^{-\frac{1}{2}} = \frac{6.6833 \times 10^{-5}}{1.06593}$$

$$= 6.2699 \times 10^{-5}.$$

One beauty of the iterative method is that minor errors correct themselves; an error in r_i (unless so gross that it destroys convergence) at worst slows down convergence (and may even speed it up if made judiciously). The final result is

$$r_4 = \frac{6.6833 \times 10^{-5}}{1.06615} = 6.28686 \times 10^{-5},$$

$$r_5 = \frac{6.6833 \times 10^{-5}}{1.06616} = 6.2686 \times 10^{-5}.$$

The last digit is uncertain.

It might be noted that, if only the result 6.27×10^{-5} were required, the whole operation could be carried out very rapidly by hand or on the slide rule.

• **PROBLEM 7-3**

The equation $e^{at} - at - b = 0$ occurs in heat transfer. Solve, given $a = .4$, $b = 9$. Use the graphical approach.

Solution: Here, write $e^{at} = at + b$ and graph the two standard curves $y = e^{at}$ and $y = at + b$ in the figure shown. From the graph, it is immediately clear that there are two roots, a negative root nearly equal to the value $-9/.4 = -22.5$, and a positive root (the physically significant one) which is a little greater than 6. (For $t = 6$, the values of $e^{.4t}$ and $.4t + 9$ are 11.023 and 11.4 respectively.)

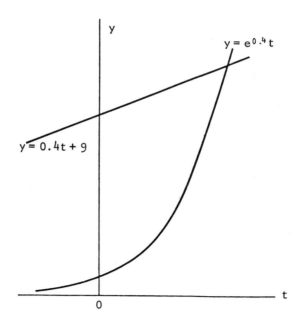

● **PROBLEM 7-4**

Consider the equation $x = (x^3 + 1)/4$. Find by trial and error method the approximations $x_i = T(x_i)$ to eight decimal places fo $i = 0,\ldots.7$ such that x is a real number.

Solution: This equation is of the form $T(x) = (x^3 + 1)/4$. By trial and error, it is found that $0 < T(0)$, while $1/2 > T(1/2)$. Thus the equation must have a solution between 0 and 1/2. Taking $x_0 = 0$ as the initial guess and using fixed-point arithmetic truncated to eight decimal places, the following approximations are obtained:

$x_0 = 0$,

$x_1 = T(x_0) = 0.2500\ 0000$,

$x_2 = T(x_1) = 0.2539\ 0625$,

$x_3 = T(x_2) = 0.2540\ 9223$,

$x_4 = T(x_3) = 0.2541\ 0123$,

$x_5 = T(x_4) = 0.2541\ 0166$,

$x_6 = T(x_5) = 0.2541\ 0168$,

$x_7 = T(x_6) = 0.2541\ 0168$.

• **PROBLEM 7-5**

Find the larger root of the equation $f(x) = x^2 - 3.6 \log_{10} x - 2.7 = 0$, correct to four decimal places (assume 3.6 and 2.7 are exact numbers). Use the graph to get an initial value. Solve the problem by an alternative method.

Solution: It can be ascertained from a graph, or otherwise, that the equation has two real roots, one between 0.1 and 0.2 and the other, the desired root, between 1.9 and 2. From

$$f'(x) = 2x - \frac{3.6 \log e}{x},$$

gives $f'(2) = 3.2$. Using $x = x - \frac{f(x)}{f'}$, (1)

one gets $\phi(x) = x - \frac{x^2 - 3.6 \log x - 2.7}{3.2}$;

whence, starting with $r_0 = 2$, one finds $r_1 = 1.93$, $r_2 = 1.9310$. Further substitution produces no change in the fourth decimal place of the root.

Alternative Solution: Forms other than those derived from (1) can be used for iteration. For example, the given equation can be solved for x^2 and then both sides divided by x to yield $x = (3.6 \log x + 2.7)/x$, whence $\phi(x) = (3.6 \log x + 2.7)/x$, and $\phi'(2) = -0.6$. Hence $\phi(x)$ is suitable for iteration; again beginning with $r_0 = 2$, one finds $r_1 = 1.9$, $r_2 = 1.95$, $r_3 = 1.92$, $r_4 = 1.94$, $r_5 = 1.93$, $r_6 = 1.932$, $r_7 = 1.931$, $r_8 = 1.9310$.

The rate of convergence is much more rapid in solution 1 than in solution 2. In the first case, $|\phi'(x)| \approx 0.05$ in the neighborhood of the root; in the second, $|\phi'(x)| \approx 0.6$.

• **PROBLEM 7-6**

Find the solution to the equation.

$y' = y^2$, $y \in E^1$, $y(0) = 1$.

Solution:

$$y_0(t) \equiv 1,$$

$$y_1(t) = 1 + \int_0^t 1 d\tau = 1 + t.$$

$$y_2(t) = 1 + \int_0^t [1 + \tau]^2 d\tau = 1 + t + t^2 + \frac{1}{3}t^3,$$

$$y_3(t) = 1 + t + t^2 + t^3 + \frac{2}{3}t^4 + \frac{1}{3}t^5 + \frac{1}{9}t^6 + \frac{1}{63}t^7.$$

215

The exact solution is

$$y(t) = \frac{1}{1-t} = \sum_{i=0}^{\infty} t^i.$$

● **PROBLEM 7-7**

Find more exactly the root of

$$\sin x + 2 \sin y = 1, \qquad 2 \sin 3x + 3 \sin 3y = 0.3$$

in the neighborhood of $x/(\frac{1}{6}\pi) = 2.1$.

Solution:

$x/(\frac{1}{6}\pi)$	$x°$	$\sin y$ $=\frac{1}{2}(1-\sin x)$	$Y=\phi_1(x)$	$\sin 3y$ $= 0.1 - \frac{2}{3}\sin 3x$	$Y=\phi_2(x)$	$\phi_2(x) - \phi_1(x)$
2.0	60°	.0670	.0671	.1	.0334	−.0337 240
2.05	61½°	.0606	.0607	.1523	.0510	−.0097 238
2.1	63°	.0545	.0545	.2043	.0686	+.0141 233
2.15	64½°	.0487	.0487	.2556	.0862	+.0374 231
2.2	66°	.0432	.0432	.3060	.1037	+.0605

and inverse interpolation then gives the required solution, approximately $x = 2.070 \, (\frac{1}{6}\pi)$.

• PROBLEM 7-8

Find more accurately the solution of the equations

$$xy(2x^2 - y^2) + 16(x + y) = 48, \qquad x^2 + y^2 = 16,$$

in the neighborhood of $x = 1.8$, $y = 3.6$.

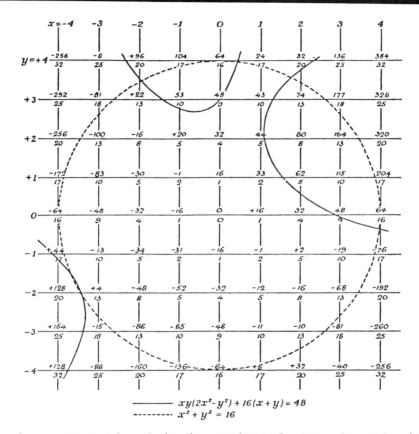

Fig. 1

Solution: From Fig. 1 it is estimated that the solution lies between $x = 1.7$ and 1.9, and between $y = 3.5$ and 3.7. The improvement of this solution can be carried out by the following procedure. A set of values of f_1 and f_2 is first evaluated for $y = 3.5$, 3.6, and 3.7, $x = 1.6$ to 2.0, this range of x being taken in order to provide enough values to check by differences. The x-contours and y-contours drawn using these points are shown in Fig. 2; those for $y = 3.5$ and 3.6 already enclose the point $f_1 = f_2 = 0$; but some points for $y = 3.7$ have been calculated to check the spacing of the y-contours and to show the curvature of the x-contours. The (f_1, f_2) point for $x = 1.85$, $y = 3.55$ is also shown. The solution estimated from these contours was $x = 1.84$, $y = 3.55$.

Values of (f_1, f_2) are now calculated for these sets of values $x = 1.83$, 1.84 and $y = 3.55$, 3.56, and the results plotted on a larger scale. For this small range of x and

217

y, the contours can be taken as straight and equally spaced to the accuracy of the plot. Inverse interpolation in x and y is required to determine the values of x and y to give $f_1 = f_2 = 0$, and this is most easily done by measurement. The values obtained can be checked by calculating f_1 and f_2 for them.

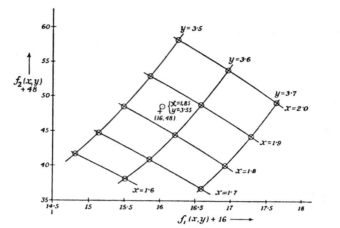

Fig. 2

Notes: (i) As in Fig. 1 the functions plotted in Fig. 2 are the left-hand sides of the two equations, namely $f_1(x, y) + 48$ and $f_2(x, y) + 16$.

(ii) A convenient way of carrying out the final interpolation is as follows. Consider the interpolation between the two x-contours, say $x = x_0$ and x_1, and let $x = x_0 + p(x_1 - x_0)$ be the interpolated value required. Lay a ruler on the (f_1, f_2) diagram so that its edge passes through the point $f_1 = f_2 = 0$; rotate it about this point and move it in the direction of its length until it cuts both x-contours at exact graduations on the scale, at a convenient interval (say 5, 10, or 20 units of the scale graduation); then the value of the fraction p of the x-interval between the contours can be read off directly.

(iii) This method of plotting is not satisfactory for the preliminary location of roots, since (x, y) is not in general a single-valued function of (f_1, f_2), so that two or more x-contours and two or more y-contours may pass through each point in a region of the (f_1, f_2) plane. If this occurs, and it will occur if the equations have more than one solution, the (f_1, f_2) diagram becomes complicated and its interpretation needs considerable care.

• **PROBLEM 7-9**

Solve the equations

$$\sin x + 2 \sin y = 1,$$

$$2 \sin 3x + 3 \sin 3y = 0.3.$$

Solution: It is most conveneint here to solve the first

equation for sin y, then from this to calculate sin 3y either from the formula

$$\sin 3y = \sin y (3 - 4 \sin^2 y)$$

or by use of inverse sine and sine tables, and then to evaluate

$$f_2(x) = 2 \sin 3x + 3 \sin 3y - 0.3$$

for these values of sin 3y and the corresponding values of sin 3x. The work is conveniently arranged in tabular form.

	sin y $= \frac{1}{2}(1-\sin x)$	sin 3y	sin 3x	$f_2(x)$
0	0.5	+1	0	+2.70
$1(\frac{1}{6}\pi) = 30°$	0.25	0.688	1	3.76
$2(\frac{1}{6}\pi) = 60°$	0.067	0.200	0	+0.30
$3(\frac{1}{6}\pi) = 90°$	0	0	-1	-2.30
$4(\frac{1}{6}\pi) = 120°$	0.067	0.200	0	+0.30
$5(\frac{1}{6}\pi) = 150°$	0.25	0.688	1	3.76
$6(\frac{1}{6}\pi) = 180°$	0.5	1	0	+2.70
$7(\frac{1}{6}\pi) = 210°$	0.75	+0.562	-1	-0.61
$8(\frac{1}{6}\pi) = 240°$	0.933	-0.450	0	-1.65
$9(\frac{1}{6}\pi) = 270°$	1	-1	1	-1.30
$10(\frac{1}{6}\pi) = 300°$	0.933	-0.450	0	-1.65
$11(\frac{1}{6}\pi) = 330°$	0.75	+0.562	-1	-0.61
$12(\frac{1}{6}\pi) = 360°$	0.5	1	0	+2.70

Two decimals are adequate to locate the roots approximately. A graph drawn from these values, or even inspection of the table without actually drawing a graph, shows that there are roots in the neighborhood of $x/(\frac{1}{6}\pi) = 2.1, 3.9, 6.8,$ and 11.2.

The approximate solutions so determined can be improved by tabulation at smaller intervals and inverse interpolation, or by an iterative process. If both the equations can be solved for one variable in terms of the other, say for y in terms of x:

$$y = \phi_1(x) \text{ for the first equation,}$$

$y = \phi_2(x)$ for the second equation, then it may be more convenient to evaluate $\phi_1(x) - \phi_2(x)$ as a function of x and interpolate for the zero of this function.

NEWTON'S METHOD

● **PROBLEM 7-10**

Approximate the positive root of the equation

$$x^3 + \sqrt{3}x^2 - 2x - 2\sqrt{3} = 0 \tag{1}$$

by Newton's method. The general expression for Newton's method is $x^{(n+1)} - x^{(n)} = \delta^{(n+1)} = -\dfrac{f(x^{(n)})}{f'(x^{(n)})}$ where the superscript n denotes values obtained on the nth iteration and n + 1 indicates values to be found on the (n + 1)th interation.

Solution: If one sets

$$f(x) = x^3 + \sqrt{3}x^2 - 2x - 2\sqrt{3},$$

then

$$f(0) = -2\sqrt{3} < 0$$

$$f(1) = 1 + \sqrt{3} - 2 - 2\sqrt{3} = -1 - \sqrt{3} < 0$$

$$f(2) = 8 + 4\sqrt{3} - 4 - 2\sqrt{3} = 4 + 2\sqrt{3} > 0,$$

and a root of (1) lies between 1 and 2. Newton's formula has the form

$$x^{(n+1)} = x^{(n)} - \frac{[x^{(n)}]^3 + \sqrt{3}[x^{(n)}]^2 - 2x^{(n)} - 2\sqrt{3}}{3[x^{(n)}]^2 + 2\sqrt{3}x^{(n)} - 2}. \tag{2}$$

Approximating $\sqrt{3}$ by 1.7, setting $x^{(1)} = 2.0$, and rounding to one decimal place, one has from (2) that

$$x^{(2)} \sim 1.6$$

$$x^{(3)} \sim 1.4$$

$$x^{(4)} \sim 1.4,$$

where the symbol ~ is used to designate an approximate value. Since $x^{(3)} = x^{(4)}$, the Newton iteration formula will yield no new value of x and the approximation which results is x ~ 1.4. The exact solution if $x = \sqrt{2}$ ~ 1.414213562.

● **PROBLEM 7-11**

Describe Newton's method of approximating the solution of a nonlinear system of equations. Apply it to solve

$$x_1^2 + x_1 x_2 + x_2 - 1 = 0$$
$$e^{x_1} + 2x_2^2 - 3 = 0 \tag{1}$$

The initial approximation is $\bar{x} = (3,2)$.

Solution: We shall write down Newton's method in the form of an algorithm.

The system $\bar{F}(\bar{x}) = \bar{0}$ is a nonlinear system, and $\bar{x}$ is the initial approximation

$$\bar{x} = (x_1, \ldots, x_n)$$

INPUT ⟶ OUTPUT

n - number of equations
$\bar{x}$ - initial approximation
T - tolerance
N - maximum number of iterations

1. $\bar{x} = (x_1, \ldots, x_n)$ approximate or exact solution, or
2. information that the number of iterations was exceeded.

Step 1: Set $k = 1$
Step 2: For $k \leq N$, do Steps 3 to 8
Step 3: Compute $\bar{F}(\bar{x})$
Step 4: Compute $I(\bar{x})_{i,j} = \left(\dfrac{\partial f_i}{\partial x_j}(\bar{x})\right)$
$1 \leq i \leq n$, $1 \leq j \leq n$

Step 5: Solve the n×n linear system

$$I(\bar{x}) \Delta \bar{x} = -\bar{F}(\bar{x}) \tag{2}$$

Step 6: Set $\bar{x} = \bar{x} + \Delta \bar{x}$
Step 7: If $||\Delta \bar{x}|| \leq T$, then OUTPUT $= \bar{x}$, and the procedure is completed successfully. STOP.
Step 8: Set $k = k+1$
Step 9: Number of iterations exceeded. STOP.

The Jacobian matrix $I(\bar{x})_{i,j}$ is

$$I(\overline{x}) = \begin{pmatrix} \frac{\partial f_1(\overline{x})}{\partial x_1} & \frac{\partial f_1(\overline{x})}{\partial x_2} & \cdots & \frac{\partial f_1(\overline{x})}{\partial x_n}, \\ \frac{\partial f_2(\overline{x})}{\partial x_1} & \frac{\partial f_2(\overline{x})}{\partial x_2} & \cdots & \frac{\partial f_2(\overline{x})}{\partial x_n}, \\ \vdots & \vdots & & \vdots \\ \frac{\partial f_n(\overline{x})}{\partial x_1} & \frac{\partial f_n(\overline{x})}{\partial x_2} & \cdots & \frac{\partial f_n(\overline{x})}{\partial x_n} \end{pmatrix} \quad (3)$$

From (1) and (3), we get

$$I(\overline{x}) = \begin{pmatrix} 2x_1+x_2 & x_1+1 \\ e^{x_1} & 4x_2 \end{pmatrix} \quad (4)$$

Since the initial approximation is $\overline{x} = (3,2)$, we get

$$\begin{pmatrix} 8 & 4 \\ e^3 & 8 \end{pmatrix} \begin{pmatrix} \Delta x_1 \\ \Delta x_2 \end{pmatrix} = -\begin{pmatrix} 16 \\ 25.085536 \end{pmatrix} \quad (5)$$

Solving (5), we get

$$\Delta x_1 = 1.75 \qquad \Delta x_2 = -7.5$$

Thus,

$$x_1 = 4.75 \qquad x_2 = -5.5$$

The results are shown in the following table.

k	$x_1^{(k)}$	$\Delta x_1^{(k)}$	$x_2^{(k)}$	$\Delta x_2^{(k)}$
0	3		2	
		1.75		-7.5
1	4.75		-5.5	
		-1.030534		2.456023
2	3.719465		-3.043976	
		-1.021031		1.281390
3	2.698433		-1.762585	
		-.802283		.851970
4	1.896149		-.910614	

Note that we can write (1) in the form

$$x_1 + x_2 - 1 = 0$$
$$e^{x_1} + 2x_2^2 - 3 = 0 \qquad x_1 \neq -1 \qquad (6)$$

or

$$e^{x_1} + 2x_1^2 - 4x_1 - 1 = 0 \qquad (7)$$

We can solve (7) using Newton's method for a single equation, and use $x_2 = 1 - x_1$ to obtain the approximations of x_2.

● **PROBLEM 7-12**

Find the positive real roots of the function

$$f(x) = x^4 - 8.6x^3 - 35.51x^2 + 464.4x - 998.46$$

by Newton's method.

For the simple root, choose $x_0 = 7.0$ as the initial guess and the final magnitude of δ be less than 10^{-6}. For a possible multiple root, choose $x_0 = 4.0$ as the initial guess and $|\delta|$ on the final interation be less in magnitude than $\varepsilon = 10^{-6}$.

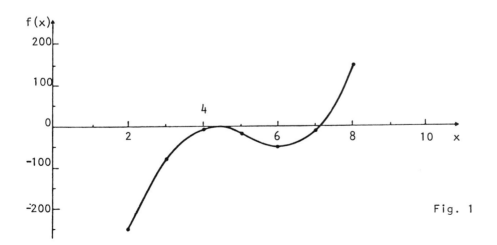

Fig. 1

Solution: A sketch of this function is shown in Fig. 1. This sketch is based on a tabulation of $f(x)$ on the interval $0 \leq x \leq 10$ using a very coarse interval of 1 unit.

It appears that there may be a multiple root (and a tangent point) near x = 4 and a simple root between x = 7 and x = 8. The root(s) near x = 4 could also be two closely spaced real roots if the function crosses the axis, or perhaps no real root(s) at all if the function does not touch the axis (the tabulation is too coarse to tell for sure).

Begin by finding the simple root between x = 7 and x = 8. Newton's method should be suitable for this root. The general expression for Newton's method can be written as

$$x^{(n+1)} - x^{(n)} = \delta^{(n+1)} = -\frac{f(x^{(n)})}{f'(x^{(n)})}$$

where the superscript n denotes values obtained on the nth iteration and n + 1 indicates values to be found on the (n + 1)th iteration. A flow chart of the algorithm is shown in Fig. 2.

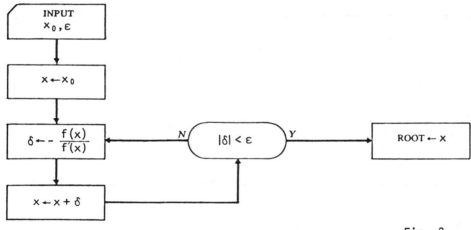

Newton's method.   Fig. 2

Note, the algorithm is terminated when the magnitude of the computed change in the value of the root, δ, is less than some predetermined quantity ε.

Take the first derivative

$$f'(x) = 4x^3 - 25.8x^2 = 71.02x + 464.4$$

Now applying the algorithm of Newton's method,

$$f(7.0) = -36.4500, \quad f'(7.0) = 75.0600$$

$$\delta = \frac{-36.4500}{75.0600} = 0.485612$$

$$x = x + \delta = 7.0 + 0.485612 = 7.485612$$

The second iteration is

$f(7.485612) = 20.6451,$ $f'(7.485612) = 164.891$

$$\delta = -\frac{20.6451}{164.891} = -0.125205$$

$x = x + \delta = 7.485612 - 0.125205 = 7.36041$

The next three iterations yield

$x = 7.34857,$ $\delta = -0.118 \times 10^{-1}$

$x = 7.34847,$ $\delta = -0.102 \times 10^{-3}$

$x = 7.34847,$ $\delta = -0.758 \times 10^{-8}$

The root has now been located accurately. Note that the value of δ decreases very rapidly. For most simple roots, the value of δ on any iteration is of the order of $|\delta|^2$ from the previous iteration. This holds for the present problem. Newton's method is thus said to have a convergence rate which is quadratic for most functions.

Now turn to the possible multiple root. A finer tabulation of $f(x)$ near $x = 4$ reveals that $f(x)$ becomes very small but never changes sign. Since the function is simple to differentiate choose the modified Newton's method in anticipation of a multiple root. Thus define a new function $u(x) = \frac{f(x)}{f'(x)}$. Apply Newton's method to $u(x)$:

$$x^{(n+1)} - x^{(n)} = \delta^{(n+1)} = \frac{u(x^{(n)})}{u'(x^{(n)})}$$

where

$$u'(x) = \frac{(f'(x))^2 - f(x)f''(x)}{(f'(x))^2}.$$

It is necessary to supply subprograms to compute $f(x)$ and the derivatives

$f'(x) = 4x^3 - 25.8x^2 - 71.02x + 464.4,$

$f''(x) = 12x^2 - 51.6x - 71.02$

Choosing $x_0 = 4.0$ as the initial guess, and asking that $|\delta|$ on the final iteration be less in magnitude than $\varepsilon = 10^{-6}$:

$f(4.0) = -3.42,$ $f'(4.0) = 23.52,$ $f''(4.0) = -85.42$

Now

$$u(4) = \frac{f(4)}{f'(4)} = \frac{-3.42}{23.52} = -0.145408$$

$$u'(4) = 1 - \frac{f(4)f''(4)}{(f'(4))^2} = 1 - \frac{(-3.42)(-85.42)}{(23.52)^2}$$

$$= 0.471906$$

and

$$\delta = -\frac{u(4)}{u'(4)} = -\frac{-0.145408}{0.471906} = 0.308129$$

$$x = x + \delta = 4.0 + 0.308129 = 4.308129$$

Three more iterations yield

$x = 4.300001,$ $\quad \delta = -0.812 \times 10^{-2}$

$x = 4.300000,$ $\quad \delta = -0.807 \times 10^{-5}$

$x = 4.300000,$ $\quad \delta = -0.660 \times 10^{-9}$

Hence the multiple roots have been found very accurately in four iterations. Note that the convergence rate of the modified Newton's method for this multiple root is quadratic.

For comparison, one can also find the multiple root by using the conventional Newton's method, with the same initial guess of $x = 4.0$ and the same ε of 10^{-6}:

$$f(4.0) = -3.42, \quad f'(4.0) = 23.52$$

so

$$\delta = -\frac{f(4.0)}{f'(4.0)} = -\frac{-3.42}{23.52} = 0.145408$$

$$x = x + \delta = 4.0 + 0.145408 = 4.145408$$

The next four iterations produce

$x = 4.22138,$ $\quad \delta = 0.075974$

$x = 4.26033,$ $\quad \delta = 0.038952$

$x = 4.28007,$ $\quad \delta = 0.019740$

$x = 4.29001,$ $\quad \delta = 0.009939$

The method is obviously converging very slowly, with each δ only about one-half of the magnitude of the preceding δ. This convergence rate is termed linear. In all, 19 iterations are necessary to obtain

$x = 4.300000,$ $\quad \delta = 0.612 \times 10^{-6}$

The advantages of the modified Newton's method for multiple roots are obvious.

• **PROBLEM 7-13**

Solve the equation $f(x) = x^3 - 4x + 1$ by Newton's Method.

Solution: Given $f(x) = x^3 - 4x + 1$. Now $f(0) = 1$ and $f(0.5) = -0.875$. Thus there is a root to the equation

$f(x) = 0$ in the interval $[0, 0.5]$. Letting $x_0 = 0$, one gets

$$x_1 = x_0 - f(x_0)/f'(x_0) = 0 - 1/(-4) = 0.25.$$

Similarly,

$$x_2 = 0.2540\ 9836,$$
$$x_3 = 0.2541\ 0168,$$
$$x_4 = 0.2541\ 0168.$$

Thus convergence has been attained.

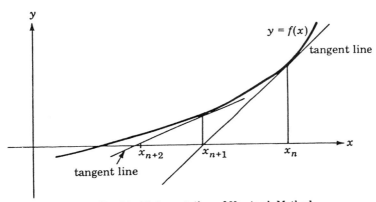

Graphical Interpretation of Newton's Method.

● **PROBLEM 7-14**

Use the correction method of Newton to solve $\tan x = x$.

Solution: Consider the transcendental equation $\tan x = x$. Solving this equation is equivalent to the determination of the zeros of the function $f(x) = \tan x - x$. From a tangent table arranged in terms of the angle in radians, one obtains

$x =$	4.49	4.50
$\tan x =$	4.4223	4.6373
$f =$	-0.0677	0.1373

The unknown value x therefore lies between the two abscissas $x_0 = 4.49$ and $X_0 = 4.50$. The Correction formula

$$x_1 = x_0 - \frac{f(x_0)}{f(X_0) - f(x_0)} h,$$

yields:

$$\underline{x_1} = 4.49 - \frac{0.0677}{0.2050} 0.01 = \underline{4.4933}.$$

• **PROBLEM 7-15**

Calculate the square root x of a given number ($a = 2$) by using the correction method of Newton.

Solution: To calculate the square root x of a given number a, one has to solve the quadratic equation $x^2 - a = 0$; i.e., one must find the zero of the quadratic function $f(x) = x^2 - a$. Because $f'(x) = 2x$, the Newton formula

$$\left(x_1 = x_0 - \frac{f(x_0)}{f'(x_0)} \right)$$

yields for the correction of an approximating value x_0 of the root:

$$x_1 = x_0 - \frac{x_0^2 - a}{2x_0} = \frac{1}{2}\left(x_0 + \frac{a}{x_0}\right). \tag{1}$$

This computational method is very convenient and can be executed by desk-calculator without noting down intermediate results. One performs the division $q = a/x_0$ on the machine and then calculates the arithmetic mean of x_0 and q. The geometric mean of these two is

$$\sqrt{x_0 q} = \sqrt{x_0 \frac{a}{x_0}} = \sqrt{a},$$

i.e., the desired root x. Since the arithmetic mean of two numbers $\geq$ the geometric mean, one sees that the approximation x_1 will containly be too large:

$$x_1 \geq \sqrt{a}.$$

Dividing this inequality by a, one obtains

$$\frac{x_1}{a} \geq \frac{1}{\sqrt{a}} \quad \text{or} \quad \frac{a}{x_1} \leq \sqrt{a}.$$

The number (a/x_1) -- which, according to (1), will be needed anyway in the calculation of the next approximant x_2 -- is therefore a lower bound root. Thus, the simple algorithm (1) gives a succession of upper bounds for the root, and these bounds approach each other increasingly, closely.

For the square root of 2, a slide rule furnishes the first approximating value x_0. Then

$$\frac{a}{x_0} = \frac{2}{1.41} = 1.4185.$$

The last decimal place was rounded up, not down, in order to ensure that the arithmetic mean (1) does not become too small, and will therefore not lose its property as an upper bound.

$$x_1 = \frac{1}{2}(1.41 + 1.4185) = 1.4143 \qquad \text{(rounded up)}$$

$$\frac{a}{x_1} = 1.4141 \qquad \text{(rounded down)}$$

$$1.4141 \le \sqrt{2} \le 1.4143.$$

A more accurate execution of the division yields

$$\frac{a}{x_1} = 1.4141272 \qquad \text{(rounded up)}$$

Hence,

$$x_2 = \frac{1}{2}\left(x_1 + \frac{a}{x_1}\right) = 1.4142136,$$

$$\frac{a}{x_2} = 1.4142135,$$

$$\underline{1.4142135} \le \sqrt{2} \le \underline{1.4142136}.$$

● **PROBLEM 7-16**

Solve the nonlinear system of equations

$$x^2 - y^2 = 1, \quad x^2 + y^2 = 4 \quad \text{by Newton's method.}$$

Use graphical means, for initial approximations.

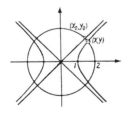

Fig. 1

Intersection of circle and hyperbola.

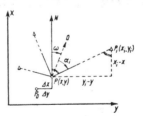

Resectioning.

Fig. 2

Solution: This can be accomplished in the elementary way and without the help of approximation methods. By addition and subtraction of the two equations a solution is immediately found

$$x = \sqrt{2.5} = \underline{1.5811}, \qquad y = \sqrt{1.5} = \underline{1.2247}. \qquad (1)$$

Now apply the Newton method. Beginning with the approximating values x_0, y_0, set $x = x_0 + \Delta x$, $y = y_0 + \Delta y$ and linearize

$$x^2 - y^2 = (x_0 + \Delta x)^2 - (y_0 + \Delta y)^2 \sim x_0^2 - y_0^2$$
$$+ 2x_0 \Delta x - 2y_0 \Delta y$$

$$x^2 + y^2 = (x_0 + \Delta x)^2 + (y_0 + \Delta y)^2 \sim x_0^2 + y_0^2$$
$$+ 2x_0 \Delta x + 2y_0 \Delta y.$$

The linear equations for the corrections therefore read

$$2x_0 \Delta x - 2y_0 \Delta y = 1 - (x_0^2 - y_0^2)$$
$$2x_0 \Delta x_0 + 2y_0 \Delta y = 4 - (x_0^2 + y_0^2) \qquad (2)$$

The initial approximations can best be found graphically (Fig. 1). In fact, the problem concerns the intersections of the hyperbola $x^2 - y^2 = 1$ with the circle $x^2 + y^2 = 4$. Whenever a hyperbola is involved, one can replace it in first approximation by its asymptotes. In this case, the hyperbola is equi-axial and the equation of one of the asymptotes is $y = x$.

Inserting this into the equation of the circle, obtain the approximating point

$$x_0 = y_0 = \sqrt{2} = 1.4.$$

The linear system of equations (2) then assumes the form

$$2.8 \Delta x - 2.8 \Delta y = 1$$
$$2.8 \Delta x + 2.8 \Delta y = 0.08$$

with the solutions $\Delta x = 0.193$, $\Delta y = -0.164$, which, in turn, gives improved approximation point

$$x_1 = 1.593, \qquad y_1 = 1.236.$$

The next Newton step yeilds

$$\underline{x_2 = 1.5812}, \qquad \underline{y_2 = 1.2248}.$$

NEWTON-RAPHSON METHOD

● **PROBLEM 7-17**

Using the Newton-Raphson method estimate the size of a population using capture/recapture data. Denote the number of individuals caught n times by f_n ($n = 1, 2, \ldots$). An estimate of the population size is given by $\hat{P}$, where

$$\ln(\hat{P}) - \ln(\hat{P} - \sum_n f_n) = (\sum_n n f_n)/\hat{P}. \qquad (1)$$

The following data, collected by Istanbul University ecology students at Abant Lake in May 1982, relate to a population of lizards:

n	1	2	3	≥ 4
f_n	91	29	1	0

As the initial value, let $\hat{P}_0 = 300$. Carry up to three iterations.

Solution: Equation (1) may be written as

$$f(\hat{P}) = 0,$$

where

$$f(\hat{P}) = \ln(\hat{P}) - \ln(\hat{P} - \Sigma f_n) - (\Sigma n f_n)/\hat{P},$$

and

$$f'(\hat{P}) = \frac{1}{\hat{P}} - \frac{1}{\hat{P} - \Sigma f_n} + (\Sigma n f_n)/\hat{P}^2.$$

Note that

$$\Sigma f_n = 91 + 29 + 1 = 121,$$

and

$$\Sigma n f_n = 1 \times 91 + 2 \times 29 + 3 \times 1 = 152.$$

An improved estimate for $\hat{P}_0 = 300$ is given by

$$\hat{P}_1 = 300 - (5.703\ 78 - 5.187\ 39 - 0.506\ 67)$$
$$\div (0.003\ 33 - 0.005\ 59 + 0.001\ 69)$$
$$= 317.$$

A second approximation is given by

$$\hat{P}_2 = 317 - (5.758\ 90 - 5.278\ 11 - 0.479\ 50)$$
$$\div (0.003\ 15 - 0.005\ 10 + 0.001\ 51)$$
$$= 320.$$

A third approximation is given by

$$\hat{P}_3 = 320 - (5.768\ 32 - 5.293\ 30 - 0.475\ 00)$$
$$\div (0.003\ 13 - 0.005\ 03 + 0.001\ 48)$$
$$= 320,$$

which is the estimate of the size of the lizard population.

● **PROBLEM 7-18**

Use the two-dimensional Newton-Raphson Method to solve the two simultaneous non-linear equations

$$x^2 + (y - 4)^2 - 9 = 0,$$
$$(x - 4)^2 + y^2 - 9 = 0,$$

using the initial values $x_0 = 2.4$ and $y_0 = 2.6$ ($\varepsilon = 0.001$).

Solution: Note that

$$\frac{\partial f_1}{\partial x} = 2x, \quad \frac{\partial f_2}{\partial x} = 2(x-4);$$

$$\frac{\partial f_1}{\partial y} = 2(y-4), \quad \frac{\partial f_2}{\partial y} = 2y.$$

At the first iteration therefore, one needs to solve the simultaneous linear equations

$$4.8h_0 - 2.8k_0 = 1.28,$$

$$-3.2h_0 + 5.2k_0 = -0.32.$$

One finds that $h_0 = 0.36$ and $k_0 = 0.16$ so that a better approximation is given by $x_1 = 2.76$, $y = 2.76$.

When the operation is repeated, the following results are obtained:

n	x_n	y_n
2	2.7089	2.7089
3	2.7071	2.7071
4	2.7071	2.7071

A solution is therefore $x = y = 2.7071$ (a result which can be verified geometrically).

• **PROBLEM 7-19**

Find the root of $x \tan x = \frac{1}{2}$ which lies between $x = 0.6$ and 0.7 by the Newton-Raphson process. Start with $x_0 = 0.6$.

Solution: There are several forms in which the equation can be written, for example:

$$f(x) \equiv x \tan x - \tfrac{1}{2} = 0; \quad f(x) \equiv 2x - \cot x = 0;$$

$$f(x) \equiv 2x \sin x - \cos x = 0.$$

The third of these is adopted, as it gives the most convenient formula for $f'(x)$, namely

$$f'(x) = 2x \cos x + 3 \sin x.$$

Starting with $x_0 = 0.6$, $\sin x_0 = 0.5646$, $\cos x_0 = 0.8253$, one will have

$$f(x_0) = 2x_0 \sin x_0 - \cos x_0 = -0.1478$$
$$f'(x_0) = 2x_0 \cos x_0 + 3 \sin x_0 = +2.6842$$

$$f(x_0)/f'(x_0) = -0.0551,$$

$$x_1 = 0.6 + 0.0551$$

$$= 0.6551,$$

$x_1 = 0.655$, $\sin x_1 = 0.609159$ $\cos x_1 = 0.793048$

$$\left. \begin{array}{l} f(x_1) = 2x_1 \sin x_1 - \cos x_1 = +0.004950 \\ f'(x_1) = 2x_1 \cos x_1 + 3 \sin x_1 = 2.86637 \end{array} \right\},$$

$$f(x_1)/f'(x_1) = +0.001727,$$
$$x_2 = 0.655 - 0.001727$$
$$= 0.653273,$$

$x_2 = 0.653273$, $\sin x_2 = 0.607788$, $\cos x_2 = 0.794099$

$$\left. \begin{array}{l} f(x_2) = 2x_2 \sin x_2 - \cos x_2 = +0.0000077 \\ f'(x_2) = 2x_2 \cos x_2 + 3 \sin x_2 = 2.86097 \end{array} \right\},$$

$$f(x_2)/f'(x_2) = +0.0000027,$$
$$x_3 = 0.653270.$$

Note that: (i) The first approximation $x_0 = 0.6$ is a rough one and four-figure values of $\sin x$, $\cos x$ are adequate at at this stage; more figures are used later when the accuracy of x_n has been improved.

(ii) For the second stage of the iteration, x_1 is taken as 0.655 instead of the value 0.6551 obatined from the first stage. It is not to be expected that the fourth decimal of this value will be correct, and the rounded value $x_1 = 0.655$ enables tables with interval $\delta x = 0.001$ to be used without interpolation. For the third stage, however, interpolation in the tables is necessary.

● **PROBLEM 7-20**

Find the real root of

$$2x - \cos x - 1 = 0,$$

by the Newton-Raphson method. Use graphical means for the first approximation ($\varepsilon = 0.0001$).

Solution: Find the first approximation by locating graphically the point of intersection of

$$y_1 = 2x - 1,$$
$$y_2 = \cos x.$$

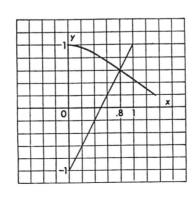

Now

$f(x) = 2x - \cos x - 1,$

$f'(x) = 2 + \sin x.$

Then

i	x_i	$f(x_i)$	$f'(x_i)$
0	.8	−.096702	2.717356
1	.835588	.000443	2.741699
2	.835427	.000067	2.741583
3	.835403		

• **PROBLEM 7-21**

Solve the non-linear equation

$$f(x) = e^x - 3x = 0 \qquad (1)$$

using different initial estimates and methods to examine the convergence properties of the various combinations. The Newton-Raphson, secant, and bisection methods operate directly on equation (1) and the Newton-Raphson method also requires the derivative $f'(x) = e^x - 3$. Equation (1) is rearranged to give the simple iteration function $x = e^x/3$ for the fixed-point method.

(a) Sketch the curves (i) $y = e^x - 3x$, and (ii) $y = e^x/3$ with $y = x$, and use them to obtain approximate values for the roots of $f(x)$ (a calculator will be useful here).

(b) Attempt to locate the exact roots by the fixed-point and Newton-Raphson methods starting from the initial estimates $x_0 = -0.1, 0.5, 0.6, 0.8, 1.0, 1.2, 1.5, 1.6$, and 2.0. Interpret the results noting in particular whether the iteration converges or diverges, to which roots convergence occurs, and whether it is monotonic or oscillatory. Compare and contrast the two methods.

(c) Repeat (b) with the secant and bisection methods using each of the suggested initial estimates x_0 with a second initial estimate $x_0^* = 0.7$. Discuss the results as suggested in (b).

Solution: (a) (i) The derivative, $y' = e^x - 3$, so $y' < 0$ when $x < 1.0986$ and $y' > 0$ when $x > 1.0986$.

(ii) $y' = e^x/3$ and $y' < 1$ when $x < 1.0986$ and $y' > 1$ when $x > 1.0986$.

(b) The fixed-point method produces convergence to the lower root $0.619\ 061$ when $x_0 < 1.51213$ (the upper root), and diverges when $x_0 > 1.51213$. From 20 to 40 iterations are needed to achieve convergence depending on the closeness of the initial estimate to the root. Newton's method converges to the lower root when $x_0 < 1.0986$, the turning point in the curve, and to the upper root when $x_0 > 1.0986$. Less than ten cycles are required for convergence.

(c) The secant method gives convergence in four to nine cycles for all initial estimates, the root found depending on the slope of the secant. Positive slopes produce estimates in the vicinity of the upper root and negative slopes give iterates that tend to converge to the lower root. Notice also that the root found depends on the order in which the initial estimates are entered.

With the bisection method, convergence occurs to a single root (here ~ 20 cycles) bracketed by the initial estimates. If these estimates bracket an even number of roots then the iterative sequence cannot start and if they bracket an odd number, then the root to which convergence occurs depends upon the location of the original estimates.

● **PROBLEM 7-22**

Find by the Newton-Raphson method the real root of

$$3x - \cos x - 1 = 0.$$

It is graphically found that the approximate value of the root is 0.61.

<u>Solution:</u>

$$f(x) = 3x - \cos x - 1,$$

$$f'(x) = 3 + \sin x.$$

Hence

$$h_1 = -\frac{3(0.61) - \cos(0.61) - 1}{3 + \sin(0.61)} = -\frac{0.010}{3.57}$$

$$= -0.00290.$$

$$\therefore a_1 = 0.61 - 0.0029 = 0.6071$$

$$h_2 = -\frac{3(0.6071) - \cos(0.6071) - 1}{3 + \sin(0.6071)}$$

$$= 0.00000381.$$

$$\therefore a_2 = 0.60710381.$$

This result also is true to its last figure.

Note that the root is obtained to a higher degree of accuracy and with less labor by this method than can be achieved by the regula falsi method.

• **PROBLEM 7-23**

Solve $x^3 - 4x + 1 = 0$ by using the Newton-Raphson method.

Solution: A plot of $y = x^3 - 4x + 1$ shows that the equation $x^3 - 4x + 1 = 0$ has roots between -3 and -2, 0 and 1, and 1 and 2. First, find the larger root.

$$F(x) = x^3 - 4x + 1$$

$$F'(x) = 3x^2 - 4$$

Using $x_0 = 2$, $F(2) = 2^3 - 4(2) + 1 = 1$

$$F'(2) = 3(2^2) - 4 = 8$$

$$x_1 = x_0 - \frac{F(x_0)}{F'(x_0)} = 2 - \frac{1}{8} = 1.875$$

$$x_2 = 1.863$$

$$x_3 = 1.863$$

Using $x_0 = 0$, $F(0) = 1$

$$F'(0) = -4$$

$$x_1 = 0 - \frac{1}{-4} = .25$$

$$x_2 = .25 + .0041 = .2541$$

$$x_3 = .2541$$

Using $x_0 = -3$, $F(-3) = 16$

$$F'(-3) = 23$$

$$x_1 = -2.305$$

$$x_2 = -2.038$$

$$x_3 = -2.116$$

$$x_4 = -2.116$$

Since this is one type of problem that can be solved using a direct method, solve this using a standard method for cubics. The equation $x^3 + ax + b = 0$ where $b^2/4 + a^3/27 < 0$ has as solutions

$$x_k = \pm 2\sqrt{\frac{-a}{3}} \cos\left(\frac{\phi}{3} + 120°k\right), \qquad k = 0, 1, 2$$

where

$$\cos \phi = \sqrt{\frac{b^2}{4} \div \frac{(-a)^3}{27}}$$

In this problem a = -4, b = 1,

$$\cos \phi = \sqrt{\frac{\frac{1}{4}}{4^3/27}} = \frac{\sqrt{27}}{16} = \frac{3\sqrt{3}}{16}$$

$$\phi = 71.04°$$

Since b > 0 use the negative sign and $\phi/3 = 23.7°$.

$$x_0 = -\frac{4}{\sqrt{3}} \cos 23.7° = -2.12$$

$$x_1 = -\frac{4}{\sqrt{3}} \cos 143.7° = +\frac{4}{\sqrt{3}} \cos 36.3° = 1.86$$

$$x_2 = -\frac{4}{\sqrt{3}} \cos 263.7° = +\frac{4}{\sqrt{3}} \cos 83.7° = +.254$$

These calculations are done with a slide rule. Although the solution is straightforward, it should be fairly apparent that this method would be difficult to use in a computer program. In addition it will, of course, only solve cubic equations.

• **PROBLEM** 7-24

Correctly to four decimal places, use the Newton-Raphson method to determine the roots of the equation $f(x) = x^3 - 10x + 2 = 0$.

The Newton-Raphson (iterative) formula is $x_{n+1} = x_n - \frac{f(x_n)}{f'(x_n)}$, where x_n is the nth approximation of the desired root and $f'(x)$ is the derivative of $f(x)$ with respect to x. In this problem $f'(x) = 3x^2 - 10$.

Solution: Since the function is a third-degree polynomial, one knows that it has three zeros and that the equation has three roots. Also, from Descartes' rule one knows that there is one negative root and either two or zero positive roots. The determination of the approximate values of the three roots can be done by making a table of x vs f(x), and watching for sign changes, and/or by graphing the function f(x). The easiest way is probably to rewrite the equation $f(x) = 0$ as $f_1(x) = x^3 = f_2(x) = 10x - 2$. Then after graphing each of these functions on the same axes, the three intersection points of $f_1 = x^3$ and $f_2 = 10x - 2$ will be the desired approximate roots of the original equation $f(x) = 0$. The three values obtained are $x_i = -3, 0,$ and 3; these values are the three initial approximations to use in the Newton-Raphson formula.

(a) $x_1 = -3$, $x_2 = -3 - \frac{f(-3)}{f'(-3)} = -3 - \frac{(-3)^3 - 10(-3) + 2}{3(-3)^2 - 10}$

$$= -3 - \frac{5}{17} = -3 - 0.3 = -3.3.$$

$$x_3 = -3.3 - \frac{f(-3.3)}{f'(-3.3)} = -3.3 - \frac{-0.937}{22.67} = -3.3 + 0.04$$

$$= -3.26$$

$$x_4 = -3.26 - \frac{f(-3.26)}{f'(-3.26)} = -3.26 + 0.00210 = -3.2579$$

In most cases the correction term in the Newton-Raphson formula, $-f(x_n)/f'(x_n)$, will be accurate to 2m decimal places where m is the number of zeros in the correction term between the decimal and the first non-zero digit. This was used in the calculations above, and will be used below similarly.

(b) $x_1 = 0$, $x_2 = 0 - \frac{2}{-10} = 0.2$;

$$x_3 = 0.2 - \frac{0.008}{-9.88} = 0.2 + 0.00081 = 0.2008$$

(c) $x_1 = 3$, $x_2 = 3 - \frac{-1}{17} = 3 + 0.06 = 3.06$;

$$x_3 = 3.06 - \frac{0.052616}{18.0908} = 3.06 - 0.00291$$

$$x_3 = 3.0571$$

The sum of the roots in this problem is $-a_2 = 0$, and the product of the roots is $(-1)^3 a_0 = -2$. Checking the above results, one gets $\Sigma r_i = -3.2579 + 0.2008 + 3.0571 = 0$ and $\pi r_i = (-3.2579)(0.2008)(3.0571) = -1.9999$. Note that as soon as one root is determined the other two roots can be determined by using the known results regarding the sum and the product of roots; but it is perhaps better to save at least one of the relationships for a check.

• **PROBLEM 7-25**

For the equation $f(x) = e^{-x} + x^2 - 10 = 0$:

(a) Determine the approximate location of all of its real roots.

(b) Determine the value of each positive root correctly to eight significant digits.

Solution: (a) Rewriting the given equation as $f_1(x) = e^{-x} = 10 - x^2 = f_2(x)$, it is very easy to graph each of the functions $f_1(x)$ and $f_2(x)$ separately and observe the intersections of the two curves. See the figure.

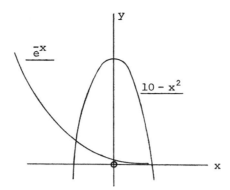

There are two roots of the given equation; one is positive and one is negative. Since the curve $f_1 = e^{-x}$ approaches the +x axis so rapidly as x increases, the positive root is approximately where the parabola crosses the +x axis; that is, an approximate value of the positive root is $x = \sqrt{10} \doteq 3.16$. The negative root is near $x = -2$. A little calculating will show it to be about $x = -1.87$.

(b) Using the Newton-Raphson formula

$$x_{n+1} = x_n - \frac{f(x_n)}{f'(x_n)},$$

where x_n is the nth approximation of the desired root and $f' = df/dx$.

Starting with $x_1 = 3.16$, one gets $x_2 = 3.16 - \frac{f(3.16)}{f'(3.16)}$

$$x_2 = 3.16 - \frac{0.0424 + 9.9856 - 10}{-0.0424 + 6.32} = 3.16 - \frac{0.0280}{6.2776} =$$

$$3.16 - 0.0045 = 3.1555$$

(This correction term is probably correct to four decimal places since the correction term starts with two zeros)

$$x_3 = 3.1555 - \frac{-0.00020266}{6.26838} = 3.1555 + 0.0000323$$

$$= 3.1555323$$

THE METHOD OF FALSE POSITION (REGULA FALSI)

• **PROBLEM 7-26**

Solve the quadratic equation by the method of false position.

Regula Falsi: $\quad c = \dfrac{bf(a) - af(b)}{f(a) - f(b)}$

$f(x) = x^2 - 2 = 0.$

Solution: Begin by trying

$a = 1.5, \quad f(a) = 0.25;$

$b = 1.4, \quad f(b) = -0.04.$

The new point is

$c = \{0.25 \times 1.4 - (-0.04) \times 1.5\}/\{0.25 - (-0.04)\}$

$= 1.4138.$

Now repeat the process with

$a = 1.5, \qquad f(a) = 0.25;$

$b = 1.4138, \qquad f(b) = -0.0012.$

The new point is

$c = \{0.25 \times 1.4138 - (-0.0012) \times 1.5\}/\{0.25 - (-0.0012)\} = 1.4142.$

When the process is repeated, the same answer (to four decimal places) is obtained. The solution has been attained.

● **PROBLEM 7-27**

Find the smallest positive root of

$$F(x) = \tan x + \tanh x = 0$$

using the initial value $x = 2.36$.

Solution: Since $\tanh x$ tends rapidly to 1 for $x > 0$, it is obvious that the roots will be close to the points at which $\tan x = -1$, or $x = 3\pi/4, 7\pi/4, \ldots$ Therefore, start near the point $x = 2.36$, and the calculation by repeated linear interpolation proceeds as follows--

x	F(x)	
2.36	−0.01009	$m_0 = 0.02000/0.01 = 2.000$
2.37	+0.00991	$x_1 = 2.36 - (-0.01009/2.000)$
		$= 2.36505$
2.36505	+0.000059	$m_1 = 0.01015/0.00505 = 2.001$
		$x_2 = 2.36505 - 0.000059/2.001$
		$= 2.365021$
2.365021	+0.0000001	$x_3 = 2.365020_5$

This carries the calculation of the root to the limit possible with 6D tables of $\tan x$ and $\tanh x$. The last stage has little significance.

Alternatively one can tabulate $F(x)$ as follows--

x	F(x)			
2.35	−0.030495			
		20403		
2.36	−0.010092		−401	
		20002		15
2.37	+0.009910		−386	
		19616		
2.38	+0.029526			

By inverse interpolation $X = 2.365021$, which is correct to 6D (D means decimal).

• **PROBLEM 7-28**

Find the positive root of $\sqrt[3]{x-1} - \cos x = 0$ correct to five significant figures by the method of false position. Then, review the solution for possible defects, if any; any; continue your iteration for a suitable solution.

Solution: 1. It is readily determined that there is only one positive root and that it is slightly greater than unity. We have $f(1) = -0.54$, $f(1.2) = 0.22$; hence the root is between $a = 1$ and $b = 1.2$. Use

$$r_0 = \frac{af(b) - bf(a)}{f(b) - f(a)}, \tag{1}$$

or

$$r_0 = a - \frac{b - a}{f(b) - f(a)} f(a). \tag{2}$$

and find $r_0 = 1.1$. Since $f(1.1) = 0.01$, regard 1.1 as a new b; using the new b and the old $a = 1$, find $r_1 = 1.098$. Since $f(1.098) = 0.00566$, 1.098 is also a new b. Continue in this fashion—it will take quite a number of additional steps—to get the answer 1.0957 correct to five significant figures. Note that as the work proceeds, more and more significant figures in the values of $f(x)$ will be needed.

2. The preceding method of solution is poor on two counts, the rate of convergence to the root is quite slow and the length of the interval within which the root is known to lie does not become arbitrarily small so that one does not have a good measure of the precision of the approximation at any step. To overcome these defects, proceed as follows. Take, as before, $a = 1$, $b = 1.2$ to find $r_0 = 1.1$. Since $f(1.1) = 0.01$, the root is between 1.00 and 1.10 and probably closer to 1.1. As a conservative guess, take the new $a = 1.06$ and the new $b = 1.10$. Using these values, obtain the approximation $r_1 = 1.096$.

Since $f(1.096) = 0.00073$, 1.096 is a new b and choose 1.095 as a new a. Since $f(1.095) = -0.00176$, 1.095 is indeed a new a. If it had turned out that $f(1.095)$ were positive, 1.095 would have been a new and better b than 1.096. Another step yields $f(1.0957) = -0.00001$, $f(1.0958) = 0.00023$. Hence, not only is the root between 1.0957 and 1.0958, but the former is certainly correct to five significant figures.

If in the right-hand member of (1) or (2) one replaces a by x to obtain

$$\phi(x) = \frac{xf(b) - bf(x)}{f(b) - f(x)} = x - \frac{b - x}{f(b) - f(x)} f(x), \tag{3}$$

then for b close to the root r, $\phi(x)$ is suitable for iteration. In the problem, start with b = 1.2 and r_0 = 1. Obtain on successive substitutions into (3), r_1 = 1.14, r_2 = 1.09, r_3 = 1.097, r_4 = 1.0955, r_5 = 1.0957, with no further change in the fourth decimal place.

• **PROBLEM 7-29**

Solve $x^3 - 9x + 1$ by the rule of false position.

Solution: Use $x_1 = 2$ and $x_2 = 4$, then

$$F(x_1) = (+2)^3 - 9(2) + 1 = +8 - 18 + 1 = -9$$

$$F(x_2) = (4)^3 - 9(4) + 1 = 64 - 36 + 1 = 29$$

A good approximation of the root is found with the formula

$$x_3 = \frac{x_1 F(x_2) - x_2 F(x_1)}{F(x_2) - F(x_1)}$$

$$x_3 = \frac{2(29) - 4(-9)}{29 - (-9)} = \frac{58 + 36}{38} = \frac{94}{38} = 2.47368$$

$x_4 = 2.73989$

$x_5 = 2.86125$

$x_6 = 2.91107$

$x_7 = 2.93816$

$x_8 = 2.94104$

$x_9 = 2.94214$

$x_{10} = 2.94256$

$x_{11} = 2.94278$

$x_{12} = 2.94281$

Thus ten iterations after the initial guesses give six significant figures.

To find the root between zero and one using $x_1 = -1$ and $x_2 = 1$

$x_3 = .125000$

$x_4 = .109827$

$x_5 = .111265$

Here only three approximations to achieve the same precision are necessary.

OTHER ROOTFINDING METHODS

• PROBLEM 7-30

Find the roots of $x^4 - 5x^3 + 8x^2 - 3x - 3 = 0$ by the root squaring method.

(1)	1	5	8	3	−3
	1	2.5 1	6.4 1	9	9
		−1.6	−3.0	48	
			−0.6		
(2)	1	0.9 1	2.8 1	5.7 1	9
	1	8.1 1	7.84 2	3.249 3	8.1 2
		−5.6	−10.26	−0.504	
			0.18		
(4)	1	2.5 1	−2.24 2	2.745 3	8.1 2
	1	6.25 2	5.0176 4	7.53503 6	6.561 3
		4.48	−13.7250	0.03629	
			0.0162		
(8)	1	1.073 3	−8.6912 4	7.57132 6	6.561 3
	1	1.15133 6	7.55370 9	5.73249 13	4.30467 7
		0.17382	−16.24805	0.00011	
			0.00001		
(16)	1	1.32515 6	−8.69434 9	5.73260 13	4.30467 7
	1	1.75602 12	7.55915 19	3.28627 27	1.85302 15
		0.01739	−15.19311	*	
			*		
(32)	1	1.77341 12	−7.63396 19	3.28627 27	1.85302 15
	1	3.14498 24	5.82773 39	1.07996 55	3.43368 30
		0.00015	−11.65581	*	
			*		
(64)	1	3.14513 24	−5.82808 39	1.07996 55	3.43368 30

Solution: The root squaring tabulations are shown in the accompanying table. Stop at the 64th power equation because after that point all coefficients with the exception of the coefficients of x^2 will be merely the squares of the previous corresponding coefficients.

The appearance of negative signs in the coefficients of x^2 indicates the presence of imaginary conjugate roots.

One obtains from the table that $R_1 = |r_1|^{64} = 3.1451 \times 10^{24}$; whence $\log|r_1| = 0.38278$ and $|r_1| = 2.4142$. Next, since r_2 and r_3 are apparently imaginary roots, put $|r_2| = |r_3| = r$ and get, using

244

$$R_g + R_{g+1} = B_g/B_{g-1},$$

$$R_g R_{g+1} = B_{g+1}/B_{g-1},$$

$$R_2 R_3 = |r_2 r_3|^{64} = (r^2)^{64} = 1.0800 \times 10^{55}/3.1451 \times 10^{24};$$

$\log r^2 = 0.47712$ and $r^2 = 3.0000$.

Finally,

$$R_1 R_2 R_3 R_4 = |r_1 r_2 r_3|^{64} |r_4|^{64} = 3.4337 \times 10^{30};$$

$\log|r_4| = 9.61722 - 10$, $|r_4| = 0.4142$.

A graph or substitution into the original equation shows that the two real roots are $r_1 = 2.4142$, $r_4 = -0.4142$. If one puts $r_2 = u + iv$, $r_3 = u - iv$, then

$$r_1 + r_2 + r_3 + r_4 = 5 = 2.4142 + (u + iv) + (u - iv) - 0.4142,$$

so that $u = 1.5000$. Then $v = \sqrt{r^2 - u^2} = \sqrt{3 - 2.25} = 0.8660$ and $r_2 = 1.5000 + 0.8660i$, $r_3 = 1.5000 - 0.8660i$. The roots are correct as far as they are written; the roots are actually $1 \pm \sqrt{2}$, $(3 \pm \sqrt{3}i)/2$.

● **PROBLEM 7-31**

Find x_3 with the Secant method if:

$x^3 - 9x + 1 = 0$ using $x_1 = 3$, $x_2 = 4$.

Solution: The Secant method is:

$$\frac{x_1 f(x_2) - x_2 f(x_1)}{f(x_2) - x_2}$$

$$x_3 = \frac{3(29) - 4(1)}{29 - 4} = \frac{83}{25} = 3.32$$

• **PROBLEM 7-32**

Using Wegstein's method:

$$\bar{x}_n = qx_{n-1} + (1-q)x_n,$$

solve $F(x) = x^2 - 4 = 0$ using $x = x + x^2 - 4$ and $x_0 = 6$.

Solution: $x_1 = 6 + 36 - 4 = 38$

$x_2 = 1478$

$a = \dfrac{1478 - 38}{38 - 6} = 45$

$q = \dfrac{45}{44} = 1.0227273$

$\bar{x}_2 = \dfrac{45}{44}(38) + (1 - \dfrac{45}{44})(1478) = 5.272688$

$x_3 = f(\bar{x}_2) = 29.073927$

$a = \dfrac{f(\bar{x}_2) - f(x_1)}{\bar{x}_2 - x_1} = \dfrac{29.073927 - 1478}{5.272688 - 38} = 44.312$

$q = \dfrac{a}{a-1} = 1.0231$

$\bar{x}_3 = 4.7226580$

$x_4 = f(\bar{x}_3) = 23.026157$

$x_5 = 7.251968$

$x_6 = 3.6954468$

$x_7 = 2.2756761$

$x_8 = 2.0198933$ $\qquad q_8 = 1.2286599$

$x_9 = 2.0002670$ $\qquad q_9 = 1.2463955$

$x_{10} = 2.0000005$ $\qquad q_{10} = 1.2497485$

Note that convergence is slow but that the iteration does converge.

• **PROBLEM 7-33**

Consider Halley's formula:

$$x_{n+1} = x_n - \frac{F(x_n)}{F'(x_n) - \frac{F''(x_n) \cdot F(x_n)}{2F'(x_n)}}$$

Solve $x^3 - x - 10 = 0$ using $x_0 = 4$.

Solution: $F(x) = x^3 - x - 10$, $F'(x) = 3x^2 - 1$, $F''(x) = 6x$, and

$$x_{i+1} = x_i - \frac{x_i^3 - x_i - 10}{3x_i^2 - 1 - \frac{6x_i(x_i^3 - x_i - 10)}{2(3x_i^2 - 1)}}$$

Using $x_0 = 4$,

$x_1 = 2.5394656$

$x_2 = 2.3104300$

$x_3 = 2.3089074$

$x_4 = 2.3089073$

$x_5 = 2.3089073$

It appears that in the solution of cubic equations this method could very well be used along with synthetic division to obtain one real root and the reduced quadratic in a very short time.

• **PROBLEM 7-34**

Compute $p(8)$, where $p(x) = 2x^3 + x + 7$ by Horner's Rule.

Solution:

	2	0	1	7	
		16	128	1,032	
	2	16	129	1,039	$p(8) = 1,039$.

Horner's rule for evaluating a polynomial of degree n,

$$p(x) = a_0 x^n + a_1 x^{n-1} + \ldots + a_{n-1} x + a_n,$$

at a point z, is described by the recursive formula:

$$b_0 = a_0, \quad b_i = a_i + z \cdot b_{i-1}$$

$(i = 1, 2, \ldots, n)$, $b_n = p(z)$.

If the intermediate b_i are of no interest, then, in most programming languages, the algorithm can be described without subscripts for the b_i, such as in the flowchart in the Fig. shown and the corresponding Algol-fragment:

    b: = a[0];

    for i: = 1 step 1 until n do

    b: = a[i] + z * b;

(The symbol: = is read "is given the value of.")

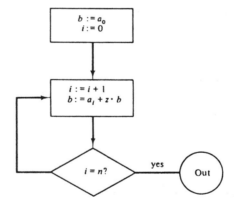

● **PROBLEM 7-35**

Solve $x^3 - 9x + 1 = 0$ for the root between $x = 2$ and $x = 4$ by interval halving or bisection.

Solution: Since $F(x) = x^3 - 9x + 1$ is a polynomial it is continuous,

$$F(2) = -9$$

$$F(4) = 29$$

and one is now sure that a root exists between $x = 2$ and $x = 4$. Now

$$x_2 = \frac{x_0 + x_1}{2} \text{ or } x_2 = \frac{2+4}{2} = 3$$

$F(3) = 1$, so the next approximation is made by checking to see that $F(2) \cdot F(3) = (-9)(1) < 0$. So

$x_2 = 3$

$x_3 = 2.5$

$x_4 = 2.75$

$x_5 = 2.875$

$x_6 = 2.9375$

$x_7 = 2.96875$

$x_8 = 2.953125$

$x_9 = 2.9453125$

$x_{10} = 2.9414063$

$x_{11} = 2.9433594$

$x_{12} = 2.9423829$

$x_{13} = 2.9428712$

$x_{14} = 2.9427492$

$x_{15} = 2.9428102$

$x_{16} = 2.9428407$

● **PROBLEM 7-36**

$f(x) = 2x^3 - 5x^2 + 3x - 1$; evaluate this function at $x_n = 2$, using synthetic division.

Solution: This is essentially the case

$$f(x) = a_3 x^3 + a_2 x^2 + ax + a_0$$

where

$a_3 = 2$

$a_2 = -5$

$a_1 = 3$

$a_0 = -1$.

This is a mth-degree polynomial in x, so that $m = 3$, and

$$b_m = b_3 = a_3 = 2$$

$$b_2 = a_2 + b_3 x_n = -5 + (2)(2) = -1$$

$$b_1 = a_1 + b_2 x_n = 3 + (-1)(2) = 1$$

$$b_0 = a_0 + b_1 x_n = -1 + (1)(2) = 1 = f(2).$$

This is merely synthetic division in a new form, where the numbers b_i represent the last line in the synthetic division

```
2|    2    -5    3    -1
           4    -2    2
      2    -1    1    |1|
      b₃   b₂   b₁   b₀.
```

Now, the name "synthetic division" comes from the fact that the numbers b_3, b_2, b_1 (or b_m, mb_{m-1}, ... b_2, b_1 in the general case) represent the coefficients of another polynomial which is the quotient of $f(x)$ divided by $x - x_n$.

This can be shown easily enough by actually performing the following long division:

$$\begin{array}{r}
a_3 x^2 \\
x - x_n \overline{\smash{\big)}\, a_3 x^3 + a_2 x^2 + a_1 x + a_0} \\
\underline{a_3 x^3 - a_3 x_n x^2} \\
a_3 x_n x^2 + a_2 x^2 \\
\underbrace{}\\
(a_2 + a_3 x_n) x^2
\end{array}$$

At this point, stop and substitute b_3 and b_2:

$$b_3 = a_3$$

$$b_2 = a_2 + b_3 x_n$$

and $b_2 x^2 = (a_2 + a_3 x_n) x^2$;

$$\begin{array}{r}
b_3 x^2 + b_2 x + b_1 \\
x - x_n \overline{\big)\, a_3 x^3 + a_2 x^2 + a_1 x + a_0} \\
\underline{a_3 x^3 - b_3 x_n x^2 } \\
b_2 x^2 \\
\underline{b_2 x^2 - b_2 x_n x } \\
\underbrace{(a_1 + b_2 x_n) x}_{} \\
b_1 x \\
\underline{b_1 x - b_1 x_n} \\
\underbrace{a_0 + b_1 x_n}_{} \\
\text{Remainder } b_0
\end{array}$$

One sees here that b_3, b_2, and b_1 are the coefficients of the quotient. In the general case, the quotient is a function $h(x)$ equal to

$$h(x) = b_m x^{m-1} + b_{m-1} x^{m-2} + \ldots + b_2 x + b_1,$$

which is only of $(m - 1)$ degree, one less than the original function. The quantity b_0 is not part of the quotient, but actually represents the remainder.

The remainder can be made part of the quotient by dividing it by the divisor. That is, if a function $f(x)$ is divided by $(x - x_n)$ one obtains

$$\frac{f(x)}{(x - x_n)} = h(x) + \frac{b_0}{x - x_n}.$$

Returning to the above problem, one gets

$$h(x) = b_3 x^2 + b_2 x + b_1$$

$$= 2x^2 - x + 1,$$

and if it is checked, it is found that

$$\frac{2x^3 - 5x^2 + 3x - 1}{x - 2} = \underbrace{2x^2 - x + 1}_{\text{quotient}} + \underbrace{\frac{1}{x - 2}}_{\text{remainder}}.$$

Multiplying both sides of the general equation by $(x - x_n)$ gives

$$f(x) = (x - x_n)h(x) + b_0.$$

Differentiating,

$$f'(x) = (x - x_n)h'(x) + h(x).$$

$$f'(x_n) = h(x_n).$$

• **PROBLEM 7-37**

Consider the equation $f(x) = 2x^5 - 9x^4 - 4x^3 + 71x^2 - 126x + 90 = 0$.

(a) With the aid of Descartes' rule of signs list the possibilities regarding positive, negative, and complex roots of this equation.

(b) Determine the sum and the product of the five roots of this equation. (An nth-degree polynomial equation always has n roots).

Solution: (a) To apply Descartes' rule, first count the sign changes in the coefficients of f(x). Going from left to right, there is a change between +2 and -9, between -4 and +71, between +71 and -126, and between -126 and +90, for a total of four sign changes in f(x). According to the rule it is known that there are either 4, 2, or 0 real positive roots. When the coefficients in f(x) are all real (as in the given function), then the "lost" real roots show up two at a time as conjugate complex numbers.

Next investigate possible negative roots by checking the sign changes in f(-x); this gives

$$f(-x) = -2x^5 - 9x^4 + 4x^3 + 71x^2 + 126x + 90.$$

There is only one sign change (between -9 and +4), so there is one real negative root. Realizing that there must be a total of five roots, one gets the following three possibilities:

(i) 4 positive, 1 negative, and 0 complex roots, or

(ii) 2 positive, 1 negative, and 2 complex roots, or

(iii) 0 positive, 1 negative, and 4 complex roots.

(b) For the polynomial equation $y(x) = x^n + a_{n-1}x^{n-1} + a_{n-2}x^{n-2} + \ldots + a_1x + a_0 = 0$, the following results are known regarding the sum and the product of its n roots:

$$\Sigma r_i = -a_{n-1} \quad \text{and} \quad \Pi r_i = (-1)^n a_0.$$

It is important to note, and remember, that the coefficient of x^n above is one. For the problem being considered, one therefore, first, divides by the 2 (the coefficient of $x^n = x^5$) and obtains

$$y(x) = x^5 - 4.5x^4 - 2x^3 + 35.5x^2 - 63x + 45.$$

Therefore, $\Sigma r_i = 4.5$ and $\Pi r_i = -45$.

The actual roots of the given equation $f(x) = 0$ are

$$r_i = 3, \ 2.5, \ -3, \ 1+i, \ 1-i.$$

It can be seen that the sum and the product of the roots does agree with the above results in part (b). Case (ii) in part (a) happens to be the correct one of the three possibilities determined there.

● **PROBLEM 7-38**

Apply the steepest descent algorithm to the function $F(x) = x_1^2 + \alpha x_2^2$ with $\alpha > 0$. Illustrate the result for $\alpha = 100$.

Solution: The function $F(x) = x_1^2 + \alpha x_2^2$, with $\alpha > 0$, has a global minimum at $x = 0$. Its gradient is linear,

$$\nabla F = [2x_1, \ 2\alpha x_2]^T$$

Therefore, one could determine at once its unique critical point from the system

$$2x_1 = 0$$

$$2\alpha x_2 = 0$$

But, consider steepest descent instead, to make a point. This requires determining the minimum of the function

$$g(t) = F(x - t\nabla F(x))$$

253

$$= F(x_1(1 - 2t), x_2(1 - 2\alpha t))$$

Setting $g'(t) = 0$ gives the equation

$$0 = 2(x(1 - 2t))(-2) + \alpha 2(x(1 - 2\alpha t))(-2\alpha)$$

whose solution is $t^* = \frac{1}{2}(x_1^2 + \alpha^2 x_2^2)/(x_1^2 + \alpha^3 x_2^2)$. Hence, if $x = [x_1, x_2]^T$ is the current guess, then

$$\frac{x_1 x_2 (\alpha - 1)}{x_1^2 + \alpha^3 x_2^2} [\alpha^2 x_2, - x_1]^T$$

is the next guess.

Now take x in the specific form $c(\alpha, \pm 1)^T$. Then the next guess becomes

$$c \frac{\alpha - 1}{\alpha + 1} [\alpha, \mp 1]^T$$

i.e., the error is reduced by the factor $(\alpha - 1)/(\alpha + 1)$.

For $\alpha = 100$, and $x^{(0)} = [1, 0.01]^T$, one gets, after 100 steps of steepest descent, the point

$$x^{(100)} = \left(\frac{\alpha - 1}{\alpha + 1}\right)^{100} [1, 0.01]^T = [.135 \ldots, 0.00135 \ldots]^T$$

which is still less than $\frac{7}{8}$ of the way from the first guess to the solution.

• **PROBLEM 7-39**

Solve by Muller's method:

$$x + 2y - 4 = 0$$

$$x - y - 1 = 0,$$

using (x_0, y_0) as $(4,4)$ and $K = 1$.

Solution: $G(x,y) = (x + 2y - 4)^2 + (x - y - 1)^2$

$$\frac{\partial G}{\partial x} = 2(x + 2y - 4) + 2(x - y - 1)$$

$$\frac{\partial G}{\partial y} = 4(x + 2y - 4) - 2(x - y - 1)$$

By substitution into the following formulas:

$$x_1 = x_0 - \frac{K\frac{\partial G}{\partial x}}{\sqrt{\left(\frac{\partial G}{\partial x}\right)^2 + \left(\frac{\partial G}{\partial y}\right)^2}}$$

$$y_1 = y_0 - \frac{K\frac{\partial G}{\partial y}}{\sqrt{\left(\frac{\partial G}{\partial x}\right)^2 + \left(\frac{\partial G}{\partial y}\right)^2}})$$

one gets

X	Y	H
3.6095	3.0794	1.
3.1990	2.1675	1.
2.7478	1.2751	1.
2.1476	.47526	1.
2.2009	.97241	.5
2.0880	.91891	.25
2.0408	1.0347	.125
2.0113	.97959	.0625
2.0104	1.0108	.03125
2.0035	.99680	.015625

CHAPTER 8

LEAST SQUARES – CURVE FITTING

LEAST - SQUARES APPROXIMATION

● PROBLEM 8-1

Given the following noisy data:

x	2.10	6.22	7.17	10.52	13.68
$f(x)$	2.90	3.83	5.98	5.71	7.74

Fit a straight line to this data by using least squares.

<u>Solution</u>: The general least-squares equations are

$$\begin{bmatrix} n & \sum x_i & \sum x_i^2 & \cdots & \sum x_i^\ell \\ \sum x_i & \sum x_i^2 & \sum x_i^3 & \cdots & \sum x_i^{\ell+1} \\ \sum x_i^2 & \sum x_i^3 & \sum x_i^4 & \cdots & \sum x_i^{\ell+2} \\ \vdots & & & & \vdots \\ \sum x_i^\ell & \sum x_i^{\ell+1} & \sum x_i^{\ell+2} & \cdots & \sum x_i^{2\ell} \end{bmatrix} \begin{bmatrix} a_0 \\ a_1 \\ a_2 \\ \vdots \\ a_\ell \end{bmatrix}$$

$$= \begin{bmatrix} \sum f(x_i) \\ \sum x_i f(x_i) \\ \sum x_i^2 f(x_i) \\ \vdots \\ \sum x_i^\ell f(x_i) \end{bmatrix}$$

For 5 data points and $\ell = 1$ (a first degree polynomial), the least-squares equations are

$$\begin{bmatrix} 5 & \sum_{i=1}^{5} x_i \\ \sum_{i=1}^{5} x_i & \sum_{i=1}^{5} x_i^2 \end{bmatrix} \begin{bmatrix} a_0 \\ a_1 \end{bmatrix} = \begin{bmatrix} \sum_{i=1}^{5} f(x_i) \\ \sum_{i=1}^{5} x_i f(x_i) \end{bmatrix}$$

Each element in these equations can now be computed:

$$\sum_{i=1}^{5} x_i = 2.10 + 6.22 + 7.17 + 10.52 + 13.68 = 39.69$$

$$\sum_{i=1}^{5} x_i^2 = (2.10)^2 + (6.22)^2 + (7.17)^2 + (10.52)^2 + (13.68)^2 = 392.3201$$

$$\sum_{i=1}^{5} f(x_i) = 2.90 + 3.83 + 5.98 + 5.71 + 7.74 = 26.16$$

$$\sum_{i=1}^{5} x_i f(x_i) = (2.10)(2.90) + (6.22)(3.83) + (7.17)(5.98) + (10.52)(5.71) + (13.68)(7.74) = 238.7416$$

The set of equations is

$$\begin{bmatrix} 5 & 39.69 \\ 39.69 & 392.3201 \end{bmatrix} \begin{bmatrix} a_0 \\ a_1 \end{bmatrix} = \begin{bmatrix} 26.16 \\ 238.7416 \end{bmatrix}$$

Gauss-Jordan elimination yields

$$a_0 = 2.038392, \quad a_1 = 0.4023190$$

The required straight line is thus

$$g(x) = 2.038392 + 0.4023190 x$$

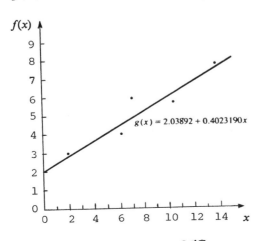

Fig. 1

The data and the straight line approximation are shown in Fig. 1. The straight line appears to provide a good approximation to the data.

● **PROBLEM 8-2**

Find the least squares straight line approximation for the following data.

x	0.0	0.2	0.4	0.6	0.8
$f(x)$	0.9	1.9	2.8	3.3	4.2

Solution: If one takes $n = 1$, $\psi_0(x) \equiv 1$ and $\psi_1(x) \equiv x$, the functions Φ_a defined by

$$\Phi_a(x) = \sum_{r=0}^{n} a_r \psi_r(x),$$

will all be straight lines. To find the least squares straight line one must solve the normal equations

$$a_0 \sum_{i=0}^{N} \psi_0(x_i) \psi_r(x_i) + \ldots + a_n \sum_{i=0}^{N} \psi_n(x_i) \psi_r(x_i) = \sum_{i=0}^{N} f(x_i) \psi_r(x_i),$$

which in this case are

$$a_0(N+1) + a_1 \sum_{i=0}^{N} x_i = \sum_{i=0}^{N} f(x_i)$$

$$a_0 \sum_{i=0}^{N} x_i + a_1 \sum_{i=0}^{N} x_i^2 = \sum_{i=0}^{N} f(x_i) \cdot x_i.$$

For the above data, these equations are

$$5a_0 + 2a_1 = 13.1$$

$$2a_0 + 1.2a_1 = 6.84$$

with solution $a_0 = 1.02$, $a_1 = 4$. The required straight line is therefore $1.02 + 4x$.

● **PROBLEM 8-3**

Standard weights x are hung on a spring, and the corresponding lengths y are measured. It is known (Hooke's Law) that the lengths should follow a linear relation

$$F(x) = \alpha + \beta x.$$

A. Determine, according to the principle of least squares, estimates a and b for α and β.

B. Apply the results to the numerical data:

x_i	y_i
2	7.32
4	8.24
6	9.20
8	10.19
10	11.01
12	12.05

Solution: A. Here pairs of observed values (x_1, y_1), (x_2, y_2), ..., (x_n, y_n) are given. The theoretical values for y_1, ..., y_n are $\alpha + \beta x_1$, $\alpha + \beta x_2$, ..., $\alpha + \beta x_n$, respectively. Hence, the sum of squared differences between observed and theoretical values is formed as

$$S(\alpha, \beta) = \sum_{i=1}^{n} (y_i - \alpha - \beta x_i)^2.$$

$S(\alpha, \beta)$ is a function of α and β; consequently, if it is to be minimized, it is required that

$$\frac{\partial S}{\partial \alpha} = \frac{\partial S}{\partial \beta} = 0.$$

The required estimates a and b will thus be the solutions of the two linear equations found by equating

$$\frac{\partial S}{\partial \alpha} = -2 \sum_{i=1}^{n} (y_i - \alpha - \beta x_i)$$

and

$$\frac{\partial S}{\partial \beta} = -2 \sum_{i=1}^{n} x_i (y_i - \alpha - \beta x_i)$$

to zero. Note that the x_i and y_i are not variables, but are observed numbers. Simplifying the resulting equations gives

$$\left. \begin{array}{l} \sum_{i=1}^{n} y_i = na + b \sum_{i=1}^{n} x_i, \\[2mm] \sum_{i=1}^{n} x_i y_i = a \sum_{i=1}^{n} x_i + b \sum_{i=1}^{n} x_i^2. \end{array} \right\} \quad (1)$$

This gives a line $f(x) = a + bx$ which estimates the true line $F(x) = \alpha + \beta x$. Note that this is not the true line $F(x) = \alpha + \beta x$; there is absolutely no way, using mathematics whereby one can obtain theoretical constants α and β from physical observations (x_i, y_i); one can only obtain estimates of α and β. Thus, for any index i, one gets three values: y_i, the observed value; $F(x_i) = \alpha + \beta x_i$, the theoretical (and unknown) value; and $f(x_i) = a + bx_i$, the least-squares estimate of $F(x_i)$.

B. Now form the following table

x_i	x_i^2	y_i	$x_i y_i$
2	4	7.32	14.64
4	16	8.24	32.96
6	36	9.20	55.20
8	64	10.19	81.52
10	100	11.01	110.10
12	144	12.05	144.60
Σ 42	364	58.01	439.02

Then Equations (1) become

$$6a + 42b = 58.01, \qquad 42a + 364b = 439.02,$$

and these are readily solved to yield $a = 6.3733333$, $b = .4707143$. Consequently, a straight line is fitted to the given data as

$$f(x) = 6.3733333 + .4707143x,$$

and this line may reasonably be used as an estimator of the true value $F(x)$ for any x in the range $2 \leq x \leq 12$.

• **PROBLEM 8-4**

A. Find the least-squares estimator for the true curve

$$F(x) = \alpha + \beta x + \gamma x^2,$$

given data (x_i, y_i) for $i = 1, 2, \ldots, n$.

B. Apply the results to the case

x	y
2	3.07
4	12.85
6	31.47
8	57.38
10	91.29

Solution: For the ith observation, the y-deviation between observed and theoretical values is

$$y_i - \alpha - \beta x_i - \gamma x_i^2,$$

and the sum of squared deviations is

$$S = \sum_{i=1}^{n} (y_i - \alpha - \beta x_i - \gamma x_i^2)^2.$$

Then

$$\frac{\partial S}{\partial \alpha} = -2 \sum_{i=1}^{n} (y_i - \alpha - \beta x_i - \gamma x_i^2),$$

$$\frac{\partial S}{\partial \beta} = -2 \sum_{i=1}^{n} x_i (y_i - \alpha - \beta x_i - \gamma x_i^2),$$

$$\frac{\partial S}{\partial \gamma} = -2 \sum_{i=1}^{n} x_i^2 (y_i - \alpha - \beta x_i - \gamma x_i^2).$$

Equating these three partial derivatives to zero, one obtains a, b, and c (the estimates of α, β, γ) as solutions of the equations

$$\left. \begin{array}{l} \sum y = na + b \sum x + c \sum x^2, \\ \sum xy = a \sum x + b \sum x^2 + c \sum x^3, \\ \sum x^2 y = a \sum x^2 + b \sum x^3 + c \sum x^4. \end{array} \right\} \quad (1)$$

Here, for simplicity, the range of summation is omitted.

B. In this case, since there is an odd number of items, matters are greatly simplified by introducing $X = (x - 6)/2$. Then form

x	y	xy	x^2y
-2	3.07	-6.14	12.28
-1	12.85	-12.85	12.85
0	31.47	.00	.00
1	57.38	57.38	57.38
2	91.29	182.58	365.16
$\sum$ 0	196.06	220.97	447.67

Also, $\sum x^2 = 10$, $\sum x^3 = 0$, $\sum x^4 = 34$, and Equations (1) give

$$196.06 = 5a + 10c,$$
$$220.97 = 10b,$$
$$447.67 = 10a + 34c.$$

Hence
$$a = 31.276286,$$
$$b = 22.097,$$
$$c = 3.967857,$$

and the estimator is

$$f(X) = 31.276286 + 22.097X + 3.967857X^2.$$

In terms of the original variables, this becomes

$$f(x) = .695999 - .855071x + .991964x^2,$$

and can now be used for estimating values. In particular, note that $y_2 = 3.07$, $f_2 = 2.953713$, whereas the true value F_2 is not known. Note also that the sum $\sum_{i=1}^{5}(y_i - f_i)^2$ is less for the estimator f_i found than for any other estimator.

● **PROBLEM 8-5**

Fit the following data, given the values of the weights, with the expression

$$\log y = \log k + n \log x$$

x	172	210	320	400
y	66	80	100	120
w	4356	6400	10000	14400

For the logarithmic function it can be shown that if the weights of y are equal, the weights of log y are given by

$$w_f = \frac{y^2}{M^2},$$

where $M = .43429$.

Solution: For some data more weight may be given to some observations than others, in which case the best fit is that for which the sum of the weighted squares of the residuals is a minimum:

$$\sum wR^2 = w_1 R_1^2 + w_2 R_2^2 + \ldots + w_n R_n^2, \tag{1}$$

where w_i is a set of weights. To minimize the function (1), set the partial derivatives equal to zero to obtain a set of weighted normal equations.

For fitting the data having weights w_j with a quadratic:

$$y = a_0 + a_1 x + a_2 x^2, \tag{2}$$

the weighted residuals are

$$w_j R_j = w_j (a_0 + a_1 x_j + a_2 x_j^2 - y_j), \tag{3}$$

and the sum of the weighted squares of the residuals is

$$\sum w_j R_j^2 = \sum w_j (a_0 + a_1 x_j + a_2 x_j^2 - y_j)^2, \tag{4}$$

the summation being over j from 0 to m for the m observations. Taking the partial derivative of (4) with respect to a_i and setting each equal to zero yields the weighted normal equations which may be transformed into

$$\begin{aligned}
(\sum w_j) a_0 + (\sum w_j x_j) a_1 + (\sum w_j x_j^2) a_2 &= \sum w_j y_j, \\
(\sum w_j x_j) a_0 + (\sum w_j x_j^2) a_1 + (\sum w_j x_j^3) a_2 &= \sum w_j x_j y_j, \\
(\sum w_j x_j^2) a_0 + (\sum w_j x_j^3) a_1 + (\sum w_j x_j^4) a_2 &= \sum w_j x_j^2 y_j,
\end{aligned} \tag{5}$$

and from which a_i can be solved to obtain the fit (2). To determine log k and n, write the equation

$$Y = K + nX,$$

where

$$Y = \log y, \quad K = \log k, \quad X = \log x,$$

so that the data is changed to

X	2.23553	2.32222	2.50515	2.60206
Y	1.81954	1.90309	2.00000	2.07918
w	4356	6400	10000	14400

The weighted normal equations [see (5)] are easily calculated:

$$35156K + 87121.34n = 70045.88,$$
$$87121.34K + 216538.92n = 174011.92,$$

the solution of which yields

$$K = 0.33346, \quad n = .66944.$$

The equation is

$$\log y = .33346 + .66944 \log x \; .$$

● **PROBLEM 8-6**

Use harmonic analysis to obtain the trigonometric approximation to

$x°$	−150	−120	−90	−60	−30	0	30	60	90	120	150	180
y	10	16	18	24	38	32	16	5	−7	−13	−14	−5

The table below gives the values for sin kx and cos kx for 12 observations.

TABLE: Harmonic Analysis - 12 Observations

x	cos x	cos 2x	cos 3x	cos 4x	cos 5x	cos 6x
0	1	1.0	1	1.0	1	1
30	.866025	0.5	0	−0.5	−.866025	−1
60	.500000	−0.5	−1	−0.5	.500000	1
90	0	−1.0	0	1.0	0	−1
120	−.500000	−0.5	1	−0.5	−.500000	1
150	−.866025	0.5	0	−0.5	.866025	−1
180	−1	1.0	−1	1.0	−1	1

x	sin x	sin 2x	sin 3x	sin 4x	sin 5x
30	.500000	.866025	1	.866025	.500000
60	.866025	.866025	0	−.866025	−.866025
90	1	0	−1	0	1
120	.866025	−.866025	0	.866025	−.866025
150	.500000	−.866025	1	−.866025	.500000

Compare your results to the given values to get an idea on how good the approximation is.

Solution: Dividing the interval from $-\pi$ to π into 2k equal parts means

$$x_j = \frac{j}{k}\pi.$$

Now seek to minimize the residual equation
$$\Sigma R^2 = \sum_{j=-k}^{k-1}\left[y_j - \sum_{n=0}^{m}(a_n \cos nx_j + b_n \sin nx_j)\right]^2.$$
Take the partial derivatives with respect to a_n and b_n and set them equal to zero to obtain the normal equations. With the aid of trigonometric identities, obtain expressions for a_n and b_n.

Define
$$F_j = f(x_j) + f(-x_j), \quad (j = 1, \ldots, k-1),$$
$$G_j = f(x_j) - f(-x_j), \quad (j = 1, \ldots, k-1),$$
$$F_0 = f(0) \text{ and } F_k = f(x_k).$$

Then
$$a_0 = \frac{1}{2k}\sum_{j=0}^{k}F_j,$$

$$a_r = \frac{1}{k}\sum_{j=-k}^{k-1}y_j\cos rx_j = \frac{1}{k}\sum_{j=0}^{k}F_j\cos rx_j,$$

$$a_k = \frac{1}{2k}\sum_{j=-k}^{k-1}F_j\cos kx_j,$$

$$b_r = \frac{1}{k}\sum_{j=-k}^{k-1}y_j\sin rx_j = \frac{1}{k}\sum_{j=1}^{k-1}G_j\sin rx_j,$$

$(r = 1, \ldots, k-1)$. The computation may be arranged in a schematic for $k = 6$, that is, the interval $(-\pi, \pi)$ divided into twelve parts.

Thus, for $k = 6$ (i.e. the interval $(-\pi, \pi)$ is divided into twleve parts)

$$a_0 = \frac{1}{12}(F_0 + F_1 + \ldots + F_6),$$

$$a_1 = \frac{1}{6}[F_0 - F_6 + \frac{1}{2}\sqrt{3}(F_1 - F_5) + \frac{1}{2}(F_2 - F_4)],$$

$$a_2 = \frac{1}{6}[F_0 - F_3 + F_6 + \frac{1}{2}(F_1 - F_2 - F_4 + F_5)],$$

$$a_3 = \frac{1}{6}(F_0 - F_2 + F_4 - F_6),$$

$$a_4 = \frac{1}{6}[F_0 + F_3 + F_6 - \frac{1}{2}(F_1 + F_2 + F_4 + F_5)],$$

$$a_5 = \frac{1}{6}[F_0 - F_6 - \frac{1}{2}\sqrt{3}(F_1 - F_5) + \frac{1}{2}(F_2 - F_4)],$$

$$a_6 = \frac{1}{12}(F_0 - F_1 + F_2 - F_3 + F_4 - F_5 + F_6)$$

and

$$b_1 = \frac{1}{6}[G_3 + \frac{1}{2}(G_1 + G_5) + \frac{1}{2}\sqrt{3}(G_2 + G_4)],$$

$$b_2 = \frac{1}{6}[\frac{1}{2}\sqrt{3}(G_1 + G_2 - G_4 - G_5)],$$

$$b_3 = \frac{1}{6}(G_1 - G_3 + G_5),$$

$$b_4 = \frac{1}{6}[\frac{1}{2}\sqrt{3}(G_1 - G_2 + G_4 - G_5)],$$

$$b_5 = \frac{1}{6}[G_3 + \frac{1}{2}(G_1 + G_5) - \frac{1}{2}\sqrt{3}(G_2 + G_4)].$$

Hence,

x	F	G				
0	32		$a_0 = 10$			
30	54	-22	$a_1 = 16.705$		$b_1 = 14.928$	
60	29	-19	$a_2 = 4.167$		$b_2 = 1.732$	
90	11	-25	$a_3 = 1.833$		$b_3 = -3.500$	
120	3	-29	$a_4 = -.500$		$b_4 = -1.155$	
150	-4	-24	$a_5 = -.038$		$b_5 = -1.072$	
180	-5		$a_6 = -.167$			

$$y = 10 + 16.705 \cos x + 4.167 \cos 2x$$
$$+ 1.833 \cos 3x - .5 \cos 4x$$
$$- .038 \cos 5x - .167 \cos 6x$$
$$+ 14.928 \sin x + 1.732 \sin 2x - 3.5 \sin 3x$$
$$- 1.155 \sin 4x - 1.072 \sin 5x.$$

Check:

$x°$	-150	-120	-90	-60	-30	0	30	60	90	120	150	180
y (given)	10	16	18	24	38	32	16	5	-7	-13	-14	-5
y (calc.)	10.89	16.25	18.17	23.42	37.45	31.17	15.44	4.42	-6.83	-12.75	-13.11	-4.33

● **PROBLEM 8-7**

Fit a cubic to the data

x	-4	-2	-1	0	1	3	4	6
y	-35.1	15.1	15.9	8.9	.1	.1	21.1	135

using the method of least squares.

Solution: Consider the cubic

$$y = a_0 + a_1 x + a_2 x^2 + a_3 x^3.$$

It is required to obtain the coefficients a_i ($i = 0$, 1, 2, 3).

The sums

$$S_0 = m \qquad\qquad t_0 = \sum_{j=1}^{m} y_j$$

$$S_1 = \sum_{i=1}^{m} x_i \quad \text{and} \quad t_1 = \sum_{j=1}^{m} x_j y_j$$

$$\vdots \qquad\qquad\qquad \vdots$$

$$S_k = \sum_{i=1}^{m} x_i^k \qquad\qquad t_k = \sum_{j=1}^{m} x_j^k y_j$$

which are required to construct the normal equations, can be arranged into a schematic:

x^0	x	x^2	x^3	x^4	x^5	x^6	y	xy	x^2y	x^3y
1	x_1	x_1^2	x_1^3	x_1^4	x_1^5	x_1^6	y_1	$x_1 y_1$	$x_1^2 y_1$	$x_1^3 y_1$
1	x_2	x_2^2	x_2^3	x_2^4	x_2^5	x_2^6	y_2	$x_2 y_2$	$x_2^2 y_2$	$x_2^3 y_2$
1	x_3	x_3^2	x_3^3	x_3^4	x_3^5	x_3^6	y_3	$x_3 y_3$	$x_3^2 y_3$	$x_3^3 y_3$
.										
.										
1	x_m	x_m^2	x_m^3	x_m^4	x_m^5	x_m^6	y_m	$x_m y_m$	$x_m^2 y_m$	$x_m^3 y_m$
$S_0 = m$	S_1	S_2	S_3	S_4	S_5	S_6	k_0	k_1	k_2	k_3

The normal equations can then be written in this manner:

$$S_0 a_0 + S_1 a_1 + S_2 a_2 + S_3 a_3 = k_0$$

$$S_1 a_0 + S_2 a_1 + S_3 a_2 + S_4 a_3 = k_1$$

$$S_2 a_0 + S_3 a_1 + S_4 a_2 + S_5 a_3 = k_2$$

$$S_3 a_0 + S_4 a_1 + S_5 a_2 + S_6 a_3 = k_3.$$

Hence, first obtain the required sums:

x^0	x	x^2	x^3	x^4	x^5	x^6	y	xy	x^2y	x^3y
1	-4	16	-64	256	-1024	4096	-35.1	140.4	-561.6	2246.4
1	-2	4	-8	16	-32	64	15.1	-30.2	60.4	-120.8
1	-1	1	-1	1	-1	1	15.9	-15.9	15.9	-15.9
1	0	0	0	0	0	0	8.9	0	0	0
1	1	1	1	1	1	1	.1	.1	.1	.1
1	3	9	27	81	243	729	.1	.3	.9	2.7
1	4	16	64	256	1024	4096	21.1	84.4	337.6	1350.4
1	6	36	216	1296	7776	46656	135.0	810.0	4860.0	29160.0
8	7	83	235	1907	7987	55643	161.1	989.1	4713.3	32622.9

The normal equations are

$$8a_0 + 7a_1 + 83a_2 + 235a_3 = 161.1$$

$$7a_0 + 83a_1 + 235a_2 + 1907a_3 = 989.1$$

$$83a_0 + 235a_1 + 1907a_2 + 7987a_3 = 4713.3$$

$$235a_0 + 1907a_1 + 7987a_2 + 55643a_3 = 32622.9.$$

Solving by Crout's method, one obtains the values of a_i,

a_0	a_1	a_2	a_3
9.011039	-8.966140	-1.000093	.999074.

The cubic is given by

$$.999x^3 - 1.000x^2 - 8.966x + 9.011 = 0.$$

A check on the given points shows a good fit.

x	-4	-2	-1	0	1	3	4	6
y (GIVEN)	-35.1	15.1	15.9	8.9	.1	.1	21.1	135
y (CALC.)	-35.1	15.0	16.0	9.0	.0	.1	21.1	135
DIFF.	0	.1	-.1	-.1	.1	0	0	0

• **PROBLEM** 8-8

Consider the function $F(x) = x_n = 10 + \frac{n-1}{5}$, $n = 1, \ldots, 6$. Find the minimum coefficients using the least square approximation method and the coefficient matrix

$$A = \begin{bmatrix} 6 & 63 & 662.2 \\ 63 & 662.2 & 6,967.8 \\ 662.2 & 6,967.8 & 73,393.5664 \end{bmatrix} \quad (1)$$

Solution: Given approximate values $f_n \approx f(x_n)$ with

$$x_n = 10 + \frac{n-1}{5} \quad n = 1, \ldots, 6,$$

one can reason that this data can be adequately represented by a parabola. Accordingly, it can be chosen that

$$\phi_1(x) = 1 \qquad \phi_2(x) = x \qquad \phi_3(x) = x^2.$$

From the coefficient matrix A it follows that $\|A\|_\infty \approx 8 \cdot 10^4$. On the other hand, with

$$x = \begin{bmatrix} 10.07 \\ -2 \\ 0.099 \end{bmatrix} \quad \text{one gets} \quad Ax = \begin{bmatrix} -0.02 \\ -0.18 \\ -1.28 \end{bmatrix}$$

Hence, from the inequality

$$\|Ax\| \geq \frac{\|x\|}{\|A^{-1}\|} \quad \text{(for all nonvectors x)}$$

one gets

$$1.28 = \|Ax\|_\infty \geq \frac{10.07}{\|A^{-1}\|_\infty} \quad \text{or} \quad \|A^{-1}\|_\infty \geq 7.8$$

Therefore the condition number of A is

$$\text{cond}(A) = \|A\|_\infty \|A^{-1}\|_\infty \geq 10^5$$

Actually, the condition number of A is much larger than 10^5, as the following specific results show. Choose

$$f(x) = 10 - 2x + \frac{x^2}{10}$$

and use exact data,

$$f_n = f(x_n) \quad n = 1, \ldots, 6$$

Then, since $f(x)$ is a polynomial of degree 2, $F^*(x)$ should be $f(x)$ itself; therefore the results will be

$$c_1^* = 10 \quad c_2^* = -2 \quad c_3^* = 0.1$$

Using Gaussian elimination to solve (1) for this case on the CDC 6500 produces the result

$$c_1^* = 9.9999997437 \ldots \quad c_2^* = -1.9999999511 \ldots$$
$$c_3^* = 0.0999999976 \ldots$$

so that 14-decimal-digit floating-point arithmetic for this 3 x 3 system gives only about 8 correct decimal digits. If one rounds the (3, 3) entry of A to 73,393.6 and repeat the calculation, the computed answer turns out to be an astonishing

$$c_1^* = 6.035 \ldots \quad c_2^* = -1.243 \ldots$$
$$c_3^* = 0.0639 \ldots$$

Similarly, if all calculations are carried out in seven-decimal-digit floating-point arithmetic, the results are

$$c_1^* = 8,492 \ldots \quad c_2^* = -1.712 \ldots$$
$$c_3^* = 0.0863 \ldots$$

● **PROBLEM 8-9**

Find the best linear fit to the data

$$x_1 = 1 \quad f_1 = 1$$
$$x_2 = 2 \quad f_2 = 3$$
$$x_3 = 3 \quad f_3 = 1$$

if best is interpreted in the Chebyshev sense:

$$\min_{a,b} \max_{i=1,2,3} |ax_i + b - f_i|.$$

Solution: It is geometrically obvious and easily proved that the line in question is such that the residuals r_1, r_2, r_3 where

$$r_i = ax_i + b - f_i$$

are equal in magnitude but alternate in sign. Thus the best fit is given by

$$y = 2.$$

● **PROBLEM 8-10**

A. Fit a parabola $y = a_0 + a_1x + a_2x^2$ to the points $(-2,9)$, $(-1,6)$, $(0,3)$, $(1,-1)$, $(2,-2)$, $(3,-3)$, $(5,-1)$, $(7,3)$.

B. Develop a method of averages to solve the above problem.

Solution: A. Group the first three, the second three, and the last two points; the centroids of the three groups are, respectively, $(-1,6)$, $(2,-2)$, $(6,1)$. The equation of the parabola through these three points is $y = (198 - 265x + 41x^2)/84$.

The "true" values of y, corresponding to the given values of x, are computed from the equation to be $446/42$, $252/42$, $99/42$, $-13/42$, $-84/42$, $-114/42$, $-51/42$, $176/42$. Hence, the respective errors are $68/42, 0$, $-27/42$, $29/42$, 0, $12/42$, $-9/42$, $50/42$. The sum of the errors is $123/42$.

B. A method of averages can be developed as follows. By substituting X_i for x in $y = a_0 + a_1x + a_2x^2 + \ldots + a_m x^m$, obtain the corresponding "true" value of y, namely, $Y_i = a_0 + a_1X_i + a_2X_i^2 + \ldots + a_m X_i^m$. The error at this point is

$$e_i = Y_i - Y_i = a_0 + a_1X_i + \ldots + a_m X_i^m - Y_i;$$

the sum of the errors for all the points is

$$\sum e_i = \sum (a_0 + a_1X_i + \ldots + a_m X_i^m - Y_i), \text{ where,}$$

1 and n are to be understood as the limits of summation. By setting $\sum e_i = 0$, one obtains only one condition for the determination of the $m + 1$ parameters $a_0, a_1, \ldots, a_m$; therefore divide the points $(X_1, Y_1), (X_2, Y_2), \ldots, (X_n, Y_n)$, into $m + 1$ groups

$$(X_{11}, Y_{11}), \quad (X_{12}, Y_{12}), \ldots, (X_{1,a}, Y_{1,a});$$
$$(X_{21}, Y_{21}), \quad (X_{22}, Y_{22}), \ldots, (X_{2,b}, Y_{2,b});$$
$$\ldots\ldots\ldots\ldots\ldots\ldots\ldots\ldots\ldots\ldots\ldots\ldots$$
$$(X_{m+1,1}, Y_{m+1,1}), (X_{m+1,2}, Y_{m+1,2}), \ldots,$$
$$(X_{m+1,k}, Y_{m+1,k});$$

and set the sum of the errors in each group equal to zero. The resulting $m + 1$ equations can be solved for the $m + 1$ parameters. If these equations are inconsistent, a regrouping is necessary.

If the points are grouped as above, the three equations obtained are

$$3a_0 - 3a_1 + 5a_2 = 18$$

$$3a_0 + 6a_1 + 14a_2 = -6$$

$$2a_0 + 12a_1 + 74a_2 = 2,$$

hence $a_0 = 522/255$, $a_1 = -803/255$, $a_2 = 123/255$. The required equation is $y = (522 - 803x + 123x^2)/255$, which is in close agreement with the previously found solution.

• **PROBLEM 8-11**

Determine the parameters in $y = a(b + x)^c$ so that its graph is a good fit for the points $(-2.5, 17.90)$, $(-1.5, 4.70)$, $(-0.5, 2.35)$, $(0.5, 1.45)$, $(1.5, 1.05)$, $(2.5, 0.80)$.

Solution: The given points are plotted on ordinary graph paper and from the hyperbolic shape of the curve, it can be concluded that a and b are positive and c is negative. The given equation yields $\log y = \log a + c \log(b + x)$. Make the substitutions $y^* = \log y$, $x^* = \log(b + x)$, $a^* = \log a$, so that $y^* = a^* + cx^*$. The value of a^* is to be found, the values of y^* (for the given points) can be found, but the values of x^* cannot be found because b is unknown. Hence the first job is to find a good estimate for b. This can be done in several ways. Since $x + b = 0$ is a vertical asymptote of the required curve, the diagram tells that b is approximately 3. Try $b = 2.8, 3, 3.2$ and form the following table:

TABLE

x	$x + 2.8$	$x + 3$	$x + 3.2$	y
−2.5	0.3	0.5	0.7	17.90
−1.5	1.3	1.5	1.7	4.70
−0.5	2.3	2.5	2.7	2.35
0.5	3.3	3.5	3.7	1.45
1.5	4.3	4.5	4.7	1.05
2.5	5.3	5.5	5.5	0.80

If the points $(x + 2.8, y)$, $(x + 3, y)$, $(x + 3.2, y)$ are plotted on log log paper, or if the points $(\log(x + 2.8), \log y)$, $(\log(x + 3), \log y)$, $(\log(x + 3.2), \log y)$ are plotted on ordinary graph paper, it will be seen that the points $(x + 3.2, y)$ or $(\log(x + 3.2), \log y)$ almost lie on a straight line, whereas the others do not. Hence take $b = 3.2$ and proceed.

An approximate value for b may also be found as follows. Plot the given points on ordinary cross-section paper as before and draw a smooth, good fitting curve. Let (x_i, y_i), $i = 1, 2, \ldots$, be arbitrary points on the curve so that $y_i = a(b + x_i)^c$. Choose four points so that $y_1:y_2 = y_3:y_4$, then $(b + x_1)/(b + x_2) = (b + x_3)/(b + x_4)$; hence, $b = (x_2 x_3 - x_1 x_4)/(x_1 - x_2 - x_3 + x_4)$. If several such tetrads are chosen, then the average of all the b's calculated from these tetrads will give a fairly precise result.

The end result is $y = 10.5(3.2 + x)^{-1.5}$.

• **PROBLEM** 8-12

The set $\{1, x, x^3\}$ is not a Chebyshev set on $[-1, 1]$, since the function $0.1 + (-1).x + 1.x^3$ has three zeros belonging to $[-1, 1]$. However, these three functions are linearly independent on $[-1, 1]$.

Now, find the least squares approximation to a function f on a finite point set $\{x_0, x_1, \ldots, x_N\}$ by a function of the form Φ_a defined by

$$\Phi_a(x) = \sum_{r=0}^{n} a_r \psi_r(x),$$

where it is assumed that $\{\psi_0, \ldots, \psi_n\}$, with $n < N$, is a Chebyshev set on some interval $[a,b]$ which contains all the x_i.

Solution: One requires the minimum of

$$E(a_0, \ldots, a_n) = \sum_{i=0}^{N} (f(x_i) - \Phi_a(x_i))^2$$

over all values of $a_0, \ldots, a_n$. A necessary condition for E to have a minimum is $\partial E/\partial a_r = 0$ for $r = 0, 1, \ldots, n$ at the minimum. This gives

$$-2 \sum_{i=0}^{N} [f(x_i) - (a_0 \psi_0(x_i) + \ldots + a_n \psi_n(x_i))] \psi_r(x_i) = 0$$

which yields the equations

$$a_0 \sum_{i=0}^{N} \psi_0(x_i) \psi_r(x_i) + \ldots + a_n \sum_{i=0}^{N} \psi_n(x_i) \psi_r(x_i)$$

$$= \sum_{i=0}^{N} f(x_i) \psi_r(x_i), \qquad (1)$$

for $0 \leq r \leq n$. These $n + 1$ linear equations in the $n + 1$ unknowns $a_0, \ldots, a_n$ are called the normal equations. It is now stated, without proof, that these linear equations have a unique solution if there is no set of numbers $b_0, \ldots, b_n$ (except $b_0 = b_1 = \ldots = b_n = 0$) for which

$$\sum_{r=0}^{n} b_r \psi_r(x_i) = 0, \qquad 0 \leq i \leq N.$$

Since $N > n$ and the ψ_r form a Chebyshev set, this condition is satisfied. It is for this reason that one uses functions ψ_r which form a Chebyshev set. One has seen so far that a necessary condition for a minimum of $E(a_0, \ldots, a_n)$ is that the a_r satisfy the normal equations (1), since it is known that these equations have a unique solution. It still remains to show that the solution of these equations does, in fact, provide the minimum. To see this, let $a_0^*, \ldots, a_n^*$ denote the solution of (1). First, consider the case where there are only two functions ψ_0 and ψ_1. Consider the difference

$$E(a_0^* + \delta_0, a_1^* + \delta_1) - E(a_0^*, a_1^*)$$

$$= \sum_{i=0}^{N} [f(x_i) - (a_0^* + \delta_0)\psi_0(x_i) - (a_1^* + \delta_1)\psi_1(x_i)]^2$$

$$- \sum_{i=0}^{N} [f(x_i) - a_0^* \psi_0(x_i) - a_1^* \psi_1(x_i)]^2$$

$$= \sum_{i=0}^{N} [\delta_0 \psi_0(x_i) + \delta_1 \psi_1(x_i)]^2 - 2\delta_0 \sum_{i=0}^{N} \psi_0(x_i) [f(x_i)$$

$$- a_0^* \psi_0(x_i) - a_1^* \psi_1(x_i)]$$

$$- 2\delta_1 \sum_{i=0}^{N} \psi_1(x_i) [f(x_i) - a_0^* \psi_0(x_i) - a_1^* \psi_1(x_i)].$$

The last two summations are both zero, since a_0^* and a_1^* satisfy the normal equations (1) for the case $n = 1$.

Therefore
$$E(a_0^* + \delta_0, a_1^* + \delta_1) - E(a_0^*, a_1^*) =$$

$$\sum_{i=0}^{N} [\delta_0 \psi_0(x_i) + \delta_1 \psi_1(x_i)]^2. \qquad (2)$$

The right side of (2) can only be zero if
$$\delta_0 \psi_0(x_i) + \delta_1 \psi_1(x_i) = 0$$
for $0 \leq i \leq N$, which cannot be so, unless $\delta_0 = \delta_1 = 0$, since ψ_0 and ψ_1 form a Chebyshev set. Thus the right side of (2) is always strictly positive unless $\delta_0 = \delta_1 = 0$, showing that $E(a_0, a_1)$ has indeed a minimum when $a_0 = a_0^*$, $a_1 = a_1^*$. For a general value of $n < N$, one may similarly show that
$$E(a_0^* + \delta_0, \ldots, a_n^* + \delta_n) - E(a_0^*, \ldots, a_n^*)$$
$$= \sum_{i=0}^{N} [\delta_0 \psi_0(x_i) + \ldots + \delta_n \psi_n(x_i)]^2,$$
showing that the minimum value of $E(a_0, \ldots, a_n)$ occurs when $a_0 = a_0^*, \ldots, a_n = a_n^*$.

● **PROBLEM 8-13**

A. Compute the minimal sum of squared deviations in the numerical data

x_i	y_i
2	7.32
4	8.24
6	9.20
8	10.19
10	11.01
12	12.05

B. Illustrate by an appropriate method for $n \to +\infty$ the significant figures in (A). Compute the T-distribution for α and β if the errors ε_i in $y_i = F(x_i) + \varepsilon_i$ follow the normal probability distribution, also find the confidence interval.

<u>Solution</u>: A. In this case, the following table is formed

TABLE

x_i	y_i	$f(x_i)$	$y_i - f_i$
2	7.32	7.314762	+.005238
4	8.24	8.256191	−.016191
6	9.20	9.197619	+.002381
8	10.19	10.139048	+.050952
10	11.01	11.080476	−.070476
12	12.05	12.021905	+.028095

As a check, one should have
$$\Sigma(y_i - f_i) = \Sigma(y_i - a - bx_i) = 0;$$
actually, it comes to $-.000001$. Accumulating the sum of squares, one obtains
$$m = \Sigma(y_i - f_i)^2$$
$$= .00864756.$$
B. Now this evaluation involves little work, since $n = 6$; however, for larger n, it is advantageous to simply note that
$$m = \Sigma(y_i - a - bx_i)^2$$
$$= \Sigma(y_i - a - bx_i)y_i - a\Sigma(y_i - a - bx_i)$$
$$- b\Sigma x_i(y_i - a - bx_i),$$
and both the second and third terms vanish, in virtue of the equations determining a and b; then
$$m = \Sigma y_i^2 - a\Sigma y_i - b\Sigma x_i y_i$$
$$= 576.3787 - 6.3733333(58.01) - 439.02(.4707143)$$
$$= .00864328.$$
Note that, while seven decimals were used in b, b was muliplied by 439.02; so the round-off error in the seventh digit of b moves up to the fifth decimal in m; this leaves one with only three-figure accuracy, $m = .00864$, despite the retention of seven figures in b. This illustrates the fact that it may often be necessary to retain ten or eleven figures in a and b in order to compute m; too often, the estimating line is given as
$$f(x) = 6.373 + .471x$$
(after all, it is claimed, that's one more decimal than in the data); then it is computed
$$m = \Sigma y_i^2 - a\Sigma y_i - b\Sigma x_i y_i$$
$$= -.09745,$$
thus obtaining m, a sum of squares, as a negative number, very different from the value $.00864$.

The way in which m is used is the following: if the errors ε_i in $y_i = F(x_i) + \varepsilon_i$ follow the normal probability distribution, then the following quantities are computed:
$$T_\alpha = t_{n-2}\sqrt{\frac{m}{n-2}\left[\frac{1}{n} + \frac{\bar{x}^2}{\Sigma(x_i - \bar{x})^2}\right]} ,$$
$$T_\beta = t_{n-2}\sqrt{\frac{m}{(n-2)[\Sigma(x_i - \bar{x})^2]}} .$$

Here $\bar{x} = \frac{1}{n}\sum x_i$, and t_{n-2} is a numerical constant tabulated as "Student's t"; tables give t_4, for 95% fiducial limits, as 2.776. Using this value in the present problem, one gets

$$T_\alpha = .03978,$$
$$T_\beta = .010397.$$

Consequently, there is a 95% probability that β lie in the range

$$b - .010397 < \beta < b + .010397,$$

and that α lie in the range

$$a - .03978 < \alpha < a + .03978.$$

• **PROBLEM 8-14**

Assume a heavy object is dropped from a height of 1100 feet. Letting x stand for the time, in seconds, after the object is dropped, one can solve for the equation of motion using elementary physics as $f(x) = 1100 - 16x^2$. Using this equation one can then calculate the height above the ground during the first four seconds:

Time (seconds)	Height (feet)
0	1100
1	1084
2	1036
3	956
4	844

Rather than employ this analytical method, suppose this experiment is actually performed and the height, to the nearest ten feet, is measured. Suppose then, this gives the following table of measured values:

x_i	$f(x_i)$
$x_0 = 0$	$f(x_0) = 1100$
$x_1 = 1$	$f(x_1) = 1080$
$x_2 = 2$	$f(x_2) = 1040$
$x_3 = 3$	$f(x_3) = 960$
$x_4 = 4$	$f(x_4) = 840$

Now fit a least-squares curve to the measured data, and compare it with the values obtained from the equation.

Solution: Knowing where the data originated, one suspects that the best fit will be obtained with a parabola, and so one looks for a second-degree polynomial $p_2(x)$ as a solution. To achieve this, let

$$g_0(x) = 1 \qquad g_1(x) = x \qquad g_2(x) = x^2$$

so that from
$$p_m(x) = a_m g_m(x) + a_{m-1} g_{m-1}(x) + \ldots + a_1 g_1(x) + a_0 g_0(x) \tag{1}$$
(where the functions $g_m(x), \ldots, g_0(x)$ are assumed to be some known functions of x), one seeks a curve of the form
$$p_2(x) = a_2 x^2 + a_1 x + a_0,$$
as the solution. Hence m = 2, and since one has five given tabulated points,
$$n + 1 = 5$$
$$n = 4.$$
Thus, one will have m + 1 = 3 equations to solve for the three unknowns a_2, a_1, and a_0.
From
$$\alpha_{kj} = \sum_{i=0}^{n} g_k(s_i) g_j(x_i), \quad \begin{array}{l} \text{for } k = 0,1,2, \ldots, m \\ j = 0,1,2, \ldots, m, \end{array} \tag{2}$$
the coefficients α_{kj} of a_k in the jth equation are found as follows:
$$\alpha_{00} = \sum_{i=0}^{4} g_0(x_i) g_0(x_i) = 1 + 1 + 1 + 1 + 1 = 5,$$
since $g_0(x) = 1$ for any x. Similarly,
$$\alpha_{01} = \alpha_{10} = \sum_{i=0}^{4} g_1(x_i) g_0(x_i) = x_0 + x_1 + x_2 + x_3 + x_4$$
$$= 0 + 1 + 2 + 3 + 4 = 10;$$
$$\alpha_{02} = \alpha_{20} = \sum_{i=0}^{4} g_2(x_i) g_0(x_i) = \sum_{i=0}^{4} x_i^2 \cdot 1$$
$$= x_0^2 + x_1^2 + x_2^2 + x_3^2 + x_4^2$$
$$= 0^2 + 1^2 + 2^2 + 3^2 + 4^2$$
$$= 30;$$
$$\alpha_{11} = \sum_{i=0}^{4} g_1(x_i) g_1(x_i) = \sum_{i=0}^{4} x_i \cdot x_i = \sum_{i=0}^{4} x_i^2$$
$$= x_0^2 + x_1^2 + x_2^2 + x_3^2 + x_4^2$$
$$= 30;$$
$$\alpha_{12} = \alpha_{21} = \sum_{i=0}^{4} g_2(x_i) g_1(x_i) = \sum_{i=0}^{4} x_i^2 \cdot x_i = \sum_{i=0}^{4} x_i^3$$
$$= 0^3 + 1^3 + 2^3 + 3^3 + 4^3$$
$$= 100;$$
$$\alpha_{22} = \sum_{i=0}^{4} g_2(x_i) g_2(x_i) = \sum_{i=0}^{4} x_i^2 \cdot x_i^2 = \sum_{i=0}^{4} x_i^4$$
$$= 0^4 + 1^4 + 2^4 + 3^4 + 4^4$$
$$= 354;$$

Using the α's, give the three equations

$$\alpha_{20}a_2 + \alpha_{10}a_1 + \alpha_{00}a_0 = \sum_{i=0}^{n} f(x_i) g_0(x_i)$$

$$\alpha_{21}a_2 + \alpha_{11}a_1 + \alpha_{01}a_0 = \sum_{i=0}^{n} f(x_i) g_1(x_i)$$

$$\alpha_{22}a_2 + \alpha_{12}a_1 + \alpha_{02}a_0 = \sum_{i=0}^{n} f(x_i) g_2(x_i).$$

Evaluating the summations on the right, and substituting the values for the α's, obtain the following three equations in three unknowns:

$$30a_2 + 10a_1 + 5a_0 = 5020$$

$$100a_2 + 30a_1 + 10a_0 = 9400$$

$$354a_2 + 100a_1 + 30a_0 = 27320.$$

Solving these equations for the three unknowns gives the results:

$$a_2 = -17.14285714$$

$$a_1 = 4.573428573$$

$$a_0 = 1097.714286,$$

so that the polynomial fitted to the data has the form (accurate to ten significant digits)

$$p_2(x) = -17.14285714x^2 + 4.573428573x + 1097.714286 \qquad (3)$$

It is interesting to compare this equation with the original data:

x_i	Accurate Value if Determined from the Correct Equation $f(x) = 1100 - 16x^2$	Inaccurate Value as "Measured" to the Nearest 10 feet	Value Obtained from the Least-Squares Fitted Curve [Eq. (3)]
0	1100	1100	1097.7
1	1084	1080	1085.1
2	1036	1040	1038.3
3	956	960	957.1
4	844	840	841.7

As one can see, the least-squares curve fitted to the slightly inaccurate starting data is, first of all, a reasonable equation for the data, and moreover tends to balance out the randomly introduced errors. At four of the five given points the least-squares curve is closer to the true value than the original tabulated point was; only at one point, where the given tabulated value happened to be exact, is the least-squares curve worse than the measured value. Thus it has been started with five given points, and has been developed to an equation which can be used to calculate other values in the range of the original table. It is not justified to think, of course, that the least-squares curve will be a good approximation at, say, $x = 8$.

● **PROBLEM 8-15**

Find the best linear fit (in the least squares sense) to the data

$$x_1 = 1 \quad f_1 = 1$$

$$x_2 = 2 \quad f_2 = 3$$

$$x_3 = 3 \quad f_3 = 1$$

Use both the direct solution of the normal equations and the orthogonalization method.

<u>Solution:</u> (1) The following notation

$$f = \begin{pmatrix} 1 \\ 3 \\ 1 \end{pmatrix}, \quad Q = \begin{pmatrix} 1 & 1 \\ 1 & 2 \\ 1 & 3 \end{pmatrix}$$

gives

$$Q'Q = \begin{pmatrix} 3 & 6 \\ 6 & 14 \end{pmatrix}, \quad (Q'Q)^{-1} = \frac{1}{6}\begin{pmatrix} 14 & -6 \\ -6 & 3 \end{pmatrix},$$

$$Q'f = \begin{pmatrix} 5 \\ 10 \end{pmatrix}, \quad c = \frac{1}{6}\begin{pmatrix} 14 & -6 \\ -6 & 3 \end{pmatrix}\begin{pmatrix} 5 \\ 10 \end{pmatrix} = \begin{pmatrix} 5/3 \\ 0 \end{pmatrix},$$

so that $y = 5/3$ is the best approximation.

(2) Alternatively take $Q'Q = LL'$ where

$$L = \begin{bmatrix} \sqrt{3} & 0 \\ 2\sqrt{3} & \sqrt{2} \end{bmatrix}$$ and solve first $Ly = \begin{bmatrix} 5 \\ 10 \end{bmatrix}$ getting

$$y = \begin{bmatrix} 5/\sqrt{3} \\ 0 \end{bmatrix}$$ and then $L'c = y$ getting

$$c = \begin{bmatrix} 5/3 \\ 0 \end{bmatrix}$$ as before.

(3) An analytic solution to the problem is as follows: Assume $y = ax + b$ to be the linear fit. Then

$$E = (a + b - 1)^2 + (2a + b - 3)^2 + (3a + b - 1)^2$$

which gives

$$\frac{\partial E}{\partial a} = 28a + 12b - 20,$$

$$\frac{\partial E}{\partial b} = 12a + 6b - 10.$$

Solving

$$7a + 3b = 5 \text{ and } 6a + 3b = 5$$

one finds $a = 0$, $b = 5/3$, as before.

(4) Consider the over-determined system

$$x + y = 1$$
$$x + 2y = 3$$
$$x + 3y = 1$$

The factorization $Q = \Phi U$ is:

$$\begin{bmatrix} 1 & 1 \\ 1 & 2 \\ 1 & 3 \end{bmatrix} = \begin{bmatrix} 1/\sqrt{3} & -1/\sqrt{2} \\ 1/\sqrt{3} & 0 \\ 1/\sqrt{3} & 1/\sqrt{2} \end{bmatrix} \begin{bmatrix} \sqrt{3} & 2\sqrt{3} \\ 0 & \sqrt{2} \end{bmatrix}$$

and find first

$$y = \begin{bmatrix} 1/\sqrt{3}, & 1/\sqrt{3}, & 1/\sqrt{3} \\ -1/\sqrt{2}, & 0, & 1/\sqrt{2} \end{bmatrix} \begin{bmatrix} 1 \\ 3 \\ 1 \end{bmatrix} = \begin{bmatrix} 5/\sqrt{3} \\ 0 \end{bmatrix}$$

and then solve $Uc = y$, i.e.,

$$\begin{bmatrix} \sqrt{3}, & 2\sqrt{3} \\ 0, & \sqrt{2} \end{bmatrix} c = \begin{bmatrix} 5/\sqrt{3} \\ 0 \end{bmatrix}$$

giving $y' = [5/3, 0]$, as before.

● **PROBLEM 8-16**

Use the method of least squares to fit a function of the form $f^*(x) = c_0 + c_1 x$ to the following measured data. All the weights are equal to one.

x	1	3	4	6	7
$f(x)$	−2.1	−0.9	−0.6	0.6	0.9

Solution: Consider the following theorem:

When $\phi_0, \phi_1, \ldots, \phi_n$ are linearly independent, the least-squares approximation problem has a unique solution,

$$f^* = \sum_{j=0}^{n} c_j^* \phi_j, \qquad (1)$$

where the coefficients c_j^* satisfy the so-called normal equations

$$\sum_{j=0}^{n} (\phi_j, \phi_k) c_j^* = (f, \phi_k), \quad (k = 0, 1, 2, \ldots, n) \qquad (2)$$

The solution is characterized by the orthogonality property that $f^* - f$ is orthogonal to all ϕ_j, ($j = 0, 1, \ldots, n$).

An important special case is when $\phi_0, \phi_1, \ldots, \phi_n$ form an orthogonal system; then the coefficients are computed more simply by the formula

$$c_j^* = \frac{(f, \phi_j)}{(\phi_j, \phi_j)}, \quad (j = 0, 1, \ldots, n).$$

The coefficients c_j^* are called the orthogonal coefficients (or occasionally, Fourier coefficients).

Using the notation of the theorem, gives

$$\phi_0(x) = 1, \quad \phi_1(x) = x.$$

The normal equations,

$$\sum_{j=0}^{n} (\phi_j, \phi_k) c_j^* = (f, \phi_k), \quad (k = 0, 1, 2, \ldots, n)$$

are, then:

$$(\phi_0, \phi_0) c_0 + (\phi_1, \phi_0) c_1 = (f, \phi_0)$$

$$(\phi_0, \phi_1) c_0 + (\phi_1, \phi_1) c_1 = (f, \phi_1).$$

The table of values of f on the net can be considered as a column vector, which is denoted by tab(f),

$$\text{tab}(f) = [f(x_0), f(x_1), \ldots, f(x_m)]^T.$$

(T denotes transpose; that is, the column vector is written as the transpose of a row vector.)

The necessary inner products are scalar products of the vectors

$$\text{tab}(\phi_0) = \begin{pmatrix} 1 \\ 1 \\ 1 \\ 1 \\ 1 \end{pmatrix}, \quad \text{tab}(\phi_1) = \begin{pmatrix} 1 \\ 3 \\ 4 \\ 6 \\ 7 \end{pmatrix}, \quad \text{tab}(f) = \begin{pmatrix} -2.1 \\ -0.9 \\ -0.6 \\ 0.6 \\ 0.9 \end{pmatrix}$$

Thus, for example, $(f, \phi_1) = -2.1 \cdot 1 - 0.9 \cdot 3 - 0.6 \cdot 4 + 0.6 \cdot 6 + 0.9 \cdot 7 = 2.7$. One gets

$$5c_0 + 21c_1 = -2.1$$

$$21c_0 + 111c_1 = 2.7$$

Check:

with the result $c_0 = -2.542$, $c_1 = 0.5053$.

$$\text{tab}(f^* - f) = \text{tab}(c_0 + c_1 x - f)$$

$$= (0.063, -0.126, 0.079, -0.110, 0.095)^T.$$

Notice that the sum of the components of tab $(f^* - f)$ is 0.001. Without round-off error, the sum would have been 0.

• **PROBLEM 8-17**

Using the least squares method, find a polynomial of degree two which fits the data of Table 1.

Table 1

i	0	1	2	3
x_i	0	.5	1	1.5
y_i	0	.6	1	1.5

Solution: In the most general form the problem can be stated as follows: a discrete set of points is given

$$(x_0,y_0), (x_1,y_1), \ldots, (x_M,y_M),$$

or briefly

$$(x_i,y_i), \quad i = 0,1,\ldots,M.$$

We have to find a polynomial of degree $n < M$,

$$P_n(x) = \sum_{\ell=0}^{n} a_\ell x^\ell,$$

in such a way that the value of

$$G = \sum_{i=0}^{M} (y_i - P_n(x_i))^2$$

is a minimum; otherwise, we choose the constant coefficients $a_0, a_1, \ldots, a_n$ to minimize G. For G to be minimized, it is necessary that

$$\frac{\partial G}{\partial a_\ell} = 0, \text{ for } \ell = 0,1,\ldots,n.$$

Hence,

$$0 = \frac{\partial G}{\partial a_\ell} = -2 \sum_{i=0}^{M} y_i x_i^\ell + 2 \sum_{i=0}^{M} P_n(x_i) x_i^\ell$$

$$= -2 \sum_{i=0}^{M} y_i x_i^\ell + 2 \sum_{k=0}^{n} a_k \sum_{i=0}^{M} x_i^{\ell+k}$$

We obtain the system of n+1 equations with n+1 unknowns, $a_0,\ldots,a_n$, called the normal equations.

$$\sum_{k=0}^{n} a_k \sum_{i=0}^{M} x_i^{\ell+k} = \sum_{i=0}^{M} y_i x_i^\ell, \quad \ell = 0,\ldots,n.$$

For this problem n=2, M=3. There are three normal equations,

$$4a_0 + 3a_1 + 3.5a_2 = 3.1$$

$$3a_0 + 3.5a_1 + 4.5a_2 = 3.55$$

$$3.5a_0 + 4.5a_1 + 6.125a_2 = 4.525$$

The solution to this system is

$$a_0 = .015, \quad a_1 = 1.13, \quad a_2 = -.1$$

and the second degree polynomial is

$$P_2(x) = -.1x^2 + 1.13x + .015$$

The graph of the polynomial is shown in the following figure.

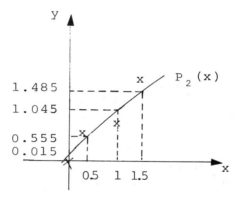

The points taken from Table 1 are denoted by x. We see that the function $P_2(x)$ of second degree fits the points from Table 1 fairly well.

Table 2

i	0	1	2	3
x_i	0	.5	1	1.5
y_i	0	.6	1	1.5
$P_2(x_i)$	.015	.555	1.045	1.485
$|y_i - P_2(x_i)|$	.015	.045	.045	.015

The method of least squares leads to the error

$$\sum_{i=0}^{3} |y_i - P_2(x_i)|^2 = .0045$$

One cannot find the second degree polynomial for which the error would be less than .0045.

DATA SMOOTHING

• **PROBLEM** 8-18

Obtain a smoother set of values to (x_i, y_i) given below by a cubic fit to five consecutive points.

x	0	1	2	3	4	5	6	7	8	9	10	11	12	13	14
y	0	3	7	9	12	13	14	14	13	11	10	11	10	8	7

x	15	16	17	18	19
y	6	6	7	5	6

Solution: The formulas for the new values of y_i, for a third-degree polynomial fit to five equidistant points are

$$y_0^* = \frac{1}{70}(69y_0 + 4y_1 - 6y_2 + 4y_3 - y_4),$$

$$y_1^* = \frac{1}{35}(2y_0 + 27y_1 + 12y_2 - 8y_3 + 2y_4),$$

$$y_2^* = \frac{1}{35}(-3y_0 + 12y_1 + 17y_2 + 12y_3 - 3y_4), \quad (1)$$

$$y_3^* = \frac{1}{35}(2y_0 - 8y_1 + 12y_2 + 27y_3 + 2y_4),$$

$$y_4^* = \frac{1}{70}(-y_0 + 4y_1 - 6y_2 + 4y_3 + 69y_4).$$

The midpoint formula y_2^* is most frequently used when smoothing the data by fitting a cubic to five points, for it can apply to any portion of the data for which $h = 1$.

The formulas for y_0^*, y_1^*, and y_2^* of (1) are used to calculate these new values. Then the midpoint formula, y_2^*, is used until y_{17}^* is reached. The formulas for y_3^* and y_4^* give y_{18}^* and y_{19}^*.

x	y	y*
0	0	-.1
1	3	3.3
2	7	6.5
3	9	9.5
4	12	11.6
5	13	13.3
6	14	13.9
7	14	14.0
8	13	12.8
9	11	11.1
10	10	10.4
11	11	10.6
12	10	9.9
13	8	8.3
14	7	6.8
15	6	6.1
16	6	6.3
17	7	6.1
18	5	5.6
19	6	5.9

• **PROBLEM** 8-19

Smooth the following data.

x	0.0	0.1	0.2	0.3	0.4	0.5	0.6	0.7	0.8
f(x)	2	3	5	7	11	13	17	19	23

x	0.9	1.0
f(x)	29	31

Solution: There is not a unique solution to this problem. Although the function f is clearly not linear in x, one could argue that each set of three consecutive points may reasonably be approximated by a straight line. For each point, except the first and last, one can therefore obtain a smoothed value by using

$$y_0 = (f_{-1} + f_0 + f_1)/3.$$

To smooth the values corresponding to x = 0.0 and x = 1.0, use

$$y_1 = f_1 - \frac{1}{6}\delta^2 f_0 \quad \text{(where } \delta^2 f_0 \text{ denotes the second difference } f_1 - 2f_0 + f_{-1}\text{)}.$$

and

$$y_{-1} = f_{-1} - \frac{1}{6}\delta^2 f_0$$

respectively. Denoting the smoothed value at x by y(x), one obtains the following (to 1 decimal place).

x	0.0	0.1	0.2	0.3	0.4	0.5	0.6	0.7
y(x)	1.8	3.3	5.0	7.7	10.3	13.7	16.3	19.7

0.8	0.9	1.0
23.7	27.7	31.7

APPROXIMATE DIFFERENTIATION

● **PROBLEM 8-20**

Find a functional representation for the data

x	1	2	3	4	5	6	7
y	.5	1.6	2.4	3.2	3.9	4.6	5.2

by the method of differential correction. Hint: choose to have three constants.

Solution: A formula which is to relate the two variables x and y and which will have a number of undertermined constants is needed (assumed to be three, a, b, c, in this case). Symbolically, write

$$y = f(x,a,b,c). \tag{1}$$

This formula is to be a good fit to the data (x_i, y_i)
$(i = 1, \ldots, m)$. The residuals are given by

$$R_1 = f(x_1, a, b, c) - y_1,$$
$$R_2 = f(x_2, a, b, c) - y_2, \qquad (2)$$
$$\vdots$$
$$R_m = f(x_m, a, b, c) - y_m,$$

where y_i $(i = 1, \ldots, m)$ are the given (observed) values from the original data. Make a plot of the given data and from it determine approximate values to the constants, a, b, c, and call them a_0, b_0, c_0; or simply pick a form for f based on past experience perhaps and then make an educated guess on some initial values for the parameters. It is desired to correct these approximate values by some incremental amount α, β, γ such that

$$a = a_0 + \alpha,$$
$$b = b_0 + \beta, \qquad (3)$$
$$c = c_0 + \gamma$$

will yield better values and the formula will fit the data.

If one substitutes the values in (3) into the residuals in (2) and transpose the y_i,

$$R_i + y_i = f(x_i, a_0 + \alpha, b_0 + \beta, c_0 + \gamma). \qquad (4)$$

Expand the right-hand side by Taylor's theorem for a function of several variables to obtain the set of equations

$$R_i + y_i = f(x_i, a_0, b_0, c_0) + \alpha \left(\frac{\partial f_i}{\partial a}\right)_0 + \beta \left(\frac{\partial f_i}{\partial b}\right)_0$$
$$+ \gamma \left(\frac{\partial f_i}{\partial c}\right)_0 \qquad (5)$$

+ higher order terms in α, β, γ,

where

$\left(\dfrac{\partial f_i}{\partial u}\right)_0 \equiv$ the value of the partial derivative $\partial f/\partial u$ at $x = x_i$,

$a = a_0$, $b = b_0$, and $c = c_0$

$\equiv f_{iu}.$

A first approximation is obtained from
$$y^* = f(x, a_0, b_0, c_0),$$
so that
$$f(x_i, a_0, b_0, c_0) = y_i^*$$
and these first approximations can be put into (5). For simplicity let
$$r_i = y_i^* - y_i.$$

If one ignores the high-order terms, one then has a set of residual equations of the form
$$R_i = \alpha \left(\frac{\partial f_i}{\partial a}\right)_0 + \beta \left(\frac{\partial f_i}{\partial b}\right)_0 + \gamma \left(\frac{\partial f_i}{\partial c}\right)_0 + r_i,$$

which are linear in α, β, γ. Thus one may determine the corrections by the method of least squares. Now minimize
$$\sum R_i^2 = g(\alpha, \beta, \gamma)$$

from which the normal equations are
$$(\sum f_{ia}^2)\alpha + (\sum f_{ia}f_{ib})\beta + (\sum f_{ia}f_{ic})\gamma + \sum f_{ia}r_i = 0,$$
$$(\sum f_{ia}f_{ib})\alpha + (\sum f_{ib}^2)\beta + (\sum f_{ib}f_{ic})\gamma + \sum f_{ib}r_i = 0, \quad (6)$$
$$(\sum f_{ic}f_{ia})\alpha + (\sum f_{ic}f_{ib})\beta + (\sum f_{ic}^2)\gamma + \sum f_{ic}r_i = 0.$$

A plot of the function shows a logarithmic tendency with something added. A simple expression is
$$y = ax + b \log x.$$

Since $y = .5$ when $x = 1$ and $\log x$ (at 1) $= 0$, choose $a_0 = .5$ and guess $b_0 = 1$; the first approximation is
$$y^* = .5x + \log x.$$

Furthermore,
$$\frac{\partial y}{\partial a} = x \quad \text{and} \quad \frac{\partial y}{\partial b} = \log x.$$
Then calculate

x	y	y^*	r	$\frac{\partial y}{\partial a}$	$\frac{\partial y}{\partial b}$
1	.5	.5	0	1	0
2	1.6	1.30103	−.29897	2	.30103
3	2.4	1.97712	−.42288	3	.47712
4	3.2	2.60206	−.59794	4	.60206
5	3.9	3.19897	−.70103	5	.69897
6	4.6	3.77815	−.82185	6	.77815
7	5.2	4.34510	−.85490	7	.84510

Using (6), calculate the normal equations,

$$140\alpha + 18.52111\beta = 18.67889,$$

$$18.5211\alpha + 2.48901\beta = 2.50376,$$

with $D = 5.42988$, from which one gets

$$\alpha = .002, \quad \beta = .842,$$

and

$$a = .5 + .022 = .522,$$

$$b = 1 + .842 = 1.842;$$

the function then becomes

$$y = .522x + 1.842 \log x \ .$$

Although the correction β is large, it turns out that in this problem no improvement in the degree of accuracy is made by finding another set of corrections, using the foregoing expression as a second approximation. There are times, though, when this may be necessary, and it is simple to do if a and b are not involved in the partial derivatives, for then one simply calculates new values for r_i and the constant terms in the normal equations.

The answer yields a set of residuals for which

$$\sum r_i^2 = .0038.$$

• **PROBLEM 8-21**

Show that the solution to a least squares problem is not necessarily unique by discussing the system

$$x + y = 1, \quad 2x + 2y = 0, \quad -x - y = 2.$$

Solution:

$$E = (x + y - 1)^2 + (2x - 0)^2 + (-x + 3y - 2)^2.$$

$$\frac{1}{2} \cdot \frac{\partial E}{\partial x} = 6x - 2y - 1; \quad \frac{1}{2} \cdot \frac{\partial E}{\partial y} = -2x + 10y - 7.$$

Hence $x = 1/14$, $y = 5/7$ for a minimum which is $1/14$.

• **PROBLEM 8-22**

Determine the parameters in $y = ae^{bx}$ (e is not a parameter) so that the graph of the curve is a good fit for the points (0.5, 0.58), (1.0, 0.68), (2.0, 0.90), (4.0, 1.65), (9.0, 7.45); by computational means, and by graphical means, using two significant figures.

Solution: To fit a curve to a set of points (x_1, y_1), $(x_2, y_2), \ldots, (x_n, y_n)$, where the equation of the curve is not of the form $y = P(x)$, $P(x)$ a polynomial in x, entails the graph of the points plus some knowledge of the underlying physical law may indicate the suitability of an equation of the type $y = ae^{bx}$. The parameters involved may usually be determined in at least one of three ways: the problem can be reduced to a previous case by an appropriate change of variables; one or more special devices can be used, some of which are illustrated below.

By reduction to previous type: Take the natural logarithm of both sides of the given equation; then $\ln y = \ln a + bx$. Put $y^* = \ln y$, $a^* = \ln a$; then $y^* = a^* + bx$. From the given data, one finds the corresponding values of ln Y, namely, -0.545, -0.386, -0.105, 0.501, 2.011. The problem reduces to fitting a line to the points $(0.5, -0.545)$, $(1.0, -0.386)$, $(2.0, -0.105)$, $(4.0, 0.501)$, $(9.0, 2.011)$. (It can be verified that these points lie approximately on a line by plotting them on ordinary graph paper or by plotting the original points on semi-log paper.) The line $y^* = -0.697 + 0.299x$ was found by the method of centroids, whence $a = 0.498$. The final answer, using two significant figures, is $y = 0.50e^{0.30x}$.

It is to be carefully noted that if the equation of the straight line $y^* = a^* + bx$ is found by the method of least squares, it is generally not true that the values of the parameters so determined will minimize the sum of the squares of the errors in the equation whose graph fits the original data.

Solution by special device: Differentiate $y = ae^{bx}$ to obtain $y' = abe^{bx}$ so that $y'/y = b$. The values of y are approximately the respective Y's, the corresponding values of y' are thus needed. Note that if $P_0: (x_0, y_0)$, $P_1: (x_1, y_1)$, $P_2: (x_2, y_2)$ are three points on or near a curve with $x_0 < x_1 < x_2$, the slope of the curve at or near P_1 should be some sort of weighted average of the slopes of the chords P_0P_1 and P_1P_2. Taking as the weights the distances $x_2 - x_1$ and $x_1 - x_0$, respectively, the slope at or near P_1 is approximately

$$\frac{1}{x_2 - x_0}\left[\frac{x_2 - x_1}{x_1 - x_0}(y_1 - y_0) + \frac{x_1 - x_0}{x_2 - x_1}(y_2 - y_1)\right]. \quad (1)$$

Using this formula to estimate the derivative at each of the three interior points of the given data, and then evaluating b, one obtains the three values 0.30, 0.30, 0.36. Their arithmetic average, 0.32, gives a reasonable estimate for b. Using this value for b, and substituting the given values of X and Y for x and y in $a = y/e^{bx}$, obtain five values for a; 0.49, 0.49, 0.47, 0.46, 0.42. These average out to 0.47.

Hence, the final equation is $y = 0.47e^{0.32x}$, a result fairly close to the previous one.

Note that if $x_1 - x_0 = x_2 - x_1$, then (1) becomes $(y_2 - y_0)/(x_2 - x_0)$, the slope of the line P_0P_2.

Solution by graphing: Plot the given points carefully on graph paper and draw a smooth curve that seems to be a good fit. Extend the curve until it meets the y-axis. The y-intercept gives the value of a in the equation, but this value should be averaged in with the value obtained as now described to even out drawing inaccuracy. Select five points, more or less equally spaced, on the graph and draw tangents to the curve at these points, by inspection, but with the aid of a straightedge. The slopes of the tangents yield the values of y'; as above, calculate the values of a and b. Average this a with the y-intercept. Essentially the same result as before should be obtained.

● **PROBLEM** 8-23

Make a least-squares fit of the form

$$y = c_1 e^{c_2 x}$$

to the points (0, 1), (1, 2), and (2, 6), using as an initial estimate $c_1 \approx 1$, $c_2 \approx 1$.

Solution: Let $k_1 = 1$, $k_2 = 1$. Then

$$f(x, k_1, k_2) = e^x$$

$$\frac{\partial f}{\partial c_1} = e^{c_2 x}, \quad \text{so} \quad \frac{\partial f(x, k_1, k_2)}{\partial c_1} = e^x$$

$$\frac{\partial f}{\partial c_2} = xc_1 e^{c_2 x}, \quad \text{so} \quad \frac{\partial f(x, k_1, k_2)}{\partial c_2} = xe^x$$

Hence the formula desired to be used for the first step is

$$y = e^x + E_1 e^x + E_2 x e^x$$

In order to apply the method of least squares fit, combine the term e^x with y, giving

$$y - e^x = E_1 e^x + E_2 x e^x$$

Now one can write the equations as

$$\begin{pmatrix} e^0 & 0 \\ e^1 & e^1 \\ e^2 & 2e^2 \end{pmatrix} \begin{pmatrix} E_1 \\ E_2 \end{pmatrix} = \begin{pmatrix} 1 - e^0 \\ 2 - e \\ 6 - e^2 \end{pmatrix}$$

or, approximately (since e = 2.718),

$$\begin{pmatrix} 1 \\ 2.718 \\ 7.389 \end{pmatrix} \begin{pmatrix} 0 \\ 2.718 \\ 14.778 \end{pmatrix} \begin{pmatrix} E_1 \\ E_2 \end{pmatrix} = \begin{pmatrix} 0 \\ -.718 \\ -1.389 \end{pmatrix}$$

Upon multiplying by the transpose, obtain

$$62.987 E_1 + 116.585 E_2 = -12.216$$

$$116.585 E_1 + 225.781 E_2 = -22.479$$

$$E_1 = -.218, \quad E_2 = .0132$$

So that the next estimate is

$$c_1 = .782, \quad c_2 = 1.0132$$

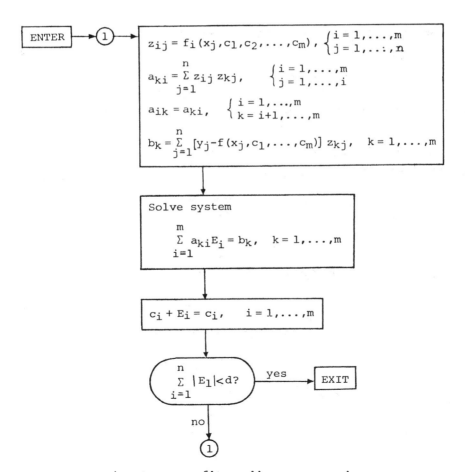

Least-square fit-nonlinear expressions

More careful calculation gives for this and succeeding steps:

Step	k_1	k_2	E_1	E_2
1	1	1	-.218312	.0131631
2	.781687	1.01316	-.000921	.0040536
3	.780766	1.01721	-.000313	.0002056
4	.780435	1.01742		

The flow chart in the figure shown describes the above iteration scheme. In this flow chart the functions f_1, f_2, etc., are defined by

$$f_1(x, c_1, c_2, \ldots, c_m) = \frac{\partial f(x, c_1, c_2, \ldots, c_m)}{\partial c_1}$$

$$f_2(x, c_1, c_2, \ldots, c_m) = \frac{\partial f(x, c_1, c_2, \ldots, c_m)}{\partial c_2}$$

$$\vdots$$

$$f_m(x, c_1, c_2, \ldots, c_m) = \frac{\partial f(x, c_1, c_2, \ldots, c_m)}{\partial c_m}$$

• **PROBLEM 8-24**

Consider a function $f(x) = \sin \pi x$ defined and continuous on the interval $[0,1]$. Using the least squares method find a polynomial of degree two approximating this function.

Solution: Suppose $f(x)$ is a function continuous on the interval $[a,b]$, $f \in C[a,b]$. We have to find a polynomial of degree n, $P_n(x)$, which will minimize

$$G(a_0, a_1, \ldots, a_n) = \int_a^b [f(x) - P_n(x)]^2 dx$$

where

$$P_n(x) = a_n x^n + a_{n-1} x^{n-1} + \ldots + a_1 x + a_0$$

We have to find real coefficients $a_0, a_1, \ldots, a_n$ which will minimize G. A necessary condition for the numbers $a_0, \ldots, a_n$ to minimize G is that

$$\frac{\partial G}{\partial a_i} = 0, \quad i = 0, 1, \ldots, n$$

Thus, we have to solve the normal equations, that is, the system of n+1 linear equations

$$\sum_{k=0}^{n} a_k \int_a^b x^{i+k} dx = \int_a^b x^i f(x) dx, \quad i = 0, 1, \ldots, n$$

with n+1 unknowns $a_0, a_1, \ldots, a_n$. Whenever $f(x) \in C[a,b]$ the normal equations have a unique solution. In this problem we are looking for the second degree polynomial

$$P_2(x) = a_2 x^2 + a_1 x + a_0$$

Thus,

$$a_0 \int_0^1 1 dx + a_1 \int_0^1 x dx + a_2 \int_0^1 x^2 dx = \int_0^1 \sin\pi x dx$$

$$a_0 \int_0^1 x dx + a_1 \int_0^1 x^2 dx + a_2 \int_0^1 x^3 dx = \int_0^1 x \sin\pi x dx$$

$$a_0 \int_0^1 x^2 dx + a_1 \int_0^1 x^3 dx + a_2 \int_0^1 x^4 dx = \int_0^1 x^2 \sin\pi x dx$$

Since $\int_0^1 dx = 1$, $\int_0^1 x dx = \frac{1}{2}$,

$$\int_0^1 x^2 dx = \frac{1}{3}, \quad \int_0^1 x^3 dx = \frac{1}{4}, \quad \int_0^1 x^4 dx = \frac{1}{5}$$

We get

$$a_0 + \frac{1}{2} a_1 + \frac{1}{3} a_2 = \frac{2}{\pi}$$

$$\frac{1}{2} a_0 + \frac{1}{3} a_1 + \frac{1}{4} a_2 = \frac{1}{\pi}$$

$$\frac{1}{3} a_0 + \frac{1}{4} a_1 + \frac{1}{5} a_2 = \frac{\pi^2 - 4}{\pi^3}$$

Solving the system of linear equations, we find

$$a_0 = \frac{12\pi^2 - 120}{\pi^3} \approx -.0504655$$

$$a_1 = \frac{720 - 60\pi^2}{\pi^3} \approx 4.122512$$

$$a_2 = \frac{60\pi^2 - 720}{\pi^3} \approx -4.122512$$

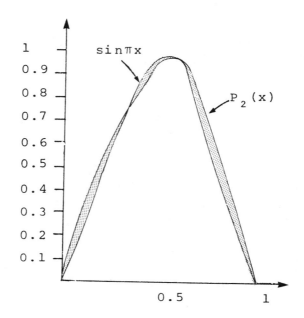

The polynomial of degree two best approximating $f(x) = \sin\pi x$ is given by

$$P_2(x) = -4.122512x^2 + 4.122512x - .0504655$$

and the error is

$$\int_0^1 [\sin\pi x + 4.122512x^2 - 4.122512x + .0504655]dx$$

CHAPTER 9

NUMERICAL DIFFERENTIATION

GENERAL APPLICATIONS OF NUMERICAL DIFFERENTIATION

● **PROBLEM 9-1**

Find the slope of the empirical function in the table shown at the point $x = 60$ using

$$f'_{-2} = \frac{1}{12h}(-25f_{-2} + 48f_{-1} - 36f_0 + 16f_1 - 3f_2).$$

The difference table

x	$f(x)$	Δf	$\Delta^2 f$	$\Delta^3 f$
60	63 620			
		−1536		
61	62 084		− 97	
		−1633		−8
62	60 451		−105	
		−1738		−6
63	58 713		−111	
		−1849		−5
64	56 864		−116	
		−1965		0
65	54 899		−116	
		−2081		−1
66	52 818		−117	
		−2198		
67	50 620			

Solution: According to

$$f'_{-2} = \frac{1}{12h}(-25f_{-2} + 48f_{-1} - 36f_0 + 16f_1 - 3f_2),$$

the slope is −1491. One can also use the finite-difference method. Third differences are effectively constant, and they are relatively small anyway. Hence, according to the finite-difference method:

297

$$f'(x) = \Delta f(x) - \frac{1}{2}\Delta^2 f(x) + \frac{1}{3}\Delta^3 f(x),$$

$$f'(60) = \Delta f(60) - \frac{1}{2}\Delta^2 f(60) + \frac{1}{3}\Delta^3 f(60)$$

$$= -1536 - \frac{1}{2}(-97) + \frac{1}{3}(-8)$$

$$= -1490.$$

● **PROBLEM 9-2**

Find the slope of the empirical function in the table shown at the point x = 61.5.

The difference table

x	f(x)	Δf	Δ²f	Δ³f
60	63 620			
		−1536		
61	62 084		−97	
		−1633		−8
62	60 451		−105	
		−1738		−6
63	58 713		−111	
		−1849		−5
64	56 864		−116	
		−1965		0
65	54 899		−116	
		−2081		−1
66	52 818		−117	
		−2198		
67	50 620			

Solution: Make use of formulae
$$f'(x) = [\ln(1+\Delta)](f(x)) = \Delta f(x) - \frac{1}{2}\Delta^2 f(x) + \frac{1}{3}\Delta^3 f(x) - \ldots$$
and
$$f(x+r) = (1+\Delta)^r f(x)$$
Hence,
$$f'(61.5) = (\ln(1+\Delta))f(61.5)$$

$$= (\ln(1+\Delta))(1+\Delta)^{\frac{1}{2}}f(61)$$

$$= (\Delta - \frac{1}{2}\Delta^2 + \frac{1}{3}\Delta^3 - \ldots)(1 + \frac{1}{2}\Delta - \frac{1}{8}\Delta^2 + \ldots)f(61)$$

$$= (\Delta - \frac{1}{24}\Delta^3 + \ldots)f(61)$$

$$= \Delta f(61) - \frac{1}{24}\Delta^3 f(61)$$

$$= -1633 - \frac{1}{24}(-6)$$

$$= -1633.$$

• **PROBLEM 9-3**

Values of $\tan^{-1}(e^x)$ are given in the table shown. Calculate the first and second derivatives of the function at $x = 1.2$.

Table: Values of $\tan^{-1}(e^x)$

x	$\tan^{-1}(e^x)$
1.0	1.218 283
1.1	1.249 462
1.2	1.278 244
1.3	1.304 726
1.4	1.329 023

Solution: Explicit formulae for these two derivatives are somewhat complicated. Use formula

$$f'_0 = \frac{1}{12h}(f_{-2} - 8f_{-1} + 8f_1 - f_2),$$

to calculate the first derivative:

$$f'(1.2) = \frac{1}{12 \times 0.1}(1.218283 - 8 \times 1.249462 + 8 \times 1.304726 - 1.329023)$$

$$= 0.276143,$$

and this answer happens to be correct in all six places.

The second derivative (the slope of the slope curve) may be obtained via formula

$$f''_0 = \frac{1}{12h^2}(-f_{-2} + 16f_{-1} - 30f_0 + 16f_1 - f_2)$$

$$f''(1.2) = \frac{1}{12 \times (0.1)^2}(-1.218283 + 16 \times 1.249462 - 30 \times 1.278244 + 16 \times 1.304726 - 1.329023)$$

$$= -0.23015$$

The actual value is -0.23021; so the answer is correct to four decimal places.

TRUNCATION ERROR USING TAYLOR'S FORMULA

• **PROBLEM** 9-4

Given the Lagrange differentiation formula (derivative of Lagrange polynomials at grid points)

$$F'(J) = \Sigma F(K) * DER(LC(N,K; X(J))), K\varepsilon //0, N//,$$

$$+ E'(N; X(J)) \quad (1)$$

and its truncation error

$$E'(N; X(J)) = DER ** (N + 1) F(2(X(J))) * (-1)**J * J!$$

$$* (N - J)! * H ** N/(N + 1)!.$$

Derive a Lagrange polynomial which passes through the three points $(X(J), F(J))$, $J\varepsilon //0, 2//$ and evaluate the derived polynomial at $X(0)$.

Note: Formula (1) has the form

$$F'(J) = \Sigma F(K) * A(K,H), K\varepsilon //0, N//, + T(F,H),$$

where the $A(K,H)$ depends only on the assumed (fixed) uniform grid and the fixed $X(J)$, and the truncation error $T(F,H)$ depends only on H and the choice of function whose derivative is being estimated at $X(J)$.

<u>Solution</u>: To facilitate the thorough understanding of this problem and the ones that follow, Weierstrass' (Uniform) Approximation Theorem and certain mathematical notations are included as a handy guide to these problems that suggest studies of number systems used by computers.

The existence of polynomials that define estimates to $F(x)$ on a specified interval is assured by Weierstrass' (Uniform) Approximation Theorem:

Let $F(x)$ be continuous on $[A,B]$ and choose any $E > 0$; then there exists a polynomial $P(N,X)$ (of sufficiently high degree N) for which

$$|F(x) - P(N,X)| \leq E, \text{ for each } X \in [A,B].$$

Notations:

The asterisk (*) denotes the operation multiplication, the double asterisk (**) indicates exponentiation, brackets [ ] denote closed intervals and braces { } denote certain sets.

The much used sets of integers

$$P, P+1, P+2,\ldots,P+N = R$$

$$P, P+1, P+2,\ldots$$

$$P, P+Q, P+2*Q, P+3*Q,\ldots,P+N*Q = R$$

are written

$$//P,R//, \quad //P,\infty//, \quad //P,Q,R//,$$

respectively, where Q may be a negative integer.

Use this	For
(A + B)/C + 2 * F	$\dfrac{a+b}{c} + 2f$
ΣA(K) * X ** K, K ∈ //0, N//	$\sum_{k=0}^{n} a_k x^k$
PROD(X - X(J)), J ∈ //0, N//	$(x - x_0)(x - x_1) \ldots (x - x_n)$
ORDB(H ** 2)	order function: $O(h^2)$
DER(F(A)) = F'(A)	$f'(a)$
DER ** (2) F(A) = F''(A)	$f''(a)$
DER ** (K) F(A)	$f^{(k)}(a)$
LC	Lagrange coefficient

Special functions (operators)

    Δ: Forward difference operator

    ∇: Backward difference operator.

    δ: Central difference operator.

  [...]F: Divided difference operator.

    μ: Averaging operator.

    ∫: Integral operator.

The symbol ∈ means in or is an element of.

Taylor's Formula: Truncation error

P(N;X,C)

    = ΣDER ** (M) F(C) * (X - C) ** M/M!, M ∈ //0,N//,   (2)

where

    DER ** (0) F(C) ≡ F(C)

and, for

$$M > 0, \text{DER}**(M)F(C)$$

is the Mth derivative of F(X) evaluated at X = C.
For each X in [A,B], there exists a number
Z = Z(X) between X and C (X ≤ Z ≤ C or C ≤ Z ≤ X)
such that

$$F(X) = P(N;X,C) + R(N;X,C), \qquad (3)$$

where the truncation error (Lagrange Form for the remainder) is

R(N;X,C) = F(X) - P(N;X,C)

$$= \text{DER}**(N+1)F(Z) * (X-C)**(N+1)/(N+1)!. \quad (4)$$

The equation (3) with (4) substituted is called Taylor's formula in one variable with the Lagrange Form for the remainder.

Consider now the problem.

If N = 2 and J = 0, then (1) becomes

F'(0) = F(0) * DER((X - X(1)) * (X - X(2))/

((X(0) - X(1)) * (X(0) - X(2)))), at X = X(0),

+ F(1) * DER((X - X(0)) * (X - X(2))/

((X(1) - X(0)) * (X(1) - X(2)))), at X = X(0),

+ F(2) * DER((X - X(0)) * (X - X(1))/

((X(2) - X(0)) * (X(2) - X(1)))), at X = X(0),

+ F'''(Z(0)) * (X(0) - X(1)) * (X(0) - X(2))/3!

= F(0) * ((X(0) - X(1)) + (X(0) - X(2)))/

((X(0) - X(1)) * (X(0) - X(2)))

+ F(1) * (X(0) - X(2))/((X(1) - X(0)) *

(X(1) - X(2)))

+ F(2) * (X(0) - X(1))/((X(2) - X(0)) *

(X(2) - X(1)))

+ F'''(Z(0)) * (X(0) - X(1)) * (X(0) - X(2))/3!

$$= F(0) * (-3 * H)/(2 * H ** 2) + F(1) * (-2 * H)/$$

$$(-H ** 2)$$

$$+ F(2) * (-H)/(2 * H ** 2) + F'''(Z(0)) *$$

$$(2 * H ** 2)/3!$$

and evaluating the derived polynomial at X(0) one obtains

$$F'(0) = 1/(2 * H) * (-3 * F(0) + 4 * F(1) - F(2))$$

$$+ (1/3) * H ** 2 * F'''(Z(0)).$$

• PROBLEM 9-5

Find the truncation error for the estimation of F'(0) by the first divided difference [x(0), x(1)]F with known values of N = 1, K = 0, and J = 1. Use the formula

$$F'(X(0)) \doteq [X(0),X(1)]F = \Delta F(0)/H =$$

$$(F(1) - F(0))/H.$$

Assume that a uniform grid with mesh H is used; the formula has been picked to estimate a derivative at a gridpoint. The basic idea of the method is to replace in the estimated values of F(X) and its derivatives at grid points with equivalent values given by Taylor's Formula to obtain

(Estimate) = (Function being estimated)

- (Truncation error).

<u>Solution</u>: Write the divided difference in the form

$$(F(1) - F(0))/H = F'(0) - T(F,H). \qquad (1)$$

First, observe that F'(0) is needed, so expand F(1) about X(0) using

$$F(X(J)) = P(N;X(J),X(K)) + R(N;X(J),X(K))$$

$$= \Sigma DER ** (M)F(X(K)) * (X(J) - X(K)) ** M/M!,$$

$$M \in //0,N//,$$

$$+ DER ** (N + 1)F(Z(J))$$

$$* (X(J) - X(K)) ** (N + 1)/(N + 1)!$$

(where $Z(J) = Z(X(J))$ is known only to be between $X(J)$ and $X(K)$), with $N = 1$, $K = 0$, and $J = 1$

$$F(1) = F(0) + H * F'(0) + .5 * H ** 2 * F''(Z(1)).$$

Substitute this in the left member of (1) to obtain

$$(F(1) - F(0))/H = ((F(0) + H * F'(0)$$
$$+ .5*H**2*F''(Z(1)))-F(0))/H$$
$$= F'(0)+.5 * H * F''(Z(1)).$$

Thus the truncation error for the estimation of $F'(0)$ by the first divided difference $[X(0),X(1)]F$ is

$$T(F,H) = F'(0) - (F(1) - F(0))/H$$
$$= -.5*H*F''(Z(1)).$$

The formula has order 1 and degree 1.

Note: The exponent of the highest power of H which is a factor of the truncation error is called the order of the formula. The truncation error $T(F,H)$ of a formula of order P is at most of the order $H ** P$ written $T(F,H) = ORDB(H ** P)$.

A formula is of degree M for a specified grid when $DEG(T) = M$, if $T(F,Grid) \equiv 0$, on the interval of interest, for $F(X)$ any polynomial of degree M or less, and $T(F,Grid) \not\equiv 0$ for $F(X)$ equal to some polynomial of degree $M + 1$. The symbols $T(F,Grid)$ and $T(F,H)$ can be used interchangeably to denote truncation error for a formula.

• **PROBLEM** 9-6

Find the **truncation** error for the estimate

$$F''(1) \doteq 2 * [X(0),X(1),X(2)]F = \Delta ** (2)F(0)/H ** 2$$
$$= (F(2) - 2 * F(1) + F(0))/H ** 2. \qquad (1)$$

Solution: Write

$$F(0) = F(1) - H * F'(1) + .5 * H ** 2 * F''(1)$$
$$-H ** 3/6 * F'''(Z(0))$$
$$F(1) = F(1)$$
$$F(2) = F(1) + H * F'(1) + .5 * H **2 * F''(1)$$
$$+ H ** 3/6 * F'''(Z(2)),$$

and compute

$$F(2) - 2*F(1) + F(0) = H**2*F''(1) - H**3/6 * (F'''(Z(0)) - F'''(Z(2))),$$

so that the truncation error for the estimate (1) is

$$T(F,H) = F''(1) - (F(2) - 2*F(1) + F(0))/H**2$$

$$= H/6 * (F'''(Z(0)) - F'''(Z(2))).$$

The formula has order 1 and degree at least 2.

● **PROBLEM 9-7**

> Two of the methods for obtaining formulas to estimate derivatives in terms of known values of F(X) and/or known values of derivatives of F(X) are as follows: The first method consists in picking a formula that the user has reason to believe may give a usable estimate to the required derivative. The second method consists in picking a form for the formula and requiring that a set of conditions be satisfied. This leads to a linear system which defines the coefficients of the form.
>
> Find the coefficients in the estimate of F'(1),
>
> $$F'(1) \doteq A*F(0) + B*F'(0) + C*F''(0) \qquad (1)$$
>
> so that the corresponding formula is exact for any polynomial of degree two or less.

Solution: For a uniform grid, representations of truncation error, written T(F,H), are usually functions of higher derivatives of F(X) at mysterious points times powers of the mesh length H. The constant coefficient and the power of H are the primary measures of the quality of a formula, since the nature of the derivatives involved is usually unknown.

Write

$$T(F,H) = F'(1) - (A*F(0) + B*F'(0) + C*F''(0))$$

and require that

$$T((X - X(1))**M, H) = 0, \text{ for each } M \in \{0,2\}.$$

That is,

$$\begin{bmatrix} 1 & 0 & 0 \\ -H & 1 & 0 \\ H**2 & -2*H & 2 \end{bmatrix} * \begin{bmatrix} A \\ B \\ C \end{bmatrix} = \begin{bmatrix} 0 \\ 1 \\ 0 \end{bmatrix},$$

which has solution $[A\ B\ C]^t = [0\ 1\ H]^t$. The differentiation formula corresponding to (1) is then

$$F'(1) = F'(0) + H * F''(0) + T(F,H). \quad (2)$$

Since the formula must be exact for the second degree Taylor Polynomial $P(2;X,X(0))$, which agrees with $F(X)$, $F'(X)$, and $F''(X)$ at $X(0)$, then the estimating function in (2) is $P'(2;X(1),X(0))$. But $P'(2;X(1),X(0))$ for $F(X)$ is just $P(1;X(1),X(0))$ for $F'(X)$, which has truncation error

$$T(F,H) = R(1;X(1),X(0)) = F'(1) - (F'(0) + H * F''(0))$$

$$= H ** 2 * F'''(Z(1))/2!.$$

In general, if $N \geq K$, then DER $**$ $(K)P(N;X(1),X(0))$ for $F(X)$ is just $P(N - K; X(1), X(0))$ for DER $**$ $(K)F(X)$, and one has the formula

$$T(F,H) = DER * (K)F(1)$$
$$- (P(N - K;X(1),X(0)) \text{ for DER } ** (K)F(X))$$

$$= R(N - K;X(1),X(0)) \text{ for DER } ** (K)F(X)$$

$$= H ** (N - K + 1)/(N - K + 1)!$$
$$* DER ** (N + 1)F(Z(1)).$$

The order of the formula is $(N - K + 1)$, while its degree in F is always N (DER stands for derivative).

● **PROBLEM** 9-8

Suppose one has reason to believe that the derivative of the interpolation polynomial $P(X)$ characterized by

$$P(X(J)) = F(J), J \in //0,1//, \text{ and } P'(X(0))$$

$$= F'(0) \quad (1)$$

will give a usable estimate to $F'(X)$ at $X(1)$. Compute the coefficients of $F(0)$, $F(1)$, and $F'(0)$ in the formula

$$T(F,H) = F'(1) - (A * F(0) + B * F(1)$$

$$+ C * F'(0)) \quad (2)$$

Solution: It is required that (2) be exact for any polynomial of degree two or less, (i.e., the linear functional $T(F,H)$ in (2) is exact for the function $F = P$ if $T(P,H) = 0$). In particular, one requires that

$$T((X - X(1)) ** M, H) = 0, \text{ for each } M \in //0,2//. \quad (3)$$

That is,
$$T(1,H) = 0 - (A * 1 + B * 1 + C * 0) = 0$$

$$T(X - X(1), H) = 1 - (A * (X(0) - X(1))$$
$$+ B * (X(1) - X(1)) + C * 1) = 0$$

$$T((X - X(1)) ** 2, H) = 2 * (X(1) - X(1)) - (A * (X(0)$$
$$- X(1)) ** 2 + B * (X(1) - X(1)) ** 2$$
$$+ C * 2 * (X(0) - X(1))) = 0,$$

or

$$\begin{bmatrix} 1 & 1 & 0 \\ -H & 0 & 1 \\ H**2 & 0 & -2*H \end{bmatrix} * \begin{bmatrix} A \\ B \\ C \end{bmatrix} = \begin{bmatrix} 0 \\ 1 \\ 0 \end{bmatrix},$$

which has solution $[A \; B \; C]^t = [-2/H \; 2/H \; -1]^t$.
Thus, the differentiation formula

$$F'(1) = -2/H * F(0) + 2/H * F(1)$$
$$- F'(0) + T(F,H) \quad (4)$$

is exact [$T(F,H) = 0$] whenever $F(X)$ is any polynomial of degree two or less. In particular, (4) is exact for the interpolation polynomial $P(X)$ characterized by (1). Now find a representation of the truncation error for (4). Using Taylor's Method, write

$$F(0) = F(1) - H * F'(1) + H**2/2 * F''(1)$$
$$- H**3/6 * F'''(Z(0))$$

$$F(1) = F(1)$$

$$F'(0) = F'(1) - H * F''(1) + H**2/2 * F'''(Z1(0))$$

and compute $-2/H * F(0) + 2/H * F(1) - F'(0)$

$$= F'(1) - H**2/6 * (3 * F'''(Z1(0)) - 2 * F'''(Z(0))).$$

Thus, the truncation error for the formula (4) is

$$T(F,H) = F'(1) - (-2/H * F(0) + 2/H * F(1) - F'(0))$$
$$= H**2/6 * (3 * F'''(Q) - 2 * F'''(R)), \quad (5)$$

where Q and R are generally unknown numbers between $X(0)$ and $X(1)$. From (5) it is again clear that $T(F,H) = 0$ if $F(X)$ is any polynomial of degree two or less. The formula (4) has order 2 and degree 2.

DIFFERENTIATION BY LAGRANGE'S FORMULA, NEWTON'S FORMULA AND STIRLING'S FORMULA

• **PROBLEM 9-9**

Given the values

x	0	1	3	4	5	7	9
y	150	108	0	-54	-100	-144	-84

Use Lagrange's formula to find the value of the derivative of y at x = 6.

Solution: The Lagrangian formula is

$$L'(x) = P(x) \sum_{i=0}^{n} y_i \sigma_i R_i^{-1}$$

where

$L'(x)$ is the approximation of

$$\frac{dy}{dx} = f'(x), \text{ and}$$

$P(x)$ is the product of the terms in the principal diagonal of the square array

$$\begin{array}{ccccc} x - x_0 & x_0 - x_1 & x_0 - x_2 & \cdots & x_0 - x_n \\ x_1 - x_0 & x - x_1 & x_1 - x_2 & \cdots & x_1 - x_n \\ x_2 - x_0 & x_2 - x_1 & x - x_2 & \cdots & x_2 - x_n \\ \vdots & & & & \\ x_n - x_0 & x_n - x_1 & x_n - x_2 & \cdots & x - x_n \end{array}$$,

$$\sigma_i = \sum_{j=0}^{n} (x - x_j)^{-1} \quad j \neq i,$$

R_i: the products of elements on each row, i.e.,

$$R_i = (x_i - x_0) \cdots (x - x_i) \cdots (x_i - x_n)$$
$$= (x - x_i) P_i(x_i)$$

For x = 6, the array of $s - s_i$ and $s_i - s_j$ after being augmented by R_i and y_i/R_i becomes

$S_i - S_0$	$S_i - S_1$	$S_i - S_2$	$S_i - S_3$	$S_i - S_4$	$S_i - S_5$	$S_i - S_6$	R_i	Y_i/R_i
6	-1	-3	-4	-5	-7	-9	22680	.006613757
1	5	-2	-3	-4	-6	-8	-5760	-.01875
3	2	3	-1	-2	-4	-6	864	0
4	3	1	2	-1	-3	-5	-360	.150
5	4	2	1	1	-2	-4	320	-.3125
7	6	4	3	2	-1	-2	2016	-.071428571
9	8	6	5	4	2	-3	-51840	.001620370

So P(6) = 540 and one may start with the R_i values. If the values are left in fractional form, the calculations for this problem can be performed without the use of a calculating machine.

σ_i	$y_i\sigma_i R_i^{-1}$
$\frac{7}{10}$	$\frac{1}{216}$
$\frac{2}{3}$	$-\frac{1}{80}$
$\frac{8}{15}$	0
$\frac{11}{30}$	$\frac{11}{200}$
$-\frac{2}{15}$	$\frac{1}{24}$
$\frac{28}{15}$	$-\frac{2}{15}$
$\frac{8}{3}$	$\frac{7}{360}$

$$y'(6) = P(6) \sum_{i=0}^{n} y_i\sigma_i R_i^{-1} = 540(-\frac{23}{540}) = -23.$$

• **PROBLEM 9-10**

Let $x_i = x_0 + ih$ for $i = 0, 1, 2$. Here x_0 and h are specified real numbers. Let $f \in C^3[x_0, x_2]$. Determine differentiation formulas for $f'(x_0)$, $f'(x_1)$, and $f'(x_2)$.

Use the Lagrange coefficient polynomials to solve:

$$L_i(x) = \frac{(x-x_0)(x-x_1)\cdots(x-x_{i-1})(x-x_{i+1})\cdots(x-x_n)}{(x_i-x_0)(x_i-x_1)\cdots(x_i-x_{i-1})(x_i-x_{i+1})\cdots(x_i-x_n)}$$

Solution: According to the following theorem:

If $f \in C^{n+1}[a,b]$ and $x_0, x_1, \ldots, x_n$ are distinct points of $[a,b]$, then

$$f'(x_k) = \sum_{j=0}^{n} L'_j(x_k) f(x_j) + \overbrace{\prod_{\substack{j=0 \\ j \neq k}}^{n} (x_k - x_j) \frac{f^{(n+1)}(\xi_k)}{(n+1)!}}^{R'(x_k)}$$

(the remainder formula for Lagrange interpolation on $n+1$ points is $R'(x_k)$)

for $k = 0, 1, \ldots, n$, where

$$\min(x_0, x_1, \ldots, x_n) < \xi_k < \max(x_0, x_1, \ldots, x_n),$$

one gets

$$f'(x_k) = L'_0(x_k) f_0 + L'_1(x_k) f_1 + L'_2(x_k) f_2 + R'(x_k)$$

for k = 0, 1, 2. The Lagrange coefficient polynomials on the three points are

$$L_0(x) = \frac{(x-x_1)(x-x_2)}{(x_0-x_1)(x_0-x_2)} = \frac{[(x-x_0)-h][(x-x_0)-2h]}{2h^2}$$

$$L_1(x) = \frac{(x-x_0)(x-x_2)}{(x_1-x_0)(x_1-x_2)} = \frac{[(x-x_0)][(x-x_0)-2h]}{-h^2}$$

$$L_2(x) = \frac{(x-x_0)(x-x_1)}{(x_2-x_0)(x_2-x_1)} = \frac{[(x-x_0)][(x-x_0)-h]}{2h^2}$$

Thus one gets

$$L_0'(x) = \frac{1}{2h^2}\{[(x-x_0)-2h]+[(x-x_0)-h]\}$$

$$L_1'(x) = -\frac{1}{h^2}\{[(x-x_0)-2h]+(x-x_0)\}$$

$$L_2'(x) = \frac{1}{2h^2}\{[(x-x_0)-h]+(x-x_0)\}$$

This gives, for k = 0,

$$f'(x_0) = -\frac{3}{2h}f_0 + \frac{2}{h}f_1 - \frac{1}{2h}f_2 + R'(x_0) \qquad (1)$$

where

$$R'(x_0) = \frac{(x_0-x_1)(x_0-x_2)}{3!}f^{(3)}(\xi_0) = \frac{2h^2}{3!}f^{(3)}(\xi_0) \qquad (2)$$

A better form for Eq. (1) is

$$f'(x_0) = \frac{1}{2h}(-3f_0 + 4f_1 - f_2) + \frac{h^2}{3}f^{(3)}(\xi_0) \qquad (3)$$

In a similar manner one obtains the formulas for $f'(x_1)$ and $f'(x_2)$.

$$f'(x_1) = \frac{1}{2h}(-f_0 + f_2) - \frac{h^2}{6}f^{(3)}(\xi_1) \qquad (4)$$

$$f'(x_2) = \frac{1}{2h}(f_0 - 4f_1 + 3f_2) + \frac{h^2}{3}f^{(3)}(\xi_2) \qquad (5)$$

Usually in numerical differentiation the function f(x) will be tabulated or available at a number of equally spaced points. Then, for a given point any one of the formulas (3) to (5) could be used. The most commonly used of the three formulas is

$$f'(x_1) = \frac{1}{2h}(-f_0 + f_2) - \frac{h^2}{6} f^{(3)}(\xi_1)$$

This formula requires only two evaluations of $f(x)$. In addition, if $f^{(3)}(x)$ is nearly constant in the interval $[x_0, x_2]$, then the truncation error using this formula is about half that for the other formulas.

• **PROBLEM 9-11**

Find the maximum and minimum values of the function

x	0	1	2	3	4	5
y	0	.25	0	2.25	16.00	56.25

by using Stirling's formula for numerical differentiation (terminate after the sixth difference). Take $x_0 = 2$.

Solution: The maximum and minimum values of a function are obtained by setting first derivative equal to zero and solving for the variable. The process can be applied to a tabulated function. Using Stirling's formula and writing the result as a polynomial in u, one gets

$$y'(u) = a_5 u^5 + a_4 u^4 + a_3 u^3 + a_2 u^2 + a_1 u + a_0, \quad (1)$$

where

$$a_5 = \frac{1}{120} \Delta^6 y_{-3},$$

$$a_4 = \frac{1}{24} m_5,$$

$$a_3 = \frac{1}{36}(6\Delta^4 y_{-2} - \Delta^6 y_{-3}),$$

$$a_2 = \frac{1}{8}(4m_3 - m_5),$$

$$a_1 = \frac{1}{180}(180\Delta^2 y_{-1} - 15\Delta^4 y_{-2} + 2\Delta^6 y_{-3}),$$

$$a_0 = \frac{1}{30}(30m_1 - 5m_3 + m_5);$$

(2)

where m_r are the arithmetic means of vertically adjacent differences of the difference table.

In solving the equation obtained by setting Formula 1 equal to zero, one may be faced with the problem of finding the roots of an algebraic equation whose coefficients are approximate numbers. This will require more elaborate techniques. Otherwise, the value of x may be found from (1) and (2) by equating

$$x = x_0 + hu$$

where h is the increment of x.

For this problem, calculate a central difference table with the horizontal line through the central value having the entries:

| y_0 | m_1 | $\Delta^2 y_{-1}$ | m_3 | $\Delta^4 y_{-2}$ | ... , |

and the line just below with the entries

| m_0 | Δy_0 | m_2 | $\Delta^3 y_{-1}$ | m_4 | ... , |

where

$$m_{2i-1} = \tfrac{1}{2}(\Delta^{2i-1} y_{-i} + \Delta^{2i-1} y_{-i+1}), \quad (i = 1,2,3...)$$

so that

$$i = 1: \quad m_1 = \tfrac{1}{2}(\Delta y_{-1} + \Delta y_0)$$

$$i = 2: \quad m_3 = \tfrac{1}{2}(\Delta^3 y_{-2} + \Delta^3 y_{-1}), \text{ etc.,}$$

and

$$m_{2i} = \tfrac{1}{2}(\Delta^{2i} y_{-i} + \Delta^{2i} y_{-i+1}), \quad (i = 1,2,3...)$$

so that

$$i = 0: \quad m_0 = \tfrac{1}{2}(y_0 + y_1),$$

$$i = 1: \quad m_2 = \tfrac{1}{2}(\Delta^2 y_{-1} + \Delta^2 y_0), \text{ etc.}$$

Hence:

For $x_0 = 2$, since $\Delta^5 y = \Delta^6 y = 0$,

$$y'(u) = a_3 u^3 + a_2 u^2 + a_1 u + a_0$$

$$= \tfrac{1}{6}(6)u^3 + \tfrac{1}{2}(6)u^2 + \tfrac{1}{12}(24)u + \tfrac{1}{30}(0)$$

$$= u^3 + 3u^2 + 2u.$$

The solutions of $y'(u) = 0$ are $u = 0, -1, -2$. Then, since $h = 1$ there are maxima and minima at

$$x = x_0 + u = 2 + u_i = 2,1,0.$$

The values of the function can now be read out of the table.

z	y	Δy	Δ²y	Δ³y	Δ⁴y
0	0				
		.25			
1	.25		−.50		
		−.25		3	
2	0		2.50		6
		1.00		6	
		2.25		9	
3	2.25		11.50		6
		13.75		15	
4	16.00		26.50		
		40.25			
5	56.25				

To determine whether there is a maximum or a minimum at the resulting value of x, inspect the table of values. It is not necessary to be concerned with the second derivative. Thus in the problem it is easily seen that the values at x = 0 and x = 2 are minima values.

• **PROBLEM 9-12**

Find the value of the derivative of y = f(x) at x = 1.03 using the given central difference table with $x_0 = 1.5$.

In the table, m_r are simply the arithmetic means of the vertically adjacent differences and are calculated after the differences have been evaluated. The arithmetic means of the odd and even order differences are defined by the following formulas:

$$m_{2i-1} = \frac{1}{2}(\Delta^{2i-1}y_{-i} + \Delta^{2i-1}y_{-i+1}) \qquad (i = 1,2,3,\ldots)$$

and

$$m_{2i} = \frac{1}{2}(\Delta^{2i}y_{-i} + \Delta^{2i}y_{-i+1}). \qquad (i = 0,1,2,\ldots)$$

In this case all the differences above the fifth are zero.

Solution: To obtain the values of the first derivative, first derive a formula by differentiating any of the classical interpolation formulas. From differential calculus

$$\frac{dy}{dx} = \frac{dy}{du} \cdot \frac{du}{dx}.$$

Now $uh = x - x_0$, where h is the interval length, so that

and

$$\frac{du}{dx} = \frac{1}{h}$$

$$\frac{dy}{dx} = \frac{1}{h}\frac{dy}{du}.$$

Central Difference Table

x	y	Δy	$\Delta^2 y$	$\Delta^3 y$	$\Delta^4 y$	$\Delta^5 y$
$x_{-5} = 1.0$	$y_{-5} = -3.00000$					
		$\Delta y_{-5} = 137372$				
$x_{-4} = 1.1$	$y_{-4} = -1.62628$		$\Delta^2 y_{-4} = 40840$			
		$\Delta y_{-4} = 178212$				
$x_{-3} = 1.2$	$y_{-3} = .15584$		$\Delta^2 y_{-3} = 51460$			
		$\Delta y_{-3} = 229672$			2160	
$x_{-2} = 1.3$	$y_{-2} = 2.45256$		$\Delta^2 y_{-2} = 64240$	12780		$\Delta^5 y_{-4} = 240$
		$\Delta y_{-2} = 293912$			2400	
$x_{-1} = 1.4$	$y_{-1} = 5.39168$		$\Delta^2 y_{-1} = 79420$	15180		$\Delta^5 y_{-3} = 240$
		$\Delta y_{-1} = 373332$			2640	
$x_0 = 1.5$	$y_0 = 9.12500$		$\Delta^2 y_{-1} = 97240$	17820		$\Delta^5 y_{-2} = 240$
		$m_1 = 421952$		$m_3 = 192600 \; \Delta^4 y_{-2} = 2880$		$m_5 = 240$
$x_1 = 1.6$	$m_0 = 11.47786$		$m_2 = 107590 \; \Delta^3 y_{-1} = 20700$		$m_4 = 3000 \; \Delta^5 y_{-2} = 240$	
		$\Delta y_0 = 470572$			3120	
	$y_1 = 13.83072$		117940	23820		$\Delta^5 y_{-2} = 240$
		588512			3360	
$x_2 = 1.7$	$y_2 = 19.71584$		141760	27180		$\Delta^6 y_{-1} = 240$
		730272			3600	
$x_3 = 1.8$	$y_3 = 27.01856$		168940	30780		
		$\Delta y_3 = 899212$				
$x_4 = 1.9$	$y_4 = 36.01068$		199720			
		$\Delta y_4 = 1098932$				
$x_5 = 2.0$	$y_5 = 47.00000$					

Now consider Newton's formula

$$y = y_0 + u\Delta y_0 + \binom{u}{2}\Delta^2 y_0 + \ldots + \binom{u}{i}\Delta^i y_0$$

$$+ \ldots + \binom{u}{n}\Delta^n y_0$$

and differentiate it with respect to x;

$$\frac{dy}{dx} = \frac{1}{h} [\Delta y_0 + D_2(u) \Delta^2 y_0 + \ldots$$
$$+ D_k(u) \Delta^k y_0 + \ldots + D_n(u) \Delta^n y_0],$$

where $D_k(u)$ is defined by

$$D_k(x) = \frac{d}{dx} \binom{x}{k}$$

from which

$$D_1(x) = 1$$

and

$$D_{i+1}(x) = \frac{1}{i+1} [(x-i) D_i(x) + (x-i+1) b_i(x)] \qquad (1)$$

$$i = 1, 2, 3, \ldots$$

with

$$b_i(x) = \frac{1}{i} \binom{x}{i-1}$$

The calculations can be arranged in the following schematic:

u	b	D
u−1	$b_1(u)$	$D_1(u)$
u−2	$b_2(u)$	$D_2(u)$
u−3	$b_3(u)$	$D_3(u)$
⋮	⋮	⋮
u−i	$b_i(u)$	$D_i(u)$
		$D_{i+1}(u)$

From the data $h = .1$, and at $x = 1.03$, $u = .3$. Now evaluate the derivatives $D_i(u)$ according to Formula 1 and the preceding schematic. At the same time list the differences $\Delta^i y_0$ from the table. The value of the derivative is then the sum of the products of the entries in the last two columns divided by h. The work is arranged in the following calculation sheet:

$$b_{i+1} = \frac{1}{i+1} b_i(u-i+1),$$

$$D_{i+1} = \frac{1}{i+1} [(u-i) D_i + (u-i+1) b_i],$$

$$y' = \frac{1}{h} \Sigma D_i \Delta^i y_0.$$

i	$u-i$	b_i	D_i	$\Delta^i y_0$
0	.3			
1	−.7	1	1	1.37372
2	−1.7	.150000	−.200000	.40840
3	−2.7	−.035000	.078333	.10620
4	−3.7	.014875	−.038000	.02160
5			.020087	.00240

$$y' = \boxed{12.99586}.$$

● **PROBLEM 9-13**

Using Table 1, find

$$\cos 1.76 = \left[\frac{d}{dx} \sin x\right]_{x=1.76},$$

by Gregory-Newton forward interpolation and Stirling formulas.

Table 1

x	$y = \sin x$	Δy	$\Delta^2 y$	$\Delta^3 y$	$\Delta^4 y$
1.70	0.9916 6481				
		−27 7504			
1.72	0.9888 8977		−3 9555		
		−31 7059		128	
1.74	0.9857 1918		−3 9427		14
		−35 6486		142	
1.76	0.9821 5432		−3 9285		16
		−39 5771		158	
1.78	0.9781 9661		−3 9127		17
		−43 4898		175	
1.80	0.9738 4763		−3 8952		14
		−47 3850		189	
1.82	0.9691 0913		−3 8763		15
		−51 2613		204	
1.84	0.9639 8300		−3 8559		18
		−55 1172		222	
1.86	0.9584 7128		−3 8337		14
		−58 9509		236	
1.88	0.9525 7619		−3 8101		
		−62 7610			
1.90	0.9463 0009				

The interval in this table is $h = 0.02$.

Solution: The Gregory-Newton formula for forward interpolation is

$$y = y_0 + u^{[1]} \Delta y_0 + \frac{1}{2!} u^{[2]} \Delta^2 y_0$$

$$+ \ldots + \frac{1}{m!} u^{\{m\}} \Delta^m y_0$$

where
$$u^{[k]} = u(u-1)(u-2)\ldots(u-k+1).$$

If one lets $x_0 = 1.76$ and uses the Gregory-Newton formula to approximate $y = \sin x$, one has

$$y \cong P_m(u) = y_0 + u\Delta y_0 + \frac{u(u-1)}{2!}\Delta^2 y_0 + \frac{u(u-1)(u-2)}{3!}\Delta^3 y_0$$

$$+ \frac{u(u-1)(u-2)(u-3)}{4!}\Delta^4 y_0 + \frac{u(u-1)(u-2)(u-3)(u-4)}{5!}\Delta^5 y_0 + \ldots$$

$$\left[\frac{dy}{dx}\right]_{x=1.76} \cong \left[\frac{1}{h}\frac{dP_m(u)}{du}\right]_{u=0}$$

$$= \frac{1}{h}(\Delta y_0 - \frac{1}{2}\Delta^2 y_0 + \frac{1}{3}\Delta^3 y_0 - \frac{1}{4}\Delta^4 y_0 + \frac{1}{5}\Delta^5 y_0 + \ldots) \qquad (1)$$

$$\left[\frac{dy}{dx}\right]_{x=1.76} = \cos 1.76 \cong \frac{1}{0.02}(-0.00395771 + 0.000195635$$

$$+ 0.000000583 - 0.000000035 - 0.000000002 + \ldots)$$

$$\cos 1.76 = -0.1978855 + 0.00978175 + 0.00002915$$

$$- 0.00000175 - 0.00000010 = -0.18807645$$

Since Δy_0 is accurate only to six significant figures, it is clear that dy/dx need be good only to six decimal places. The correct answer is

$$\cos 1.76 = -0.18807680$$

and hence the above calculation is well within the limit of error which might arise from the round-off error in the tabulated values.

The first approximation to the derivative is given by

$$\left(\frac{dy}{dx}\right)_0 \cong \frac{\Delta y_0}{h} = -0.1978855 \qquad (2)$$

the first term in the series. Note that it is off in the second decimal place.

If one uses Stirling's formula instead of the Gregory-Newton interpolation formula, the result should be better, since the polynomial then extends approximately the same distance on each side of x_0. Stirling's formula is

$$y = y_0 + u\,\mu\delta y_0 + \frac{1}{2!} u^2 \delta^2 y_0 + \frac{1}{3!} u(u^2 - 1)\mu\delta^3 y_0$$

$$+ \frac{1}{4!} u^2(u^2 - 1) \delta^4 y_0 + \ldots$$

$$+ \frac{1}{(2k-1)!} u(u^2 - 1)(u^2 - 4)$$

$$\ldots [u^2 - (k-1)^2]\mu\delta^{2k-1} y_0$$

$$+ \frac{1}{(2k)!} u^2(u^2 - 1)(u^2 - 4)$$

$$\ldots [u^2 - (k-1)^2] \delta^{2k} y_0 + \ldots$$

where

$$\mu\delta^{2k-1} y_0 = \frac{1}{2}(\delta^{2k-1} y_{\frac{1}{2}} + \delta^{2k-1} y_{-\frac{1}{2}})$$

is termed a mean difference.

Hence,

$$y \cong y_0 + u \frac{\Delta y_{-1} + \Delta y_0}{2} + \frac{1}{2} u^2 \Delta^2 y_{-1}$$

$$+ \frac{1}{6} u(u^2 - 1) \frac{\Delta^3 y_{-2} + \Delta^3 y_{-1}}{2}$$

$$+ \frac{1}{24} u^2(u^2 - 1) \Delta^4 y_{-2}$$

$$+ \frac{1}{120} u(u^2 - 1)(u^2 - 4) \frac{\Delta^5 y_{-3} + \Delta^5 y_{-2}}{2} \quad (3)$$

Therefore

$$\left[\frac{dy}{dx}\right]_{x=x_0} \cong \frac{1}{h}\left[\frac{dP_m(u)}{du}\right]_{u=0} = \frac{1}{h}\left(\frac{\Delta y_{-1} + \Delta y_0}{2}\right.$$

$$\left. - \frac{1}{6} \frac{\Delta^3 y_{-2} + \Delta^3 y_{-1}}{2} + \frac{1}{30} \frac{\Delta^5 y_{-3} + \Delta^5 y_{-2}}{2}\right) \quad (4)$$

and hence

$$\left[\frac{dy}{dx}\right]_{x=x_0} = \cos 1.76 = -0.18806425 - 0.00001250 = 0.18807675$$

the fifth difference term being too small to affect the
result. This answer is correct to seven decimal places.

If Stirling's formula is terminated with the first difference, one obtains as a first approximation for the
derivative the equation

$$\left[\frac{dy}{dx}\right]_{x=x_0} \cong \frac{\Delta y_{-1} + \Delta y_0}{2h} = \frac{1}{h}\mu\delta y_0 \tag{5}$$

Since this yields the value of -0.18806425 for the
derivative, it differs from the true value in the fifth
decimal place. Thus this approximation is very much
better than that given by Eq. (2).

APPROXIMATE DIFFERENTIATION

• **PROBLEM** 9-14

Table 1 gives the values of $f(x) = e^x$ to four decimal places for certain x's in [1.0, 1.2]. Using

$$f'(x_1) = \frac{1}{2h}(-f_0 + f_2) - \frac{h^2}{6}f'''(\xi_1),$$

approximate $(d/dx)e^x$ at $x = 1.1$, with $h = 0.1, 0.05$, and 0.01. Analyze the error in using this approximation for each h, since values of $f(x)$ are usually not exact owing to experimental error and/or round-off errors.

Note: The approximate value of $f'(x_1)$ is given by

$$\bar{f}'(x_1) = \frac{1}{2h}(-y_0 + y_2)$$

while the exact value is

$$f'(x_1) = \frac{1}{2h}(-y_0 - \varepsilon_0 + y_2 + \varepsilon_2) - \frac{h^2}{6}f^{(3)}(\xi_1)$$

Thus,

$$\left|f'(x_1) - \bar{f}'(x_1)\right| = \left|\frac{1}{2h}(-\varepsilon_0 + \varepsilon_2) - \frac{h^2}{6}f^{(3)}(\xi_1)\right|$$

for some $\xi_1 \in (x_0, x_2)$. The effect of the inherent error is $(-\varepsilon_0 + \varepsilon_2)/2h$.

In a typical problem some bound on $|\varepsilon_0|$ and $|\varepsilon_2|$ is available. If ε is such a bound, then the inherent error is bounded by

$$\left|\frac{1}{2h}(-\varepsilon_0 + \varepsilon_2)\right| \leq \frac{|\varepsilon_0| + |\varepsilon_2|}{2h} \leq \frac{\varepsilon}{h}$$

Notice that the bound on the inherent error increases as h decreases. To ensure an accurate result, one must take h small enough so that the magnitude of the truncation error

$$\frac{h^2}{6}|f^{(3)}(\xi)|$$

will be small. But if h is small relative to the errors inherent in the data, then the inherent error in the calculated result may be quite large.

Table 1

x	$f(x)$
1.00	2.7183
1.05	2.8577
1.09	2.9743
1.10	3.0042
1.11	3.0344
1.15	3.1582
1.20	3.3201

Solution: Since Table 1 gives the values of e^x to four decimal places, it contains rounding errors. A reasonable assumption is that these errors are bounded by $\varepsilon = .00005$. The approximations to f'(1.1) are given in Table 2.

Table 2

h	Approximation to $f'(1.1)$	Bound on inherent error	Bound on truncation error
.10	3.009	.0005	$.00167 f'''(\xi_1)$
.05	3.005	.0010	$.00043 f'''(\xi_2)$
.01	3.005	.0050	$.00002 f'''(\xi_3)$

From the last two entries in the table one would probably conclude that 3.005 was a reasonable approximation to f(x). The exact derivative of $f(x) = e^x$ is $f'(x) = e^x$. Thus, from the table of e^x, the correct value of f'(1.1) is 3.0042 to four decimal places. The actual error is considerably smaller than the error bound.

● **PROBLEM 9-15**

Find the approximate values of f'(2) for the function $f(x) = \sqrt{x}$ by using

$$f'(x) = \lim_{h \to 0} \frac{f(x+h) - f(x-h)}{2h} \tag{1}$$

for h = 1, .5, .1, .05, .01, .005, .001, .0005 and .0001. Calculate to 4 decimal digits. Compute the upper and lower bounds on the error, for h = 0.1.

__Solution:__ Since $f(x) = \sqrt{x}$, one gets

$$f'(2) \doteq \frac{\sqrt{2+h} - \sqrt{2-h}}{2h}$$

from formula (1).

h	$\sqrt{2+h}$	$\sqrt{2-h}$	Approx. Value $f'(2)$	Approx. Error
1	1.7320	1.0000	0.3660	−0.012447
0.5	1.5811	1.2247	0.3564	−0.002847
0.1	1.4491	1.3784	0.3535	0.000053
0.05	1.4317	1.3964	0.3530	0.000553
0.01	1.4177	1.4106	0.3550	−0.001447
0.005	1.4159	1.4124	0.3500	0.003553
0.001	1.4145	1.4138	0.3500	0.003553
0.0005	1.4143	1.4140	0.3000	0.053553
0.0001	1.4142	1.4141	0.5000	−0.146447

Table. Approximate Values of $f'(2)$ as a Function h.

The table gives the approximate value of $f'(2)$ for several different values of h. These were calculated from the above formula using fixed-point arithmetic with numbers truncated to four decimal places. The exact value is $1/(2\sqrt{2}) \doteq 0.353\,553$.

Notice here that although theoretically

$$\lim_{h \to 0} \frac{f(x+h) - f(x-h)}{2h} = f'(x),$$

in practice round-off error is important and the smallest value of h does not necessarily give the best approximation. A bound on the error is

$$|E| = \left| \frac{f^{(3)}(\xi)}{3!} \prod_{\substack{k=0 \\ k \neq 1}}^{2} (x_1 - x_k) \right| = \frac{3h^2}{6.8 \xi^{5/2}}.$$

For example, if $h = 0.1$, this gives

$$|E| \leq \frac{10^{-2}}{16(1.9)^{5/2}} \doteq 1.26 \cdot 10^{-4}.$$

Actually, one also gets

$$|E| \geq \frac{10^{-2}}{10(2.1)^{5/2}} \doteq 9.78 \cdot 10^{-5}.$$

This is inconsistent with the observed error of $5.3 \cdot 10^{-5}$, but but the discrepancy is caused by round-off. A more accurate computation with $h = 0.1$ gives

$$f'(2) \doteq 0.353\,664.$$

The error is now about $-1.11 \cdot 10^{-4}$, which is consistent with the upper and lower bounds computed for $|E|$.

• **PROBLEM 9-16**

Find the approximate values of f'(.800) for f(x) = cosx and for h = .001, .002, .005, .01, .02, .05, .1, .2. Use the data in Table 1.

Table 1

x	cosx	x	cosx
.600	.82534	.801	.69599
.700	.76484	.802	.69527
.750	.73169	.805	.69311
.780	.71091	.810	.68950
.790	.70385	.820	.68222
.795	.70028	.850	.65998
.798	.69814	.900	.62161
.799	.69742	1.000	.54030

Solution: From the formula

$$f'(x) = \lim_{h \to 0} \frac{f(x+h) - f(x-h)}{2h}$$

we obtain

$$f'(.800) \approx \frac{f(.800+h) - f(.800-h)}{2h}$$

For h = .001, we get

$$f'(.800) \approx \frac{f(.801) - f(.799)}{.002} = -.715$$

The results are shown in Table 2.

Table 2

h	.001	.002	.005	.01	.02	.05	.1	.2
f'(.800)	-.715	-.7175	-.717	-.7175	-.71725	-.7171	-.71615	-.7126

The exact value is

$$f'(.800) = -.717356091$$

The values in Table 2 yield a fairly good approximation of f'(.800).

• **PROBLEM 9-17**

In Problem 9-10, using Lagrange's formula, we derived the following

$$f'(x_0) = \frac{1}{2h}[-3f(x_0) + 4f(x_0+h) - f(x_0+2h)]$$

$$+ \frac{h^2}{3} f^{(3)}(\eta_0) \qquad h \neq 0 \tag{1}$$

where $x_0 < \eta_0 < (x_0+2h)$, and η_0 is dependent upon x_0. Similarly,

$$f'(x_0) = \frac{1}{2h}[-f(x_0-h)+f(x_0+h)] - \frac{h^2}{6} f^{(3)}(\eta_1) \quad h \neq 0 \quad (2)$$

where η_1 is dependent upon x_0 and $(x_0-h) < \eta_1 < (x_0+h)$.

Using the data in the following table, approximate $f'(1.85)$, setting $h = .1$ and $h = -.1$.

x	f(x)
1.65	4.21175
1.75	5.90625
1.85	7.79075
1.95	9.87725
2.05	12.17775

Solution: Setting $h = .1$ in (1), we obtain

$$f'(1.85) \approx \frac{1}{.2}[-3f(1.85)+4f(1.95)-f(2.05)]$$

$$= 19.795 \quad (3)$$

Setting $h = -.1$ in the same formula, we find

$$f'(1.85) \approx -\frac{1}{.2}[-3f(1.85)+4f(1.75)-f(1.65)]$$

$$= 19.795 \quad (4)$$

Similarly, setting $h = .1$ in (2), we obtain

$$f'(1.85) \approx \frac{1}{.2}[-f(1.75)+f(1.95)]$$

$$= 19.855 \quad (5)$$

Since we know that the function in the table is $f(x) = 2x^3 - x^2 + 3x - 7$, we can compute its exact derivative

$$f'(1.85) = (6x^2-2x+3)\Big|_{x=1.85} = 19.835 \quad (6)$$

In eq.(3) and (4), the error was .04, while in eq.(5) the error was .02. Formula (2) leads to more accurate results than does formula (1). In this case, the errors can be easily computed because $f^{(3)}(\eta_0) = 12$, and the last term becomes $4h^2$ in (1) and $-2h^2$ in (2).

Since the table contains the values of $f(x)$ for five points, we can use the even more accurate five-point formula

$$f'(x_0) = \frac{1}{12h}[f(x_0-2h)-8f(x_0-h)+8f(x_0+h)-f(x_0+2h)]$$

$$+\frac{h^4}{30} f^{(5)}(\eta) \quad (7)$$

Since $f^{(5)}(\eta) = 0$, this formula leads to the accurate result, 19.835, which was evaluated in (6). You can double check it using the data from the table.

• **PROBLEM 9-18**

Let $f(x) = e^x$. Approximate $f'(2.2)$ with error

$$\frac{|hf''(\zeta)|}{2} \leq \frac{|h|}{2} e^{x_0+h} \tag{1}$$

where $x_0 < \zeta < (x_0+h)$. Set $h = .1, .01, .001$, and $x_0 = 2.2$.

Solution: For small values of h to approximate $f'(x_0)$, we can use the difference quotient

$$\frac{f(x_0+h)-f(x_0)}{h} \tag{2}$$

In this case the error is bounded by $\frac{h}{2}$ S, where S is a bound of $f''(x)$ for $x \in [x_0, x_0+h]$

$$f'(x_0) = \frac{f(x_0+h)-f(x_0)}{h} + \frac{h}{2} f''(\zeta) \tag{3}$$

For $h > 0$, this formula is known as the forward-difference formula; for $h < 0$, the backward-difference formula. The results are shown in the following table. The quotient is equal to

$$\frac{f(x_0+h)-f(x_0)}{h} = \frac{e^{2.2+h} - e^{2.2}}{h}$$

h	f(2.2+h)	$\frac{f(2.2+h)-f(2.2)}{h}$	$\frac{|h|}{2} e^{x_0+h}$
0	9.025013499	–	–
.1	9.974182455	9.491689560	.498709123
.01	9.115716393	9.070289400	.045578582
.001	9.034043027	9.029528000	.004517022
.0001	9.025916046	9.025470000	.000451296

• **PROBLEM 9-19**

Using a table of $f(x) = e^x$ correct to four decimal places, find the approximate value of $f'(1.15)$ for various h from .01 to .08, with .01 increments, and discuss the results.

h	Approximation to $f'(1.15)$	Error
.08	3.1613	−.0031
.07	3.1607	−.0025
.06	3.1600	−.0018
.05	3.1590	−.0008
.04	3.1588	−.0006
.03	3.1583	−.0001
.02	3.1575	.0008
.01	3.1550	.0032

Solution: The exact value of $f'(1.15)$ to four decimal places is 3.1582. For $h = .01, .02, \ldots, .08$ and by using

$$f'(x_1) = \frac{1}{2h}(-f_0 + f_2) - \frac{h^2}{6}f^{(3)}(\xi_1)$$

the approximations of the accompanying table were found. Observe that in this table the total error decreases in magnitude as h decreases to .03. Then the error increases as h decreases.

Numerical differentiation is particularly useful when the function $f(x)$ involved can be evaluated quite accurately. In such a case the truncation error can be made small without introducing a large inherent error.

● **PROBLEM 9-20**

Find the series expansion that gives y as a function of x in the neighborhood of $x = 0$ when

$$\frac{dy}{dx} = x^2 + y^2 \tag{1}$$

with boundary condition $y = 0$ for $x = 0$.

Solution: From Eq. (1) it is seen that $y' = 0$ at $x = 0$. Therefore, since $y = 0$ and its first derivative is zero, y will be quite small for x small. As a first approximation, assume that $y = 0$. Then by

$$u_2(x) = y_0 + \int_{x_0}^{x} f(x, u_1(x))\, dx,$$

letting $u_1(x) = 0$,

326

$$u_2(x) = \int_0^x x^2 dx = \tfrac{1}{3}x^3$$

This is the second-order approximation to the solution of Eq. (1). The third-order approximation, given by

$$u_i(x) = y_0 + \int_{x_0}^x f(x, u_{i-1}(x))\, dx,$$

is then

$$u_3(x) = \int_0^x \left(x^2 + \frac{x^6}{9}\right) dx = \tfrac{1}{3}x^3 + \tfrac{1}{63}x^7$$

Likewise the higher-order approximations are

$$u_4(x) = \int_0^x \left(x^2 + \tfrac{1}{9}x^6 + \tfrac{2}{189}x^{10} + \tfrac{1}{3,969}x^{14}\right) dx$$

$$= \tfrac{1}{3}x^3 + \tfrac{1}{63}x^7 + \tfrac{2}{2,079}x^{11} + \tfrac{1}{59,535}x^{15}$$

and

$$u_5(x) = \int_0^x \left(x^2 + \tfrac{1}{9}x^6 + \tfrac{2}{189}x^{10} + \tfrac{13}{14,553}x^{14} + \cdots\right) dx$$

$$= \tfrac{1}{3}x^3 + \tfrac{1}{63}x^7 + \tfrac{2}{2,079}x^{11} + \tfrac{13}{218,295}x^{15} + \cdots$$

If the series is terminated after the third term and used to approximate y to four decimal places, then using the first neglected term as an approximation of the truncation error,

$$\tfrac{13}{218,295}x^{15} \leq 0.00005$$

or

$$15 \log x \leq \log \frac{(0.00005)(218,295)}{13}$$

$$x \leq 0.988$$

Since negative values can be used as well as positive ones, the polynomial

$$\tfrac{1}{3}x^3 + \tfrac{1}{63}x^7 + \tfrac{2}{2,079}x^{11}$$

represents y correct to four decimal places for the range $-0.988 \leq x \leq 0.988$.

CHAPTER 10

NUMERICAL INTEGRATION

GENERAL APPLICATIONS OF NUMERICAL INTEGRATION

• **PROBLEM** 10-1

The mean errors for a certain gun at a range of 3000 yards are

$$n_x = 8.3 \text{ yds.}, \quad n_y = 4.6 \text{ yds.}$$

If 30 shots are fired at the side of a house 12 yds. wide and 6 yds. high at a distance of 3000 yds,

(a) How many hits may be expected?

(b) What is the chance of hitting a door 6 ft. × 3 ft. in the lower right-hand corner of the side of the house?

Solution: (a) If the gun is accurately aimed at the geometric center of the side of the house, any shot will be a hit if it passes within 6 yards of the central vertical line and within 3 yards of the central horizontal line. Hence

$$x = 6 \text{ yds.}, \quad y = 3 \text{ yds.};$$

and

$$h_x x = \frac{0.5642x}{n_x} = \frac{0.5642 \times 6}{8.3} = 0.407,$$

$$h_y y = \frac{0.5642y}{n_y} = \frac{0.5642 \times 3}{4.6} = 0.368.$$

From the probability table

$$P_x = 0.435, \quad P_y = 0.397.$$

The chance of a hit for each shot is therefore

$$P = P_x P_y = 0.435 \times 0.397 = 0.173.$$

For 30 shots the number of hits would probably be

$$30 \times 0.173 = 5.2$$

or 5, say.

(b) To find the probability that the door would be hit during the bombardment, assume that the gun is aimed at the geometric center of the side of the house, as in (a). Then the door will be hit if a shot strikes within the rectangle bounded by the lines $x = 5$, $x = 6$, $y = -1$, $y = -3$. The chance of hitting the door at each shot is therefore

$$P = P_x \cdot P_y = \frac{h_x}{\sqrt{\pi}} \int_5^6 e^{(-h_x^2 x^2)} dx \cdot \frac{h_y}{\sqrt{\pi}} \int_1^3 e^{(-h_y^2 y^2)} dy$$

$$= \left[\frac{1}{\sqrt{\pi}} \int_0^6 e^{-(h_x x)^2} d(h_x x) - \frac{1}{\sqrt{\pi}} \int_0^5 e^{-(h_x x)^2} d(h_x x) \right]$$

$$\times \left[\frac{1}{\sqrt{\pi}} \int_0^3 e^{-(h_y y)^2} d(h_y y) - \frac{1}{\sqrt{\pi}} \int_0^1 e^{-(h_y y)^2} d(h_y y) \right]$$

Hence the two values of $h_x x$ to be used in the probability table are

$$\frac{0.5642}{8.3} \times 6 = 0.407$$

and

$$\frac{0.5642}{8.3} \times 5 = 0.340,$$

for which the probabilities are $P_{x_6} = 0.435/2$, $P_{x_5} = 0.369/2$. Therefore

$$P_x = \frac{0.435}{2} - \frac{0.369}{2} = \frac{0.066}{2}.$$

Likewise, the two values of $h_y y$ are

$$\frac{0.5642}{4.6} \times 3 = 0.368,$$

$$\frac{0.5642}{4.6} \times 1 = 0.1226.$$

The corresponding probabilities are found from the table to be

$$P_{y_3} = \frac{0.397}{2},$$

$$P_{y_1} = \frac{0.138}{2}.$$

Hence $P_y = 0.397/2 - 0.138/2 = 0.159/2$, and finally

$$P = P_x \times P_y = \frac{0.066 \times 0.159}{4} = 0.0026.$$

The door will be hit unless every one of the 30 shots misses it. The chance that any shot will miss it is

$$1 - 0.0026 = 0.9974.$$

The chance that every one of the 30 shots misses is therefore $(0.9974)^{30} = 0.9249$. The chance of a hit is therefore

$$1 - 0.9249 = 0.0751.$$

The door would probably be hit once out of the every $1/0.0026 = 380$ shots.

● **PROBLEM 10-2**

Correctly to nine decimal places, evaluate

$$I = \int_0^{0.2} e^{-x^2} dx.$$

Solution: This integral can be evaluated quite easily by replacing the integrand by the first few terms of its Taylor's series, and then integrating term by term. One knows that

$$e^{-x} = 1 - x + x^2/2! - x^3/3! + x^4/4! - x^5/5!$$

$$\cdots +$$

Therefore,

$$e^{-x^2} = 1 - x^2 + x^4/2 - x^6/6 + x^8/24 - x^{10}/120 + \ldots$$

and $I = \int_0^{0.2} e^{-x^2} dx = \int_0^{0.2} (1 - x^2 + x^4/2 - x^6/6 + x^8/24 - x^{10}/120 + \ldots) dx$

$$= (x - x^3/3 + x^5/2(5) - x^7/6(7) + x^9/24(9) - x^{11}/120(11) + \ldots) \Big|_0^{0.2}$$

$$= 0.2 - (0.2)^3/3 + (0.2)^5/10 - (0.2)^7/42 + (0.2)^9/216 - (0.2)^{11}/1320 + \ldots$$

$= 0.2 - 0.002\ 666\ 666\ 67 + 0.000\ 032$

$-0.000\ 000\ 304\ 76 + 0.000\ 000\ 002\ 37$

$-0.000\ 000\ 000\ 02 + \ldots$

$I = 0.197\ 365\ 031$

It is noted that the sixth, and last, term above does not affect the answer to the nine decimal places being determined. Thus the sixth term becomes the first term being neglected and, one knows that the truncation error is less than 2 in the 11th decimal place.

Integrals such as the one above are sometimes called non-elementary integrals when, as is the case here, the quantity following the integral sign is not the differential of any elementary function. That is, the integration can not be performed in the usual fashion as for integrands which have inverses, in the integration sense. The above integral is extensively tabulated, though, since similar integrals are important in probability studies.

● **PROBLEM 10-3**

Express the integral from x_0 to x_2 in terms of the three ordinates and three derivatives at the three equally-spaced points x_0, x_1, x_2.

The required formula is of the type

$$\int_{x_0}^{x_2} y\, dx = A_0 y_0 + A_1 y_1 + A_2 y_2 + B_0 y_0' + B_1 y_1' + B_2 y_2'.$$

Solution: Set $x_0 = x_1 - h$, $x_2 = x_1 + h$, and let $y = (x - x_1)^k$, $k = 0, 1, \ldots, 5$. The six equations that result are, after obvious factors have been removed:

$$2h = A_0 + A_1 + A_2,$$

$$0 = -A_0 h \quad + A_2 h + B_0 + B_1 + B_2,$$

$$\frac{2h^2}{3} = A_0 h \quad + A_2 h - 2B_0 \quad + 2B_2,$$

$$0 = -A_0 h \quad + A_2 h + 3B_0 \quad + 3B_2,$$

$$\frac{2h^2}{5} = A_0 h \quad + A_2 h - 4B_0 \quad + 4B_2,$$

$$0 = -A_0 h \quad + A_2 h + 5B_0 \quad + 5B_2,$$

From the second, fourth, and six equations it follows that

$$A_0 = A_2, \qquad B_0 = -B_2, \qquad B_1 = 0.$$

The remaining three equations are thereby reduced to

$$2h = 2A_0 + A_1$$

$$\frac{2h}{3} = 2A_0 \quad - 4B_0$$

$$\frac{2h}{5} = 2A_0 \quad - 8B_0$$

whence $B_0 = \dfrac{h^2}{15}$, $A_0 = \dfrac{7h}{15}$, $A_1 = \dfrac{16h}{15}$,

and the final formula is

$$\int_{x_0}^{x_2} y\, dx = \frac{h}{15}(7y_0 + 16y_1 + 7y_2) + \frac{h^2}{15}(y_0' - y_2').$$

● **PROBLEM 10-4**

Assume

$$\int_{x_0}^{x_2} f(x)\,dx = A_0 f_0 + A_1 f_1 + A_2 f_2 + R$$

where $x_i = x_0 + ih$, $i = 0, 1, 2$. Determine A_0, A_1, A_2, independent of $f(x)$ and such that $R = 0$ if $f(x)$ is any polynomial of degree ≤ 2.

Solution: The calculation is simplified by writing the polynomial

$$p(x) = a_0 x^2 + a_1 x + a_2$$

in the form $b_0(x - x_0)^2 + b_1(x - x_0) + b_2$ and observing that if one chose the A_i so that $R = 0$ for $f(x) = 1$, $x - x_0$ and $(x - x_0)^2$ then $R = 0$ for $f(x) = p(x)$. Using these three special polynomials one obtains three equations in the three unknown A_i:

$$A_0 + A_1 + A_2 = 2h$$
$$A_1 + 2A_2 = 2h$$
$$A_1 + 4A_2 = \frac{8}{3}h$$

hence $A_0 = \frac{h}{3}$, $A_1 = \frac{4h}{3}$, $A_2 = \frac{h}{3}$ and

$$\int_{x_0}^{x_2} f(x)\,dx = \frac{h}{3}[f_0 + 4f_1 + f_2] + R$$

where $R = 0$ if $f(x)$ is any polynomial of degree ≤ 2. Actually in this case $R = 0$ for any polynomial of degree ≤ 3.

● **PROBLEM 10-5**

Let S be the square whose vertices are $(0,0)$, $(1,0)$, $(1,1)$, $(0,1)$, and whose interior is denoted by R. Find a numerical solution of the associated plateau problem if $f(x,y) = x - 3y$.

Solution: The numerical approach will center about minimizing functional

333

$$J = \iint_{R+S} \sqrt{1 + u_x^2 + u_y^2} \, dA \qquad (1)$$

subject to the boundary condition

$$f(x,y) = x - 3y. \qquad (2)$$

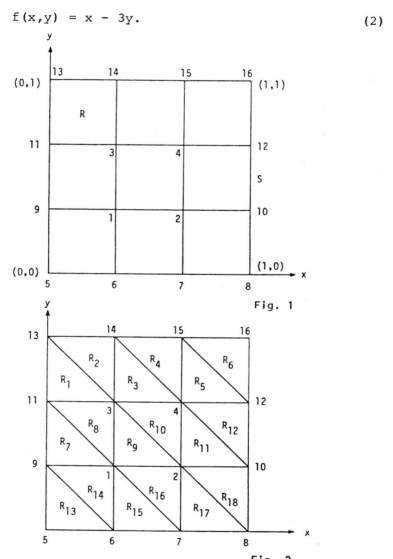

Fig. 1

Fig. 2

For this purpose, take $h = \frac{1}{3}$ and consturct R_h and S_h. It will be convenient, in the present development, to adjoin to S_h the vertices (0,0), (1,0), (1,1), and (0,1). Number the points of R_h with 1, 2, 3, 4, and those of S_h with 5, 6, 7, ..., 16, as shown in Figure 1. Next triangulate, in any fashion, each subsquare shown in Figure 1, so that R + S is thereby divided into 18 mutually disjoint, subtriangular regions, one possible arrangement of which is shown in Figure 2. Notice that the process of

334

triangularization introduces no new grid points. Number these triangular regions

$$R_1, R_2, \ldots, R_{18},$$

in any order, and let the boundary of each R_i be denoted by S_i, $i = 1, 2, \ldots, 18$.

Now, note that (1) can be rewritten as

$$J = \iint_{R_1+S_1} \sqrt{1 + u_x^2 + u_y^2}\, dA + \iint_{R_2+S_2} \sqrt{1 + u_x^2 + u_y^2}\, dA$$

$$+ \ldots + \iint_{R_{18}+S_{18}} \sqrt{1 + u_x^2 + u_y^2}\, dA,$$

and consider first, $I_1 = \iint_{R_1+S_1} \sqrt{1 + u_x^2 + u_y^2}\, dA.$

In order to approximate I_1, find the right angle vertex of S_1, which is the point numbered 11 in Figure 2, and at it approximate u_x and u_y by using function values only at other points of S_1. Thus,

$$u_x\big|_{11} \sim (u_3 - u_{11})/(1/3), \quad u_y\big|_{11} \sim (u_{13} - u_{11})/(1/3),$$

which, from (2) implies that

$$u_x\big|_{11} \sim \frac{u_3 - (-2)}{1/3} = 3u_3 + 6, \quad u_y\big|_{11} \sim \frac{-3 - (-2)}{1/3} = -3.$$

Thus, I_1 can be approximated by

$$I_1^* = \frac{1}{18} \cdot \sqrt{1 + (3u_3+6)^2 + (-3)^2}.$$

Consider next $I_2 = \iint_{R_2+S_2} \sqrt{1 + u_x^2 + u_y^2}\, dA.$

In order to approximate I_2, fix the right angle vertex of S_2, which is the point number 14 in Figure 2 and at it approximate u_x and u_y by using function values only at other points of S_2. Thus,

$$u_x\big|_{14} \sim \frac{u_{14} - u_{13}}{1/3} = \frac{-\frac{8}{3} - (-3)}{1/3} = 1$$

$$u_y\big|_{14} \sim \frac{u_{14} - u_3}{1/3} = \frac{-\frac{8}{3} - u_3}{1/3} = -8 - 3u_3,$$

and, as an approximation to I_2, take

$$I_2^* = \frac{1}{18} \cdot \sqrt{1 + (1)^2 + (-8 - 3u_3)^2}.$$

Proceed in the indicated fashion until each integral

$$I_i = \iint_{R_i + S_i} \sqrt{1 + u_x^2 + u_y^2} \, dA, \quad i = 1, 2, \ldots, 18$$

is approximated by an I_i^*. In each I_i^*, u_x and u_y are approximated at the right angle vertex of S_i by means of function values only at points of S_i.

One next approximates J by J_{18}, where $J_{18} = \sum_{i=1}^{18} I_i^*$.

Note immediately, that J_{18} is a function only of u_1, u_2, u_3, u_4. As an approximation to the minimum of J, take the minimum of J_{18} at the points of R_h, and these are found, by solving the system

$$\frac{\partial J_{18}}{\partial u_i} = 0, \quad i = 1, 2, 3, 4,$$

to be $u_1 = -2/3$, $u_2 = -1/3$, $u_3 = -5/3$, $u_4 = -4/3$.

Interestingly enough, these values coincide with the exact values of the solution $u = x - 3y$ of the given problem at the points of R_h.

NUMERICAL INTEGRATION: SIMPSON'S RULE, TRAPEZOIDAL RULE, WEDDLE'S RULE, AND ROMBERG'S METHOD

• **PROBLEM 10-6**

Use Simpson's rule and Weddle's rule to find the value of

$$\int_0^6 y \, dx$$

for the function tabulated.

x	0	1	2	3	4	5	6
y	0	-45	-496	-2541	-8184	-19525	-37320

Solution: Simpson's rule states that, if n is even,

$$\int_{x_0}^{x_n} y \, dx \doteq \frac{h}{3}(y_0 + 4y_1 + 2y_2 + 4y_3 + 2y_4 + \cdots$$
$$+ 4y_{n-1} + y_n)$$
$$= \frac{h}{3} \sum_{i=0}^{n} c_i y_i \quad \text{where}$$

i	0	1	2	3	4	5	...	$n-2$	$n-1$	n
c_i	1	4	2	4	2	4	...	2	4	1

and h is the increment and $\doteq$ stands for "estimate." This is the easiest of the quadrature formulas to apply. In its application there must be an odd number of points, and it is quite accurate for small values of h.

Weddle's rule states that

$$\int_{x_0}^{x_n} y \, dx \doteq .3h(y_0 + 5y_1 + y_2 + 6y_3 + y_4 + 5y_5$$
$$+ 2y_6 + 5y_7 + y_8 + 6y_9 + y_{10} + 5y_{11}$$
$$+ \cdots + 2y_{n-6} + 5y_{n-5} + y_{n-4} + 6y_{n-3} + y_{n-2}$$
$$+ 5y_{n-1} + y_n)$$
$$= .3h \sum_{i=0}^{n} c_i y_i, \quad \text{where}$$

i	0	1	2	3	4	5	6	7	8	9	10	11	...	n
c_i	1	5	1	6	1	5	2	5	1	6	1	5		1

The coefficients may best be remembered in groups of six:

First group	1, 5, 1, 6, 1, 5
All interior groups	2, 5, 1, 6, 1, 5
Last group	2, 5, 1, 6, 1, 5, 1

Weddle's rule requires at least seven consecutive values of the function and uses them in multiples of six if more than the first seven are required. In general, it is more accurate than Simpson's rule.

In this case h = 1.

x	y	$c_i(S)$	$c_i(W)$
0	0	1	1
1	−45	4	5
2	−496	2	1
3	−2541	4	6
4	−8184	2	1
5	−19525	4	5
6	−37320	1	1

$$I_S = \tfrac{1}{3} \sum c_i(S) y_i = -47708.0$$

$$I_W = .3 \sum c_i(W) y_i = -47728.8$$

This is a tabulation of the function

$$y(x) = x^6 - 9x^5 - 9x^4 - 9x^3 - 9x^2 - 10x,$$

and the integral from $x = 0$ to $x = 6$ correct to one decimal place is -47733.9. Thus Simpson's rule gives a value that is in error by .054% and Weddle's rule gives a value that is in error by .011%.

● **PROBLEM 10-7**

The solution of the initial value problem

$$y' = 1 - 2xy$$

$$y(0) = 0$$

is

$$y(x) = e^{-x^2} \int_0^x e^{u^2} du.$$

Use Simpson's rule to calculate values of $y(x)$ at $x = 0, 0.1, 0.2, \ldots, 0.8$.

Solution:

x	u	u^2	e^{u^2}	$\int_0^x e^{u^2} du$	e^{-x^2}	y
0	0	0	1	0	1	0
	0.05	0.0025	1.00250			
0.1	0.1	0.01	1.01005	0.10033	0.99005	0.0993

	0.15	0.0225	1.02276			
0.2 0.2	0.04	1.04081	0.20269	0.96079	0.1947	
	0.25	0.0625	1.06449			
0.3 0.3	0.09	1.09417	0.30923	0.91393	0.2826	
	0.35	0.1225	1.13032			
0.4 0.4	0.16	1.17351	0.42237	0.85214	0.3599	
	0.45	0.2025	1.22446			
0.5 0.5	0.25	1.28403	0.54495	0.77880	0.4244	
	0.55	0.3025	1.35324			
0.6 0.6	0.36	1.43333	0.68045	0.69768	0.4747	
	0.65	0.4225	1.52577			
0.7 0.7	0.49	1.63232	0.83326	0.61263	0.5105	
	0.75	0.5625	1.75505			
0.8 0.8	0.64	1.89648	1.00907	0.52729	0.5321	

● **PROBLEM 10-8**

Find the value of the integral

$$I = \int_1^{2.2} \int_1^{2.2} \frac{dx\,dy}{xy},$$

by repeated application of Weddle's rule and Simpson's rule. The increments for x and y are $h = .3$ and $k = .2$ respectively.
Find the approximate errror committed by Weddle's and Simpson's rules if the true value of the integral is o.62167.

Solution: Let $z = f(x,y)$ be a function of two independent variables x and y and let its values be given at equally spaced intervals, x_i, (i = 0, ..., n), of length h and y_i, (i = 0, ..., m), of length k. Then since $dx = h\,du$ and $dy = k\,dv$, one gets

$$I = \int_{x_0}^{x_n} \int_{y_0}^{y_m} z(x,y)\,dy\,dx = hk \int_0^n \int_0^m z(u,v)\,dv\,du. \quad (1)$$

If the values of $Z_{ij}(x_i, y_i)$ are exhibited in a rectangular array of the form:

TABLE: Function of two variables

	y_0	y_1	...	y_m
x_0	z_{00}	z_{01}		z_{0m}
x_1	z_{10}	z_{11}		z_{1m}
.	.	.		.
.	.	.		.
.	.	.		.
x_n	z_{n0}	z_{n1}		z_{nm}

then the value of I in Formula 1 may be found by applying the following rule:

Rule: The value of the double integral may be found by applying to each horizontal row (or to each vertical column) any quadrature formula employing equidistant ordinates. Then, to the results thus obtained for the rows (or columns), again apply a similar formula.

The value of the double integral can thus be found by repeated application of Simpson's rule, Weddle's rule, or any other quadrature formula.

For this problem, let the values of the integrand be given as shown in the table which has h = .3 and k = .2:

y \ x	1.0	1.2	1.4	1.6	1.8	2.0	2.2	A_x	$c_i(S)$
1.0	1.00000	.83333	.71429	.62500	.55556	.50000	.45455	.788463	1
1.3	.76923	.64103	.54945	.48077	.42735	.38462	.34965	.606513	4
1.6	.62500	.52083	.44643	.39063	.34722	.31250	.28409	.492790	2
1.9	.52632	.43860	.37594	.32895	.29240	.26316	.23923	.414983	4
2.2	.45455	.37879	.32468	.28409	.25253	.22727	.20661	.358393	1
$c_i(W)$	1	5	1	6	1	5	1		

$I \doteq .62184$, where $\doteq$ means approximate.

Apply Weddle's rule to each row to obtain

$$A_x = (.06) \Sigma [c_i(W)] \text{(entry)},$$

which have been entered in a column for each x. Then apply Simpson's rule to the column of A_x to obtain

$$I \doteq .1 \Sigma [c_i(s)] A_x = .62184.$$

The true value of the integral is

$$I = (\ln 2.2)(\ln 2.2) = .62167.$$

Thus the error is -.00017.

• **PROBLEM 10-9**

Find the value of the integral

$$I = \int_4^{5.2} \int_2^{3.2} \frac{dydx}{xy}$$

by repeated application of Weddle's rule and Simpson's rule. The increments for x and y are h = 0.2 and k = 0.3 respectively.

Solution: Given h = 0.2, k = 0.3, compute the following table of values of z = 1/xy.

x \ y	4.0	4.2	4.4	4.6	4.8	5.0	5.2
2.0	0.125000	0.119048	0.113636	0.108696	0.104167	0.100000	0.096154
2.3	0.108696	0.103520	0.0988142	0.0945180	0.0905797	0.0869565	0.0836120
2.6	0.096154	0.0915751	0.0874126	0.0836120	0.0801282	0.0769231	0.0739645
2.9	0.0862069	0.0821018	0.0783699	0.0749625	0.0718391	0.0689655	0.0663130
3.2	0.078125	0.0744048	0.0710227	0.0679348	0.0651042	0.0625000	0.0600962

Applying Weddle's Rule to each horizontal row, one will have

A = 0.06 [0.125000 + 5(0.119048) + 0.113636 + 6(0.108696)

+ 0.104167 + 5(0.100000) + 0.096154] = 0.131182,

A_1 = 0.114072, A_2 = 0.100909, A_3 = 0.090470,

A_4 = 0.081989.

Now applying Simpson's Rule to the A's, one gets

I = 0.1 [0.131182 + 4(0.114072) + 2(0.100909)

+ 4(0.090470) + 0.081989] = 0.123316.

The true value of this integral is

$$\int_4^{5.2} \int_2^{3.2} \frac{dydx}{xy} = \ln 1.3 \times \ln 1.6 = 0.123321,$$

the error is therefore

E = 0.123321 - 0.123316 = 0.000005.

• **PROBLEM 10-10**

Given the following function tabulated at evenly spaced intervals:

(x)	0	1	2	3	4	5	6	7	8	9
f(x)	0	0.5687	0.7909	0.5743	0.1350	-0.1852	-0.1802	0.0811	0.2917	0.3031

Evaluate

$$\int_0^9 f(x)\,dx$$

using suitable trapezoidal and Simpson's methods. Compare the results obtained from the various methods.

Solution: The trapezoidal rule gives

$$\bar{I}_A = \frac{1}{2}\left[f(0) + f(9) + 2\sum_{j=1}^{8} f(j)\right] = 2.22785$$

Simpson's rule cannot be applied directly, since the number of panels (9) is odd. However, one might wonder what would happen if Simpson's rule were used for the first 8 panels and the trapezoidal rule for the last. This gives

$$\bar{I}_B = \frac{1}{3}\left[f(0) + f(8) + 4\sum_{\substack{j=1\\j\text{ odd}}}^{7} f(j) + 2\sum_{\substack{j=2\\j\text{ even}}}^{6} f(j)\right]$$

$$+ \frac{1}{2}\left[f(8) + f(9)\right] = 1.97957 + 0.29740 = 2.27697$$

Presumably the most accurate results can be obtained by using a fourth order method throughout the region. A formula which is of the same error order as the standard Simpson's rule, but which uses three panels, is often called Simpson's 3/8 rule and is given by

$$\int_{x_{j-2}}^{x_{j+1}} f(x)\,dx \approx \frac{3\Delta x}{8}[f(x_{j-2}) + 3f(x_{j-1})$$

$$+ 3f(x_j) + f(x_{j+1})]$$

The integral over the entire interval can be evaluated by using Simpson's 1/3 rule for the first 6 panels and Simpson's 3/8 rule for the last 3 panels. This gives

$$\bar{I}_C = \frac{1}{3}\{f(0) + f(6) + 4[f(1) + f(3) + f(5)]$$
$$+ 2[f(2) + f(4)]\}$$
$$+ \frac{3}{8}[f(6) + 3f(7) + 3f(8) + f(9)]$$
$$= 1.83427 + 0.46548 = 2.29975$$

Accepting $\bar{I}_C$ as the most accurate value, note that the trapezoidal rule value $\bar{I}_A$ is considerably different (and presumably less accurate). The value $\bar{I}_B$, which was obtained by using Simpson's rule over all but one panel, and the trapezoidal rule over that panel, appears to be reasonably accurate. While this combination of methods with different error orders is not aesthetically pleasing, it will usually produce fairly good results, particularly for small mesh sizes and gently varying functions where the trapezoidal rule portion contributes relatively little to the total and/or is reasonably accurate. However, since the combination of Simpson's 1/3 and 3/8 rules is as simple to compute as any other method, there would seldom be any reason to choose another technique.

• **PROBLEM 10-11**

Use the trapezoidal and Simpson's rules to evaluate
$$\int_0^2 x^2 dx, \quad \int_0^2 e^x dx, \quad \int_0^4 x^3 dx$$
Compare with the exact values.

Solution: The trapezoidal rule is

$$\int_a^b f(x)dx \approx \frac{b-a}{2}[f(b)+f(a)]$$

Hence,
$$\int_0^2 x^2 dx \approx f(2) + f(0) = 4$$

$$\int_0^2 e^x dx \approx e^2 + 1 = 8.39$$

$$\int_0^4 x^3 dx \approx \frac{4}{2} \cdot 4^3 = 128$$

Simpson's rule states that

$$\int_{x_0}^{x_n} f(x)\,dx = \tfrac{h}{3}[f(x_0)+4f(x_1)+2f(x_2)+\ldots+4f(x_{n-1})+f(x_n)]$$

which leads to

$$\int_0^2 x^2\,dx \approx \tfrac{1}{3}[f(0)+4f(1)+f(2)] = \tfrac{4+4}{3} = \tfrac{8}{3}$$

$$\int_0^2 e^x\,dx \approx \tfrac{1}{3}[e^0+4e+e^2] = 6.4207$$

$$\int_0^4 x^3\,dx \approx \tfrac{1}{3}[f(0)+4f(1)+2f(2)+4f(3)+f(4)]$$

$$= \tfrac{4+16+108+64}{3} = 64$$

The exact values are

$$\int_0^2 x^2\,dx = \tfrac{8}{3}, \quad \int_0^2 e^x\,dx = e^2-1, \quad \int_0^4 x^3\,dx = 64$$

Note that we used the increment $h = 1$. From these examples, it is obvious that Simpson's rule is more accurate than the trapezoidal rule.

• **PROBLEM 10-12**

Consider the integral equation

$$u(x) = \tfrac{5x}{6} - \tfrac{1}{9} + \tfrac{1}{3}\int_0^1 (t+x)u(t)\,dt.$$

Evaluate by Simpson's rule, taking $n = 3$ or $h = \tfrac{1}{2}$. Check the result.

Solution: The integral will be evaluated by Simpson's rule, taking $n = 3$ or $h = \tfrac{1}{2}$. Then

$$u(x) = \tfrac{5x}{6} - \tfrac{1}{9} + \tfrac{1}{3}\cdot\tfrac{1}{3}\cdot\tfrac{1}{2}\Big[(t_1+x)u(t_1)$$

$$+4(t_2+x)u(t_2)+(t_3+x)u(t_3)\Big]. \tag{1}$$

Since (1) must hold for all values of x from 0 to 1, it holds for $x = t_1, t_2, t_3$. Hence from (1) one gets

$$u(t_1) = \tfrac{5t_1}{6} - \tfrac{1}{9} + \tfrac{1}{18}\Big[2t_1 u(t_1) + 4(t_2+t_1)u(t_2)$$

$$+ (t_3+t_1)u(t_3)\Big],$$

344

$$u(t_2) = \frac{5t_2}{6} - \frac{1}{9} + \frac{1}{18} \Big[(t_1 + t_2)u(t_1) + 4(2t_2)u(t_2)$$
$$+ (t_3 + t_2)u(t_3) \Big] ,$$

$$u(t_3) = \frac{5t_3}{6} - \frac{1}{9} + \frac{1}{18} \Big[(t_1 + t_3)u(t_1) + 4(t_2 + t_3)u(t_2)$$
$$+ 2t_3 u(t_3) \Big] .$$

Now putting $t_1 = 0$, $t_2 = \frac{1}{2}$, $t_3 = 1$ and writing $u(t_i) = u_i$, results in

$$u_1 = -\frac{1}{9} + \frac{1}{18} \Big[2u_2 + u_3 \Big]$$

$$u_2 = \frac{5}{12} + \frac{1}{18} \Big[\frac{1}{2}u_1 + 4u_2 + \frac{3}{2}u_3 \Big]$$

$$u_3 = \frac{5}{6} + \frac{1}{18} \Big[u_1 + 6u_2 + 2u_3 \Big].$$

Clearing of fractions and transposing the u's to the left side of the equations yields

$$36u_1 - 4u_2 - 2u_3 = -4$$
$$-u_1 + 28u_2 - 3u_3 = 11$$
$$-2u_1 - 12u_2 + 32u_3 = 26.$$

On solving these equations by determinants (Cramer's rule), or otherwise, one finds $u_1 = 0$, $u_2 = \frac{1}{2}$, $u_3 = 1$. Then substituting in (1) these values of the u's and putting $t_1 = 0$, $t_2 = \frac{1}{2}$, $t_3 = 1$, one gets

$$u(x) = \frac{5x}{6} - \frac{1}{9} + \frac{1}{18} \Big[0 + 4\Big(\frac{1}{2} + x\Big)\Big(\frac{1}{2}\Big) + (1 + x)(1) \Big] = x .$$

This result can be checked by substituting it in the integrand of the original equation and performing the integration. Thus, putting $u(t) = t$, one has

$$u(x) = \frac{5x}{6} - \frac{1}{9} + \frac{1}{3} \int_0^1 (t + x)t \, dt = \frac{5x}{6} - \frac{1}{9}$$
$$+ \frac{1}{3} \Big[\frac{t^3}{3} + \frac{xt^2}{2} \Big]_0^1$$
$$x = \frac{5x}{6} - \frac{1}{9} + \frac{1}{9} + \frac{x}{6} = x$$

Simpson's rule gives the exact result because the integrand is a second-degree polynomial in t.

● **PROBLEM 10-13**

Evaluate
$$\int_0^1 \frac{\log x}{(1 - x^2)^{\frac{1}{2}}} dx.$$

Apply Simpson's rule using an interval of $[0,1]$ with h = 0.25.

Solution: The integrand is infinite at the lower limit, and its derivative is infinite at the upper limit. However, the integral is conveniently rewritten as

$$\int_0^{1/2} \left\{ \frac{1}{(1-x^2)^{1/2}} - 1 \right\} \log x \, dx + \int_0^{1/2} \log x \, dx$$

$$+ \int_{1/2}^1 \left\{ \frac{\log x}{(1-x^2)^{1/2}} - (1/2) \sin^{-1} x \right\} dx + (1/2) \int_{1/2}^1 \sin^{-1} x \, dx$$

so that in the first integral the integrand is finite when x = 0, and in the third the derivative is finite when x = 1. Apply Simpson's rule to these integrals, using an interval of 0.25 and the following values of the integrands—

x	$\left\{\frac{1}{(1-x^2)^{1/2}} - 1\right\}\log x$	x	$\frac{\log x}{(1-x^2)^{1/2}} - (1/2)\sin^{-1} x$
0.0	0	0.50	−1.0622
0.25	−0.0455	0.75	−1.8590
0.50	−0.1072	1.00	−0.7854

The estimates are −0.0242 and −0.4403 respectively.

The second and fourth integrals are respectively

$$\left[x \log x - x \right]_0^{1/2} = -0.8466$$

$$\frac{1}{2} \left[x \sin^{-1} x - (1-x^2)^{1/2} \right]_{1/2}^1 = +0.2215$$

Hence the estimate of the original integral is

$$-0.0242 - 0.8466 - 0.4403 + 0.2215$$

or − 1.0896

The correct value is $-\frac{1}{2} \pi \log_e 2$ or −1.0888 to 4 decimals. Note that if the behavior of the derivative of the integrand near x = 1 is ignored, a double application of Simpson's rule gives −1.0821, in error by 0.6 per cent. An alternative treatment is to evaluate the equivalent integral

$$-\int_0^1 (1/x) \sin^{-1} x\, dx,$$

obtained by integrating by parts, in which the integrand is always finite but its derivative again infinite when x = 1. If this is ignored, and Simpson's rule used with four intervals, the answer is -1.1026.

• **PROBLEM 10-14**

Consider the trapezoidal rule, Simpson's rule, and Weddle's rule to compute

$$\int_0^{.6} x^4\, dx$$

using h = .1.

Solution: Calculate

x	0	.1	.2	.3	.4	.5	.6
y	0	.0001	.0016	.0081	.0256	.0625	.1296

$I_T = \frac{.1}{2}$ [0 + 2(.0001) + 2(.0016) + 2(.0081) + 2(.0256)

    + 2(.0625) + .1296

I_T = .01627

$I_S = \frac{.1}{3}$ [0 + 4(.0001) + 2(.0016) + 4(.0081) + 2(.0256)

    + 4(.0625) + .1296

I_S = .01556

$I_W = \frac{3h}{10}$ [$y_0 + 5y_1 + y_2 + 6y_3 + y_4 + 5y_5 + y_6$]

(Weddle's rule)

NOTE: This rule only fits polynomials of degree five or less, since Weddle's rule require the data groups of six intervals. Thus for 12 intervals:

$I_W = \frac{3h}{10}$ [$y_0 + 5y_1 + y_2 + 6y_3 + y_4 + 5y_5 + 2y_6 + 5y_7$

    $+ y_8 + 6y_9 + y_{10} + 5y_{11} + y_{12}$]

$$I_W = \frac{3(.1)}{10} [0 + 5(.0001) + .0016 + 6(.0081) + .0256$$

$$I_W = .015552$$

The exact value of the integral is

$$\int_0^{.6} x^4 \, dx = .015552$$

Notice that the Weddle's rule gives exact results, since the function is only fourth-degree and Weddle's rule gives exact results for quintics.

• **PROBLEM 10-15**

Using Romberg integration with an absolute convergence criterion of

$$\varepsilon = 1.0 \times 10^{-6},$$

evaluate

$$I = \int_0^{0.8} e^{-x^2} \, dx$$

Correct to five decimal places.

Solution: Romberg algorithm:

This numerical integration technique is based on the use of the trapezoidal rule combined with the Richardson extrapolation. In order to apply this extrapolation, it is necessary to know the general form of the error terms for the trapezoidal rule. Considering this, the trapezoidal rule may be written as

$$I = \frac{\Delta x}{2} [f(a) + f(b) + 2 \sum_{j=1}^{n-1} f(a + j\Delta x)]$$

$$+ C(\Delta x)^2 + D(\Delta x)^4 + E(\Delta x)^6 + \ldots \quad (1)$$

where C, D, E, etc. are functions of $f(x)$ and its derivatives, but are not functions of Δx. The terms involving the odd powers of Δx have vanished from the error.

Let $\bar{I} = \frac{\Delta x}{2} [f(a) + f(b) + 2 \sum_{j=1}^{n-1} f(a + j\Delta x)]$ \quad (2)

Equation (1) may then be rearranged in the form

$$\bar{I} = I - C(\Delta x)^2 - D(\Delta x)^4 - E(\Delta x)^6 - \ldots \qquad (3)$$

Consider now the use of two different mesh sizes, Δx_1 and Δx_2. If one denotes the values of $\bar{I}$ corresponding to Δx_1 and Δx_2 as $\bar{I}_1$ and $\bar{I}_2$ respectively, then from (3),

$$\bar{I}_1 = I - C(\Delta x_1)^2 - D(\Delta x_1)^4 - E(\Delta x_1)^6 - \ldots \qquad (4)$$

$$\bar{I}_2 = I - C(\Delta x_2)^2 - D(\Delta x_2)^4 - E(\Delta x_2)^6 - \ldots \qquad (5)$$

Now assume $\Delta x_1 = 2\Delta x_2$. Then (4) becomes, in terms of Δx_2,

$$\bar{I}_1 = I - 4C(\Delta x_2)^2 - 16D(\Delta x_2)^4 - 64E(\Delta x_2)^6 - \ldots \qquad (6)$$

Now multiply equation (5) by 4, subtract (6), and divide by 3:

$$\frac{4\bar{I}_2 - \bar{I}_1}{3} = I + 4D(\Delta x_2)^4 + 20E(\Delta x_2)^6 + \ldots \qquad (7)$$

The error term involving $(\Delta x)^2$ has vanished and (7) thus furnishes an approximation to the integral which is of the order $(\Delta x_2)^4$. Extrapolation of this type is termed Richardson extrapolation. (By inserting the expressions for $\bar{I}_1$ and $\bar{I}_2$ into (7), also find that Simpson's rule has been rediscovered. Now evaluate $\bar{I}_3$, where $\Delta x_3 = \Delta x_2/2$, and extrapolate $\bar{I}_2$ and $\bar{I}_3$, obtain

$$\frac{4\bar{I}_3 - \bar{I}_2}{3} = I + 4D(\Delta x_3)^4 + 20E(\Delta x_3)^6 + \ldots \qquad (8)$$

Between (7) and (8), the term in $(\Delta x)^4$ may be eliminated to furnish an estimate to I which is accurate to $\mathcal{O}(\Delta x)^6$.

Thus for each new evaluation of an $\bar{I}$, one more term in the error can be eliminated by extrapolation. This systematic procedure is called Romberg integration.

In order to describe the algorithm in detail, adopt a new notation. The trapezoidal rule estimates of the integral will be denoted as

$$T_{1,k} = \frac{\Delta x}{2}\left[f(a) + f(b) + 2\sum_{j=1}^{\ell} f(a + j\Delta x)\right] \qquad (9)$$

where $\Delta x = (b-a)/2^{k-1}$ and $\ell = 2^{k-1} - 1$. The number of panels involved in $T_{1,k}$ is 2^{k-1}.

Thus

$$T_{1,1} = \frac{b-a}{2}\left[f(a) + f(b)\right]$$

$$T_{1,2} = \frac{b-a}{4}\left[f(a) + f(b) + 2f\left(a + \frac{b-a}{2}\right)\right]$$

$$T_{1,3} = \frac{b-a}{8}\left[f(a) + f(b) + 2f\left(a + \frac{b-a}{4}\right)\right.$$
$$\left. + 2f\left(a + \frac{b-a}{2}\right) + 2f\left(\frac{3(b-a)}{4}\right)\right] \quad \text{etc.}$$

Note that

$$T_{1,2} = \frac{T_{1,1}}{2} + \frac{b-a}{2}\left[f\left(a + \frac{b-a}{2}\right)\right]$$

$$T_{1,3} = \frac{T_{1,2}}{2} + \frac{b-a}{4}\left[f\left(a + \frac{b-a}{4}\right)\right.$$
$$\left. + f\left(a + \frac{3(b-a)}{4}\right)\right]$$

etc.

This means that each succeeding trapezoidal rule approximation can be obtained from the preceding approximation without having to recompute f(x) at any of the points where it has already been computed.

The extrapolation is carried out according to

$$T_{\ell,k} = \frac{1}{4^{\ell-1} - 1}\left(4^{\ell-1}T_{\ell-1,k+1} - T_{\ell-1,k}\right) \quad (10)$$

For example, for $\ell = 2$,

$$T_{2,1} = \frac{1}{3}\left(4T_{1,2} - T_{1,1}\right)$$

$$T_{2,2} = \frac{1}{3}\left(4T_{1,3} - T_{1,2}\right)$$

(These extrapolations each eliminate the $O(\Delta x)^2$ error term.) Now for $\ell = 3$,

$$T_{3,1} = \frac{1}{15}\left(16T_{2,2} - T_{2,1}\right)$$

(This extrapolation eliminates the $O(\Delta x)^4$ error term.) These results can conveniently be arranged in tabular form.

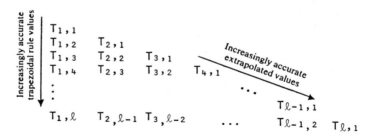

The extrapolated values along the diagonal will converge to the correct answer much more rapidly than the trapezoidal rule values in the first column.

The Romberg algorithm yields the following table:

0.61092

0.64632   0.65812

0.65485   0.65770   0.65767

0.65697   0.65767   0.65767   0.65767

0.65749   0.65767   0.65767   0.65767   (0.65767)

The table has been rounded to 5 decimal places. The encircled value is actually 0.6576691 as compared to the exact value, which can be found from tables of the error function to be 0.6576698. Using 16 panels (17 functional evaluations), 6 digits (5 decimal places if rounded) of the exact answer have been reproduced. Note that the best trapezoidal rule value (the bottom entry in the first column) is accurate to only three digits (2 decimal places if rounded). This function is well suited to Romberg integration.

● **PROBLEM 10-16**

Find the integral of cos x from 0 to 0.8 by Simpson's rule, using h = 0.1.

Solution: The values of cos x and the corresponding multipliers in Simpson's rule are tabulated below:

x	cos x	Multiplier
0	1.00000000	1
0.1	0.99500417	4
0.2	0.98006658	2
0.3	0.95533649	4
0.4	0.92106099	2
0.5	0.87758256	4
0.6	0.82533562	2
0.7	0.76484219	4
0.8	0.69670671	1

The multiplication and summation is performed in a continuous operation on the calculating machine, and the result is then multiplied by h/3. In this manner one finds the integral to be 0.71735649. The error term when evaluated with $y^{(4)} = 1$ has the value -0.00000045. The actual error is -0.00000040.

● **PROBLEM** 10-17

Compute the value of the definite integral

$$\int_4^{5.2} \ln x \, dx$$

by Simpson's rule and Weddle's rule using h = 0.2. Find the approximate error committed by Simpson's rule and Weddle's rule.

Solution: Divide the interval of integration into six equal parts each of width 0.2. Hence h = 0.2. The values of the function y = ln x are next computed for each point of subdivision. These values are given in the table below.

x	ln x	x	ln x
4.0	1.38629436	4.8	1.56861592
4.2	1.43508453	5.0	1.60943791
4.4	1.48160454	5.2	1.64865863
4.6	1.52605630		

(a) By Simpson's rule

$$I_S = \frac{0.2}{3} [3.03495299 + 4(4.57057874) + 2(3.05022046)]$$
$$= 1.82784726.$$

(b) By Weddle's rule

$$I_W = (0.3)(0.2)[3.03495299 + 5(3.04452244)$$
$$+ 3.05022046 + 6(1.52605630)] = 1.82784741.$$

The true value of the integral is

$$I = \int_4^{5.2} \ln x \, dx = \left[x(\ln x - 1) \right]_4^{5.2} = 1.82784741.$$

First consider Simpson's rule. Let the function f(x)

be continuous in the interval under consideration and has as many continuous derivatives as required. Then write

$$F(x) = \int_a^x f(x)\,dx, \quad F'(x) = f(x), \quad F''(x) = f'(x), \text{ etc.}$$

The definite integral from $x_0 - h$ to $x_0 + h$ is

$$I = \int_{x_0-h}^{x_0+h} f(x)\,dx = F(x_0 + h) - F(x_0 - h).$$

Simpson's rule yields

$$I_S = \frac{h}{3}\left[f(x_0 - h) + 4f(x_0) + f(x_0 + h)\right]$$

and the difference between the true value of the integral and the solution obtained by Simpson's rule is the inherent error

$$E_S = I - I_S.$$

Hence the errors are

$$E_S = 0.00000015 = 15 \times 10^{-8}, \quad E_W = 0.$$

● **PROBLEM 10-18**

Find the integral of $\cos x$ from $x = 0$ to $x = 0.8$ using $h = 0.1$. Solve by trapezoidal rule. How can the accuracy of the trapezoidal rule be improved? Use the formula

$$\int_{x_{n-1}}^{x_n} y\,dx = \frac{h}{24}[-y_{n-2} + 13y_{n-1} + 13y_n - y_{n+1}]$$

$$+ \frac{11 y^{(4)} h^5}{720}.$$

<u>Solution</u>: The required values of $\cos x$ are tabulated below:

x	cos x
0	1.00000000

0.1	0.99500417
0.2	0.98006658
0.3	0.95533649
0.4	0.92106099
0.5	0.87758256
0.6	0.82533562
0.7	0.76484219
0.8	0.69670671

The value of the integral, as given by

$$\int_{x_{n-1}}^{x_n} y\, dx = \frac{h}{2}(y_{n-1} + y_n) - \frac{y'' h^3}{12},$$

is easily obtained and proves to be 0.716758196.

Calculating E, using the fact that $|y''| \leq 1$ for the interval in question, one finds that

$$E < \frac{8(0.1)^3}{12} = 0.00067,$$

which casts doubt on the last six figures of the result. In actual fact, the error is about 0.00059789.

By the addition of a relatively simple correction term the accuracy of the trapezoidal rule can be materially improved. For this purpose, use the formula

$$\int_{x_{n-1}}^{x_n} y\, dx$$

$$= \frac{h}{24}[-y_{n-2} + 13y_{n-1} + 13y_n - y_{n+1}] + \frac{11 y^{(4)} h^5}{720}$$

which is found by undetermined coefficients. Assuming $n = 1, 2, \ldots, m$ and adding, one derives the equation

$$\int_{x_0}^{x_m} y\, dx = h\left[\frac{1}{2} y_0 + y_1 + y_2 + \ldots + y_{m-1} + \frac{1}{2} y_m\right]$$

$$+ \frac{h}{24}[-y_{-1} + y_1 + y_{m-1} - y_{m+1}] + \frac{11 m y^{(4)} h^5}{720}$$

The second term on the right in this equation is the correction term.

• **PROBLEM** 10-19

Find by Simpson's rule the value of

$$I = \int_{-1}^{1} \sqrt{(1-x^2)(2-x)}\, dx,$$

for h = 0.5, 0.2 and 0.1, correct to six decimal digits. What can be suggested to make a better approximation within the means of Simpson's rule?

Solution: The values of the integrand are given in the table below.

x	y	x	y
−1	0	0.1	1.371496
−0.9	0.742294	0.2	1.314534
−0.8	1.003992	0.3	1.243756
−0.7	1.173456	0.4	1.159310
−0.6	1.289961	0.5	1.060660
−0.5	1.369307	0.6	0.946573
−0.4	1.419859	0.7	0.814248
−0.3	1.446720	0.8	0.657267
−0.2	1.453272	0.9	0.457165
−0.1	1.441874	1	0
0	1.414214		

The correct value of the given integral to five significant figures is found from the table of elliptic integrals to be

$$I = 2.2033.$$

Simpson's Rule gives the following values for different values of h:

(a) I = 2.0914 for h = 0.5. Percentage error = 5.1%

(b) I = 2.1751 for h = 0.2. Percentage error = 1.28%.

(c) I = 2.1934 for h = 0.1. Percentage error = 0.42%.

It will be observed that when the interval of integration was divided into 20 subintervals the error was nearly a half of one per cent, which is less than slide-rule accuracy. Inasmuch as the tabular values are all correct to six or seven figures, the errors in the results found above are due entirely to the inherent inaccuracy of Simpson's Rule. The trouble with this problem lies in the fact that the integrand cannot be approximated closely by a polynomial near the end points of the range of integration, for at these points the slope of the integrand is infinite. A better approximation can be obtained by using horizontal parabolas for the regions near the ends of the interval. Thus, for the regions at the left end of the interval, one has

$$I_1 = (2/3)(0.2 \times 1.003992) = 0.1338656;$$

and for the region at the right end, one has

$$I_2 = (2/3)(0.2 \times 0.657267) = 0.0876356.$$

Then the application of Simpson's rule to the region from $x = -0.8$ to $x = 0.8$ gives

$$I_3 = 1.9780924.$$

The sum of these is

$$I_1 + I_2 + I_3 = I = 2.1996,$$

the percentage error of which is 0.17 per cent.

• **PROBLEM 10-20**

With the trapezoidal rule, evaluate the integral

$$I = \int_1^2 \frac{dx}{x}$$

using (a) two intervals, (b) four intervals, and (c) eight intervals.

Solution: The trapezoidal rule evaluates integrals such as

$$\int_a^{a+h} f(x)\,dx$$

by replacing the function $f(x)$ with a straight line, with the resulting area thus being a trapezoid with an area equal to

$$h \frac{f(a) + f(a+h)}{2}.$$

If the rule is used n times for a given interval, the rule becomes

$$\int_a^{a+nh} f(x)\,dx = (h/2)(f_0 + 2f_1 + 2f_2 + \ldots + 2f_{n-1} + f_n)$$

where

$$f_i = f(a + ih).$$

All of the f_i's except for the first and last ones are multiplied by two because the inside ordinates are used twice as the areas of the various trapezoids are added together. Applying the trapezoidal rule to the above problems one gets

(a) With 2 intervals,

$$I = (h/2)(f_1 + 2f_{1.5} + f_2) = \left[\left(\tfrac{1}{2}\right)/2\right](1 + 2/1.5 + 1/2)$$

$$= 0.708$$

(b) With 4 intervals,

$$I = (h/2)(f_1 + 2f_{1.25} + 2f_{1.5} + 2f_{1.75} + f_2)$$

$$I = \left[(1/4)/2\right](1 + 2/1.25 + 2/1.5 + 2/1.75 + 1/2)$$

$$= 0.697$$

(c) With 8 intervals,

$$I = (h/2)(f_1 + 2f_{1.125} + 2f_{1.25} + \ldots + 2f_{1.75}$$

$$+ 2f_{1.875} + f_2)$$

$$I = [(1/8)/2](1 + 2/1.125 + 2/1.25 + 2/1.375$$

$$+ 2/1.5 + 2/1.625 + 2/1.75 + 2/1.875 + \tfrac{1}{2})$$

$$I = 0.694$$

The procedure above gives some feeling for the accuracy of numerical integration formulas. That is, by watching the change in the answer each time one reduces the size of h (which, of course, simultaneously increases the number of intervals), one will often have some idea of the accuracy of the latest answer. For example, in this problem the answers are, respectively, 0.708, 0.697, and 0.694; this certainly suggests that the correct answer to two decimal places is probably 0.69 and to three decimal places it is perhaps 0.694 or 0.693. The exact answer for the integral above is $I = \ln 2 \doteq 0.6931$.

● **PROBLEM 10-21**

Use Simpsons's Rule with two, four, and ten intervals to evaluate

$$\int_1^2 \frac{1}{x}\,dx = \log 2.$$

Solution: With two intervals, one obtains

$$\frac{.5}{3} [1 + 2.66667 + .5] = \frac{4.16667}{6} = .69444.$$

With four intervals, one obtains

$$\frac{.25}{3} [1 + 3.2 + 1.33333 + 2.28571 + .5] =$$

$$\frac{8.31904}{12} = .69325.$$

With ten intervals (using a table of reciprocals),

$$1 + .5 = 1.5$$
$$4(.9090909 + .7692308 + .6666667$$
$$+ .5882352 + .5263158) = 13.8381580$$
$$2(.8333333 + .7142857 + .625 + .5555556) = \underline{5.4563492}$$
$$20.7945072$$

$$\text{Result} = \frac{20.7945072}{30} = .693150.$$

Since log 2 = .693147, one sees that the result approaches the true value quite rapidly as the number of intervals is increased.

If one assumes that the given data or the function appearing in

$$\int_a^b f_x \, dx$$

can be represented by a sextic polynomial, then Stirling's formula gives

$$f_x = f_0 + \frac{x}{1!} \mu \delta f_0 + \frac{x^2}{2!} \delta^2 f_0 + \frac{x(x^2 - 1)}{3!} \mu \delta^3 f_0$$

$$+ \frac{x^2(x^2 - 1)}{4!} \delta^4 f_0 + \frac{x(x^2 - 1)(x^2 - 4)}{5!} \mu \delta^5 f_0$$

$$+ \frac{x^2(x^2-1)(x^2-4)}{6!} \delta^6 f_0.$$

Integrating from -3 to $+3$, note that the second, fourth, and sixth terms yield even functions of x, and so do not contribute to the integral. Hence

$$\int_{-3}^{3} f_x \, dx = \left[f_0 x + \frac{x^3}{6} \delta^2 f_0 + \frac{1}{4!}\left(\frac{x^5}{5} - \frac{x^3}{3}\right) \delta^4 f_0 \right.$$

$$\left. + \frac{1}{6!}\left(\frac{x^7}{7} - x^5 + \frac{4}{3}x^3\right) \delta^6 f_0 \right]_{-3}^{3}$$

$$= 6 f_0 + 9 \delta^2 f_0 + \frac{33}{10} \delta^4 f_0 + \frac{41}{140} \delta^6 f_0.$$

Sixth differences are, in general, very small; so an error of only

$$\frac{1}{140} \delta^6 f_0$$

is introduced if the coefficient 41 in this expression is replaced by the more manageable coefficient 42. Then

$$\int_{-3}^{3} f_x \, dx = \frac{3}{10} [20 + 30 \delta^2 + 11 \delta^4 + \delta^6] f_0.$$

Simplifying the operator, one uses

$$\delta^2 = E - 2 + E^{-1};$$

then

$$20 + \delta^2[30 + \delta^2(11 + \delta^2)] = (E^3 + E^{-3}) + 5(E^2 + E^{-2})$$

$$+ (E + E^{-1}) + 6.$$

Hence $\int_{-3}^{3} f_x \, dx = \frac{3}{10} [6 f_0 + (f_1 + f_{-1}) + 5(f_2 + f_{-2})$

$$+ (f_3 + f_{-3})].$$

By moving the range three units to the right, this formula can be written as

$$\int_0^6 f(x)\,dx = \frac{3}{10}[6f_3 + (f_2 + f_4) + 5(f_1 + f_5)$$

$$+ (f_0 + f_6)]. \tag{1}$$

If the range is 0 to 6h, one obtains the most general form of (1) known as Weddle's rule,

$$\int_0^{6h} f_x\,dx = \frac{3h}{10}[6f_3 + (f_2 + f_4) + 5(f_1 + f_5)$$

$$+ (f_0 + f_6)], \tag{2}$$

where again one writes $f_0, f_1, \ldots, f_6$, rather than $f_0, f_h, \ldots, f_{6h}$.

The main disadvantage of Weddle's rule is that the number of intervals must be a multiple of 6.

● **PROBLEM** 10-22

Consider the following data which comes from a fourth-degree polynomial:

x	-2	-1	0	1	2	3	4
f(x)	21	1	-1	-3	1	41	171

(a) Evaluate

$$I_a = \int_{-2}^{2} f(x)\,dx,$$

using Simpson's rule.

(b) Evaluate

$$I_b = \int_{-2}^{4} f(x)\,dx,$$

using Weddle's rule.

(c) In each of the above two cases explain whether or not the answers obtained should be exactly the same as the actual definite integrals

$$I = \int_c^d f(x)\,dx.$$

Solution: (a) Applied to this problem, Simpson's rule is

$$I_a = h/3[1f(-2) + 4f(-1) + 2f(0) + 4f(1) + 1f(2)]$$

$$= (1/3)(21 + 4 - 2 - 12 + 1) = 4$$

(b) Applied to this problem, Weddle's rule is

$$I_b = (3h/10)\Big[1f(-2) + 5f(-1) + 1f(0) + 6f(1) + 1f(2)$$

$$+5f(3) + 1f(4)\Big]$$

$$I_b = (3/10)(21 + 5 - 1 - 18 + 1 + 205 + 171)$$

$$= (3/10)(384) = 115.2$$

(c) When $f(x)$ is a polynomial, the derivation of Simpson's rule shows that it will give exact results if $f(x)$ is third degree or less. The derivation of Weddle's rule shows that it gives exact results if $f(x)$ is fifth degree or less. Since the problem states that the data for $f(x)$ came from a fourth-degree polynomial, the value of I_a above cannot be expected to be the same as the actual corresponding integral, but the value of I_b above should be exactly the same as the corresponding integral.

Using a polynomial interpolation formula, one can find

the polynomial from which the given data comes; it is

$$f(x) = x^4 - x^3 - x^2 - x - 1.$$

The actual values obtained by integration are

$$\int_{-2}^{2} f\,dx = (52/15) \doteq 3.47$$

and

$$\int_{-2}^{4} f\,dx = 115.2.$$

The interval of differencing, h, is too large in part (a) for Simpson's rule to give a very accurate result.

● PROBLEM 10-23

Use Simpson's rule with four intervals to evaluate each of the following integrals:

(a)
$$I_a = \int_{0}^{2} \frac{x+1}{x+2}\,dx$$

(b)
$$I_b = \int_{0}^{\pi} \sin^2 x\,dx$$

<u>Solution</u>: Applied to this problem Simpson's rule is

$$I_a = \int_{0}^{2} \frac{x+1}{x+2}\,dx = h/3[1f(0) + 4f(0.5) + 2f(1) + 4f(1.5) + 1f(2)]$$

$$I_a = (\tfrac{1}{2})/3 \,[1(1/2) + 4(1.5/2.5) + 2(2/3)$$

$$+ 4(2.5/3.5) + 1(3/4)]$$

$$I_a = (1/6)(1/2 + 12/5 + 4/3 + 20/7 + 3/4) = 7.8405/6$$

$$= 1.307$$

The exact value of the integral is $I = 2 - \ln 2 = 1.307$.

(b) Applied to this problem Simpson's rule is

$$I_b = \int_0^\pi \sin^2 x \, dx = h/3[1f(0) + 4f(\pi/4) + 2f(\pi/2)$$

$$+ 4f(3\pi/4) + 1f(\pi)]$$

$$I_b = (\pi/4)/3[1(0) + 4(\sqrt{2}/2)^2 + 2(1)^2 + 4(\sqrt{2}/2)^2 + 1(0)]$$

$$I_b = (\pi/12)(0 + 2 + 2 + 2 + 0) = \pi/2$$

This answer given by Simpson's rule just happens to be exactly the same as the actual value of the integral. One would not expect this to happen since the integrand is obviously not a polynomial of third degree or less.

OTHER METHODS FOR NUMERICAL INTEGRATION

• **PROBLEM 10-24**

Illustrate the quadrature formulas of Guass by evaluating the integral

$$I = \int_1^2 \frac{dx}{x} \, .$$

<u>Solution</u>: The quadrature formulas due to Gauss are the most accurate in ordinary use. In seeking numerical integration formulas of the greatest possible accuracy,

Gauss found that the values of the independent variable x used to compute various values of f(x), the ordinates, should not be equidistant from one another. This is in contrast to the trapezoidal rule, and the formulas of Simpson, Weddle, and others which use intervals of equal width.

An integral of the form

$$I = \int_a^b f(x)\,dx$$

can be put into a more convenient form for Gauss's method by changing the variable of integration from x to u by the transformation

$$x = (b-a)u + (a+b)/2$$

This gives $u = -\frac{1}{2}$ for $x = a$ and $u = \frac{1}{2}$ for $x = b$. Then noting that $dx = (b-a)du$, the integral becomes

$$I = \int_a^b f(x)\,dx = (b-a)\int_{-\frac{1}{2}}^{\frac{1}{2}} \Phi(u)\,du.$$

Then the integration formula can be written as

$$I = (b-a)\int_{-\frac{1}{2}}^{\frac{1}{2}} \Phi(u)\,du = \sum_{i=1}^{n} R_i \Phi(u_i),$$

where n is the number of ordinates to be used in approximating the integral, and the u_i are the n values of the new independent variable u where the ordinates are to be computed.

As noted above, the u_i are not equally spaced but they are always symmetrical with respect to $u = 0$. Extensive tables are available for values of n up to at least 16, with values of u_i and R_i given (in some places) to at least 15 decimal places. The Gauss method will be illustrated by evaluating the given integral with $n = 4$. The following values may be found in various tables:

$$u_1 = -0.430568 = -u_4$$

$$u_2 = -0.169991 = -u_3$$

$$R_1 = 0.173927 = R_4$$

$$R_2 = 0.325073 = R_3$$

For this problem, one gets

$$a = 1, \quad b = 2$$

and

$$x = (b-a)u + (a+b)/2 = u + 1.5.$$

Therefore,

$$\Phi(u) = 1/(u + 1.5)$$

and

$$I = (b-a)[R_1\Phi(u_1) + R_2\Phi(u_2) + R_3\Phi(u_3) + R_4\Phi(u_4)]$$

$$R_1\Phi(u_1) = (0.173927)/(-0.430568 + 1.5),$$

$$R_2\Phi(u_2) = (0.326073)/(-0.169991 + 1.5)$$

$$R_3\Phi(u_3) = (0.326073)/(0.169991 + 1.5),$$

$$R_4\Phi(u_4) = (0.173927)/(0.430568 + 1.5)$$

Then

$$I = \int_1^2 \frac{dx}{x} = (2-1) \sum_{i=1}^{4} R_i\Phi(u_i) = 0.693146.$$

The exact value of the integral is $I = \ln 2 \doteq 0.693147$. Ordinarily, the Gauss quadrature formulas would be used only when the data available for $f(x)$ is very accurate and the desired results are needed very accurately.

• **PROBLEM 10-25**

Obtain an approximate formula (correct to eight decimal places) for ln x by integrating

$$\ln x = \int_1^x \frac{dt}{t}$$

using Gauss' three-point formula. Use the formula for $x = 2$.

Solution: By the transformation

$$u = \frac{t - 1}{x - 1}$$

or

$$t = (x - 1)u + 1$$

this integral becomes

$$\ln x = (x - 1) \int_0^1 \frac{du}{xu + (1 - u)}$$

Therefore, by Gauss' three-point formula

$$\ln x \cong \frac{x - 1}{18} \left(\frac{5}{0.11270167x + 0.88729833} + \frac{8}{0.5x + 0.5} \right.$$

$$\left. + \frac{5}{0.88729833x + 0.11270167} \right)$$

This formula may be written

$$\ln x = \frac{z - 2}{18} \left[\frac{16}{z} + \frac{5z}{y(z - y)} \right]$$

where

$$z = x + 1$$

and

$$y = 0.11270167x + 0.88729833$$

For $x = 2$, one gets $z = 3$, $y = 1.11270167$, and therefore

$$\ln 2 = \frac{1}{18}\left[\frac{16}{3} + \frac{15}{(1.11270167)(1.88729833)}\right]$$

$$= 0.69312169$$

This result is in error by only three units in the fifth decimal place.

• **PROBLEM 10-26**

Find the Euler equation for the functional

$$\int_0^1 [xy^3 - (y')^2 + 3xyy']dx.$$

Solution: The Euler equation of this functional is

$$(3xy^2 + 3xy') - \frac{d}{dx}(-2y' + 3xy) = 0,$$

or equivalently,

$$2y'' - 3y + 3xy^2 = 0,$$

from

$$J = \int_a^b F(x,y,y')dx \tag{1}$$

$$\frac{\partial F}{\partial y} - \frac{d}{dx}\frac{\partial F}{\partial y'} = 0 \tag{2}$$

Euler differential equation (2) derived from the functional form (1) is, in general, a nonlinear, second order, ordinary differential equation. Although such equations are, in general, very difficult to solve, still they seem to be more viable analytically than the functionals from which they are derived. Nevertheless, numerically it is so often easier to approximate a solution of the original variational problem than to solve the problem defined by (2), that the common approach is to examine applied problems which are usually stated in terms of (2) by returning to their primitive, variational formulation.

CHAPTER 11

LINEAR SYSTEMS

VECTOR SPACES, MATRICES

● PROBLEM 11-1

Interpret the problem of solving the equations

$$2x_1 + 2x_2 + 3x_3 = 3$$
$$4x_1 + 7x_2 + 7x_3 = 1$$
$$-2x_1 + 4x_2 + 5x_3 = -7$$

as that of determining x_1, x_2 and x_3 such that

$$\begin{bmatrix} 3 \\ 1 \\ -7 \end{bmatrix} = x_1 \begin{bmatrix} 2 \\ 4 \\ -2 \end{bmatrix} + x_2 \begin{bmatrix} 2 \\ 7 \\ 4 \end{bmatrix} + x_3 \begin{bmatrix} 3 \\ 7 \\ 5 \end{bmatrix}. \tag{1}$$

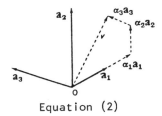

Equation (2)

Solution: One seeks x_1, x_2 and x_3 such that the vector on the left may be expressed as a linear combination of the other three vectors.

If one has any three vectors $\vec{a}_1$, $\vec{a}_2$ and $\vec{a}_3$ in three-dimensional space, then

$$\vec{v} = \alpha_1 \vec{a}_1 + \alpha_2 \vec{a}_2 + \alpha_3 \vec{a}_3 \tag{2}$$

is also a vector in the space. By varying α_1, α_2 and

α_3, one obtains all possible vectors in the space unless $\vec{a}_1$, $\vec{a}_2$ and $\vec{a}_3$ lie in a common plane. This result may be visualized from the parallelogram law for adding vectors. If $\vec{a}_1$, $\vec{a}_2$ and $\vec{a}_3$ lie in a common plane, that is, if they are coplanar, one can only generate vectors in this plane in (2). In the accompanying figure, all vectors lie in the plane of the paper and no matter how one varies the scalars α_1, α_2 and α_3 one cannot move $\vec{v}$ out of that plane.

If $\vec{a}_1$, $\vec{a}_2$ and $\vec{a}_3$ have a common direction, that is, if they are collinear, one can only generate vectors with this direction in (2). Call the set of all vectors generated by (2) the space spanned by $\vec{a}_1$, $\vec{a}_2$ and $\vec{a}_3$. In (1) the three vectors on the right are not coplanar or collinear and any three-dimensional vector can be expressed in terms of them. If the three vectors were coplanar, it would only be possible to obtain a solution if the vector on the left were in their common plane.

In the general case, one seeks $x_1, \ldots, x_n$ such that

$$\vec{b} = x_1\vec{a}_1 + x_2\vec{a}_2 + \ldots + x_n\vec{a}_n$$

where $\vec{a}_1, \vec{a}_2, \ldots, \vec{a}_n$ are the columns of the coefficient matrix. The vector $\vec{b}$ must be in the space spanned by $\vec{a}_1, \ldots, \vec{a}_n$, if there is to be a solution of the equations, that is, if the equations are to be consistent.

If, for the special case $\vec{b} = \vec{0}$, there is a solution other than $\vec{x} = \vec{0}$ it can be said that the columns $\vec{a}_1, \ldots, \vec{a}_n$ are linearly dependent. If $m = n$ it follows that the columns are linearly dependent if, and only if, A is singular.

Suppose that $\vec{a}_1$, $\vec{a}_2$ and $\vec{a}_3$ in three-dimensional space are linearly dependent and

$$\vec{0} = x_1\vec{a}_1 + x_2\vec{a}_2 + x_3\vec{a}_3$$

where at least one of x_1, x_2 and x_3 is non-zero. Assuming that $x_1 \neq 0$ one gets

$$\vec{a}_1 = \left(\frac{-x_2}{x_1}\right)\vec{a}_2 + \left(\frac{-x_3}{x_1}\right)\vec{a}_3.$$

Thus $\vec{a}_1$, $\vec{a}_2$ and $\vec{a}_3$ are coplanar as $\vec{a}_2$ and $\vec{a}_3$ have at least one common plane and $\vec{a}_1$, a linear combination of $\vec{a}_2$ and $\vec{a}_3$, must lie in such a plane. If $x_1 = 0$, the same argument will be used with $\vec{a}_1$ and $\vec{a}_2$ or $\vec{a}_1$ and $\vec{a}_3$ interchanged.

• **PROBLEM 11-2**

Show that the space R^n (comprised of n-tuples of real numbers $x_1, \ldots, x_n$) is a vector space over the field R of real numbers. The operations are addition of n-tuples, i.e., $(x_1, \ldots, x_n) + (y_1, y_2, \ldots, y_n) = (x_1 + y_1, x_2 + y_2, \ldots, x_n + y_n)$, and scalar multiplication, $\alpha(x_1, x_2, \ldots, x_n) = (\alpha x_1, \alpha x_2, \ldots, \alpha x_n)$ where $\alpha \in R$.

Solution: Any set that satisfies the axioms for a vector space over a field is known as a vector space. One must show that R^n satisfies the vector space axioms. The axioms fall into two distinct categories:
A) the axioms of addition for elements of a set
B) the axioms involving multiplication of vectors by elements from the field.
1) Closure under addition
By definition, $(x_1, x_2, \ldots, x_n) + (y_1, y_2, \ldots, y_n)$
$= (x_1 + y_1, x_2 + y_2, \ldots, x_n + y_n)$.
Now, since $x_1, y_1, x_2, y_2, \ldots, x_n, y_n$ are real numbers, the sums of $x_1 + y_1, x_2 + y_2, \ldots, x_n + y_n$ are also real numbers. Therefore, $(x_1 + y_1, x_2 + y_2, \ldots, x_n + y_n)$ is also an n-tuple of real numbers; hence, it belongs to R^n.

2) Addition is commutative

The numbers $x_1, x_2, \ldots, x_n$ are the coordinates of the vector $(x_1, x_2, \ldots, x_n)$, and $y_1, y_2, \ldots, y_n$ are the coordinates of the vector $(y_1, y_2, \ldots, y_n)$.

Show $(x_1, x_2, \ldots, x_n) + (y_1, \ldots, y_n) = (y_1, \ldots, y_n)$

$+ (x_1, \ldots, x_n)$. Now, the coordinates $x_1 + y_1$, $x_2 + y_2$, $\ldots$, $x_n + y_n$ are sums of real numbers. Since real numbers satisfy the commutativity axiom, $x_1 + y_1 = y_1 + x_1$, $x_2 + y_2 = y_2 + x_2, \ldots, x_n + y_n = y_n + x_n$.

Thus, $(x_1, x_2, \ldots, x_n) + (y_1, y_2, \ldots, y_n)$

$= (x_1 + y_1, x_2 + y_2, \ldots, x_n + y_n)$ (by definition)

$= (y_1 + x_1, y_2 + x_2, \ldots, y_n + x_n)$ (by commutativity of real numbers)

$= (y_1, y_2, \ldots, y_n) + (x_1, x_2, \ldots, x_n)$ (by definition)

It has been shown that n-tuples of real numbers satisfy the commutativity axiom for a vector space.

3) Addition is associative: $(a+b) + c = a + (b+c)$.
Let $(x_1, x_2, \ldots, x_n)$, $(y_1, y_2, \ldots, y_n)$ and $(z_1, z_2, \ldots, z_n)$ be three points in R^n.

Now, $((x_1, x_2, \ldots, x_n) + (y_1, y_2, \ldots, y_n))$
$+ (z_1, z_2, \ldots, z_n)$

$= (x_1 + y_1, x_2 + y_2, \ldots, x_n + y_n) + (z_1, z_2, \ldots, z_n)$

$= ((x_1 + y_1) + z_1, (x_2 + y_2) + z_2, \ldots, (x_n + y_n) + z_n)$. (1)

The coordinates $(x_i + y_i) + z_i$ $(i = 1, \ldots, n)$ are real numbers. Since real numbers satisfy the associativity axiom, $(x_i + y_i) + z_i = x_i + (y_i + z_i)$. Hence, (1) may be rewritten as

$(x_1 + (y_1 + z_1), x_2 + (y_2 + z_2), \ldots, x_n + (y_n + z_n))$

$= (x_1, x_2, \ldots, x_n) + (y_1 + z_1, y_2 + z_2, \ldots, y_n + z_n)$

$= (x_1, x_2, \ldots, x_n) + ((y_1, y_2, \ldots, y_n) + (z_1, z_2, \ldots, z_n))$.

4) Existence and uniqueness of a zero element

The set R^n should have a member $(a_1, a_2, \ldots, a_n)$ such that for any point $(x_1, \ldots, x_n)$ in R^n, $(x_1, x_2, \ldots, x_n)$
$+ (a_1, a_2, \ldots, a_n) = (x_1, x_2, \ldots, x_n)$. The point $\underbrace{(0, 0, 0, \ldots, 0)}_{n \text{ zeros}}$, where 0 is the unique zero of the real

number system, satisfies this requirement.

5) Existence and uniqueness of an additive inverse.

Let $(x_1, x_2, \ldots, x_n) \in R^n$. An additive inverse of $(x_1, x_2, \ldots, x_n)$ is an n-tuple $(a_1, a_2, \ldots, a_n)$ such that $(x_1, x_2, \ldots, x_n) + (a_1, a_2, \ldots, a_n) = (0, 0, \ldots, 0)$. Since $x_1, x_2, \ldots, x_n$ belong to the real number system, they have unique additive inverses $(-x_1), (-x_2), \ldots, (-x_n)$.

Consider $(-x_1, -x_2, \ldots, -x_n) \in R^n$.

$(x_1, x_2, \ldots, x_n) + (-x_1, -x_2, \ldots, -x_n)$
$= ((x_1 + (-x_1)), x_2 + (-x_2), \ldots, x_n + (-x_n))$
$= (0, 0, \ldots, 0)$.

Now, turn to the axioms involving scalar multiplication.

6) Closure under scalar multiplication.

By definition, $\alpha(x_1, x_2, \ldots, x_n) = (\alpha x_1, \alpha x_2, \ldots, \alpha x_n)$ where the coordinates αx_i are real numbers. Hence, $(\alpha x_1, \alpha x_2, \ldots, \alpha x_n) \in R^n$.

7) Associativity of scalar multiplication.

Let α, β be elements of R. Show that

$(\alpha \beta)(x_1, x_2, \ldots, x_n) = \alpha(\beta x_1, \beta x_2, \ldots, \beta x_n)$.

Since α, β and $x_1, x_2, \ldots, x_n$ are real numbers,
$(\alpha \beta)(x_1, x_2, \ldots, x_n) = \alpha(\beta(x_1, x_2, \ldots, x_n))$
$= \alpha(\beta x_1, \beta x_2, \ldots, \beta x_n)$.

8) The first distributive law

Show that $\alpha(x+y) = \alpha x + \alpha y$ where x and y are vectors in R^n and $\alpha \in R$.

$\alpha[(x_1, x_2, \ldots, x_n) + (y_1, y_2, \ldots, y_n)]$
$= \alpha[(x_1 + y_1), (x_2 + y_2), \ldots, (x_n + y_n)]$
$= (\alpha(x_1 + y_1), \alpha(x_2 + y_2), \ldots, \alpha(x_n + y_n))$ \hfill (2)
(by definition of scalar multiplication).

Since each coordinate is a product of a real number and the sum of two real numbers, $\alpha(x_i + y_i) = \alpha x_i + \alpha y_i$.

Hence, (2) becomes $[(\alpha x_1 + \alpha y_1), (\alpha x_2 + \alpha y_2), \ldots, (\alpha x_n + \alpha y_n)]$

$= (\alpha x_1, \alpha x_2, \ldots, \alpha x_n) + (\alpha y_1, \alpha y_2, \ldots, \alpha y_n)$

$= \alpha x + \alpha y$.

9) The second distributive law

Show that $(\alpha + \beta)x = \alpha x + \beta x$ where $\alpha, \beta \in R$ and x is a vector in R^n. Since $\alpha + \beta$ is also a scalar,

$$(\alpha + \beta)(x_1, x_2, \ldots, x_n)$$
$$= ((\alpha + \beta)x_1, (\alpha + \beta)x_2, \ldots, (\alpha + \beta)x_n) \quad (3)$$

Since α, β, x_i are all real numbers, then $(\alpha + \beta)x_i = \alpha x_i + \beta x_i$. Therefore, (3) becomes

$((\alpha x_1 + \beta x_1), (\alpha x_2 + \beta x_2), \ldots, (\alpha x_n + \beta x_n))$

$= (\alpha x_1, \alpha x_2, \ldots, \alpha x_n) + (\beta x_1, \beta x_2, \ldots, \beta x_n)$

$= \alpha(x_1, x_2, \ldots, x_n) + \beta(x_1, x_2, \ldots, x_n)$.

10) The existence of a unit element from the field. It is required that there exist a scalar in the field R, call it "1", such that $1(x_1, x_2, \ldots, x_n) = (x_1, x_2, \ldots, x_n)$.

Now the real number 1 satisfies this requirement.

Since a set defined over a field that satisfies (1) - (10) is a vector space, the set R^n of n-tuples is a vector space when equipped with the given operations of addition and scalar multiplication.

• **PROBLEM 11-3**

Show that V, the set of all functions from a set $S \neq \phi$ to the field R, is a vector space under the following operations: if $f(s)$ and $g(s) \in V$, then $(f + g)(s) = f(s) + g(s)$. If c is a scalar from R, then $(cf)(s) = cf(s)$.

Solution: Since the points of V are functions, V is called a function space. Because the field is R, V is the space of real-valued functions defined on S. Also, since addition of real numbers is commutative, $f(s) + g(s) = g(s) + f(s)$.

(Here f(s) and g(s) are real numbers, the values of the functions f and g at the point s ∈ S). Since addition in R is associative, f(s) + [g(s) + h(s)] = [f(s) + g(s)] + h(s) for all s ∈ S. Hence, addition of functions is associative. Next, the unique zero vector is the zero function which assigns to each s ∈ S the value 0 ∈ R. For each f in V, let (-f) be the function given by

(-f)(s) = -f(s).

Then f + (-f) = 0 as required.

Next, verify the scalar axioms. Since multiplication in R is associative,

(cd) f(s) = c (df(s)).

The two distributive laws are: c (f + g)(s) = cf(s) + cg(s) and (c + d) f(s) = cf(s) + df(s). They hold by the properties of the real numbers. Finally, for 1 ∈ R, 1f(s) = f(s).

Thus, V is a vector space. If, instead of R one had considered a field F, one would have had to verify the above axioms using the general properties of a field (any set that satisfies the axioms for a field).

• PROBLEM 11-4

Does the space V of potential functions of the nth degree form a vector space over the field of real numbers? Addition and scalar multiplication are as defined below:

i)   f(x, y) + g(x, y) = (f + g)(x, y) for f, g ∈ V.

ii) c [f(x, y)] = cf(x, y).

Solution: A potential function is any twice differentiable function V(x, y) that satisfies the second-order, partial differential equation

$$\frac{\partial^2 V}{\partial^2 x} + \frac{\partial^2 V}{\partial y^2} = 0 \qquad (1)$$

Examples of functions that satisfy (1) are 1) V = k (a constant); 2) V = x; 3) V = y; 4) V = xy.

Note that if V_1 and V_2 are potential functions,

$$\frac{\partial^2 V_1}{\partial x^2} + \frac{\partial^2 V_1}{\partial y^2} + \frac{\partial^2 V_2}{\partial x^2} + \frac{\partial^2 V_2}{\partial y^2} = 0 + 0 = 0,$$

their sum is again a potential function. (This gives closure under addition.)

If V is a potential function, then so is cV where c is a scalar. To show this, observe that

$$\frac{\partial^2 cV}{\partial x^2} = c\frac{\partial^2 V}{\partial x^2} \text{ and } \frac{\partial^2 cV}{\partial y^2} = c\frac{\partial^2 V}{\partial y^2}.$$

Then, $\frac{\partial^2 cV}{\partial x^2} + \frac{\partial^2 cV}{\partial y^2} = c\left[\frac{\partial^2 V}{\partial x^2} + \frac{\partial^2 V}{\partial x^2}\right] = c(0) = 0.$

Finally, the set of potential functions V is a subset of the vector space of all real valued functions G. Since V is closed under addition and scalar multiplication, it is a subspace of G. Thus, it is a vector space in its own right.

To find the dimension of the nth degree potential function, let

$$V(x, y) = a_0 x^n + a_1 x^{n-1} y + a_2 x^{n-2} y^2 + \ldots + a_{n-2} x^2 y^{n-2}$$

$$+ a_{n-1} xy^{n-1} + a_n y^n.$$

Then, $\frac{\partial^2 V}{\partial x^2} + \frac{\partial^2 V}{\partial y^2} = 0$ is given by

$$a_0 n(n-1) x^{n-2} + a_1(n-1)(n-2) x^{n-3} y + \ldots + 3(2) a_{n-3} xy^{n-3}$$

$$+ 2a_{n-2} y^{n-2} + 2a_2 x^{n-2} + 3(2) a_3 x^{n-3} y + \ldots$$

$$+ (n-1)(n-2) a_{n-1} xy^{n-3} + n(n-1) a_n y^{n-2} = 0.$$

$$[n(n-1) a_0 + 2a_2] x^{n-2} + [(n-1)(n-2) a_1 + 3(2) a_3] x^{n-3} y + \ldots$$

$$+ [(n-1)(n-2) a_{n-1} + 3(2) a_{n-3}] xy^{n-3} + [n(n-1) a_n + 2a_{n-2}] y^{n-2}$$

$$= 0.$$

One sees that, from the coefficient of x^{n-2},

$$a_2 = \frac{-n(n-1)}{2(1)} a_0. \tag{1}$$

Similarly, for $x^{n-3} y$, $a_3 = \frac{-(n-1)(n-2)}{3(2)} a_1. \tag{2}$

The general recurrence relationship is, for $x^{n-k} y^{(k-2)}$,

$$a_k = \frac{-(n-(k+2))(n-(k+1)) a_{k-2}}{k(k-1)}. \tag{3}$$

Now, $\dfrac{\partial^2 V}{\partial x^2} + \dfrac{\partial^2 V}{\partial y^2} = 0$ only if the coefficients of $x^r y^{n-r+2}$ are equal to zero. Hence,

$$n(n-1)a_0 + 2a_2 = (n-1)(n-2)a_1 + 3(2)a_3 = \ldots$$

$$= (n-1)(n-2)a_{n-1} + 3(2)a_{n-3} = n(n-1)a_n + 2a_{n-2} = 0.$$

But from (1) and (2), if n and a_0 are given, then a_2, a_4, $\ldots a_{2k}$ can be calculated using the recurrence relation (3). Similarly, if a_1 is given, then a_1, a_3, $\ldots$, a_{2k-1} can be calculated using (3).

Hence, every potential function of degree n can be expressed as a linear combination of two potential functions of degree n, one for a_0 and one for a_1.

Thus, potential functions of degree n form a space of dimension 2.

● **PROBLEM 11-5**

Show that the set of semi-magic squares of order 3 x 3 form a vector space over the field of real numbers with addition defined as: $a_{ij} + b_{ij} = (a + b)_{ij}$ for i, j = 1, ..., 3.

Solution: First, consider an example of a magic square.

$$\begin{bmatrix} 4 & 9 & 2 \\ 3 & 5 & 7 \\ 8 & 1 & 6 \end{bmatrix} \quad (1)$$

One sees that in (1)

i) every row has the same total, T;

ii) every column has the same total, T;

iii) the two diagonals have the same total, again T.

A square array of numbers that satisfies i) and ii) but not iii) is called a semi-magic square. For example, from (1)

$$\begin{bmatrix} 6 & 2 & 7 \\ 8 & 4 & 3 \\ 1 & 9 & 5 \end{bmatrix}$$

is semi-magic. Let M be the set of 3 x 3 semi-magic squares. Notice that when two semi-magic squares are added, the result is a semi-magic square. For example, suppose m_1 and $m_2 \in M$ have row (and column) sums of T_1 and T_2 respectively; then each row and column in $m_1 + m_2$ will have a sum of $T_1 + T_2$. Now, for m_1, m_2 and $m_3 \in M$

i) $m_1 + m_2 = m_2 + m_1$.

This follows from the commutativity property of the real numbers.

ii) $(m_1 + m_2) + m_3 = m_1 + (m_2 + m_3)$

Again, since the elements of m_1, m_2 and m_3 are real numbers and real numbers obey the associative law, semi-magic squares are associative with respect to addition.

iii) Existence of a zero element

$$\begin{bmatrix} a_{11} & a_{12} & a_{13} \\ a_{21} & a_{22} & a_{23} \\ a_{31} & a_{32} & a_{33} \end{bmatrix} + \begin{bmatrix} 0 & 0 & 0 \\ 0 & 0 & 0 \\ 0 & 0 & 0 \end{bmatrix} = \begin{bmatrix} a_{11} & a_{12} & a_{13} \\ a_{21} & a_{22} & a_{23} \\ a_{31} & a_{32} & a_{33} \end{bmatrix}$$

The 3 x 3 array with zeros everywhere is a semi-magic square.

iv) Existence of an additive inverse.

If one replaces every element in a semi-magic square with its negative, the result is a semi-magic square. Adding one obtains the zero semi-magic square.

Next, let α be a scalar from the field of real numbers.

Then, αm_1 is still a semi-magic square. The two distributive laws hold, and $(\alpha \beta) m_1 = \alpha (\beta m_1)$ from the properties of the real numbers. Finally, $1(m_1) = m_1$, i.e., multiplication of a vector in M by the unit element from R leaves the vector unchanged.

Thus, M is a vector space.

• **PROBLEM 11-6**

Consider the vector space $C[0,1]$ of all continuous functions defined on $[0,1]$. If $f \in C[0,1]$, show that

$$\left(\int_0^1 f^2(x)\,dx\right)^{1/2}$$

defines a norm on all elements of this vector space.

Solution: Since f is continuous, f^2 is continuous. If a function is continuous, it is integrable; thus,

$$\left(\int_0^1 f^2(x)\,dx\right)^{1/2} = \|f\|$$

is well-defined.

Let V be a vector space. A norm on V is a nonnegative number associated with every $v \in V$, denoted $\|v\|$, such that

a) $\quad \|v\| > 0 \quad$ and $\quad \|v\| = 0 \iff v = 0$.

b) $\quad \|kv\| = |k|\,\|v\| \quad$ for k, a scalar.

c) $\quad \|v+w\| \leq \|v\| + \|w\|$, for $w \in V$.

It is necessary to show that

$$\left(\int_0^1 f^2(x)\,dx\right)^{1/2}$$

satisfies conditions a) - c) above.

a) If $f(x) \in C[0,1]$, then $f^2(x) \geq 0$ on $[0,1]$. But this implies that

$$\left(\int_0^1 f^2(x)\,dx\right)^{1/2} \geq 0.$$

If

$$\|f\| = \left(\int_0^1 f^2(x)\,dx\right)^{1/2} = 0,$$

then $f^2(x) = 0$.

This suggests that $f(x) = 0$ on $[0,1]$ since a nonnegative function continuous over the interval $[0,1]$ can have zero integral over $[0,1]$ only if that function is identically zero on $[0,1]$. So,

$$\|f\| = 0$$

if and only if $f \equiv 0$.

b) Let k be a scalar from the field over which $C[0,1]$ is defined. Then,

$$\|kf\| = \left[\int_0^1 [kf(x)]^2 \, dx\right]^{1/2} = |k| \left[\int_0^1 f^2(x) \, dx\right]^{1/2}$$

$$= |k| \, \|f\|.$$

c) First, show that $\|f+g\|$, $f,g \in C[0,1]$ is well-defined. Now,

$$\|f+g\| = \left(\int_0^1 [f(x) + g(x)]^2 \, dx\right)^{1/2}$$

exists since sums and squares of continuous functions in $C[0,1]$ are also in $C[0,1]$ and, therefore, integrable over $[0,1]$. Note also that

$$fg = \frac{1}{2}[(f+g)^2 - f^2 - g^2]$$

is also in $C[0,1]$ and integrable over $[0,1]$. It is now possible to show

$$\|f+g\| \leq \|f\| + \|g\|.$$

$$\|f+g\| \leq \|f\| + \|g\|$$

and $\quad \|f+g\|^2 \leq (\|f\| + \|g\|)^2.$ \hfill (1)

But,

$$\|f+g\|^2 = \int_0^1 [f(x) + g(x)]^2 \, dx$$

$$= \int_0^1 f^2(x)\,dx + 2\int_0^1 f(x)g(x)\,dx + \int_0^1 g^2(x)\,dx \qquad (2)$$

and

$$(\|f\| + \|g\|)^2 = \|f\|^2 + 2\|f\|\,\|g\| + \|g\|^2$$

$$= \int_0^1 f^2(x)\,dx + 2\left(\int_0^1 f^2(x)\,dx\right)^{\frac{1}{2}}\left(\int_0^1 g^2(x)\,dx\right)^{\frac{1}{2}}$$

$$+ \int_0^1 g^2(x)\,dx. \qquad (3)$$

Comparing (2) and (3), one sees that (1) can hold if and only if

$$\int_0^1 f(x)g(x)\,dx \leq \left(\int_0^1 f^2(x)\,dx\right)^{\frac{1}{2}}\left(\int_0^1 g^2(x)\,dx\right)^{\frac{1}{2}} \qquad (4)$$

By the properties of the absolute value functions,

$$\int_0^1 f(x)g(x)\,dx \leq \int_0^1 |f(x)| \cdot |g(x)|\,dx.$$

It can be proved that

$$\int_0^1 |f(x)| \cdot |g(x)|\,dx \leq \|f\|\,\|g\|.$$

Let λ be a real variable and form

$$\int_0^1 [|f(x)| + \lambda|g(x)|]^2\,dx \geq 0.$$

But,

$$\int_0^1 [|f(x)| + \lambda|g(x)|]^2\,dx =$$

$$\lambda^2 \int_0^1 g^2(x)\,dx + 2\lambda \int_0^1 |f(x)||g(x)|\,dx + \int_0^1 f^2(x)\,dx$$

is a nonnegative quadratic polynomial in λ. Hence, it has no real roots and its discriminant is nonpositive; i.e.,

$$4[\int_0^1 |f(x)||g(x)|\,dx]^2 - 4\int_0^1 g^2(x)\,dx \int_0^1 f^2(x)\,dx \leq 0$$

which implies

$$\int_0^1 |f(x)||g(x)|\,dx \leq \left(\int_0^1 f^2(x)\,dx\right)^{\frac{1}{2}} \left(\int_0^1 g^2(x)\,dx\right)^{\frac{1}{2}},$$

as was to be shown.

Thus,

$$\left(\int_0^1 f^2(x)\,dx\right)^{\frac{1}{2}}$$

defines a norm on $C[0,1]$. It is known as the Euclidean norm.

● **PROBLEM 11-7**

Define the polynomial of degree 2 which passes through the three points $(-H, F(-H))$, $(0, F(0))$, $(H, F(H))$. What is the solution of the linear system thus obtained? Use Cramer's rule to solve.

Solution: It is required that the polynomial form

$$P(X; A, B, C) = A * X ** 2 + B * X + C$$

satisfy the conditions

$$P(-H; A, B, C) = F(-H)$$
$$P(0; A, B, C) = F(0)$$
$$P(H; A, B, C) = F(H),$$

to obtain the linear system

$$(-H)**2 * A + (-H) * B + C = F(-H)$$

$$(0)**2 * A + (0) * B + C = F(0)$$

$$(H)**2 * A + (H) * B + C = F(H).$$

The corresponding matrix equation is

$$\begin{bmatrix} H**2 & -H & 1 \\ 0 & 0 & 1 \\ H**2 & H & 1 \end{bmatrix} * \begin{bmatrix} A \\ B \\ C \end{bmatrix} = \begin{bmatrix} F(-H) \\ F(0) \\ F(H) \end{bmatrix}.$$

The solution of the linear system is

$$(A,B,C) = ((F(H) - 2 * F(0) + F(-H))/(2 * H**2),$$

$$(F(H) - F(-H))/(2 * H), F(0)).$$

The polynomial P (X;A,B,C), with the A, B, and C of the solution (A,B,C) substituted, agrees with F(X) at the specified points X = -H, 0 and H.

• **PROBLEM 11-8**

Define the coefficients A, B, and C of the linear form (here, a functional: real-valued function of the function F)

$$T(F) = \int F(X), \text{ over } X \in (-H,H), \tag{1}$$

$$- (A * F(-H) + B * F(0) + C * F(H))$$

$$T(F) = \int_{-H}^{H} F(X)\,dX - (AF(-H) + BF(0) + CF(H))$$

subject to the conditions

$$T(F) = 0, \text{ for } F(X) \equiv 1.$$

$$T(F) = 0, \text{ for } F(X) = X.$$

$$T(F) = 0, \text{ for } F(X) = X**2.$$

$$T(F) = 0, \text{ for } F(X) = X^2.$$

Solution: With $F(X) \equiv 1$, the condition $T(F) = 0$ becomes

$$\int 1, \text{ over } X \in (-H, H), = A * 1 + B * 1 + C * 1$$

or $A + B + C = 2 * H$.

The condition $T(F) = 0$, for $F(X) = X$ becomes

$$\int X, \text{ over } X \in (-H, H), = A * (-H) + B * (0) + C * (H),$$

or

$$-H * A + H * C = H ** 2/2 - (-H) ** 2/2 = 0.$$

Also, the condition $T(F) = 0$, for $F(X) = X ** 2$ becomes

$$T(F) = 0 \text{ for } F(X) = X^2$$

$$\int X ** 2, \text{ over } X \in (-H, H), = A * (-H) ** 2 + B * 0$$

$$+ C * H ** 2,$$

or $H ** 2 * (A + C) = H ** 3/3 - (-H) ** 3/3$

$$= 2/3 * H ** 3.$$

Then, the coefficients A, B, and C are defined by the linear system

$$A + B + C = 2 * H$$
$$-A \quad\quad + C = 0 \tag{2}$$
$$A \quad\quad + C = 2/3 * H.$$

The solution of (2) is $(A, B, C) = (H/3, 4 * H/3, H/3)$. The form (1) with the solution (A, B, C) substituted defines Simpson's Rule, where $T(F)$ denotes the truncation error of the estimate

$$\int F(X), \text{ over } X \in (-H, H), \doteq H/3 * (F(-H) + 4 * F(0)$$

$$+ F(H)).$$

• PROBLEM 11-9

Use Crout's method to solve the following set of simultaneous equations:

$$2x_1 + 6x_2 + 2x_3 = -1$$

$$x_1 + x_2 + 4x_3 = 8$$

$$3x_1 - x_2 - 6x_3 = 5$$

Solution: Crout's method of solving simultaneous equations is a condensation of the method of Gauss.

A set of simultaneous linear equations with n equations and n unknowns can be written as

$$\sum_{j=1}^{n} a_{ij} x_j = c_i \qquad (i = 1, 2, \ldots, n)$$

or as AX = C in matrix notation. Crout's method starts by forming an (n) × (n+1) augmented matrix A_1 which is identical to the original coefficient matrix A, except that A_1 has an (n+1)th column consisting of the c_i's.

Next, an auxiliary (n) × (n+1) matrix B is formed according to the following rules:

1. The b_{ij} elements are found in the following order: elements of the first column are found first, then the remaining elements of the first row; next to be found are the remaining elements of the second column, etc.

2. The first column of B is identical to the first column of A_1. The other elements of the first row of B are obtained by dividing the corresponding elements of A_1 by a_{11}.

3. Each element either on or below the principal diagonal of B equals the corresponding element of A_1 minus the sum of the products of elements in its row and corresponding elements in its column. The elements in these products are previously computed elements of B.

4. The elements to the right of the principal diagonal of B are found as in rule 3 above except there is a final division by b_{ii}, the corresponding diagonal element of B.

Using the matrix B, the unknown x_j's are determined by the following rules:

1. The unknowns are found in reverse order:
   $x_n, x_{n-1}, \ldots, x_2, x_1$.

2. The unknown x_n is equal to the corresponding (last) element of the last column of B.

3. Each other unknown x_j equals the corresponding element of the last column of B minus the sum of the products of elements in its row in B and corresponding x_j's already computed. These computations require only elements to the right of B's principal diagonal.

The given problem's solution will now be indicated using the rules above. Below the solution, sample calculations will be indicated. The solution may, of course, be checked by substituting back into the original equations.

$$A_1 = \begin{bmatrix} 2 & 6 & 2 & -1 \\ 1 & 1 & 4 & 8 \\ 3 & -1 & -6 & 5 \end{bmatrix}$$

$$B = \begin{bmatrix} 2 & 3 & 1 & -0.5 \\ 1 & -2 & -1.5 & -4.25 \\ 3 & -10 & -24 & 1.50 \end{bmatrix}$$

$$X = \begin{bmatrix} 4 & -2 & 1.5 \end{bmatrix}$$

Sample Calculations:

$(j > 1) \quad b_{1j} = a_{1j}/a_{11} = a_{1j}/2$

$b_{32} = -1 - (3)(3) = -10$

$b_{23} = [4 - (1)(1)]/(-2) = -1.5$

$b_{34} = [5 - (3)(-0.5) - (-10)(-4.25)]/(-24) = 1.50$

$x_2 = -4.25 - (-1.5)(1.5) = -2$

• **PROBLEM 11-10**

Use Crout's method to solve the system with complex coefficients, composed of 4 variables and 4 equations:

x_1	x_2	x_3	x_4	k
8.342	−9.012	3.345	−4.518	65.65
7.130	1.132	−1.248	−3.362	−1.810
9.123	4.567	−2.222	8.041	−87.30
1.071	5.432	7.444	−2.111	6.500
3.789	−2.421	7.342	3.467	−34.25
−1.242	7.321	−2.181	−7.182	−35.45
−4.142	8.042	3.732	−2.111	−42.63
−3.181	−2.131	−7.801	−2.932	14.50

where the second number in each group refers to the imaginary part. Thus

$$a_{11} = 8.342 + 7.130i.$$

Solution: If the coefficients of the system are complex numbers, the solutions will be complex numbers. The method of solution is the same, except that some scheme must be devised for recording the real and imaginary part at each step.

The division by a complex number may be carried out by a multiplication.

$$\frac{1}{A} = \frac{a}{a^2 + b^2} - \frac{b}{a^2 + b^2} i$$

where

$$A = a + bi.$$

The method may therefore be carried out by writing the complex number in two-element form in the matrix and augmenting the derived matrix. Consider the system

$$\begin{pmatrix} x & y & z & k \\ a_{11} + b_{11}i & a_{12} + b_{12}i & a_{13} + b_{13}i & a_{14} + b_{14}i \\ a_{21} + b_{21}i & a_{22} + b_{22}i & a_{23} + b_{23}i & a_{24} + b_{24}i \\ a_{31} + b_{31}i & a_{32} + b_{32}i & a_{33} + b_{33}i & a_{34} + b_{34}i \end{pmatrix}$$

When numbers are involved, this system is conveniently written on two lines for each element instead of two columns.

The derived matrix is

$$\begin{pmatrix} A_{11} + B_{11}i & A_{12} + B_{12}i & A_{13} + B_{13}i & A_{14} + B_{14}i \\ A_{21} + B_{21}i & A_{22} + B_{22}i & A_{23} + B_{23}i & A_{24} + B_{24}i \\ A_{31} + B_{31}i & A_{32} + B_{32}i & A_{33} + B_{33}i & A_{34} + B_{34}i \end{pmatrix}$$

where the elements are obtained in the regular manner. However, to accomplish this step it is necessary to record additional data.

Supplement the derived matrix both on the left and right. In the left supplementary matrix form the sum of the squares of the real numbers and the reciprocals of each element of the principal diagonal. Thus

$$\begin{Vmatrix} A_{ii}^2 + B_{ii}^2 & \dfrac{1}{A_{ii} + B_{ii}i} \\ \hline A_{11}^2 + B_{11}^2 & \dfrac{A_{11}}{LS_{11}} - \dfrac{B_{11}}{LS_{11}}i \\ A_{22}^2 + B_{22}^2 & \dfrac{A_{22}}{LS_{21}} - \dfrac{B_{22}}{LS_{21}}i \\ A_{33}^2 + B_{33}^2 & \dfrac{A_{33}}{LS_{31}} - \dfrac{B_{33}}{LS_{31}}i \end{Vmatrix}$$

where LS_{ij} are the elements of the left supplementary matrix.

In the right supplementary matrix are recorded the sums of the products which form the numerators of the elements to the right of the principal diagonal before each is divided by the corresponding diagonal elements. Thus in the problem

$$\begin{vmatrix} RS_{23} & RS_{24} \\ & RS_{34} \end{vmatrix}$$

and

$$RS_{23} = (a_{23} + b_{23}i) - (A_{21} + B_{21}i)(A_{13} + B_{13}i)$$

$$= (a_{23} - A_{21}A_{13} + B_{21}B_{13}) + (b_{23} - B_{21}A_{13} - A_{21}B_{13})i,$$

$RS_{24} = (a_{24} + b_{24}i) - (A_{21} + B_{21}i)(A_{14} + B_{14}i)$

$RS_{34} = (a_{34} + b_{34}i) - (A_{31} + B_{31}i)(A_{14} + B_{14}i)$
$\qquad + (A_{32} + B_{32}i)(A_{24} + B_{34}i)$

$\qquad = a_{34} - A_{31}A_{14} + B_{31}B_{14} - A_{32}A_{24} + B_{32}B_{34}$
$\qquad + (b_{34} - B_{31}A_{14} - A_{31}B_{14} - B_{32}A_{24} - A_{32}B_{34})i.$

The derived matrix and solution are as follows:

A^2+B^2	$\dfrac{1}{A+Bi}$	x_1	x_2	x_3	x_4	k	Check		RS	
				Derived Matrix						
120.43	.069268 −.059204	8.342 7.130	−.55722 .61196	.15781 −.28448	−.5120 .03460	4.4403 −4.0121	4.5288 −3.6500			
106.41	.096851 −.004190	9.123 1.071	10.306 .44587	−.34279 .97257	1.2269 −.23533	−12.6339 4.2675	−10.7497 5.0047	−3.9664 9.8703	12.749 −1.8783	−132.106 38.347	−113.016 46.7846
121.90	.089603 −.013213	3.789 −1.242	−1.0697 4.3102	10.9227 1.6107	.32911 −1.28346	−3.10756 4.51272	−1.77845 3.22921		5.6621 −13.4889	−41.2120 44.2863	−24.6270 32.4075
115.84	.050184 −.078192	−4.142 −3.181	3.78735 −1.3688	5.2576 −12.6300	5.81342 9.05788	7.1094 −3.4450	−6.1094 −3.4450			−10.12549 −84.42388	−4.3122 −75.3656
		−.5507 −.4159	−5.2309 2.0754	3.6537 −3.4781	−7.1094 −3.4450	SOLUTION					

The second number in each group is the imaginary part. Thus $x_1 = -.5507 - .4159i$.

• **PROBLEM 11-11**

Using Crout Reduction, solve

$$A * X = \begin{bmatrix} 1 & 1 & 1 \\ 0 & 0 & 1 \\ 1 & -1 & 1 \end{bmatrix} * \begin{bmatrix} X(3) \\ X(2) \\ X(1) \end{bmatrix} = \begin{bmatrix} -1 \\ 1 \\ 3 \end{bmatrix} = B.$$

Solution: The Linear system that is required to be solved, is, for each $I \in \{1,N\}$,

$$\Sigma A(I,J) * X(J), J \in \{1,N\}, = B(I), \quad [A * X = B], \quad (1)$$

$$(A|B): \begin{matrix} A(1,1) & A(1,2) & \ldots & A(1,N) & B(1) \\ A(2,1) & A(2,2) & \ldots & A(2,N) & B(2) \\ \vdots & \vdots & & \vdots & \vdots \\ A(N,1) & A(N,2) & \ldots & A(N,N) & B(N) \end{matrix}$$

Compute elements of a new matrix "B," with the same dimension as (A|B), and store "B" in the storage cells originally occupied by (A|B).

$$\text{"B"}: \begin{matrix} A1(1,1) & A1(1,2) & \ldots & A1(1,N) & B1(1) \\ C(2,1) & A1(2,2) & \ldots & A2(2,N) & B2(2) \\ C(3,1) & C(3,2) & \ldots & A3(3,N) & B3(3) \\ \vdots & \vdots & & \vdots & \vdots \\ C(N,1) & C(N,2) & \ldots & ANM1(N,N) & BN(N) \end{matrix}$$

The elements of "B" are computed one block at a time in the sequence indicated by the circled numbers in the blocks of (2):

"B":
$$\begin{array}{|c|c|c|c|c|c|c|} \hline A(1,1) & A1(1,2) & A1(1,3) & A1(1,4) & ② \; A1(1,N) & B1(1) \\ \hline C(2,1) & A1(2,2) & A2(2,3) & A2(2,4) & ④ \; A2(2,N) & B2(2) \\ \hline C(3,1) & C(3,2) & A2(3,3) & A3(3,4) & ⑥ \; A3(3,N) & B3(3) \\ \vdots \; ① & \vdots \; ③ & \vdots \; ⑤ & & & \vdots \\ C(N,1) & C(N,2) & C(N,3) & \cdots & & BN(N) \\ \end{array}$$
(2)

Block (1): $A(1,1) = A(1,1)$, and $C(K,1) = A(K,1)$,

for $K \in \{2,N\}$

Block (2): $A1(1,K) = A(1,K)/A(1,1)$,

for $K \in \{2,N\}$,

and $B1(1) = B(1)/A(1,1)$.

Block (3): $A1(2,2) = A(2,2) - C(2,1) * A1(1,2)$

$C(K,2) = A(K,2) - C(K,1) * A1(1,2)$,

for $K \in \{3,N\}$.

Block (4): $A2(2,K) = (A(2,K) - C(2,1) * A1(1,K))/A1(2,2)$,

for $K \in \{3,N\}$,

$B2(2) = (B(2) - C(2,1) * B1(1))/A1(2,2)$.

Block (5): $A2(3,3) = A(3,3) - C(31) * A1(1,3)$

$ - C(3,3) * A2(2,3)$

$C(K,3) = A(K,3) - C(K,1) * A1(1,3)$

$ - C(K,2) * A2(2,3)$,

for $K \in \{4,N\}$.

Block (6): $A3(3,K) = (A(3,K) - C(3,1) * A1(1,K)$

$ - C(3,2) * A2(2,K))/A2(3,3)$,

for $K \in \{4,N\}$,

$B3(3) = (B(3) - C(3,1) * B1(1)$

$ - C(3,2) * B2(2))/A2(3,3)$.

To complete the Jth column, Block $(2 * J - 1)$:
Compute

$AJM1(J,J) = A(J,J) - C(J,1) * A1(1,J) - C(J,2) * A2(2,J)$

$ - \ldots - C(J,J-1) * AJM1(J-1,J)$,

and for each $K \in \{J + 1, N\}$,

$$C(K,J) = A(K,J) - C(K,1) * A1(1,J) - C(K,2) * A2(2,J)$$

$$- \ldots - C(K,J-1) * AJM1(J-1,J).$$

To complete the Ith row, Block (2 * I): Compute for each $K \in \{I + 1, N\}$,

$$AI(I,K) = 1/AIM1(I,I) * (A(I,K) - C(I,1) * A1(1,K)$$

$$- C(I,2) * A2(2,K) - \ldots$$

$$- C(I,I-1) * AIM1(I-1,K)),$$

and

$$BI(I) = 1/AIM1(I,I) * (B(I) - C(I,1) * B1(1) -$$

$$C(I,2) * B2(2)$$

$$- \ldots - C(I,I-1) * BIM1(I-1)).$$

When the calculation of "B" is complete, the elements $C(I,J)$ of "B" below the principal diagonal are no longer needed, and these storage cells are available for use as workspace when executing the back-substitution to solve

$$X(1) + A1(1,2) * X(2) + A1(1,3) * X(3) + \ldots$$

$$+ A1(1,N) * X(N) = B1(1)$$

$$X(2) + A2(2,3) * X(3) + \ldots + A2(2,N) * X(N) = B2(2)$$

$$\vdots \qquad\qquad (3)$$

$$X(N-1) + ANM1(N-1,N) * X(N) = BNM1(N-1)$$

$$X(N) = BN(N),$$

and thus obtain a solution of (1).

Thus,

$$(A|B): \quad \begin{matrix} 1 & 1 & 1 & -1 \\ 0 & 0 & 1 & 1 \\ 1 & -1 & 1 & 3 \end{matrix}$$

Compute Blocks (1), (2), and (3).

$$\text{"B"}: \quad \begin{matrix} 1 & 1 & 1 & -1 \\ 0 & 0 & & \\ 1 & -2 & & \end{matrix}$$

Since the number in A1(2,2) position is zero, one cannot execute the operations indicated in Block (4). Thus, a row or column interchange is in order. For a row interchange rewrite (A|B) and the incomplete "B" with their second and third rows interchanged.

$$(A|B): \quad \begin{matrix} 1 & 1 & 1 & -1 \\ 1 & -1 & 1 & 3 \\ 0 & 0 & 1 & 1 \end{matrix}$$

M = 1. (M = 1 indicates that a row or a column interchange has been executed.)

$$\text{"B"}: \quad \begin{matrix} 1 & 1 & 1 & -1 \\ 1 & -2 & & \\ 0 & 0 & & \end{matrix} \quad M = 1.$$

Continue with the calculation of "B." Note that no recalculation of elements of "B" is required after a row interchange.

$$\text{"B"}: \quad \begin{matrix} 1 & 1 & 1 & -1 \\ 1 & -2 & 0/(-2) & 4/(-2) \\ 0 & 0 & 1 & 1/1 \end{matrix} \quad M = 1.$$

Then (3) becomes

$$X(3) + (1) * X(2) + (1) * X(1) = (-1)$$

$$X(2) + (0) * X(1) = (-2)$$

393

$$X(1) = (1),$$

where

$X(1) = 1$, $X(2) = -2$,

$X(3) = 0$.

and

$$DET(A) = (-1) ** M * (1) * (-2) * (1) = 2,$$

where A is the matrix of coefficients before the row interchange, and DET is the determinant.

● **PROBLEM 11-12**

Upper-triangulate the system below:

$$x_2 + 2x_3 = 1$$
$$x_1 + x_2 + x_3 = 0$$
$$2x_1 - x_2 = 5$$

Solution:

(a) Interchange the first and third equations.

$$2x_1 - x_2 = 5$$
$$x_1 + x_2 + x_3 = 0$$
$$x_2 + 2x_3 = 1 \ .$$

(b) Multiply the first equation by $\frac{1}{2}$.

$$x_1 - \frac{1}{2}x_2 = \frac{5}{2}$$
$$x_1 + x_2 + x_3 = 0$$
$$x_2 + 2x_3 = 1$$

(c) Add -1 times the first equation to the second equation.

$$x_1 - \frac{1}{2}x_2 = \frac{5}{2}$$

$$\tfrac{3}{2}x_2 + x_3 = -\tfrac{5}{2}$$

$$x_2 + 2x_3 = 1$$

(d) Multiply the second equation by $\tfrac{2}{3}$.

$$x_1 - \tfrac{1}{2}x_2 = \tfrac{5}{2}$$

$$x_2 + \tfrac{2}{3}x_3 = -\tfrac{5}{3}$$

$$x_2 + 2x_3 = 1$$

(e) Add -1 times the second equation to the third equation.

$$x_1 - \tfrac{1}{2}x_2 = \tfrac{5}{2}$$

$$x_2 + \tfrac{2}{3}x_3 = -\tfrac{5}{3}$$

$$\tfrac{4}{3}x_3 = \tfrac{8}{3}$$

(f) Multiply the third equation by $\tfrac{3}{4}$.

$$x_1 - \tfrac{1}{2}x_2 = \tfrac{5}{2}$$

$$x_2 + \tfrac{2}{3}x_3 = -\tfrac{5}{3}$$

$$x_3 = 2$$

Note that in this procedure one has done more than just upper-triangulate the original system; one has made the diagonal elements $a_{ii} = 1$.

While it is always possible to upper-triangulate a system of linear equations, it is not always possible to upper-triangulate a system and have the diagonal elements $a_{ii} = 1$. This is because a system of equations having no solutions or infinitely many solutions must necessarily upper-triangulate into a form having one or more of the diagonal elements equal to zero. However, a system of equations having a unique solution must necessarily upper-triangulate into a system having no zero elements on the diagonal, and hence the diagonal elements can be made to be 1 by multiplying an equation by a nonzero constant.

• **PROBLEM 11-13**

Let $f(x)$ be a given function defined on the distinct points $x_0, x_1, \ldots, x_n$. It is desired to find a polynomial

$$P(x) = b_0 + b_1 x + b_2 x^2 + \ldots + b_n x^n \tag{1}$$

such that

$$P(x_i) = f(x_i) \qquad i = 0, 1, \ldots, n \tag{2}$$

Set up the problem as a system of simultaneous linear equations.

Solution: Here the unknown quantities are $b_0, b_1, \ldots, b_n$. By using Eq. (1), Eq. (2) becomes

$$b_0 + b_1 x_i + b_2 x_i^2 + \ldots + b_n x_i^n = f(x_i)$$

$$i = 0, \ldots, n$$

This system can be rewritten in the form

$$a_{11} x_1 + a_{12} x_2 + \ldots + a_{1n} x_n = b_1$$

$$a_{21} x_1 + a_{22} x_2 + \ldots + a_{2n} x_n = b_2$$

$$\cdots\cdots\cdots\cdots\cdots\cdots\cdots$$

$$a_{n1} x_1 + a_{n2} x_2 + \ldots + a_{nn} x_n = b_n$$

The a_{ij}'s and b_i's are known real numbers, and the x_i's are unknown real numbers to be determined so that each of the n equations is satisfied by these x_i's.

Hence

$$1 b_0 + x_0 b_1 + x_0^2 b_2 + \ldots + x_0^n b_n = f(x_0)$$

$$1 b_0 + x_1 b_1 + x_1^2 b_2 + \ldots + x_1^n b_n = f(x_1)$$

$$\cdots\cdots\cdots\cdots\cdots\cdots\cdots\cdots\cdots\cdots$$

$$1 b_0 + x_n b_1 + x_n^2 b_2 + \ldots + x_n^n b_n = f(x_n)$$

This is a system of $n + 1$ simultaneous linear equations in $n + 1$ unknowns.

● **PROBLEM 11-14**

Solve equation A × X = B, where

$$X = \begin{bmatrix} x_1 \\ x_2 \end{bmatrix}, \quad A = \begin{bmatrix} a_{1,1} & a_{1,2} \\ a_{2,1} & a_{2,2} \end{bmatrix}, \quad B = \begin{bmatrix} b_1 \\ b_2 \end{bmatrix},$$

for x_1 and x_2.

Solution: First, calculate the inverse of A:

$$\begin{bmatrix} a_{1,1} & a_{1,2} & 1 & 0 \\ a_{2,1} & a_{2,2} & 0 & 1 \end{bmatrix}$$

Now to produce unity in the 1,1 position, divide the first row by $a_{1,1}$.

$$\begin{bmatrix} 1 & \dfrac{a_{1,2}}{a_{1,1}} & \dfrac{1}{a_{1,1}} & 0 \\ a_{2,1} & a_{2,2} & 0 & 1 \end{bmatrix}$$

Next, generate a zero in the 2,1 position by subtracting from the second row $a_{2,1}$ times the first row.

$$\begin{bmatrix} 1 & \dfrac{a_{1,2}}{a_{1,1}} & \dfrac{1}{a_{1,1}} & 0 \\ 0 & \dfrac{a_{1,1}a_{2,2} - a_{1,2}a_{2,1}}{a_{1,1}} & -\dfrac{a_{2,1}}{a_{1,1}} & 1 \end{bmatrix}$$

Generate a 1 in the 2,2 position by making the appropriate division in the second row.

$$\begin{bmatrix} 1 & \dfrac{a_{1,2}}{a_{1,1}} & \dfrac{1}{a_{1,1}} & 0 \\ 0 & 1 & -\dfrac{a_{2,1}}{a_{1,1}a_{2,2} - a_{1,2}a_{2,1}} & \dfrac{a_{1,1}}{a_{1,1}a_{2,2} - a_{1,2}a_{2,1}} \end{bmatrix}$$

Generate a zero in the 1,2 position by both multiplying row 2 by $a_{1,2}$, dividing by $a_{1,1}$, and then subtracting from the first row.

$$\begin{bmatrix} 1 & 0 & \dfrac{a_{2,2}}{a_{1,1}a_{2,2} - a_{1,2}a_{2,1}} & -\dfrac{a_{1,2}}{a_{1,1}a_{2,2} - a_{1,2}a_{2,1}} \\ \\ 0 & 1 & -\dfrac{a_{2,1}}{a_{1,1}a_{2,2} - a_{1,2}a_{2,1}} & \dfrac{a_{1,1}}{a_{1,1}a_{2,2} - a_{1,2}a_{2,1}} \end{bmatrix}$$

Therefore,

$$A^{-1} = \begin{bmatrix} \dfrac{a_{2,2}}{a_{1,1}a_{2,2} - a_{1,2}a_{2,1}} & -\dfrac{a_{1,2}}{a_{1,1}a_{2,2} - a_{1,2}a_{2,1}} \\ \\ -\dfrac{a_{2,1}}{a_{1,1}a_{2,2} - a_{1,2}a_{2,1}} & \dfrac{a_{1,1}}{a_{1,1}a_{2,2} - a_{1,2}a_{2,1}} \end{bmatrix}$$

The solution immediately follows by using the inverse

$$X = \begin{bmatrix} x_1 \\ x_2 \end{bmatrix}$$

$$= \begin{bmatrix} \dfrac{a_{2,2}}{a_{1,1}a_{2,2} - a_{1,2}a_{2,1}} & -\dfrac{a_{1,2}}{a_{1,1}a_{2,2} - a_{1,2}a_{2,1}} \\ \\ -\dfrac{a_{2,1}}{a_{1,1}a_{2,2} - a_{1,2}a_{2,1}} & \dfrac{a_{1,1}}{a_{1,1}a_{2,2} - a_{1,2}a_{2,1}} \end{bmatrix} \times \begin{bmatrix} b_1 \\ b_2 \end{bmatrix} \quad (1)$$

Equation (1) may also be written in the form of Eqs. (2) and (3):

$$x_1 = \dfrac{a_{2,2}b_1 - a_{1,2}b_2}{D}, \qquad x_2 = \dfrac{a_{1,1}b_2 - a_{2,1}b_1}{D} \quad (2)$$

$$D = a_{1,1}a_{2,2} - a_{1,2}a_{2,1} \qquad (3)$$

Check these solutions against the ones obtained by the method of determinants

$$x_1 = \frac{\begin{vmatrix} b_1 & a_{1,2} \\ b_2 & a_{2,2} \end{vmatrix}}{D} = \frac{a_{2,2}b_1 - a_{1,2}b_2}{D} \qquad (4)$$

$$x_2 = \frac{\begin{vmatrix} a_{1,1} & b_1 \\ a_{2,1} & b_2 \end{vmatrix}}{D} = \frac{a_{1,1}b_2 - a_{2,1}b_1}{D} \qquad (5)$$

One sees that Eq. (2), by matrix method, agrees with Eqs. (4) and (5).

MATRIX NORMS AND APPLICATIONS

● **PROBLEM 11-15**

Find the norm of the three-dimensional vector

$$u = (-3, 2, 1)$$

and the distance between the points $(-3, 2, 1)$ and $(4, -3, 1)$.

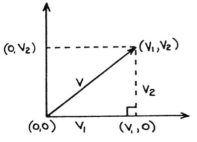

Fig. 1

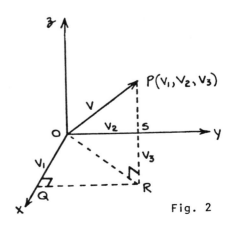

Fig. 2

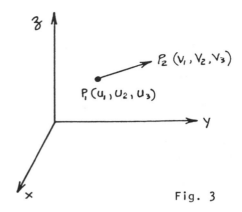

Fig. 3

Solution: The length of a vector is called its norm and is denoted by $\|v\|$. If one lets v be a vector from the origin to the point (v_1, v_2) in R^2, one sees that the norm of v is

$$\sqrt{v_1^2 + v_2^2} :$$

This follows from the theorem of Pythagoras since

$$\|v^2\| = v_1^2 + v_2^2,$$

and hence,

$$\|v\| = \sqrt{v_1^2 + v_2^2}.$$

Next, consider a vector from the origin $(0,0,0)$ to the point v_1, v_2, v_3 in R^3.

Here, two applications of the Pythagorean theorem are necessary to find $\|v\|$. From the figure, ORP is a right triangle. Hence,

$$\|v\|^2 = OP^2 = OR^2 + RP^2. \tag{1}$$

But now, the vector OR in the x y plane is itself the hypotenuse of a right triangle whose other sides are OQ and OS. Hence,

$$OR^2 = OQ^2 + OS^2 \tag{2}$$

Substituting (2) into (1),

$$OP^2 = OQ^2 + OS^2 + RP^2$$

$$\|v\|^2 = v_1^2 + v_2^2 + v_3^2$$

$$\|v\| = \sqrt{v_1^2 + v_2^2 + v_3^2}. \qquad (3)$$

The given vector is $u = (-3,2,1)$. Using formula (3),

$$\|u\| = \sqrt{(-3)^2 + (2)^2 + (1)^2} = \sqrt{14}.$$

To find the distance between two points P_1, P_2 in R^3, proceed as follows:

The distance d is the norm of the vector P_1P_2. But

$$P_1P_2 = (v_1 - u_1, v_2 - u_2, v_3 - u_3)$$

and, using (3),

$$\|P_1P_2\| = \sqrt{(v_1 - u_1)^2 + (v_2 - u_2)^2 + (v_3 - u_3)^2} \qquad (4)$$

The given points are

$$u = (-3,2,1)$$

and

$$v = (4,-3,1).$$

Hence,

$$\|v - u\| = \sqrt{(4 - (-3))^2 + ((-3) - (2))^2 + (1-1)^2}$$

$$= \sqrt{74}.$$

Note the following points:

i) in n-dimensional space, (n-1) applications of the Pythagorean theorem will yield the norm of a vector v as

$$\|v\| = \sqrt{v_1^2 + v_2^2 + \ldots + v_n^2}$$

where

$$v = (v_1, v_2, \ldots, v_n).$$

ii) $\|v-u\| = \|u-v\|$,

that is, the distance between two points is a symmetric function of its arguments.

● **PROBLEM 11-16**

Find the row and column norms of the matrix

$$A = \begin{bmatrix} 1 & -1 & 3 \\ -5 & 2 & 1 \\ 1 & 3 & -1 \end{bmatrix}$$

Solution: To find the row norm, examine the sums of the absolute values of the row elements of A. These are 5, 8, 5 respectively. The maximum of these sums is 8; so

$$\|A\|_r = 8.$$

The sums of the absolute values of the column elements are 7, 6, 5, respectively. Thus

$$\|A\|_c = 7.$$

● **PROBLEM 11-17**

Find the norms of

$$x = \begin{bmatrix} 1 \\ -2 \\ 3 \end{bmatrix}$$

and

$$y = \begin{bmatrix} 0 \\ 2 \\ 3 \end{bmatrix},$$

and show that two different vectors (x,y) have the same norm such that x = y.

Solution: The norms of

$$x = \begin{bmatrix} 1 \\ -2 \\ 3 \end{bmatrix} \quad \text{and} \quad y = \begin{bmatrix} 0 \\ 2 \\ 3 \end{bmatrix}$$

are

$$\|x\|_1 = 6, \qquad \|y\|_1 = 5,$$

$$\|x\|_2 = \sqrt{14}, \qquad \|y\|_2 = \sqrt{13},$$

$$\|x\|_\infty = \|y\|_\infty = 3.$$

Two different vectors may have the same norm, so that

$$\|x\| = \|y\| \Rightarrow x = y.$$

However, from condition $\|x\| > 0$, unless $x = 0$, and $\|0\| = 0$,

$$\|x - y\| = 0 \Rightarrow x - y = 0 \Rightarrow x = y.$$

It can be said that a sequence of n-dimensional vectors

$$\{x_m\}_{m=0}^{\infty}$$

converges to a vector α if, for $i = 1, 2, \ldots, n$, the sequence formed from the ith elements of the x_m converges to the ith element of α. One writes

$$\lim_{m \to \infty} x_m = \alpha.$$

• **PROBLEM 11-18**

Consider the equations

$$\begin{bmatrix} 33 & 25 & 20 \\ 20 & 17 & 14 \\ 25 & 20 & 17 \end{bmatrix} x = \begin{bmatrix} 78 \\ 51 \\ 62 \end{bmatrix},$$

which have solution

$$x = [1\ 1\ 1]^T.$$

Use the iterative correction process from residual vectors ($r_0 = Ax_0 - b$) to improve the computed approximate solution.

Solution: If one uses the simple elimination method with arithmetic accuracy to only two significant decimal digits except in a double-length accumulator, one obtains

$$\begin{bmatrix} 33 & 25 & 20 & : & 78 \\ 0 & 1.8 & 1.8 & : & 3.4 \\ 0 & 0 & 0.79 & : & 0.80 \end{bmatrix}.$$

The first three columns form U' and the last column is the reduced right side

$$(L')^{-1}b,$$

where the multipliers are elements of

$$L' = \begin{bmatrix} 1 & 0 & 0 \\ 0.61 & 1 & 0 \\ 0.76 & 0.56 & 1 \end{bmatrix}.$$

The computed approximation to x is

$$x_0 = [1.1\ \ 0.89\ \ 1.0]^T.$$

One now finds, working to four significant digits, that

404

$$r_0 = [0.5500 \quad 0.1300 \quad 0.3000]^T$$

which, when treated by the multipliers as above, becomes

$$(L')^{-1} r_0 = [0.5500 \quad -0.2055 \quad -0.002920]^T .$$

On solving

$$U' \delta x_0 = -(L')^{-1} r_0 ,$$

one obtains

$$\delta x_0 = [-0.1026 \quad 0.1105 \quad 0.003696]^T ,$$

so that

$$x_1 = x_0 + \delta x_0 = [0.9974 \quad 1.000 \quad 1.004]^T .$$

Notice that

$$\|x - x_0\|_\infty = 0.11,$$

whereas

$$\|x - x_1\|_\infty = 0.004,$$

showing that there has been a considerable improvement in the approximation.
A second application of the process yields

$$\delta x_1 = [0.002737 \quad -0.0001538 \quad -0.004034]^T$$

so that if $x_2 = x_1 + \delta x_1$,

$$\|x - x_2\|_\infty < 0.0002.$$

● **PROBLEM 11-19**

Show that $f(X) = 2x_1 - 3x_2 + 6$ is continuous at $P = (0,-4)$.

Solution: The function $f(X)$ is defined for all $X \in E^2$.
$$|f(X) - f(P)| = |2(x_1 - 0) - 3(x_2 + 4)|$$
One must show that given an arbitrary $\varepsilon > 0$, there exists a $\delta > 0$ such that if
$$[(x_1 - 0)^2 + (x_2 + 4)^2]^{\frac{1}{2}} < \delta$$

then
$$|2(x_1 - 0) - 3(x_2 + 4)| < \varepsilon$$

Note that
$$|2(x_1 - 0) - 3(x_2 + 4)| \leq 2|x_1 - 0| + 3|x_2 + 4|$$

Let $\delta = \varepsilon/5$. Then $\|X - P\| < \delta$ implies that

$$[(x_1 - 0)^2]^{\frac{1}{2}} < \frac{\varepsilon}{5} \qquad [(x_2 + 4)^2]^{\frac{1}{2}} < \frac{\varepsilon}{5}$$

so that
$$2|x_1 - 0| + 3|x_2 + 4| < \varepsilon$$
This completes the proof.

• **PROBLEM 11-20**

Suppose that $n = 3$ and

$$P_m = \left[1 - \frac{1}{m}, \; 2 + \frac{1}{m}, \; \cos\frac{1}{m}\right]$$

Find $\lim_{m \to \infty} P_m$.

Note that the $\lim_{m \to \infty} P_m = P$ means that for each $\varepsilon > 0$ there exists an integer M (which often depends on ε) such that

$$\|P_m - P\| < \varepsilon$$

whenever $m > M$.

Solution: The sequence $\{P_m\}$ is made up of three sequences of real numbers, one for each component of P_m. The limits for these sequences are easily determined.

$$\lim_{m\to\infty} \left(1 - \frac{1}{m}\right) = 1$$

$$\lim_{m\to\infty} \left(2 + \frac{1}{m}\right) = 2 \tag{1}$$

$$\lim_{m\to\infty} \cos \frac{1}{m} = 1$$

Thus, one expects that the limit for the sequence will exist and will be $P = (1,2,1)$. It remains to prove this.

Let $\varepsilon > 0$ be arbitrary. Then for the three limits in Eqs. (1) there exist integers M_1, M_2, M_3, respectively, such that

$$\left| \left(1 - \frac{1}{m}\right) - 1 \right| < \frac{\varepsilon}{3^{\frac{1}{2}}} \quad \text{if } m > M_1$$

$$\left| \left(2 + \frac{1}{m}\right) - 2 \right| < \frac{\varepsilon}{3^{\frac{1}{2}}} \quad \text{if } m > M_2 \tag{2}$$

$$\left| \cos \frac{1}{m} - 1 \right| < \frac{\varepsilon}{3^{\frac{1}{2}}} \quad \text{if } m > M_3$$

Set M equal to the largest of the three integers, M_1, M_2, M_3. Then, if $m > M$, all the inequalities (2) are satisfied. Hence,

$$\|P_m - P\| < \left[\left(\frac{\varepsilon}{3^{\frac{1}{2}}}\right)^2 + \left(\frac{\varepsilon}{3^{\frac{1}{2}}}\right)^2 + \left(\frac{\varepsilon}{3^{\frac{1}{2}}}\right)^2\right]^{\frac{1}{2}} = \varepsilon$$

Thus,
$$\lim_{m \to \infty} P_m = P.$$

MATRIX FACTORIZATION

● **PROBLEM 11-21**

Factorize

$$A = \begin{bmatrix} 1 & 2 & 3 & -1 \\ 2 & -1 & 9 & -7 \\ -3 & 4 & -3 & 19 \\ 4 & -2 & 6 & -21 \end{bmatrix}$$

<u>Solution</u>: Matrix Factorization Method:

The elimination process for n equations in n unknowns

$$Ax = b \tag{1}$$

may be described as a factorization method. One seeks factors L (lower triangular with units on the diagonal) and U (upper triangular) such that

$$A = LU,$$

when (1) becomes

$$LUx = b.$$

First, find c by solving the lower triangular equations

$$Lc = b$$

by forward substitution. One gets

$$1 \cdot c_1 = b_1$$
$$\ell_{21} c_1 + 1 \cdot c_2 = b_2$$
$$\ell_{31} c_1 + \ell_{32} c_2 + 1 \cdot c_3 = b_3$$

$$\vdots$$

$$\ell_{n_1} c_1 + \ell_{n_2} c_2 + \ldots + 1 \cdot c_n = b_n.$$

From the first equation, c_1 is found, then c_2 from the second and so on. Secondly find x by back substitution in the triangular equations

$$Ux = c.$$

The solution of (1) is this vector x, since

$$Ax = (LU)x = L(Ux) = Lc = b.$$

It can be seen from the above that, given any method of calculating factors L and U such that A = LU, one can solve the equations. One way of finding the factors L and U is to build them up from submatrices. Use this method to describe how an n × n matrix A may be factorized in the form

$$A = LDV \qquad (2)$$

$$D = \begin{bmatrix} d_1 & & & & 0 \\ & d_2 & & & \\ & & \cdot & & \\ & & & \cdot & \\ 0 & & & & d_n \end{bmatrix}$$

is a diagonal matrix with $d_i \neq 0$, $i = 1, 2, \ldots, n$, L is a lower triangular matrix with units on the diagonal as before and V is an upper triangular matrix with units on the diagonal. It will be shown that, under certain circumstances, such a factorization exists and is unique. In this factorization call the diagonal elements of D the pivots.

Let A_k denote the kth leading submatrix of A. By this one means the submatrix consisting of the first k rows and columns of A. Note that $A_1 = [a_{11}]$ and $A_n = A$.

Suppose that for some value of k, one already has a factorization

$$A_k = L_k D_k V_k$$

of the form (2), where L_k, D_k and V_k are matrices of the appropriate kinds. Now seek L_{k+1}, D_{k+1} and V_{k+1} such that

$$A_{k+1} = L_{k+1} D_{k+1} V_{k+1} .$$

Write

$$A_{k+1} = \begin{bmatrix} A_k & c_{k+1} \\ r_{k+1}^T & a_{k+1,k+1} \end{bmatrix},$$

where

$$c_{k+1} = \begin{bmatrix} a_{1,k+1} \\ \vdots \\ a_{k,k+1} \end{bmatrix}$$

and

$$r_{k+1}^T = [a_{k+1,1} \quad a_{k+1,2} \quad \cdots \quad a_{k+1,k}] .$$

Seek an element d_{k+1} and vectors

$$l_{k+1}^T = [\ell_{k+1,1} \; \ell_{k+1,2} \; \cdots \; \ell_{k+1,k}], \quad v_{k+1} = \begin{bmatrix} v_{1,k+1} \\ \vdots \\ v_{k,k+1} \end{bmatrix}$$

such that

$$A_{k+1} = \begin{bmatrix} A_k & c_{k+1} \\ r_{k+1}^T & a_{k+1,k+1} \end{bmatrix} = \begin{bmatrix} L_k & 0 \\ l_{k+1}^T & 1 \end{bmatrix} \begin{bmatrix} D_k & 0 \\ 0^T & d_{k+1} \end{bmatrix} \begin{bmatrix} V_k & v_{k+1} \\ 0^T & 1 \end{bmatrix} .$$

On multiplying out the right side and equating blocks, it is found that one requires

$$A_k = L_k D_k V_k \qquad (3)$$

$$c_{k+1} = L_k D_k v_{k+1} \qquad (4)$$

$$r_{k+1}^T = l_{k+1}^T D_k V_k \qquad (5)$$

$$a_{k+1,k+1} = l_{k+1}^T D_k v_{k+1} + d_{k+1}. \qquad (6)$$

Equation (3) is satisfied by hypothesis. One can solve (4) for v_{k+1} by forward substitution in the lower triangular equations

$$(L_k D_k) v_{k+1} = c_{k+1}. \qquad (7)$$

Use forward substitution in the triangular equations

$$l_{k+1}^T (D_k V_k) = r_{k+1}^T \qquad (8)$$

to find l_{k+1}^T. Having found v_{k+1} and l_{k+1}^T, use (6) to determine d_{k+1}.

Thus, this is a practical method of factorizing A since

$$A_1 = [a_{11}] = [1][a_{11}][1] = L_1 D_1 V_1$$

and the process only involves solving triangular equations. The L_k, D_k and V_k are successively determined for $k = 1,2,\ldots,n$ and are leading submatrices of L, D and V.

Considering the foregoing explanations,

$$A_1 = [1]. \qquad L_1 = [1], \qquad D_1 = [1], \qquad V_1 = [1].$$

$$A_2 = \begin{bmatrix} 1 & 2 \\ 2 & -1 \end{bmatrix}, \quad L_2 = \begin{bmatrix} 1 & 0 \\ 2 & 1 \end{bmatrix}, \quad D_2 = \begin{bmatrix} 1 & 0 \\ 0 & -5 \end{bmatrix}$$

$$V_2 := \begin{bmatrix} 1 & 2 \\ 0 & 1 \end{bmatrix}.$$

$$A_3 = \begin{bmatrix} 1 & 2 & 3 \\ 2 & -1 & 9 \\ -3 & 4 & -3 \end{bmatrix}, \quad L_3 = \begin{bmatrix} 1 & 0 & 0 \\ 2 & 1 & 0 \\ -3 & -2 & 1 \end{bmatrix},$$

$$D_3 = \begin{bmatrix} 1 & 0 & 0 \\ 0 & -5 & 0 \\ 0 & 0 & 12 \end{bmatrix} \quad V_3 = \begin{bmatrix} 1 & 2 & 3 \\ 0 & 1 & -\frac{3}{5} \\ 0 & 0 & 1 \end{bmatrix}.$$

$A_4 = A;$

$$(L_3 D_3) v_4 = \begin{bmatrix} 1 & 0 & 0 \\ 2 & -5 & 0 \\ -3 & 10 & 12 \end{bmatrix} v_4 = c_4 = \begin{bmatrix} -1 \\ -7 \\ 19 \end{bmatrix},$$

$$\therefore v_4 = \begin{bmatrix} -1 \\ 1 \\ \frac{1}{2} \end{bmatrix};$$

$$l_4^T (D_3 V_3) = l_4^T \begin{bmatrix} 1 & 2 & 3 \\ 0 & -5 & 3 \\ 0 & 0 & 12 \end{bmatrix} = r_4^T = [4 \quad -2 \quad 6]$$

$$\therefore l_4^T = [4 \quad 2 \quad -1];$$

$$d_4 = a_{4,4} - l_4^T D_3 v_4$$

$$= -21 - \begin{bmatrix} 4 & 2 & -1 \end{bmatrix} \begin{bmatrix} -1 \\ -5 \\ 6 \end{bmatrix}$$

$$= -21 + 20 = -1.$$

Thus

$$A = \begin{bmatrix} 1 & & & 0 \\ 2 & 1 & & \\ -3 & -2 & 1 & \\ 4 & 2 & -1 & 1 \end{bmatrix} \begin{bmatrix} 1 & & & 0 \\ & -5 & & \\ & & 12 & \\ 0 & & & -1 \end{bmatrix}$$

$$\begin{bmatrix} 1 & 2 & 3 & -1 \\ & 1 & -\frac{3}{5} & 1 \\ & & 1 & \frac{1}{2} \\ 0 & & & 1 \end{bmatrix} .$$

The process will fail if, and only if, any of the pivots d_k become zero. This can only happen if one of the leading submatrices is singular.

• **PROBLEM 11-22**

Factorize the following matrix A in the form

$$A = L(DV).$$

$$A = \begin{bmatrix} 1 & 2 & 3 & -1 \\ 2 & -1 & 9 & -7 \\ -3 & 4 & -3 & 19 \\ 4 & -2 & 6 & -21 \end{bmatrix} .$$

Solution: The first row of DV must be the same as the first row of A, since

[1st row of A] = [1st row of L](DV)

$$= [1 \quad 0 \quad \ldots \quad 0](DV)$$
$$= [\text{1st row of DV}].$$

The first column of L is calculated from

$$\begin{bmatrix} \text{1st} \\ \text{column} \\ \text{of A} \end{bmatrix} = L \begin{bmatrix} \text{1st} \\ \text{column} \\ \text{of DV} \end{bmatrix}$$

$$= L \begin{bmatrix} 1 \\ 0 \\ \vdots \\ 0 \end{bmatrix} = \begin{bmatrix} \text{1st} \\ \text{column} \\ \text{of L} \end{bmatrix}.$$

One now gets

$$\begin{bmatrix} 1 & 0 & 0 & 0 \\ 2 & 1 & 0 & 0 \\ -3 & \ell_{32} & 1 & 0 \\ 4 & \ell_{42} & \ell_{43} & 1 \end{bmatrix} \begin{bmatrix} 1 & 2 & 3 & -1 \\ 0 & u_{22} & u_{23} & u_{24} \\ 0 & 0 & u_{33} & u_{34} \\ 0 & 0 & 0 & u_{44} \end{bmatrix}$$

The second row of DV is calculated using

$$[\text{2nd row of A}] = [\text{2nd row of L}](DV).$$

For the:

(2,2) element, $-1 = 4 + u_{22}$, $\therefore$ $u_{22} = -5$;

(2,3) element, $9 = 6 + u_{23}$, $\therefore$ $u_{23} = 3$;

(2,4) element, $-7 = -2 + u_{24}$, $\therefore$ $u_{24} = -5$.

One now gets

$$\begin{bmatrix} 1 & 0 & 0 & 0 \\ 2 & 1 & 0 & 0 \\ -3 & \ell_{32} & 1 & 0 \\ 4 & \ell_{42} & \ell_{43} & 1 \end{bmatrix} \begin{bmatrix} 1 & 2 & 3 & -1 \\ 0 & -5 & 3 & -5 \\ 0 & 0 & u_{33} & u_{34} \\ 0 & 0 & 0 & u_{44} \end{bmatrix}$$

The second column of L is calculated using

$$\begin{bmatrix} \text{2nd} \\ \text{column} \\ \text{of A} \end{bmatrix} = L \begin{bmatrix} \text{2nd} \\ \text{column} \\ \text{of DV} \end{bmatrix}.$$

For the:

$\quad$ (3,2) element, $\quad 4 = -6 - 5\ell_{32}, \quad \therefore \ell_{32} = -2;$

$\quad$ (4,2) element, $\quad -2 = 8 - 5\ell_{42}, \quad \therefore \ell_{42} = 2.$

One now gets

$$\begin{bmatrix} 1 & 0 & 0 & 0 \\ 2 & 1 & 0 & 0 \\ -3 & -2 & 1 & 0 \\ 4 & 2 & \ell_{43} & 1 \end{bmatrix} \begin{bmatrix} 1 & 2 & 3 & -1 \\ 0 & -5 & 3 & -5 \\ 0 & 0 & u_{33} & u_{34} \\ 0 & 0 & 0 & u_{44} \end{bmatrix}.$$

The third column of DV is claculated using

$$\begin{bmatrix} \text{3rd row of A} \end{bmatrix} = \begin{bmatrix} \text{3rd row of L} \end{bmatrix}(DV).$$

For the:

$\quad$ (3,3) element, $-3 = -9 - 6 + u_{33}, \quad \therefore u_{33} = 12;$

$\quad$ (3,4) element, $19 = 3 + 10 + u_{34}, \quad \therefore u_{34} = 6.$

The third column of L and the fourth row of DV are similarly computed. One obtains

$$A = \begin{bmatrix} 1 & 0 & 0 & 0 \\ 2 & 1 & 0 & 0 \\ -3 & -2 & 1 & 0 \\ 4 & 2 & -1 & 1 \end{bmatrix} \begin{bmatrix} 1 & 2 & 3 & -1 \\ 0 & -5 & 3 & -5 \\ 0 & 0 & 12 & 6 \\ 0 & 0 & 0 & -1 \end{bmatrix}$$

● **PROBLEM 11-23**

Factorize the coefficient matrix A using a compact elimination method with partial pivoting and working to three significant digits.

$$A = \begin{bmatrix} 0.5 & 1.1 & 3.1 \\ 2.0 & 4.5 & 0.36 \\ 5.0 & 0.96 & 6.5 \end{bmatrix}$$

Solution: The (1,1) element of DV may be any of the elements in the first column of A and thus

$$w_1 = 0.5, \qquad w_2 = 2.0, \qquad w_3 = 5.0.$$

Choose w_3 as the pivot u_{11} and therefore interchange rows 1 and 3. For the first column of L, one gets

$$\ell_{21} = \frac{w_2}{u_{11}} = \frac{2.0}{5.0} = 0.400, \qquad \ell_{31} = \frac{w_1}{u_{11}} = \frac{0.5}{5.0} = 0.100.$$

One obtains

$$A' = \begin{bmatrix} 5.0 & 0.96 & 6.5 \\ 2.0 & 4.5 & 0.36 \\ 0.5 & 1.1 & 3.1 \end{bmatrix} = L(DV)$$

$$= \begin{bmatrix} 1 & 0 & 0 \\ 0.400 & ① & ⓞ \\ 0.100 & x & ① \end{bmatrix} \begin{bmatrix} 5.00 & 0.960 & 6.50 \\ 0 & x & x \\ 0 & ⓞ & x \end{bmatrix}.$$

The possible values of the (2,2) element of DV are

$$w_2 = a'_{22} - \ell_{21}u_{12} = 4.5 - (0.400)(0.960) \simeq 4.12;$$

$$w_3 = a'_{32} - \ell_{31}u_{12} = 1.1 - (0.100)(0.960) \simeq 1.00.$$

Choose w_2 as pivot u_{22}. Thus no interchanges are required and

$$\ell_{32} = \frac{w_3}{u_{22}} = \frac{1.00}{4.12} \simeq 0.243.$$

Completing the calculations, one obtains

$$u_{23} \simeq -2.24, \qquad u_{33} \simeq 2.99.$$

● **PROBLEM** 11-24

If A is a symmetric band matrix of width 2m + 1 show that the matrix L occurring in the Cholesky factorization A = LL' is also a band matrix.

Solution: The factorization of an (m+1) × (m+1) matrix which can be regarded as a band matrix of width 2m + 1 gives (in general) full triangular factors, which can be regarded as band matrices.

Suppose it has been established, for n ≥ m + 1 that an n × n band matrix of width 2m + 1 can be factorized into matrices of the same type. Consider now the problem for an (n+1) × (n+1) matrix. Use the notation

$$A = LU$$

where

$$L = \begin{bmatrix} L_{n-1} & 0 \\ x' & \ell_{nn} \end{bmatrix}, \quad U = \begin{bmatrix} U_{n-1} & y \\ 0 & u_{nn} \end{bmatrix}$$

and

$$\begin{bmatrix} A_{n-1} & u \\ v' & a_{nn} \end{bmatrix} = \begin{bmatrix} L_{n-1}U_{n-1}, & L_{n-1}y \\ x'U_{n-1}, & x'y + \ell_{nn}u_{nn} \end{bmatrix}.$$

Therefore,

$$u = L_{n-1}y \qquad \text{and} \qquad v = xU_{n-1}$$

$$x = vU_{n-1}^{-1}, \qquad y = L_{n-1}^{-1}u .$$

$$\ell_{nn}u_{nn} = a_{nn} - x'y .$$

Also, assume that all the leading submatrices of A are non-singular.

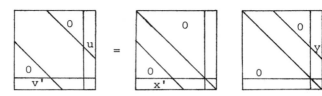

Fig. 1

The induction hypothesis validates the diagram in Figure 1. It remains to be shown that the vectors x',y will each have n-m initial zeros. These are obtained as

$$x' = v'U_{n-1}^{-1} , \qquad y = L_{n-1}^{-1}u$$

and since v,u have n-m initial zeros, so have x,y.

ITERATIVE METHODS

● **PROBLEM 11-25**

Make a series of three iterations to achieve an acceptable solution to the set of simultaneous linear equations:

$$-x_1 + 5x_2 - 2x_3 - 3 = 0$$
$$x_1 + x_2 - 4x_3 + 9 = 0$$
$$4x_1 - x_2 + 2x_3 - 8 = 0$$

Use the initial approximation $\vec{x}^{(0)} = (0,0,0)$.

Solution: Given

$$-x_1 + 5x_2 - 2x_3 - 3 = 0 \qquad (1)$$
$$x_1 + x_2 - 4x_3 + 9 = 0 \qquad (2)$$
$$4x_1 - x_2 + 2x_3 - 8 = 0 \qquad (3)$$

That there is some advantage in rearranging these equations is not obvious at this moment; the advantage will become apparent later. See Eqs. (4), (5), and (6).

$$4x_1 - x_2 + 2x_3 - 8 = 0 \qquad (4)$$

$$-x_1 + 5x_2 - 2x_3 - 3 = 0 \qquad (5)$$

$$x_1 + x_2 - 4x_3 + 9 = 0 \qquad (6)$$

In rearranging the equations, note that the magnitude of the first coefficient in the first equation is the largest of all the coefficient magnitudes in that equation. In the second equation, the magnitude of the second coefficient is the largest. In any set of equations, this procedure is continued such that the magnitude of the coefficient of the kth term in the kth equation is the largest coefficient magnitude in that equation.

Then Array (7) represents the Eqs. (4), (5), and (6):

$$\begin{array}{cccc} 4 & -1 & 2 & -8 \\ -1 & 5 & -2 & -3 \\ 1 & 1 & -4 & 9 \end{array} \qquad (7)$$

Obviously, one could solve the set of equations represented by Array (7) by a number of different techniques: elimination, substitution, determinants, and matrices.

As shown in the following outline, note that in Step 1 each of the variables is set at some initial value. Having no special insight into the solution of the problem, the programmer might initially set each variable at zero. Then in Step 2, first x_1 is calculated using the first row of Array (7) or from Eq. (4). In turn x_2 is calculated using the second row of the array and using the current values of the other two variables x_1 and x_2. See (b) of Step 2. As shown in (c), x_3 is then calculated. Step 3 is an iteration of Step 2. As these iterations repeat, x_1 is always calculated using the equation represented by the first row of the array that represents the family of equations; x_2 is always calculated from the second row; and so forth. As each variable is computed in its turn, most recently computed values of other variables are used.

By the particular iteration procedure under discussion, the steps are repeated as illustrated below until sufficient accuracy has been achieved.

Step 1: Initialize Variables

$$x_1 = 0 \qquad (8)$$

$$x_2 = 0 \qquad (9)$$

$$x_3 = 0 \qquad (10)$$

Step 2:

(a) $\quad x_1 = \dfrac{x_2 - 2x_3 + 8}{4} = \dfrac{8}{4} = 2 \qquad (11)$

(from 1st row of Array (7) or from Eq. (4))

(b) $\quad x_2 = \dfrac{x_1 + 2x_3 + 3}{5} = \dfrac{2 + 3}{5} = 1 \qquad (12)$

(from 2nd row)

(c) $\quad x_3 = \dfrac{-x_1 - x_2 - 9}{-4} = \dfrac{-2 - 1 - 9}{-4} = 3 \qquad (13)$

(from 3rd row)

Step 3:

(a) $\quad x_1 = \dfrac{x_2 - 2x_3 + 8}{4} = \dfrac{1 - 6 + 8}{4} = .75 \qquad (14)$

(from 1st row)

(b) $\quad x_2 = \dfrac{x_1 + 2x_3 + 3}{5} = \dfrac{.75 + 6 + 3}{5} = 1.95 \qquad (15)$

(from 2nd row)

(c) $\quad x_3 = \dfrac{-x_1 - x_2 - 9}{-4} = \dfrac{-.75 - 1.95 - 9}{-4} = 2.925 \qquad (16)$

(from 3rd row)

Step 4:

(a) $\quad x_1 = \dfrac{x_2 - 2x_3 + 8}{4} = \dfrac{1.95 - 2 \times 2.925 + 8}{4}$

$$= 1.025 \tag{17}$$

(b) $$x_2 = \frac{x_1 + 2x_3 + 3}{5} = \frac{1.025 + 2 \times 2.925 + 3}{5}$$

$$= 1.975 \tag{18}$$

(c) $$x_3 = \frac{-x_1 - x_2 - 9}{-4} = \frac{-1.025 - 1.975 - 9}{-4}$$

$$= 3.000 \tag{19}$$

The exact solution to the set of equations represented by Array (7) is given by Eqs. (20), (21), and (22).

$$x_1 = 1 \tag{20}$$

$$x_2 = 2 \tag{21}$$

$$x_3 = 3 \tag{22}$$

It is interesting to note in this problem how the solution converges as the iteration continues. On the other hand, one must not assume that—independent of the set of equations and independent of the iteration technique—the solution will always converge.

● **PROBLEM 11-27**

Using the iterative technique approximate the solution of the linear system.

$$5x_1 - x_2 + x_3 = 7 \tag{1}$$

$$x_1 - 10x_2 - 10x_3 = 1 \tag{2}$$

$$-x_1 + 4x_2 + 5x_3 = 0 \tag{3}$$

Use the initial approximation

$$\bar{x}^{(0)} = (0,0,0) \tag{4}$$

Carry out the calculation to the fourth digit.

Solution: We shall approximate the solution of the n×n linear system $A\bar{x} = \bar{a}$. Starting with the initial approximation $\bar{x}^{(0)}$, we generate the sequence of approximations

$$\left\{\bar{x}^{(\ell)}\right\}_{\ell=0}^{\infty} \tag{5}$$

which converges to some $\bar{x}$.

For this purpose, the system $A\bar{x} = \bar{a}$ will be written in the form

$$\bar{x} = T\bar{x} + \bar{t} \tag{6}$$

which is equivalent.

Here, T is an n×n matrix and $\bar{t}$ is a vector, $\bar{t} = (t_1, \ldots, t_n)$.

Starting with $\bar{x}^{(0)}$, the consecutive approximations are obtained from

$$\bar{x}^{(k)} = T\bar{x}^{(k-1)} + \bar{t} \tag{7}$$

$$k = 1, 2, \ldots.$$

To find the representation (6) of the systems (1)-(3), we write (1)-(3) in the form

$$x_1 = \frac{1}{5}x_2 - \frac{1}{5}x_3 + \frac{7}{5} \tag{8}$$

$$x_2 = \frac{1}{10}x_1 - x_3 - \frac{1}{10} \tag{9}$$

$$x_3 = \frac{1}{5}x_1 - \frac{4}{5}x_2 \tag{10}$$

Thus,

$$T = \begin{pmatrix} 0 & .2 & -.2 \\ .1 & 0 & -1 \\ .2 & -.8 & 0 \end{pmatrix}, \quad \bar{t} = \begin{pmatrix} 1.4 \\ -.1 \\ 0 \end{pmatrix} \tag{11}$$

and

$$\bar{x}^{(1)} = T\bar{x}^{(0)} + \bar{t} = \bar{t} = \begin{pmatrix} 1.4 \\ -.1 \\ 0 \end{pmatrix} \tag{12}$$

$$\bar{x}^{(2)} = T\bar{x}^{(1)} + \bar{t} = \begin{pmatrix} 0 & .2 & -.2 \\ .1 & 0 & -1 \\ .2 & -.8 & 0 \end{pmatrix} \begin{pmatrix} 1.4 \\ -.1 \\ 0 \end{pmatrix} + \begin{pmatrix} 1.4 \\ -.1 \\ 0 \end{pmatrix}$$

$$= \begin{pmatrix} 1.38 \\ .04 \\ .36 \end{pmatrix} \tag{13}$$

In a similar manner, we obtain other approximations shown in the following table.

k	0	1	2	3	4	5	6	7	8
x_1	0	1.4	1.38	1.336	1.287	1.254	1.217	1.191	1.163
x_2	0	-.1	.04	-.322	-.210	-.491	-.401	-.622	-.545
x_3	0	0	.36	.244	.525	.426	.644	.564	.736

Note that not every linear system can be handled in this manner. For example, the solution of the system

$$x_1 + x_2 + x_3 = 2$$

$$c = \begin{bmatrix} \frac{3}{5} \\ \frac{25}{11} \\ \frac{-11}{10} \\ \frac{15}{8} \end{bmatrix}$$

For the initial approximation $\vec{x}^{(0)} = (0,0,0,0)^t$, generate $\vec{x}^{(1)}$ by:

$$x_1^{(1)} = \qquad \frac{1}{10}x_2^{(0)} - \frac{1}{5}x_3^{(0)} \qquad + \frac{3}{5}$$

$$x_2^{(1)} = \frac{1}{11}x_1^{(0)} \qquad + \frac{1}{11}x_3^{(0)} - \frac{3}{11}x_4^{(0)} + \frac{25}{11}$$

$$x_3^{(1)} = -\frac{1}{5}x_1^{(0)} + \frac{1}{10}x_2^{(0)} \qquad + \frac{1}{10}x_4^{(0)} - \frac{11}{10}$$

$$x_4^{(1)} = \qquad -\frac{3}{8}x_2^{(0)} + \frac{1}{8}x_3^{(0)} \qquad + \frac{15}{8}$$

$x_1^{(1)} = .6000,$

$x_2^{(1)} = 2.2727,$

$x_3^{(1)} = -1.1000,$

$x_4^{(1)} = 1.8750.$

Additional iterates, $\vec{x}^{(k)} = (x_1^{(k)}, x_2^{(k)}, x_3^{(k)}, x_4^{(k)})^t$, are generated in a similar manner and are presented in the table shown.

Table

k	0	1	2	3	4	5	6	7	8	9	10
$x_1^{(k)}$	0.0000	.6000	1.0473	.9326	1.0152	.9890	1.0032	.9981	1.0006	.9997	1.0001
$x_2^{(k)}$	0.0000	2.2727	1.7159	2.0533	1.9537	2.0114	1.9922	2.0023	1.9987	2.0004	1.9998
$x_3^{(k)}$	0.0000	−1.1000	−.8052	−1.0493	−.9681	−1.0103	−.9945	−1.0020	−.9990	−1.0004	−.9998
$x_4^{(k)}$	0.0000	1.8750	.8852	1.1309	.9739	1.0214	.9944	1.0036	.9989	1.0006	.9998

$$2x_1 + x_2 + 3x_3 = 7$$

$$x_1 + 4x_2 + 5x_3 = 7$$

cannot be approximated using this method. This is because the sequence of matrices $T, T^2, T^3, T^4, \ldots$ where

$$T = \begin{pmatrix} 0 & -1 & -1 \\ -2 & 0 & -3 \\ -.2 & -.8 & 0 \end{pmatrix}$$

does not converge.

● **PROBLEM 11-27**

Consider the algorithm:

Let $P_0 = (p_{10}, p_{20}, \ldots, p_{n_0})$ be a first approximation to a solution of a system $X = G(X)$. Generate the sequence $\{P_m\}$ recursively by the relation

$$P_m = G(P_{m-1}) \qquad m = 1, 2, \ldots$$

Investigate the possible convergence of the sequence $\{P_m\}$ to a solution of $X = G(X)$.

Draw the flow chart and apply this algorithm to the system given below, with $P_0 = (0,0)$.

$$x_1 = \frac{x_1}{4} + \frac{x_2}{8} + \frac{11}{8}$$

$$x_2 = \frac{-x_1}{8} + \frac{x_2}{4} + 1$$

Solution: Here

$$g_1(x_1, x_2) = \frac{x_1}{4} + \frac{x_2}{8} + \frac{11}{8}$$

$$g_2(x_1, x_2) = \frac{-x_1}{8} + \frac{x_2}{4} + 1$$

The exact solution is given by $x_1 = 2$, $x_2 = 1$. The iteration gives

$$P_1 = [g_1(0,0), g_2(0,0)] = \left(\frac{11}{8}, 1\right)$$

$$P_2 = [g_1\left(\frac{11}{8}, 1\right), g_2\left(\frac{11}{8}, 1\right)] = \left(\frac{59}{32}, \frac{69}{64}\right)$$

$$P_3 = [g_1\left(\frac{59}{32}, \frac{69}{64}\right), g_2\left(\frac{59}{32}, \frac{69}{64}\right)] = \left(\frac{1,009}{512}, \frac{133}{128}\right)$$

$$P_4 = \left(\frac{4,091}{2,048}, \frac{4,151}{4,096}\right)$$

Because the exact solution is known, one can compute the errors in the approximations provided by P_1, P_2, P_3, P_4. The values in the table shown have been rounded to three decimal places.

i	$|p_{1i} - 2|$	$|p_{2i} - 1|$
0	2.000	1.000
1	.875	.000
2	.094	.078
3	.006	.039
4	.002	.011

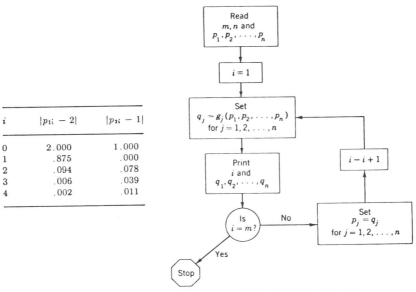

Flow chart for Algorithm

● **PROBLEM 11-28**

Apply the Seidel method of iteration to the following system:

$$x_1 = \frac{x_1}{4} + \frac{x^2}{8} + \frac{11}{8}$$

$$x_2 = \frac{-x_1}{8} + \frac{x^2}{4} + 1$$

Take $P_0 = (0,0)$.

Solution: Consider the algorithm. Let $P_0 = (P_{10}, P_{20}, \ldots, P_{no})$ be a first approximation to a solution of a system

425

$X = G(X)$. Then, generate the sequence $\{P_m\}$ recursively by the relation

$$P_m = G(P_{m-1}) \qquad M = 1, 2, \ldots$$

This describes a procedure where, given

$$P_0 = (p_{10}, p_{20}, \ldots, p_{n0})$$

compute $P_m = (p_{1m}, p_{2m}, \ldots, p_{nm})$ by $P_m = G(P_{m-1})$. Thus

$$p_{1,m} = g_1(p_{1,m-1}, p_{2,m-1}, \ldots, p_{n,m-1})$$

$$p_{2,m} = g_2(p_{1,m-1}, p_{2,m-1}, \ldots, p_{n,m-1})$$

$$\ldots\ldots\ldots\ldots\ldots\ldots\ldots\ldots\ldots\ldots\ldots$$

$$p_{n,m} = g_n(p_{1,m-1}, p_{2,m-1}, \ldots, p_{n,m-1})$$

Normally one would expect that $p_{1m}, p_{2m}, \ldots, p_{nm}$ would be calculated sequentially. Thus, at the time p_{2m} is being calculated, the value of p_{1m} is known. If the iterative procedure is converging, one would expect that p_{1m} is closer to the first component of the solution vector than is $p_{1,m-1}$. Thus it would seem natural to compute p_{2m} by the formula

$$p_{2m} = g_2(p_{1m}, p_{2,m-1}, p_{3,m-1}, \ldots, p_{n,m-1})$$

Repetition of this reasoning leads to the following formulas for $P_m = (p_{1m}, p_{2m}, \ldots, p_{nm})$:

$$p_{1,m} = g_1(p_{1,m-1}, p_{2,m-1}, \ldots, p_{n,m-1})$$

$$p_{2,m} = g_2(p_{1,m}, p_{2,m-1}, \ldots, p_{n,m-1})$$

$$p_{3,m} = g_3(p_{1,m}, p_{2,m}, p_{3,m-1}, \ldots, p_{n,m-1})$$

$$\ldots\ldots\ldots\ldots\ldots\ldots\ldots\ldots\ldots\ldots\ldots$$

$$p_{n,m} = g_n(p_{1,m}, p_{2,m}, p_{3,m}, \ldots, p_{n-1,m}, p_{n,m-1})$$

That is, $p_{i,m}$ is computed by using $p_{1,m}$, $p_{2,m}$, ..., $p_{i-1,m}$ and $p_{i,m-1}$, $p_{i+1,m-1}$, ..., $p_{n,m-1}$. This modification of algorithm (1) is called Seidel's method of iteration.

In many problems of general interest it turns out that the Seidel iteration converges somewhat more rapidly than the ordinary iteration. However, there are several additional reasons for using the Seidel iteration.

The Seidel iteration is somewhat easier to implement on a computer than algorithm (1). This can be seen by studying the flow chart. In addition, the Seidel iteration requires less storage space. In many problems if algorithm (1) produces a convergent sequence, then so does the Seidel iteration, and vice versa.

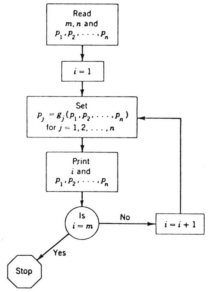

Flow chart for Seidel iteration.

Thus

$$p_{1,m} = \frac{p_{1,m-1}}{4} + \frac{p_{2,m-1}}{8} + \frac{11}{8}$$

$$p_{2,m} = -\frac{p_{1,m}}{8} + \frac{p_{2,m-1}}{4} + 1$$

The Seidel iteration gives

m = 1: $\qquad p_{1,1} = \frac{11}{8}$

$$P_{2,1} = \frac{-\left(\frac{11}{8}\right)}{8} + 0 + 1 = \frac{53}{64}$$

or

$$P_1 = \left(\frac{11}{8}, \frac{53}{64}\right)$$

$m = 2$:
$$P_{1,2} = \frac{\frac{11}{8}}{4} + \frac{\frac{53}{64}}{8} + \frac{11}{8} = \frac{933}{512}$$

$$P_{2,2} = -\frac{933}{512 \times 8} + \frac{53}{64 \times 4} + 1 = \frac{4,011}{4,096}$$

Using the approximation $P_2 = (1.8223, .97925)$ continue.

$m = 3$:
$$P_{1,3} = \frac{1.8223}{4} + \frac{.97925}{8} + \frac{11}{8} = 1.9530$$

$$P_{2,3} = -\frac{1.9530}{8} + \frac{.97925}{4} + 1 = 1.00068$$

or $P_3 = (1.9530, 1.0007)$. A similar calculation gives
$P_4 = (1.9883, 1.0016)$.

An error table is given below.

i	$|p_{1i} - 2|$	$|p_{2i} - 1|$
0	2.000	1.000
1	0.875	.172
2	.178	.021
3	.147	.001
4	.012	.002

● **PROBLEM** 11-29

Solve the equations

$$\begin{bmatrix} 5 & -1 & -1 & -1 \\ -1 & 10 & -1 & -1 \\ -1 & -1 & 5 & -1 \\ -1 & -1 & -1 & 10 \end{bmatrix} \vec{x} = \begin{bmatrix} -4 \\ 12 \\ 8 \\ 34 \end{bmatrix}$$

by the Jacobi iterative method. Stop at six iterations starting with $\vec{x}_0 = 0$ and work to three decimal places. (The computer solution is $\vec{x} = [1 \quad 2 \quad 3 \quad 4]^T$.)

Solution: Consider the problem of solving n non-singular equations in n unknowns

$$Ax = b. \tag{1}$$

If E and F are n x n matrices such that

$$A = E - F, \tag{2}$$

(2) is called a splitting of A. For such a splitting, (1) may be written as

$$Ex = Fx + b.$$

This form of the equations suggest an iterative procedure

$$Ex_{m+1} = Fx_m + b, \quad m = 0, 1, 2, \ldots, \tag{3}$$

for arbitrary x_0. If the sequence is to be uniqely defined for a given x_0, it is required that E be non-singular, when

$$x_{m+1} = E^{-1} Fx_m + E^{-1} b.$$

It can be said that the sequence $\{x_m\}_{m=0}^{\infty}$ converges if

$$||E^{-1}F|| < 1. \tag{4}$$

It can also be seen that, in this case, $\{x_m\}$ converges to the solution vector x.

The important question is the suitable choice of E and F. If E = A and F = 0 one achieves nothing. Repeatedly solve (3) in order to determine x_{m+1}; it would be necessary to use one of the elimination methods unless E is of a particularly simple form. If E is not of a simpler form than A, the work involved for each iteration will be identical to that required in solving the original equations (1) by an elimination method from the start.

The simplest choice of E is a diagonal matrix, usually the diagnoal of A provided all the diagonal elements are non-zero. Obtain the splitting

$$A = D - B,$$

where D is diagonal with non-zero diagonal elements and B has zeros on its diagonal. The relation (3) becomes

$$x_{m+1} = D^{-1} B x_m + D^{-1} b, \quad m = 0, 1, 2, \ldots \quad (5)$$

This is known as the Jacobi iterative method. D^{-1} is simply the diagonal matrix whose diagonal elements are the inverses of those of D. The method is convergent if

$$||D^{-1} B|| < 1.$$

This condition is certainly satisfied if the matrix A is a strictly diagonally dominant matrix, that is, a matrix such that

$$|a_{ii}| > \sum_{\substack{j=1 \\ j \neq i}}^{n} |a_{ij}| \quad \text{for all} \quad i = 1, 2, \ldots, n.$$

$D^{-1} B$ has off-diagonal elements a_{ij}/a_{ii} and zeros on the diagonal, so that for a strictly diagonally dominant matrix A,

$$||D^{-1} B||_\infty = \max_{1 \leq i \leq n} \sum_{\substack{j=1 \\ j \neq i}}^{n} \left|\frac{a_{ij}}{a_{ii}}\right| = \max_{1 \leq i \leq n} \frac{1}{|a_{ii}|} \sum_{\substack{j=1 \\ j \neq i}}^{n} |a_{ij}| < 1.$$

Consider now the above problem.

The coefficient matrix is diagonally dominant and, therefore, the Jacobi method is convergent. The equations (5) are:

$$x_{m+1} = \begin{bmatrix} 0 & 0.2 & 0.2 & 0.2 \\ 0.1 & 0 & 0.1 & 0.1 \\ 0.2 & 0.2 & 0 & 0.2 \\ 0.1 & 0.1 & 0.1 & 0 \end{bmatrix} x_m + \begin{bmatrix} -0.8 \\ 1.2 \\ 1.6 \\ 3.4 \end{bmatrix}$$

and

$$||D^{-1}B||_\infty = 0.6.$$

Table 1 shows six successive iterates starting with $\vec{x}_0 = 0$ and working to three decimal places.

TABLE 1

m	0	1	2	3	4	5
x_m	0	−0.800	0.440	0.716	0.883	0.948
	0	1.200	1.620	1.840	1.929	1.969
	0	1.600	2.360	2.732	2.880	2.948
	0	3.400	3.600	3.842	3.929	3.969

CHAPTER 12

MATRIX EIGENVALUES AND EIGENVECTORS

EIGENVALUES AND EIGENVECTORS OF A MATRIX

● PROBLEM 12-1

(1) Define an eigenvalue.

(2) Show that if u and v are eigenvectors of a linear operator f which belong to λ and if a is a real number, then (a) u + v and (b) au are also eigenvectors of f which belong to λ.

Solution: (1) A real number, λ, is an eigenvalue of f if and only if there exists a non-zero vector u in V such that $f(u) = \lambda u$.

Thus, an eigenvalue is a number which acts as a scalar multiple of some non-zero vector to give its f-image. If λ is an eigenvalue and $f(u) = \lambda u$, then u is called an eigenvector of f belonging to λ. The set of eigenvectors of f which belong to a given eigenvalue, λ, constitutes a subspace of R^n which is called the eigenspace of λ.

Let P be the graph point of an eigenvector u which belongs to an eigenvalue λ of an operator f on R^2. If Q is the graph point of λu, then $\overline{OP}$ and $\overline{OQ}$ are collinear. The eigenvalue, λ, signifies an extension, contraction or reversal in direction of its eigenspace according to whether its value is greater than 1, between 0 and 1 or less than 0. (See accompanying figure.)

Extension
$\lambda > 1$

Contraction
$0 < \lambda < 1$

Reversal of direction
$\lambda < 0$

(2a) $f(u+v) = f(u) + f(v) = \lambda u + \lambda v = \lambda(u+v)$. Thus, u+v are also eigenvectors of f which belong to λ.

(b) $f(au) = af(u) = a(\lambda u) = a(\lambda u)$. Thus, each scalar multiple au is an eigenvector.

● **PROBLEM 12-2**

Find the real eigenvalues of the matrix,

$$A = \begin{bmatrix} -2 & -1 \\ 5 & 2 \end{bmatrix}$$

Solution: Let T be a linear transformation and A its matrix with respect to a given basis. Then λ is an eigenvalue if

$$AX = \lambda X, \tag{1}$$

where X is a non-zero vector. Choosing R^n as the underlying vector space, $T: R^n \to R^n$, (1) becomes

$$\begin{bmatrix} a_{11} & a_{12} & \cdots & a_{1n} \\ a_{21} & & & \vdots \\ \vdots & & & \vdots \\ a_{n1} & a_{n2} & \cdots & a_{nn} \end{bmatrix} \begin{bmatrix} x_1 \\ x_2 \\ \vdots \\ x_n \end{bmatrix} = \lambda \begin{bmatrix} x_1 \\ x_2 \\ \vdots \\ x_n \end{bmatrix}. \tag{2}$$

Expanding (2),

$$a_{11}x_1 + a_{12}x_2 + \cdots + a_{1n}x_n = \lambda x_1 \tag{3}$$

$$a_{21}x_1 + a_{22}x_2 + \cdots + a_{2n}x_n = \lambda x_2$$

$$\vdots \qquad \vdots \qquad \qquad \vdots$$

$$a_{n1}x_1 + a_{n2}x_2 + \cdots + a_{nn}x_n = \lambda x_n$$

Rewriting (3),

$$(\lambda - a_{11})x_1 - a_{12}x_2 - \cdots - a_{1n}x_n = 0 \tag{4}$$

$$-a_{21}x_1 + (\lambda - a_{22})x_2 - \cdots - a_{2n}x_n = 0$$

$$\vdots$$

$$-a_{n1}x_1 - a_{n2}x_2 - \ldots + (\lambda-a_{nn})x_n = 0.$$

The set of linear homogeneous equations (4) can be expressed in matrix form as:

$$\begin{bmatrix} \lambda & 0 & \ldots & 0 \\ 0 & \lambda & \ldots & 0 \\ \vdots & & & \\ 0 & 0 & \ldots & \lambda \end{bmatrix} \begin{bmatrix} x_1 \\ x_2 \\ \vdots \\ x_n \end{bmatrix} - \begin{bmatrix} a_{11} & a_{12} & \ldots & a_{1n} \\ a_{21} & & & \\ \vdots & & & \\ a_{n1} & & \ldots & a_{nn} \end{bmatrix} \begin{bmatrix} x_1 \\ x_2 \\ \vdots \\ x_n \end{bmatrix} = \begin{bmatrix} 0 \\ 0 \\ \vdots \\ 0 \end{bmatrix}$$

or,

$$[\lambda I - A][X] = [0]. \quad (5)$$

Recall now that a set of n linear homogeneous equations in n unknowns can have a non-trivial solution only if

$$\det[\lambda I - A] = 0.$$

The equation $\det[\lambda I - A] = 0$ is an nth degree polynomial in λ and its roots provide the eigenvalues of A and, thus, of T. By the Fundamental Theorem of Algebra there are n such roots in the complex field.

Form the matrix

$$\lambda I - A = \lambda \begin{bmatrix} 1 & 0 \\ 0 & 1 \end{bmatrix} - \begin{bmatrix} -2 & -1 \\ 5 & 2 \end{bmatrix}$$

$$= \begin{bmatrix} \lambda + 2 & 1 \\ -5 & \lambda - 2 \end{bmatrix}$$

Take its determinant to obtain the characteristic polynomial of A:

$$f(\lambda) = \det(\lambda I - A)$$

$$= \det \begin{vmatrix} \lambda + 2 & 1 \\ -5 & \lambda - 2 \end{vmatrix}$$

$$= (\lambda + 2)(\lambda - 2) + 5$$
$$= \lambda^2 + 1$$

The eigenvalues of A must, therefore, satisfy the quadratic equation $\lambda^2 + 1 = 0$. Since the only solutions to this equation are the imaginary numbers $\lambda = i$ and $\lambda = -i$. A has no real eigenvalues.

● **PROBLEM 12-3**

Determine the eigenvalues and eigenvectors of the matrix

$$A = \begin{pmatrix} 5 & 1 & -1 \\ 1 & 3 & -1 \\ -1 & -1 & 3 \end{pmatrix}.$$

Solution: Many problems in engineering and other fields lead to the matrix eigenvalue problem which can be written algebraically as $AX = \lambda X$. In this matrix equation A is an (n×n) square matrix, X is an (n×1) column vector representing the eigenvectors, and λ is a scalar parameter representing the eigenvalues. This equation can be rewritten as

$$(A - \lambda I)X = 0,$$

and this matrix equation represents a set of n simultaneous, homogeneous equations in the n unknown x_i's. The only possibility for this set of equations to have a nontrivial solution is for the determinant of the coefficient matrix to be zero. That is, one must have

$$|A - \lambda I| = 0$$

for a nontrivial solution.

Therefore, for the given problem one gets

$$|A - \lambda I| = \begin{vmatrix} 5-\lambda & 1 & -1 \\ 1 & 3-\lambda & -1 \\ -1 & -1 & 3-\lambda \end{vmatrix} = 0.$$

Expanding this determinant, and after multiplying through by -1, one gets the following characteristic equation:

$$\lambda^3 - 11\lambda^2 + 36\lambda - 36 = 0.$$

It can be checked that $\lambda_i = 6, 3,$ and 2 satisfy this equation, so one can factor it as

$$(\lambda-6)(\lambda-3)(\lambda-2) = 0,$$

and the eigenvalues for the given matrix A are

$$\lambda_1 = 6, \qquad \lambda_2 = 3, \qquad \lambda_3 = 2.$$

The corresponding eigenvectors are found by, respectively, substituting these values of λ back into the basic equation $AX = \lambda X$. For example, for $\lambda_1 = 6$, one gets $AX = 6X$ or

$$5x_1 + 1x_2 - 1x_3 = 6x_1$$
$$1x_1 + 3x_2 - 1x_3 = 6x_2$$
$$-1x_1 - 1x_2 + 3x_3 = 6x_3$$

or

$$-1x_1 + x_2 - x_3 = 0$$
$$1x_1 - 3x_2 - x_3 = 0$$
$$-x_1 - x_2 - 3x_3 = 0$$

Rewriting these equations once more:

$$-1x_1/x_3 + 1x_2/x_3 = 1$$
$$1x_1/x_3 - 3x_2/x_3 = 1$$
$$-1x_1/x_3 - 1x_2/x_3 = 3$$

and the solution of these 3 nonindependent equations can be easily found to be

$$x_1/x_3 = -2$$

and

$$x_2/x_3 = -1.$$

The solution is, of course, not unique and the eigenvectors are determined only up to a constant of proportionality. In this case write the answer as

$$X_1 = c_1\{-2,-1,1\}.$$

The other two eigenvectors can be found in like manner from the equations $AX = 3X$ and $AX = 2X$. The final results are

$$\lambda_1 = 6, \; X_1 = c_1\{-2,-1,1\};$$
$$\lambda_2 = 3, \; X_2 = c_2\{1,-1,1\};$$
$$\lambda_3 = 2, \; X_3 = c_3\{0,1,1\}.$$

Besides substituting these results back into $AX = \lambda X$ for a check, it is also known that $\Sigma\lambda_i = \Sigma a_{ii}$ = trace of A and $\pi\lambda_i = |A|$. Using the above, one gets

$$\Sigma\lambda_i = 11 = \Sigma a_{ii}$$

and

$$\pi\lambda_i = 36 = |A|.$$

Also, when the matrix A is symmetric, as above, it is also known that the X_i's must be orthogonal to each other. For the above results one gets:

$$(-2)(1) + (-1)(-1) + (1)(1) = 0,$$

$$(-2)(0) + (-1)(1) + (1)(1) = 0,$$

and

$$(1)(0) + (-1)(1) + (1)(1) = 0.$$

Thus they are all orthogonal to each other.

● **PROBLEM 12-4**

By iteration, determine the dominant eigenvalue of matrix

$$A = \begin{pmatrix} 2 & -1 & 0 \\ -1 & 2 & -1 \\ 0 & -1 & 2 \end{pmatrix}.$$

(Note: The dominant eigenvalue is the numerically largest eigenvalue). Use the initial vector

$$(1,1,1)^T$$

and compute up to three decimal places.

Solution: The iterative procedure for determining an eigenvalue of a square matrix A consists of starting with a (reasonable) initial approximation to the eigenvector X_1, (call this value Z_0), and consecutively performing the following products:

$$AZ_0 = Z_1, \quad AZ_1 = Z_2, \quad AZ_2 = Z_3, \quad \ldots, \quad AZ_{n-1} = AZ_n.$$

In general, it is found that Z_i is a better approximation to X_1 than is Z_{i-1}. Also, if the procedure is continued until Z_n is equal to a constant times Z_{n-1}, to a certain number of decimal places, then Z_n will be a good approximation for X_1 and the constant just referred to will be a good approximation for the dominant eigenvalue. Although there are exceptional cases where this iterative method does not work, it usually does. Ordinarily, the selection of an initial value for Z_0 is not very critical.

A slightly different numerical procedure to that just

described is usually to be preferred. It consists of scaling each product, the Z_i's, so that the vector premultiplied by A will always have one of its elements equal to 1; this slightly different vector will be called Y_i. Not only does this make the calculations simpler, but the factor taken out of Z_i to form Y_i is the current approximation of the eigenvalue λ_i.

Taking $Z_0 = \{1,1,1\}$, in the given problem, one gets

$$AZ_0 = \begin{pmatrix} 2 & -1 & 0 \\ -1 & 2 & -1 \\ 0 & -1 & 2 \end{pmatrix} \begin{pmatrix} 1 \\ 1 \\ 1 \end{pmatrix} = \begin{pmatrix} 1 \\ 0 \\ 1 \end{pmatrix} = Z_1 = 1 \begin{pmatrix} 1 \\ 0 \\ 1 \end{pmatrix} = 1Y_1$$

$$AY_1 = \begin{pmatrix} 2 & -1 & 0 \\ -1 & 2 & -1 \\ 0 & -1 & 2 \end{pmatrix} \begin{pmatrix} 1 \\ 0 \\ 1 \end{pmatrix} = \begin{pmatrix} 2 \\ -2 \\ 2 \end{pmatrix} = 2 \begin{pmatrix} 1 \\ -1 \\ 1 \end{pmatrix} = 2Y_2$$

Similarly,

$AY_2 = \{3,-4,3\} = 3\{1,-4/3,1\} = 3Y_3$

$AY_3 = \{10/3,-14/3,10/3\} = 10/3\{1,-1.4,1\} = 10/3\ Y_4$

$AY_4 = \{3.4,-4.8,3.4\} = 3.4\{1,-1.412,1\} = 3.4Y_5$

$AY_5 = \{3.412,-4.824,3.412\} = 3.412\{1,-1.414,1\}$
$\qquad = 3.412Y_6$

$AY_6 = \{3.414,-4.828,3.414\} = 3.414\{1,-1.414,1\}$
$\qquad = 3.414Y_7$

Since $Y_6 = Y_7$ to three decimal places, one gets $\lambda_1 = 3.414$, probably correct to three decimal places. And

$$X_1 = c_1\{1,-1.414,1\}.$$

Since it is known that $\Sigma\lambda_i = \Sigma a_{ii}$ and $\pi\lambda_i = |A|$, the other two eigenvalues can be found as follows:

$$\Sigma\lambda_i = 3.414 + \lambda_2 + \lambda_3 = 6$$

and

$$\pi\lambda_i = 3.414\lambda_2\lambda_3 = 4.$$

The simultaneous solution of these two equations is $\lambda_2 = 2.000$ and $\lambda_3 = 0.586$.

The exact values of the eigenvalues for the above A can easily be found to be $\lambda_i = 2 + \sqrt{2},\ 2,\ 2 - \sqrt{2}$.

The answers obtained are thus seen to be correct to three decimal places. Matrices similar to the given

A, including larger ones, actually appear in the solution of engineering problems; e.g., in vibration problems.

● **PROBLEM 12-5**

Find the eigenvalues and the corresponding eigenvectors of A where

$$A = \begin{bmatrix} 0 & \frac{1}{2} \\ \frac{1}{2} & 0 \end{bmatrix}$$

Solution: An eigenvalue of A is a scalar λ such that $AX = \lambda X$ for some non-zero vectors X. This may be converted to $(\lambda I - A)X = 0$ which implies $\det(\lambda I - A) = 0$, the characteristic equation. The roots of this equation yield the required eigenvalues.

$$\lambda I - A = \begin{bmatrix} \lambda & 0 \\ 0 & \lambda \end{bmatrix} - \begin{bmatrix} 0 & \frac{1}{2} \\ \frac{1}{2} & 0 \end{bmatrix} = \begin{bmatrix} \lambda & -\frac{1}{2} \\ -\frac{1}{2} & \lambda \end{bmatrix}$$

$$\det(\lambda I - A) = \lambda^2 - \frac{1}{4}.$$

Then, the characteristic equation is $\lambda^2 - \frac{1}{4} = 0$ and the eigenvalues are $\lambda_1 = \frac{1}{2}$ and $\lambda_2 = -\frac{1}{2}$. Substitute $\lambda = \frac{1}{2}$ in the equation $(\lambda I - A)x = 0$ to obtain the corresponding eigenvectors.

$$(\tfrac{1}{2} I - A) x = 0$$

$$\begin{bmatrix} \frac{1}{2} & -\frac{1}{2} \\ -\frac{1}{2} & \frac{1}{2} \end{bmatrix} \begin{bmatrix} x_1 \\ x_2 \end{bmatrix} = \begin{bmatrix} 0 \\ 0 \end{bmatrix}$$

or

$$\tfrac{1}{2} x_1 - \tfrac{1}{2} x_2 = 0$$

$$-\tfrac{1}{2} x_1 + \tfrac{1}{2} x_2 = 0$$

or,
$$x_1 - x_2 = 0.$$

Thus,

$$X_1 = \begin{bmatrix} 1 \\ 1 \end{bmatrix}$$

is an eigenvector of A associated with the eigenvalue

$$\lambda_1 = \frac{1}{2}.$$

Now, let $\lambda = -\frac{1}{2}$. Then,

$$\begin{bmatrix} -\frac{1}{2} & -\frac{1}{2} \\ -\frac{1}{2} & -\frac{1}{2} \end{bmatrix} \begin{bmatrix} x_1 \\ x_2 \end{bmatrix} = \begin{bmatrix} 0 \\ 0 \end{bmatrix}$$

therefore,

$$-\frac{1}{2} x_1 - \frac{1}{2} x_2 = 0 ;$$

$$-\frac{1}{2} x_1 - \frac{1}{2} x_2 = 0$$

or,

$$x_1 + x_2 = 0$$

Then,

$$X_2 = \begin{bmatrix} 1 \\ -1 \end{bmatrix}$$

is an eigenvector of A associated with the eigenvalue

$$\lambda_2 = -\frac{1}{2}.$$

If one lets $L: R^2 \to R^2$ be defined by

$$L(X) = AX = \begin{bmatrix} 0 & \frac{1}{2} \\ \frac{1}{2} & 0 \end{bmatrix} \begin{bmatrix} x_1 \\ x_2 \end{bmatrix}$$

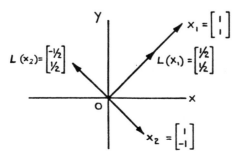

then the accompanying figure shows that X_1 and $L(X_1)$ are parallel and that X_2 and $L(X_2)$ are parallel also.

This illustrates the fact that if X is an eigenvector of A, then X and AX are parallel.

● **PROBLEM 12-6**

Let $A = \begin{bmatrix} 3 & 1 \\ 4 & 3 \end{bmatrix}$. Find an eigenvector for A.

Solution: Then:

$$f(\lambda) = \begin{vmatrix} 3 - \lambda & 1 \\ 4 & 3 - \lambda \end{vmatrix} = (3 - \lambda)^2 - 4.$$

The zeros of f (and hence the eigenvalues of A) are given by

$$\lambda_1 = 5, \quad \lambda_2 = 1.$$

An eigenvector corresponding to λ_1 can be obtained by solving the equation above with $\lambda = \lambda_1$:

$$\begin{bmatrix} -2 & 1 \\ 4 & -2 \end{bmatrix} \begin{bmatrix} x_1 \\ x_2 \end{bmatrix} = \vec{0},$$

$$-2x_1 + x_2 = 0,$$
$$4x_1 - 2x_2 = 0.$$

This system is dependent, so x_2 may be chosen arbitrarily. Choosing $x_2 = 2$, one finds that (1,2) is an eigenvector corresponding to λ_1.

● **PROBLEM 12-7**

Find the eigenvalues and eigenvectors for the matrix

$$\begin{pmatrix} 0 & 1 & 0 \\ 0 & 0 & 1 \\ -8 & -12 & -6 \end{pmatrix}$$

Solution: Set

$$\begin{vmatrix} -\lambda & 1 & 0 \\ 0 & -\lambda & 1 \\ -8 & -12 & -6 - \lambda \end{vmatrix} = \vec{0}$$

and obtain the characteristic equation

$$-\lambda^3 - 6\lambda^2 - 12\lambda - 8 = \vec{0}$$

which has the roots

$$\lambda_1 = -2, \quad \lambda_2 = -2, \quad \lambda_3 = -2.$$

To find the eigenvectors, set

$$\begin{pmatrix} 2 & 1 & 0 \\ 0 & 2 & 1 \\ -8 & -12 & -4 \end{pmatrix} \begin{pmatrix} x_1 \\ x_2 \\ x_3 \end{pmatrix} = \vec{0}.$$

Since the coefficient matrix has rank 2, there is only one linearly independent eigenvector. It turns out to be:

$$\begin{pmatrix} 1 \\ -2 \\ 4 \end{pmatrix}$$

Since all roots are the same, no more eigenvectors can be obtained. In this case one gets a triple eigenvalue, -2, and only one eigenvector (which might be considered an eigenvector of multiplicity 3):

$$\begin{pmatrix} 1 \\ -2 \\ 4 \end{pmatrix}$$

• **PROBLEM 12-8**

Find the eigenvalues and eigenvectors for the matrix

$$\begin{pmatrix} 3 & 2 & 4 \\ 1 & 4 & 4 \\ -1 & -2 & -2 \end{pmatrix}$$

Solution: To determine the eigenvalues, set

$$\begin{vmatrix} 3-\lambda & 2 & 4 \\ 1 & 4-\lambda & 4 \\ -1 & -2 & -2-\lambda \end{vmatrix} = \vec{0}.$$

Expanding, obtain the characteristic equation

$$-\lambda^3 + 5\lambda^2 - 8\lambda + 4 = 0$$

which has the roots

$$\lambda_1 = 1, \quad \lambda_2 = \lambda_3 = 2.$$

To find the eigenvector corresponding to λ_1, set

$$\begin{pmatrix} 3 - \lambda_1 & 2 & 4 \\ 1 & 4 - \lambda_1 & 4 \\ -1 & -2 & -2 - \lambda_1 \end{pmatrix} \begin{pmatrix} x_1 \\ x_2 \\ x_3 \end{pmatrix} = \vec{0}$$

or

$$\begin{pmatrix} 2 & 2 & 4 \\ 1 & 3 & 4 \\ -1 & -2 & -3 \end{pmatrix} \begin{pmatrix} x_1 \\ x_2 \\ x_3 \end{pmatrix} = \vec{0}.$$

The coefficient matrix has rank 2, so this system has one linearly dependent vector solution. Hence, the eigenvector corresponding to λ_1 is

$$\begin{pmatrix} 1 \\ 1 \\ -1 \end{pmatrix}$$

To find the eigenvector corresponding to λ_2, set

$$\begin{pmatrix} 3 - \lambda_2 & 2 & 4 \\ 1 & 4 - \lambda_2 & 4 \\ -1 & -2 & -2 - \lambda_2 \end{pmatrix} \begin{pmatrix} x_1 \\ x_2 \\ x_3 \end{pmatrix} = \vec{0}$$

or

$$\begin{pmatrix} 1 & 2 & 4 \\ 1 & 2 & 4 \\ -1 & -2 & -4 \end{pmatrix} \begin{pmatrix} x_1 \\ x_2 \\ x_3 \end{pmatrix} = \vec{0}.$$

The coefficient matrix has rank 1, so this system has two linearly independent vector solutions,

$$\begin{pmatrix} -4 \\ 0 \\ 1 \end{pmatrix} \quad \text{and} \quad \begin{pmatrix} -2 \\ 1 \\ 0 \end{pmatrix}.$$

The root λ_3, being the same as λ_2, has the same eigenvectors. Hence there is a single root, 1, with one eigenvector, and a double root, 2, with two eigenvectors.

● **PROBLEM 12-9**

A surprisingly large number of problems in physics and engineering reduce to the following mathematical problem: Given a square matrix A, find a nonzero vector x and a constant λ such that

$$A\vec{x} = \lambda \vec{x}.$$

That is, find a vector $\vec{x}$ such that $A\vec{x}$ is simply a multiple of itself. One can rewrite this equation as

$$A\vec{x} - \lambda \vec{x} = \vec{0}$$

or

$$(A - \lambda I)\vec{x} = \vec{0}.$$

With the aid of the above information, find the eigenvalues and eigenvectors for the matrix

$$\begin{pmatrix} 1 & 3 \\ 2 & 2 \end{pmatrix}$$

Solution: To find the eigenvalues, set

$$\begin{vmatrix} 1 - \lambda & 3 \\ 2 & 2 - \lambda \end{vmatrix} = \vec{0}.$$

Expanding, one obtains the characteristic equation

$$(1 - \lambda)(2 - \lambda) - 6 = \lambda^2 - 3\lambda - 4 = \vec{0}.$$

This factors into

$$(\lambda - 4)(\lambda + 1) = \vec{0}$$

so the eigenvalues are

$$\lambda_1 = 4, \quad \lambda_2 = -1.$$

To find the eigenvector corresponding to λ_1, set

$$\begin{pmatrix} 1 - \lambda_1 & 3 \\ 2 & 2 - \lambda_1 \end{pmatrix} \begin{pmatrix} x_1 \\ x_2 \end{pmatrix} = \vec{0}$$

or

$$\begin{pmatrix} -3 & 3 \\ 2 & -2 \end{pmatrix} \begin{pmatrix} x_1 \\ x_2 \end{pmatrix} = \vec{0}$$

443

Since there are two unknowns and the coefficient matrix has rank 1, these equations have one linearly independent vector solution $\vec{U}_1$, and all other solutions are multiples of this one. One sees by inspection that the vector

$$\begin{pmatrix} 1 \\ 1 \end{pmatrix}$$

is a solution, and hence is an eigenvector corresponding to λ_1. All other solutions are of the form

$$c_1 \begin{pmatrix} 1 \\ 1 \end{pmatrix}$$

where c_1 is an arbitrary constant. Hence, the eigenvector is really determined only up to an arbitrary constant multiple.

To find the eigenvector corresponding to λ_2, set

$$\begin{pmatrix} 1 - \lambda_2 & 3 \\ 2 & 2 - \lambda_2 \end{pmatrix} \begin{pmatrix} x_1 \\ x_2 \end{pmatrix} = \vec{0}$$

or

$$\begin{pmatrix} 2 & 3 \\ 2 & 3 \end{pmatrix} \begin{pmatrix} x_1 \\ x_2 \end{pmatrix} = \vec{0}$$

Again there is one linearly independent solution vector. One sees by inspection that

$$\begin{pmatrix} 3 \\ -2 \end{pmatrix}$$

is a solution. All solutions are of the form

$$c_2 \begin{pmatrix} 3 \\ -2 \end{pmatrix}$$

where c_2 is an arbitrary constant.

Hence the eigenvalues are

$$4, \quad -1$$

and the corresponding eigenvectors are

$$\begin{pmatrix} 1 \\ 1 \end{pmatrix} \quad \text{and} \quad \begin{pmatrix} 3 \\ -2 \end{pmatrix}$$

(Ordinarily, the arbitrary constant multiple is ignored when writing an eigenvector.)

• **PROBLEM 12-10**

The characteristic values of the matrix

$$A = \begin{bmatrix} 8 & 2 & -2 \\ 3 & 3 & -1 \\ 24 & 8 & -6 \end{bmatrix}$$

are 2 and 1. Find the characteristic vectors.

Solution: Let W_1 denote the set of characteristic vectors that belong to 2. A vector

$$V = (x_1, x_2, x_3)$$

is in W_1 if and only if $VA = 2V$; that is, $V[A - 2I] = 0$ or,

$$[x_1, x_2, x_3] \begin{bmatrix} 6 & 2 & -2 \\ 3 & 1 & -1 \\ 24 & 8 & -8 \end{bmatrix} = [0, 0, 0]$$

or

$$6x_1 + 3x_2 + 24x_3 = 0$$
$$2x_1 + x_2 + 8x_3 = 0 \qquad (1)$$
$$-2x_1 - x_2 - 8x_3 = 0.$$

Thus, W_1 is the solution space of (1). The system (1) is clearly equivalent to

$$2x_1 + x_2 + 8x_3 = 0.$$

Since

$$x_2 + 8x_3 = -2x_1$$

and

$$2x_1 + x_2 = -8x_3,$$

it follows that

$$v_1 = (1, -2, 0)$$

and

$$v_2 = (0, -8, 1);$$

$[v_1, v_2]$ is a basis of W_1. The vectors v_1 and v_2 are characteristic vectors belonging to 2, and every characteristic vector belonging to 2 is a linear combination of v_1 and v_2.

Let W_2 denoted the set of characteristic vectors that belong to 1. A vector $V = (x_1, x_2, x_3)$ belongs to W_2 if and only if $VA = V$; that is,

$$v[A - 1I] = 0,$$

or,

$$[x_1 x_2 x_3] \begin{bmatrix} 7 & 2 & -2 \\ 3 & 2 & -1 \\ 24 & 8 & -7 \end{bmatrix} = [0,0,0]$$

or,

$$7x_1 + 3x_2 + 24x_3 = 0$$
$$2x_1 + 2x_2 + 8x_3 = 0$$
$$-2x_1 - x_2 - 7x_3 = 0$$

Solving this system gives $x_1 = 3$, $x_2 = 1$, $x_3 = -1$ as one uncomplicated solution. Thus, $v_3 = (3,1,-1)$ is a basis of W_2. The vector v_3 is a characteristic vector belonging to 1, and every characteristic vector belonging to 1 is a scalar multiple of v_3.

• **PROBLEM 12-11**

Given

$$A = \begin{bmatrix} 1 & 0 & -1 & 0 \\ 0 & 1 & 0 & -1 \\ 1 & 0 & -1 & 0 \\ 0 & 1 & 0 & -1 \end{bmatrix}$$

and

$$B = \begin{bmatrix} 4 & -1 & -1 & 0 \\ -1 & 4 & 0 & -1 \\ 1 & 0 & 2 & -1 \\ 0 & 1 & -1 & 2 \end{bmatrix},$$

find a characteristic vector common to A and B.

Solution: If $AB = BA$, then A and B have a characteristic vector in common.

Let λ_1 be a characteristic root of A and let $x_1, \ldots, x_k$ be a basis for the null space of $\lambda_1 I_n - A$. Then, any non-zero linear combination of $x_1, \ldots, x_k$ is a characteristic vector corresponding to λ_1. Conversely, any characteristic vector of A corresponding to λ_1 can be expressed as a linear combination of $(x_1, x_2, \ldots, x_k)$. Suppose that $Bx_i \neq 0$, $i = 1, \ldots, k$, then

$$A(Bx_i) = BAx_i$$
$$= B(\lambda_1 x_i)$$
$$= \lambda_1 (Bx_i) ;$$

thus,

$$Bx_i \in Sp(x_1, \ldots, x_k), \quad i = 1, \ldots, k .$$

Let X be the matrix whose jth column is x_j, $j = 1, \ldots, k$. Therefore, the system of equations,

$$Xy_i = Bx_i , \quad i = 1, \ldots, k, \qquad (1)$$

is the matrix whose ith column is y_i, $i = 1, \ldots, k$. The relation (1) can be written compactly as

$$XY = BX . \qquad (2)$$

Let μ_1 be any characteristic root of Y and $z = (\alpha_1, \ldots, \alpha_k)$, a characteristic vector of Y corresponding to μ_1. Then, by (2),

$$B(Xz) = XYz$$

and since

$$Yz = \mu_1 z,$$
$$B(Xz) = \mu_1 (Xz) .$$

Now,

$$Xz = \sum_{t=1}^{k} \alpha_t x_t \neq 0$$

because $z = (\alpha_1, \ldots, \alpha_k) \neq 0$ and $x_1, \ldots, x_k$ are linearly independent. Thus, Xz is a characteristic vector of B. But,

$$Xz = \sum_{t=1}^{k} \alpha_t x_t \in (x_1, \ldots, x_k)$$

and, therefore, Xz is a characteristic vector of A. Now, find the product of AB and BA.

$$AB = \begin{bmatrix} 3 & -1 & -3 & 1 \\ -1 & 3 & 1 & -3 \\ 3 & -1 & -3 & 1 \\ -1 & 3 & 1 & -3 \end{bmatrix} = BA.$$

Thus, AB = BA. Therefore, A and B have a common characteristic vector. The matrix A is clearly singular and, therefore, $\det(I\lambda - A) = 0$ implies $\lambda = 0$ (since $\det A = 0$). Since the rank of $\lambda_1 I_4 - A$ is 2, the dimension of its null space is $4 - 2 = 2$. The vector $x_1 = (1,0,1,0)$ and $x_2 = (0,1,0,1)$ form a basis for this null space

$$Bx_1 = (3,-1,1,3); \quad Bx_2 = (-1,3,-1,3)$$

Thus,
$$Bx_1 = 3x_1 - x_2 \text{ and } Bx_2 = -x_1 + 3x_2 .$$

Let
$$X = \begin{bmatrix} 1 & 0 \\ 0 & 1 \\ 1 & 0 \\ 0 & 1 \end{bmatrix}$$

and
$$Y = \begin{bmatrix} 3 & -1 \\ -1 & 3 \end{bmatrix} .$$

Then, as in (2),
$$XY = BX.$$

To find the characteristic root of Y, solve the equation, $\det(\lambda I - Y) = 0$. Then,

$$\det \begin{vmatrix} \lambda-3 & 1 \\ 1 & \lambda-3 \end{vmatrix} = 0$$

or,
$$(\lambda-3)^2 - 1 = 0$$
$$(\lambda-4)(\lambda-2) = 0.$$

Therefore, $\mu_1 = 4$ is the characteristic root of Y, and $z = (1,-1)$ is the corresponding characteristic vector. Then,

$$Xz = \begin{bmatrix} 1 & 0 \\ 0 & 1 \\ 1 & 0 \\ 0 & 1 \end{bmatrix} \begin{bmatrix} 1 \\ -1 \end{bmatrix} = \begin{bmatrix} 1 \\ -1 \\ 1 \\ -1 \end{bmatrix}$$

Thus, $Xz = (1,-1,1,-1)$ is a characteristic vector common to A and B.

• **PROBLEM 12-12**

Show how to evaluate $f'_n(\lambda)$ and $f'_n(\lambda)$ for specific λ, where $f_n(\lambda)$ is the characteristic polynomial of the triple diagonal n × n matrix

$$A = [\ldots, c_r, a_r, b_r, \ldots].$$

Do the same when $f_n(\lambda)$ is the characteristic polynomial of a Hessenberg matrix.

Evaluate the characteristic polynomials of

$$A = \begin{vmatrix} 0 & 10 & 0 & 0 & 0 \\ 10 & 0 & 10 & 0 & 0 \\ 0 & 10 & 0 & 10 & 0 \\ 0 & 0 & 10 & 0 & 10 \\ 0 & 0 & 0 & 10 & 0 \end{vmatrix},$$

$$B = \begin{vmatrix} 0 & 1 & 0 & 0 & 0 \\ 3 & 0 & 1 & 0 & 0 \\ 4 & 3 & 0 & 1 & 0 \\ 5 & 4 & 3 & 0 & 1 \\ 6 & 5 & 4 & 3 & 0 \end{vmatrix}$$

in the interesting ranges.

Solution: Write
$$f_0(\lambda) = 1, \quad f_1(\lambda) = a_1 - \lambda$$
and then use
$$f_r(\lambda) = (a_r - \lambda) f_{r-1}(\lambda) - b_{r-1} c_r f_{r-2}(\lambda)$$
for $r = 2, 3, \ldots, n$.
The corresponding formulas for the derivatives are:
$$f'_0(\lambda) = 0, \quad f'_1(\lambda) = -1$$
$$f'_r(\lambda) = (a_r - \lambda) f'_{r-1}(\lambda) - b_{r-1} c_r f'_{r-2}(\lambda) - f_{r-1}(\lambda)$$

and these can be computed at the same time as the $f_r(\lambda)$. The characteristic values of A are

$$0 + 20 \cos r\, \pi/6, \quad r = 1,2,3,4,5.$$

i.e.,

$$\pm 10\sqrt{3},\ \pm 10, 0.$$

The characteristic polynomial of B is

$$-\lambda^5 + * + 12\lambda^3 + 12\lambda^2 - 17\lambda - 18$$

$$= -(\lambda + 1)(\lambda + 2)(\lambda^3 - 3\lambda^2 - 5\lambda + 9)$$

and the characteristic values are

$$-2,\ -1.9459970,\ -1,\ 1.2520004,\ 3.6939948.$$

Since tr B = 0, the term in λ^4 being absent, some idea of the errors in the solution is given by the actual sum of the characteristic values which is

$$4 \times 10^{-7}.$$

● **PROBLEM** 12-13

Find the eigenvalues of m'm:

(1) $m = \begin{bmatrix} 1 & 0 \\ 0 & 0 \end{bmatrix}$

(2) $m = \begin{bmatrix} 0 & 1 \\ -1 & 0 \end{bmatrix}$

(3) $m = \begin{bmatrix} 1 & 1 \\ 0 & 1 \end{bmatrix}$.

Solution: The map $X \to mX = Y$ maps the unit sphere in x-space onto an ellipsoid in y space, and the squares of the semi-axes of the image are equal to the eigenvalues of m'm. If one has the map $X \to mX$ and puts $mX = Y$, then the x's which are mapped into $Y'Y = 1$ satisfy $X'm'mX = 1$. They lie on a quadric surface in the original space.

When the eigenvalues are all positive, the locus is an ellipsoid.

(1) $m = \begin{bmatrix} 1 & 0 \\ 0 & 0 \end{bmatrix}$

The locus is X'm'mX = 1 (in this case, the pair of lines $x_1 = \pm 1$). These map onto the two points $(\pm 1, 0)$. m'm has eigenvalues 1 and 0.

(2) $\quad m = \begin{bmatrix} 0 & 1 \\ -1 & 0 \end{bmatrix}$

This is a rotation. The unit circle is rotated through a right angle. m'm = I. The unit circle is invariant.

(3) $\quad m = \begin{bmatrix} 1 & 1 \\ 0 & 1 \end{bmatrix}$

This is a shear. The unit circle becomes an ellipse

$$x^2 - 2x_1 x_2 + 2x_2^2 = 1.$$

The major axis is in the direction $(2, -1 + \sqrt{5})$ and is of length $1 + \sqrt{5}$. The minor axis is in the direction $(2, -1 - \sqrt{5})$ and is of length $-1 + \sqrt{5}$.

$$mm' = \begin{bmatrix} 1 & 1 \\ 1 & 2 \end{bmatrix}.$$

The eigenvalues are $\frac{1}{2}(3 \pm \sqrt{5})$. The ellipse

$$x_1^2 + 2x_1 x_2 + 2x_2^2 = 1$$

is mapped onto the unit circle.

● **PROBLEM 12-14**

Find the eigenvalues and an orthonormal basis for the eigenspace of A where,

$$A = \begin{bmatrix} 1 & 2 & 0 \\ 2 & 1 & 0 \\ 0 & 0 & 3 \end{bmatrix}.$$

Solution: A is a symmetric matrix. An important result concerning symmetric matrices is that all eigenvalues of a symmetric matrix are real. Form the matrix

$$(\lambda I - A) = \begin{bmatrix} \lambda & 0 & 0 \\ 0 & \lambda & 0 \\ 0 & 0 & \lambda \end{bmatrix} - \begin{bmatrix} 1 & 2 & 0 \\ 2 & 1 & 0 \\ 0 & 0 & 3 \end{bmatrix}$$

$$= \begin{bmatrix} \lambda-1 & -2 & 0 \\ -2 & \lambda-1 & 0 \\ 0 & 0 & \lambda-3 \end{bmatrix}.$$

Now, expanding along the third column yields

$$\det(\lambda I - A) = (\lambda-3) \begin{vmatrix} \lambda-1 & -2 \\ -2 & \lambda-1 \end{vmatrix}$$

$$= (\lambda-3)(\lambda^2 - 2\lambda + 1 - 4)$$

$$= (\lambda-3)(\lambda-3)(\lambda+1).$$

The characteristic equation of A is $(\lambda-3)^2(\lambda+1) = 0$ and, therefore, the eigenvalues of A are $\lambda = 3$ and $\lambda = -1$.

Now find the eigenvectors corresponding to $\lambda_1 = 3$. To do this, solve

$$(\lambda I - A)X = 0 \text{ for } X \text{ with } \lambda = 3.$$

$$(3I - A)X = 0$$

or,

$$\begin{bmatrix} 2 & -2 & 0 \\ -2 & 2 & 0 \\ 0 & 0 & 0 \end{bmatrix} \begin{bmatrix} x_1 \\ x_2 \\ x_3 \end{bmatrix} = \begin{bmatrix} 0 \\ 0 \\ 0 \end{bmatrix}.$$

This is equivalent to

$$2x_1 - 2x_2 = 0$$

$$-2x_1 + 2x_2 = 0$$

or

$$x_1 - x_2 = 0.$$

Solving this system gives

$$x_1 = s, \quad x_2 = s, \quad x_3 = t.$$

Therefore,

$$X = \begin{bmatrix} s \\ s \\ t \end{bmatrix} = \begin{bmatrix} s \\ s \\ 0 \end{bmatrix} + \begin{bmatrix} 0 \\ 0 \\ t \end{bmatrix} = s \begin{bmatrix} 1 \\ 1 \\ 0 \end{bmatrix} + \begin{bmatrix} 0 \\ 0 \\ 1 \end{bmatrix}$$

or,

$$X_1 = \begin{bmatrix} 1 \\ 1 \\ 0 \end{bmatrix} \quad X_2 = \begin{bmatrix} 0 \\ 0 \\ 1 \end{bmatrix}$$

Note that x_1 and x_2 are orthogonal to each other since $x_1 \cdot x_2 = 0$. Next, normalize x_1 and x_2 to obtain the unit orthogonal solutions by replacing x_i with

$$\frac{x_i}{|x_i|}.$$

Since $|x_1| = \sqrt{2}$ and $|x_2| = 1$,

$$u_1 = \begin{bmatrix} 1/\sqrt{2} \\ 1/\sqrt{2} \\ 0 \end{bmatrix} ; \quad u_2 = \begin{bmatrix} 0 \\ 0 \\ 1 \end{bmatrix}$$

and they form a basis for the eigenspace corresponding to $\lambda = 3$. To find the eigenvectors corresponding to $\lambda = -1$, solve $(\lambda I - A)X = 0$ for X with $\lambda = -1$.

$(-1I - A)X = 0$ or,

$$\begin{bmatrix} -2 & -2 & 0 \\ -2 & -2 & 0 \\ 0 & 0 & -4 \end{bmatrix} \begin{bmatrix} x_1 \\ x_2 \\ x_3 \end{bmatrix} = 0 \quad .$$

Carrying out the indicated matrix multiplication,

$$-2x_1 - 2x_2 = 0$$
$$-2x_1 - 2x_2 = 0$$
$$-4x_3 = 0$$

or,

$$x_1 + x_2 = 0$$
$$x_3 = 0 \quad .$$

Solving this system gives

$$x_3 = \begin{bmatrix} 1 \\ -1 \\ 0 \end{bmatrix} \quad .$$

Now, normalize x_3 to obtain the unit orthogonal solution. Thus,

$$u_3 = \begin{bmatrix} 1/\sqrt{2} \\ -1/\sqrt{2} \\ 0 \end{bmatrix}$$

forms a basis for the eigenspace corresponding to $\lambda = -1$. Since

$$u_1 \cdot u_3 = 0$$

and
$$u_2 \cdot u_3 = 0,$$
$\{u_1, u_2, u_3\}$ is an orthonormal basis of R^3. In general, if A is a symmetric $n \times n$ matrix, then the eigenvectors of A contain an orthonormal basis of R^n.

● **PROBLEM** 12-15

Find a basis for the eigenspace of
$$A = \begin{bmatrix} 3 & -2 & 0 \\ -2 & 3 & 0 \\ 0 & 0 & 5 \end{bmatrix}.$$

Solution: If λ is an eigenvalue of A, then the solution space for the system of equations $(\lambda I - A)X = 0$ is called the eigenspace of A corresponding to λ, and the non-zero vectors in the eigenspace are called the eigenvectors of A corresponding to λ.

Form the matrix

$$\lambda I - A = \lambda \begin{bmatrix} 1 & 0 & 0 \\ 0 & 1 & 0 \\ 0 & 0 & 1 \end{bmatrix} - \begin{bmatrix} 3 & -2 & 0 \\ -2 & 3 & 0 \\ 0 & 0 & 5 \end{bmatrix}$$

$$= \begin{bmatrix} \lambda-3 & 2 & 0 \\ 2 & \lambda-3 & 0 \\ 0 & 0 & \lambda-5 \end{bmatrix}.$$

$$\det(\lambda I - A) = \det \begin{vmatrix} \lambda-3 & 2 & 0 \\ 2 & \lambda-3 & 0 \\ 0 & 0 & \lambda-5 \end{vmatrix}$$

$$= (\lambda-5) \begin{vmatrix} \lambda-3 & 2 \\ 2 & \lambda-3 \end{vmatrix}$$

$$= (\lambda-5)[(\lambda-3)^2 - 4]$$

$$= (\lambda-5)[\lambda^2 - 6\lambda + 9 - 4]$$

$$= (\lambda-5)[\lambda^2 - 6\lambda + 5]$$

$$= (\lambda-5)(\lambda-5)(\lambda-1)$$

$$= (\lambda-5)^2(\lambda-1).$$

The characteristic equation of A is $(\lambda-5)^2(\lambda-1) = 0$, so that the eigenvalues of A are $\lambda = 1$ and $\lambda = 5$.

By definition,

$$X = \begin{bmatrix} x_1 \\ x_2 \\ x_3 \end{bmatrix}$$

is an eigenvector of A corresponding to λ if and only if x is a non-trivial solution of $(\lambda I - A)X = 0$. Thus,

$$\begin{bmatrix} \lambda-3 & 2 & 0 \\ 2 & \lambda-3 & 0 \\ 0 & 0 & \lambda-5 \end{bmatrix} \begin{bmatrix} x_1 \\ x_2 \\ x_3 \end{bmatrix} = \begin{bmatrix} 0 \\ 0 \\ 0 \end{bmatrix}. \quad (1)$$

If $\lambda = 5$, then equation (1) becomes

$$\begin{bmatrix} 2 & 2 & 0 \\ 2 & 2 & 0 \\ 0 & 0 & 0 \end{bmatrix} \begin{bmatrix} x_1 \\ x_2 \\ x_3 \end{bmatrix} = \begin{bmatrix} 0 \\ 0 \\ 0 \end{bmatrix}.$$

Solving this system yields $x_1 = -s$, $x_2 = s$, $x_3 = t$, where s and t are any scalars. Thus, the eigenvectors of A corresponding to $\lambda = 5$ are the non-zero vectors of the form

$$X = \begin{bmatrix} -s \\ s \\ t \end{bmatrix} = \begin{bmatrix} -s \\ s \\ 0 \end{bmatrix} + \begin{bmatrix} 0 \\ 0 \\ t \end{bmatrix} = s\begin{bmatrix} -1 \\ 1 \\ 0 \end{bmatrix} + t\begin{bmatrix} 0 \\ 0 \\ 1 \end{bmatrix}$$

Since

$$\begin{bmatrix} -1 \\ 1 \\ 0 \end{bmatrix}$$

and

$$\begin{bmatrix} 0 \\ 0 \\ 1 \end{bmatrix}$$

are linearly independent, they form a basis for the eigenspace corresponding to $\lambda = 5$. If $\lambda = 1$, then equation (1) becomes

$$\begin{bmatrix} -2 & 2 & 0 \\ 2 & 2 & 0 \\ 0 & 0 & -4 \end{bmatrix} \begin{bmatrix} x_1 \\ x_2 \\ x_3 \end{bmatrix} = \begin{bmatrix} 0 \\ 0 \\ 0 \end{bmatrix}.$$

Solving this system yields $x_1 = t$, $x_2 = t$, $x_3 = 0$; t is any scalar. Thus, the eigenvectors corresponding to $\lambda = 1$ are non-zero vectors of the form:

$$X = \begin{bmatrix} t \\ t \\ 0 \end{bmatrix} = t \begin{bmatrix} 1 \\ 1 \\ 0 \end{bmatrix}$$

so that

$$\begin{bmatrix} 1 \\ 1 \\ 0 \end{bmatrix}$$

is a basis for the eigenspace corresponding to $\lambda = 1$.

● **PROBLEM** 12-16

Find the dominant characteristic value and vector for

$$\begin{bmatrix} 133 & 6 & 135 \\ 44 & 5 & 46 \\ -88 & -6 & -90 \end{bmatrix}$$

by the power method.

Find also the other characteristic values and vectors.

<u>Solution</u>: For a symmetric $n \times n$ matrix H, the eigenvectors span the space, that is, any n-dimensional vector V can be written as

$$V = c_1 X_1 + c_2 X_2 + c_3 X_3 + \ldots + c_n X_n \tag{1}$$

Note that if V is a good estimate of one of the eigenvectors X_i, then the constant c_i associated with that eigenvector will be considerably larger in magnitude than the other c's. Now suppose one forms the product HV. From (1) obtain

$$HV = c_1 HX_1 + c_2 HX_2 + c_3 HX_3 + \ldots + c_n HX_n \tag{2}$$

but

$$HX_1 = \lambda_1 X_1, \quad HX_2 = \lambda_2 X_2, \text{ etc.}$$

Thus (2) can be written as

$$HV = c_1 \lambda_1 X_1 + c_2 \lambda_2 X_2 + c_3 \lambda_3 X_3 + \ldots + c_n \lambda_n X_n \tag{3}$$

If one again premultiplies by H, (3) becomes

$$H(HV) = c_1 \lambda_1^2 X_1 + c_2 \lambda_2^2 X_2 + c_3 \lambda_3^2 X_3 + \ldots \tag{4}$$

$$+ c_n \lambda_n^2 x_n$$

Each succeeding multiplication by H increases by one the power to which the eigenvalues are raised, and it should be apparent that the term involving the eigenvalue which is largest in magnitude will eventually dominate the right side of the equation. This is the principle behind the power method. If the largest eigenvalue in magnitude (called the dominant eigenvalue) is denoted as λ_1 and the next largest as λ_2, then the ratio

$$r = \frac{|\lambda_2|}{|\lambda_1|} \tag{5}$$

defined as the dominance ratio, is clearly of fundamental importance in how rapidly the term involving λ_1 overwhelms all other terms. If V is a good estimate of the eigenvector associated with λ_1, then this will also accelerate the process, since the corresponding value of c will be large in magnitude compared with the other c's.

In practice, the power method is applied as follows. Denote the first estimate of the eigenvector as V_0. (All elements of V_0 are usually taken as 1 unless there is a better estimate available.) Then calculate

$$HV_0 = V_1 \tag{6}$$

Now normalize the vector V_1 by making one of its elements equal to 1. This is done by dividing each of its elements by any one element (usually chosen as the largest in magnitude to ensure best accuracy). Thus

$$V_1 = p_1 V_1' \tag{7}$$

where p_1 is a constant equal to the element which has been normalized to 1, and V_1' is the normalized vector found by dividing each element of V_1 by p_1. Now form the product

$$HV_1' = V_2 = p_2 V_2' \tag{8}$$

and continue this process as many times as desired. As the number of iterations increases, p approaches the largest eigenvalue in magnitude of H, and V' approaches its associated eigenvector.

Considering the above problem, the solution is given below.

The characteristic pairs are

$$45, \ [-3, \ -1, \ 2]'$$
$$2, \ [-3, \ -2, \ 3]'$$
$$1, \ [-2, \ -1, \ 2]'.$$

● PROBLEM 12-17

Consider the matrix

$$A = \begin{bmatrix} 1 & 2 \\ 1 & 1 \end{bmatrix}$$

and the initial values

$$x_0 = \begin{bmatrix} 1 \\ 1 \end{bmatrix}$$

$$x_1 = Ax_0$$

$$x_2 = Ax_1$$

$$x_3 = Ax_2$$

$$x_4 = Ax_3$$

By the power method, find an approximate eigenvector of A.

Solution: Applying the power method to the matrix,

$$A = \begin{bmatrix} 1 & 2 \\ 1 & 1 \end{bmatrix},$$

take

$$x_0 = \begin{bmatrix} 1 \\ 1 \end{bmatrix}.$$

Then

$$x_1 = Ax_0 = \begin{bmatrix} 3 \\ 2 \end{bmatrix},$$

$$x_2 = Ax_1 = \begin{bmatrix} 7 \\ 5 \end{bmatrix},$$

$$x_3 = Ax_2 = \begin{bmatrix} 17 \\ 12 \end{bmatrix},$$

$$x_4 = Ax_3 = \begin{bmatrix} 41 \\ 29 \end{bmatrix}.$$

Hence

$$\lambda_1 \doteq \frac{41}{17} = 2.4118 \quad \text{or} \quad \lambda_1 \doteq \frac{29}{12} = 2.4167.$$

In fact, $\lambda_1 = 1 + \sqrt{2} = 2.4142$. An approximate eigenvector is given by

$$u = \frac{x_4}{(2.4118)^4} = \begin{bmatrix} 1.2118 \\ 0.8571 \end{bmatrix}.$$

The ratio $u_1/u_2 = 1.2118/0.8571 = 1.4138$, while in fact one should have

$$u_1/u_2 = \sqrt{2} = 1.4142.$$

• **PROBLEM 12-18**

Find the dominant eigenvalue and corresponding eigenvector of

$$\begin{bmatrix} -3 & 8 & 9 \\ 8 & -2 & 4 \\ 9 & 4 & 2 \end{bmatrix}$$

Use the power method, and show the effects of shifting to increase the convergence rate. Use an initial guess of

$$V = \begin{bmatrix} 1 \\ 1 \\ 1 \end{bmatrix}$$

and an absolute convergence criterion of 10^{-5}.

Solution: The straight power method, without relaxation, and with an initial guess of

$$V = \begin{bmatrix} 1 \\ 1 \\ 1 \end{bmatrix}$$

and an absolute convergence criterion of 10^{-5} on the eigenvalue, requires 124 iterations to produce the following results:

$$\lambda_1 = 13.25761, \qquad X_1 = \begin{bmatrix} 0.9199474 \\ 0.7445194 \\ 1 \end{bmatrix}$$

From the large number of iterations required, it should be apparent that $|\lambda_1| \approx |\lambda_2|$, or in other words that the dominance ratio is close to unity. A strategy of shifting all of the eigenvalues by a constant may be helpful in this situation, particularly if λ_1 and λ_2 have opposite signs. Simply by adding a constant to the main diagonal elements of the matrix, the eigenvalues are all shifted by this constant. Attempting to accelerate the slow convergence in the present problem by adding 5 to each main diagonal element gives the resulting matrix

$$\begin{bmatrix} 2 & 8 & 9 \\ 8 & 3 & 4 \\ 9 & 4 & 7 \end{bmatrix}$$

Again using an initial estimate of all 1's for the eigenvector, the power method applied to this matrix with the same convergence criterion as before requires only 14 iterations to yield

$$\lambda_1 = 18.25760, \qquad X_1 = \begin{bmatrix} 0.9199485 \\ 0.7445199 \\ 1 \end{bmatrix}$$

Substracting 5 from λ_1 yields a value of 13.25760 which is virtually identical to that obtained before the shift with 124 iterations. The eigenvectors are also in good agreement.

The enormous increase in convergence rate which was observed in this case cannot be expected in general when applying shifting. (The exact values of the first two eigenvalues of the original matrix are 13.25761 and -12.70709. The shift makes these eigenvalues 18.25761 and -7.70709 respectively. The dominance ratio thus changes from 0.95848 to 0.42213, which is obviously extremely beneficial to the power method.) However, shifting is certainly worth trying if slow convergence of the power method is encountered, and shifting strategies for such other methods as the LR and QR algorithms are almost indispensible.

• **PROBLEM** 12-19

A 4 × 4 matrix is known to have characteristic values near 20, 10, 5, 1. The power method will therefore enable one to find the root near 20. Show that the same method, when applied to a suitably shifted matrix $A - \alpha I$, will enable the characteristic value near 1 to be determined.

Solution: Taking $\alpha = \frac{1}{2}(20 + 5)$, the roots of the new matrix $A - \alpha I$ are

$$-\frac{23}{2}, \ -\frac{15}{2}, \ -\frac{5}{2}, \ \frac{15}{2}$$

and there is a separation factor of 15/23 which makes the power method quite attractive.

• **PROBLEM** 12-20

If λ_i are the eigenvalues of the square matrix A, prove that the eigenvalues of At are $\lambda_i t$.

Solution: Let A be the n×n matrix,

$$A = \begin{bmatrix} a_{11} & a_{12} & \cdots & a_{1n} \\ a_{21} & a_{22} & \cdots & a_{2n} \\ \vdots & & & \vdots \\ a_{n1} & a_{n2} & \cdots & a_{nn} \end{bmatrix} \quad (a)$$

Then

$$At = \begin{bmatrix} a_{11}t & a_{12}t & \cdots & a_{1n}t \\ a_{21}t & a_{22}t & \cdots & a_{2n}t \\ \vdots & & & \vdots \\ a_{n1}t & a_{n2}t & \cdots & a_{nn}t \end{bmatrix} \quad (b)$$

The eigenvalues of (a) are numbers λ, such that:

$$Ax = \lambda x,$$

where x is an n×1 column matrix,

$$x = \begin{bmatrix} x_1 \\ x_2 \\ \vdots \\ x_n \end{bmatrix}.$$

It must be shown that the eigenvalues of At are numbers λ, such that

$$Atx = \lambda tx.$$

Since λ is an eigenvalue of A, it must satisfy the characteristic equation:

$$\det(A - \lambda I) = 0, \qquad (c)$$

where I is the identity matrix. Now consider the determinant,

$$\det(At - \lambda tI). \qquad (d)$$

This may be rewritten as

$$\det[t(A - \lambda I)] = t^n \det(A - \lambda I).$$

But, from (c), $\det(A - \lambda I) = 0$. Therefore, (d) is equal to zero, i.e., λt is an eigenvalue of At.

● **PROBLEM 12-21**

Let

$$A_1 = \begin{bmatrix} 2 & 1 & 0 \\ 1 & 3 & 1 \\ 0 & 1 & 4 \end{bmatrix} \qquad (1)$$

Solve by QR algorithm.

Solution: Given a matrix A, there is a factorization

$$A = QR$$

with R upper triangular and Q orthogonal. With A real, both Q and R can be chosen real. Assume that A is real. Let $A_1 = A$, and define a sequence of matrices A_m, Q_m, and R_m by

$$A_m = Q_m R_m \qquad A_{m+1} = R_m Q_m \qquad m = 1, 2, \ldots \qquad (2)$$

If A satisfies any one of a number of different assumptions on its form, then the sequence $\{A_m\}$ will converge to either a triangular matrix with the eigenvalues of A on its diagonal or to a near-triangular matrix from which the eigenvalues can be easily calculated. In this form the convergence is usually slowly convergent; and a technique known as shifting is used to accelerate the convergence.

Before applying (2) to the problem, a few properties of (2) will be derived. From (2)

$$R_m = Q_m^T A_m,$$

and thus

$$A_{m+1} = Q_m^T A_m Q_m \qquad (3)$$

A_{m+1} is orthogonally similar to A_m, and thus by induction to $A_1 = A$. From (3),

$$A_{m+1} = Q_m^T \ldots Q_1^T A_1 Q_1 \ldots Q_m \qquad (4)$$

Introduce the matrices P_m and U_m by

$$P_m = Q_1 \ldots Q_m \qquad U_m = R_m \ldots R_1 \qquad (5)$$

Then from (4)

$$A_{m+1} = P_m^T A_1 P_m \qquad m \geq 1 \qquad (6)$$

The matrix P_m is orthogonal, and U_m is upper triangular. Derive a further relation involving P_m and U_m.

$$P_m U_m = Q_1 \ldots Q_m R_m \ldots R_1$$

$$= Q_1 \ldots Q_{m-1} A_m R_{m-1} \ldots R_1$$

From (4) with m replacing m + 1,

$$Q_1 \ldots Q_{m-1} A_m = A_1 Q_1 \ldots Q_{m-1}$$

Using this,

$$P_m U_m = A_1 Q_1 \ldots Q_{m-1} R_{m-1} \ldots R_1 = A_1 P_{m-1} U_{m-1}$$

Since $P_1 U_1 = Q_1 R_1 = A_1$, induction on the last statement implies

$$P_m U_m = A_1^m \qquad m \geq 1 \qquad (7)$$

For the above problem the eigenvalues are

$$\lambda_1 = 3 + \sqrt{3} \doteq 4.7321 \qquad \lambda_2 = 3.0$$

$$\lambda_3 = 3 - \sqrt{3} \doteq 1.2679$$

The iterates A_m do not converge rapidly, and only a few are given to indicate the qualitative behavior of the convergence.

$$A_2 = \begin{bmatrix} 3.0000 & 1.0954 & 0 \\ 1.0954 & 3.0000 & -1.3416 \\ 0 & -1.3416 & 3.0000 \end{bmatrix}$$

$$A_3 = \begin{bmatrix} 3.7059 & .9558 & 0 \\ .9558 & 3.5214 & .9738 \\ 0 & .9738 & 1.7727 \end{bmatrix}$$

$$A_7 = \begin{bmatrix} 4.6792 & .2979 & 0 \\ .2979 & 3.0524 & .0274 \\ 0 & .0274 & 1.2684 \end{bmatrix}$$

$$A_8 = \begin{bmatrix} 4.7104 & .1924 & 0 \\ .1924 & 3.0216 & -.0115 \\ 0 & -.0115 & 1.2680 \end{bmatrix}$$

$$A_9 = \begin{bmatrix} 4.7233 & .1229 & 0 \\ .1229 & 3.0087 & .0048 \\ 0 & .0048 & 1.2680 \end{bmatrix}$$

$$A_{10} = \begin{bmatrix} 4.7285 & .0781 & 0 \\ .0781 & 3.0035 & -.0020 \\ 0 & -.0020 & 1.2680 \end{bmatrix}$$

The elements in the (1,2) position decrease geometrically with a ratio of about .64 per iterate, and those in the (2,3) position decrease with a ratio of about .42 per iterate. The value in the (3,3) position of A_{15} will be 1.2679, which is correct to five places.

● PROBLEM 12-22

Let A be a symmetric, tridiagonal n×n matrix

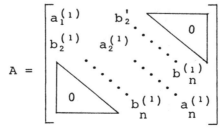

Write the QL algorithm to find the eigenvalues of A.

Solution: INPUT: n-dimension of the matrix A

$a_1^{(1)}, a_2^{(1)}, \ldots, a_n^{(1)}, b_1^{(1)}, \ldots, b_n^{(1)}$ - elements of A

M - maximum number of iterations

OUTPUT: $\lambda_1, \ldots, \lambda_n$ - eigenvalues of A, or a way to split A, or information that the maximum number of iterations was exceeded.

Step 1: Set k = 1

MOVE = 0

Step 2: For k ≤ M, do Steps 3-8

Step 3: If $b_2^{(k)} \approx 0$, then set $\lambda = a_1^{(k)}$ + MOVE

OUTPUT: λ
set n = n-1
$a_1^{(k)} = a_2^{(k)}$
for ℓ = 2,...,n,

set $a_\ell^{(k)} = a_{\ell+1}^{(k)}$
$b_\ell^{(k)} = b_{\ell+1}^{(k)}$

If $b_\ell^{(k)} \approx 0$ for 3 ≤ ℓ ≤ (n-1), then

OUTPUT: (split into
$a_1^{(k)}, \ldots, a_{\ell-1}^{(k)}, b_2^{(k)}, \ldots, b_{\ell-1}^{(k)}$
and
$a_\ell^{(k)}, \ldots, a_n^{(k)}, b_{\ell+1}^{(k)}, \ldots, b_n^{(k)}$, MOVE)

STOP

If $b_n^{(k)} \approx 0$, then set
$$\lambda = a_n^{(k)} + \text{MOVE}$$

OUTPUT: λ

set $n = n-1$

Step 4: $\quad \beta = -a_1^{(k)} - a_2^{(k)}$

$$\gamma = a_1^{(k)} a_2^{(k)} - \left[b_2^{(k)}\right]^2$$
$$\delta = (\beta^2 - 4\gamma)^{1/2}$$

if $\beta > 0$, then set
$$\alpha_1 = \frac{-2\gamma}{\beta + \delta}$$
$$\alpha_2 = -\frac{\beta + \delta}{2}$$

if $\beta \leq 0$, then set
$$\alpha_1 = \frac{\delta - \beta}{2}$$
$$\alpha_2 = \frac{2\gamma}{\delta - \beta}$$

If $n = 2$, then set $\quad \lambda_1 = \alpha_1 + \text{MOVE}$
$$\lambda_2 = \alpha_2 + \text{MOVE}$$

OUTPUT: λ_1, λ_2

STOP

Choose t in such a way that
$$|t - a_1^{(k)}| = \min(|\alpha_1 - a_1^{(k)}|, |\alpha_2 - a_1^{(k)}|)$$

Step 5: $\quad \text{MOVE} = \text{MOVE} + t$

Step 6: For $\ell = 1,\ldots,n$, set
$$d_\ell = a_\ell^{(k)} - t$$

Step 7: Set $x_n = b_n^{(k)}$
$$y_n = d_n$$
for $\ell = n, n-1, \ldots, 2$

$$\text{set } f_\ell = (y_\ell^2 + [b_\ell^{(k)}]^2)^{1/2}$$
$$g_{\ell-1} = \frac{y_\ell}{f_\ell}$$
$$h_{\ell-1} = \frac{b_\ell^{(k)}}{f_\ell}$$
$$g_\ell = g_{\ell-1} x_\ell + h_{\ell-1} d_{\ell-1}$$
$$y_{\ell-1} = g_{\ell-1} d_{\ell-1} - h_{\ell-1} x_\ell$$

if $\ell \neq 2$, then set
$$x_{\ell-1} = g_{\ell-1} b_{\ell-1}^{(k)}$$
$$r_\ell = h_{\ell-1} b_{\ell-1}^{(k)}$$

Note that in Step 7 we compute matrices K_ℓ, $A_{n-\ell+2}^{(k)}$, $A_{n-\ell+1}^{(k)}$, such that

$$A_{n-\ell+2}^{(k)} = K_\ell A_{n-\ell+1}^{(k)}$$

where

$$A_{n-\ell+2}^{(k)} = \begin{bmatrix} d_1 & b_2^k & & & & & & & 0 \\ b_2^k & \ddots & \ddots & & & & & & \\ 0 & \ddots & \ddots & \ddots & & & & & \\ & \ddots & \ddots & b_{\ell-2} & d_{\ell-2} & b_{\ell-1} & & & \\ & & 0 & & & & & & \\ & & & x_{\ell-1} & y_{\ell-1} & 0 & \ddots & & \\ & & & r_\ell & q_\ell & f_\ell & \ddots & 0 & \\ 0 & & & & \ddots & \ddots & \ddots & \ddots & \\ & & & & & r_n & q_n & & f_\ell \end{bmatrix}$$

$$A_n^{(k)} = \begin{bmatrix} y_1 & & & & & & & \\ q_2 & f_2 & & & & & 0 & \\ r_3 & q_3 & f_3 & & & & & \\ & \ddots & \ddots & \ddots & & & & \\ & & \ddots & \ddots & \ddots & & & \\ 0 & & & & & & & \\ & & & & r_n & q_n & f_n & \end{bmatrix}$$

Step 8: Calculate $A^{(k+1)}$

Set $f_1 = y_1$

$a_n^{(k+1)} = h_{n-1} q_n + g_{n-1} f_n$

$b_n^{(k+1)} = h_{n-1} f_{n-1}$

for $\ell = n-1, n-1, \ldots, 2$

set $a_\ell^{(k+1)} = h_{\ell-1} q_\ell + g_{\ell-1} g_\ell f_\ell$

$b_\ell^{(k+1)} = h_{\ell-1} f_{\ell-1}$

set $a_1^{(k+1)} = g_1 y_1$

$k = k+1$

Step 9:   OUTPUT: (Maximum number of iterations M exceeded)
          STOP.

EIGENVALUES AND EIGENVECTORS OF LINEAR OPERATORS

● **PROBLEM** 12-23

The matrix of the transformation, $(x,y)T = (y,x)$, is

$$A = \begin{bmatrix} 0 & 1 \\ 1 & 0 \end{bmatrix}.$$

Find the eigenvalues, the eigenvectors, and also the diagonal matrix of T.

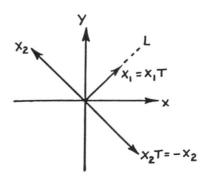

Solution: To find eigenvalues, set $AX = \lambda X$. Then,

$$[A - \lambda I] = \begin{bmatrix} 0 & 1 \\ 1 & 0 \end{bmatrix} - \begin{bmatrix} \lambda & 0 \\ 0 & \lambda \end{bmatrix}$$

$$= \begin{bmatrix} -\lambda & 1 \\ 1 & -\lambda \end{bmatrix}.$$

Therefore, the characteristic equation $\det(A - \lambda I) = 0$ is $\lambda^2 - 1 = 0$, and the eigenvalues are $\lambda = 1$ and $\lambda = -1$. To find an eigenvector corresponding to $\lambda = 1$, solve

$$X(A - 1I) = 0$$

for X. Thus,

$$\begin{bmatrix} x_1 \\ x_2 \end{bmatrix} - \begin{bmatrix} -1 & 1 \\ 1 & -1 \end{bmatrix} = \begin{bmatrix} 0 \\ 0 \end{bmatrix}$$

or,

$$-x_1 + x_2 = 0$$
$$x_1 - x_2 = 0$$

or,

$$x_1 - x_2 = 0.$$

Thus,

$$x_1 = \begin{bmatrix} 1 \\ 1 \end{bmatrix}$$

is an eigenvector corresponding to $\lambda = 1$.

To find an eigenvector corresponding to $\lambda = -1$, solve

$X[A - (-1)I] = 0$ for X. Hence,

$$\begin{bmatrix} x_1 \\ x_2 \end{bmatrix} \begin{bmatrix} 1 & 1 \\ 1 & 1 \end{bmatrix} = \begin{bmatrix} 0 \\ 0 \end{bmatrix}$$

or,
$$x_1 + x_2 = 0$$

or,
$$x_1 + x_2 = 0$$

$$x_1 + x_2 = 0.$$

So, in general, $x_1 = -x_2$. One specific solution is:
$$x_1 = -1, \quad x_2 = +1.$$

Therefore,
$$x_2 = \begin{bmatrix} -1 \\ 1 \end{bmatrix}$$

is an eigenvector corresponding to $\lambda = -1$.

Since there are two distinct eigenvalues and a two-dimensional vector space, the eigenvectors form a basis for R^2. Relative to this basis, T has the diagonal matrix,

$$B = \begin{bmatrix} 1 & 0 \\ 0 & -1 \end{bmatrix},$$

whose diagonal entries are the eigenvalues. T is now seen to be the reflection across the line

$$L = \{a(1,1): a \in R\}$$

in the given figure.

● **PROBLEM 12-24**

Find the eigenvalues and a basis for each of the eigenspaces of the linear operator $T: P_2 \to P_2$ defined by:

$$T(a + bx + cx^2) = (3a - 2b) + (-2a + 3b)x + (5c)x^2.$$

Solution: The matrix of T with respect to the standard basis

$$B = \{1, x, x^2\}$$

is

$$A = \begin{bmatrix} 3 & -2 & 0 \\ -2 & 3 & 0 \\ 0 & 0 & 5 \end{bmatrix}.$$

The eigenvalues of T are the eigenvalues of A. One forms the matrix

$$\lambda I - A = \begin{bmatrix} \lambda & 0 & 0 \\ 0 & \lambda & 0 \\ 0 & 0 & \lambda \end{bmatrix} - \begin{bmatrix} 3 & -2 & 0 \\ -2 & 3 & 0 \\ 0 & 0 & 5 \end{bmatrix}$$

$$= \begin{bmatrix} \lambda-3 & 2 & 0 \\ 2 & \lambda-3 & 0 \\ 0 & 0 & \lambda-5 \end{bmatrix}.$$

Expanding along the third column yields

$$\det(\lambda I - A) = (\lambda-5) \begin{vmatrix} \lambda-3 & 2 \\ 2 & \lambda-3 \end{vmatrix}$$

$$= (\lambda-5)[(\lambda-3)^2 - 4]$$

$$= (\lambda-5)[\lambda^2 - 6\lambda + 5]$$

$$= (\lambda-5)(\lambda-5)(\lambda-1)$$

$$= (\lambda-5)^2(\lambda-1).$$

The characteristic equation of A is $(\lambda-5)^2(\lambda-1) = 0$ so that the eigenvalues of A are $\lambda = 1$ and $\lambda = 5,5$. If $\lambda = 1$, then $(1I - A)u = 0$
or

$$\begin{bmatrix} -2 & 2 & 0 \\ 2 & 2 & 0 \\ 0 & 0 & -4 \end{bmatrix} \begin{bmatrix} u_1 \\ u_2 \\ u_3 \end{bmatrix} = \begin{bmatrix} 0 \\ 0 \\ 0 \end{bmatrix}.$$

Solving this system yields $u_1 = t$, $u_2 = t$, $u_3 = 0$. Thus, the eigenvectors corresponding to $\lambda = 1$ are the non-zero vectors of the form

$$u = \begin{bmatrix} t \\ t \\ 0 \end{bmatrix} = t \begin{bmatrix} 1 \\ 1 \\ 0 \end{bmatrix},$$

so that

$$\begin{bmatrix} 1 \\ 1 \\ 0 \end{bmatrix}$$

is a basis for the eigenspace corresponding to $\lambda = 1$.
If $\lambda = 5$, then

$$(\lambda I - A)u = 0$$

or,

$$\begin{bmatrix} 2 & 2 & 0 \\ 2 & 2 & 0 \\ 0 & 0 & 0 \end{bmatrix} \begin{bmatrix} u_1 \\ u_2 \\ u_3 \end{bmatrix} = \begin{bmatrix} 0 \\ 0 \\ 0 \end{bmatrix}.$$

Solving this system gives $u_1 = -s$, $u_2 = s$, $u_3 = t$.
Thus, the eigenvectors of A corresponding to $\lambda = 5$ are the non-zero vectors of the form:

$$u = \begin{bmatrix} -s \\ s \\ t \end{bmatrix} = \begin{bmatrix} -s \\ s \\ 0 \end{bmatrix} + \begin{bmatrix} 0 \\ 0 \\ t \end{bmatrix} = s \begin{bmatrix} -1 \\ 1 \\ 0 \end{bmatrix} + t \begin{bmatrix} 0 \\ 0 \\ 1 \end{bmatrix}$$

Hence, the eigenspace of A corresponding to $\lambda = 5$ has the basis $\{u_1, u_2\}$ and that corresponding to $\lambda = 1$ has the basis $\{u_3\}$ where

$$u_1 = \begin{bmatrix} -1 \\ 1 \\ 0 \end{bmatrix} \quad u_2 = \begin{bmatrix} 0 \\ 0 \\ 1 \end{bmatrix} \quad u_3 = \begin{bmatrix} 1 \\ 1 \\ 0 \end{bmatrix}.$$

These matrices are the coordinate matrices with respect to B of

$$P_1 = -1 + x, \quad P_2 = x^2, \quad P_3 = 1 + x.$$

Thus, $\{-1 + x, x^2\}$ is a basis for the eigenspace of T corresponding to $\lambda = 5$, and $\{1 + x\}$ is a basis for the eigenspace corresponding to $\lambda = 1$.

• **PROBLEM 12-25**

(a) Let T_1 be the linear operator on R^2 defined by $T_1(xy) = (x'y')$ if and only if

$$\begin{bmatrix} 1 & 1 \\ 0 & 2 \end{bmatrix} \begin{bmatrix} x \\ y \end{bmatrix} = \begin{bmatrix} x' \\ y' \end{bmatrix}$$

Is T_1 diagonalizable?

(b) Let T be the linear operator on R^2 such that:

$T(xy) = (x, y')$ if and only if

$$\begin{bmatrix} 3 & -1 \\ -1 & 3 \end{bmatrix} \begin{bmatrix} x \\ y \end{bmatrix} = \begin{bmatrix} x' \\ y' \end{bmatrix}.$$

Show that TT_1 is not diagnoalizable.

(c) Let T_2 be the linear operator on R^2 defined by

$T_2(xy) = (x', y')$ if and only if

$$\begin{bmatrix} 1 & 100 \\ 0 & 2 \end{bmatrix} \begin{bmatrix} x \\ y \end{bmatrix} = \begin{bmatrix} x' \\ y' \end{bmatrix}.$$

Show that $T + T_2$ is not diagonalizable.

<u>Solution</u>: (a) The matrix of T_1 is $A = \begin{bmatrix} 1 & 1 \\ 0 & 2 \end{bmatrix}$.

$$\det[\lambda I - A] = \det \begin{vmatrix} \lambda-1 & -1 \\ 0 & \lambda-2 \end{vmatrix} = (\lambda-1)(\lambda-2).$$

Therefore, the characteristic values for T_1 and $\lambda = 1$ and $\lambda = 2$. Thus, T_1 has two distinct real roots; T_1 is diagonalizable. The characteristic vectors for T_1 are obtained by solving the equations $[1I - A]X = 0$ and $[2I - A]X = 0$, or,

$$\begin{bmatrix} 0 & -1 \\ 0 & -1 \end{bmatrix} \begin{bmatrix} x_1 \\ x_2 \end{bmatrix} = \begin{bmatrix} 0 \\ 0 \end{bmatrix}$$

and

$$\begin{bmatrix} 1 & -1 \\ 0 & 0 \end{bmatrix} \begin{bmatrix} x_1 \\ x_2 \end{bmatrix} = \begin{bmatrix} 0 \\ 0 \end{bmatrix}$$

Thus,

$$X_1 = \begin{bmatrix} 1 \\ 0 \end{bmatrix} \quad \text{and} \quad X_2 = \begin{bmatrix} 1 \\ 1 \end{bmatrix}.$$

(b) The matrix of TT_1 with respect to the standard basis is

$$\begin{bmatrix} 3 & -1 \\ -1 & 3 \end{bmatrix} \begin{bmatrix} 1 & 1 \\ 0 & 2 \end{bmatrix} = \begin{bmatrix} 3 & 1 \\ -1 & 5 \end{bmatrix}$$

then,

$$\det(\lambda I - TT_1) = \det \begin{vmatrix} \lambda-3 & -1 \\ 1 & \lambda-5 \end{vmatrix}$$

$$= (\lambda-3)(\lambda-5) + 1$$

$$= (\lambda-4)^2.$$

Therefore, $\lambda = 4$ is the only characteristic value of TT_1. Since nullity $(4I - TT_1)$ = nullity

$$\begin{bmatrix} 1 & -1 \\ 1 & -1 \end{bmatrix} = 1,$$

conclude that TT_1 is not diagonalizable.

(c) The matrix of $T + T_2$ with respect to the standard basis is

$$\begin{bmatrix} 3 & -1 \\ -1 & 3 \end{bmatrix} + \begin{bmatrix} 1 & 100 \\ 0 & 2 \end{bmatrix} = \begin{bmatrix} 4 & 99 \\ -1 & 5 \end{bmatrix}.$$

$$\det[\lambda I - (T + T_2)] = \begin{bmatrix} \lambda-4 & -99 \\ 1 & \lambda-5 \end{bmatrix}$$

$$= (\lambda-4)(\lambda-5) + 99$$

$$= \lambda^2 - 9\lambda + 119.$$

Using the quadratic formula, conclude that $T + T_2$ has no real characteristic values. Therefore, $T + T_2$ is not diagonalizable.

APPLICATIONS OF THE SPECTRAL THEOREM

● **PROBLEM** 12-26

Find a orthogonal matrix P such that $P^{-1}AP$ is a diagonal matrix B where,

$$A = \begin{bmatrix} 3 & 1 \\ 1 & 3 \end{bmatrix}.$$

Solution: Recall that the transpose A^t of A is the matrix obtained from A by interchanging the rows and columns of A. It is said that A is symmetric if $A^t = A$.

For symmetric matrices there is the following theorem:

If A is symmetric, there is an invertible matrix P such that $B = P^{-1}AP$ is a diagonal matrix.

This theorem is usually called the Spectral Theorem; it tells, in particular, that if a matrix is symmetric, its characteristic polynomial must have only real roots. If A is symmetric, one may actually find an orthogonal matrix P such that $P^{-1}AP$ is diagonal. Recall that an orthogonal matrix is a matrix whose columns are orthonormal.

Now, consider the given matrix

$$A = \begin{bmatrix} 3 & 1 \\ 1 & 3 \end{bmatrix}.$$

A is symmetric. Form the matrix

$$(\lambda I - A) = \begin{bmatrix} \lambda & 0 \\ 0 & \lambda \end{bmatrix} - \begin{bmatrix} 3 & 1 \\ 1 & 3 \end{bmatrix}$$

$$= \begin{bmatrix} \lambda-3 & -1 \\ -1 & \lambda-3 \end{bmatrix}.$$

Then,

$$\det(\lambda I - A) = \det \begin{matrix} \lambda-3 & -1 \\ -1 & \lambda-3 \end{matrix}$$

$$= (\lambda-3)^2 - 1$$

$$= \lambda^2 - 6\lambda + 9 - 1$$

$$= \lambda^2 - 6\lambda + 8$$

$$= (\lambda-4)(\lambda-2).$$

The characteristic equation of A is $(\lambda-4)(\lambda-2) = 0$ so that the characteristic values are $\lambda = 4$ and $\lambda = 2$. If $\lambda = 4$, then

$$4I - A = \begin{bmatrix} 4 & 0 \\ 0 & 4 \end{bmatrix} - \begin{bmatrix} 3 & 1 \\ 1 & 3 \end{bmatrix} = \begin{bmatrix} 1 & -1 \\ -1 & 1 \end{bmatrix}.$$

Now find the characteristic vectors (or eigenvectors).

$$(4I - A)X = 0$$

$$\begin{bmatrix} 1 & -1 \\ -1 & 1 \end{bmatrix} \begin{bmatrix} x_1 \\ x_2 \end{bmatrix} = 0$$

$x_1 - x_2 = 0$, $-x_1 + x_2 = 0$ Thus, $x_1 = \begin{bmatrix} t \\ t \end{bmatrix}$

or, $x_1 - x_2 = 0$.

where $t \in R$ are the eigenvectors. Clearly, $\begin{bmatrix} 1 \\ 1 \end{bmatrix}$ is a basis for the space. Since the norm of $\{1,1\}$ is $(\{1,1\})^{\frac{1}{2}} = \sqrt{2}$, normalize $\{1,1\}$ to obtain

$$x_1 = \begin{bmatrix} 1/\sqrt{2} \\ 1/\sqrt{2} \end{bmatrix}.$$

This is an orthonormal basis for the null space of $4I - A$. Now let $\lambda = 2$. Then,

$$2I - A = \begin{bmatrix} 2 & 0 \\ 0 & 2 \end{bmatrix} - \begin{bmatrix} 3 & 1 \\ 1 & 3 \end{bmatrix} = \begin{bmatrix} -1 & -1 \\ -1 & -1 \end{bmatrix}.$$

Thus, $(4I - A)X = 0$,

$$\begin{bmatrix} -1 & -1 \\ -1 & -1 \end{bmatrix} \begin{bmatrix} x_1 \\ x_2 \end{bmatrix} = \begin{bmatrix} 0 \\ 0 \end{bmatrix}$$

or,

$-x_1 - x_2 = 0$

or, $x_1 + x_2 = 0$.

$-x_1 - x_2 = 0$

Clearly, $\begin{bmatrix} -1 \\ 1 \end{bmatrix}$ is a basis for the space. If normalized, $u_2 = \begin{bmatrix} -1/\sqrt{2} \\ 1/\sqrt{2} \end{bmatrix}$ is an orthonormal basis for the null space of $2I - A$.

Observe that $u_1 \cdot u_2 = 0$ so that u_1 and u_2 are an orthonormal basis for R^2. Thus, in general, if A is symmetric and λ_1 and λ_2 are distinct characteristic values of A, the corresponding characteristic vectors x_1, x_2 must be orthogonal. Now construct an orthogonal matrix P whose columns are orthonormal. Thus,

$$P = \begin{bmatrix} 1/\sqrt{2} & -1/\sqrt{2} \\ 1/\sqrt{2} & 1/\sqrt{2} \end{bmatrix}.$$

Since P is an orthogonal matrix, $P^{-1} = P^t$. Therefore,

$$P^{-1} = \begin{bmatrix} 1/\sqrt{2} & 1/\sqrt{2} \\ -1/\sqrt{2} & 1/\sqrt{2} \end{bmatrix}.$$

Then, $B = P^{-1}AP$ is a diagonal matrix.

$$B = \begin{bmatrix} 1/\sqrt{2} & 1/\sqrt{2} \\ -1/\sqrt{2} & 1/\sqrt{2} \end{bmatrix} \begin{bmatrix} 3 & 1 \\ 1 & 3 \end{bmatrix} \begin{bmatrix} 1/\sqrt{2} & -1/\sqrt{2} \\ 1/\sqrt{2} & 1/\sqrt{2} \end{bmatrix}$$

$$= \begin{bmatrix} 1/\sqrt{2} & 1/\sqrt{2} \\ -1/\sqrt{2} & 1/\sqrt{2} \end{bmatrix} \begin{bmatrix} 4/\sqrt{2} & -2/\sqrt{2} \\ 4/\sqrt{2} & 2/\sqrt{2} \end{bmatrix}$$

$$= \begin{bmatrix} 2+2 & -1+1 \\ -2+2 & 1+1 \end{bmatrix}$$

$$= \begin{bmatrix} 4 & 0 \\ 0 & 2 \end{bmatrix}.$$

• **PROBLEM 12-27**

Reduce the following quadratic equation to standard form:

$$ax^2 + bxy + cy^2 = d.$$

Solution: $ax^2 + bxy + cy^2 = d.$ \hfill (1)

Put

$$A = \begin{bmatrix} a & b/2 \\ b/2 & c \end{bmatrix} \quad \text{and} \quad \bar{u} = \begin{bmatrix} x \\ y \end{bmatrix}.$$

A is symmetric and

$$A\bar{u} \cdot \bar{u} = \begin{bmatrix} ax + b/2\ y \\ b/2\ x + cy \end{bmatrix} \cdot \begin{bmatrix} x \\ y \end{bmatrix}$$

$$= ax^2 + bxy + cy^2,$$

so one can rewrite equation (1) as

$$Au \cdot u = d. \qquad (2)$$

Since A is symmetric, apply the spectral theorem to obtain an orthogonal matrix P and diagonal matrix B such that $B = P^{-1}AP$. Solve this for A to obtain $A = PBP^{-1}$. Substituting this in equation (2) gives

$$PBP^{-1}u \cdot u = d. \qquad (3)$$

For any matrix A and any u and v, one gets $Au \cdot v = u \cdot A^t v$. Using this to shift P across the dot product yields

$$PBP^{-1}u \cdot u = BP^{-1}u \cdot P^t u.$$

Since P is orthogonal, it is known that $P^t = P^{-1}$. Hence, write equation (3) as

$$BP^{-1}u \cdot P^{-1}u = d. \qquad (4)$$

One can also write

$$B = \begin{bmatrix} b_1 & 0 \\ 0 & b_2 \end{bmatrix} \quad \text{and} \quad P^{-1}u = \begin{bmatrix} x_1 \\ y_1 \end{bmatrix}$$

Therefore, equation (3) can now be expressed as

$$\begin{bmatrix} b_1 & 0 \\ 0 & b_2 \end{bmatrix} \begin{bmatrix} x_1 \\ y_1 \end{bmatrix} \cdot \begin{bmatrix} x_1 \\ y_1 \end{bmatrix} = d$$

which, after carrying out these products, gives

$$b_1 x_1^2 + b_2 y_1^2 = d. \qquad (5)$$

Thus, relative to the $x_1 y_1$ coordinate system, rewrite equation (1) in the form (4), in which no xy terms appears. Equation (4) can easily be graphed in the $x_1 y_1$ system and, thereby, results in the graph of equation (1).

This problem illustrates the application of the spectral theorem to a geometric problem.

CHAPTER 13

LINEAR PROGRAMMING

BASIC PROBLEM

● **PROBLEM** 13-1

Three products are processed through three different operations. The time (in minutes) required per unit of each product, the daily capacity of the operations (in minutes per day) and the profit per unit sold of each product (in dollars), are as given in the accompanying table.

The zero times indicate that the product does not require the given operation. It is assumed that all units produced are sold. Moreover, the given profits per unit are net values that result after all pertinent expenses are deducted. Formulate a linear programming model that fits the daily production for the three products.

Operation	Time per unit (minutes)			Operation capacity (minutes/day)
	Product 1	Product 2	Product 3	
1	1	2	1	430
2	3	0	2	460
3	1	4	0	420
Profit/unit ($)	3	2	5	

Solution: The main elements of a mathematical model are (1) the variables or unknowns, (2) the objective function, and (3) the constraints. The variables are immediately identified as the daily number of units to be manufactured of each product. Let x_1, x_2, and x_3 be the number of daily units produced of products 1, 2, and 3. Because of the assumption that all units produced are sold, the total profit x_0 (in dollars) for the three products is $x_0 = 3x_1 + 2x_2 + 5x_3$.

The constraints of the problem must ensure that the total processing time required by all produced units does not exceed the daily capacity of each operation. These are expressed as

Operation 1: $1x_1 + 2x_2 + 1x_3 \leq 430$

Operation 2: $3x_1 + 0x_2 + 2x_3 \leq 460$

Operation 3: $1x_1 + 4x_2 + 0x_3 \leq 420$

Because it is nonsensical to produce negative quantities, the additional nonnegativity constraints $x_1 \geq 0$, $x_2 \geq 0$, and $x_3 \geq 0$ must be added.

The linear programming model is now summarized as follows.

Maximize $x_0 = 3x_1 + 2x_2 + 5x_3$

subject to

$$x_1 + 2x_2 + x_3 \leq 430$$
$$3x_1 + 2x_3 \leq 460$$
$$x_1 + 4x_2 \leq 420$$
$$x_1, x_2, x_3 \geq 0$$

● **PROBLEM 13-2**

One of the most successful applications of linear programming deals with the determination of optimal feed mix for meeting the nutritional needs of an animal or a broiler at the least cost. The model assumes the availability of certain feedstuffs or ingredients from which the feed is mixed. The nutritive content of each ingredient is known. The constraints of the model include (1) daily nutrient requirements of the animal or broiler and (2) physical or nonnutritive limitations such as supply, feed texture, or ability of feed to be pelleted. The objective is to minimize the total cost of a given batch size of the feed mix such that the nutritive and physical constraints are satisfied.

Assume the required daily batch of the feed mix is 100 pounds. The diet must contain

1. At least 0.8% but not more than 1.2% calcium.

2. At least 22% protein.

3. At most 5% crude fiber.

Assume further that the main ingredients (or feedstuffs) used include limestone (calcium carbonate), corn, and soybean meal. The nutritive content of these ingredients is summarized in the accompanying table. Formulate a linear programming model using the above information.

Ingredient	Pounds per pound of ingredient			Cost ($) per pound
	Calcium	Protein	Fiber	
Limestone	0.380	0.00	0.00	0.0164
Corn	0.001	0.09	0.02	0.0463
Soybean meal	0.002	0.50	0.08	0.1250

Solution: Let x_1, x_2, and x_3 be the amounts (in pounds) of limestone, corn, and soybean meal used in producing the feed mix batch of 100 pounds. Then the linear programming model becomes

$$\text{minimize } x_0 = 0.0164x_1 + 0.0463x_2 + 0.1250x_3$$

subject to

$$x_1 + x_2 + x_3 = 100$$
$$0.380x_1 + 0.001x_2 + 0.002x_3 \leq 0.012 \times 100$$
$$0.380x_1 + 0.001x_2 + 0.002x_3 \geq 0.008 \times 100$$
$$0.09 x_2 + 0.50 x_3 \geq 0.22 \times 100$$
$$0.02 x_2 + 0.08 x_3 \leq 0.05 \times 100$$
$$x_1, x_2, x_3 \geq 0$$

The first constraint specifies the batch size. The second and third constraints specify the maximum and minimum requirements of calcium. The minimum requirement of protein is given by the fourth constraint, while the fifth constraint stipulates the maximum requirement of crude fiber.

● **PROBLEM 13-3**

The product mix problem is most naturally modeled as a mathematical programming problem. Let

X_i = quantity of product i, i = 1, 2, ..., n, produced in the period

b_k = amount of resource k, k = 1, 2, ..., K, available during the period

a_{ik} = number of units of resource k required to produce one unit of product i

U_i = maximum sales potential of product i in the period

L_i = required minimum production level of product i in the period

r_i = revenue, net of variable selling expense, from selling one unit of product i

c_i = unit variable cost of producing a unit of product i

Assume that $(r_i - c_i)X_i$ is the contribution to overheads and profits resulting from the production and sale of X_i units of product i, where one supposes that all production of product i up to U_i units can be sold in the period.

Furthermore, assume that production of X_i units of product i will use up $a_{ik}X_i$ units of resource k. Write up a mathematical programming model to maximize the total contribution to overhead and profit, Z, from all products.

Solution: Mathematically, one wishes to choose $X_1, X_2, \ldots, X_n$ to maximize

$$Z = \sum_{i=1}^{n} (r_i - c_i)X_i \qquad (a)$$

subject to

$$\sum_{i=1}^{n} a_{ik}X_i \leq b_k, \qquad (k = 1, 2, \ldots, K) \qquad (b)$$

$$X_i \leq U_i, \qquad (i = 1, 2, \ldots, n) \qquad (c)$$

$$X_i \geq L_i, \qquad (L_i \geq 0,\ i = 1, 2, \ldots, n) \qquad (d)$$

The left-hand side of (b) is the total amount of resource k required by the production program. The lower bound constraints, (d), occur when there is a prior commitment to deliver a given amount, L_i, of a product i in the period, or when management decides to produce at least that much of the product, regardless of the economic consequences.

● **PROBLEM 13-4**

In a given factory there are three machines M_1, M_2, M_3 used in making two products P_1, P_2. One unit of P_1 occupies M_1 5 minutes, M_2 3 minutes, and M_3 4 minutes. The corresponding figures for one unit of P_2 are: M_1 1 minute, M_2 4 minutes, M_3 3 minutes. The net profit per unit of P_1 produced is 30 dollars, and for P_2 20 dollars (independent of wether the machines are used to full capacity or not). What production plan gives the most profit?

Hint: Assume that x_1 units of P_1 and x_2 units of P_2 are produced per hour.

Solution: The problem is to maximize

$$f = 30x_1 + 20x_2$$

subject to the constraints

$$5x_1 + x_2 \leq 60 \quad \text{(for } M_1\text{)}$$

$$3x_1 + 4x_2 \leq 60 \quad \text{(for } M_2\text{)}$$

$$4x_1 + 3x_2 \leq 60 \quad \text{(for } M_3\text{)}$$

$$x_1 \geq 0$$

$$x_2 \geq 0$$

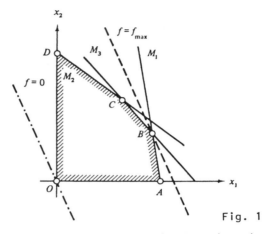

Fig. 1

The problem is illustrated geometrically in Figure 1. The first of the inequalities above means that the point (x_1,x_2) must lie to the left of or on the line AB whose equation is $5x_1 + x_2 = 60$. The other inequalities can be interpreted in a similar way. The point (x_1,x_2) must satisfy all of these inequalities. Thus it must lie within or on the boundary of the pentagon OABCD. The value of the function to be maximized is proportional to the distance between (x_1,x_2) and the dashed line $f(x_1,x_2) = 0$; it clearly takes on its largest value at the vertex B. Every vertex is the intersection of two sides of the pentagon. Thus at the vertices, one must have equality in (at least) two of the inequalities. At the vertex B, equality holds in the inequality for M_1 and the inequality for M_3. In this way, one gets two equations for determining the two unknowns x_1 and x_2; the solution is $x_1 = 120/11$, $x_2 = 60/11$. The maximal profit is thus $4,800/11 = 436.36$ dollars/hour. M_1 and M_3 are to be used continuously, while M_2 is used only

$$\frac{3 \cdot 120 + 4 \cdot 60}{11} = 54.55 \text{ min/hour}.$$

● **PROBLEM 13-5**

A manufacturer produces a line of household products fabricated from sheet metal. To illustrate his production planning problem, assume that he makes only four products and that his production system consists of five production centers: stamping, drilling, assembly, finishing (painting and printing), and packaging. For a given month, he must decide how much of each product to manufacture, and to aid in this decision, he has assembled the data shown in Tables 1 and 2. Furthermore, he knows that only 2000 square feet of the type of sheet metal used for products 2 and 4 will be available during the month. Product 2 requires 2.0 square feet per unit and product 4 uses 1.2 square feet per unit.

Formulate this as a linear programming model.

Table 1   Production Data

DEPARTMENT	PRODUCTION RATES IN HOURS PER UNIT				PRODUCTION HOURS AVAILABLE
	PRODUCT 1	PRODUCT 2	PRODUCT 3	PRODUCT 4	
Stamping	0.03	0.15	0.05	0.10	400
Drilling	0.06	0.12	—	0.10	400
Assembly	0.05	0.10	0.05	0.12	500
Finishing	0.04	0.20	0.03	0.12	450
Packaging	0.02	0.06	0.02	0.05	400

Table 2   Product Data

PRODUCT	NET SELLING PRICE/UNIT	VARIABLE COST/UNIT	SALES POTENTIAL	
			MINIMUM	MAXIMUM
1	$10	$6	1000	6000
2	25	15	—	500
3	16	11	500	3000
4	20	14	100	1000

Solution: Let X_i be the number of units of product i produced in the month and Z be the total contribution to profit and overhead. The problem is to choose nonnegative X_1, X_2, X_3, and X_4 to maximize

$$Z = 4X_1 + 10X_2 + 5X_3 + 6X_4$$

subject to

(1) constraints on production time (e.g., machine-hours):

$0.03X_1 + 0.15X_2 + 0.05X_3 + 0.10X_4 \leq 400$ (Stamping)

$0.06X_1 + 0.12X_2 \qquad\qquad + 0.10X_4 \leq 400$ (Drilling)

$0.05X_1 + 0.10X_2 + 0.05X_3 + 0.12X_4 \leq 500$ (Assembly)

$0.04X_1 + 0.20X_2 + 0.03X_3 + 0.12X_4 \leq 450$ (Finishing)

$0.02X_1 + 0.06X_2 + 0.02X_3 + 0.05X_4 \leq 400$ (Packaging)

(2) constraint on sheet metal availability:

$2.0X_2 + 1.2X_4 \leq 2000$

(3) constraints on minimum production and maximum sales:

$$1000 \leq X_1 \leq 6000$$
$$0 \leq X_2 \leq 500$$
$$500 \leq X_3 \leq 3000$$
$$100 \leq X_4 \leq 1000$$

The problem was solved by the simplex algorithm to obtain the information described in the following paragraphs:

The optimal production program is $X_1^* = 5500$, $X_2^* = 500$, $X_3^* = 3000$, and $X_4^* = 100$. We are to make as much of products 2 and 3 as we can sell in the period. Product 4 is produced only because of the lower bound constraint $X_4 \geq 100$. This program uses all of the capacity in stamping and drilling. The unused capacity in the remaining production centers is 13 hours in assembly, 28 hours in finishing, and 195 hours in packaging. Only 1120 square feet of sheet metal are used, leaving 880 square feet unused. The maximum profit is $42,600.

By examining values of the dual variables, one learns that it would not be desirable to produce more of product 4 (at the expense of reducing production of some other product) unless its unit profit were at least $1.78 higher, or $7.78. The marginal value of an additional hour of stamping capacity is $22.22 and of drilling capacity, $55.56. If one could sell one more unit of product 3, one could increase the profit by $3.89 (accounting for the fact that some other product would have to be cut back). The ability to sell one more unit of product 2 would yield no more profit.

• **PROBLEM 13-6**

Consider the three-stage system shown in the accompanying figure. There is one product produced by each stage. The product of stage 1 is used in the production of products at stages 2 and 3. The product of stage 2 is used in the production of the product at stage 3 and also can be sold directly to customers. The product of stage 3 is sold.

Let

X_j = units of production at stage j

P_j = capacity of stage j in units of product

a_{ij} = number of units of product at stage i required to produce one unit at stage j

c_j = unit variable production cost incurred at stage j

r_j = revenue from the sale of product $j, j = 2, 3$

L_j = minimum sales requirements for product $j, j = 2, 3$

U_j = maximum sales possible for product $j, j = 2, 3$

Z = contribution to profit and overhead for the period.

Assume that all production of products 2 and 3 up to U_2 and U_3, respectively, can be sold. Formulate a balance equation (product mix) model to maximize contribution to profit and overhead for the period.

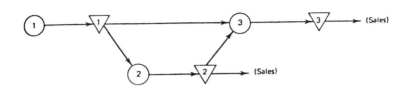

Figure: Production System.

Solution: The programming model becomes:

Maximize

$$Z = r_3 X_3 + r_2(X_2 - a_{23} X_3) - c_1 X_1 - c_2 X_2 - c_3 X_3$$

$$= (r_3 - a_{23} r_2 - c_3) X_3 + (r_2 - c_2) X_2 - c_1 X_1$$

subject to

$X_j \le P_j$, $(j = 1, 2, 3)$

$X_1 = a_{12} X_2 + a_{13} X_3$

$X_2 \ge a_{23} X_3$

$L_2 \le X_2 - a_{23} X_3 \le U_2$

$L_3 \le X_3 \le U_3$

An alternative to this model results from defining additional variables S_2 and S_3, representing the sales of products 2 and 3. Then, maximize

$$Z = r_2 S_2 + r_3 S_3 - c_1 X_1 - c_2 X_2 - c_3 X_3$$

subject to

$X_j \le P_j$, $(j = 1, 2, 3)$

$X_1 = a_{12} X_2 + a_{13} X_3$

$X_2 = a_{23} X_3 + S_2$

$X_3 = S_3$

$L_2 \le S_2 \le U_2$

$L_3 \le S_3 \le U_3$

The formulation completes the material balance equation

at the second stage and adds an inventory balance equation for the third stage. There is no advantage to this approach in the single time period case. However, it can be the only way to model certain product mix problems in the multiple period situation.

SIMPLEX METHOD

• **PROBLEM 13-7**

Maximize $x_0 = 4x_1 + 3x_2$

subject to

$$2x_1 + 3x_2 \leq 6 \tag{1}$$

$$-3x_1 + 2x_2 \leq 3 \tag{2}$$

$$2x_2 \leq 5 \tag{3}$$

$$2x_1 + x_2 \leq 4 \tag{4}$$

$$x_1, x_2 \geq 0 \tag{5}$$

(a) Solve graphically, (b) Given the standard form of the preceding linear programming model

Maximize

$$x_0 = 4x_1 + 3x_2 + 0S_1 + 0S_2 + 0S_3 + 0S_4$$

subject to

$$2x_1 + 3x_2 + S_1 = 6$$

$$-3x_1 + 2x_2 + S_2 = 3$$

$$0x_1 + 2x_2 + S_3 = 5$$

$$2x_1 + x_2 + S_4 = 4$$

$$x_1, x_2, S_1, S_2, S_3, S_4 \geq 0.$$

Solve with the aid of the simplex method.

Solution: (a) The idea of the method is to plot the (feasible) solution space, which is defined as the space enclosed by constraints ① through ⑤. The optimum solution is the point (in the solution space) which maximizes the value of the objective function x_0. The solution space is plotted in the (x_1, x_2)-plane, as shown in Figure 1. The nonnegativity restrictions ⑤ specify that the feasible solutions must lie in the first quadrant defined by $x_1 \geq 0$ and $x_2 \geq 0$. Each of the constraints ① through ④ will be plotted first with (≤) replaced by (=), thus yielding simple straight-line equations. The region in which each constraint holds is indicated by an arrow on its associated straight line. The resulting (feasible) solution space is given in the figure shown by the area ABCDE. Constraint ③ is redundant because it can be deleted without affecting the solution space.

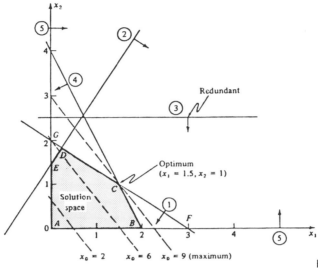

Fig. 1

Every point within or on the boundaries of the area ABCDE satisfies all the constraints, namely, ① through ⑤. Thus, the optimum solution is the point in ABCDE which yields the largest (maximum) value of x_0. Figure 1 plots the objective function $x_0 = 4x_1 + 3x_2$ for $x_0 =$ 2, 6, and 9 to show the direction in which x_0 increases. The figure shows that if x_0 is increased beyond 9, the objective function will not pass through any feasible points, which means $x_0 = 9$ is the maximum. The value $x_0 = 9$ is determined by moving the line $4x_1 + 3x_2 = x_0$ parallel to itself in the direction of increasing x_0, one sees that the maximum value of x_0 occurs where the objective function passes through point C, whose coordinates are $x_1 = 1.5$ and $x_2 = 1$. Substituting these values into the objective function gives $x_0 = 4x_1 + 3x_2 = 4(1.5) + 3(1) = 9$.

An interesting observation is that the maximum value of x_0 always occurs at one of the corner points A, B, C, D, and E of the solution space. The choice of a specific corner point as the optimum depends on the slope of the objective function. It can be verified graphically that the changes in the objective function given in the accompanying table produce the new optimum (maximum) corner points shown.

The problem reveals that the search for the optimum is reduced to considering only a finite number of feasible points, namely, the corner points of the solution space. Mathematically, a corner point is known as an extreme point. Thus, after all extreme points are determined,

the optimum is the (feasible) extreme point that yields
the best value of the objective function.

Objective function	Optimum corner point	Optimum solution
$x_0 = 10x_1 + x_2$	B	($x_1 = 2$, $x_2 = 0$, $x_0 = 20$)
$x_0 = x_1 + 20x_2$	D	($x_1 = 3/13$, $x_2 = 24/13$, $x_0 = 483/13$)
$x_0 = -4x_1 + 2x_2$	E	($x_1 = 0$, $x_2 = 3/2$, $x_0 = 3$)
$x_0 = -x_1 - x_2$	A	($x_1 = 0$, $x_2 = 0$, $x_0 = 0$)

(b) The simplex method starts from a basic feasible
solution (or extreme point), which for this problem must
have two (out of six) variables equal to zero. An obvious
starting basic feasible solution is obtained by setting
$x_1 = x_2 = 0$. This immediately gives (with no further
calculations) the all-slack starting solution as $S_1 = 6$,
$S_2 = 3$, $S_3 = 5$, and $S_4 = 4$. The slack variables yield
an obvious starting solution in this case because (1)
their constraint coefficients form an identity matrix
in which the diagonal elements are ones and all the
remaining elements are zeros and (2) the right-hand-side
constants of the equations are always nonnegative (property of the standard form).

A convenient way for recording the information about the
starting solution is the following tableau.

Basic	x_0	x_1	x_2	S_1	S_2	S_3	S_4	Solution	
x_0	①	-4	-3	0	0	0	0	0	x_0-equation
S_1	0	2	3	①	0	0	0	6	S_1-equation
S_2	0	-3	2	0	①	0	0	3	S_2-equation
S_3	0	0	2	0	0	①	0	5	S_3-equation
S_4	0	2	1	0	0	0	①	4	S_4-equation

Notice that for the purpose of the tableau, the objective
function is expressed in equation form as

$$x_0 - 4x_1 - 3x_2 - 0S_1 - 0S_2 - 0S_3 - 0S_4 = 0$$

The Basic column labels the (current) basic variables
S_1, S_2, S_3, and S_4. The Solution column gives the
current values of the basic variables; namely, $S_1 = 6$,
$S_2 = 3$, $S_3 = 5$, and $S_4 = 4$. Each equation in the tableau
will be identified by the associated variable in the basic
column. The current non-basic variables are those that
do not appear in the basic column and their values are
zero, that is, $x_1 = x_2 = 0$. The values assigned to the
variables yield $x_0 = 0$ as shown in the Solution column.

The special arrangement of the above tableau provides
useful information. The columns associated with the
starting basic variables (S_1, S_2, S_3, and S_4 in this
problem) always appear immediately to the left of the
Solution column and their constraints coefficients
constitute an identity matrix.

For convenience, the graphical representation of the

solution space is given in Figure 2. The above starting solution corresponds to point A in the solution space.

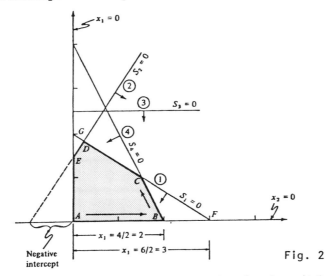

Fig. 2

The next step is to determine a new basic feasible solution (extreme point) with an improved value of the objective function. The simplex method does this by selecting a current nonbasic variable to be increased above zero providing its coefficient in the objective function has the potential to improve the value of x_0. Since an extreme point (in this problem) must have two zero nonbasic variables, one of the current basic variables must be made nonbasic at zero level providing the new solution is feasible. The current nonbasic variable to be made basic is usually called the entering variable and is determined by the optimality condition. The current basic variable to be made nonbasic is called the leaving variable and is determined by the feasibility condition.

The optimality condition stipulates that the entering variable be selected as the nonbasic variable having a negative coefficient in the x_0-equation of the tableau. This step follows since it is equivalent to selecting a positive coefficient in the original (maximization) objective function. In the above tableau, both x_1 and x_2 have negative coefficients and thus each qualifies as an entering variable. However, the variable with the most negative coefficient is usually selected, since it is likely (but not certain) to yield the largest improvement in the objective value. Thus x_1 is the entering variable.

Next, the feasibility condition is developed. Figure 2 shows that if x_1, the entering variable, is increased, the solution must move from point A to point B. This follows since B is the only feasible extreme point that can be reached from the current point A by increasing x_1. In other words, point B is now the new (feasible) extreme point solution. Interestingly, the nonbasic variables at B are $x_2 = 0$ and $S_4 = 0$. By comparison with point A where x_1 and x_2 are nonbasic, one can see that S_4 takes the place of x_1 and hence S_4 is the leaving variable.

Before showing how the leaving variable S_4 can be determined directly from the tableau, a geometric interpretation is given as follows. Determine the intercepts of all the constraints with the nonnegative direction of the x_1-axis (x_1 is the entering variable). Then the constraint with the smallest nonnegative intercept defines the leaving variable.

In Figure 2, only constraints ① and ④ intersect the nonnegative direction of the x_1-axis. Constraint ② intersects the negative direction and ③ is parallel to the x_1-axis. This result can be identified directly from the tableau since both ① and ④ have positive constraint coefficients under x_1 (namely, 2 and 2), while ② and ③ have negative and zero coefficients (namely, -3 and 0). The general conclusion then is that if the constraint coefficient under the entering variable is negative or zero, the corresponding constraint does not intersect the nonnegative direction of the axis defining the entering variable and hence will have no effect on feasibility.

Figure 2 shows that the intercepts of constraints ① and ④ with the nonnegative x_1-axis are given by $AB = 4/2 = 2$ and $AF = 6/2 = 3$. Since AB is the smaller of AB and AF, the value of x_1 is $AB = 2$. Since at B the current basic variable S_4 (associated with constraint ④) becomes zero, it follows that S_4 is the leaving variable.

The same information may be secured directly from the tableau by taking ratios of the values in the Solution column to the positive constraint coefficients under x_1. The basic variable associated with the minimum ratio (that is, smallest intercept) is the leaving variable. This is summarized below

Basic	Solution	x_1	Ratios	
S_1	6	2	$6/2 = 3$	
S_2	3	-3	—	
S_3	5	0	—	
S_4	4	2	$4/2 = 2$	minimum

After determining the entering and the leaving variables, the next step is to modify the tableau so that the Solution column will directly give the new values of x_0 and the (new) basic variables S_1, S_2, S_3, and x_1. This result is achieved by applying the Gauss-Jordan (or row operations) method, which eliminates (actually, substitutes out) the entering variable from all the equations of the tableau except the one associated with the leaving variable. To illustrate the method, the last tableau is repeated here for convenience.

Basic	x_0	x_1	x_2	S_1	S_2	S_3	S_4	Solution
x_0	①	-4	-3	0	0	0	0	0
S_1	0	2	3	①	0	0	0	6
S_2	0	-3	2	0	①	0	0	3
S_3	0	0	2	0	0	①	0	5
S_4	0	[2]	1	0	0	0	①	4

The coefficient "2" under the entering variable (designated by □) corresponding to the minimum ratio is called the pivot element. The S_4-equation associated with the leaving variable is called the pivot-equation. The first step in the Gauss-Jordan elimination is to divide the pivot equation by the pivot element and replace S_4 as a basic variable by the entering variable x_1. The pivot equation appears in the new tableau as

Basic	x_0	x_1	x_2	S_1	S_2	S_3	S_4	Solution
x_0								
S_1								
S_2								
S_3								
x_1	0	①	1/2	0	0	0	1/2	4/2 = 2

The elimination of x_1 from all but the pivot equation is accomplished by adding a proper multiple of the new pivot equation (whose pivot element is now equal to 1) to each of the other equations. The detailed row operations are given as follows.

1. Multiply the new pivot equation by +4 and add it to the old x_0-equation to obtain the new one [where (+4) is the negative of the element (-4) under x_1 in the x_0-equation].

```
            Old x0-equation:         (1   -4   -3   0   0   0   0    0)
+ (+4) x (new pivot equation):       (0    4    2   0   0   0   2    8)
            = new x0-equation:       (1    0   -1   0   0   0   2    8)
```

2. Multiply the new pivot equation by -2 and add it to the S_1-equation.

```
            Old S1-equation:         (0    2    3   1   0   0   0    6)
+ (-2) x (new pivot equation):       (0   -2   -1   0   0   0  -1   -4)
            = new S1-equation:       (0    0    2   1   0   0  -1    2)
```

3. Multiply the new pivot equation by +3 and add it to the S_2-equation.

```
            Old S2-equation:         (0   -3    2   0   1   0   0    3)
+ (+3) x (new pivot equation):       (0    3  3/2   0   0   0  3/2   6)
            = new S2-equation:       (0    0  7/2   0   1   0  3/2   9)
```

4. Leave the S_2-equation unchanged since its coefficient under x_1 is already zero.

The result of these calculations is the following tableau.

Basic	x_0	x_1	x_2	S_1	S_2	S_3	S_4	Solution
x_0	①	0	-1	0	0	0	2	8
S_1	0	0	②	①	0	0	-1	2
S_2	0	0	7/2	0	①	0	3/2	9
S_3	0	0	2	0	0	①	0	5
x_1	0	①	1/2	0	0	0	1/2	2

This shows that the new basic solution is $x_1 = 2$, $S_1 = 2$, $S_2 = 9$, and $S_3 = 5$ with $x_2 = S_4 = 0$ and $x_0 = 8$. This is point B in Figure 2.

The optimality condition shows that x_2 is the entering variable since it has a (only) negative coefficient in the x_0-equation. The leaving variable is determined by taking the ratios as given below, which show that S_1 leaves the solution.

Basic	Solution	x_2	Ratios	
S_1	2	2	$2/2 = 1$	minimum
S_2	9	7/2	$9/(7/2) = 18/7$	
S_3	5	2	$5/2 = 2.5$	
x_1	2	1/2	$2/(1/2) = 4$	

Given x_2 is the entering variable and S_1 is the leaving variable, one finds that the pivot equation is the S_1-equation with its pivot element equal to 2. The new tableau is obtained from the immediately preceding one as follows.

1. Divide the S_1-equation by 2 to obtain the new pivot equation.
2. Multiply the new pivot equation by +1 and add it to the old x_0-equation.
3. Multiply the new pivot equation by $-7/2$ and add it to the old S_2-equation.
4. Multiply the new pivot equation by -2 and add it to the old S_3-equation.
5. Multiply the new pivot equation by $-1/2$ and add it to the old x_1-equation.

These steps result in the following tableau.

Basic	x_0	x_1	x_2	S_1	S_2	S_3	S_4	Solution
x_0	①	0	0	1/2	0	0	3/2	9
x_2	0	0	①	1/2	0	0	$-1/2$	1
S_2	0	0	0	$-7/4$	①	0	13/4	11/2
S_3	0	0	0	-1	0	①	1	3
x_1	0	①	0	$-1/4$	0	0	3/4	3/2

Since all the coefficients on the left-hand side of the x_0-equation are nonnegative, no further improvements are possible and the current solution is optimum. This yields $x_1 = 1.5$, $x_2 = 1$, and $x_0 = 9$, which is point C in Figure 2.

Note that only three feasible extreme points are encountered before the optimum is determined. Thus, it is not necessary to enumerate all five feasible points (A, B, C, D, and E) let alone the remaining ten infeasible or nonexisting extreme points. The problem gives some idea of the computational efficiency of the simplex method.

The only difference between maximization and minimization occurs in the optimality condition: In minimization the entering variable is the one with the most positive coefficient in the objective function. The feasibility condition remains the same since it depends on the constraints and not the objective function.

• **PROBLEM 13-8**

Consider the linear programming problem

maximize $x_0 = 3x_1 + 2x_2 + 5x_3$

subject to

$$x_1 + 2x_2 + x_3 \le 430$$
$$3x_1 + 2x_3 \le 460$$
$$x_1 + 4x_2 \le 420$$

$x_1, x_2, x_3 \ge 0$, using the simplex method.

Solution: The problem is expressed in tableau form as follows.

Starting Tableau:

Basic	x_0	x_1	x_2	x_3	S_1	S_2	S_3	Solution
x_0	①	-3	-2	-5	0	0	0	0
S_1	0	1	2	1	①	0	0	430
S_2	0	3	0	2	0	①	0	460
S_3	0	1	4	0	0	0	①	420

First iteration: x_3 is the entering variable. By taking ratios,

Current basic solution	Ratios to coefficients of x_3
$S_1 = 430$	$430/1 = 430$
$S_2 = 460$	$460/2 = 230 \leftarrow S_2 = 0, x_3 = 230$
$S_3 = 420$	$420/0 = -$

S_2 becomes the leaving variable. The new tableau is thus given by

Basic	x_0	x_1	x_2	x_3	S_1	S_2	S_3	Solution
x_0	①	9/2	-2	0	0	5/2	0	1150
S_1	0	-1/2	2	0	①	-1/2	0	200
x_3	0	3/2	0	①	0	1/2	0	230
S_3	0	1	4	0	0	0	①	420

Second Iteration: x_2 is the entering variable. By taking ratios,

Current basic solution	Ratios to coefficients of x_2
$S_1 = 200$	$200/2 = 100 \leftarrow S_1 = 0, x_2 = 100$
$x_3 = 230$	$230/0 = -$
$S_3 = 420$	$420/4 = 105$

S_1 leaves the solution. The new tableau is

Basic	x_0	x_1	x_2	x_3	S_1	S_2	S_3	Solution
x_0	①	4	0	0	1	2	0	1350
x_2	0	-1/4	①	0	1/2	-1/4	0	100
x_3	0	3/2	0	①	0	1/2	0	230
S_3	0	2	0	0	-2	1	①	20

This is optimal since all the coefficients in the x_0-equation are nonnegative. The optimal solution is $x_1 = 0$, $x_2 = 100$, $x_3 = 230$, $S_1 = 0$, $S_2 = 0$, $S_3 = 20$, and $x_0 = 1350$.

• **PROBLEM 13-9**

Consider the problem

minimize $x_0 = 4x_1 + x_2$

subject to

$3x_1 + x_2 = 3$

$4x_1 + 3x_2 \geq 6$

$x_1 + 2x_2 \leq 3$

$x_1, x_2 \geq 0$

Use the simplex method to solve.

Hint: Optimality condition: In minimization the entering variable is the one with the most positive coefficient in the objective function. The feasibility condition is the same as for the maximization problem since it depends on the constraints and not the objective function.

Solution: The standard form of the problem is as follows.
Minimize $x_0 = 4x_1 + x_2$
subject to

$3x_1 + x_2 = 3$

$4x_1 + 3x_2 - S_2 = 6$

$x_1 + 2x_2 + S_3 = 3$

$x_1, x_2, S_2, S_3 \geq 0$

The first and second equations do not provide obvious starting basic feasible variables, but S_3 can be used as a starting basic variable for the third equation. By adding artificial variables to the first and second equations and modifying the objective function accordingly, one reduces the above standard form to minimize $x_0 = 4x_1 + x_2 + MR_1 + MR_2$

subject to

$3x_1 + x_2 + R_1 = 3$

$4x_1 + 3x_2 - S_2 + R_2 = 6$

$x_1 + 2x_2 + S_3 = 3$

$x_1, x_2, S_2, S_3, R_1, R_2 \geq 0$

where M is a very large positive value.

The tableau form of the problem is

Basic	x_0	x_1	x_2	S_2	R_1	R_2	S_3	R.H.S.
x_0	①	−4	−1	0	−M	−M	0	0
R_1	0	3	1	0	①	0	0	3
R_2	0	4	3	−1	0	①	0	6
S_3	0	1	2	0	0	0	①	3

The objective equation must be expressed in terms of the nonbasic variables only to reflect the proper worth per unit of each variable.

In general, the tableau is in the proper form if each coefficient of the starting basic variables in the x_0-equation is zero. In the above tableau, this is equivalent to substituting out R_1 and R_2 in the x_0-equation. The result can be accomplished by using the following row operation.

new x_0-equation = old x_0-equation + M × R_1-equation + M × R_2-equation

The new x_0-equation produces the following tableau, which is now ready for the application of optimality condition.

Basic	x_0	x_1	x_2	S_2	R_1	R_2	S_3	Solution
x_0	①	−4 + 7M	−1 + 4M	−M	0	0	0	9M
R_1	0	3	1	0	①	0	0	3
R_2	0	4	3	−1	0	①	0	6
S_3	0	1	2	0	0	0	①	3

Noting that this is a minimization problem, the entering variable has the largest positive coefficient in the x_0-equation. Thus,

First Iteration: Introduce x_1 and drop R_1.

Basic	x_0	x_1	x_2	S_2	R_1	R_2	S_3	Solution
x_0	①	0	$\frac{1 + 5M}{3}$	−M	$\frac{4 - 7M}{3}$	0	0	4 + 2M
x_1	0	①	1/3	0	1/3	0	0	1
R_2	0	0	5/3	−1	−4/3	①	0	2
S_3	0	0	5/3	0	−1/3	0	①	2

Second Iteration: Introduce x_2 and drop R_2.

Basic	x_0	x_1	x_2	S_2	R_1	R_2	S_3	Solution
x_0	①	0	0	1/5	$8/5 - M$	$-1/5 - M$	0	18/5
x_1	0	①	0	1/5	3/5	$-1/5$	0	3/5
x_2	0	0	①	$-3/5$	$-4/5$	3/5	0	6/5
S_3	0	0	0	1	1	-1	①	0

Third Iteration: Introduce S_2 and drop S_3.

Basic	x_0	x_1	x_2	S_2	R_1	R_2	S_3	Solution
x_0	①	0	0	0	$7/5 - M$	$-M$	$-1/5$	18/5
x_1	0	①	0	0	2/5	0	$-1/5$	3/5
x_2	0	0	①	0	$-1/5$	0	3/5	6/5
S_2	0	0	0	①	1	-1	1	0

This gives the optimal solution, $x_1 = 3/5$, $x_2 = 6/5$, and $x_0 = 18/5$.

● **PROBLEM 13-10**

Consider the following problem.

   Maximize $x_0 = 3x_1 + 5y + 2x_3$

subject to

   $x_1 + 2y + 2x_3 \leq 14$

   $2x_1 + 4y + 3x_3 \leq 23$

   $0 \leq x_1 \leq 4, \quad 2 \leq y \leq 5, \quad 0 \leq x_3 \leq 3$

Solve by the simplex method.

Solution: Since y has a positive lower bound, it must be substituted at its lower bound. Let $y = x_2 + 2$, then $0 \leq x_2 \leq 5 - 2 = 3$ and the problem becomes

Starting tableau:

Basic	x_0	x_1	x_2	x_3	S_1	S_2	Solution
x_0	①	-3	-5	-2	0	0	10
S_1	0	1	2	2	①	0	10
S_2	0	2	4	3	0	①	15

First Iteration:

Select x_2 as the entering variable ($z_2 - c_2 = -5$). Thus

$$\vec{a}^2 = \begin{pmatrix} 2 \\ 4 \end{pmatrix} > \vec{0}$$

and

   $\theta_1 = \min\{10/2, 15/4\} = 3.75$

Since all $a_i^2 > 0$, it follows that $\theta_2 = \infty$. Consequently,

$\theta = \min\{3.75, \infty, 3\} = 3$.

Because $\theta = u_2, x_2$ is substituted at its upper limit but it remains nonbasic. Thus putting $x_2 = u_2 - x_2' = 3 - x_2'$, the new tableau becomes

Basic	x_0	x_1	x_2'	x_3	S_1	S_2	Solution
x_0	①	-3	5	-2	0	0	25
S_1	0	1	-2	2	①	0	4
S_2	0	2	-4	3	0	①	3

Second Iteration:

Select x_1 as the entering variable $(z_1 - c_1 = -3)$. Thus,

$$\vec{a}^1 = \begin{pmatrix} 1 \\ 2 \end{pmatrix}$$

$\theta_1 = \min\{4/1, 3/2\} = 3/2$, corresponding to S_2

$\theta_2 = \infty$

Hence, $\theta = \min\{3/2, \infty, 4\} = 3/2$. Since $\theta = \theta_1$, introduce x_1 and drop S_2. This yields

Basic	x_0	x_1	x_2'	x_3	S_1	S_2	Solution
x_0	①	0	-1	5/2	0	3/2	59/2
S_1	0	0	0	1/2	①	$-1/2$	5/2
x_1	0	①	-2	3/2	0	1/2	3/2

Third Iteration:

Select x_2' as the entering variable. Since

$$\vec{a}^2 = \begin{pmatrix} 0 \\ -2 \end{pmatrix} \leq 0$$

$\theta_1 = \infty$

$\theta_2 = \left| \dfrac{4 - 3/2}{-(-2)} \right| = \dfrac{5}{4}$, corresponding to x_1

Thus, $\theta = \min\{\infty, 5/4, 3\} = 5/4$. Since $\theta = \theta_2$, introduce x_2' into basis and drop x_1 then substitute it out at its upper bound $(4 - x_1')$. Thus, by removing x_1 and introducing x_2', the tableau becomes

Basic	x_0	x_1	x_2'	x_3	S_1	S_2	Solution
x_0	①	$-1/2$	0	7/4	0	5/4	115/4
S_1	0	0	0	1/2	①	$-1/2$	5/2
x_2'	0	$-1/2$	①	$-3/4$	0	$-1/4$	$-3/4$

Now, by substituting for $x_1 = 4 - x_1'$, the final tableau becomes

Basic	x_0	x_1'	x_2'	x_3	S_1	S_2	Solution
x_0	①	1/2	0	7/4	0	5/4	123/4
S_1	0	0	0	1/2	①	−1/2	5/2
x_2'	0	1/2	①	−3/4	0	−1/4	5/4

which is now optimal and feasible.

The optimal solution in terms of the original variables x_1, x_2, and x_3 is found as follows: Since $x_1' = 0$, it follows that $x_1 = 4$. Also, since $x_2' = 5/4$, $x_2 = 3 - 5/4 = 7/4$ and $y = 2 + 7/4 = 15/4$. Finally x_3 equals 0. These values yield $x_0 = 123/4$ as shown in the optimal tableau.

● PROBLEM 13-11

Maximize

$$f = 2x_1 + 2x_2 + 3x_3$$

subject to the constraints

$$x_1 + x_3 \leq 1$$

$$x_2 + x_3 \leq 1$$

$$x_i \geq 0, \text{ for } i = 1,2,3$$

Solve by the aid of graphical depiction.

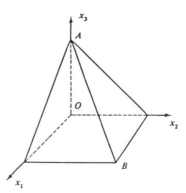

Solution: The feasible vectors form a four-sided pyramid in (x_1, x_2, x_3)-space; see Figure. Introduce the slack variables x_4, x_5 and take $\{x_1, x_2, x_3\}$ as right-hand variables. Notice that the equations $x_4 = 0$ and $x_5 = 0$ also mean sides of a pyramid. This gives a feasible vector, since $x_1 = x_2 = x_3 = 0$ (the point O in the figure) satisfies the constraints. Then

$$x_4 = 1 - \underline{x_1} - \underline{x_3}, \qquad \Delta x_1 = 1, \; \Delta f = 2$$

$$x_5 = 1 - \underline{x_2} - \underline{x_3}, \qquad \Delta x_2 = 1, \; \Delta f = 2$$

$$f = 2x_1 + 2x_2 + 3x_3, \quad \Delta x_3 = 1, \; \Delta f = 3 \text{ (largest)}$$

$$x_i \geq 0 \text{ for } i = 1,2,3,4,5$$

Notice also that x_3, which gave the largest increase in f, in underlined in two places; x_3 is to be exchanged with either x_4 or x_5. In such a situation one can choose either one -say, x_4. (Then one comes to the point A.)

$$x_3 = 1 - x_1 - x_4, \quad \Delta x_2 = 0, \; \Delta f = 0$$

$$x_5 = \underline{x_1} - x_2 + x_4$$

$$f = 3 - x_1 + 2x_2 - 3x_4.$$

The point A is a degenerate feasible vector; x_5 is also zero at this point. Four sides of a pyramid go through A. Since x_2 is the only right-hand coordinate which has a positive coefficient in f, exchange x_2 and x_5. Thus, in this step, one remains at the point A. (Recall that in nondegenerate cases, one step in the iteration takes one to a neighboring basic point on the boundary of the admissible region.)

$$x_2 = x_1 + x_4 - x_5, \quad \Delta x_1 = 1, \; \Delta f = 1$$

$$x_3 = 1 - \underline{x_1} - x_4$$

$$f = 3 + x_1 - x_4 - 2x_5.$$

Exchange x_1 and x_3. (One goes from A to B.)

$$x_1 = 1 - x_3 - x_4$$

$$x_2 = 1 - x_3 - x_5$$

$$f = 4 - x_3 - 2x_4 - 2x_5.$$

The maximum criterion is fulfilled. $f_{max} = 4$. The optimal solution is the point $(1,1,0,0,0)$.

● **PROBLEM** 13-12

Maximize

$$f = x_1 - x_2$$

under the constraints

$$x_3 = -2 + 2x_1 - x_2$$

$$x_4 = 2 - x_1 + 2x_2$$

$$x_5 = 5 - x_1 - x_2$$

$$x_i \geq 0 \quad (i = 1,2,3,4,5)$$

Solve by the "M-technique" based on the use of "artificial" variables.

Solution: x_1 and x_2 cannot be used as right-hand variables, since, when $x_1 = x_2 = 0$, x_3 is negative. It is not immediately obvious which pair of variables suffice as right-hand variables. In such a case, one can modify the problem by introducing a new artificial variable x_6, defined by the equation

$$x_3 = -2 + 2x_1 - x_2 + x_6,$$

with the constraint $x_6 \geq 0$. If one now takes x_1, x_2, x_3 as right-hand variables, then x_4, x_5, x_6 are all nonnegative when the right-hand variables are set equal to zero. Thus one has found a feasible vector for an extended (six-dimensional) problem. In order for this problem to correspond, finally, to the original problem, one must make sure that the artificial variable (here x_6) is zero in the final solution. To accomplish this, modify the function to be maximized by adding a term with a very large negative coefficient, say, $-M$, in front of the artificial variable (M is assumed to be much larger than all other numbers in the computations). Thus a positive value of an artificial variable will tend to make the function to be maximized quite small; in this way, the value of the artificial variable becomes zero in the final solution. Here, the modified problem becomes:

Maximize

$$\bar{f} = x_1 - x_2 - Mx_6 = (1 + 2M)x_1 - (1 + M)x_2 - Mx_3 - 2M$$

under the constraints

$$x_6 = 2 - \underline{2x_1} + x_2 + x_3, \quad \Delta x_1 = 1, \; \Delta\bar{f} = 1 + 2M$$

$$x_4 = 2 - x_1 + 2x_2$$

$$x_5 = 5 - x_1 - x_2$$

$$x_i \geq 0, \; (i = 1,2,3,4,5,6).$$

Exchange x_1 and x_6:

$$x_1 = \frac{2 + x_2 + x_3 - x_6}{2}$$

$$x_4 = \frac{2 + 3x_2 \underline{-x_3} + x_6}{2}, \quad \Delta x_3 = 2, \; \Delta\bar{f} = 1$$

$$x_5 = \frac{8 - 3x_2 - x_3 + x_6}{2}$$

$$\bar{f} = \frac{2 - x_2 + x_3 + x_6}{2} - Mx_6.$$

Now the artificial variable has become a right-hand variable. If one takes away x_6 and sets the rest of the right-hand variables equal to zero, then one gets a feasible vector to the original (five-dimensional) problem. The artificial variable is no longer needed in the computations. After two iteratons (x_3 is exchanged with x_4, and thereafter x_2 with x_5), the maximum criterion is satisfied and one finds $f_{max} = 3$ for $(x_1, x_2, x_3, x_4, x_5) = (4,1,5,0,0)$. In more difficult situations, it may be necessary to introduce several artificial variables.

There can be many optimal solutions to a given problem, but of course f_{max} is uniquely determined. Such a situation is indicated in the calculations when one or more of the right-hand variables is absent in the expression for f when the maximum criterion is satisfied.

DUALITY

● **PROBLEM 13-13**

Minimize $x_0 = 2x_1 + x_2$

subject to

$3x_1 + x_2 \geq 3$

$4x_1 + 3x_2 \geq 6$

$x_1 + 2x_2 \leq 3$

$x_1, x_2 \geq 0$,

using the Dual Simplex method.

Solution: By putting the constraints in the form ($\leq$) and adding the slack variables, the corresponding starting tableau is given by

Basic	x_0	x_1	x_2	S_1	S_2	S_3	Solution
x_0	①	-2	-1	0	0	0	0
S_1	0	-3	-1	①	0	0	-3
S_2	0	-4	-3	0	①	0	-6
S_3	0	1	2	0	0	①	3

The starting solution is $S_1 = -3$, $S_2 = -6$, and $S_3 = 3$. This solution is infeasible since at least one basic variable is negative. In the meantime, the x_0-equation is optimal since all its coefficients are nonpositive (minimization problem). This problem is typical of the type that can be handled by the dual simplex method.

As in the regular simplex method, the method of solution is based on the optimality and feasibility conditions. The optimality condition guarantees that the solution remains always optimal, while the feasibility condition forces the basic solutions toward the feasible space.

Feasibility Condition: The leaving variable is the basic variable having the most negative value. (Break ties arbitrarily.) If all the basic variables are nonnegative the process ends and the feasible (optimal) solution is reached.

Optimality Condition: The entering variable is selected from among the nonbasic variables as follows. Take the ratios of the left-hand side coefficients of the x_0-equation to the corresponding coefficients in the equation associated with the leaving variable. Ignore the ratios associated with positive or zero denominators. The entering variable is the one with the smallest ratio if the problem is minimization, or the smallest absolute value of the ratios if the problem is maximization. (Break ties arbitrarily.) If all the denominators are zero or positive, the problem has no feasible solution.

After selecting the entering and the leaving variables, row operations are applied as usual to obtain the next iteration of the solution.

The leaving variable in the above tableau is S_2 $(= -6)$ since it has the most negative value. For the entering variable, the ratios are given by

Variable	x_1	x_2	S_1	S_2	S_3
x_0-equation	−2	−1	0	0	0
S_2-equation	−4	−3	0	1	0
Ratios	1/2	1/3	−	−	−

The entering variable is x_2 since it corresponds to the smallest ratio 1/3. By applying row operations as usual, one finds the new tableau

Basic	x_0	x_1	x_2	S_1	S_2	S_3	Solution
x_0	①	−2/3	0	0	−1/3	0	2
S_1	0	−5/3	0	①	−1/3	0	−1
x_2	0	4/3	①	0	−1/3	0	2
S_3	0	−5/3	0	0	2/3	①	−1

The new solution is still optimal but infeasible ($S_1 = -1$, $S_3 = -1$). If S_1 is arbitrarily selected to leave the solution, x_1 becomes the entering variable. This gives

Basic	x_0	x_1	x_2	S_1	S_2	S_3	Solution
x_0	①	0	0	−2/5	−1/5	0	12/5
x_1	0	①	0	−3/5	1/5	0	3/5
x_2	0	0	①	4/5	−3/5	0	6/5
S_3	0	0	0	−1	1	①	0

which is now optimal and feasible.

The application of the dual simplex method is especially useful in sensitivity analysis. This occurs, for example, when a new constraint is added to the problem after the optimal solution is obtained. If this constraint is not satisfied by the optimal solution, the problem remains optimal but it becomes infeasible. The dual simplex method is then used to clear the infeasibility in the problem.

• **PROBLEM 13-14**

Find a solution to the following problem by solving its dual:

$$\text{Minimize } 9x_1 + 12x_2 + 15x_3 \tag{1}$$

subject to

$$2x_1 + 2x_2 + x_3 \geq 10$$

$$2x_1 + 3x_2 + x_3 \geq 12 \tag{2}$$

$$x_1 + x_2 + 5x_3 \geq 14$$

$$x_1 \geq 0, \ x_2 \geq 0, \ x_3 \geq 0. \tag{3}$$

Solution: The dual to (1), (2) and (3) is

$$\text{Maximize } 10y_1 + 12y_2 + 14y_3 \tag{4}$$

subject to

$$2y_1 + 2y_2 + y_3 \leq 9$$

$$2y_1 + 3y_2 + y_3 \leq 12 \tag{5}$$

$$y_1 + y_2 + 5y_3 \leq 15$$

$$y_1 \geq 0, \ y_2 \geq 0, \ y_3 \geq 0. \tag{6}$$

Solve this problem using the simplex method. Converting the inequalities to equalities by adding slack variables

$$2y_1 + 2y_2 + y_3 + y_4 = 9$$

$$2y_1 + 3y_2 + y_3 + y_5 = 12$$

$$y_1 + y_2 + 5y_3 + y_6 = 15.$$

This may be rewritten as:

$$y_1 \begin{pmatrix} 2 \\ 2 \\ 1 \end{pmatrix} + y_2 \begin{pmatrix} 2 \\ 3 \\ 1 \end{pmatrix} + y_3 \begin{pmatrix} 1 \\ 1 \\ 5 \end{pmatrix} + y_4 \begin{pmatrix} 1 \\ 0 \\ 0 \end{pmatrix} + y_5 \begin{pmatrix} 0 \\ 1 \\ 0 \end{pmatrix}$$

$$+ y_6 \begin{pmatrix} 0 \\ 0 \\ 1 \end{pmatrix} = \begin{pmatrix} 9 \\ 12 \\ 15 \end{pmatrix}. \tag{7}$$

Since there are three equations in six unknowns, one can obtain a basic feasible solution by setting any three of the y_i's to zero. Let

$$v_1 = \begin{pmatrix} 2 \\ 2 \\ 1 \end{pmatrix}, \quad v_2 = \begin{pmatrix} 2 \\ 3 \\ 1 \end{pmatrix}, \quad v_3 = \begin{pmatrix} 1 \\ 1 \\ 5 \end{pmatrix}, \quad v_4 = \begin{pmatrix} 1 \\ 0 \\ 0 \end{pmatrix},$$

$$v_5 = \begin{pmatrix} 0 \\ 1 \\ 0 \end{pmatrix}, \quad v_6 = \begin{pmatrix} 0 \\ 0 \\ 1 \end{pmatrix}$$

and

$$b = \begin{pmatrix} 9 \\ 12 \\ 15 \end{pmatrix}.$$ Let $d_1 = 10$, $d_2 = 12$, $d_3 = 14$. Setting $y_1 = y_2 = y_3 = 0$, a basic feasible solution is $y_4 = 9$, $y_5 = 12$, $y_6 = 15$. Start the simplex algorithm with v_4, v_5 and v_6 in the basis.

	v_1	v_2	v_3	b	v_4	v_5	v_6
v_4	2	2	1	9	1	0	0
v_5	2	3	1	12	0	1	0
v_6	1	1	5	15	0	0	1
d	-10	-12	-14	0	0	0	0

To increase the value of D, one must choose another vector for the basis. Choose the column which has the most negative value in the row labelled 'd'. Here v_3 is the chosen column. Next, decide which vector in the basis to discard. The rule here is to choose that row for which the ratio of b to v_3 is the smallest. Here the ratios are 9/1, 12/1 and 15/5. Hence replace v_6 by v_3, by pivoting on 5. The process of conversion is carried out by first converting the 5 to a 1 and then using elementary row operations to reduce every other element under v_3 to zero. Thus one obtains the new tableau

	v_1	v_2	v_3	b	v_4	v_5	v_6
v_4	9/5	9/5	0	6	1	0	-1/5
v_5	9/5	14/5	0	9	0	1	-1/5
v_3	1/5	1/5	1	3	0	0	1/5
d	-36/5	-46/5	0	42	0	0	14/5

The value 42 was obtained by using $b = c_1y_1 + c_2y_2 +$

$c_3y_3 + c_4y_4 + c_5y_5 + c_6y_6$ where the y_i are the coefficients of the vectors in the basis. Here the basis vectors are v_4, v_5, v_6. The coefficients are, from (4), $c_1 = 10$, $c_2 = 12$, $c_3 = 14$, $c_4 = c_5 = c_6 = 0$ and the y_i are obtained from the column labelled b. Hence

$$D = c_4(6) + c_5(9) + c_3(3) = 0(6) + 0(9) + 14(3) = 42.$$

Since the row d still contains negative entries repeat the above procedure by pivoting on 14/5. One obtains

	v_1	v_2	v_3	b	v_4	v_5	v_6
v_4	9/14	0	0	3/14	1	-9/14	-1/14
v_2	9/14	1	0	45/14	0	5/14	-1/14
v_3	1/14	0	1	33/14	0	-1/14	3/14
d	-9/7	0	0	501/7	0	23/7	15/7

Next, pivot on the first element in the first column, i.e., 9/14. The final simplex tableau is

	v_1	v_2	v_3	b	v_4	v_5	v_6
v_1	1	0	0	1/3	14/9	-1	-1/9
v_2	0	1	0	3	-1	1	0
v_3	0	0	1	7/3	-1/9	0	2/9
d	0	0	0	72	2	2	2

Thus the solution to the maximization problem (the dual of the given problem) is

$$y_1 = 1/3, \quad y_2 = 3, \quad y_3 = 7/3$$

with value 72. It is known from duality theory that if a program has an optimal feasible point then so does its dual and both programs have the same value. The dual of (4), (5), (6) is (1), (2), (3). Thus the solution to the minimization problem has value 72. But what are the values of x_1, x_2, x_3 at this optimum point? From the Complementary Slackness Theorem of duality theory, these values are the values of the slack variables in the final simplex tableau of the dual problem. Thus $x_1 = 2$, $x_2 = 2$, $x_3 = 2$ is the required solution.

● **PROBLEM 13-15**

Solve the following problem in integer programming:

Find non-negative integers X_{ij} which will

Minimize $200 x_{11} + 300 x_{12} + 250 x_{21} + 100 x_{22}$

$+ 250 x_{31} + 250 x_{32}$

subject to

$$X_{11} + X_{21} + X_{31} \geq 30$$
$$X_{12} + X_{22} + X_{32} \geq 20$$
$$-X_{11} - X_{12} \geq -20$$
$$-X_{21} - X_{22} \geq -20$$
$$-X_{31} - X_{32} \geq -20$$

Solution: Since some of the constraints are negative one cannot apply the simplex algorithm to the given problem. However, the dual, which is a maximization problem, is receptive to the simplex technique. The dual is

Maximize $30y_1 + 20y_2 - 20y_3 - 20y_4 - 20y_5$

subject to:

$$y_1 - y_3 \leq 200$$
$$y_2 - y_3 \leq 300$$
$$y_1 - y_4 \leq 250$$
$$y_2 - y_4 \leq 100$$
$$y_1 - y_5 \leq 250$$
$$y_2 - y_5 \leq 250$$
$$y_1, y_2, \ldots, y_5 \geq 0$$

Convert the inequalities to equalities by adding six slack variables

$y_6, y_7, y_8, y_9, y_{10}, y_{11}$.

An initial basic feasible solution is obtained by setting $y_1 = y_2 = y_3 = y_4 = y_5 = 0$. The simplex tableau is then

	v_1	v_2	v_3	v_4	v_5	v_6	v_7	v_8	v_9	v_{10}	v_{11}	b
v_6	1	0	-1	0	0	1	0	0	0	0	0	200
v_7	0	1	-1	0	0	0	1	0	0	0	0	300
v_8	1	0	0	-1	0	0	0	1	0	0	0	250
v_9	0	1	0	-1	0	0	0	0	1	0	0	100
v_{10}	1	0	0	0	-1	0	0	0	0	1	0	250
v_{11}	0	1	0	0	-1	0	0	0	0	0	1	250
d	-30	-20	20	20	20	0	0	0	0	0	0	0

Since -30 is the largest negative entry and 200/1 is the minimum positive ratio, replace v_6 by v_1 in the basis on the left. Thus, pivot on the first element in the first column to obtain

	v_1	v_2	v_3	v_4	v_5	v_6	v_7	v_8	v_9	v_{10}	v_{11}	b
v_1	1	0	-1	0	0	1	0	0	0	0	0	200
v_7	0	1	-1	0	0	0	1	0	0	0	0	300
v_8	0	0	1	-1	0	-1	0	1	0	0	0	50
v_9	0	(1)	0	-1	0	0	0	0	1	0	0	100
v_{10}	0	0	1	0	-1	-1	0	0	0	1	0	50
v_{11}	0	1	0	0	-1	0	0	0	0	0	0	250
d	0	-20	-10	20	20	30	0	0	0	0	0	6,000

Now, replace v_9 by v_2 to obtain a new basic feasible solution. Pivoting on the encircled element:

	v_1	v_2	v_3	v_4	v_5	v_6	v_7	v_8	v_9	v_{10}	v_{11}	b
v_1	1	0	-1	0	0	1	0	0	0	0	0	200
v_7	0	0	-1	1	0	0	1	0	-1	0	0	200
v_8	0	0	1	-1	0	-1	0	1	0	0	0	50
v_2	0	1	0	-1	0	0	0	0	1	0	0	100
v_{10}	0	0	(1)	0	-1	-1	0	0	0	1	0	50
v_{11}	0	0	0	1	-1	0	0	0	-1	0	1	50
d	0	0	-10	0	20	30	0	0	20	0	0	8,000

The only negative entry is under v_3. Notice that the ratios of b to v_8 and of b to v_{10} are both the same, i.e., 50/1. By the rules of the simplex algorithm choose either v_8 or v_{10} for liquidation. Choose to replace v_{10} by v_3 and hence pivot on the encircled element. The next tableau is

	v_1	v_2	v_3	v_4	v_5	v_6	v_7	v_8	v_9	v_{10}	v_{11}	b
v_1	1	0	0	0	-1	0	0	0	0	1	0	250
v_7	0	0	0	1	-1	-1	1	0	1	1	0	250
v_8	0	0	0	-1	1	0	0	1	0	-1	0	0
v_2	0	1	0	-1	0	0	0	0	1	0	0	100
v_3	0	0	1	0	-1	-1	0	0	0	1	0	50
v_{11}	0	0	0	1	-1	0	0	0	-1	0	1	50
d	0	0	0	0	10	20	0	0	20	10	0	8,500

Since there are no more negative entries in the last row, the algorithm has converged to a solution. The maximum feasible solution is 8,500. Therefore the minimum feasible solution is also 8,500. At this value, the values of the X_{ij} are read off from the slack variable values in the last row. Thus, $X_{11} = 20$, $X_{12} = 0$, $X_{21} = 0$, $X_{22} = 20$, $X_{31} = 10$, $X_{32} = 0$. Note that the main body of the simplex tableau consisted only of ones and zeros. This is a characteristic feature of integer programs. It makes pivoting easier and also ensures that the solution values will be integers.

OTHER OPTIMIZATION PROBLEMS

● **PROBLEM** 13-16

Find (ZZ,XX) where

$$XX = (XX(1), XX(2), XX(3), XX(4), XX(5))$$

is in S, and ZZ = F(XX) is the minimum value of

$$Z = F(X) = -2*X(1) - X(2) + 0*X(3) + 0*X(4) + 0*X(5) \quad (1)$$

as X = (X(1),X(2),X(3),X(4),X(5)) varies over the set S defined by

$$X(1) + 3*X(2) + X(3) \qquad\qquad = 6 \qquad (2a)$$

$$X(1) +X(2) + X(4) \qquad = 3 \qquad (2b)$$

$$X(1) -X(2) + X(5) = 2 \qquad (2c)$$

$$X(1) \geq 0, \ X(2) \geq 0, \ X(3) \geq 0, \ X(4) \geq 0, \ \text{and} \ X(5) \geq 0. \quad (2d)$$

It is known that the vertices (extreme points) of S correspond to vectors X with at most K nonzero components, where K is the number of equations in (2). Since the minimum feasible solution is a vertex of S, then one only needs to look at vertices of S. Thus, examine five-tuples that satisfy (2) and that have at least two components equal zero.

Solution: Step 1: Begin by choosing X(1) = X(2) = 0 in (2a,b,c). Then, the equations become

$$X(3) = 6$$
$$X(4) = 3$$
$$X(5) = 2,$$

and trivially obtain the feasible solution (element of S)

$$X = (0,0,6,3,2).$$

Since (i) the components of this feasible solution are nonnegative, (ii) at least two components are zero, and (iii) the "columns" of (2a,b,c) that correspond to nonzero components are linearly independent, then this feasible solution is a basic feasible solution. Basic feasible solutions are vertices (or corners) of S. The basic feasible solution just obtained gives the objective function F(X) the value

$$F(X) = -2*(0) - (0) + 0*(6) + 0*(3) + 0*(2) = 0.$$

Step 2: Note that F(X), (1), is reduced more by a unit of X(1) than by a unit of X(2); so choose X(1) as large as possible, consistent with (2a,b,c,d), and leave X(2) = 0. With X(2) = 0, (2a,b,c) becomes

$$X(1) + X(3) = 6$$
$$X(1) + X(4) = 3$$
$$X(1) + X(5) = 2.$$

Choose X(3) = 0, or X(4) = 0, or X(5) = 0, have X(1) as large as possible and still have (2) satisfied.

X(3) = 0, and X(1) = 6 implies X(4) = -3 < 0.

X(4) = 0, and X(1) = 3 implies X(5) = -1 < 0.

X(5) = 0, and X(1) = 2 implies X(3) = 4, and X(4) = 1.

Therefore, with X(2) = 0, choose X(5) = 0. Then (2a,b,c) becomes

$$X(1) + X(3) = 6$$
$$X(1) + X(4) = 3$$
$$X(1) = 2,$$

which can be solved by Jordan Elimination to obtain

$$X(1) = 2$$
$$X(3) = 4$$
$$X(4) = 1.$$

Thus, a new basic feasible solution is

$$X = (2, 0, 4, 1, 0).$$

which gives the objective function the value

$$F(X) = -2*(2) - (0) + 0*(4) + 0*(1) + 0*(0) = -4.$$

This is indeed less than F(X) for the first basic feasible solution.

It is convenient to replace (2a,b,c) by an equivalent linear system obtained by a Gauss Elimination:

$$4*X(2) + X(3) \qquad - X(5) = 4 \qquad (3a)$$

$$2*X(2) \qquad\qquad + X(4) - X(5) = 1 \qquad\qquad (3b)$$

$$X(1) - X(2) \qquad\qquad + X(5) = 2. \qquad\qquad (3c)$$

"Cover" the X(2) and X(5)-columns, and read-off the potentially nonzero components of the new basic feasible solution. From (3c)

$$X(1) = 2 + X(2) - X(5).$$

Thus, the objective function (1) can be written

$$F(X) = -2*(2 + X(2) - X(5)) - X(2) \qquad\qquad (4)$$

$$= -3*X(2) + 2*X(5) - 4.$$

Notice, if X(2) = X(5) = 0, then F(X) = -4, as claimed.

Step 3: From (4) see that F(X) can be reduced by increasing X(2), and keeping X(5) = 0. With X(5) = 0, (3a,b,c) becomes

$$4*X(2) + X(3) \qquad\qquad = 4$$

$$2*X(2) \qquad\qquad + X(4) = 1$$

$$X(1) - X(2) \qquad\qquad = 2.$$

Set X(1) = 0, or X(3) = 0, or X(4) = 0, and find the largest "allowable" X(2).

X(3) = 0 and X(2) = 1 implies X(4) = -1 < 0.

X(4) = 0 and X(2) = 1/2 implies X(3) = 2, and X(1) = 5/2.

X(1) = 0 and X(2) = -2 implies X(2) < 0.

The only choice consistent with (3a,b,c), (2d), and X(5) = 0 is the choice X(4) = 0. So "keep" (3b) and eliminate X(2) from the other equations to obtain

$$+ X(3) - 2*X(4) + X(5) = 2 \qquad\qquad (4a)$$

$$X(2) \qquad + .5*X(4) - .5*X(5) = 1/2 \qquad\qquad (4b)$$

$$X(1) \qquad + .5*X(4) + .5*X(5) = 5/2. \qquad\qquad (4c)$$

Thus, a new basic feasible solution is

$$X = (5/2, 1/2, 2, 0, 0).$$

From (4b)

$$X(2) = .5*(1 - X(4) + X(5)),$$

so the objective function (4) can be written

$$F(X) = -3*.5*(1 - X(4) + X(5)) + 2*X(5) - 4 \qquad\qquad (5)$$

$$= 3/2*X(4) + 1/2*X(5) - 11/2.$$

When $X(4) = X(5) = 0$, then $F(X) = -11/2$.

Step 4: Examine (5). Since the coefficients of $X(4)$ and $X(5)$ in $F(x)$ are positive, then no nonnegative choice for $X(4)$ and/or $X(5)$ will reduce $F(x)$. Thus, the minimum feasible solution of the linear programming problem is (ZZ,XX) where

   XX = (5/2, 1/2, 2, 0, 0)

and

   $F(XX) = -2*(5/2) - (1/2) + 0*(2) + 0*(0) + 0*(0)$

   $= -11/2$.

● **PROBLEM 13-17**

Find the coefficients A and B of the linear polynomial

   $P(T) = A*T + B$ (1a)

so that the graph of P(T), a line, comes "close" to the points

   (.2,.1),   (.3,.2),   (.4,.2),

and

   $|P(.2) - .1| \leq M$

   $|P(.3) - .2| \leq M$ (1b)

   $|P(.4) - .2| \leq M$

for the constant M as small as possible. That is, no line passes through the three points, but the line that comes closest to the three points in the sense that the maximum error at the specified points is a minimum, is desired.

What is the linear programming model that corresponds to this problem? Choose an initial basic feasible solution.

Solution: Note:

   $|W| \leq E$ is equivalent to $-E \leq W \leq E$

           is equivalent to $W \leq E$ and $-W \leq E$.

Thus, the conditions (1b) may be written in the equivalent form(s)

    $P(.2) - .1 \leq M$ $-M + P(.2) \leq .1$

    $-P(.2) + .1 \leq M$ $-M - P(.2) \leq -.1$

    $P(.3) - .2 \leq M$ $-M + P(.3) \leq .2$

$$-P(.3) + .2 \leq M \quad \text{, or} \quad -M - P(.3) \leq -.2 \qquad (2)$$

$$P(.4) - .2 \leq M \qquad -M + P(.4) \leq .2$$

$$-P(.4) + .2 \leq M \qquad -M - P(.4) \leq -.2.$$

The linear programming problem associated with the second set of inequalities in (2), after slack variables have been introduced, may be written in the following form.

Find (ZZ,XX) where

$$XX = (MM, AA, BB, CC1, CC2, CC3, CC4, CC5, CC6)$$

is in S, and $ZZ = F(XX)$ is the minimum value of

$$Z = F(X) = 1*M + 0*A + 0*B + 0*C1 + 0*C2 + 0.*C3$$
$$+ 0*.C4 + 0.C5 + 0.C6 \qquad (3)$$

as $X = (M,A,B,C1,C2,C3,C4,C5,C6)$ varies over the set S defined by

$$-M + .2*A + B + C1 \qquad\qquad\qquad = .1$$
$$-M - .2*A - B \qquad + C2 \qquad\qquad = -.1$$
$$-M + .3*A + B \qquad\qquad + C3 \qquad = .2 \qquad (4a)$$
$$-M - .3*A - B \qquad\qquad\qquad + C4 \qquad = -.2$$
$$-M + .4*A + B \qquad\qquad\qquad\qquad + C5 \qquad = .2$$
$$-M - .4*A - B \qquad\qquad\qquad\qquad\qquad + C6 = -.2$$

$$M \geq 0,\ C1 \geq 0,\ C2 \geq 0,\ C3 \geq 0,\ C4 \geq 0,\ C5 \geq 0,\ C6 \geq 0,\ (4b)$$

but A and B are not restricted in sign.

Three components of X are to be assigned the value zero. One cannot pick $M = A = B = 0$, for then $C2 = -.1 < 0$. Anyway, $M \geq 0$ and M is to be minimized, so it is not surprising that $M = 0$ will not work. Natural choices for this problem are $(A = 0, B = 0, C4 = 0)$, and $(A = 0, B = 0, C6 = 0)$, so that $M = .2$ [the largest constant term in (4a)]. With $A = 0$, $B = 0$, and $C4 = 0$, (4a) becomes

$$-M + C1 \qquad\qquad\qquad = .1$$
$$-M \qquad + C2 \qquad\qquad = -.1$$
$$-M \qquad\qquad + C3 \qquad = .2$$
$$-M \qquad\qquad\qquad\qquad = -.2$$
$$-M \qquad\qquad\qquad\qquad + C5 \qquad = .2$$
$$-M \qquad\qquad\qquad\qquad\qquad + C6 = -.2$$

Apply Gauss Elimination to the M-column of (4a) and "keep" the fourth row, to obtain

$$
\begin{aligned}
.5*A + 2*B + C1 \quad\quad\quad - C4 \quad\quad &= .3 \\
.1*A \quad\quad\quad + C2 \quad - C4 \quad\quad &= .1 \\
.6*A + 2*B \quad\quad\quad + C3 - C4 \quad\quad &= .4 \\
M + .3*A + \quad B \quad\quad\quad\quad - C4 \quad\quad &= .2 \\
.7*A + 2*B \quad\quad\quad\quad - C4 + C5 \quad &= .4 \\
-.1*A \quad\quad\quad\quad\quad - C4 \quad\quad + C6 &= 0.
\end{aligned}
\tag{5}
$$

Thus, a first basic feasible solution, a vertex of S, is ["cover" the A,B, and C4-columns in (5)]

$$X = (M, A, B, C1, C2, C3, C4, C5, C6)$$
$$= (.2, 0, 0, .3, .1, .4, 0, .4, 0).$$

The objective function F(X) for this vertex has the value

$$Z = F(X) = M = .2.$$

But, using the fourth equation in (5) one can write F(X) in the equivalent form

$$Z = F(X) = M = -.3*A - B + C4 + .2,$$

and again see at A = B = C4 = 0 that F(X) = .2.

Solving the system, the minimum feasible solution is

$$XX = (1/40, 1/2, 1/40, 0, 1/20, 1/20, 0, 0, 1/20).$$

The minimum value of the objective function is

$$F(XX) = 1/40 = MM \text{ (in XX)}.$$

The desired polynomial P(T) = A*T + B is then

$$P(T) = 1/2*T + 1/40.$$

The line Y = P(T) is the same distance from each of the three specified points. In particular, comparing ordinates, one gets

$$P(.2) - .1 = 5/40 - 4/40 = 1/40 \ (= M)$$
$$P(.3) - .2 = 7/40 - 8/40 = -1/40 \ (= -M)$$
$$P(.4) - .2 = 9/40 - 8/40 = 1/40 \ (= M).$$

• **PROBLEM 13-18**

Assume there are three factories (F_1, F_2 and F_3) supplying goods to three warehouses (W_1, W_2, W_3). The amounts available in each factory, the amounts needed in each factory and the costs of shipping from factory i to warehouse j are given in the table.

Find the minimum cost of satisfying warehouse demands given that any factory may supply to any warehouse.

Source / Destination	F_1	F_2	F_3	Units demanded
W_1	$.90	$1.00	$1.00	5
W_2	$1.00	$1.40	$ .80	20
W_3	$1.30	$1.00	$ .80	20
units available	20	15	10	45

Solution: This is a transportation problem. It may be solved by iteration, i.e., start with a solution and then use it to find more nearly optimal solutions.

An initial solution may be found by finding the box that has the lowest value in both its row and column. Place in that box the lower of demand or supply requirements. Next find the next lower value and repeat the placing of units shipped according to demand and supply requirements. One thus obtains

Source / Destination	F_1	F_2	F_3	units demanded
W_1	.90 / 5	$1.00 / 0	$1.00 / 0	5
W_2	1.00 / 10	1.40 / 0	.80 / 10	20
W_3	1.30 / 5	1.00 / 15	.80 / 0	20
units available	20	15	10	45

The total cost of this program is

$(.90)(5) + (1.00)(10) + (1.30)(5) + (1.00)(15)$

$+ (.80)(10) = \$44.00$.

Now check whether this solution is optimal. Pick a box with zero entry, say $W_1 F_2$. This means that F_2 supplies no goods to W_1. The cost of F_2 directly supplying W_1 is $1.00 per unit. But F_2 is also indirectly supplying W_1 since by supplying W_3 it allows F_1 to supply W_1. What is this indirect cost?

The cost of shipping one unit from F_2 to W_1 by this indirect route is:

+ $1.00 charge for shipping from F_2 to W_3

− $1.30 every unit F_2 sends to W_3 saves the cost of supplying W_3 from F_1

+ $0.90 charge for shipping from F_1 to W_1

+ $0.60

The indirect cost, i.e., the cost currently incurred is $0.60 whereas the direct cost is $1.00. Thus the current solution is cheaper.

The other zero boxes may be evaluated in a comparable manner. For example, the indirect shipment from F_2 to W_2 is the charge from F_2 to W_3 ($1.00) less the W_3F_1 charge ($1.30) plus the W_2F_1 charge ($1.00) = 70¢. Again, this is less than the cost of direct shipment ($1.40), so the current indirect route should be continued. In this way one obtains the following table:

Unused route	Cost of direct route	Cost of indirect route
W_1F_2	$1.00	$0.60
W_1F_3	$1.00	$0.70
W_2F_2	$1.40	$0.70
W_3F_3	$0.80	$1.10

By using W_3F_3 a saving of $.30 per unit can be obtained. But how many units can be shipped? The answer is the minimum number in any of the connections of the indirect route which must supply units for the transfer. This is five units, from box W_3F_1. Thus ship 5 units by the direct route W_3F_3; since F_3 produces only 10 units, this imposes a reduction in the W_2F_3 box to 5. The new pattern is shown in the table below:

Source Destination	F_1		F_2		F_3		Units demanded
W_1	.90	5	1.00	0	1.00	0	5
W_2	1.00	15	1.40	0	.80	5	20
W_3	1.30	0	1.00	15	.80	5	20
units available	20		15		10		45

Once again compare the cost of using the direct route to the cost of using the indirect route.

Unused route	Cost of using direct route	Cost of using indirect route
W_3F_1	$1.30	$1.00
W_1F_2	$1.00	$0.90
W_2F_2	$1.40	$1.00
W_1F_3	$1.00	$0.70

In every case the cost of using the indirect route is less than the cost of the direct route, indicating that the shipment costs are being minimized. The total cost of shipment from factories to warehouses is:

(5)($0.90) + (15)($1.00) + (15)($1.00) + 5($0.80)

+ 5($0.80) = $42.50.

CHAPTER 14

FOURIER TRANSFORMS

PROPERTIES OF FOURIER SERIES

● PROBLEM 14-1

Develop the definition of the Fourier transform of a function f(x) by extending the definition of the Fourier series of f to the case where the discrete spectrum of Fourier coefficients becomes a continuous spectrum.

Solution: Let f be representable by its Fourier series on $(-L, L)$. Then f can be written as

$$f(x) = \sum_{n=-\infty}^{\infty} C_n e^{i \frac{n\pi x}{L}} \quad (-L < x < L) \tag{1}$$

where C_n are the complex Fourier coefficients of f given by

$$C_n = \frac{1}{2L} \int_{-L}^{L} f(x) e^{-i \frac{n\pi x}{L}} dx. \tag{2}$$

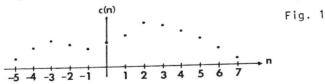

Fig. 1

To emphasize the functional dependence of C_n on n, write $C(n) = C_n$. The function $C(n)$ is called the Fourier spectrum of $f(x)$ and a typical example is plotted in Figure 1 (for convenience, $f(x)$ is assumed real and even so that the $C(n)$ are real and the graph may be made). Now make the substitutions

$$k = \frac{n\pi}{L}, \quad \left(\frac{L}{\pi}\right) C_n = C_L(k)$$

where k is called the wave number of the nth term (or kth term) in the Fourier series expansion of f. Then equations (1) and (2) may be written as

$$C_L(k) = \frac{1}{2\pi} \int_{-L}^{+L} f(x) e^{-ikx} dx,$$

$$f(x) = \frac{Lk}{\pi} \sum_{-\infty}^{\infty} C_L(k) e^{ikx} \Delta k \qquad (3)$$

since

$$\Delta k = \frac{\pi}{L} \Delta n = \frac{\pi}{L}.$$

With the change of scale from n to k the Fourier spectrum may be plotted versus wave number as in Figure 2.

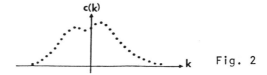

Fig. 2

Evidently as L approaches infinity, the wave number spectrum approaches a continuous spectrum, i.e., trigonometric functions of all wave numbers must be summed to represent f. Hence, as $L \to \infty$ (the function is no longer periodic) the sum in the second equation of (3) becomes an integral since $\Delta k \to 0$ and write

$$C(k) = \lim_{L \to \infty} C_L(k) = \frac{1}{2\pi} \int_{-\infty}^{\infty} f(x) e^{-ikx} dx \qquad (4)$$

and

$$f(x) = \int_{-\infty}^{\infty} C(k) e^{ikx} dk. \qquad (5)$$

With a slight change in notation, formulas (4) and (5) become the Fourier transformation. Thus, define a new function F by

$$F(k) = \sqrt{2\pi}\, C(-k)$$

so that the formulas now read

$$F(k) = \frac{1}{\sqrt{2\pi}} \int_{-\infty}^{\infty} f(x) e^{ikx} dx \qquad (6)$$

$$f(x) = \frac{1}{\sqrt{2\pi}} \int_{-\infty}^{\infty} F(k) e^{-ikx} dx \qquad (7)$$

F(k) is known as the Fourier transform of the function

f(x) and, conversely, f(x) is called the inverse Fourier transform of F(k). To emphasize the interpretation of the Fourier transform as an operator operating on f(x), write

$$F(k) = \Phi\{f(x)\}$$

where Φ denotes the operation on f described in equation (6). Finally, note that (6) is merely a definition of F(k), just as (2) is a definition of the coefficients C_n, so that there is no question about the validity of this formula. That is, if f(x) is integrable and the integral in (6) converges then F(k) exists. On the other hand, there is a question about whether the original function f(x) can be retrieved by the formula (7) since this formula was obtained by a limiting process on the Fourier series of f and there is a question as to whether this series represents f. The theory surrounding the convergence of the integral in (7) to f(x) is very similar to that of the convergence of Fourier series.

● **PROBLEM 14-2**

Prove the following properties of Fourier transforms:

a) The Fourier transform of f(x) exists if f is absolutely integrable over $(-\infty, +\infty)$,

b) If f(x) is real valued then

$$F(-k) = F^*(k)$$

where $F^*(k)$ is the complex conjugate of F(k).

Solution: a) By definition, if f is absolutely integrable over $(-\infty, +\infty)$, then

$$\int_{-\infty}^{\infty} |f(x)| dx \tag{1}$$

exists. The Fourier transform of f is given by

$$F(k) = \frac{1}{\sqrt{2\pi}} \int_{-\infty}^{\infty} f(x) e^{ikx} dx. \tag{2}$$

Now, recalling that $|e^{iy}| = 1$ for all y,

$$|f(x) e^{ikx}| = |f(x)|$$

so that

$$\frac{1}{\sqrt{2\pi}} \int_{-\infty}^{\infty} |f(x) e^{ikx}| dx = \frac{1}{\sqrt{2\pi}} \int_{-\infty}^{\infty} |f(x)| dx,$$

and it is known that this second integral exists. Thus,

$$f(x)e^{ikx}$$

is absolutely integrable over $(-\infty, \infty)$ and is therefore integrable over $(-\infty, \infty)$. Hence, $F(k)$ exists.

b) This proof is immediate. From (2)

$$F(-k) = \frac{1}{\sqrt{2\pi}} \int_{-\infty}^{\infty} f(x) e^{-ikx} dx \qquad (3)$$

and, recalling the identity $e^{iy} = \cos y + i \sin y$,

$$F^*(k) = \left(\frac{1}{\sqrt{2\pi}} \int_{-\infty}^{\infty} f(x) e^{ikx} dx \right)^*$$

$$= \left(\frac{1}{\sqrt{2\pi}} \int_{-\infty}^{\infty} f(x) \cos kx \, dx + i \frac{1}{\sqrt{2\pi}} \int_{-\infty}^{\infty} f(x) \sin kx \, dx \right)^*$$

$$= \frac{1}{\sqrt{2\pi}} \int_{-\infty}^{\infty} f(x) \cos kx \, dx - i \frac{1}{\sqrt{2\pi}} \int_{-\infty}^{\infty} f(x) \sin kx \, dx$$

$$(f \text{ real})$$

$$= \frac{1}{\sqrt{2\pi}} \int_{-\infty}^{\infty} f(x) \cos(-kx) dx + i \frac{1}{\sqrt{2\pi}} \int_{-\infty}^{\infty} f(x) \sin(-kx) dx$$

$$= \frac{1}{\sqrt{2\pi}} \int_{-\infty}^{\infty} f(x) (\cos(-kx) + i \sin(-kx)) dx$$

$$= \frac{1}{\sqrt{2\pi}} \int_{-\infty}^{\infty} f(x) e^{-ikx} dx. \qquad (4)$$

In the third step the facts that $\cos(-y) = \cos y$ and $\sin(-y) = -\sin y$ were used. Equating (3) and (4) yields

$$F(-k) = F^*(k), \text{ for real } f(x).$$

• **PROBLEM 14-3**

a) Prove the attenuation property of Fourier transforms:

$$\Phi\{f(x)\ e^{ax}\} = F(k-ai)$$

where

$$F(k) = \Phi\{f(x)\}.$$

b) Prove the shifting property of Fourier transforms:

$$\Phi\{f(x-a)\} = e^{ika}\ F(k).$$

c) Prove the derivative properties of Fourier transforms:

$$\Phi\{f'(x)\} = -ik\Phi\{f(x)\}$$

$$\Phi\{f''(x)\} = -k^2\Phi\{f(x)\}.$$

Solution: a) The Fourier transform of $f(x)$ is defined by

$$F(k) = \Phi\{f(x)\} = \frac{1}{\sqrt{2\pi}} \int_{-\infty}^{\infty} f(x)\ e^{ikx}\ dx. \qquad (1)$$

The Fourier transform of $g(x) = f(x)\ e^{ax}$ is then

$$G(k) = \Phi\{f(x)\ e^{ax}\} = \frac{1}{\sqrt{2\pi}} \int_{-\infty}^{\infty} f(x)\ e^{ax}\ e^{ikx}\ dx$$

$$= \frac{1}{\sqrt{2\pi}} \int_{-\infty}^{\infty} f(x)\ e^{(a+ik)x}\ dx$$

$$= \frac{1}{\sqrt{2\pi}} \int_{-\infty}^{\infty} f(x)\ e^{i(k-ia)x}\ dx. \qquad (2)$$

Now make the change of variable $r = k - ia$ in (2) to give

$$\Phi\{f(x)\ e^{ax}\} = \frac{1}{\sqrt{2\pi}} \int_{-\infty}^{\infty} f(x)\ e^{irx}\ dx = F(r) = F(k-ia).$$

b) Suppose f(x) is shifted a length a to the right. Then the Fourier transform of this new function f(x-a) is

$$\Phi\{f(x-a)\} = \frac{1}{\sqrt{2\pi}} \int_{-\infty}^{\infty} f(x-a) e^{ikx} dx \qquad (3)$$

Now make the substitution $x' = x-a$. Then (3) becomes

$$\Phi\{f(x-a)\} = \frac{1}{\sqrt{2\pi}} \int_{-\infty}^{\infty} f(x') e^{ik(x'+a)} dx'$$

$$= e^{ika} \frac{1}{\sqrt{2\pi}} \int_{-\infty}^{\infty} f(x') e^{ikx'} dx' = e^{ika} F(k).$$

c) Suppose that $\Phi\{f'(x)\}$ exists. Then by definition (1),

$$\Phi\{f'(x)\} = \frac{1}{\sqrt{2\pi}} \int_{-\infty}^{\infty} f'(x) e^{ikx} dx. \qquad (4)$$

Integrating (4) by parts gives

$$\Phi\{f'(x)\} = \frac{1}{\sqrt{2\pi}} f(x) e^{ikx} \Big|_{-\infty}^{\infty}$$

$$- \frac{ik}{\sqrt{2\pi}} \int_{-\infty}^{\infty} f(x) e^{ikx} dx. \qquad (5)$$

If the Fourier transform of f(x) exists, this usually implies that $f(x) \to 0$ as $x \to \pm\infty$ (this is sometimes not the case, but then f(x) can be treated as a distribution so that the derivative formula is then still valid. Thus, the first term in (5) is zero and

$$\Phi\{f'(x)\} = -ik \, \Phi\{f(x)\}. \qquad (6)$$

If
$$\Phi\{f''(x)\}$$

exists, it is given by

$$\Phi\{f''(x)\} = \frac{1}{\sqrt{2\pi}} \int_{-\infty}^{\infty} f''(x) e^{ikx} dx.$$

Again integrating by parts yields

$$\Phi\{f''(x)\} = \frac{1}{\sqrt{2\pi}} \left. f'(x) e^{ikx} \right|_{-\infty}^{\infty}$$

$$- \frac{ik}{\sqrt{2\pi}} \int_{-\infty}^{\infty} f'(x) e^{ikx} dx. \tag{7}$$

For reasons mentioned above, it is expected that $f'(x) \to 0$ as $x \to \pm\infty$ so that (7) gives

$$\Phi\{f''(x)\} = -ik \, \Phi\{f'(x)\}.$$

Using (6) gives $\Phi\{f''(x)\} = -k^2 \, \Phi\{f(x)\}$.

The obvious extension is made by using $f^{(n-1)}(x) = g(x)$ in the formula of (6) to give

$$\Phi\{g'(x)\} = -ik \, \Phi\{g(x)\} \quad \text{or} \quad \Phi\{f^{(n)}(x)\} = -ik \, \Phi\{f^{(n-1)}(x)\},$$

which yields upon iteration $\quad \Phi\{f^{(n)}(x)\} = (-ik)^n \, \Phi\{f(x)\}$.

• **PROBLEM 14-4**

a) Prove that if the functions $g(x)$ and $F(k)$ are absolutely integrable on $(-\infty, +\infty)$ and that the Fourier inversion integral for $f(x)$ is valid for all x except possibly at a countably infinite number of points, then

$$\int_{-\infty}^{\infty} F(k) \, G(-k) \, dk = \int_{-\infty}^{\infty} f(x) \, g(x) \, dx \tag{1}$$

where

$$F(x) = \Phi\{f(x)\} \quad , \quad G(k) = \Phi\{g(x)\}.$$

This is known as the second Parseval theorem of Fourier transform theory.

b) From the above equation (1), prove the first Parseval theorem of Fourier transform theory,

$$\int_{-\infty}^{\infty} |F(k)|^2 \, dk = \int_{-\infty}^{\infty} |f(x)|^2 \, dx \tag{2}$$

Solution: a) The Fourier transform of a function f(x) is defined by

$$F(k) = \frac{1}{\sqrt{2\pi}} \int_{-\infty}^{\infty} f(x) e^{ikx} dx \qquad (3)$$

so that by definition

$$G(-k) = \frac{1}{\sqrt{2\pi}} \int_{-\infty}^{\infty} g(x) e^{-ikx} dx. \qquad (4)$$

Therefore

$$\int_{-\infty}^{\infty} F(k) G(-k) dk$$

$$= \int_{-\infty}^{\infty} F(k) dk \int_{-\infty}^{\infty} \frac{1}{\sqrt{2\pi}} g(x) e^{-ikx} dx. \qquad (5)$$

Now $F(k)$ and $g(x)$ are absolutely convergent on $(-\infty, +\infty)$, that is, the integrals

$$\int_{-\infty}^{\infty} |F(k)| dk, \qquad \int_{-\infty}^{\infty} |g(x)| dx$$

are convergent, so that

$$\int_{-\infty}^{\infty} F(k) e^{-ikx} dx, \qquad \int_{-\infty}^{\infty} g(x) e^{-ikx} dx$$

are absolutely convergent (since

$$|F(k) e^{-ikx}| = |F(k)| |e^{-ikx}| = |F(k)|$$

and

$$|g(x) e^{-ikx}| = |g(x)|).$$

Hence, the order of integration in (5) may be interchanged giving

$$\int_{-\infty}^{\infty} F(k)G(-k)dk$$

$$= \int_{-\infty}^{\infty} g(x)dx \frac{1}{\sqrt{2\pi}} \int_{-\infty}^{\infty} F(k)e^{-ikx}dk. \qquad (6)$$

Since the Fourier inversion integral is valid,

$$\frac{1}{\sqrt{2\pi}} \int_{-\infty}^{\infty} F(k)e^{-ikx} dk = f(x) \qquad (7)$$

and using this result in (6) gives the second Parseval theorem:

$$\int_{-\infty}^{\infty} F(k)G(-k)dk = \int_{-\infty}^{\infty} g(x)f(x)dx. \qquad (8)$$

The validity of (8) is insured even if the Fourier inversion integral for $f(x)$ has a countably infinite number of discrepancies with $f(x)$ since this will not affect the equality of the integrals

$$\int_{-\infty}^{\infty} g(x)(f(x))dx$$

and

$$\int_{-\infty}^{\infty} g(x) \frac{1}{\sqrt{2\pi}} \int_{-\infty}^{\infty} F(k)e^{-ikx} dk\, dx.$$

b) The first Parseval theorem is a corollary to the second Parseval theorem stated in equation (8) which follows by letting $f(x) = g(x)$ so that $F(k) = G(k)$ and recalling that (assuming $f = g$ real) $G(-k) = G^*(k)$ where $G^*(k)$ is the complex conjugate of $G(k)$. Noting that

$$G(k)G^* = |G(k)|^2$$

and using these results in (8) gives

$$\int_{-\infty}^{\infty} G(k)G^*(k)\,dk = \int_{-\infty}^{\infty} [g(x)]^2\,dx$$

or

$$\int_{-\infty}^{\infty} |G(k)|^2\,dk = \int_{-\infty}^{\infty} |g(x)|^2\,dx.$$

• **PROBLEM 14-5**

Find the sums of the series

$$1 + \frac{\cos x}{p} + \frac{\cos 2x}{p^2} + \ldots + \frac{\cos nx}{p^n} + \ldots, \tag{1}$$

$$\frac{\sin x}{p} + \frac{\sin 2x}{p^2} + \ldots + \frac{\sin nx}{p^n} + \ldots,$$

where p is a real constant with absolute value greater than 1.

Solution: The two series (1) converge for all x. Consider the series

$$\left(1 + \frac{\cos x}{p} + \frac{\cos 2x}{p^2} + \ldots\right) + i\left(\frac{\sin x}{p} + \frac{\sin 2x}{p^2} + \ldots\right)$$

$$= 1 + \frac{e^{ix}}{p} + \frac{e^{2ix}}{p^2} + \ldots.$$

One has

$$1 + \frac{z}{p} + \frac{z^2}{p^2} + \ldots = \frac{1}{1-(z/p)} = \frac{p}{p-z} = F(z),$$

since the series on the left is a geometric series, which converges for $|z/p| < 1$. Therefore

$$F(e^{ix}) = \frac{p}{p-e^{ix}} = \frac{p}{(p-\cos x) - i\sin x}$$

$$= p\,\frac{(p-\cos x) + i\sin x}{(p-\cos x)^2 + \sin^2 x}$$

$$= p\,\frac{(p-\cos x) + i\sin x}{p^2 - 2p\cos x + 1},$$

and

$$\frac{p(p-\cos x)}{p^2 - 2p\cos x + 1} = 1 + \frac{\cos x}{p} + \frac{\cos 2x}{p^2} + \ldots$$

528

$$+ \frac{\cos nx}{p^n} + \ldots$$

$$\frac{p \sin x}{p^2 - 2p \cos x + 1} = \frac{\sin x}{p} + \frac{\sin 2x}{p^2} + \ldots + \frac{\sin nx}{p^n} + \ldots$$

for all x.

FOURIER SERIES EXPANSION

• **PROBLEM 14-6**

Draw a graph of the function

$$f(x) = \begin{cases} 0 & \text{if } -2 < x < 0 \\ 1 & \text{if } 0 < x < 2 \end{cases}$$

where the period of the function is 4.

Determine the Fourier coefficients and write the Fourier series corresponding to the function.

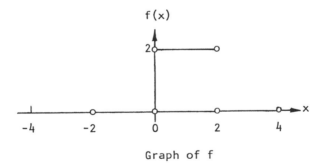

Graph of f

Solution: A periodic function is one whose behavior is recurrent. That is, there is a certain number L, called the period of the function, such that the function repeats itself over every interval of length L, $f(x+L) = f(x)$ for all $x \in R$.

The definition of the Fourier series of a periodic function f of period 2L is given by

$$f(x) \sim \frac{a_0}{2} + \sum_{n=1}^{\infty} \left[a_n \cos\left(\frac{n\pi x}{L}\right) + b_n \sin\left(\frac{n\pi x}{L}\right) \right]$$

where the Fourier coefficients a_n and b_n are

$$a_n = \frac{1}{L} \int_{-L}^{L} f(x) \cos \frac{n\pi x}{L} \, dx$$

$$n = 0, 1, 2, \ldots$$

$$b_n = \frac{1}{L} \int_{-L}^{L} f(x) \sin \frac{n\pi x}{L} dx$$

For this function $L = 2$.

$$a_n = \frac{1}{2} \int_{-2}^{2} f \cos\left(\frac{n\pi x}{2}\right) dx = \frac{1}{2} \int_{-2}^{0} 0 \cdot \cos\left(\frac{n\pi x}{2}\right) dx$$

$$+ \frac{1}{2} \int_{0}^{2} 1 \cdot \cos\left(\frac{n\pi x}{2}\right) dx$$

$$= \left(\frac{1}{2}\right)\left(\frac{2}{n\pi}\right) \left[\sin\left(\frac{n\pi x}{2}\right)\right]_{0}^{2} = 0 \qquad \text{if } n \neq 0$$

$$a_0 = \frac{1}{2} \int_{-2}^{2} f \, dx = \frac{1}{2} \int_{0}^{2} dx = 1$$

$$b_n = \frac{1}{2} \int_{-2}^{2} f \sin\left(\frac{n\pi x}{2}\right) dx = \frac{1}{2} \int_{0}^{2} 1 \cdot \sin\left(\frac{n\pi x}{2}\right) dx$$

$$= \frac{1}{n\pi}(1 - \cos n\pi) = \frac{1}{n\pi}\begin{cases} 2 & \text{if } n \text{ is odd} \\ 0 & \text{if } n \text{ is even} \end{cases}$$

Therefore,

$$b_{2n-1} = \frac{2}{(2n-1)\pi}, \qquad b_{2n} = 0 \qquad \text{if } n \in N$$

The Fourier series may be expressed as

$$f(x) \sim \frac{1}{2} + \frac{1}{\pi} \sum_{n=1}^{\infty} \frac{1 - \cos n\pi}{n} \sin\left(\frac{n\pi x}{2}\right)$$

or, if the zero coefficients are omitted,

$$f(x) \sim \frac{1}{2} + \frac{2}{\pi} \sum_{n=1}^{\infty} \frac{1}{2n-1} \sin\left[\frac{(2n-1)\pi x}{2}\right]$$

The function is undefined at points $x = 0, \pm 2, \pm 4, \ldots$ If the series converges at these jump discontinuities it can't converge to f.

Three types of convergence will be discussed: Pointwise Convergence, Uniform Convergence and Convergence in the Mean. Each has its own advantages in the determination of when a function's Fourier series adequately represents it. For instance, the Uniform Consequence Theorem has the advantage that if certain conditions are satisfied, then the Fourier series of f will converge to f at each point where f is defined.

However, these conditions on f are very stringent. On the other hand, f's Fourier series will be found to converge in the mean for all but the most pathological of functions. However, this is a statistical type of convergence in that it might not be possible to say that the Fourier series of f at x_0 converges to $f(x_0)$ for any x_0. Finally, Point-wise Convergence will be seen to be somewhere in between the other two, both in adequacy of representation and stringency of conditions on f. However, Convergence in the Mean is the most widely used for physical applications.

• **PROBLEM 14-7**

(a) Find the Fourier series for the function

$$f(x) = \begin{cases} -\cos x & \text{if } -\pi < x < 0 \\ \cos x & \text{if } 0 < x < \pi \end{cases}$$

Sketch the graph of the function, to aid the calculations.

(b) Find the convergence at all jump discontinuities.

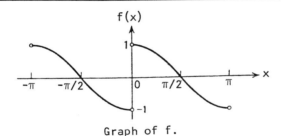

Graph of f.

Solution: a) Since f is odd one can construct a sine series representation for the function. In this case $L = \pi$.

$$b_n = \frac{2}{\pi} \int_0^\pi \cos x \sin nx \, dx$$

$$= \left(\frac{2}{\pi}\right)\left(\frac{1}{2}\right) \int_0^\pi [\sin(1+n)x - \sin(1-n)x] \, dx$$

$$= \frac{1}{\pi} \left[\frac{-1}{1+n} \cos(1+n)x + \frac{1}{1-n} \cos(1-n)x \right]_0^\pi$$

$$= \left[\frac{2n}{(n^2 - 1)\pi} \right] [1 + \cos n\pi] \quad \text{if } n \neq 1$$

$$= \left[\frac{2n}{(n^2 - 1)\pi} \right] \begin{cases} 0 & \text{if } n \text{ is odd} \\ 2 & \text{if } n \text{ is even} \end{cases}$$

$$b_{2n} = \frac{8n}{(4n^2 - 1)\pi}$$

$$b_1 = \frac{2}{\pi} \int_0^\pi \cos x \sin x \, dx = 0$$

Therefore, the series is

$$f(x) \sim \frac{8}{\pi} \sum_{n=1}^\infty \frac{n \sin 2nx}{4n^2 - 1}$$

(b) A jump discontinuity in the function exists at $x = 0$. If the odd periodic extension is considered for f, then jump discontinuities exists at $x = n\pi$, $n \in Z$. At each discontinuity the convergence is zero.

• **PROBLEM 14-8**

Let $f(x)$ be a real valued function of one variable, $f: R \to R$, with a period $2c$ (i.e., $f(x+2c) = f(x)$ for all $x \in R$). Define the Fourier series of f. Assume that the series converges uniformly.

---

y= FOURIER SERIES OF f

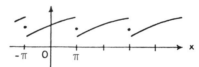

PERIODIC EXTENSION OF A FUNCTION
DEFINED BETWEEN $-\pi$ AND π. Fig. 1

Solution: Consider the trigonometric series

$$\frac{a_0}{2} + \sum_{n=1}^\infty \left(a_n \cos \frac{\pi n x}{c} + b_n \sin \frac{\pi n x}{c} \right)$$

and assume that it converges uniformly for all $x \in R$. Then it represents a function of x, i.e.,

$$f(x) = \frac{a_0}{2} + \sum_{n=1}^\infty \left(a_n \cos \frac{\pi n x}{c} + b_n \sin \frac{\pi n x}{c} \right) \quad (1)$$

It can be shown that

$$a_n = \frac{1}{c} \int_{-c}^{c} f(x) \cos \left(\frac{\pi n x}{c} \right) dx,$$

$$b_n = \frac{1}{c} \int_{-c}^{c} f(x) \sin \left(\frac{\pi n x}{c} \right) dx. \quad (2)$$

These relations will be proven now. To find the b_n, multiply each term of (1) by $\sin(k\pi x/c)\,dx$ and integrate from $-c$ to c. Thus (1) becomes

$$\int_{-c}^{c} f(x)\sin\frac{k\pi x}{c}\,dx = \frac{1}{2}a_0 \int_{-c}^{c} \sin\frac{k\pi x}{c}\,dx$$

$$+ \sum_{n=1}^{\infty}\left[a_n \int_{-c}^{c} \cos\frac{n\pi x}{c} \sin\frac{k\pi x}{c}\,dx \right.$$

$$\left. + b_n \int_{-c}^{c} \sin\frac{n\pi x}{c} \sin\frac{k\pi x}{c}\,dx \right] \qquad (3)$$

Note that

$$\int_{-a}^{b}\left[\sum_{n=1}^{\infty} g(x)\right]dx$$

is interchanged with

$$\sum_{n=1}^{\infty}\left[\int_{-a}^{b} g(x)\,dx\right].$$

This is valid if a series converges uniformly, as assumed in this problem. To evaluate (3) note that the set of functions

$$(\sin x,\ \cos x,\ \sin 2x,\ \cos 2x,\ \ldots,\ \sin kx,\ \cos kx,\ \ldots)$$

is an orthogonal set. This means that for any two members of the set,

$$\int_{a}^{b} \sin mx \cos mx\,dx = 0, \qquad (4)$$

$$\int_{a}^{b} \sin mx \sin nx\,dx = 0 \qquad (m \neq n), \qquad (5)$$

$$\int_a^b \cos mx \cos nx\, dx = 0 \qquad (m \neq n), \qquad (6)$$

$$\int_a^b [\sin mx]^2\, dx \neq 0, \qquad (7)$$

$$\int_a^b [\cos mx]^2\, dx \neq 0. \qquad (8)$$

Using (4) to (7), (3) is reduced to

$$\int_{-c}^{c} f(x) \sin \frac{k\pi x}{c}\, dx = b_k \int_{-c}^{c} \sin^2 \frac{k\pi x}{c}\, dx \qquad (9)$$

From (3) one determines b_k in the following manner. First evaluate

$$\int_{-c}^{c} \sin^2 \frac{k\pi x}{c}\, dx \qquad (10)$$

Now (10) is not zero. Using the trigonometric identities, it will equal the number c. Then (9) may be rewritten as

$$b_k = \frac{1}{c} \int_{-c}^{c} f(x) \sin \frac{k\pi x}{c}\, dx, \quad k = 1,2,3,\ldots \qquad (11)$$

Thus the coefficients b_n in (1) are

$$b_n = \frac{1}{c} \int_{-c}^{c} f(x) \sin \frac{n\pi x}{c}\, dx, \quad n = 1,2,3,\ldots \qquad (12)$$

Now, find the coefficient a_n of the cosine terms in (1). Therefore, multiplying (1) by $\cos \frac{n\pi x}{c}$ and integrating from $-c$ to c,

$$\int_{-c}^{c} f(x) \cos \frac{n\pi x}{c} dx = \frac{1}{2} a_0 \int_{-c}^{c} \cos \frac{n\pi x}{c} dx$$

$$+ \sum_{n=1}^{\infty} \left[\int_{-c}^{c} \cos \frac{n\pi x}{c} \cos \frac{k\pi x}{c} dx \right.$$

$$+ \left. \int_{-c}^{c} \cos \frac{n\pi x}{c} \sin \frac{k\pi x}{c} dx \right].$$

Using facts (4) - (8), and $k \neq 0$,

$$\int_{-c}^{c} f(x) \cos \frac{n\pi x}{c} dx = a_n \int_{-c}^{c} \left[\cos \frac{n\pi x}{c} \right]^2 dx.$$

Solving for a_n,

$$a_n = \frac{1}{c} \int_{-c}^{c} f(x) \cos \frac{n\pi x}{c} dx \qquad (13)$$

where

$$c = \int_{-c}^{c} \left[\cos \frac{n\pi x}{c} \right]^2 dx.$$

Finally, determine a_0. If $k = 0$ in (1),

$$\int_{-c}^{c} f(x) dx = \frac{1}{2} a_0 \int_{-c}^{c} dx + \sum_{n=1}^{\infty} \left[a_n \int_{-c}^{c} \cos \frac{n\pi x}{c} dx \right.$$

$$\left. + b_n \int_{-c}^{c} \sin \frac{n\pi x}{c} dx \right] \qquad (14)$$

Now

$$\int_{-c}^{c} \cos \frac{n\pi x}{2} dx = 0 \text{ for } n \geq 1$$

and

$$\int_{-c}^{c} \sin \frac{n\pi x}{c} dx = \frac{2c}{n\pi} \left[-\cos \frac{n\pi x}{c} \right]_{-c}^{c} = 0, \text{ for } n \geq 1.$$

535

Thus, (14) is reduced to

$$\int_{-c}^{c} f(x)\,dx = \frac{1}{2} a_0 \int_{-c}^{c} dx$$

$$\int_{-c}^{c} f(x)\,dx = \frac{1}{2} a_0 (2c)$$

and hence

$$a_0 = \frac{1}{c} \int_{-c}^{c} f(x)\,dx.$$

Thus

$$f(x) \sim \frac{1}{2} a_0 + \sum_{n=1}^{\infty} \left[a_n \cos \frac{n\pi x}{c} + b_n \sin \frac{n\pi x}{c} \right]$$

with a_0, a_n and b_n given by (2). Thus, the coefficients of (1) have been found for the case in which the series there represents some function and converges uniformly to it.

As was done for functions of period 2π, the series (1) will be used to motivate the definition of the Fourier series of any function $f(x)$ of period $2c$. Thus, this series is defined as the series in (1) with coefficients given in (2) but is written as

$$f(x) \sim \frac{a_0}{2} + \sum_{n=1}^{\infty} \left[a_n \cos \frac{n\pi x}{c} + b_n \sin \frac{n\pi x}{c} \right] \qquad (15)$$

since it is not known in general whether this series is convergent to $f(x)$ at any point. It should be noted that the limits of integration in (2) may be a and $a+2c$ where a is any number such that $(a,a+2c)$ is contained within the domain of definition of f. If f is defined only for some interval $(a,a+2c)$ with some specific (a), then these limits must be used, i.e.,

$$a_n = \frac{1}{c} \int_{a}^{a+2c} f(x) \cos\left(\frac{\pi n x}{c}\right) dx,$$

$$b_n = \frac{1}{c} \int_{a}^{a+2c} f(x) \sin\left(\frac{\pi n x}{c}\right) dx. \qquad (16)$$

Of course if f is defined only on $(-c,c)$ then (16) is used with $a = -c$.

When the phenomena of a particular problem are periodic in time with period T, then the variable x is usually replaced by t, and the following substitutions are usually made:

$$\frac{\pi}{c} \to \frac{2\pi}{T} = \omega, \quad A_n = \sqrt{a_n^2 + b_n^2},$$

$$\phi_n = \arctan \frac{b_n}{a_n} \quad (n > 0). \tag{17}$$

The series (15) then becomes

$$f(t) \sim \frac{a_0}{2} + \sum_{n=1}^{\infty} A_n \cos(n\omega t - \phi_n). \tag{18}$$

In any case, always remember that a Fourier series can only be associated with a periodic function. In addition, in a case where the function f is only defined on a certain interval (a, a+2c), (16) rather than (12) and (13) must be used. Then (15) represents the "periodic extension" of f to the whole real axis. That is, except for some very pathological functions, the series in (15) will converge to

$$L_0 = \frac{1}{2} \left[\lim_{x \to x_0^+} f(x) + \lim_{x \to x_0^-} f(x) \right]$$

in its domain of definition, say [a, a+2c] (if these limits are equal the series converges to $f(x_0)$). But then by the periodicity of (15), at all points $x_0 + 2nc$, the series will converge to L_0 and will thus have the same graph between a+2nc and a+2(n+1)c as f does in [a, a+2c]. This is illustrated in the figure.

● **PROBLEM 14-9**

Determine the Fourier series of the function

$$f(x) = \begin{cases} x, & 0 < x < \pi \\ 0, & \pi < x < 2\pi \end{cases}. \tag{1}$$

Solution: Recall the most general definition of the Fourier series of a function f. That is, the Fourier series of a periodic function f of period 2c defined on the interval D is given by

$$f(x) \sim \frac{a_0}{2} + \sum_{n=1}^{\infty} \left[a_n \cos\left(\frac{\pi n x}{c}\right) + b_n \sin\left(\frac{\pi n x}{c}\right) \right] \tag{2}$$

where

$$a_n = \frac{1}{c} \int_a^{a+2c} f(x) \cos\left(\frac{\pi n x}{c}\right) dx \qquad (3)$$

$$b_n = \int_a^{a+2c} f(x) \sin\left(\frac{\pi n x}{c}\right) dx, \qquad (4)$$

where a is any number such that the interval $(a, a+2c)$ is contained in D. In the case at hand, the function in (1) is defined only on the interval $(0, 2\pi)$ and is not periodic within this interval so $2c = 2\pi$ and (2) represents the periodic extension of f. Therefore, in equations (2) - (4) the substitutions

$$2c = 2\pi, \quad a = 0, \quad f(x) = \begin{cases} x, & x \in (0, \pi) \\ 0, & x \in (\pi, 2\pi) \end{cases}$$

are made. For a_n, $n > 0$, this results in

$$a_n = \frac{1}{\pi} \int_0^{2\pi} f(x) \cos nx \, dx = \frac{1}{\pi} \int_0^{\pi} x \cos nx \, dx$$

$$= \frac{1}{\pi} \left[\frac{x}{n} \sin nx + \frac{\cos nx}{n^2} \right] \Big|_0^{\pi}$$

$$= \frac{1}{n^2 \pi} (\cos n\pi - 1).$$

For $n = 2, 4, 6, \ldots$ $\cos n\pi = 1$ and hence $a_n = 0$, and for $n = 1, 3, 5, \ldots$ $\cos n\pi = -1$ and

$$a_n = \frac{-2}{n^2 \pi},$$

n = odd. Similarly, to evaluate b_n, the same substitutions are made yielding

$$b_n = \frac{1}{\pi} \int_0^{2\pi} f(x) \sin nx \, dx = \frac{1}{\pi} \int_0^{\pi} x \sin nx \, dx$$

$$= \frac{1}{\pi} \left[\frac{\sin nx}{n^2} - \frac{x \cos nx}{n} \right] \Big|_0^{\pi}$$

$$= \frac{-\pi \cos n\pi}{\pi n} \quad \text{for all } n = 1, 2, 3, \ldots$$

$$= \frac{-\pi(-1)^n}{n} = \frac{(-1)^{n+1}}{n}.$$

Finally,

$$a_0 = \frac{1}{\pi} \int_0^{2\pi} f(x)\,dx = \frac{1}{\pi} \int_0^{\pi} x\,dx = \left.\frac{x^2}{2\pi}\right|_0^{\pi} = \frac{\pi}{2}.$$

So far, the following results have been obtained:

$$a_0 = \frac{\pi}{2}$$

$$a_n = \frac{-2}{n^2 \pi}, \quad n = 1, 3, 5\ldots$$

$$b_n = \frac{(-1)^{n+1}}{n}, \quad n = 1, 2, 3, \ldots.$$

To make a_n valid for all n, ensure that n is odd by replacing n by $(2n-1)$. Finally, one obtains

$$a_n = \frac{-2}{(2n-1)^2 \pi}, \quad n = 1, 2, 3, \ldots.$$

Now, referring back to the Fourier expression,

$$f(x) \sim \frac{a_0}{2} + \sum_{n=1}^{\infty} a_n \cos nx + b_n \sin nx$$

and substituting back their respective values,

$$f(x) \sim \frac{\pi}{4} + \sum_{n=1}^{\infty} \frac{-2}{(2n-1)^2 \pi} \cos(2n-1)x + \frac{(-1)^{n+1}}{n} \sin nx$$

or

$$f(x) \sim \frac{\pi}{4} - \frac{2}{\pi}\left(\cos x + \frac{\cos 3x}{3^2} + \frac{\cos 5x}{5^2} + \ldots\right) + \sin x$$

$$- \frac{\sin 2x}{2} + \frac{\sin 3x}{3} \ldots.$$

The symbol "$\sim$" is to be read "has the Fourier series" since convergence questions have not been dealt with. In fact, this series is uniformly convergent which can be seen by applying the Weierstrass M-test.

• **PROBLEM 14-10**

Determine the Fourier series of the function given by

$$\begin{cases} f(x) = x^2, \quad x \in (-\pi, \pi) \\ \\ f(x+2\pi) = f(x), \text{ all } x \end{cases} \quad (1)$$

Solution: The most general definition of the Fourier series of a function f which is periodic with period 2c and defined on the interval D is given by

FOURIER SERIES OF f(x)

$$f(x) \sim \frac{a_0}{2} + \sum_{n=1}^{\infty} \left[a_n \cos\left(\frac{\pi n x}{c}\right) + b_n \sin\left(\frac{\pi n x}{c}\right) \right] \qquad (2)$$

where

$$a_n = \int_a^{a+2c} f(x) \cos\left(\frac{\pi n x}{c}\right) dx \qquad (3)$$

$$b_n = \int_a^{a+2c} f(x) \sin\left(\frac{\pi n x}{c}\right) dx. \qquad (4)$$

Here, a is any number such that $(a, a+2c) \subseteq D$. For the function in (1), D is the whole real axis, $2c = 2\pi$ and since any a can be used, the most convenient is $a = -\pi$. Then

$$a_0 = \frac{1}{\pi} \int_{-\pi}^{\pi} x^2 dx = \frac{2\pi^2}{3} \qquad (5)$$

$$a_n = \frac{1}{\pi} \int_{-\pi}^{\pi} x^2 \cos nx\, dx \qquad (6)$$

Integrating (6) by parts twice gives

$$a_n = \frac{1}{\pi}\left[\frac{x^2}{n}\sin nx\right]\Big|_0^{\pi} - \frac{2}{n\pi}\int_{-\pi}^{\pi} x \sin nx\, dx$$

$$= 0 + \frac{2}{\pi n^2}[x \cos nx]\Big|_{-\pi}^{\pi} - \frac{2}{\pi n^2}\int_{-\pi}^{\pi} \cos nx\, dx$$

540

$$= \frac{4}{n^2} [\cos n\pi] - \frac{2}{\pi n^3} (\sin nx) \Big|_{-\pi}^{\pi}$$

$$= \frac{4}{n^2} (-1)^n . \qquad (n > 0) \qquad (7)$$

Finally,

$$b_n = \frac{1}{\pi} \int_{-\pi}^{\pi} x^2 \sin nx \, dx.$$

Note that the integrand, $F(x) = x^2 \sin nx$, is an odd function of x, that is

$$F(-x) = (-x)^2 \sin(-nx) = -x^2 \sin nx = -F(x). \quad \text{Thus}$$

$$b_n = \frac{1}{\pi} \int_{-\pi}^{0} x^2 \sin nx \, dx + \frac{1}{\pi} \int_{0}^{\pi} x^2 \sin nx \, dx$$

$$= \frac{-1}{\pi} \int_{0}^{\pi} y^2 \sin ny \, dy + \frac{1}{\pi} \int_{0}^{\pi} x^2 \sin nx \, dx = 0 \qquad (8)$$

where the change of variables $x = -y$ was made in the first integral of (8). Using these values of a_n and b_n in (2) gives

$$f(x) \sim \frac{\pi^2}{3} + \sum_{n=1}^{\infty} (-1)^n \left(\frac{4}{n^2}\right) \cos nx. \qquad (9)$$

The graph of this series is shown in the figure. Note that even if f had been defined only on $(-\pi, \pi)$, it would have the same Fourier series (9) and this series would have the same graph. Thus, the series would represent a periodic extension of the values of x^2 in the interval $(-\pi, \pi)$ if it converged (it does by the Weierstrass M-test).

• **PROBLEM 14-11**

A horizontal beam L ft. long is simply supported at its ends. Using trigonometric series find the equation of the elastic curve of the beam if it carries a load of P lb., c ft. from one end. (Neglect the weight of the beam).

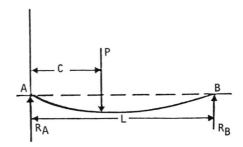

Solution: Many expansions of functions in Fourier series deal with trigonometric series with a period 2π. However, in most engineering and physics applications one requires an expansion of a given function over an interval different from π or 2π. Then one stretches or compresses this interval $(-\pi,\pi)$ by means of a transformation of variables to suit the circumstances.

Considering the above problem, first evaluate the reaction forces R_A and R_B by finding the moments with respect to points A and B. Taking counterclockwise moments as positive,

$$R_B L - Pc = 0.$$

$$P(L-c) - R_A L = 0.$$

Hence,

$$R_A = \frac{P}{L}(L-c),$$

and

$$R_B = \frac{Pc}{L}.$$

Take an arbitrary point, x ft. away from A; then the bending moment of the constraint forces with respect to this point are given by:

$$M_1 = -R_A x = -\frac{P}{L}(L-c)x, \quad 0 < x < c$$

$$M_2 = -R_B(L-x) = -\frac{Pc}{L}(L-x), \quad c < x < L.$$

Now, using the Fourier half-range sine series, the bending moment can be expressed, in the interval $(0,L)$, as

$$M = \sum_{n=1}^{\infty} b_n \sin nx,$$

where

$$b_n = \frac{2}{L} \int_0^L (M_1 + M_2) \sin \frac{n\pi x}{L} dx$$

$$= \frac{2}{L} \int_0^c - \frac{P}{L}(L-c)x \sin \frac{n\pi x}{L} dx$$

$$+ \frac{2}{L} \int_c^L - \frac{Pc}{L}(L-x) \sin \frac{n\pi x}{L} dx.$$

$$b_n = \frac{2}{L} \left\{ -\frac{P}{L}(L-c) \int_0^c x \sin \frac{n\pi x}{L} dx - \frac{Pc}{L} \int_c^L (L-x) \sin \frac{n\pi x}{L} dx \right\}$$

$$= \frac{2}{L} \left\{ \frac{P}{L}(c-L) \left[-\frac{Lx}{n\pi} \cos \frac{n\pi x}{L} + \frac{L^2}{n^2\pi^2} \sin \frac{n\pi x}{L} \right]_0^c \right.$$

$$- \frac{Pc}{L} \left[-\frac{L^2}{n\pi} \cos \frac{n\pi x}{L} + \frac{Lx}{n\pi} \cos \frac{n\pi x}{L} - \frac{L^2}{n^2\pi^2} \sin \frac{n\pi x}{L} \right]_c^L \bigg\}$$

$$= \frac{2}{L} \left[-\frac{Pc^2}{n\pi} \cos \frac{n\pi c}{L} + \frac{PLc}{n^2\pi^2} \sin \frac{n\pi c}{L} + \frac{PLc}{n\pi} \cos \frac{n\pi c}{L} \right.$$

$$- \frac{PL^2}{n^2\pi^2} \sin \frac{n\pi c}{L} + (-1)^n \frac{PLc}{n\pi}$$

$$+ (-1)^{n+1} \frac{PLc}{n\pi} - \frac{PLc}{n\pi} \cos \frac{n\pi c}{L} + \frac{Pc^2}{n\pi} \cos \frac{n\pi c}{L}$$

$$- \frac{PLc}{n^2\pi^2} \sin \frac{n\pi c}{L} \right].$$

$$b_n = -\frac{2PL}{n^2\pi^2} \sin \frac{n\pi c}{L}.$$

Hence,

$$M = -\frac{2PL}{\pi^2} \sum_{n=1}^{\infty} \frac{\sin \frac{n\pi c}{L}}{n^2} \sin \frac{n\pi x}{L}.$$

Now, recall the differential equation of the elastic curve of the beam:

$$EI \frac{d^2y}{dx^2} = M,$$

where E (lb./ft^2) is Young's modulus, I is the moment of inertia (ft.^{4}) of the cross-sectional area of the beam with respect to the neutral axis, and M is the bending moment.

Hence, combining the two equations,

$$EI \frac{d^2y}{dx^2} = -\frac{2PL}{\pi^2} \sum_{n=1}^{\infty} \frac{\sin \frac{n\pi c}{L}}{n^2} \sin \frac{n\pi x}{L},$$

or,

$$\frac{d^2y}{dx^2} = -\frac{2PL}{EI\pi^2} \sum_{n=1}^{\infty} \frac{\sin \frac{n\pi c}{L}}{n^2} \sin \frac{n\pi x}{L}.$$

Since $dy/dx \simeq 0$ and $y = 0$ at $x = 0$, double successive integration yields the desired equation of the bend:

$$y = 2 \frac{PL}{EI\pi^2} \sum_{n=1}^{\infty} \frac{\sin \frac{n\pi c}{L}}{n^2} \left(-\frac{L}{n\pi}\right)\left(\frac{L}{n\pi}\right) \sin \frac{n\pi x}{L}$$

$$= \frac{2PL^3}{EI\pi^4} \sum_{n=1}^{\infty} \frac{\sin \frac{n\pi c}{L}}{n^4} \sin \frac{n\pi x}{L}.$$

As an alternative derivation, consider the following facts. The work done in bending a piece of beam, and hence the potential energy stored in the piece, is given by:

$$\Delta w = \frac{1}{2} M \Delta s/R,$$

where Δs is the length of this element, and R is the radius of curvature of the bend, which can be approximated as $R \simeq 1/d^2y/dx^2$.

$$\frac{ds}{R^2} \simeq \frac{d^2y}{dx^2}.$$

Thus

$$W = \frac{EI}{2} \int_0^L \left(\frac{d^2y}{dx^2}\right)^2 dx.$$

Again the equation of the elastic curve is desired to be represented as:

$$y = \sum_{n=1}^{\infty} b_n \sin \frac{n\pi x}{L}.$$

Hence,

$$\frac{d^2y}{dx^2} = -\frac{\pi^2}{L^2} \sum_{n=1}^{\infty} b_n n^2 \sin \frac{n\pi x}{L}.$$

$$\left(\frac{d^2y}{dx^2}\right)^2 = \frac{\pi^4}{L^4} \sum_{n=1}^{\infty} b_n^2 n^4 \sin^2 \frac{n\pi x}{L}$$

$$+ \frac{\pi^4}{L^4} \sum_{n=1}^{\infty} \sum_{m=n-1}^{\infty} b_n n^2 \sin \frac{n\pi x}{L} \left(b_m m^2 \sin \frac{n\pi x}{L}\right),$$

where upon integration from zero to L the second sum vanishes ($n \neq m$). One has:

$$\int_0^L \left(\frac{d^2y}{dx^2}\right)^2 dx = \frac{\pi^4}{L^4} \sum_{n=1}^{\infty} b_n^2 n^4 \int_0^L \sin^2 \frac{n\pi x}{L} dx,$$

where

$$\int_0^L \sin^2 \frac{n\pi x}{L} dx = \frac{L}{2}.$$

Hence,

$$W = \frac{EI\pi^4}{4L^3} \sum_{n=1}^{\infty} b_n^2 n^4.$$

It is known that the potential energy, w, is imparted due to the weight p at x = c, producing a deflection y_c at that point, where

$$y_c = \sum_{n=1}^{\infty} b_n \sin \frac{n\pi c}{L}.$$

Now, considering the load p acting through an infintesimal displacement dy_c, it does an amount of work pdy_c, equal to the infintesimal increase in dw, which, according to the relation obtained, is a function of b_n only. Thus,

$$\frac{EI\pi^4}{4L^3} \sum_{n=1}^{\infty} 2n^4 b_n db_n = p \sum_{n=1}^{\infty} \sin \frac{n\pi c}{L} db_n.$$

This holds if and only if $b_n \neq 0$, which leads to:

$$\frac{EI\pi^4}{4L^3} n^4 b_n = p \sin \frac{n\pi c}{L}.$$

That is,

$$b_n = \frac{2L^3 p}{EI\pi^4} \frac{\sin \frac{n\pi c}{L}}{n^4}.$$

Hence the Fourier half-range expansion, with values of b_n above, becomes:

$$y = \frac{2pL^3}{EI\pi^4} \sum_{n=1}^{\infty} \frac{\sin \frac{n\pi c}{L}}{n^4} \sin \frac{n\pi x}{L},$$

as obtained before.

● **PROBLEM 14-12**

Expand the function f(x), defined by

$$f(x) = \begin{cases} \cos \frac{\pi x}{\ell} & \text{for} \quad 0 \leq x \leq \frac{\ell}{2}, \\ 0 & \text{for} \quad \frac{\ell}{2} < x \leq \ell \end{cases}$$

in cosine series.

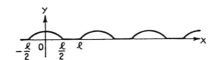

Solution: The figure shows the graph of f(x) and its even extension onto the interval $[-\ell, 0]$, together with its subsequent periodic extension (with period 2ℓ) onto the whole x-axis. The convergence criterion can be applied everywhere.

For $\ell/2 < x \le \ell$, one gets $f(x) = 0$, so that

$$a_0 = \frac{2}{\ell}\int_0^\ell f(x)\, dx = \frac{2}{\ell}\int_0^{\ell/2} \cos\frac{\pi x}{\ell}\, dx = \frac{2}{\pi},$$

$$a_n = \frac{2}{\ell}\int_0^\ell f(x)\cos\frac{\pi nx}{\ell}\, dx = \frac{2}{\ell}\int_0^{\ell/2} \cos\frac{\pi x}{\ell}\cos\frac{\pi nx}{\ell}\, dx.$$

Making the substitution $\pi x/\ell = t$ yields

$$a_n = \frac{2}{\pi}\int_0^{\pi/2} \cos t \cos nt\, dt = \frac{1}{\pi}\int_0^{\pi/2} [\cos(n+1)t + \cos(n-1)t]\, dt,$$

hence

$$a_1 = \frac{1}{\pi}\int_0^{\pi/2}(\cos 2t + 1)\, dt = \frac{1}{\pi}\left[\frac{\sin 2t}{2} + t\right]_{t=0}^{t=\pi/2} = \frac{1}{2},$$

$$a_n = \frac{1}{\pi}\left[\frac{\sin(n+1)t}{n+1} + \frac{\sin(n-1)t}{n-1}\right]_{t=0}^{t=\pi/2} \quad (n>1).$$

Therefore, for odd $n > 1$

$$a_n = 0,$$

while, for even n

$$a_n = -\frac{2(-1)^{n/2}}{\pi(n^2-1)}, \quad b_n = 0 \quad (n = 1, 2, \ldots).$$

Thus one gets

$$\frac{1}{\pi} + \frac{1}{2}\cos\frac{\pi x}{\ell} - \frac{2}{\pi}\sum_{n=1}^\infty \frac{(-1)^n}{4n^2-1}\cos\frac{2\pi nx}{\ell}$$

$$= \begin{cases} \cos \frac{\pi x}{\ell} & \text{for } 0 \leq x \leq \frac{\ell}{2}, \\ 0 & \text{for } \frac{\ell}{2} < x \leq \ell. \end{cases}$$

This series converges on the whole x-axis to the function shown in the figure.

• **PROBLEM** 14-13

Expand $f(x) = x^2$ $(0 < x < 2\pi)$ in Fourier series.

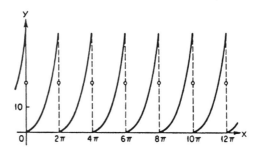

Solution: This problem resembles the case over the interval $(-\pi \leq x \leq \pi)$, but the graph of the periodic extension of $f(x)$ immediately shows the difference (see Figure). At the points of discontinuity the series converges to the arithmetic mean of the right-hand and left-hand limits, i.e., to the value $2\pi^2$. The function $f(x)$ is neither even nor odd.

Since

$$a_0 = \frac{1}{\pi} \int_0^{2\pi} x^2 \, dx = \frac{1}{\pi} \left[\frac{x^3}{3} \right]_{x=0}^{x=2\pi} = \frac{8\pi^2}{3},$$

$$a_n = \frac{1}{\pi} \int_0^{2\pi} x^2 \cos nx \, dx = -\frac{2}{\pi n} \int_0^{2\pi} x \sin nx \, dx$$

$$= \frac{2}{\pi n^2} [x \cos nx]_{x=0}^{x=2\pi} - \frac{2}{\pi n^2} \int_0^{2\pi} \cos nx \, dx = \frac{4}{n^2},$$

$$b_n = \frac{1}{\pi} \int_0^{2\pi} x^2 \sin nx \, dx$$

$$= -\frac{1}{\pi n} [x^2 \cos nx]_{x=0}^{x=2\pi} + \frac{2}{\pi n} \int_0^{2\pi} x \cos nx \, dx$$

$$= -\frac{4\pi}{n} - \frac{2}{\pi n^2} \int_0^{2\pi} \sin nx \, dx = -\frac{4\pi}{n},$$

548

one gets

$$x^2 = \frac{4\pi^2}{3} + 4(\cos x - \pi \sin x + \frac{\cos 2x}{2^2} - \frac{\pi \sin 2x}{2} + \ldots$$

$$+ \frac{\cos nx}{n^2} - \frac{\pi \sin nx}{n} + \ldots)$$

$$= \frac{4\pi^2}{3} + 4 \sum_{n=1}^{\infty} \left(\frac{\cos nx}{n^2} - \frac{\pi \sin nx}{n} \right)$$

$$= \frac{4\pi^2}{3} + 4 \sum_{n=1}^{\infty} \frac{\cos nx}{n^2} - 4\pi \sum_{n=1}^{\infty} \frac{\sin nx}{n},$$

for $0 < x < 2\pi$.

● **PROBLEM 14-14**

Expand $f(x) = |\sin x|$ in Fourier series.

Solution: This function is defined for all x, and represents a continuous, piecewise smooth, even function. Hence $f(x) = |\sin x|$ is everywhere equal to its Fourier series, which is absolutely and uniformly convergent. Its graph is shown in the figure.

$|\sin x| = \sin x$ for $0 \leq x \leq \pi$, which gives

$$a_0 = \frac{2}{\pi} \int_0^{\pi} \sin x \, dx = \frac{4}{\pi},$$

and

$$a_n = \frac{2}{\pi} \int_0^{\pi} \sin x \cos nx \, dx$$

$$= \frac{1}{\pi} \int_0^{\pi} [\sin(n+1)x - \sin(n-1)x] dx$$

$$= -\frac{1}{\pi} \left[\frac{\cos(n+1)x}{n+1} - \frac{\cos(n-1)x}{n-1} \right]_{x=0}^{x=\pi}$$

$$= -\frac{1}{\pi} \left[\frac{(-1)^{n+1} - 1}{n+1} - \frac{(-1)^{n-1} - 1}{n-1} \right] = -2 \frac{(-1)^n + 1}{\pi(n^2 - 1)},$$

for n ≠ 1, while for n = 1

$$a_1 = \frac{2}{\pi} \int_0^\pi \sin x \cos x \, dx = \frac{1}{\pi} \int_0^\pi \sin 2x \, dx = 0.$$

Moreover, $b_n = 0$ (n = 1, 2, . . .), since f(x) is even. Therefore, for all x one gets

$$|\sin x| = \frac{2}{\pi} - \frac{4}{\pi}\left(\frac{\cos 2x}{3} + \frac{\cos 4x}{15} + \frac{\cos 6x}{35} + \cdots\right).$$

• **PROBLEM 14-15**

Expand the function f(x), defined by

$$f(x) = \begin{cases} x & \text{for } 0 \le x \le \frac{\ell}{2}, \\ \ell - x & \text{for } \frac{\ell}{2} < x \le \ell, \end{cases}$$

in sine series.

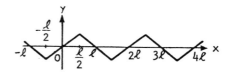

Solution: The figure shows the graph of f(x) and its odd extension onto the interval [-ℓ, 0], together with its subsequent periodic extension (with period 2ℓ) onto the whole x-axis. The convergence criterion can be applied everywhere.

In this case, one has

$$a_n = 0 \quad (n = 0, 1, 2, \ldots),$$

$$b_n = \frac{2}{\ell} \int_0^\ell f(x) \sin \frac{\pi n x}{\ell} dx$$

$$= \frac{2}{\ell} \int_0^{\ell/2} x \sin \frac{\pi n x}{\ell} dx + \frac{2}{\ell} \int_{\ell/2}^\ell (\ell - x) \sin \frac{\pi n x}{\ell} dx$$

$$(n = 1, 2, \ldots).$$

Setting πx/ℓ = t gives

$$b_n = \frac{2\ell}{\pi^2} \int_0^{\pi/2} t \sin nt \, dt + \frac{2\ell}{\pi^2} \int_{\pi/2}^\pi (\pi - t) \sin nt \, dt$$

$$= \frac{2\ell}{\pi^2}\left[-\frac{t \cos nt}{n}\right]_{t=0}^{t=\pi/2} + \frac{2\ell}{\pi^2 n}\int_0^{\pi/2} \cos nt \, dt$$

$$+ \frac{2\ell}{\pi^2} \left[- \frac{(\pi - t) \cos nt}{n} \right]_{t=\pi/2}^{t=\pi}$$

$$- \frac{2\ell}{\pi^2 n} \int_{\pi/2}^{\pi} \cos nt \, dt$$

$$= \frac{4\ell}{\pi^2 n^2} \sin \frac{\pi n}{2}.$$

Therefore

$$\frac{4\ell}{\pi^2} \left(\sin \frac{\pi x}{\ell} - \frac{1}{3^2} \sin \frac{3\pi x}{\ell} + \frac{1}{5^2} \sin \frac{5\pi x}{\ell} - \cdots \right)$$

$$= \begin{cases} x & \text{for } 0 \le x \le \frac{\ell}{2}, \\ \ell - x & \text{for } \frac{\ell}{2} < x \le \ell. \end{cases}$$

This series converges on the whole x-axis to the function shown in the figure.

● **PROBLEM 14-16**

Make a Fourier expansion of a so-called rectangular wave

$$f(x) = \begin{cases} -1, & -\pi < x < 0, \\ 1, & 0 < x < \pi. \end{cases}$$

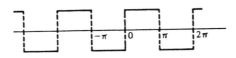

Solution: Make a periodic continuation of f(x) outside the interval $(-\pi, \pi)$ (see Fig.). f is an odd function, so $a_j = 0$ for all j.

$$b_j = -\pi^{-1} \int_{-\pi}^{0} \sin jx \, dx + \pi^{-1} \int_{0}^{\pi} \sin jx \, dx$$

$$= 2\pi^{-1} \int_{0}^{\pi} \sin jx \, dx$$

$$= 2\pi^{-1} \frac{1 - \cos j\pi}{j} = \begin{cases} 0 & \text{if } j \text{ is even}, \\ \frac{4}{j\pi} & \text{if } j \text{ is odd}. \end{cases}$$

Hence

$$f(x) = \frac{4}{\pi} \left(\sin x + \frac{\sin 3x}{3} + \frac{\sin 5x}{5} + \cdots \right).$$

Notice that the sum of the series is zero at the points where f has a jump discontinuity; this agrees with the fact that that sum should be equal to the average of the limiting values to the left and to the right of the discontintuity.

DIFFERENTIATION OF FOURIER SERIES

● **PROBLEM 14-17**

Given

$$f(x) = \begin{cases} x^2 - 1 & \text{if } x < 0 \\ x^2 + 1 & \text{if } x > 0 \end{cases}$$

discuss continuity and smoothness for the function. Compute jumps if they exist. The function f is PWC (Piece-wise continuous) with a jump discontinuity at x = 0.

<u>Solution</u>: The jump $f(0+) - f(0-)$ is 2. The derivative of f exists for all x except x = 0. Since f(0) is not defined, the definition of f' at x = 0 is not satisfied at this point, even though

$$f'_+(0) = f'_-(0) = 0 \; .$$

The jump at x = 0 for the derivative function is zero. The function f is PWS (Piece-wise smooth.) See figures 1 and 2.

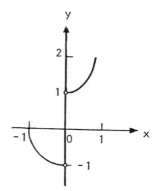

Fig. 1

The PWC Function f.

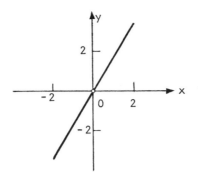

Fig. 2

The derivative of f.

● **PROBLEM 14-18**

The function is defined by

$$f(x) = \begin{cases} x^2 \sin\left(\frac{1}{x}\right) & \text{if } x \neq 0 \\ 0 & \text{if } x = 0 \end{cases}$$

(a) Find the right and left hand derivatives for f at $x = 0$

(b) Determine $f'(0)$ if it exists.

Solution: For all values of $x \neq 0$, the function has the derivative formula

$$f'(x) = 2x \sin\left(\frac{1}{x}\right) - \cos\left(\frac{1}{x}\right)$$

(a) If $x = 0$, investigate the definitions for the one-sided derivatives

$$f'_-(0) = \lim_{\substack{h \to 0 \\ h > 0}} \frac{f(0-) + h^2 \sin(1/h)}{h}$$

$$= \lim_{\substack{h \to 0 \\ h > 0}} h \sin\left(\frac{1}{h}\right) = 0$$

$$f'_+(0) = \lim_{\substack{h \to 0 \\ h > 0}} \frac{h^2 \sin(1/h) - f(0+)}{h}$$

$$= \lim_{\substack{h \to 0 \\ h > 0}} h \sin\left(\frac{1}{h}\right) = 0$$

(b) Since $f(0)$ is defined,

$$f'(0) = \lim_{h \to 0} \frac{h^2 \sin(1/h) - f(0)}{h}$$

$$= \lim_{h \to 0} h \sin\left(\frac{1}{h}\right) = 0$$

● **PROBLEM 14-19**

Differentiate the series

$$\cos x = \frac{8}{\pi} \sum_{n=1}^{\infty} \frac{n \sin 2nx}{(4n^2 - 1)}$$

and investigate the possibility of the newly formed series converging to the function - sin x.

<u>Solution</u>: By termwise differentiation one gets the series

$$\frac{16}{\pi} \sum_{n=1}^{\infty} \frac{n^2 \cos 2nx}{4n^2 - 1}$$

presumably the representation for the function -sin x. Upon investigating the limit of $(16/\pi)(n^2/(4n^2 - 1)) \cos 2nx$ as $n \to \infty$, one finds it is not zero. Therefore the new series is divergent and cannot be the convergence of -sin x.

If the function f is replaced by f' in

$$f(x) \sim \frac{a_0}{2} + \sum_{n=1}^{\infty} \left[a_n \cos\left(\frac{n\pi x}{L}\right) + b_n \sin\left(\frac{n\pi x}{L}\right) \right]$$

where the Fourier coefficients a_n and b_n are

$$a_n = \frac{1}{L} \int_{-L}^{L} f(x) \cos \frac{n\pi x}{L} dx$$
$$n = 0, 1, 2, \ldots$$
$$b_n = \frac{1}{L} \int_{-L}^{L} f(x) \sin \frac{n\pi x}{L} dx$$

with $L = \pi$, then one is assured that the series corresponding to f' converges. If f' is periodic with a period 2π and PWS on $-\pi \leq x \leq \pi$, then the corresponding Fourier series

$$\frac{a_0'}{2} + \sum_{n=1}^{\infty} (a_n' \cos nx + b_n' \sin nx)$$

where

$$a_n' = \frac{1}{\pi} \int_{-\pi}^{\pi} f' \cos nx \, dx, \qquad n \in N_0$$

$$b_n' = \frac{1}{\pi} \int_{-\pi}^{\pi} f' \sin nx \, dx, \qquad n \in N$$

converges to

$$\frac{f'(x+) + f'(x-)}{2}$$

If one adds that $f(-\pi) = f(\pi)$ and makes f a continuous function with f' PWS, then both f' and f are PWC. Coefficients

$$a_0' = 0, \quad a_n' = nb_n, \quad b_n' = -na_n$$

have been determined. The derivative f' is continuous where f'' exists. For the values of x where f'' exists,

$$f'(x) = f'(x+) = f'(x-)$$

and

$$\frac{f'(x+) + f'(x-)}{2} = f'(x)$$

or

$$f'(x) = \sum_{n=1}^{\infty} (nb_n \cos nx - na_n \sin nx) \quad .$$

The following theorem contains the results.

Theorem: Assume that f is a continuous function of period 2π on the interval $-\pi \leq x \leq \pi$ with $f(-\pi) = f(\pi)$. Let f', also a periodic function of period 2π, be PWC on the interval. Then at every point where f'' exists, f is termwise differentiable and the series converges to f'. The series

$$f(x) = \frac{a_0}{2} + \sum_{n=1}^{\infty} (a_n \cos nx + b_n \sin nx)$$

has the derivative

$$f'(x) = \sum_{n=1}^{\infty} (nb_n \cos nx - na_n \sin nx)$$

where

$$a_n = \frac{1}{\pi} \int_{-\pi}^{\pi} f \cos nx \, dx$$

$$b_n = \frac{1}{\pi} \int_{-\pi}^{\pi} f \sin nx \, dx$$

When f'' fails to exist but $f'_+(x)$ and $f'_-(x)$ exists, differentiation is valid in the sense that the series for f' converges to

$$\frac{f'(x+) + f'(x-)}{2}$$

For other types of Fourier series the theorem applies if the natural modifications are made in the theorem. If f is continuous and f' is PWC on $-L < x < L$, then where f'' exists the Fourier series for f is differentiable.

● **PROBLEM 14-20**

Consider the following operational formula for the Fourier transform:

If $f'(t) \to 0$ as $t \to \infty$, then

$$C_\alpha\{f''\} = -f'(0) + \alpha\{[f(t) \sin \alpha t]_0^\infty$$

$$- \alpha \int_0^\infty f(t) \cos \alpha t \, dt\}$$

If $f(t) \to 0$ as $t \to \infty$, then

$$C_\alpha\{f''\} = -f'(0) - \alpha^2 C_\alpha\{f\} .$$

(a) Find the Fourier sine transform for the function $f(x) = e^{-kx}$, $k > 0$, from the definition.

(b) Find the Fourier cosine transform for $f(x) = e^{-kx}$, $k > 0$, using the operational formula

$$C_\alpha\{f''\} = -f'(0) - \alpha^2 C_\alpha\{f\} .$$

Solution:

(a) $S_\alpha\{e^{-kx}\} = \int_0^\infty e^{-kt} \sin \alpha t \, dt$

$$= \left[\frac{e^{-kt}}{k^2 + \alpha^2} (-k \sin \alpha t - \alpha \cos \alpha t)\right]_0^\infty$$

$$= \frac{\alpha}{k^2 + \alpha^2}, \quad k > 0$$

(b) $C_\alpha\{e^{-kx}\}$ is to be determined by the formula

$$C_\alpha\{f''\} = -f'(0) - \alpha^2 C_\alpha\{f\}$$

If
$$f = e^{-kx}$$
$$f' = -ke^{-kx}, \quad f'(0) = -k$$
$$f'' = k^2 e^{-kx}$$
$$C_\alpha\{k^2 e^{-kx}\} = -(-k) - \alpha^2 C_\alpha\{e^{-kx}\}, \tag{1}$$

$C_\alpha\{f\}$ is a linear operator. Therefore,
$$C_\alpha\{k^2 e^{-kx}\} = k^2 C_\alpha\{e^{-kx}\}.$$

Write (1) as
$$k^2 C_\alpha\{e^{-kx}\} = k - \alpha^2 C_\alpha\{e^{-kx}\}$$
$$(k^2 + \alpha^2) C_\alpha\{e^{-kx}\} = k$$

and
$$C_\alpha\{e^{-kx}\} = \frac{k}{k^2 + \alpha^2}, \quad k > 0$$

FOURIER INTEGRALS

• **PROBLEM 14-21**

Investigate the uniform convergence of the integral

$$\int_0^\infty x e^{-tx} \, dt$$

for $0 < a \le x$, and show that $S(x) = 1$.

Solution: The convergence

$$S(x) = \lim_{q \to \infty} \int_0^q x e^{-tx} \, dt = \lim_{q \to \infty} [-e^{-tx}]_{t=0}^q = 1$$

if $x > 0$.

Using the idea expressed in the definition for uniform convergence, if $\varepsilon > 0$ one can find a P which is dependent on ε but not x, so that

$$\left| 1 - \int_0^q x e^{-tx} dt \right| = \left| 1 - (1 - e^{-qx}) \right| = e^{-qx} < \varepsilon$$

for all $q > (1/a) \ln(1/\varepsilon)$. Therefore uniform convergence follows. One should point out that as $a \to 0$, P increases without limit and the integral fails to converge uniformly for $x > 0$.

• **PROBLEM 14-22**

Evaluate the integral

$$\int_0^\infty \frac{\sin t}{t} dt$$

Solution: First, show that if

$$S(x) = \int_0^\infty \frac{e^{-xt} \sin t}{t} dt$$

with $x > 0$, then

$$\left| \frac{e^{-xt} \sin t}{t} \right| \leq e^{-xt}.$$

The integral

$$\int_0^\infty e^{-xt} dt$$

converges on any interval $0 < a \leq x < \infty$. Therefore, by the M-test

$$S(x) = \int_0^\infty \frac{e^{-xt} \sin t}{t} dt \tag{1}$$

converges uniformly for $x > 0$. Thus the derivative may be computed

$$S'(x) = -\int_0^\infty e^{-xt} \sin t \, dt \tag{2}$$

If $0 < a \leq x < \infty$, then

$$\left| e^{-xt} \sin t \right| \leq e^{-at}$$

and

$$\int_0^\infty e^{-at} \, dt$$

converges. Therefore (2) converges uniformly on $a \leq x < \infty$. Integrating the improper integral (2), one obtains

$$S'(x) = -\left[\frac{e^{-xt}(-x \sin t - \cos t)}{x^2 + 1}\right]_0^\infty = -\frac{1}{x^2 + 1}$$

Therefore,

$$S(x) = -\arctan x + C \tag{3}$$

Now from (1)

$$|S(x)| \leq \int_0^\infty \left|\frac{e^{-xt} \sin t}{t}\right| dt \leq \int_0^\infty e^{-xt} \, dt = \frac{1}{x}$$

for any $x > 0$. As $x \to \infty$, $S(x) \to 0$. In (3) as $x \to \infty$, $S(x) \to -\pi/2 + C$, and $C = \pi/2$. Formula (3) may be written

$$S(x) = -\arctan x + \frac{\pi}{2} \tag{4}$$

Since the integral of (1) is uniformly convergent,

$$\lim_{x \to 0^+} S(x) = \int_0^\infty \lim_{x \to 0^+} \left(\frac{e^{-xt} \sin t}{t}\right) dt = \int_0^\infty \frac{\sin t}{t} \, dt \tag{5}$$

Finally, as $x \to 0^+$ in (4), $S(x) \to \pi/2$. Therefore,

$$\int_0^\infty \frac{\sin t}{t} \, dt = \frac{\pi}{2} \tag{6}$$

● **PROBLEM 14-23**

(a) Draw a graph for the function

$$f(x) = \begin{cases} 0 & \text{when } x < 0 \\ x & \text{when } 0 < x < 1 \\ 0 & \text{when } x > 1 \end{cases}$$

(b) Find the Fourier integral representing f of part (a).

(c) Determine the convergence of the integral at $x = 1$.

Solution: (a)

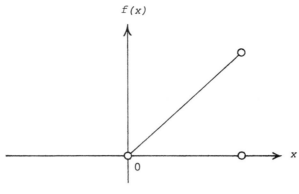

Graph of f.

(b) The integral representation of f is

$$f(x) \sim \int_0^\infty [A(\alpha) \cos \alpha x + B(\alpha) \sin \alpha x] \, d\alpha \qquad (1)$$

where

$$A(\alpha) = \frac{1}{\pi} \int_{-\infty}^\infty f(t) \cos \alpha t \, dt$$

$$= \frac{1}{\pi} \int_{-\infty}^0 0 \cdot \cos \alpha t \, dt + \frac{1}{\pi} \int_0^1 t \cos \alpha t \, dt$$

$$+ \int_1^\infty 0 \cdot \cos \alpha t \, dt$$

$$= \frac{1}{\pi} \left[\frac{1}{\alpha^2} \cos \alpha t + \frac{1}{\alpha} \sin \alpha t \right]_0^1$$

$$= \frac{1}{\pi} \left[\frac{\cos \alpha + \alpha \sin \alpha - 1}{\alpha^2} \right]$$

$$B(\alpha) = \frac{1}{\pi} \int_{-\infty}^\infty f(t) \sin \alpha t \, dt = \frac{1}{\pi} \int_0^1 t \sin \alpha t \, dt$$

$$= \frac{1}{\pi} \left[\frac{1}{\alpha^2} \sin \alpha t - \frac{t}{\alpha} \cos \alpha t \right]_0^1 = \frac{1}{\pi} \left[\frac{\sin \alpha - \alpha \cos \alpha}{\alpha^2} \right]$$

Replacing $A(\alpha)$ and $B(\alpha)$ in (1) by their computed values, one gets

$$f(x) \sim \frac{1}{\pi} \int_0^\infty \left[\frac{\cos \alpha + \alpha \sin \alpha - 1}{\alpha^2} \cos \alpha x \right.$$

$$+ \frac{\sin \alpha - \alpha \cos \alpha}{\alpha^2} \sin \alpha x \Bigg] d\alpha$$

$$= \frac{1}{\pi} \int_0^\infty \frac{\cos \alpha (1-x) + \alpha \sin \alpha (1-x) - \cos \alpha x}{\alpha^2} d\alpha$$

(c) At $x = 1$, the integral fails to converge to the function. In fact, the function is not defined at this point. The convergence at $x = 1$ is

$$\frac{f(1+) + f(1-)}{2} = \frac{0+1}{2} = \frac{1}{2}$$

• **PROBLEM 14-24**

Draw a graph of the function

$$f(x) = \begin{cases} 0 & \text{when } -\infty < x < -\pi \\ -1 & \text{when } -\pi < x < 0 \\ 1 & \text{when } 0 < x < \pi \\ 0 & \text{when } \pi < x < \infty \end{cases}$$

(a) Determine the Fourier integral for the function described.

(b) To what number does the integral found in (a) converge at $x = -\pi$?

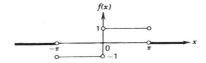

Graph of f.

Solution: (a) Since f is an odd function and PWS, use

$$f(x) \sim \int_0^\infty B(\alpha) \sin \alpha x \, d\alpha, \quad 0 < x < \infty$$

with the coefficients

$$B(\alpha) = \frac{2}{\pi} \int_0^\infty f(t) \sin \alpha t \, dt$$

for its representation.

Then

$$B(\alpha) = \frac{2}{\pi} \int_0^\pi 1 \cdot \sin \alpha t \, dt + \frac{2}{\pi} \int_\pi^\infty 0 \cdot \sin \alpha t \, dt$$

561

$$= \frac{2}{\pi}\left[-\frac{\cos \alpha t}{\alpha}\right]_0^\pi = \frac{2}{\pi\alpha}(1 - \cos \alpha\pi)$$

Therefore,

$$f(x) \sim \frac{2}{\pi}\int_0^\infty \left(\frac{1 - \cos \alpha\pi}{\alpha}\right)\sin \alpha x \, d\alpha$$

(b) Refer to the convergence theorem and to

$$\frac{f(x+) + f(x-)}{2}.$$

(The odd extension of f in the graph when represented by a sine integral will be an odd function over the entire real axis. Similarly, the even extension of f represented in (a) by a cosine integral will be even over the entire real axis.) One concludes that the integral converges to $-\frac{1}{2}$ at $x = -\pi$.

• **PROBLEM 14-25**

Solve the integral equation

$$\int_0^\infty f(x) \cos \alpha x \, dx = \begin{cases} 1 & \text{when } 0 < \alpha < \pi \\ 0 & \text{when } \pi < \alpha < \infty \end{cases}$$

Solution: An equation of the form

$$\int_0^\infty f(x)\cos \alpha x \, dx = g(\alpha)$$

is an integral equation. To solve the equation one needs to determine f. Therefore,

$$f(x) = \frac{2}{\pi}\int_0^\infty g(\alpha)\cos \alpha x \, d\alpha$$

In the problem,

$$f(x) = \frac{2}{\pi}\int_0^\pi 1 \cdot \cos \alpha x \, d\alpha = \frac{2}{\pi}\left[\frac{1}{x}\sin \alpha x\right]_0^\pi$$

$$f(x) = \frac{2}{\pi x}\sin \pi x$$

• **PROBLEM 14-26**

Show that

$$E_\alpha\{f(x - c)\} = e^{i\alpha c}E_\alpha\{f(x)\}$$

if c is a real constant.

Solution: $E_\alpha\{f(x-c)\} = \int_{-\infty}^{\infty} f(x-c)\exp(i\alpha x)\,dx$

If $t = x - c$ or $x = t + c$, then

$$E_\alpha\{f(x-c)\} = \int_{-\infty}^{\infty} f(t)\exp[i\alpha(t+c)]\,dt$$

$$= e^{i\alpha c}\int_{-\infty}^{\infty} f(t)\exp(i\alpha t)\,dt$$

$$= e^{i\alpha c}\int_{-\infty}^{\infty} f(x)\exp(i\alpha x)\,dx$$

$$E_\alpha\{f(x-c)\} = e^{i\alpha c}E_\alpha\{f(x)\}$$

This result is sometimes referred to as shifting or translating.

● **PROBLEM 14-27**

Determine a solution for the integral equation

$$f(x) = g(x) + \int_{-\infty}^{\infty} f(t)h(x-t)\,dt$$

Solution: Assume that the Fourier transforms of f, g, and h exist and are represented by $F_e(\alpha)$, $G_e(\alpha)$, and $H_e(\alpha)$, respectively. If one writes the transforms of the equation, using the convolution theorem, one gets

$$F_e(\alpha) = G_e(\alpha) + F_e(\alpha) \cdot H_e(\alpha)$$

then one obtains:

$$F_e(\alpha)[1 - H_e(\alpha)] = G_e(\alpha)$$

or

$$F_e(\alpha) = \frac{G_e(\alpha)}{1 - H_e(\alpha)}$$

If $F_e(\alpha)$ has an inverse, then

$$f(x) = \frac{1}{2\pi}\int_{-\infty}^{\infty} \frac{G_e(\alpha)}{1 - H_e(\alpha)}\exp(-i\alpha x)\,d\alpha$$

• **PROBLEM 14-28**

If $f(x,y) = xy$, $0 < x < 1$, $0 < y < 2$, determine the double series representation.

Solution: The function $f(x,y)$ satisfies the condition

$$f(-x,y) = -xy = -f(x,y)$$

$$f(x,-y) = -xy = -f(x,y)$$

Therefore, adopt the sine series representation

$$f(x,y) = \sum_{m=1}^{\infty} \sum_{n=1}^{\infty} d_{mn} \sin\left(\frac{m\pi x}{1}\right) \sin\left(\frac{n\pi x}{2}\right)$$

where

$$d_{mn} = \frac{4}{1(2)} \int_0^2 \int_0^1 xy \sin(m\pi x) \sin\left(\frac{n\pi y}{2}\right) dx\, dy$$

$$= 2 \int_0^2 \left[\frac{\sin m\pi x}{m^2 \pi^2} - \frac{x \cos m\pi x}{m\pi}\right]_0^1 y \sin\left(\frac{n\pi y}{2}\right) dy$$

$$= \frac{-2(-1)^m}{m\pi} \int_0^2 y \sin\left(\frac{n\pi y}{2}\right) dy$$

$$= \frac{-2(-1)^m}{m\pi} \left(\frac{-4(-1)^n}{n\pi}\right)$$

$$d_{mn} = \frac{8(-1)^{m+n}}{mn\pi^2}$$

• **PROBLEM 14-29**

If f' is continuous and f'' is PWC on $[0,L]$ show that

$$S_n\{f''\} = \frac{n\pi}{L}\left[f(0) - (-1)^n f(L)\right] - \frac{n^2\pi^2}{L^2} F_s(n)$$

Solution: To show this relation, replace f in

$$S_n\{f\} = \int_0^L f(x) \sin\left(\frac{n\pi x}{L}\right) dx = F_s(n), \quad n \in N$$

by f'' and integrate by parts two times.

$$S_n\{f''\} = \int_0^L f'' \sin\left(\frac{n\pi x}{L}\right) dx$$

$$= -\frac{n\pi}{L} \int_0^L f' \cos\left(\frac{n\pi x}{L}\right) dx$$

$$= \frac{n\pi}{L}\left[f(0) - (-1)^n f(L)\right] - \frac{n^2\pi^2}{L^2} F_s(n)$$

• **PROBLEM 14-30**

Determine the Fourier complex series for the function $f(x) = e^{2x}$, $-\pi < x < \pi$, given the series

$$e^{2x} \sim \sum_{n=-\infty}^{\infty} c_n e^{inx}$$

where

$$c_n = \frac{1}{2\pi} \int_{-\pi}^{\pi} f e^{-inx} dx$$

Solution: If $e^{2x} \sim \sum_{n=-\infty}^{\infty} c_n e^{inx}$ where

$$c_n = \frac{1}{2\pi}\int_{-\pi}^{\pi} e^{2x} e^{-inx} dx = \frac{1}{2\pi}\int_{-\pi}^{\pi} e^{(2-in)x} dx$$

then

$$c_n = \frac{1}{2\pi(2-in)}[e^{(2-in)x}]\Big|_{-\pi}^{\pi}$$

$$= \frac{1}{2\pi(2-in)}[e^{(2-in)\pi} - e^{-(2-in)\pi}]$$

$$= \frac{(-1)^n(2+in)\sinh 2\pi}{\pi(4+n^2)}$$

Therefore the series is

$$\frac{1}{\pi}\sum_{n=-\infty}^{\infty} \frac{(-1)^n(2+in)\sinh 2\pi e^{inx}}{4+n^2} \ .$$

Considering

$$f(x) \sim \sum_{n=-\infty}^{\infty} c_n \exp\left(\frac{2n\pi i x}{b-a}\right)$$

where

$$c_n = \frac{1}{b-a}\int_a^b f\,\exp\left(\frac{-2n\pi ix}{b-a}\right)dx,$$

employing Euler's identity, displaying the series as an isolated term plus two series, changing an index, and finally combining the two series again, one obtains by inserting c_n in the series $f(x)$,

$$f(x) \sim \frac{1}{(b-a)}\int_a^b f(t)\,dt$$

$$+ \frac{2}{(b-a)}\sum_{n=1}^{\infty}\int_a^b f(t)\cos\left[\frac{2n\pi(x-t)}{b-a}\right]dt$$

● **PROBLEM 14-31**

Let $f(x) = -\ln|2\sin(x/2)|$. This function is even and becomes infinite at $x = 2k\pi$ ($k = 0, \pm 1, \pm 2, \ldots$). Moreover, $f(x)$ is periodic, since

$$\left|2\sin\frac{x+2\pi}{2}\right| = \left|2\sin\left(\frac{x}{2}+\pi\right)\right| = \left|-2\sin\frac{x}{2}\right|$$

$$= \left|2\sin\frac{x}{2}\right|,$$

so that

$$\ln\left|2\sin\frac{x+2\pi}{2}\right| = \ln\left|2\sin\frac{x}{2}\right|.$$

The graph of $f(x)$ is shown in Fig. 1.

Prove that $f(x)$ is integrable.

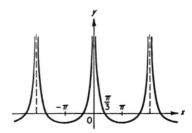

Fig. 1

Solution: It is sufficient to prove that it is integrable on the interval $[0, \pi/3]$ (see the graph of $f(x)$). Clearly, one has

$$-\int_\varepsilon^{\pi/3}\ln\left|2\sin\frac{x}{2}\right|dx = -\int_\varepsilon^{\pi/3}\ln\left(2\sin\frac{x}{2}\right)dx$$

$$= -\left[x \ln\left(2 \sin \frac{x}{2}\right)\right]_{x=\varepsilon}^{x=\pi/3}$$

$$+ \int_\varepsilon^{\pi/3} \frac{x \cos (x/2)}{2 \sin (x/2)} \, dx$$

$$= \varepsilon \ln \left(2 \sin \frac{\varepsilon}{2}\right) + \int_\varepsilon^{\pi/3} \frac{x \cos(x/2)}{2 \sin(x/2)} dx,$$

where the absolute value sign is dropped, since $2 \sin(x/2) > 1$ for $0 < x < \pi/3$. As $\varepsilon \to 0$, the quantity $\varepsilon \ln(2 \sin(\varepsilon/2))$ approaches zero, as can easily be verified by using L'Hospital's rule, while the last integral converges to the integral

$$\int_0^{\pi/3} \frac{x \cos (x/2)}{2 \sin (x/2)} \, dx,$$

which obviously has meaning, since the integrand is bounded. Thus

$$\lim_{\varepsilon \to 0} \int_\varepsilon^{\pi/3} \left| \ln \left| 2 \sin \frac{x}{2} \right| \right| dx$$

exists, i.e., $f(x)$ is integrable on the interval $[0, \pi/3]$. Moreover, $f(x)$ is absolutely integrable on the interval $[0, \pi/3]$, since it does not change sign there (see Figure).

Since $f(x)$ is even, one has

$$b_n = 0 \qquad (n = 1, 2, \ldots),$$

$$a_n = -\frac{2}{\pi} \int_0^\pi \ln\left(2 \sin \frac{x}{2}\right) \cos nx \, dx \qquad (n = 0, 1, 2, \ldots).$$

First of all, calculate the integral

$$I = \int_0^\pi \ln (2 \sin \frac{x}{2}) \, dx$$

$$= \int_0^\pi (\ln 2 + \ln \sin \frac{x}{2}) \, dx$$

$$= \pi \ln 2 + \int_0^\pi \ln \sin \frac{x}{2} \, dx.$$

Denote the last integral by Y and make the substitution $x = 2t$:

$$Y = 2\int_0^{\pi/2} \ln \sin t \, dt = 2\int_0^{\pi/2} \ln(2 \sin \tfrac{t}{2} \cos \tfrac{t}{2}) dt$$

$$= \pi \ln 2 + 2\int_0^{\pi/2} \ln \sin \tfrac{t}{2} \, dt + 2\int_0^{\pi/2} \ln \cos \tfrac{t}{2} \, dt.$$

The substitution $t = \pi - u$ gives

$$2\int_0^{\pi/2} \ln \cos \tfrac{t}{2} \, dt = 2\int_{\pi/2}^{\pi} \ln \sin \tfrac{u}{2} \, du$$

$$= 2\int_{\pi/2}^{\pi} \ln \sin \tfrac{t}{2} \, dt.$$

Therefore $Y = \pi \ln 2 + 2Y$, so that

$$Y = -\pi \ln 2.$$

This implies that $I = 0$, so that $a_0 = 0$.

Furthermore, integration by parts gives

$$a_n = -\frac{2}{\pi}\left\{ \left[\frac{\ln(2 \sin(x/2)) \sin nx}{n}\right]_{x=0}^{x=\pi} \right.$$

$$\left. - \frac{1}{n}\int_0^{\pi} \frac{\sin nx \cos(x/2)}{2 \sin(x/2)} dx \right\}$$

$$= \frac{1}{n\pi}\int_0^{\pi} \frac{\sin nx \cos(x/2)}{\sin(x/2)} dx.$$

(The first term in braces vanishes, since the indeterminacy for $x \to 0$ is easily "removed" by using L'Hospital's rule.) But

$$\sin nx \cos \tfrac{x}{2} = \tfrac{1}{2}[\sin(n+\tfrac{1}{2})x + \sin(n-\tfrac{1}{2})x],$$

and therefore

$$a_n = \frac{1}{n\pi}\int_0^{\pi} \frac{\sin(n+\tfrac{1}{2})x}{2 \sin(x/2)} dx + \frac{1}{n\pi}\int_0^{\pi} \frac{\sin(n-\tfrac{1}{2})x}{2 \sin(x/2)} dx,$$

which gives

$$a_n = \frac{1}{n} \quad (n = 1, 2, \ldots).$$

Since the function $f(x)$ is obviously differentiable for $x \neq 2k\pi$ ($k = 0, \pm 1, \pm 2, \ldots$), it follows that

$$-\ln\left|2\sin\frac{x}{2}\right| = \cos x + \frac{\cos 2x}{2} + \frac{\cos 3x}{3} + \ldots \qquad (1)$$

for $x \neq 2k\pi$ ($k = 0, \pm 1, \pm 2, \ldots$). It should be noted that for $x = 2k\pi$, both sides of (1) become infinite. Thus, in this sense, the equation (1) can be regarded as valid for all x.

Setting $x = \pi$ in (1), the familiar formula is obtained

$$\ln 2 = 1 - \frac{1}{2} + \frac{1}{3} - \frac{1}{4} + \ldots$$

CHAPTER 15

DIFFERENCE EQUATIONS

GENERAL APPLICATIONS OF DIFFERENCE EQUATIONS

● **PROBLEM 15-1**

Find the sum of n terms of the series

$$\frac{1}{1 \cdot 4} + \frac{1}{4 \cdot 7} + \frac{1}{7 \cdot 10} + \ldots$$

Solution: The sum required is

$$\sum_{x=0}^{n-1} \frac{1}{(3x+1)(3x+4)} = \sum_{x=0}^{n-1} (3x-2)^{(-2)}$$

$$= \left[\frac{(3x-2)^{(-1)}}{-3} \right]_0^n$$

$$= \frac{1}{3} \left[\frac{-1}{3n+1} + 1 \right]$$

$$= \frac{1}{3} \left[1 - \frac{1}{3n+1} \right] = \frac{n}{3n+1}.$$

● **PROBLEM 15-2**

Find the sum of n terms of the double arithmetic-geometric progression

$$1 \cdot 2 \cdot 3 + 2 \cdot 3 \cdot 3^2 + 3 \cdot 4 \cdot 3^3 + 4 \cdot 5 \cdot 3^4 + 5 \cdot 6 \cdot 3^5 + \ldots$$

Solution: $\sum_{0}^{n-1} (x+1)(x+2) 3^{x+1} = \sum_{0}^{n-1} (x+2)^{(2)} 3^{x+1}$

$$= \sum_{0}^{n-1} (x + 2)^{(2)} \Delta \frac{3^{x+1}}{2}$$

$$= \left[(x + 2)^{(2)} \frac{3^{x+1}}{2} \right]_0^n$$

$$- \sum_{0}^{n-1} 3^{x+2} (x + 2)^{(1)}.$$

The required sum then is

$$(n + 2)^{(2)} \frac{3^{n+1}}{2} - 3 - \sum_{0}^{n-1} (x + 2)^{(1)} \Delta \frac{3^{x+2}}{2}$$

$$= (n + 2)^{(2)} \frac{3^{n+1}}{2} - 3$$

$$- \left[(x + 2)^{(1)} \frac{3^{x+2}}{2} \right]_0^n + \sum_{0}^{n-1} \frac{3^{x+3}}{2}$$

$$= (n + 2)^{(2)} \frac{3^{n+1}}{2} - 3 - (n + 2)^{(1)} \frac{3^{n+2}}{2}$$

$$+ 3^2 + \frac{3^{n+3}}{4} - \frac{27}{4}$$

$$= \frac{3^{n+1}}{4} (2n^2 + 1) - \frac{3}{4}$$

$$= \frac{3}{4} \left[3^n (2n^2 + 1) - 1 \right].$$

• **PROBLEM 15-3**

Prove

$$\Delta^{-1} a^x f(x) = \frac{a^x}{a - 1} \left[1 - \frac{a}{a-1} \Delta + \left(\frac{a}{a-1} \right)^2 \Delta^2 - \cdots \right] f(x).$$

Solution: One has

$$\Delta^{-1} a^x f(x) = (E - 1)^{-1} a^x f(x).$$

Now introduce two partial enlargement operators, E_1 and

E_2 which affect only $f(x)$ and a^x respectively. Thus

$$E_1 f(x) = f(x+1), \text{ but } E_1 a^x = a^x;$$

$$E_2 a^x = a^{x+1}, \qquad \text{but } E_2 f(x) = f(x).$$

Let Δ_1 and Δ_2 be the corresponding difference operators. Then $E = E_1 E_2$ and

$$\Delta^{-1} a^x f(x) = (E_1 E_2 - 1)^{-1} a^x f(x)$$

$$= (\Delta_1 \Delta_2 + \Delta_1 + \Delta_2)^{-1} a^x f(x)$$

$$= \Delta_2^{-1} \left[1 + \frac{\Delta_1 E_2}{\Delta_2} \right]^{-1} a^x f(x)$$

$$= \left[\Delta_2^{-1} - \frac{\Delta_1}{\Delta_2^2} E_2 + \frac{\Delta_1^2}{\Delta_2^3} E_2^2 - \ldots \right] a^x f(x)$$

$$= \left[\frac{a^x}{a-1} - \frac{a^{x+1}}{(a-1)^2} \Delta_1 + \frac{a^{x+2}}{(a-1)^3} \Delta_1^2 - \ldots \right] f(x).$$

Since Δ_1 now operates only on $f(x)$, one can dispense with the subscript and conclude with the result

$$\Delta^{-1} a^x f(x) = \frac{a^x}{a-1} \left[1 - \frac{a}{a-1} \Delta + \left(\frac{a}{a-1} \right)^2 \Delta^2 - \ldots \right] f(x).$$

• **PROBLEM 15-4**

Form the differential and difference equations corresponding to the two-parameter family

$$y = ax^2 - bx.$$

Solution: One deduces
$$y' = 2ax - b,$$
$$y'' = 2a.$$

Eliminating a and b, one obtains

$$\begin{vmatrix} y & x^2 & x \\ y' & 2x & 1 \\ y'' & 2 & 0 \end{vmatrix} = 0,$$

that is, $x^2 y'' - 2xy' + 2y = 0$.

A similar procedure yields

$$\Delta y = a(2x + 1) - b,$$

$$\Delta^2 y = 2a.$$

Hence

$$\begin{vmatrix} y & x^2 & x \\ \Delta y & 2x + 1 & 1 \\ \Delta^2 y & 2 & 0 \end{vmatrix} = 0,$$

that is,

$$x^2 \Delta^2 y + x \Delta^2 y - 2x \Delta y + 2y = 0.$$

Thus the system of curves $y = ax^2 - bx$ is equivalent to the differential equation

$$x^2 D^2 y - 2x Dy + 2y = 0,$$

or to the difference equation

$$(x^2 + x) \Delta^2 y - 2x \Delta y + 2y = 0.$$

The difference equation might equally well be given in terms of E as

$$(x^2 + x) y_{x+2} - (2x^2 + 4x) y_{x+1} + (x^2 + 3x + 2) y_x = 0.$$

● **PROBLEM 15-5**

Obtain a fifth order formula of the form

$$y_{n+1} = a_1 y_n + a_2 y_{n-1} + a_3 y_{n-2} + a_4 y_{n-3}$$

$$+ h(b_0 y'_{n+1} + b_1 y'_n + b_2 y'_{n-1})$$

Express each coefficient in terms of a_2. Calculate the explicit form of the error term.

Solution: Expanding each term y_i and y'_i about t_n in the Taylor series and equating the coefficients of h^0 through h^5 to zero, one gets

$$1 = a_1 + a_2 + a_3 + a_4$$

$$1 = -a_2 - 2a_3 - 3a_4 + b_0 + b_1 + b_2$$

$$\frac{1}{2} = \frac{1}{2}(a_2 + 4a_3 + 9a_4) + b_0 - b_2$$

$$\frac{1}{6} = \frac{1}{6}(-a_2 - 8a_3 - 27a_4) + \frac{1}{2}(b_0 + b_2)$$

$$\frac{1}{24} = \frac{1}{24}(a_2 + 16a_3 + 81a_4) + \frac{1}{6}(b_0 - b_2)$$

$$\frac{1}{120} = \frac{1}{120}(-a_2 - 32a_3 - 243a_4) + \frac{1}{24}(b_0 + b_2)$$

The truncation error is given by

$$T_n = -\frac{h^6}{5!}\int_{-3}^{1} G(u)y^{(6)}(t_n + hu)\,du$$

where

$$G(u) = \begin{cases} (u-1)^5 + 5b_0(u-1)^4, & 0 \le u \le 1 \\ a_2(u+1)^5 - 5b_2(u+1)^4 + \\ \quad a_3(u+2)^5 + a_4(u+3)^5, & -1 \le u \le 0 \\ a_2(u+2)^5 + a_4(u+3)^5, & -2 \le u \le -1 \\ a_4(u+3)^5, & -3 \le u \le -2 \end{cases}$$

The coefficients a_i and b_i may be obtained as

$$306a_1 = -413a_2 + 468, \quad 34a_3 = 13a_2 - 20,$$

$$153a_4 = -5a_2 + 9, \quad 34b_0 = -a_2 + 12,$$

$$51b_1 = 31a_2 + 36, \quad 34b_2 = 37a_2 - 36$$

The value of T_n is given by

$$T_n = \frac{h^6}{5! \times 17}(-216 + 86a_2)y^{(6)}(\eta)$$

where $\quad -3 < \eta < 1$

• PROBLEM 15-6

Derive the governing finite difference equation for an interior node of a three-dimensional steady-state conduction region with generation, but with constant thermal conductivity.

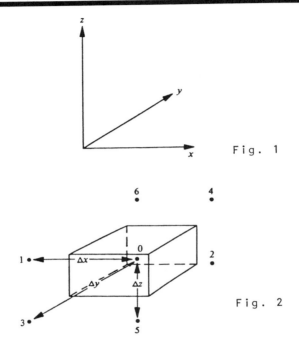

Fig. 1 coordinate system and Fig. 2 general interior node in a three-dimensional conduction region.

Solution: In a steady conduction problem, there are two different approaches used to arrive at the finite difference equation for a node. One approach is to take the governing partial differential equation such as, for constant thermal conductivity,

$$\frac{\partial^2 T}{\partial x^2} + \frac{\partial^2 T}{\partial y^2} + \frac{\partial^2 T}{\partial z^2} + \frac{q'''}{k} = 0 \qquad (1)$$

and since it must be satisfied everywhere in the conduction region, it must be satisfied at each node, say at node j, so that

$$\left.\frac{\partial^2 T}{\partial x^2}\right|_j + \left.\frac{\partial^2 T}{\partial y^2}\right|_j + \left.\frac{\partial^2 T}{\partial z^2}\right|_j + \frac{q'''_j}{k} = 0.$$

Now, the various terms, such as

$$\left.\frac{\partial^2 T}{\partial x^2}\right|_j,$$

are approximated by using the definition of the derivative, except that the spacing Δx between nodes in the x-direction does not approach zero, but remains finite. The other approach identifies the volume of material associated with each node and then applies the law of conservation of energy,

$$R_{in} + R_{gen} = R_{out} + R_{stor} \tag{2}$$

to this volume of material to derive the finite difference equation for the node. The phrase "material associated with each node" means the material is considered to be essentially at the temperature of the node. This approach is used herein since it is similar to the derivation of the governing partial differential equation when the volume of material associated with the node is a differential volume (compare with Equation (1)). Note Eq. (2) implies that the rate at which energy of all forms enters the control volume plus the rate at which energy is generated within the control volume itself is equal to the rate at which energy of all forms leaves the control volume plus the time rate of change of stored energy within the control volume itself (a control volume is a region of space across whose surface, called the control surface, both mass and energy may pass).

Figure 2 shows an interior node, labelled node 10, in a three-dimensional conduction region surrounded by six other nodes in the three coordinate directions. The lattice spacing Δx in the x-direction is the distance between any two nodes in the x-direction, and the spacings in the y- and z-directions are called Δy and Δz, respectively. The rectangular parallelopiped shown in Fig. 2 with node 0. The six nodes surrounding node 0 also have material associated with them, but are not outlined on the figure.

The spacing between nodes 1 and 0, and 0 and 2, is Δx, while between 3 and 0, and 0 and 4, it is Δy, and it is Δz between 5 and 0, and 0 and 6. Note that each of the six faces of the volume of material $\Delta x \Delta y \Delta z$ associated with node 0 lie halfway between node 0 and the adjacent node in the coordinate direction perpendicular to the face. Now, the law of conservation of energy, Eq. (2), applied to the control volume $\Delta x \Delta y \Delta z$ in the steady state gives

$$R_{in} + R_{gen} = R_{out}.$$

However, the only way energy can enter or leave a control volume which has all of its surfaces bounded by solid material is by conduction across the six faces of the control volume. Without loss of generality, it can be assumed that all of the conduction terms are into the node of interest; hence, Eq. (2) becomes

$$R_{in} + R_{gen} = 0$$

Let q_{10} be the rate at which energy is conducted into the control volume from the material around node 1, and use a similar numbering scheme for the conduction heat transfer rates from the other five nodes surrounding node 0. Then,

$$R_{in} = q_{10} + q_{20} + q_{30} + q_{40} + q_{50} + q_{60} .$$

The total rate at which energy is generated within the control volume is given by the local generation rate per unit volume times the volume of material associated with node 0. Thus

$$R_{gen} = q_0''' \Delta x \Delta x \Delta z .$$

The energy balance on node zero becomes

$$q_{10} + q_{20} + q_{30} + q_{40} + q_{50} + q_{60} + q_0''' \Delta x \Delta y \Delta z = 0.$$

Now, to relate q_{10}, for example, to the nodal temperatures, Fourier's simplified law of conduction is used. This is part of the finite difference approximation, since the conditions for the use of the simplified form of Fourier's law are not satisfied unless the finite spacings Δx, Δy, and Δz approach the infinitesimal quantities dx, dy, and dz. Fourier's simplified law of conduction for q_{10} gives

$$q_{10} = kA \frac{(T_1 - T_0)}{L} .$$

From Fig. 2, L is equal to Δx and the area A perpendicular to the heat flow is equal to $\Delta y \Delta z$; hence,

$$q_{10} = k \Delta y \Delta z \frac{(T_1 - T_0)}{\Delta x} .$$

For q_{20},

$$q_{20} = k \Delta y \Delta z \frac{(T_2 - T_0)}{\Delta x} .$$

(Note that the temperature difference is written as $T_2 - T_0$ since it is assumed that q_{20} is into the control volume.)

The other heat transfer rates are, using the same type of reasonsing as above,

$$q_{30} = k\Delta z\Delta x \frac{(T_3 - T_0)}{\Delta y} \quad , \quad q_{40} = k\Delta z\Delta x \frac{(T_4 - T_0)}{\Delta y},$$

$$q_{50} = k\Delta y\Delta x \frac{(T_5 - T_0)}{\Delta z} \quad , \quad q_{60} = k\Delta y\Delta x \frac{(T_6 - T_0)}{\Delta z}.$$

Inserting these expressions for the heat transfer rates into Eq. (2) and dividing each term by $k\Delta x\Delta y\Delta z$, the algebraic finite difference equation for the general interior node 0 is

$$\frac{T_1 + T_2 - 2T_0}{\Delta x^2} + \frac{T_3 + T_4 - 2T_0}{\Delta y^2} + \frac{T_5 + T_6 - 2T_0}{\Delta z^2}$$

$$+ \frac{q_0'''}{k} = 0 \qquad (3)$$

There will be an equation of the same type as Eq. (3) for all interior nodes, that is one for node 5, one for node 3, one for each node including the nodes that have not been shown since the entire conduction region has nodes throughout spaced Δx apart in the x-direction, Δy in the y-direction, and Δz in the z-direction. Since all these nodes are interior nodes, the finite difference equation is the same as Eq. (3) with just the subscripts changed. For instance, if the node labeled 40 is surrounded by nodes labeled 39 and 41 in the x-direction, 17 and 18 in the y-direction, and 178 and 180 in the z-direction, the finite difference equation for node 40 becomes, by comparison with Eq. (3),

$$\frac{T_{39} + T_{41} - 2T_{40}}{\Delta x^2} + \frac{T_{17} + T_{18} - 2T_{40}}{\Delta y^2} + \frac{T_{178} + T_{180} - 2T_{40}}{\Delta z^2}$$

$$+ \frac{q_{40}'''}{k} = 0.$$

If the temperatures of the different portions of the outer surface of the entire conduction region are specified and the region's shape is such that all nodes are interior nodes of the type 0, as shown in Fig. 2, then an equation of the type of Eq. (3) applies at each of the n nodes, and in the equations for the nodes that are exactly Δx or Δy or Δz from the outer surface, one or more of the temperatures appearing will be known, since at least one of the temperatures will apply to a point on the outer surface of the conduction region where the temperature has been specified. In this situation, the n linear algebraic equations are solved for the n unknown temperatures.

Equation (3) is the finite difference equation for an interior node (one which has the volume of material $\Delta x \Delta y \Delta z$ associated with it) in a steady, three-dimensional, constant thermal conductivity, conduction region, where the nodes are referenced to a rectangular cartesian coordinate system, as shown in Fig. 1.

● **PROBLEM 15-7**

Consider a portion of a conduction region, as shown in the accompanying figure. The temperatures of the surface nodes 1, 0, and 2 are not specified. Derive a finite difference equation for node zero.

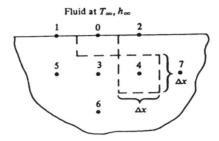

Solution: Clearly, unlike an interior node such as 4 which has the material Δx by Δx by 1 associated with it, node 0 has the material

$$\Delta x \text{ by } \frac{1}{2} \Delta x \text{ by } 1$$

associated with it as shown by dashed lines. Applying

$$R_{in} + R_{gen} = R_{out} + R_{stor},$$

to the control volume surrounding node 0, and assuming that all heat transfer rates are into the control volume,

$$R_{in} + R_{gen} = 0.$$

Energy enters the control volume by conduction from nodes 3, 1, and 2, and by convection from the fluid since node 0 is at a fluid solid interface. Assume that node 0 receives no net radiant gain from any surfaces beyond the fluid. Hence,

$$R_{in} = q_{30} + q_{10} + q_{20} + q_c,$$

where

$$q_{30} = k(\Delta x)(1) \frac{(T_3 - T_0)}{\Delta x}, \quad q_{20} = k \frac{\Delta x}{2} (1) \frac{(T_2 - T_0)}{\Delta x},$$

$$q_{10} = k \frac{\Delta x}{2} (1) \frac{(T_1 - T_0)}{\Delta x} .$$

Note that in the expressions for q_{10} and q_{20}, the area perpendicular to the heat flow is only $\frac{1}{2} \Delta x (1)$, rather than $(\Delta x)(1)$ as it is for the flow between 3 and 0. This is a consequence of the reduced amount of material associated with a noninterior node. By Newton's law of cooling,

$$q_c = h (\Delta x)(1)(T_\infty - T_0),$$

$$R_{gen} = q_0''' \frac{1}{2} (\Delta x)(\Delta x)(1) = q_0''' \frac{1}{2} \Delta x^2 .$$

Note that the reduced size of the control volume also makes itself felt in the R_{gen} term since R_{gen} is the total rate at which energy is generated within the control volume itself. After combining these results, the finite difference equation for node 0 becomes

$$T_3 + \frac{1}{2} T_2 + \frac{1}{2} T_1 + (h_\infty \Delta x T_\infty)/k - (2 + h_\infty \Delta x / k) T_0$$

$$+ q_0''' \Delta x^2 / 2k = 0 .$$

BOUNDARY VALUE PROBLEMS

● **PROBLEM 15-8**

Obtain the finite difference solution of the boundary value problem

$$-y'' - 2y' + 2y = e^{-2x}$$

$$y(0) = 1, \; y(\infty) = 0 \text{ with } h = \frac{1}{2} .$$

Solution: It can be easily verified that the analytical solution of the boundary value problem is

$$y(x) = \frac{1}{2} (e^{-(1+\sqrt{3})x} + e^{-2x})$$

The finite difference approximation to the differential equation at the nodal point $x = x_n$ is given by

$$-\frac{y_{n-1} - 2y_n + y_{n+1}}{h^2} - 2\frac{y_{n+1} - y_{n-1}}{2h} + 2y_n = e^{-2x_n}$$

The boundary conditions become

$$y_0 = 1, \qquad y_{N+1} = 0$$

For $h = 1/2$, the difference equation becomes

$$-y_{n-1} + 5y_n - 3y_{n+1} = \frac{1}{2} e^{-n}$$

Table 1 SOLUTION OF $-y''-2y'+2y=e^{-2x}$, $y(0)=1$, $y(\infty)=0$ FROM FINITE BOUNDARY CONDITIONS, $h=1/2$

y_N	$y_N=0$	Differences
$y_2 = \frac{5}{3}\left(y_1 - \frac{1}{5} - \frac{1}{10}e^{-1}\right)$	$y_1 = 0.2367879$	
		$0.415165 - 01$
$y_3 = \frac{22}{9}\left(y_1 - \frac{5}{22} - \frac{5}{44}e^{-1} - \frac{3}{44}e^{-2}\right)$	$y_1 = 0.2783046$	
		$0.891351 - 02$
$y_4 = \frac{95}{27}\left(y_1 - \frac{22}{95} - \frac{22}{190}e^{-1} - \frac{3}{38}e^{-2} - \frac{9}{190}e^{-3}\right)$	$y_1 = 0.2872182$	
		$0.204293 - 02$
$y_5 = \frac{409}{81}\left(y_1 - \frac{95}{409} - \frac{95}{818}e^{-1} - \frac{33}{409}e^{-2} - \frac{45}{818}e^{-3} - \frac{27}{818}e^{-4}\right)$	$y_1 = 0.2892611$	

The values of y_1 for a series of values of n are given in Table 1. Furthermore, with

$$\varepsilon = 10^{-8},$$

one obtains N = 16 so that $b^{(N)} = 8.5$.

• **PROBLEM 15-9**

Solve the equation

$$f(x + 2h) + f(x + h) - 12f(x) = 10x$$

in the interval (0,1) with interval of differencing h = .1 and subject to the boundary values f(0) = 0, f(1) = 50.

Solution: Denoting the ordinates at 0, .1, .2, ... by names f_0, f_1, f_2, ..., it is found that one must deal with the system of equations

$$f_1 + f_2 = 0$$

$$-12f_1 + f_2 + f_3 = 1,$$

$$-12f_2 + f_3 + f_4 = 2,$$

$$-12f_3 + f_4 + f_5 = 3,$$

$$\dots\dots\dots\dots\dots\dots$$

$$-12f_7 + f_8 + f_9 = 7,$$

$$-12f_8 + f_9 + 50 = 8.$$

This is a set of nine equations in nine unknowns and, as such, can be solved explicitly to give values of f_x for $x = 0(.1)1$.

In this particular case, one obtains

$$f_2 = -f_1,$$
$$f_3 = 1 + 13f_1,$$
$$f_4 = 1 - 25f_1,$$
$$f_5 = 14 + 181f_1,$$
$$f_6 = 2 - 481f_1,$$
$$f_7 = 171 + 2653f_1,$$
$$f_8 = -141 - 8425f_1,$$
$$f_9 = 2200 + 40261f_1.$$

Then, from $f_9 - 12f_8 = -42$, one obtains

$$141361f_1 = 3934,$$

that is,

$$f_1 = -.02782945791.$$

SOLUTION OF
$f(x + 2h) + f(x + h) - 12f(x) = 10x$

x	$f(x)$
0	0.00000
.1	−.02783
.2	+.02783
.3	+.63822
.4	+1.69574
.5	+8.96287
.6	+15.38597
.7	+97.16845
.8	+93.46318
.9	+1079.55820
1.0	50.00000

Table

The other values can now all be computed (note, however, that this method has certain disadvantages; since some of the coefficients multiplying are f_1 as large as 40000, the retention of eleven decimals in f_1 is necessary in order to have a tabulated solution good to six decimal places). The complete solution is given in the table shown. (Entries have been rounded to five decimals after being computed from the eleven-decimal value of f_1.)

• **PROBLEM 15-10**

Find the function satisfying Laplace's two-dimensional equation with the boundary conditions

$$(u = 0, x = 0); \quad (u = \tfrac{1}{2}x^2, y = 0);$$

$$(u = 8 + 2y, x = 4); \quad (u = x^2, y = 4),$$

by the method of relaxation.

Solution: Consider first the coarse network of $x = y = i$ ($i = 0,1,2,3,4$). The first values of the interior points are computed in the manner:

$u_5 = \tfrac{1}{4}(a_7 + a_3 + a_{15} + a_{11})$	$u_2 = \tfrac{1}{4}(u_3 + a_3 + u_1 + u_5)$
$u_1 = \tfrac{1}{4}(u_5 + a_3 + a_1 + a_{15})$	$u_4 = \tfrac{1}{4}(u_5 + u_1 + a_{15} + u_7)$
$u_3 = \tfrac{1}{4}(a_7 + a_5 + a_3 + u_5)$	$u_6 = \tfrac{1}{4}(a_7 + u_3 + u_5 + u_9)$
$u_7 = \tfrac{1}{4}(u_5 + a_{15} + a_{13} + a_{11})$	$u_8 = \tfrac{1}{4}(u_9 + u_5 + u_7 + a_{11})$
$u_9 = \tfrac{1}{4}(a_7 + u_5 + a_{11} + a_9)$	

The residuals R_{ij} are calculated and recorded in the left column of each block. The largest residual, $R_{22} = .7500$, is at the point P_{22}, which has a function value of $u_{22} = 4.5000$. This value is altered by the amount

$$\tfrac{1}{4} R_{22} = \tfrac{1}{4}(.7500) = .1875,$$

which is recorded in the right column of each block. This reduces R_{22} to zero and changes R_{23} to .1875, R_{12} to .1875, R_{21} to .1875, and R_{32} to .1875. After these changes have been made, the largest residual is $R_{31} = R_{33} = -.500$. Assume one chooses to work on R_{33} and change u_{33} by the amount

$$\tfrac{1}{4} R_{33} = -.125.$$

This changes R_{33} to zero, R_{23} to .0625, and R_{32} to .0625. Note that there are no residuals at the boundary points. Next, change the functional value u_{31} to make R_{31} zero, etc. The completed calculations are arranged as follows:

x / y	0		1			2			3			4
			R	u	Δu	R	u	Δu	R	u	Δu	
0	0			.5			2			4.5		8
1	0	−.25	1.625			0	3.6875		−.25	6.625		10
		0	1.5625	−.0625		.1875			0	6.5625	−.0625	
						.1250						
						.0625		.0156				
		.0156		.0039		.0001	3.7031		.0156		.0039	
		0	1.5664			.0040			0	6.5664		
						.0079		.0020				
		.0020		.0005		−.0001	3.7051		.0020		.0005	
		0	1.5669			.0004			0	6.5669		
		.0002				.0009		.0002	.0002			
						.0001	3.7053					
2	0	0	2.0625			.75	4.5	.1875	0	8.0625		12
		.1875				0	4.6875		.1875			
		.0625				.0156			.0625			
		0				0			0			
		.0039				.0020			.0039			
		0				0			0			
		.0005				.0002			.0005			
		0				0			0			
3	0	−.5	2.125			0	4.9375		−.5	9.125		14
		0	2.000	−.125		.1875			0	9.000	−.125	
						.0625						
		−.0156		−.0039		−.0625		−.0156	−.0156		−.0039	
		0	1.9961			−.0001	4.9219		0	8.9961		
						−.0040						
						−.0079	4.9199	−.0020	−.0020			
		−.0020		−.0005		.0001			0	8.9956	−.0005	
		0	1.9956			−.0004						
						−.0009		−.0002	−.0002			
		−.0002				−.0001	4.9197					
4	0		1				4			9		16

• **PROBLEM** 15-11

Solve the equation

$$y'' - 3y' - 10y = 10x,$$

in $(0,1)$ with mesh width $h = .1$, subject to the conditions $y(0) = 0$, $y(1) = 100$.

Solution: The approximating difference equation for the typical point (x_0, y_0) is obtained as

$$y_{-1}(2 + 3h) - y_0(4 + 20h^2) + y_1(2 - 3h) = 20h^2 x_0.$$

The sequence of tables which solve this problem is as follows:

Table 1   Relaxation Table

0	0	0	0	0	0	0	0	0
			6 −1	12 −2 −2	18 −3 −2	24 −2	36	60
	1	2						
0	10	20	50	80	130	220	360	600
4		5					4	4
	2		−4					
1	1			−3		−2		
			−1				−1	−1
5	13	25	45	77	130	218	363	603

Table 2   Residual Table

−2	−4	−6	−8	−10	−12	−14	−16	16982
		1014 844	−488 −408 −748 −288	−610 −510 −10	−732 68 108	166 −524 −144	584 124	62
168	336 −84	4 234						
168	−84	234	−288	−10	108	−144	124	62
0	8 93	24 70	−173			−76	24	−14
34	9	2	−5	−102				
9	−10	25	−56	24	39 5	8	−22	
		8	−14	1		−9	3	5
90	−100	80	−140	10	50	−90	30	50

This gives the set of equations

$$2.3y_0 - 4.2y_1 + 1.7y_2 = .02,$$

$$2.3y_1 - 4.2y_2 + 1.7y_3 = .04,$$

$$\cdots\cdots\cdots\cdots\cdots\cdots\cdots\cdots\cdots\cdots$$

$$2.3y_8 - 4.2y_9 + 1.7y_{10} = .18.$$

There is a single relaxation operator

$$\begin{array}{ccc} 1.7 & -4.2 & 2.3 \\ \circ\!\!-\!\!\!-\!\!\!-\!\!\bullet\!\!-\!\!\!-\!\!\!-\!\!\circ \end{array}$$
$$y_i$$

The solution is given in the two following tables (all equatins have been multiplied by 100 to avoid decimals). The first step employs the wedge operator

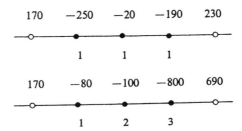

(here one writes the changes in the residuals above the line and the changes in the y_i's below the line).

Other wedge and line operators such as

```
        170   -250   -20   -190   230
        ──o────●─────●─────●──────o──
                 1     1     1

        170   -80   -100   -800   690
        ──o────●─────●─────●──────o──
                 1     2     3
```

and

are also employed. Note that in Table 2, at the second stage of relaxation, the relaxation operator (170, -420, 230) is replaced by (17, -42, 23) instead of multiplying all residuals by 10.

In connection with this problem, it should be pointed out that there is little point in carrying the solution onward to give further decimal places. To do so would

be to solve the difference equation

$$y_{-1}(2 + 3h) - y_0(4 + 20h^2) + y_1(2 - 3h) = 20h^2 x_0$$

to an ever-increasing degree of accuracy. But this difference equation is only an approximate representation of the given differential equation

$$y'' - 3y' - 10y = 10x.$$

Consequently, excessive accuracy in determining the approximate solution is unwarranted; the exact solution of the approximating difference equation is not relevant to the problem.

• PROBLEM 15-12

Find the values of the function satisfying

$$\frac{\partial U}{\partial t} = 4 \frac{\partial^2 U}{\partial x^2}$$

with boundary conditions

$U = 0$ at $x = 0$ and $x = 8$,

$U = 4x - \frac{1}{2}x$ at $t = 0$

using the direct step-by-step method. Hint: obtain the corresponding difference equation with h (the increment on x) = 1 and difference equation constant

$$r = \frac{1}{2}.$$

How can the formula be improved?

Solution: Equations of the parabolic or hyperbolic type may be solved by a direct step-by-step method to show the growth of the function, provided sufficient boundary conditions are given. Considering the parabolic equation

$$\frac{\partial u}{\partial t} = c^2 \frac{\partial^2 u}{\partial x^2},$$

one could obtain an expression for $u(x, t + k)$ in terms of $u(x,t)$, $u(x + h,t)$ and $u(x - h,t)$.

In dividing the xt-plane into a network, choose different increments on x and t, $\Delta t = k$ and $\Delta x = h$. The transformation to a difference equation yields

$$u(x, t + k) = r[u(x + h, t) + u(x - h, t)]$$
$$+ (1 - 2r)u(x,t), \qquad (1)$$

where $r = c^2 k h^{-2}$.

This expression gives the growth of the function as t increases; it is dependent on the choice of h and k and having sufficient boundary values.

Formula (1) would simplify to

$$u(x, t + k) = \frac{1}{2} [u(x + h, t) + u(x - h, t)] \qquad (2)$$

if one could choose h and k such that $r = c^2 k h^{-2}$ is $\frac{1}{2}$. Therefore choose h = 1. Then $r = 4k = \frac{1}{2}$ would yield $k = \frac{1}{8}$ making the increment on $t = \frac{1}{8}$. Now arrange the work according to the following schematic. The entries in the first row, first and last columns, are calculated from the boundary conditions, and the other entries are taken from Formula (2).

t \ x	0	1	2	3	4	5	6	7	8
0	0	3.5000	6.0000	7.5000	8.0000	7.5000	6.0000	3.5000	0
1/8	0	3.0000	5.5000	7.0000	7.5000	7.0000	5.5000	3.0000	0
2/8	0	2.7500	5.0000	6.5000	7.0000	6.5000	5.0000	2.7500	0
3/8	0	2.5000	4.625	6.0000	6.5000	6.0000	4.6250	2.5000	0
4/8	0	2.3125	4.2500	5.5625	6.0000	5.5625	4.2500	2.3125	0
5/8	0	2.1250	3.9375	5.1250	5.5625	5.1250	3.9375	2.1250	0
6/8	0	1.9688	3.6250	4.7500	5.1250	4.7500	3.6250	1.9688	0
7/8	0	1.8125	3.3594	4.3750	4.7500	4.3750	3.3594	1.8125	0
8/8	0	1.6797	3.0938	4.0547	4.3750	4.0547	3.0938	1.6797	0

Formula (2) is a simple averaging formula and could yield large errors. In fact under certain conditions this formula will lead to unstable solutions, and it is usually desirable to have $r < \frac{1}{2}$. The formula can be improved if the function U(x,t) has continuous partial derivatives with respect to x of order 6 and with respect to t of order 3. Then it can be shown that

$$U(x, t + k) = \frac{1}{6} [U(x + h, t) + 4U(x, t) + U(x - h, t)]$$

$$+ \text{ error function} \qquad (3)$$

if h and k are chosen so that $r = \frac{1}{6}$. In this problem this can be accomplished by different combinations of h and k, such as $\left(2, \frac{1}{6}\right)$, $\left(4, \frac{2}{3}\right)$, and $\left(1, \frac{1}{24}\right)$.

Improvements on the functional values may also be obtained by choosing a finer mesh, that is, smaller values for h and k. Note, however, that the formula for the calculation of the values of the function depends on h and k through r, and thus the formula will be changed whenever h and k are changed.

MOTIVATION.
PARTICULAR SOLUTIONS

● **PROBLEM** 15-13

If the coefficients A(K;N), K∈//0,M//, and B(N) are constants, A(M;N) ≠ 0, and A(0;N) ≠ 0, for each N in the index set I, (I = { N/N is an integer, N ≥ M} ≡ // M, ∞//), then find a particular solution (1):

{ y(0), Y(1), Y(2), ..., Y(M), Y(M+1), ... }

where each set of M+1 consecutive elements

{Y(N), Y(N-1), Y(N-2), ..., Y(N-M)}

satisfies (2): A(M;N) * Y(N) + A(M-1;N) * Y(N-1)

+ ... + A(0;N) * Y(N-M) = B(N) .

Solution: The particular solution (1) of (2) is obtained if one chooses as starting-values any M consecutive elements of that particular solution. With starting-values Y(J), J∈// 0,M - 1//, solve (2) for Y(N) to define, for N ∈ //M, ∞ //,

Y(N) = 1/A(M;N) * (B(N)   (3)
    - ∑ A(M - K;N) * Y(N - K), K ∈//1,M// ,).

With Q > M, and starting-values Y(J), J ∈ // Q - M, Q - 1// , define, for N ∈ //Q - 1, -1, M //,

Y(N - M) = 1/A(0;N) * (B(N)
    - ∑ A(M - K;N) * Y(N - K), K ∈ //0, M - 1//,)

and, for N ∈ // Q, ∞ // ,

Y(N) = 1/A(M;N) * (B(N)
    - ∑ A(M - K;N) * Y(N - K), K∈ //1,M //,).

Specifically, with M = 2, A(2;N) ≡ 1, A(1;N) = N, A(0;N) = -2 * N**2, and B(N) ≡ 0, then (2) becomes

Y(N) + N * Y(N - 1) - 2 * N ** 2 * Y(N - 2) = 0.

Given: J=0, 1; From (3): J=2, 3, 4, 5

J	0	1	2	3	4	5
Y(J)	1	1	6	0	192	-960

With index set I = //2,5//, and starting-values Y(0) = 1, Y(1) = 1, then the particular solution is given in in the table above.

● **PROBLEM 15-14**

Find a particular solution of Y(N) - 3 * Y(N-1) + 2 * Y(N-2) = 0, with index set I,

  I = {N/N is an integer, N ≥ M} ≡ //M,∞//,

and starting-values Y(P) = YP and (YQ) = YQ, P≠Q.

Solution: The particular solution corresponding to the table of starting-values Y(P) = YP, Y(Q) = YQ is defined by

  Y(N) = YY(N) = A * 2 ** N + B * 1 ** N   (1)

where A and B are elements of the solution of the linear system

  Y(P) = YY(P) = A * 2 ** P + B * 1 ** P = YP

  Y(Q) = YY(Q) = A * 2 ** Q + B * 1 ** Q = YQ.

That is, the particular solution of

  Y(N) - 3 * Y(N-1) + 2 * Y(N-2) = 0   (2)

with the index set I and starting-values Y(P) = YF, Y(Q) = YQ is

  Y(N) = YY(N)

   = (YP - YQ)/(2 ** P - 2 ** Q) * 2 ** N

   + (YQ * 2 ** P - YP * 2 ** Q)/(2 ** P - 2 ** Q)

     * 1 ** N.

In this problem, (1) is a general solution of (2). The order of the difference equation is 2.

● **PROBLEM 15-15**

Obtain the numerical solutions of the differential equation problem

  Y'(X) = A * Y(X), X ∈ [0,1], Y(0) = Y0,

     A = constant

defined in terms of particular solutions of linear homogeneous difference equations, where the difference equations are obtained in the following way. On the interval [0,1] choose the grid

$$X(J) = H * J, J \in //0, Ml//, \text{ with } H = 1/Ml,$$

so that $X(0) = 0$ and $X(Ml) = 1$. Solve by the half-open Lagrange Integration formula

$$DI(F;A,B) \equiv \int F(X), \text{ over } X \in (A,B),$$

$$= \int (P(N;X) + E(N;X)), \text{ over } X \in (A,B),$$

$$= \int (\sum F(K) * LC(N,K;X), K \in //0,N//,),$$
$$\text{over } X \in (A,B),$$

$$+ \int POLP(X) * DER ** (N+1)F(Z(X))/(N+1)!,$$
$$\text{over } X \in (A,B),$$

$$= \sum F(K) * DI(LC(N,K;X);A,B), K \in //0,N//, +$$
$$T(F,GRID)$$

$$= \sum F(K) * B(K), K \in //0,N//, + T(F,GRID).$$

$$T(F,GRID) = DI(POLP(X) * DER ** (N+1)F(Z(X));$$
$$A,B)/(N+1)!.$$

Solution: One has

$$Y(S) - Y(R) = \int Y'(X), \text{ over } X \in (X(R), X(S)), \quad (1)$$

$$= \int A * Y(X), \text{ over } X \in (X(R), X(S)),$$

$$\doteq A * \sum Y(K) * B(K,H), K \in //P,Q//,$$

where $R < S$, $P \leq Q$, and R, S, P, and Q are elements of

//0,M1//. The estimate (1) is motivation for the choice of difference equation

$$Y(S) - Y(R) = A * \sum Y(K) * B(K,H), \quad K \in //P,Q//. \quad (2)$$

Generally H appears only as a factor of the B(K,H). The order M of the difference equation (2) is the maximum of numbers

$$S - R, \quad Q - P, \quad |P - R|, \quad |Q - R|, \quad |P - S|,$$
$$\text{and } |Q - S|.$$

A table of starting-values will have M elements- usually including the given element Y(0) = Y0.

● **PROBLEM** 15-16

Obtain a numerical solution of the differential equation problem

$$Y"(X) = Y(X), \quad X \in [0,1], \quad Y(0) = A,$$

$$Y(1) = B \quad (1)$$

in the following way. On the interval [0,1] choose the grid

$$X(J) = H * J, J \in //0,M//, \text{ with } H = 1/M,$$

so that X(0) = 0 and X(M) = 1. Choose the estimate

$$F"(1) \doteq 2 * [X(0),X(1),X(2)]F$$
$$= \Delta ** (2)F(0)/H ** 2$$
$$= (F(2) - 2 * F(1) + F(0))/H ** 2,$$

$$(Y(K + 1) - 2 * Y(K) + Y(K - 1))/H ** 2 \doteq Y"(K). \quad (2)$$

Solution: Assume that M is chosen so large that H = 1/M is so small that when the elements

$$Y(X(J)) \equiv Y(J), J \in //K - 1, K + 1//,$$

of the range of the solution Y(X) of (1) are substituted in the left member of the estimate (2), one obtains a number which is usable as an estimate to Y"(K) and to Y(K). This is just another criterion for choosing H, but it does emphasize the fact that estimates like (2) are only used to motivate a choice of difference equation. In practice, one would probably try to choose H so small that the truncation error term associated with estimates like (2) would be "sufficiently small" - even though these error terms involve higher derivatives at mysterious points. One does not claim that a difference equation defined using such an H will have a particular solution which corresponds to a useful estimate to Y(X). The estimate, compare (2) and note (1),

$$(Y(N) - 2 * Y(N - 1) + Y(N - 2))/H ** 2 \doteq Y(N - 1)$$

suggests the second order linear homogeneous difference equation with constant coefficients

$$(Y(N) - 2 * Y(N - 1) + Y(N - 2))/H ** 2 = Y(N - 1)$$

or

$$Y(N) - (2 + H ** 2) * Y(N - 1) + Y(N - 2) = 0, \quad (3)$$

with the index set I = //2, M //, and starting-values Y(0) = A, Y(M) = B. The requirement that (3) be satisfied for the M - 1 elements of the index set defines a linear system.

Index	Equation
2	Y(2) - (2 + H ** 2) * Y(1) + Y(0) = 0
3	Y(3) - (2 + H ** 2) * Y(2) + Y(1) = 0
4	Y(4) - (2 + H ** 2) * Y(3) + Y(2) = 0
.	.
.	.
.	.
M	Y(M) - (2 + H ** 2) * Y(M - 1) + Y(M - 2) = 0.

This is M - 1 equations in the M - 1 unknowns

$$Y(J), J \in //1, M - 1//.$$

The corresponding matrix equation is

(4)
$$\begin{bmatrix} 2+H**2 & -1 & 0 & \cdots & 0 & 0 \\ -1 & 2+H**2 & -1 & \cdots & 0 & 0 \\ 0 & -1 & 2+H**2 & \cdots & 0 & 0 \\ \vdots & \vdots & \vdots & \ddots & \vdots & \vdots \\ 0 & 0 & 0 & \cdots & 2+H**2 & -1 \\ 0 & 0 & 0 & \cdots & -1 & 2+H**2 \end{bmatrix} * \begin{bmatrix} Y(1) \\ Y(2) \\ Y(3) \\ \vdots \\ Y(M-2) \\ Y(M-1) \end{bmatrix} = \begin{bmatrix} Y(0) \\ 0 \\ 0 \\ \vdots \\ 0 \\ Y(M) \end{bmatrix}$$

which has a matrix of coefficients that is symmetric and dominant diagonal. Since M is usually large, iteration methods are often used to solve (4). In the present case (3) has a unique particular solution for each choice of starting-values Y(0) = A, Y(M) = B. This particular solution

J	0	1	2	...	M − 1	M
Y(J)	A	Y(1)	Y(2)	...	Y(M − 1)	B

of (3) is used to define the set

Numerical Solution of (1)

J	0	1	2	...	M − 1	M
X(J)	0	1/M	2/M	...	(M − 1)/M	1
Y(J)	A	Y(1)	Y(2)	...	Y(M − 1)	B

which is considered a set of finite estimates to elements

X	0	1/M	2/M	...	(M − 1)/M	1
Y(X)	A	Y(1/M)	Y(2/M)	...	Y((M − 1)/M)	B

of Y(X), the solution of (1). Thus a particular solution of a difference equation may be used to define a numerical solution of a differential equation problem.

CHAPTER 16

DIFFERENTIAL EQUATIONS

APPLICATIONS OF FIRST ORDER AND SECOND ORDER DIFFERENTIAL EQUATIONS

● **PROBLEM 16-1**

A function $g(x,y)$ is homogeneous of degree n if $g(tx,ty) = t^n g(x,y)$ for all t. Determine whether the following functions are homogeneous, and, if so, find their degree: (a) $xy + y^2$, (b) $x + y \sin(y/x)^2$, (c) $x^3 + xy^2 e^{x/y}$, and (d) $x + xy$.

Solution: The function $g(x,y)$ is homogeneous of degree n if $g(tx,ty) = t^n g(x,y)$. Substitute tx, ty for x and y respectively, and using the rules of algebra, determine whether the above equation holds.

(a) $g(x,y) = xy + y^2$

$$g(tx,ty) = (tx)(ty) + (ty)^2 = t^2(xy) + t^2(y^2)$$

$$= t^2[xy + y^2] = t^2 g(x,y).$$

The given function is homogeneous of degree two.

(b) $f(x,y) = x + y \sin(y/x)^2$

$$f(tx,ty) = tx + [ty \sin(ty/tx)^2]$$

$$= tx + \left[ty \sin \frac{t^2}{t^2} \left(\frac{y^2}{x^2}\right)\right]$$

$$= tx + \left[ty \sin\left(\frac{y}{x}\right)^2\right] = t[f(x,y)].$$

The function f is homogeneous of degree one.

(c) $h(x,y) = x^3 + xy^2 e^{x/y}$

$$h(tx,ty) = t^3 x^3 + (txt^2 y^2 e^{tx/ty})$$

$$= t^3x^3 + t^3xy^2 e^{x/y} = t^3 [h(x,y)].$$

h is homogeneous of degree three.

(d) $p(x,y) = x + xy$

$p(tx,ty) = tx + (tx)(ty) = tx + t^2(xy).$

The original function $p(x,y) = x + xy$ cannot be isolated; therefore it is not homogeneous.

• PROBLEM 16-2

Give examples of homogeneous first order differential equations, i.e., equations of the form $M(x,y)dx + N(x,y)dy = 0$ such that when written as

$$\frac{dy}{dx} = f(x,y),$$

there exists a function g such that $f(x,y)$ can be expressed in the form $g(y/x)$.

Solution: Consider $(x^2 - 3y^2)dx + 2xy\, dy = 0.$

This equation is homogeneous. To show this, rewrite the equation in the form

$$\frac{dy}{dx} = f(x,y).$$

Thus,

$$\frac{dy}{dx} = \frac{3y^2 - x^2}{2xy} = \frac{3y}{2x} - \frac{x}{2y} = \frac{3}{2}(y/x) - \frac{1}{2}\left(\frac{1}{y/x}\right) = g(y/x)$$

as required.

Consider $y' = \frac{2x + y^2}{xy}$: Rewriting as $\frac{dy}{dx} = \frac{2x + y^2}{xy}$ the composite fraction can be broken down into $\frac{2x}{xy} + \frac{y^2}{xy} = \frac{2}{y} + y/x.$ Here g is not of the form $g(y/x)$ because of the term $\frac{2}{y}$. Hence $\frac{dy}{dx} = \frac{2x + y^2}{xy}$ is not a homogeneous equation. To check this, verify that $f(tx,ty) \neq f(x,y)$, i.e., $\frac{2(tx) + (ty)^2}{(tx)(ty)} \neq \frac{2x + y^2}{xy}.$

Consider $(y + \sqrt{x^2 + y^2})dx - xdy = 0.$

Rewriting as

$$\frac{dy}{dx} = \frac{y + \sqrt{x^2 + y^2}}{x}$$

596

one finds

$$\frac{dy}{dx} = \frac{y}{x} + \frac{\sqrt{x^2 + y^2}}{x} = \frac{y}{x} \pm \frac{\sqrt{x^2 + y^2}}{\sqrt{x^2}} = \frac{y}{x} \pm \sqrt{1 + (y/x)^2}$$

which is of the form $g(y/x)$. Thus the given equation is homogeneous.

● **PROBLEM 16-3**

A body of mass m, with initial velocity v_0, falls vertically. If the initial position is denoted s_0, determine the position and velocity of the body at the end of t seconds. First, assume the body is acted upon by gravity alone. Then, do the problem again assuming air resistance proportional to the square of the velocity.

Solution: Taking the positive s direction to be vertically downward, the force of gravity acting on the particle is mg. From Newton's second law

$$m \frac{dv}{dt} = mg \qquad (1)$$

which separates and integrates to be

$$v = gt + C_1 \qquad (2)$$

Evaluate C_1, using the initial condition $v(0) = v_0$. Therefore,

$$v = gt + v_0 \qquad (3)$$

Since $v = \frac{ds}{dt}$, a second integration gives

$$s = \frac{1}{2} gt^2 + v_0 t + C_2 \qquad (4)$$

The other condition, $s(0) = s$ leaves

$$s = \frac{1}{2} gt^2 + v_0 t + s_0 \qquad (5)$$

Equations (3) and (5) describe the motion for the first part of the problem.

Assume that a retarding force, proportional to the square of the velocity, say $- m k^2 v^2$, also acts upon the body. In this case Newton's second law, after dividing by m, leaves

$$\frac{dv}{dt} = g - k^2 v^2. \qquad (6)$$

Again, the variables separate. Hence,

$$\frac{dv}{g - k^2 v^2} = dt \qquad (7)$$

The integral of the left-hand side can be found in tables. Then

$$\frac{1}{k\sqrt{g}} \tanh^{-1} \frac{kv}{\sqrt{g}} = t + C_3 \tag{8}$$

or

$$v = \frac{\sqrt{g}}{k} \tanh k \sqrt{g} (t + C_3) \tag{9}$$

The condition $v(0) = (v_0)$ gives

$$C_3 = \frac{1}{k\sqrt{g}} \tanh^{-1} \frac{k v_0}{\sqrt{g}} \tag{10}$$

The solution comprised of (9) and (10) reaches a limit as t increases without bound. The hyperbolic tangent of a very large argument approaches one; therefore,

$$\lim_{t \to \infty} v = \frac{\sqrt{g}}{k} \tag{11}$$

This terminal velocity could also have been arrived at by examining equation (6). When the terminal velocity is reached, $\frac{dv}{dt} = 0$ and (6) gives

$$0 = g - k^2 v_T^2 \tag{12}$$

or $v_T = \frac{\sqrt{g}}{k}$

It remains to find $s(t)$. Equation (9), (substituting $\frac{ds}{dt}$ for v) becomes

$$\frac{ds}{dt} = \frac{\sqrt{g}}{k} \tanh k\sqrt{g} (t + C_3) \tag{14}$$

Again using tables, one finds

$$s = \frac{1}{k^2} \ln \cosh k \sqrt{g} (t + C_3) + C_2 \tag{15}$$

From $s(0) = s_0$ it can be shown that

$$C_2 = s_0 - \frac{1}{k^2} \ln \cosh k \sqrt{g} \, C_3 \tag{16}$$

The motion of a body in this resistive medium, air, is given by equations (9), (10), (15), and (16).

• **PROBLEM 16-4**

A parachutist equipped with parachute and other equipment falls from rest toward the earth. The total weight of the man plus equipment is 160 lb. Before the parachute opens, the air resistance (in pounds) is numerically equal to $\frac{1}{2}$ v, where v is the velocity (in feet per second). The parachute opens 5 sec after the fall begins; after it opens, the air resistance (in pounds) is numerically equal to $\frac{5}{8}$ v^2, where v is the velocity (in feet per second). Find the velocity of the parachutist (A) before the parachute opens, (B) after the parachute opens.

Solution: The problem requires two parts: the velocity before, and the velocity after the parachute opens.

First, find the velocity profile before the parachute opens. Use Newton's second law

$$F = (mv)' = m \frac{dv}{dt} \qquad (m \text{ is constant}).$$

Thus, find the net downward force acting on the parachutist.

His weight plus the equipment is 160 lb. and acts in a downward direction. The air-resistance is $\frac{1}{2}$ v and acts in an upward direction. Thus

$$F = mg - \frac{1}{2} v$$

where the downward direction has been taken as positive. Then

$$m \frac{dv}{dt} = mg - \frac{1}{2} v. \quad \text{Now } m = \frac{W}{g} = \frac{160}{32} = 5. \quad \text{Thus}$$

$$5 \frac{dv}{dt} = 160 - \frac{1}{2} v$$

subject to the initial condition $v(0) = 0$.

Before solving the equation, formulate the second part of the problem (after the parachute opens).

The forces now acting on the parachutist are the total weight acting downward which equals 160 lb. and the air-resistance which is now equal to $-\frac{5}{8} v_1^2$ (the sign is negative because the force is in an upward direction).

Thus $F = (mv_1)' = m \frac{dv_1}{dt} = 160 - \frac{5}{8} v_1^2$ by Newton's second law. Hence $5 \frac{dv_1}{dt} = 160 - \frac{5}{8} v_1^2$ subject to the initial condition $v_1(0) = v(5)$ in the first equation.

Proceeding to solve,

$5\dfrac{dv}{dt} = 160 - \dfrac{1}{2}v$. This is a separable equation whose solution is found by integrating

$$\dfrac{dv}{v - 320} = -\dfrac{1}{10}\,dt.$$

$$\ln(v - 320) = -\dfrac{1}{10}t + c$$

or, solving for v,

$$v = 320 + ce^{-t/10}.$$

To evaluate c, apply the initial condition $v(0) = 0$:

$$0 = 320 + c. \quad \text{Thus } c = -320.$$

Hence the solution to the first part of the problem is:

$$v = 320(1 - e^{-t/10})$$

which holds for $0 \le t \le 5$. At the end of this interval, i.e. when $t = 5$, the velocity of the parachutist is:

$$v(5) = 320(1 - e^{-1/2}) \cong 126 \text{ ft/sec}.$$

Now, proceeding to solve the second equation:

$$5\dfrac{dv_1}{dt} = 160 - \dfrac{5}{8}v_1^2.$$

Separating variables, $\dfrac{dv_1}{v_1^2 - 256} = -\dfrac{dt}{8}$.

Integrating, $\dfrac{1}{32}\ln\dfrac{v_1 - 16}{v_2 + 16} = -\dfrac{t}{8} + C_1$ (using the method of partial fractions)

or $\ln\dfrac{v_1 - 16}{v_1 + 16} = -4t + C_1$.

Taking exponents,

$$\dfrac{v_1 - 16}{v_1 + 16} = ce^{-4t}. \quad \text{Then, solving for } v_1,$$

$$v_1 = 16\dfrac{(ce^{-4t} + 1)}{1 - ce^{-4t}}.$$

To find c, apply the initial condition that $v_1(0) = v(5) \cong 126$.

Then $c = \dfrac{55}{71}e^{20}$ and so the velocity of the parachutist at any time t after his parachute has opened is given by:

$$v_1 = \frac{16\left[\frac{55}{71} e^{20-4t} + 1\right]}{1 - \frac{55}{71} e^{20-4t}} \quad ; \quad t \geq 5.$$

● **PROBLEM** 16-5

A 16-lb. weight is attached to the end of a spring suspended from a ceiling. It comes to rest in its equilibrium position, thereby stretching the spring 1 foot. The weight is then pulled down 6 inches below its equilibrium position and released at t = 0 with an initial velocity of 1 ft/sec, directed downward. Neglecting the resistance of the medium and assuming that no external forces are present, determine the amplitude, period and frequency of the resulting motion.

Solution: The general differential equation which must be solved for this type of problem is of the form

$$m \frac{d^2x}{dt^2} + kx = 0 \tag{1}$$

where m = mass, k = spring constant, x = x(t) is the displacement of the mass and is a function of t (the time elapsed).

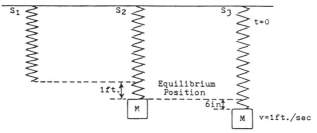

S_1: Spring, with no mass attached to it, at rest

S_2: Downward displacement is taken to be positive, upward to be negative

S_3: Mass is displaced 6 inches (note;+6inches) and given a positive initial velocity of 1ft./sec.

To solve for m and k, use F = mg. Thus the mass = $\frac{F}{g}$ =

$$\frac{16 \text{ lb}}{32 \text{ ft/sec}} = \frac{1}{2} \text{ lb/ft/sec} = \frac{1}{2} \text{ (slugs)} = m$$

Hooke's law is given by F = ks where s is the amount of elongation of the spring from its natural length. So here

16 lbs = k (1 ft)   or   k = 16 lbs/ft.

Thus $\frac{1}{2} \frac{d^2x}{dt^2} + 16x = 0$ or $\frac{d^2x}{dt^2} + 32x = 0$ \qquad (2)

One also has initial conditions, namely: $x(0) = \frac{1}{2}$ ft

$\left.\dfrac{dx}{dt}\right|_0 = 1$ ft/sec.

The auxiliary equation corresponding to equation (1) is: $m^2 + 32 = 0$. Using the quadratic formula to solve, one gets

$$\dfrac{-0 \pm \sqrt{0^2 - 4(1)(32)}}{2} = \dfrac{\pm i\sqrt{128}}{2} = \pm 4\sqrt{2}\, i = m$$

so that the solution is of the form

$$x(t) = C_1 \sin 4\sqrt{2}\, t + C_2 \cos 4\sqrt{2}\, t. \qquad (3)$$

Now solve for C_1 and C_2 by applying the initial conditions, namely

$$x(0) = \frac{1}{2} = C_1 \sin 0 + C_2 \cos 0 = C_2$$

and $\dfrac{dx}{dt} = 4\sqrt{2}\, C_1 \cos 4\sqrt{2}\,t - 4\sqrt{2}\, C_2 \sin 4\sqrt{2}\,t$

$\left.\dfrac{dx}{dt}\right|_0 = 1 = 4\sqrt{2}\, C_1$ or $C_1 = \dfrac{1}{4\sqrt{2}} = \dfrac{\sqrt{2}}{8}$.

Thus the solution of the differential equation (2) is

$$x(t) = \dfrac{\sqrt{2}}{8} \sin 4\sqrt{2}\,t + \dfrac{1}{2} \cos 4\sqrt{2}\,t. \qquad (4)$$

Now construct the right triangle ABC where $AB = \dfrac{1}{2}$, $BC = -\dfrac{\sqrt{2}}{8}$

$AC = \sqrt{\left[\dfrac{1}{2}\right]^2 + \left[-\dfrac{\sqrt{2}}{8}\right]^2}$

$AC = \dfrac{3}{8}\sqrt{2}$.

Note $\sin \theta = \dfrac{\text{Opp}}{\text{Hyp}} = \dfrac{-\dfrac{\sqrt{2}}{8}}{\dfrac{3}{8}\sqrt{2}} = -\dfrac{1}{3}$

$\cos \theta = \dfrac{\text{Adj}}{\text{Hyp}} = \dfrac{\dfrac{1}{2}}{\dfrac{3}{8}\sqrt{2}} = \dfrac{4}{3\sqrt{2}} = \dfrac{2\sqrt{2}}{3}$. $\qquad (5)$

Rewriting equation (4) (note, just multiplying by 1)

$$x(t) = \dfrac{3\sqrt{2}}{8}\left[\dfrac{1}{3}\sin 4\sqrt{2}\,t + \dfrac{2}{3}\sqrt{2}\cos 4\sqrt{2}\,t\right]$$

and noting $\cos(\alpha + \beta) = \cos\alpha \cos\beta - \sin\alpha \sin\beta$
one gets $x(t) = \frac{3\sqrt{2}}{8}(-\sin\theta \sin 4\sqrt{2}t + \cos\theta \cos 4\sqrt{2}t$

$$= \frac{3\sqrt{2}}{8} \cos(4\sqrt{2}t + \theta).$$

(6)

Here θ is determined by (5); namely

$$\tan\theta = \frac{-1}{2\sqrt{2}} \quad \text{or} \quad \theta = \tan^{-1}\left[\frac{-1}{2\sqrt{2}}\right] \approx -.35 \text{ radians.}$$

Thus equation (6) can be rewritten

$$x(t) = \frac{3\sqrt{2}}{8} \cos(4\sqrt{2}t - .35).$$

In general the solution to equation 1 is

$$x = C \cos\left[\sqrt{\frac{k}{m}} t + \theta\right]. \tag{7}$$

The constant C is defined to be the amplitude of the motion and it gives the maximum positive displacement of the mass from its equilibrium position. The motion is periodic, oscillating from $x = C$ to $x = -C$. If $x = C$, then by equation (7)

$$\sqrt{\frac{k}{m}} t + \theta = \pm 2n\pi \qquad \text{where } n = 0, 1, 2, \ldots$$
$$t > 0.$$

Transposing, one gets $t = \sqrt{\frac{m}{k}} (\pm 2n\pi - \theta).$ (8)

Now the time interval between two successive maxima is called the period of the motion. Thus, if

$$t_1 = \sqrt{\frac{m}{k}} (\pm 2n\pi - \theta) \quad \text{and} \quad t_2 = \sqrt{\frac{m}{k}} (\pm 2(n+1) - \theta)$$

$$t_2 - t_1 = \frac{2\pi}{\sqrt{\frac{k}{m}}} = \text{Period.}$$

The reciprocal of the period which gives the number of oscillations per second is called frequency of motion. Now looking at the problem, namely

$$x(t) = \frac{3\sqrt{2}}{8} \cos(4\sqrt{2}t - .35)$$

where the amplitude is $\frac{3\sqrt{2}}{8}$, the period is

$$\frac{2\pi}{\sqrt{\frac{k}{m}}} = \frac{2\pi}{\sqrt{\frac{16}{1/2}}} = \frac{\pi}{2\sqrt{2}} = \frac{\sqrt{2}\pi}{4}$$

and the frequency is $\dfrac{4}{\sqrt{2}\pi} = \dfrac{2\sqrt{2}}{\pi}$.

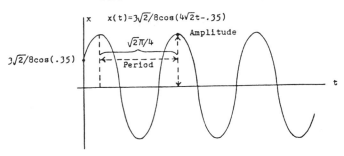

• **PROBLEM 16-6**

A hall of length L is attached to an auditorium filled with smoke at a concentration a_0. The hall is ventilated and a sufficient time has elapsed so that the concentration of smoke in the hall depends only on the distance x from the auditorium. Using the law of diffusion, determine the concentration of smoke, $y(x)$, in the hall.

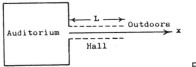

Fig.1

Solution: The law of diffusion states that the amount of smoke, $F(x)$, that passes through a unit area per unit time is proportional to the spatial rate of change of concentration, i.e.,

$$F(x) = -Dy'(x). \qquad (1)$$

The positive constant D is called the diffusion constant.

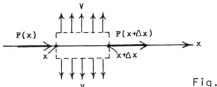

Fig.2

To find $y(x)$ consider an infinitesimal section of the hall, from x to $x + \Delta x$, as in Fig. 2. As the total amount of smoke must be conserved, one sees that

$\begin{bmatrix}\text{amount of smoke diffusing into}\\ \text{the section at x per unit time}\end{bmatrix}$

$- \begin{bmatrix}\text{amount of smoke diffusing out of the}\\ \text{section at } x + \Delta x \text{ per unit time}\end{bmatrix}$

$- \begin{bmatrix}\text{amount of smoke removed by ventila-}\\ \text{tion system per unit time}\end{bmatrix} = 0.$ \qquad (2)

Consider this ventilation system for a moment. Assume $v(x)$ is the amount of smoke removed at x and $v(x)$ is proportional to $y(x)$, the concentration at x. The total amount removed between x and $x + \Delta x$ is given by

$$\int_x^{x+\Delta x} v(z)\, dz = k \int_x^{x+\Delta x} y(z)\, dz \qquad (3)$$

where the positive constant k is called the absorption coefficient of the ventilation system. Take Δx to be of infinitesimal length--the integral can therefore be approximated by finding the concentration at the midpoint, and multiplying by Δx. Thus

$$k \int_x^{x+\Delta x} y(z)\, dz \approx k\, y\left[x + \frac{\Delta x}{2}\right] \Delta x. \qquad (4)$$

If the hall has a cross-sectional area A, express equation (2) in symbols, by applying the law of diffusion. One finds

$$-Dy'(x)\, A - [-Dy'(x + \Delta x) A] - ky\left[x + \frac{\Delta x}{2}\right]\Delta x = 0. \qquad (5)$$

Dividing through by Δx one gets

$$AD\left[\frac{y'(x+\Delta x) - y'(x)}{\Delta x}\right] - ky\left[x + \frac{\Delta x}{2}\right] = 0. \qquad (6)$$

Now shrink Δx. From (6), one obtains

$$AD \lim_{\Delta x \to 0}\left[\frac{y'(x+\Delta x) - y'(x)}{\Delta x}\right] - k \lim_{\Delta x \to 0}\left[y\left[x + \frac{\Delta x}{2}\right]\right]$$
$$= 0. \qquad (7)$$

Recognize the first term as the definition of the second derivative y". The other limit approaches $y(x)$. Thus,

$$y'' - \alpha^2 y = 0 \qquad (8)$$

where $\alpha = \left[\frac{k}{AD}\right]^{1/2}$.

The solution to equation (8), a second-order, linear equation with constant coefficients, can be expressed as

$$y = C_1 e^{-\alpha x} + C_2 e^{\alpha x}. \qquad (9)$$

The boundary conditions are assumed to be as follows: $y(0) = a_0$, and $y(L) = 0$. Substitution in (9) gives

$$a_0 = C_1 + C_2 \qquad (10)$$

$$0 = C_1 e^{-\alpha L} + C_2 e^{\alpha L} \qquad (11)$$

from which one finds that

$$C_1 = \frac{-e^{2\alpha L} a_0}{1 - e^{2\alpha L}} \tag{12}$$

and $$C_2 = \frac{a_0}{1 - e^{2\alpha L}}. \tag{13}$$

The particular solution to the boundary value problem is

$$y(x) = \frac{a_0}{1 - e^{2\alpha L}} [e^{\alpha x} - e^{2\alpha L - \alpha x}]. \tag{14}$$

The hyperbolic sine, defined as

$$\sinh t = \frac{e^t - e^{-t}}{2} \tag{15}$$

allows one to rewrite (14) as

$$y(x) = \frac{a_0 \sinh \alpha (L - x)}{\sinh \alpha L}. \tag{16}$$

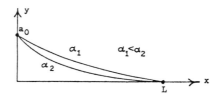

Fig.3

The solution is sketched in Fig. 3 for two values of α. One sees that the concentration drops off from the a_0 concentration in the auditorium to the smoke-free air outdoors. It is important to note that the decay rate depends mainly upon the absorption coefficient of the ventilation system. For larger values of k the concentration will drop off much more rapidly.

• **PROBLEM 16-7**

Find the solution of the suspended spring if the vibration is free and undamped.

Solution: A suspended spring is a spring hanging vertically with top end fixed and bottom end holding a mass m. Initially, the spring and mass are in equilibrium position. Then at t = 0, the spring is stretched by moving the mass to x_0, some positive distance below the equilibrium position. It is then released from rest. The initial conditions are thus:

$$x(0) = x_0, \qquad v(0) = 0 \tag{1}$$

where x(t) is the position of the mass at t, and

$v(t) = \frac{dx(t)}{dt}$ is its velocity.

If the vibration is free, then there is no external force, $F(t)$. If the vibration is undamped, then there is no air or medium resistance force. Thus, the only force acting on the system is the force of the spring, which by Hooke's law is $-kx$. By Newton's second law, the differential equation is

$$m \frac{d^2x}{dt^2} = -kx$$

or
$$\frac{d^2x}{dt^2} + \frac{k}{m} x = 0. \qquad (2)$$

Solve this by use of the characteristic equation. This equation comes from assuming a solution of the form

$$x = e^{\lambda t}.$$

Substituting in (2) one gets

$$\lambda^2 e^{\lambda t} + \frac{k}{m} e^{\lambda t} = 0,$$

or
$$\lambda^2 + \frac{k}{m} = 0,$$

$$\lambda = \pm \sqrt{\frac{-k}{m}}.$$

Since $k > 0$ and $m > 0$

$$\lambda = \pm i \sqrt{\frac{k}{m}}$$

where $i^2 = -1$. Thus the general solution is

$$x(t) = c_1 e^{i\sqrt{\frac{k}{m}} t} + c_2 e^{-i\sqrt{\frac{k}{m}} t}. \qquad (3)$$

Applying the initial conditions (1),

$$x(0) = c_1 e^0 + c_2 e^0$$

$$x_0 = c_1 + c_2. \qquad (4)$$

Also $\frac{dx}{dt}\Big|_{t=0} = i\sqrt{\frac{k}{m}} c_1 e^0 - i\sqrt{\frac{k}{m}} c_2 e^0,$

$$0 = i\sqrt{\frac{k}{m}} (c_1 - c_2).$$

Thus $c_1 = c_2$, and by (4) $c_1 = \frac{x_0}{2}$. The solution becomes

$$x(t) = \frac{x_0}{2} \left[e^{i\sqrt{\frac{k}{m}} t} + e^{-i\sqrt{\frac{k}{m}} t} \right]$$

However, $\cos\theta = \dfrac{e^{i\theta} + e^{-i\theta}}{2}$

by Euler's identity. Thus

$$x(t) = \dfrac{x_0}{2}\left[2\cos\sqrt{\dfrac{k}{m}}\,t\right],$$

$$x(t) = x_0 \cos\left(\sqrt{\dfrac{k}{m}}\,t\right). \tag{5}$$

x_0 is called the amplitude of the motion, and $\sqrt{\dfrac{k}{m}}$ the circular frequency. The natural frequency is defined as

$$f = \dfrac{1}{2\pi}\sqrt{\dfrac{k}{m}}$$

and the period is defined as $T = \dfrac{1}{f} = 2\pi\sqrt{\dfrac{m}{k}}$.

• **PROBLEM 16-8**

Solve the initial value problem

$$y'' + 2by' + y = k$$

where b and k are constants, $|b| < 1$ and $y(0) = 1$, $y'(0) = 0$.

Solution: The given equation

$$y'' + 2by' + y = k \tag{a}$$

is inhomogeneous, thus its solution will be the sum of the homogeneous solution and a particular solution.

The homogeneous equation

$$y'' + 2by' + y = 0 \tag{b}$$

has constant coefficients so assume a solution of the form $y = e^{px}$, substitute in (b) and produce the auxiliary equation

$$p^2 + 2bp + 1 = 0.$$

Solving by the quadratic formula

$$p = \dfrac{-2b \pm \sqrt{4b^2 - 4}}{2}$$

$$p = -b \pm \sqrt{b^2 - 1}.$$

Since $|b| < 1$, $b^2 - 1 < 0$ and

$$p = -b \pm i\sqrt{1 - b^2}$$

where i is the imaginary $\sqrt{-1}$.

The homogeneous solution is thus

$$y_h = c_1 e^{(-b + i\sqrt{1-b^2})x} + c_2 e^{(-b - i\sqrt{1-b^2})x} \qquad (c)$$

where c_1 and c_2 are arbitrary constants. A particular solution of (a) is seen by inspection to be

$$y_p = k.$$

The general solution is thus

$$y(x) = e^{-bx}(c_1 e^{iqx} + c_2 e^{-iqx}) + k \qquad (d)$$

where $q = \sqrt{1-b^2}$.

Now apply the initial conditions to (d)

$$y(0) = e^0 (c_1 e^0 + c_2 e^0) + k$$

$$1 = c_1 + c_2 + k \qquad (e)$$

$$y'(x) = -be^{-bx}(c_1 e^{iqx} + c_2 e^{-iqx})$$

$$+ e^{-bx}(iqc_1 e^{iqx} - iq c_2 e^{-iqx}),$$

$$y'(0) = -be^0 (c_1 e^0 + c_2 e^0) + iqe^0 (c_1 e^0 - c_2 e^0),$$

$$0 = -b(c_1 + c_2) + iq(c_1 - c_2). \qquad (f)$$

By (e)

$$c_1 = 1 - k - c_2.$$

Substituting in (f)

$$0 = -b(1 - k - c_2 + c_2) + iq(1 - k - 2c_2),$$

$$\frac{b(1-k)}{iq} = 1 - k - 2c_2,$$

$$c_2 = \frac{(1-k)}{2}\left[1 - \frac{b}{iq}\right].$$

and $\quad c_1 = (1-k)\left[1 - \frac{1}{2}\left[1 - \frac{b}{iq}\right]\right].$

The final solution is

$$y = e^{-bx}\frac{(1-k)}{2}\left[\left[1 - \frac{ib}{q}\right]e^{iqx} + \left[1 + \frac{ib}{q}\right]e^{-iqx}\right]$$

• **PROBLEM 16-9**

Consider a free damped motion of a mass m attached to a spring of force constant k. Let the coefficient of damping be a. Then show that the resulting motion of the system is completely determined by the quantity $a^2 - 4km$.

Solution: The differential equation of the system is derived from Newton's second law:

$$m \frac{d^2x}{dt^2} = -a \frac{dx}{dt} - kx$$

or $\quad m \frac{d^2x}{dt^2} + a \frac{dx}{dt} + kx = 0.$ \hfill (a)

The characteristic equation is obtained by assuming a solution of the form $x = e^{\lambda t}$. Substituting this in (a) and dividing by $e^{\lambda t}$ one obtains the characteristic equation:

$$m\lambda^2 + a\lambda + k = 0.$$

By the quadratic formula

$$\lambda = \frac{-a \pm \sqrt{a^2 - 4km}}{2m} \quad (b)$$

Consider the quantity which determines the resulting motion by

$$\Delta = a^2 - 4km.$$

Then $\lambda_1 = \frac{-a + \sqrt{\Delta}}{2m}, \qquad \lambda_2 = \frac{-a - \sqrt{\Delta}}{2m}$ \hfill (c)

Case I: $\Delta > 0$ or $a^2 > 4km$

In this case λ_1 and λ_2 are real numbers. The solution is

$$x(t) = c_1 e^{\lambda_1 t} + c_2 e^{\lambda_2 t} \quad (d)$$

where c_1 and c_2 are arbitrary constants. Note that λ_1 and λ_2 are negative. To show this, assume

$$\lambda_1 > 0;$$

then $\frac{-a + \sqrt{a^2 - 4km}}{2m} > 0,$

$$-a + \sqrt{a^2 - 4km} > 0$$

$$\sqrt{a^2 - 4km} > a > 0.$$

Squaring,

$$a^2 - 4km > a^2$$

$$-4km > 0.$$

This is a contradiction since km is positive and $-4km$ must therefore be negative. Consequently, the assumption that $\lambda_1 > 0$, is false, and $\lambda_1 < 0$. It may similarly be shown that $\lambda_2 < 0$. Thus, in solution (d)

$$\lim_{t \to \infty} \left(c_1 e^{\lambda_1 t} + c_2 e^{\lambda_2 t}\right) \to 0.$$

This is called overdamped motion.

Case II: $\Delta = 0$, $a^2 = 4km$

In this case λ_1 and λ_2 are real and equal. The general solution is

$$x(t) = c_1 e^{\lambda_1 t} + c_2 t\, e^{\lambda_2 t},$$

$$x(t) = e^{\lambda_1 t}(c_1 + c_2 t) \qquad (e)$$

where c_1 and c_2 are arbitrary constants.

Since $\lambda_1 = \dfrac{-a}{2m} < 0$,

$$\lim_{t \to \infty} e^{\lambda_1 t}(c_1 + c_2 t) = 0$$

and the system is said to be critically damped.

Case III: $\Delta < 0$, $a^2 < 4km$

In this case the roots λ_1 and λ_2 are complex:

$$\lambda_1 = \frac{-a + i\sqrt{-\Delta}}{2m}, \qquad \lambda_2 = \frac{-a - i\sqrt{-\Delta}}{2m}$$

where i is the complex number such that $i^2 = -1$.

The solution is then

$$x(t) = c_1 e^{\lambda_1 t} + c_2 e^{\lambda_2 t}, \qquad \text{or}$$

$$x(t) = e^{\frac{-a}{2m}} \left[c_1 e^{i \frac{\sqrt{-\Delta}}{2m}} + c_2 e^{-i \frac{\sqrt{-\Delta}}{2m}} \right] \qquad (f)$$

where c_1 and c_2 are arbitrary constants. By Euler's formula

$$e^{i\theta} = \cos\theta + i\sin\theta,$$

or $\quad e^{-i\theta} = \cos\theta - i\sin\theta.$

If $\theta = \frac{\sqrt{-\Delta}}{2m} t$

then (f) becomes

$$x(t) = e^{\frac{-a}{2m}} [(c_1 + c_2) \cos \theta + i(c_1 - c_2) \sin \theta]. \qquad (g)$$

If c_1 and c_2 are complex numbers they must be complex conjugates of each other. To see this, let $c_1 = \alpha + i\beta$. Then c_1 contains two arbitrary constants, α and β. Two arbitrary real constants are all that are required in the solution of a second order differential equation. Thus one may let $c_2 = \alpha - i\beta$ and still have two arbitrary constants. Hence

$$c_1 + c_2 = 2\alpha$$

and $c_1 - c_2 = 2i\beta$

Substituting into (g)

$$x(t) = e^{\frac{-a}{2m}} (2\alpha \cos \theta - 2\beta \cos \theta)$$

Let $2\alpha = A$ and $-2\beta = B$ to obtain the final form of the solution

$$x(t) = e^{\frac{-a}{2m}} \left[A \cos \left(\frac{\sqrt{-\Delta}}{2m} t \right) + B \cos \left(\frac{\sqrt{-\Delta}}{2m} t \right) \right]$$

This is damped oscillatory motion. Summarizing in a table:

If $a^2 - 4mk$ is	The Motion is
POSITIVE	OVER damped
ZERO	CRITICALLY damped
NEGATIVE	OSCILLATORY damped

● **PROBLEM 16-10**

Compute, to five decimals, $y(0.5)$, where $y(x)$ is the solution to the differential equation

$$y'' = -xy,$$

with initial conditions $y(0) = 1$, $y'(0) = 0$. Use the method of undetermined coefficients.

<u>Solution</u>: The solution cannot be simply expressed in terms of elementary functions. Substitute a series of

the form:

$$y(x) = \sum_{n=0}^{\infty} c_n x^n = c_0 + c_1 x + c_2 x^2 + \ldots$$

Differentiate twice:

$$y''(x) = \sum n(n-1) c_n x^{n-2}$$

$$= 2c_2 + 6c_3 x + 12 c_4 x^2 + \ldots$$

$$+ (m+2)(m+1) c_{m+2} x^m + \ldots$$

$$-xy(x) = -c_0 x - c_1 x^2 - c_2 x^3 - \ldots - c_{m-1} x^m - \ldots$$

Equate the coefficients of x^m in these series:

$$c_2 = 0, \quad (m+1)(m+2) c_{m+2} = -c_{m-1} \quad \text{for } m \geq 1.$$

From the initial conditions, it follows that $c_0 = 1$, $c_1 = 0$. Thus

$$c_3 = -\frac{c_0}{6} = -\frac{1}{6}, \quad c_6 = -\frac{c_3}{30} = \frac{1}{180},$$

$$c_9 = -\frac{c_6}{72} = -\frac{1}{12{,}960}, \ldots$$

$c_n = 0$, if n is not a multiple of 3.

Thus

$$y(x) = 1 - \frac{x^3}{6} + \frac{x^6}{180} - \frac{x^9}{12{,}960} + \ldots$$

$y(0.5) = 0.97925.$

The x^9-term is ignored, since it is less thant $2 \cdot 10^{-7}$.

EULER AND RUNGE - KUTTA METHODS

• **PROBLEM 16-11**

Working to only three decimal places, use Euler's method to tabulate approximate values of y(x) for x = 0, 0.1, 0.2, 0.3 for the differential equation y' = 1 + xy with an initial condition of y(0) = 1.

Solution: Consider the differential equation $dy/dx = y' = f(x,y)$ with initial condition (x_0, y_0). A rough approximation for y_1, that is the value of y at $x = x_1 =$

613

$x_0 + \Delta x$, can be found by adding to y_0 the change in y approximated by Δx times the slope y_0' at x_0. That is,

$$y_1 = y_0 + \Delta x(y_0')$$

It is noted that y' is obtainable for any value of x from the differential equation $y' = f(x,y)$. After computing y_1 and y_1', determine $y_2 = y(x = 2\Delta x)$ from

$$y_2 = y_1 + \Delta x(y_1')$$

For the nth value of y, one gets

$$y_n = y_{n-1} + \Delta x(y_{n-1}')$$

The above is the basic Euler method, but a better, modified Euler method is available by using the average value of y'. For example,

$$y_1 = y_0 + \Delta x(y_0' + y_1')/2$$

where y_1' will be available as soon as a first approximation $y_1^{(1)}$ of y_1 is found. Now consider the given problem. Superscripts are used to indicate first, second, and third approximations.

$$y' = 1 + xy, \quad x_0 = 0, \; y_0 = 1, \quad x_1 = x_0 + \Delta x = 0.1,$$

$$x_2 = 0.2, \quad x_3 = 0.3, \quad y_{n+1} = y_n + x(y_{avg}')$$

$$y_0' = 1 + (x_0)(y_0) = 1 + (0)(1) = 1$$

$$y_1^{(1)} = y_0 + \Delta x(y_0') = 1 + 0.1(1) = 1.1$$

$$y_1'^{(1)} = 1 + x_1 y_1^{(1)} = 1 + 0.1(1.1) = 1.11$$

$$y_1^{(2)} = y_0 + \Delta x(y_0' + y_1'^{(1)})/2 = 1 + 0.1(1 + 1.11)/2 = 1.106$$

$$y_1'^{(2)} = 1 + x_1 y_1^{(2)} = 1 + 0.1(1.106) = 1.111$$

$$y_1^{(3)} = y_0 + \Delta x(y_0' + y_1'^{(2)})/2 = 1 + 0.1(1 + 1.111)/2$$
$$= 1.106 = y_1$$

$$y_2^{(1)} = y_1 + \Delta x(y_1') = 1.106 + 0.1(1.111) = 1.217$$

$$y_2'^{(1)} = 1 + x_2 y_2^{(1)} = 1 + 0.2(1.217) = 1.243$$

$$y_2^{(2)} = 1.224, \quad y_2'^{(2)} = 1.245, \quad y_2^{(3)} = 1.224 = y_2, \text{ etc.}$$

The results are tabulated in the following table.

x	$y^{(1)}$	$y'^{(1)}$	$y^{(2)}$	$y'^{(2)}$	$y^{(3)}$
0	1	1			
0.1	1.1	1.11	1.106	1.111	1.106
0.2	1.217	1.243	1.224	1.245	1.224
0.3	1.348	1.404	1.356	1.407	1.357

● **PROBLEM 16-12**

Given the initial value problem

$$\frac{d^2y}{dt^2} + 2\frac{dy}{dt} + 4y = 0$$

$$y(0) = 2, \quad \frac{dy}{dt}(0) = 0$$

Solve and compare numerical solutions obtained by Euler and Runge-Kutta methods with the exact solution

$$y(t) = 2e^{-t}[\cos\sqrt{3}x + (1/\sqrt{3})\sin x].$$

in the range $0 \le t \le 5$. For Euler method, use step sizes $\Delta t = 0.1$, $\Delta t = 0.01$, $\Delta t = 0.001$, $\Delta t = 0.0001$, and for Runge-Kutta method, use $\Delta t = 0.1$.

Table 1

t	Euler				Runge-Kutta	Exact
	$\Delta t = 0.1$	$\Delta t = 0.01$	$\Delta t = 0.001$	$\Delta t = 0.0001$	$\Delta t = 0.1$	
0	2.00000	2.00000	2.00000	2.00000	2.00000	2.00000
0.5	1.35936	1.32204	1.31950	1.31784	1.31941	1.31940
1	0.185381	0.290310	0.300021	0.300537	0.301137	0.301149
1.5	− 0.429152	− 0.264378	− 0.250157	− 0.247868	− 0.248730	− 0.248709
2	− 0.400114	− 0.314608	− 0.306954	− 0.305134	− 0.306259	− 0.306245
2.5	− 0.121281	− 0.147718	− 0.148984	− 0.148551	− 0.149182	− 0.149181
3	+ 0.076168	+ 0.001436	− 0.003992	− 0.004495	− 0.004572	− 0.004579
3.5	0.107975	0.056026	0.051721	+ 0.051052	+ 0.051289	+ 0.051282
4	0.049308	0.043271	0.042095	0.041759	0.041989	0.041987
4.5	− 0.008092	0.013009	0.014018	0.014032	0.014131	0.014132
5	− 0.026704	− 0.005912	− 0.004490	− 0.004335	− 0.004342	− 0.004340

<u>Solution</u>: The problem is first converted to the first order system

$$\frac{dz}{dt} = -2z - 4y$$

$$z(0) = 0, \quad \frac{dy}{dt} = z, \quad y(0) = 2$$

Numerical solutions to this system by the Euler method for the step sizes specified above and by Runge-Kutta with $\Delta t = 0.1$ are given in Table 1 along with the exact solution $y(t) = 2e^{-t}[\cos\sqrt{3}x + (1/\sqrt{3})\sin x]$.

One purpose of this problem is to illustrate the difficulty of attempting to obtain accurate solutions with a low order method. Using a relatively coarse mesh size of $\Delta t = 0.1$, the fourth-order Runge-Kutta formula yields answers which are in error by no more than 2 x 10^{-5} at any point shown. In addition, since a relatively small number of steps (50) is required to cover the region of interest, it is reasonable to expect that the accuracy of these answers could be considerably improved by reducing the step size. However, the first-order Euler method cannot match this accuracy even with $\Delta t = 0.0001$ (50,000 steps to cover the range of interest). At some points in the interval $0 \leq t \leq 5$, an error of at least 1×10^{-3} is encountered for even the best Euler solutions, those for $\Delta t = 0.001$ and $\Delta t = 0.0001$. No results are given for smaller step sizes since the number of steps is already absurdly large compared to the number required for the fourth-order Runge-Kutta solution. It should be noted that the form of the differential equation is very simple, otherwise it is likely that roundoff error would be a serious problem for the small step size Euler solutions due to the enormous number of evaluations of $f(y,t)$ involved.

It is reasonable to conclude from this problem that the Euler method is simply too inaccurate to employ in practice.

• **PROBLEM 16-13**

Use the Runge-Kutta fourth-order equations given below to determine $y(0.1)$ and $y(0.2)$ for the differential equation $y' = x + y$, with initial condition $(0,1)$.

$$k_1 = hf(x_n, y_n)$$

$$k_2 = hf(x_n + h/2, y_n + k_1/2)$$

$$k_3 = hf(x_n + h/2, y_n + k_2/2)$$

$$k_4 = hf(x_n + h, y_n + k_3)$$

$$\Delta y = (k_1 + 2k_2 + 2k_3 + k_4)/6$$

Solution: Consider the differential equation $y' = f(x,y)$ with initial values (x_0, y_0) and with $\Delta x = h$. One wants to find y_1, the value of y at $x = x_1 = x_0 + h$, and then y_2, the value of y at $x = x_2 = x_0 + 2h$. Runge-Kutta formulas are formulas for obtaining the change in y, i.e. Δy, using various values of the function $f(x,y)$ which is in the differential equation $(y' = f)$. The most often used Runge-Kutta formulas are the fourth-order formulas (for determining $y_{n+1} = y_n + \Delta y$) given in the problem.

Starting with the initial values (x_0, y_0) these formulas would first be used (with $n = 0$) to find the four k_i's, the Δy, and then y_1 from $y_1 = y_0 + \Delta y$. After that cycle of calculations is completed, the formulas would be used again (with $n = 1$) to determine $y_2 = y_1 + \Delta y$; this would, of course, be a different value of Δy. This procedure can be continued to be used and a tabulation can be made for several values of $y(x)$. For the given problem one gets

$$y' = f(x,y) = x + y; \quad x_0 = 0, \; y_0 = 1; \quad \Delta x = h = 0.1$$

$$k_1 = hf(x_0, y_0) = 0.1(x_0 + y_0) = 0.1(0 + 1) = 0.1$$

$$k_2 = hf(x_0 + h/2, y_0 + k_1/2) = 0.1(0 + 0.05 + 1 + 0.05)$$
$$= 0.1(0.05 + 1.05) = 0.11$$

$$k_3 = hf(x_0 + h/2, y_0 + k_2/2) = 0.1(0 + 0.05 + 1 + 0.055)$$
$$= 0.1(0.05 + 1.055) = 0.1105$$

$$k_4 = hf(x_0 + h, y_0 + k_3) = 0.1(0 + 0.1 + 1 + 0.1105)$$
$$= 0.1(0.1 + 1.1105) = 0.12105$$

$$\Delta y = (k_1 + 2k_2 + 2k_3 + k_4)/6 = (0.1 + 0.22 + 0.221 + 0.12105)/6 = 0.110\,341\,7$$

$$y_1 = y(0.1) = y_0 + \Delta y = 1 + 0.110\,341\,7 \doteq 1.110\,342$$

Starting now with $x_1 = 0.1$ and $y_1 = 1.110\,342$, one gets

$$k_1 = hf(x_1, y_1) = 0.1(x_1 + y_1) = 0.1(1.210\,342)$$
$$= 0.121\,034\,2$$

$$k_2 = hf(x_1 + h/2, y_1 + k_1/2) = 0.1(0.15 + 1.170859)$$
$$= 0.132\,085\,9$$

$$k_3 = hf(x_1 + h/2, y_1 + k_2/2) = 0.1(0.15 + 1.176385)$$
$$= 0.132\,638\,5$$

$$k_4 = hf(x_1 + h, y_1 + k_3) = 0.1(0.2 + 1.2429805)$$
$$= 0.144\,298\,0$$

$$\Delta y = (k_1 + 2k_2 + 2k_3 + k_4)/6 = (0.794781)/6 = 0.132\,464$$

$$y_2 = y(0.2) = y_1 + \Delta y = 1.110\,342 + 0.132464 = 1.242\,806$$

The exact solution to the above problem is $y = 2e^x - x - 1$. The above results for y_1 and y_2 are correct to six decimal places.

● **PROBLEM 16-14**

Show that a Runge-Kutta method of order three should be partially unstable above a step size $h > 2.512/|p|$ when solving the equation $y' = py$, $p < 0$.

Solution: The solution y_r predicted by an RK method of order three is

617

$$y_r \simeq y_0 e^{rph} \simeq y_0 (1 + ph + p^2h^2/2 + p^3h^3/6)^r.$$

For the solution to be exponentially decreasing, the bracketed term must be less than 1 in magnitude. When ph = -2.512, the absolute value of this term is 0.9998. When ph = -2.513, it is 1.0004. Hence partial instability occurs when $h > 2.512/|p|$ (actually 2.512745).

● **PROBLEM 16-15**

Find the values of y corresponding to the values 0.1, 0.2, . . ., 0.5 of x which form a pointwise solution of $y' = x + y^2$ with the initial condition $y = 1$ when $x = 0$, using Runge's and Kutta's formulas. Discuss what can be done to improve the solutions obtained by Runge-Kutta formulas.

Table 1

x	y	F(x,y)		hF(X,Y)	
x_0	y_0	$F(x_0,y_0)$	k_1	$\frac{1}{2}(k_1 + k_4)$	
$x_0 + \frac{1}{2}h$	$y_0 + \frac{1}{2}k_1$	$F(x_0 + \frac{1}{2}h, y_0 + \frac{1}{2}k_1)$	k_2	$k_2 + k_3$	
$x_0 + \frac{1}{2}h$	$y_0 + \frac{1}{2}k_2$	$F(x_0 + \frac{1}{2}h, y_0 + \frac{1}{2}k_2)$	k_3	$3k = $ sum	
$x_0 + h$	$y_0 + k_3$	$F(x_0 + h, y_0 + k_3)$	k_4	$k = \frac{1}{3}$ sum	

Solution: The work, using Runge's formulas, is conveniently arranged as in Table 1.

The numbers in the first column are entered since all are known; also the next two entries in the first row. The value of k_1 is found from

$$k_1 = hF(x_m, y_m)$$
$$k_2 = hF(x_m + \tfrac{1}{2}h, y_m + \tfrac{1}{2}k_1)$$
(Runge's formulas) $\quad k_3 = hF(x_m + \tfrac{1}{2}h, y_m + \tfrac{1}{2}k_2)$
$$k_4 = hF(x_m + h, y_m + k_3),$$
$$k = \tfrac{1}{6}(k_1 + 2k_2 + 2k_3 + k_4);$$

and the next two numbers are entered in the second row. The value of k_2 is found and the next two numbers entered in the third row. Then k_3 and the next two numbers in the last row are entered. The k_i's are entered in the fourth column and then k is found as indicated in the last column.

The "sum" refers to the sum of the two entries in Table 1. The value of y_1 is $y_0 + k$; Table 1 is then continued by the same procedure until the value of the last y is found.

The entries for this problem are:

x	y	$x + y^2$	$h(x + y^2)$	
0	1	1	0.1	0.11736 848
0.05	1.05	1.1525	0.11525	0.23210 706
0.05	1.05762 5	1.16857 06	0.11685 706	0.34947 554
0.1	1.11685 706	1.34736 97	0.13473 697	0.11649 185
0.1	1.11649 185	1.34655 41	0.13465 541	0.15849 050
0.15	1.18381 956	1.55142 88	0.15514 288	0.31272 159
0.15	1.19406 329	1.57578 71	0.15757 871	0.47121 209
0.2	1.27407 056	1.82325 58	0.18232 558	0.15707 070
0.2	1.27356 255	1.82196 16	0.18219 616	0.21692 072
0.25	1.36466 063	2.11229 86	0.21122 986	0.42644 291
0.25	1.37917 748	2.15213 05	0.21521 305	0.64336 363
0.3	1.48877 560	2.51645 28	0.25164 528	0.21445 454
0.3	1.48801 709	2.51419 49	0.25141 949	0.30601 787
0.35	1.61372 684	2.95411 43	0.29541 143	0.59797 034
0.35	1.63572 281	3.02558 91	0.30255 891	0.90398 821
0.4	1.79057 600	3.60616 24	0.36061 624	0.30132 940
0.4	1.78934 649	3.60176 09	0.36017 609	0.45522 086
0.45	1.96943 453	4.32867 24	0.43286 724	0.88018 262
0.45	2.00578 011	4.47315 38	0.44731 538	1.33540 348
0.5	2.23666 187	5.50265 63	0.55026 563	0.44513 449
0.5	2.23448 098			

The table for the solution by use of Kutta's formulas is similar:

Table 2

x	y	F(x,y)		hF(x,y)	
x_0	y_0	$F(x_0, y_0)$		k_1	$k_1 + k_4$
$x_0 + \frac{1}{3}h$	$y_0 + \frac{1}{3}k_1$	$F(x_0 + \frac{1}{3}h, y_0 + \frac{1}{3}k_1)$		k_2	$3(k_2 + k_3)$
$x_0 + \frac{2}{3}h$	$y_0 - \frac{1}{3}k_1 + k_2$	$F(x_0 + \frac{2}{3}h, y_0 - \frac{1}{3}k_1 + k_2)$		k_3	sum
$x_0 + h$	$y_0 + k_1 - k_2 + k_3$	$F(x_0 + h, y_0 + k_1 - k_2 + k_3)$		k_4	$k = \frac{1}{8}(\text{sum})$

In detail:

x	y	$x + y^2$	$h(x + y^2)$	
0	1	1	0.1	0.23376 576
0.03333 333	1.03333 333	1.10111 11	0.11011 111	0.69816 846
0.06666 667	1.07677 778	1.22611 71	0.12261 171	0.93193 422
0.1	1.11250 060	1.33765 76	0.13376 576	0.11649 178
0.1	1.11649 178	1.34655 39	0.13465 539	0.31553 845
0.13333 333	1.16137 691	1.48212 97	0.14821 297	0.94102 689
0.16666 667	1.21981 962	1.65462 66	0.16546 266	1.25656 534
0.2	1.26839 686	1.80883 06	0.18088 306	0.15707 067
0.2	1.27356 245	1.82196 13	0.18219 613	0.43154 802
0.23333 333	1.33429 449	2.01367 51	0.20136 751	1.28408 925
0.26666 667	1.41419 792	2.26662 24	0.22666 224	1.71563 727
0.3	1.48105 331	2.49351 89	0.24935 189	0.31445 466
0.3	1.48801 711	2.51419 49	0.25141 949	0.60808 045
0.33333 333	1.57182 361	2.80396 28	0.28039 628	1.80255 894
0.36666 667	1.68460 689	3.20456 70	0.32045 670	2.41063 939
0.4	1.77949 702	3.56660 96	0.35666 096	0.30132 992
0.4	1.78934 703	3.60176 28	0.36017 628	0.90290 596
0.43333 333	1.90940 579	4.07916 38	0.40791 638	2.65818 288
0.46666 667	2.07720 465	4.78144 58	0.47814 458	3.56108 884
0.5	2.21975 151	5.42729 68	0.54272 968	0.44513 611
0.5	2.23448 314			

Note that the two values obtained for y_5 differ by about 2 units in the sixth decimal place. The end result is this: if y_{m+2} is obtained in two steps from y_m and y_{m+1} by means of Runge's, Kutta's, or similar formulas, and if y^*_{m+2} is obtained from y_m in one step, that is, by use of an interval of width 2h, then the value of y_{m+2} will usually be improved by adding to it the quantity

$$\frac{1}{15} (y_{m+2} - y^*_{m+2}).$$

One-half of this quantity can be added to y_{m+1} to improve its value.

The use of the correcting term is illustrated by doing the problem over again. The entries in the first eight rows of Table 3 are copied from Table 1 (Runge's method). Then compute the first four rows of Table 4 using the double interval 2h = 0.2. The value of y_2 = 1.27356 255 which was previously found is then improved by adding to it the correcting term

$$\frac{1}{15}(y_2 - y_2^*) = \frac{1}{15}(1.27356\ 255 - 1.27353\ 566)$$

$$= 0.00000\ 179.$$

The corrected value of y_2, namely 1.27356 434, is used in the next part of Table 3 and the work is continued as before.

Table 3

x	y	$x + y^2$	$h(x + y^2)$	
0	1	1	0.1	0.11736 848
0.05	1.05	1.1525	0.11525	0.23210 706
0.05	1.05762 5	1.16857 06	0.11685 706	0.34947 554
0.1	1.11685 706	1.34736 97	0.13473 697	0.11649 185
0.1	1.11649 185	1.34655 41	0.13465 541	0.15849 050
0.15	1.18381 956	1.55142 88	0.15514 288	0.31272 159
0.15	1.19406 329	1.57578 71	0.15757 871	0.47121 209
0.2	1.27407 056	1.82325 58	0.18232 558	0.15707 070
0.2	1.27356 255			
	0.00000 179 (correcting term)			
0.2	1.27356 434	1.82196 61	0.18219 661	0.21692 130
0.25	1.36466 265	2.11230 41	0.21123 041	0.42644 403
0.25	1.37917 955	2.15213 62	0.21521 362	0.64336 533
0.3	1.48877 796	2.51645 98	0.25164 598	0.21445 511
0.3	1.48801 945	2.51420 19	0.25142 019	0.30601 880
0.35	1.61372 954	2.95412 30	0.29541 230	0.59797 212
0.35	1.63572 560	3.02559 82	0.30255 982	0.90399 092
0.4	1.79057 927	3.60617 41	0.36061 741	0.30133 031
0.4	1.78934 976			
	0.00001 155 (correcting term)			
0.4	1.78936 131	3.60181 39	0.36018 139	0.45522 847
0.45	1.96945 201	4.32874 12	0.43287 412	0.88019 683
0.45	2.00579 837	4.47322 71	0.44732 271	1.33542 530
0.5	2.23668 402	5.50275 54	0.55027 554	0.44514 177
0.5	2.23450 308			

Table 3A

x	y	$x + y^2$	$h(x + y^2)$	
0	1	1	0.2	0.28277 478
0.1	1.1	1.31	0.262	0.53783 220
0.1	1.131	1.37916 10	0.27583 220	0.82060 698
0.2	1.27583 22	1.82774 78	0.36554 956	0.27353 566
0.2	1.27353 566			
0.2	1.27356 434	1.82196 61	0.36439 322	0.54364 756
0.3	1.45576 095	2.41923 99	0.48384 798	1.00318 896
0.3	1.51548 833	2.59670 49	0.51934 098	1.54683 652
0.4	1.79290 532	3.61450 95	0.72290 190	0.51561 217
0.4	1.78917 651			

• **PROBLEM 16-16**

Consider the initial value problem

$$y'' + y' + y = 1 + x, \quad y(0) = 0, \quad y'(0) = 1,$$

which, in system form, can be reformulated as

$y' = v$

$v' = 1 + x - y - v$

$y(0) = 0$, $v(0) = 1$.

Thus,

$$x_0 = 0, \quad y_0 = 0, \quad v_0 = 1$$

and

$$f(x,y,v) = 1 + x - y - v.$$

For $h = 0.5$, show how to generate y_1, v_1, y_2 and v_2 by using Kutta's Formula (1)-(5):

$$y_{i+1} = y_i + hv_i + \frac{h}{6}(m_0 + m_1 + m_2) \quad (1)$$

$$v_{i+1} = v_i + \frac{1}{6}(m_0 + 2m_1 + 2m_2 + m_3) \quad$$

$$m_0 = hf(x_i, y_i, v_i) \quad (2)$$

$$m_1 = hf(x_1 + \frac{h}{2}, y_i + \frac{v_i h}{2}, v_i + \frac{m_0}{2}) \quad (3)$$

$$m_2 = hf(x_i + \frac{h}{2}, y_i + \frac{v_i h}{2} + \frac{m_0 h}{4}, v_i + \frac{m_1}{2}) \quad (4)$$

$$m_3 = hf(x_{i+1}, y_i + v_i h + \frac{m_1 h}{2}, v_i + m_2). \quad (5)$$

Solution: First, let $i = 0$ in (2)-(5) so that

$$m_0 = (0.5)f(0,0,1) = 0$$

$$m_1 = (0.5)f(\tfrac{1}{4},\tfrac{1}{4},1) = 0$$

$$m_2 = (0.5)f(\tfrac{1}{4},\tfrac{1}{4},1) = 0$$

$$m_3 = (0.5)f(\tfrac{1}{2},\tfrac{1}{2},1) = 0.$$

From (1) it follows that

$$y_1 = 0 + (\tfrac{1}{2})(1) + \tfrac{1}{12}(0 + 0 + 0) = \tfrac{1}{2}$$

$$v_1 = 1 + \tfrac{1}{6}(0 + 2\cdot 0 + 2\cdot 0 + 0) = 1.$$

Next, let $i = 2$ in (2)-(5), so that

$$m_0 = (0.5)f(\tfrac{1}{2},\tfrac{1}{2},1) = 0$$

$$m_1 = (0.5)f(\tfrac{3}{4},\tfrac{3}{4},1) = 0$$

$m_2 = (0.5) f(\frac{3}{4}, \frac{3}{4}, 1) = 0$

$m_3 = (0.5) f(1, 1, 1) = 0.$

From (1), it follows finally that

$y_2 = \frac{1}{2} + \frac{1}{2}(1) + \frac{1}{12}(0 + 0 + 0) = 1$

$v_2 = 1 + \frac{1}{6}(0 + 2 \cdot 0 + 2 \cdot 0 + 0) = 1.$

Note, incidentally, that the approximations y_1 and y_2 coincide precisely, at the grid points, with the solution $y = x$ of the given, rather trivial, initial value problem only because second and higher order derivatives of this exact, continuous solution are all identically zero.

● **PROBLEM 16-17**

Solve the initial value problem

$y'' = (1 + t^2)y$, $y(0) = 1$, $y'(0) = 0$, $t \in [0,1]$

by the Runge-Kutta method with $h = 0.1$.

Table 1

t_n	y_n	y'_n	$y(t_n)$	$y'(t_n)$
0	1	0	1	0
0.1	1.0050167	0.100501	1.0050125	0.100501
0.2	1.0202098	0.204038	1.0202013	0.204040
0.3	1.0460407	0.313802	1.0460279	0.313808
0.4	1.0833046	0.433303	1.0832871	0.433315
0.5	1.1331710	0.566554	1.1331485	0.566574
0.6	1.1972453	0.718298	1.1972174	0.718330
0.7	1.2776552	0.894286	1.2776213	0.894335
0.8	1.3771681	1.101629	1.3771278	1.101702
0.9	1.4993498	1.349266	1.4993030	1.349372
1.0	1.6487762	1.648568	1.6487213	1.648722

Solution of $y'' = (1+t^2)y$, $y(0) = 1$, $y'(0) = 0$ by the Runge-Kutta Method with $h = 0.1$

<u>Solution</u>: For $n = 0$

$t_0 = 0$, $y_0 = 1$, $y_0' = 0$

$K_1 = \frac{h^2}{2} f(t_0, y_0) = \frac{(.1)^2}{2}(1 + 0) 1 = .005$

$K_2 = \frac{h^2}{2} f(t_0 + \frac{2}{3}h, y_0 + \frac{2}{3}h y'_0 + \frac{4}{9}K_1)$

623

$$= \frac{(.1)^2}{2} (1 + \frac{4}{9}(.1)^2) (1 + \frac{2}{3}(.1)0 + \frac{4}{9}(.005))$$

$$= .0050333827$$

$$y_1 = y_0 + hy'_0 + \frac{1}{2}(K_1 + K_2)$$

$$= 1 + 0 + \frac{1}{2}(.005 + .0050333827)$$

$$= 1.0050167$$

$$y'_1 = 0 + \frac{1}{2(.1)}(.005 + .0151001481) = 0.10050074$$

The exact solution is given by

$$y(t) = e^{t^3/2}$$

The computed solution is listed in Table 1.

● **PROBLEM** 16-18

Solve the initial value problem

$x' = y$, $x(0) = 0$

$y' = -x$, $y(0) = 1$, $t \in [0, 1]$,

by second order Runge-Kutta method with $h = 0.1$.

Table 1

t_n	x_n	y_n	$x(t_n)$	$y(t_n)$
0	0	1	0	1
0.1	0.1	0.995	0.099833	0.995005
0.2	0.1990	0.980025	0.198669	0.980067
0 3	0.296008	0.955225	0.295520	0.955336
0.4	0.390050	0.920848	0.389418	0.921061
0.5	0.480185	0.877239	0.479426	0.877583
0.6	0.565507	0.824834	0.564642	0.825336
0.7	0.645163	0.764159	0.644218	0.764842
0.8	0.718353	0.695822	0.717356	0.696707
0.9	0.784344	0.620508	0.783327	0.621610
1.0	0.842473	0.538971	0.841471	0.540302

Solution of $x' = y$, $y' = -x$, $x(0) = 0$, $y(0) = 1$
by Second Order Method with $h = 0.1$

Solution: For n = 0

$t_0 = 0$, $x_0 = 0$, $y_0 = 1$

$K_{11} = hf_1(t_0, x_0, y_0) = .1(1) = .1$

$K_{21} = hf_2(t_0, x_0, y_0) = .1(0) = 0$

$K_{12} = hf_1(t_0 + h, x_0 + K_{11}, y_0 + K_{21}) = .1(1 + 0) = .1$

$K_{22} = hf_2(t_0 + h, x_0 + K_{11}, y_0 + K_{21}) = .1(0 - .1) = -.01$

$x_1 = x_0 + \frac{1}{2}(K_{11} + K_{12}) = 0 + \frac{1}{2}(.1 + .1) = .1$

$y_1 = y_0 + \frac{1}{2}(K_{21} + K_{22}) = 1 + \frac{1}{2}(0 - .01) = 1 - .005 = .995$

For n = 1

$t_1 = .1$, $x_1 = .1$, $y_1 = .995$

$K_{11} = hf_1(t_1, x_1, y_1) = .1(.995) = .0995$

$K_{21} = hf_2(t_1, x_1, y_1) = .1(-.1) = -.01$

$K_{12} = hf_1(t_1 + h, x_1 + K_{11}, y_1 + K_{21}) = .1(.995 - .01)$
$= .0985$

$K_{22} = hf_2(t_1 + h, x_1 + K_{11}, y_1 + K_{21})$
$= .1[-(.1 + .0995)]$
$= -.01995$

$x_2 = x_1 + \frac{1}{2}(K_{11} + K_{12})$

$= .1 + \frac{1}{2}(.0995 + .0985)$

$= .1990$

$y_2 = y_1 + \frac{1}{2}(K_{21} + K_{22})$

$= .995 + \frac{1}{2}(-.01 - .01995)$

= .980025.

The exact solution is given by

x(t) = sin t, y(t) = cos t

The computed solution is listed in Table 1.

• **PROBLEM 16-19**

Solve the initial value problem

y' = t + y, y(0) = 1, t ∈ [0,1]

by classical fourth order Runge-Kutta method with h = .1.

Table 1

t_n	y_n	$y(t_n)$
0	1	1
0.1	1.11034167	1.11034184
0.2	1.24280514	1.24280552
0.3	1.39971699	1.39971762
0.4	1.58364848	1.58364940
0.5	1.79744128	1.79744254
0.6	2.04423592	2.04423760
0.7	2.32750325	2.32750542
0.8	2.65107913	2.65108186
0.9	3.01920283	3.01920622
1.0	3.43655949	3.43656366

Solution of y' = t+y, y(0) = 1, by Classical Fourth Order Method with h = 0.1

Solution: For n = 0

$t_0 = 0$, $y_0 = 1$

$K_1 = hf(t_0, y_0) = (.1)(0 + 1) = .1$

$K_2 = hf\left(t_0 + \frac{h}{2}, y_0 + \frac{K_1}{2}\right)$

$= (.1)\left[0 + \frac{.1}{2} + \left(1 + \frac{.1}{2}\right)\right] = .11$

$K_3 = hf\left(t_0 + \frac{h}{2}, y_0 + \frac{K_2}{2}\right)$

$= (.1)\left[0 + \frac{.1}{2} + \left(1 + \frac{.11}{2}\right)\right] = .1105$

$$K_4 = hf(t_0 + h, y_0 + K_3)$$
$$= (.1)[(0 + .1) + (1 + .1105)] = .12105$$
$$y_1 = 1 + \frac{1}{6}[1 + .22 + .2210 + .12105]$$
$$= 1.11034167$$

For n = 1

$$t_1 = .1, \quad y_1 = 1.11034167$$
$$K_1 = hf(t_1, y_1)$$
$$= (.1)[.1 + 1.11034167] = .121034167$$

$$K_2 = hf\left(t_1 + \frac{h}{2}, y_1 + \frac{K_1}{2}\right)$$

$$= (.1)\left[\left(.1 + \frac{.1}{2}\right) + \left(1.11034167 + \frac{1}{2}(.121034167)\right)\right]$$
$$= .132085875$$

$$K_3 = hf\left(t_1 + \frac{h}{2}, y_1 + \frac{1}{2}K_2\right)$$

$$= (.1)\left[\left(.1 + \frac{.1}{2}\right) + \left(1.11034167 + \frac{1}{2}(.132085875)\right)\right]$$
$$= .132638461$$

$$K_4 = hf(t_1 + h, y_1 + K_3)$$
$$= (.1)[(.1 + .1) + (1.11034167 + .132638461)]$$
$$= .144303013$$

$$y_2 = 1.11034167 + \frac{1}{6}[.121034167 + 2(.132085875)$$
$$+ 2(.132638461) + .144303013]$$
$$= 1.24280514$$

The exact solution is

$$y(t) = 2e^t - t - 1.$$

The computed solution is given in Table 1.

627

• **PROBLEM 16-20**

Find the values of x and y corresponding to the values t = 1.1, 1.2, . . ., 1.5 that are a point-wise solution of

$$\frac{dx}{dt} = \frac{1}{2}(y - 1)$$

$$\frac{dy}{dt} = \frac{2t}{3x}$$

with the initial condition x = 2/3, y = 3 when t = 1.

Solve by:

(a) infinite series

(b) Picard's method

(c) Runge - Kutta formulas

(d) Adams' method (use previous method's results for starting values for variables)

(e) advancing and correcting formulas A and C.

A: $y_{n+1} = y_{n-3} + \frac{4}{3}h(2y'_{n-2} - y'_{n-1} + 2y'_n)$

C: $y_{n+1} = y_{n-1} + \frac{1}{3}h(y'_{n-1} + 4y'_n + y'_{n+1})$; use Runge-Kutta solution's results for starting values for variables.

Solution: (a) Assume that a solution exists of the form

$$x = a_0 + a_1(t - 1) + a_2(t - 1)^2 + \ldots,$$

$$y = b_0 + b_1(t - 1) + b_2(t - 1)^2 + \ldots,$$

with

$$a_n = \frac{x^{(n)}(1)}{n!}, \quad b_n = \frac{y^{(n)}(1)}{n!}.$$

One gets from the first of the given differential equations

$$x^{(n)} = \frac{1}{2}y^{(n-1)}$$

for values of n greater than unity; one gets from the second equation

$$y' = \frac{2}{3}tx^{-1}$$

$$y'' = \frac{2}{3} - tx^{-2}x' + x^{-1}$$

$$y^{(3)} = \frac{2}{3} - tx^{-2}x'' + 2tx^{-3}(x')^2 - 2x^{-2}x'.$$

One finds on making use of the initial conditions that at $t = 1$,

$$x' = 1 \qquad y' = 1$$

$$x'' = \frac{1}{2} \qquad y'' = -\frac{1}{2}$$

$$x^{(3)} = -\frac{1}{4} \qquad y^{(3)} = \frac{3}{4}$$

$$x^{(4)} = \frac{3}{8}.$$

Hence

$$x = \frac{2}{3} + (t-1) + \frac{1}{4}(t-1)^2 - \frac{1}{24}(t-1)^3$$

$$+ \frac{1}{64}(t-1)^4 + \ldots ,$$

$$y = 3 + (t-1) - \frac{1}{4}(t-1)^2 + \frac{1}{32}(t-1)^3 + \ldots$$

The desired values of x and y are obtained by substituting the given values of t into the infinite series; they are listed in Table 1.

Table 1

t	x	y
1.0	0.667	3.000
1.1	0.769	3.098
1.2	0.874	3.190
1.3	0.988	3.277
1.4	1.104	3.362
1.5	1.225	3.441

(b) Put the differential equations into the forms

$$x = \frac{2}{3} + \int_1^t \frac{1}{2}(y-1)\, dt$$

$$y = 3 + \int_1^t \frac{2t}{3x}\, dt. \tag{1}$$

Use the initial value 3 for y, substitute into the integrand of the first equation of (1) and perform the integration; a first approximation for x is obtained,

$$x = t - \frac{1}{3}.$$

Similarly, substitute the initial value 2/3 for x into the integrand of the second equation of (1), integrate, and obtain

$$y = \frac{t^2}{2} + \frac{5}{2}$$

as the first approximation for y. Substituting this approximation for y into the integrand of the first equation of (1) and carrying out the integration, a second approximation for x is obtained

$$x = \frac{t^3}{12} + \frac{3t}{4} - \frac{1}{6}.$$

Substituting the first approximation for x into the integrand of the second equation of (1) and integrating, obtain a second approximation for y

$$y = 3 + \frac{2}{3}\left(t - 1 + \frac{1}{3}\ln\frac{3t-1}{2}\right).$$

The second approximations yield the following values for x and y:

t	x	y
1.0	0.667	3.000
1.1	0.769	3.098
1.2	0.877	3.192
1.3	0.991	3.283
1.4	1.112	3.371
1.5	1.240	3.458

These values agree fairly well with the first set of values. It is apparent, however, that they are beginning to diverge.

(c)   The Runge-Kutta formulas

$$k_1 = hF(x_m, y_m)$$

$$k_2 = hF(x_m + \tfrac{1}{2}h, y_m + \tfrac{1}{2}k_1)$$

$$k_3 = hF(x_m + \tfrac{1}{2}h, y_m + \tfrac{1}{2}k_2)$$

$$k_4 = hF(x_m + h, y_m + k_3),$$

$$k = \tfrac{1}{6}(k_1 + 2k_2 + 2k_3 + k_4);$$

are extended to read

$$k_{i,1} = hF_i(t_0, x_0, y_0)$$

$$k_{i,2} = hF_i(t_0 + \tfrac{1}{2}h, x_0 + \tfrac{1}{2}k_{1,1}, y_0 + \tfrac{1}{2}k_{2,1})$$

$$k_{i,3} = hF_i(t_0 + \tfrac{1}{2}h, x_0 + \tfrac{1}{2}k_{1,2}, y_0 + \tfrac{1}{2}k_{2,2})$$

$$k_{i,4} = hF_i(t_0 + h, x_0 + k_{1,3}, y_0 + k_{2,3}),$$

$$k_i = \tfrac{1}{6}(k_{i,1} + 2k_{i,2} + 2k_{i,3} + k_{i,4}).$$

Here, i takes on the values 1 and 2,

$$F_1 = \tfrac{1}{2}(y - 1), \qquad F_2 = \frac{2t}{3x},$$

and

$$x_1 = x_0 + k_1, \qquad y_1 = y_0 + k_2.$$

After x_1 and y_1 have been computed, all the appropriate subscripts are stepped up by unity and the calculations are continued as before. The generalization to m dependent variables is immediate.

The work for this problem is shown in Table 2.

Table 2

t	x	y	$k_{1,j}$ $\frac{1}{20}(y-1)$	$k_{2,j}$ $\frac{\tau}{15x}$	k_1	k_2
1	0.6667	3.0000	0.10000	0.10000	0.10244	0.09768
1.05	0.7167	3.0500	0.10250	0.09768	0.20494	0.19518
1.05	0.7179	3.0488	0.10244	0.09750	0.30738	0.29286
1.1	0.7691	3.0975	0.10488	0.09535	0.10246	0.09762
1.1	0.7691	3.0976	0.10488	0.09535	0.10721	0.09329
1.15	0.8215	3.1453	0.10727	0.09333	0.21449	0.18652
1.15	0.8227	3.1443	0.10722	0.09319	0.32170	0.27981
1.2	0.8763	3.1908	0.10954	0.09123	0.10723	0.09327
1.2	0.8763	3.1909	0.10955	0.09123	0.11179	0.08947
1.25	0.9311	3.2365	0.11183	0.08950	0.22361	0.17889
1.25	0.9322	3.2357	0.11178	0.08939	0.33540	0.26836
1.3	0.9881	3.2803	0.11402	0.08771	0.11180	0.08945
1.3	0.9881	3.2804	0.11402	0.08771	0.11617	0.08612
1.35	1.0451	3.3243	0.11622	0.08612	0.23239	0.17215
1.35	1.0462	3.3235	0.11617	0.08603	0.34856	0.25827
1.4	1.1040	3.3664	0.11832	0.08452	0.11619	0.08609
1.4	1.1043	3.3665	0.11832	0.08452	0.12040	0.08309
1.45	1.1635	3.4088	0.12044	0.08308	0.24084	0.16609
1.45	1.1645	3.4080	0.12040	0.08301	0.36124	0.24918
1.5	1.2247	3.4495	0.12248	0.08165	0.12041	0.08306
1.5	1.2247	3.4496				

(d) Replace $y_{n+1} = y_n + hF_n + \tfrac{1}{2}\Delta(hF_{n-1}) + \tfrac{5}{12}\Delta^2(hF_{n-2}) + \tfrac{3}{8}\Delta^3(hF_{n-3}) + \ldots$ by the pair of formulas

$$x_{n+1} = x_n + hF_{1,n} + \frac{1}{2}\Delta(hF_{1,n-1}) + \frac{5}{12}\Delta^2(hF_{1,n-2})$$
$$+ \ldots$$

$$y_{n+1} = y_n + hF_{2,n} + \frac{1}{2}\Delta(hF_{2,n-1}) + \frac{5}{12}\Delta^2(hF_{2,n-2})$$
$$+ \ldots,$$

for the solution of the given equations. Again, the generalization to m dependent variables is immediate. F_1 and F_2 refer to the right-hand members of the two equations

$$\frac{dx}{dt} = \frac{1}{2}(y - 1)$$

$$\frac{dy}{dt} = \frac{2t}{3x}.$$

It turns out that four starting values for t, x, and y are needed; copy these from the results of the last solution and enter them in a table. These entries are shown in the first four rows and first three columns of Table 3. Find hF_1 and hF_2 from the differential equations, calculate their differences (second differences are sufficient here) and enter them in Table 3.

Table 3

t	x	y	hF_1	$\Delta(hF_1)$	$\Delta^2(hF_1)$	hF_2	$\Delta(hF_2)$	$\Delta^2(hF_2)$
1	0.6667	3	0.10000			0.10000		
				488			−465	
1.1	0.7691	3.0976	0.10488		−21	0.09535		53
				467			−412	
1.2	0.8763	3.1909	0.10955		−20	0.09123		60
				447			−352	
1.3	0.9881	3.2804	0.11402		−16	0.08771		33
				431			−319	
1.4	1.1043	3.3666	0.11833			0.08452		
1.5	1.2247	3.4497						

Next calculate

$$x_4 = 0.9881 + 0.11402 + \frac{1}{2}(0.00447) + \frac{5}{12}(-0.00020)$$
$$= 1.1043$$

$$y_4 = 3.2804 + 0.08771 + \frac{1}{2}(-0.00352) + \frac{5}{12}(-0.00060)$$
$$= 3.3666.$$

Find $hF_{1,4}$ and $hF_{2,4}$ and then extend the difference table. Finally,

$$x_5 = 1.1043 + 0.11833 + \frac{1}{2}(0.00431) + \frac{5}{12}(-0.00016)$$

$$= 1.2247$$

$$y_5 = 3.3666 + 0.08452 + \tfrac{1}{2}(-0.00319) + \tfrac{5}{12}(0.00033)$$

$$= 3.4497.$$

(e) Modify

$$A: \quad y_{n+1} = y_{n-3} + \tfrac{4}{3}h(2y'_{n-2} - y'_{n-1} + 2y'_n)$$

$$C: \quad y_{n+1} = y_{n-1} + \tfrac{1}{3}h(y'_{n-1} + 4y'_n + y'_{n+1}),$$

to read

$$A_1: \quad x_{n+1} = x_{n-3} + \tfrac{4}{3}h(2F_{1,n-2} - F_{1,n-1} + 2F_{1,n})$$

$$C_1: \quad x_{n+1} = x_{n-1} + \tfrac{1}{3}h(F_{1,n-1} + 4F_{1,n} + F_{1,n+1}) \tag{2}$$

$$A_2: \quad y_{n+1} = y_{n-3} + \tfrac{4}{3}h(2F_{2,n-2} - F_{2,n-1} + 2F_{2,n})$$

$$C_2: \quad y_{n+1} = y_{n-1} + \tfrac{1}{3}h(F_{2,n-1} + 4F_{2,n} + F_{2,n+1}).$$

The F's are defined as above and the generalization to m equations is again immediate. There is a need for four starting values; these are copied from the Runge-Kutta solution and entered in a table. (See the first four rows of Table 4). Use $2A_1$, A_2 to predict x_4 and y_4 and find

$$x_4 = 0.6667 + \tfrac{4}{30}[2(1.0488) - 1.0955 + 2(1.1402)]$$

$$= 1.1043,$$

$$y_4 = 3.000 + \tfrac{4}{30}[2(0.9535) - 0.9123 + 2(0.8771)]$$

$$= 3.3665.$$

Table 4

t	x	y	$F_{1,j}$	$F_{2,j}$
1	0.6667	3.0000	1.0000	1.0000
1.1	0.7691	3.0976	1.0488	0.9535
1.2	0.8763	3.1909	1.0955	0.9123
1.3	0.9881	3.2804	1.1402	0.8771
1.4	1.1043	3.3664	1.1832	0.8452
1.5	1.2247	3.4495		

Substitution into the differential equations yields $F_{1,4} = 1.1832$, $F_{2,4} = 0.8452$. Next, correct the preceding values of x_4 and y_4 by use of C_1, C_2 in (2) and find

$$x_4 = 0.8763 + \frac{1}{30}[1.0955 + 4(1.1402) + 1.1832]$$

$$= 1.1043,$$

$$y_4 = 3.1909 + \frac{1}{30}[0.9123 + 4(0.8771) + 0.8452]$$

$$= 3.3664.$$

Keep these values and go on to find x_5 and y_5 in the same way. The values are shown in Table 4.

The various values obtained above can be checked from the particular solution of the differential equation $x = 2t^{3/2}/3$, $y = 2t^{1/2} + 1$. The values of x and y correct to four decimal places are:

t	x	y
1	0.6667	3.0000
1.1	0.7691	3.0976
1.2	0.8764	3.1909
1.3	0.9882	3.2804
1.4	1.1043	3.3664
1.5	1.2247	3.4495

BOUNDARY VALUE PROBLEMS

● **PROBLEM 16-21**

Consider the linear boundary-value problem

$$y'' = f(x)y' + g(x)y + h(x), \quad a \leq x \leq b \qquad (1)$$
$$y(a) = \alpha, \quad y(b) = \beta$$

Write down the linear shooting algorithm to approximate the solution of (1).

Solution: We should make some assumptions concerning the coefficient functions $f(x)$, $g(x)$, $h(x)$.

If

1. $f(x)$, $g(x)$, and $h(x)$ are continuous on $[a,b]$ and
2. $g(x) > 0$ on $[a,b]$,

then problem (1) has a unique solution.

In such a case, the linear boundary-value problem can be replaced by the two initial-value problems

$$I \begin{cases} y'' = f(x)y' + g(x)y + h(x), \quad a \leq x \leq b \\ y(a) = \alpha, \quad y'(a) = 0 \end{cases} \qquad (2)$$

and

$$\text{II} \quad \begin{cases} y'' = f(x)y' + g(x)y, & a \leq x \leq b \\ y(a) = 0, \quad y'(a) = 1 \end{cases} \quad (3)$$

If $y_1(x)$ denotes the solution of (2), and $y_2(x)$ denotes the solution of (3), then

$$y(x) = y_1(x) + y_2(x) \cdot \frac{\beta - y_1(b)}{y_2(b)}, \quad y_2(b) \neq 0 \quad (4)$$

is the unique solution of the boundary-value problem (1).

The shooting method for linear equations consists of replacement of the boundary-value problem (1) by the two initial-value problems (2) and (3). Then, substituting the solutions to (2) and (3) into (4), we obtain the solution of (1). The method is illustrated in the following figure.

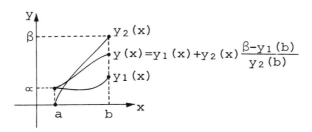

We can use any method (in our algorithm-subroutine) to find the approximations to the solutions $y_1(x)$ and $y_2(x)$. Let us utilize the classical fourth order Runge-Kutta method. In this algorithm, we shall obtain the approximations to the solution $y(x)$ of the boundary-value problem and to the derivative $y'(x)$.

Linear Shooting Algorithm with the fourth-order Runge-Kutta Method.

Problem: approximate the solution to

$$y'' = f(x)y' + g(x)y + h(x) \quad (5)$$

$$x \in [a,b]$$

$$y(a) = \alpha, \quad y(b) = \beta$$

INPUT: [a,b] - interval

α, β - boundary conditions

N - number of subintervals

λ - length of subinterval

Note: The initial interval [a,b] is divided into N subintervals of equal length $\lambda = \frac{b-a}{N}$.

OUTPUT: approximations

$Y_{1,n}$ to $y(x_n)$

$Y_{2,n}$ to $y'(x_n)$

for $n = 0,1,\ldots,N$

Step 1: Set
$$u_{1,0} = \alpha$$
$$u_{2,0} = 0$$
$$v_{1,0} = 0$$
$$v_{2,0} = 1$$

Step 2: For $n = 0,1,\ldots,N-1$, set
$$x = a + n\lambda$$
$$\ell_{1,1} = \lambda u_{2,n}$$
$$\ell_{1,2} = \lambda[f(x)u_{2,n} + g(x)u_{1,n} + h(x)]$$
$$\ell_{2,1} = \lambda[u_{2,n} + .5\ell_{1,2}]$$
$$\ell_{2,2} = \lambda[f(x+.5\lambda)(u_{2,n} + .5\ell_{1,2}) + g(x+.5\lambda)(u_{1,n} + .5\ell_{1,1})$$
$$+ h(x+.5\lambda)]$$
$$\ell_{3,1} = \lambda[u_{2,n} + .5\ell_{2,2}]$$
$$\ell_{3,2} = \lambda[f(x+.5\lambda)(u_{2,n} + .5\ell_{2,2}) + g(x+.5\lambda)(u_{1,n} + .5\ell_{2,1})$$
$$+ h(x+.5\lambda)]$$
$$\ell_{4,1} = \lambda[u_{2,n} + \ell_{3,2}]$$
$$\ell_{4,2} = \lambda[f(x+\lambda)(u_{2,n} + \ell_{3,2}) + g(x+\lambda)(u_{1,n} + \ell_{3,1})$$
$$+ h(x+\lambda)]$$

$$u_{1,n+1} = u_{1,n} + \frac{1}{6}[\ell_{1,1} + 2\ell_{2,1} + 2\ell_{3,1} + \ell_{4,1}]$$

$$u_{2,n+1} = u_{2,n} + \frac{1}{6}[\ell_{1,2} + 2\ell_{2,2} + 2\ell_{3,2} + \ell_{4,2}]$$

$$m_{1,1} = \lambda v_{2,n}$$

$$m_{1,2} = \lambda[f(x)v_{2,n} + g(x)v_{1,n}]$$

$$m_{2,1} = \lambda[v_{2,n} + .5m_{1,2}]$$

$$m_{2,2} = \lambda[f(x+.5\lambda)(v_{2,n} + .5m_{1,2}) + g(x+.5\lambda)(v_{1,n} + .5m_{1,1})]$$

$$m_{3,1} = \lambda[v_{2,n} + .5m_{2,2}]$$

$$m_{3,2} = \lambda[f(x+.5\lambda)(v_{2,n} + .5m_{2,2}) + g(x+.5\lambda)(v_{1,n} + .5m_{2,1})]$$

$$m_{4,1} = \lambda[v_{2,n} + m_{3,2}]$$

$$m_{4,2} = \lambda[f(x+\lambda)(v_{2,n} + m_{3,2}) + g(x+\lambda)(v_{1,n} + m_{3,1})]$$

$$v_{1,n+1} = v_{1,n} + \frac{1}{6}[m_{1,1} + 2m_{2,1} + 2m_{3,1} + m_{4,1}]$$

$$v_{2,n+1} = v_{2,n} + \frac{1}{6}[m_{1,2} + 2m_{2,2} + 2m_{3,2} + m_{4,2}]$$

Step 3: Set

$$Y_{1,0} = \alpha$$
$$Y_{2,0} = \frac{\beta - u_{1,N}}{v_{1,N}}$$

OUTPUT: $(a, Y_{1,0}, Y_{2,0})$

Step 4: For $n = 0, 1, \ldots, N$, set

$$Y_{1,n} = u_{1,n} + Y_{1,0} v_{1,n}$$
$$Y_{2,n} = u_{2,n} + Y_{2,0} v_{2,n}$$
$$x = a + n\lambda$$

OUTPUT: $(x_n, Y_{1,n}, Y_{2,n})$

Step 5: STOP.

Note that in this algorithm we compute two sets of numbers ℓ and m. From the ℓ's we evaluate the set of values $u_{1,n}$ and $u_{2,n}$ for $n = 0, 1, \ldots, N-1$.

From the m's we compute $v_{1,n}$ and $v_{2,n}$ for $n = 0, 1, \ldots, N-1$.

The values $u_{1,n}$ approximate $y_1(x_n)$, and $v_{1,n}$ approximate $y_2(x_n)$.

The solution of the boundary-value problem $y(x_n)$ is approximated by the set of numbers $Y_{1,n}$, and its derivative is approximated by $Y_{2,n}$.

● **PROBLEM 16-22**

Consider the nonlinear second-order boundary-value problem

$$y'' = f(x,y,y'), \quad a \leq x \leq b \tag{1}$$
$$y(a) = \alpha, \quad y(b) = \beta$$

Write an algorithm to approximate the solution of (1) which uses the shooting method for nonlinear problems.

Solution: In Problem 16-21, we found an algorithm which approximates the solution of the linear problem. In that case, the linear problem was replaced by the two initial-value problems. The situation is more complex for the nonlinear problems because their solutions cannot be expressed as a linear combination of the solutions to two initial-value problems.

The nonlinear boundary-value problem can be replaced by a sequence of initial-value problems of the form

$$y'' = f(x,y,y'), \quad a \leq x \leq b \tag{2}$$
$$y(a) = \alpha, \quad y'(a) = s_k$$

Let us denote the solution to (2) by $y(x,s_k)$. The parameter s_k introduces the consecutive corrections so that the sequence of solutions $y(x,s_k)$ approximates the solution $y(x)$ of the nonlinear boundary-value problem. The sequence (s_k) should be chosen in such a way that

$$\lim_{k \to \infty} y(b,s_k) = y(b) = \beta \qquad (3)$$

This technique of solving nonlinear problems is called the shooting method by analogy to the firing of missiles at a stationary target. The initial elevation is determined by parameter s_0. The missile travels along $y(x,s_0)$ and misses the target $\beta = y(b)$ by

$$y(b,s_0) - \beta \qquad \text{(see Fig. 1)} \qquad (4)$$

The correction is introduced (parameter is changed to t_1). This time the missile moves along $y(x,s_1)$ and gets closer to the target, but misses it by $y(b,s_1)-\beta$. The nth trajectory is the solution to the initial-value problem (see Fig. 2).

$$y'' = f(x,y,y'), \quad a \le x \le b$$
$$y(a) = \alpha, \quad y'(a) = s_n \qquad (5)$$

As the number n increases, we get closer and closer to the target. A number K can be predetermined which tells us that the accuracy is sufficient when

$$|y(b,s_n) - \beta| < K \qquad (6)$$

and the "shooting" should cease.

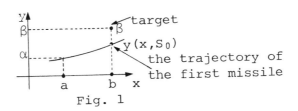
Fig. 1

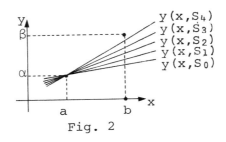
Fig. 2

In the nonlinear shooting algorithm we shall utilize the fourth-order Runge-Kutta method.

The segment [a,b] will be divided into N subsegments.

Algorithm to Approximate the Solution of the Nonlinear Boundary-Value Problem

INPUT: [a,b] - segment
         α, β - boundary conditions
         N - number of subsegments
         ACC - accuracy
         M - maximum number of iterations

OUTPUT: approximations

$$\omega_{1,n} \text{ to } y(x_n)$$
$$\omega_{2,n} \text{ to } y'(x_n)$$
$$\text{for } n = 0, 1, \ldots, N$$

or information that the maximum number of iterations M was exceeded.

Step 1: $\lambda = \dfrac{b-a}{N}$

   $k = 0$

   $s_0 = \dfrac{\beta - \alpha}{b-a}$

Step 2: For $K \leq M$, do Steps 3-9.

Step 3: Set

$$u_{1,0} = 0$$
$$u_{2,0} = 1$$
$$\omega_{1,0} = \alpha$$
$$\omega_{2,0} = s_0$$

Step 4: For $n = 0, \ldots, N-1$, do Steps 5 and 6.

Step 5: Set $x = a + n\lambda$

Step 6: $\ell_{1,1} = \lambda \omega_{2,n}$

   $\ell_{1,2} = \lambda f(x, \omega_{1,n}, \omega_{2,n})$

   $\ell_{2,1} = \lambda(\omega_{2,n} + .5\ell_{1,2})$

   $\ell_{2,2} = \lambda f(x + .5\lambda, \omega_{1,n} + .5\ell_{1,1}, \omega_{2,n} + .5\ell_{1,2})$

$$\ell_{3,1} = \lambda(\omega_{2,n} + .5\ell_{2,2})$$

$$\ell_{3,2} = \lambda f(x+.5\lambda, \omega_{1,n} + .5\ell_{2,1}, \omega_{2,n} + .5\ell_{2,2})$$

$$\ell_{4,1} = \lambda(\omega_{2,n} + \ell_{3,2})$$

$$\ell_{4,2} = \lambda f(x+\lambda, \omega_{1,n} + \ell_{3,1}, \omega_{2,n} + \ell_{3,2})$$

$$\begin{cases} \omega_{1,n+1} = \omega_{1,n} + \frac{1}{6}[\ell_{1,1} + 2\ell_{2,1} + 2\ell_{3,1} + \ell_{4,1}] \\ \omega_{2,n+1} = \omega_{2,n} + \frac{1}{6}[\ell_{1,2} + 2\ell_{2,2} + 2\ell_{3,2} + \ell_{4,2}] \end{cases}$$

$$m_{1,1} = \lambda u_{2,n}$$

$$m_{1,2} = \lambda[f_y(x, \omega_{1,n}, \omega_{2,n})u_{1,n} + f_{y'}(x, \omega_{1,n}, \omega_{2,n})u_{2,n}]$$

$$m_{2,1} = \lambda[u_{2,n} + .5m_{1,2}]$$

$$m_{2,2} = \lambda[f_y(x+.5\lambda, \omega_{1,n}, \omega_{2,n})(u_{1,n} + .5m_{1,1})$$

$$+ f_{y'}(x+.5\lambda, \omega_{1,n}, \omega_{2,n})(u_{2,n} + .5m_{2,1})]$$

$$m_{3,1} = \lambda[u_{2,n} + .5m_{2,2}]$$

$$m_{3,2} = \lambda[f_y(x+.5\lambda, \omega_{1,n}, \omega_{2,n})(u_{1,n} + .5m_{2,1})$$

$$+ f_{y'}(x+.5\lambda, \omega_{1,n}, \omega_{2,n})(u_{2,n} + .5m_{2,2})]$$

$$m_{4,1} = \lambda(u_{2,n} + m_{3,2})$$

$$m_{4,2} = \lambda[f_y(x+\lambda, \omega_{1,n}, \omega_{2,n})(u_{1,n}+m_{3,1})$$

$$+ f_{y'}(x+\lambda, \omega_{1,n}, \omega_{2,n})(u_{2,n}+m_{3,2})]$$

$$u_{1,n+1} = u_{1,n} + \frac{1}{6}[m_{1,1} + 2m_{2,1} + 2m_{3,1} + m_{4,1}]$$

$$u_{2,n+1} = u_{2,n} + \frac{1}{6}[m_{1,2} + 2m_{2,2} + 2m_{3,2} + m_{4,2}]$$

Step 7: If $|\omega_{1,N} - \beta| \leq ACC$, then do Step 8.

Step 8: For $n = 0, \ldots, N$

    set $x = a + n\lambda$

OUTPUT: $(x, \omega_{1,n}, \omega_{2,n})$

STOP

Step 9: Set $s_0 = s_0 - \left(\dfrac{\omega_{1,N} - \beta}{u_{1,N}}\right)$

    $k = k+1$

Step 10: OUTPUT: Maximum number of iterations is exceeded.

    STOP.

Note that in Steps 5 and 6 we applied the fourth-order Runge-Kutta method. The initial value $s_0 = \frac{\beta - \alpha}{b - a}$ selected in Step 1 is the slope of the straight line from (a, α) to (b, β).

● PROBLEM 16-23

Solve the boundary value problem

$$\frac{d^2y}{dx^2} + \frac{1}{4} y = 8$$

$$y(0) = 0, \quad y(10) = 0$$

by the matrix method.

Solution: This numerical approach to the solution of boundary value problems can best be illustrated by:

$$\frac{d^2y}{dx^2} + Ay = \dot{B}$$

$$y(0) = 0, \quad y(L) = 0.$$

As shown in the figure, the region $0 \le x \le L$ is first divided into $n + 1$ equally spaced intervals of length Δx.

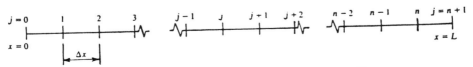

Finite difference grid for boundary value problem.

The differential equation can now be represented at the point j by

$$\frac{y_{j+1} - 2y_j + y_{j-1}}{(\Delta x)^2} + Ay_j = B$$

The derivative d^2y/dx^2 has been replaced by a central difference representation of $0(\Delta x)^2$. There are n equations of the form (1), one for each interior point of the region shown in the figure. There is also one unknown value of y for each interior point of the region. Thus one has n simultaneous linear equations in the n unknowns $y_1, y_2, \ldots, y_n$. After multiplying through each equation by $(\Delta x)^2$, this set of equations can be written as

$$y_2 - 2y_1 + (0) + A(\Delta x)^2 y_1 = B(\Delta x)^2$$
$$y_3 - 2y_2 + y_1 + A(\Delta x)^2 y_2 = B(\Delta x)^2$$
$$\vdots \qquad \qquad (2)$$
$$y_n - 2y_{n-1} + y_{n-2} + A(\Delta x)^2 y_{n-1} = B(\Delta x)^2$$
$$(0) - 2y_n + y_{n-1} + A(\Delta x)^2 y_n = B(\Delta x)^2$$

Collecting coefficients of the unknowns, and writing the set (2) in matrix form yields

$$\begin{bmatrix} \alpha & 1 & & & & \\ 1 & \alpha & 1 & & & \\ & 1 & \alpha & 1 & & \\ \hline & & & & & \\ & & & 1 & \alpha & 1 \\ & & & & 1 & \alpha \end{bmatrix} \begin{bmatrix} y_1 \\ y_2 \\ y_3 \\ \hline \\ y_{n-1} \\ y_n \end{bmatrix} = \begin{bmatrix} B(\Delta x)^2 \\ B(\Delta x)^2 \\ B(\Delta x)^2 \\ \hline \\ B(\Delta x)^2 \\ B(\Delta x)^2 \end{bmatrix}$$

where $\alpha = -2 + A(\Delta x)^2$.

Now, consider the problem.

Assume arbitrarily that $\Delta x = 1$ (10 intervals from $x = 0$ to $x = 10$). Then, following the notation where $\alpha = -2 + A(\Delta x)^2$, $\alpha = -2 + \frac{1}{4}(1)^2 = -1.75$ and the elements of the right-hand side vector are $8(1)^2 = 8$. The matrix formulation is

$$\begin{bmatrix} -1.75 & 1 & & & & & & \\ 1 & -1.75 & 1 & & & & & \\ & 1 & -1.75 & & & & & \\ & & & - & - & - & & \\ & & & & - & - & & \\ & & & & 1 & -1.75 & 1 \\ & & & & & 1 & -1.75 \end{bmatrix} \begin{bmatrix} y_1 \\ y_2 \\ y_3 \\ - \\ - \\ y_8 \\ y_9 \end{bmatrix}$$

$$= \begin{bmatrix} 8 \\ 8 \\ 8 \\ - \\ - \\ 8 \\ 8 \end{bmatrix}$$

The solution to this set, along with the exact solution given by

$$y = 32 \left[\frac{(\cos(5) - 1)}{\sin(5)} \sin(x/2) - \cos(x/2) + 1 \right]$$

is shown in Table 1. The solution is symmetric about x = 5 and only half of the region has been shown.

Table 1

x	Matrix Method	Exact
0	0	0
1	14.9384	15.3779
2	34.1422	34.8254
3	52.8105	53.5812
4	66.2761	67.0532
5	71.1173	71.9429

The accuracy of the solution can, of course, be improved by reducing Δx.

One could have taken advantage of the symmetry to reduce the number of unknowns. Since the solution is symmetric about x = 5, one could write

$$\frac{dy}{dx}(5) = 0$$

Using a backward difference representation of error order $(\Delta x)^2$ (to be consistent with the central difference representations which are also of error order $(\Delta x)^2$), this derivative can be represented as

$$\frac{3y_5 - 4y_4 + y_3}{2(1)} = 0$$

The difference representation of the differential equation written at $x = 1, 2, 3$, and 4 now provides 4 equations in the 5 unknowns y_1, y_2, y_3, y_4, and y_5. The additional equation is the difference representation of the derivative at $x = 5$ which can be rewritten as

$$(1)y_3 + (-4)y_4 + (3)y_5 = 0$$

The matrix formulation now becomes

$$\begin{bmatrix} -1.75 & 1 & & & \\ 1 & -1.75 & 1 & & \\ & 1 & -1.75 & 1 & \\ & & 1 & -1.75 & 1 \\ & & 1 & -4 & 3 \end{bmatrix} \begin{bmatrix} y_1 \\ y_2 \\ y_3 \\ y_4 \\ y_5 \end{bmatrix} = \begin{bmatrix} 8 \\ 8 \\ 8 \\ 8 \\ 0 \end{bmatrix}$$

Since this set is so small, there is no real necessity to make the set tridiagnoal. However, if one is striving for accuracy, the set would be much larger, and the advantages of having a tridiagonal coefficient matrix could be very significant. The matrix can be made tridiagonal by simply subtracting the fourth equation from the fifth equation. Thus the new matrix formulation becomes

$$\begin{bmatrix} -1.75 & 1 & & & \\ 1 & -1.75 & 1 & & \\ & 1 & -1.75 & 1 & \\ & & 1 & -1.75 & 1 \\ & & & -2.25 & 2 \end{bmatrix} \begin{bmatrix} y_1 \\ y_2 \\ y_3 \\ y_4 \\ y_5 \end{bmatrix} = \begin{bmatrix} 8 \\ 8 \\ 8 \\ 8 \\ -8 \end{bmatrix}$$

A similar method can be applied to any problem involving a gradient boundary condition, with the end result being a tridiagonal set.

• **PROBLEM 16-24**

Consider the boundary-value problem

$$y'' = F(x,y,y'), \quad a \leq x \leq b$$
$$y(a) = \alpha, \quad y(b) = \beta \tag{1}$$

Approximate the solution applying the Richardson extrapolation method. Evaluate the error of an approximation to $y''(x_0)$.

Solution: We shall briefly explain the Richardson extrapolation method. Suppose $y(x)$ has derivatives of all orders in the interval $[a,b]$. Then, $y(x)$ can be represented by a Taylor series ($h > 0$)

$$y(x_0+h) = y(x_0) + \frac{y'(x_0)}{1!}h + \ldots + \frac{y^{(n)}(x_0)}{n!}h^n + \frac{y^{(n+1)}(x')}{(n+1)!}h^{n+1} \quad (2)$$

where $x_0 < x' < (x_0+h)$.

Similarly,

$$y(x_0-h) = y(x_0) - \frac{y'(x_0)}{1!}h + \frac{y''(x_0)}{2!}h^2 - \ldots + \frac{y^{(n+1)}(x_{11})}{(n+1)!}(-h)^{n+1} \quad (3)$$

where $(x_0-h) < x_{11} < x_0$

Assume for simplicity that $y(x)$ is expanded in a fifth-degree polynomial. Adding (2) and (3), we obtain

$$\frac{1}{h^2}\left[y(x_0+h) - 2y(x_0) + y(x_0-h)\right] = y''(x_0) + \frac{h^2}{12}y^{(4)}(x_0) + \frac{h^4}{360}y^{(6)}(x_1) \quad (4)$$

$(x_0-h) < x_1 < (x_0+h)$.

Let us replace in (4) the value of h by αh, where $\alpha \neq \pm 1$

$$\frac{1}{\alpha^2 h^2}\left[y(x_0+\alpha h) - 2y(x_0) + y(x_0-\alpha h)\right] = y''(x_0) + \frac{\alpha^2 h^2}{12}y^{(4)}(x_0)$$
$$+ \frac{\alpha^4 h^4}{360}y^{(6)}(x_{11}) \quad (5)$$

From (4) and (5), we obtain

$$\frac{\alpha^2}{h^2(\alpha^2-1)}\left[y(x_0+h) - 2y(x_0) + y(x_0-h)\right]$$
$$- \frac{1}{(\alpha^2-1)\alpha^2 h^2}\left[y(x_0+\alpha h) - 2y(x_0) + y(x_0-\alpha h)\right] = y''(x_0) \quad (6)$$
$$- \frac{\alpha^2 h^4}{(1-\alpha^2)360}\left[\alpha^2 y^{(6)}(x_{11}) - y^{(6)}(x_1)\right]$$

Note that we eliminated the term involving h^2.

Solving (6) for $y''(x_0)$, we find

$$y''(x_0) = \frac{1}{\alpha^2 h^2 (\alpha^2-1)} \Big[-y(x_0+\alpha h) + \alpha^4 y(x_0+h) \\ - 2(\alpha^4-1)y(x_0) + \alpha^4 y(x_0-h) - y(x_0-\alpha h) \Big] + O(h^4) \quad (7)$$

The approximation to $y''(x_0)$ is obtained with truncation error $O(h^4)$. To obtain better accuracy, we can expand $y(x_0)$ into a seventh-degree Taylor polynomial. In such a case, the error will be $O(h^6)$.

Applying formula (7), we find first, second, etc., extrapolations to the solution of (1).

• **PROBLEM 16-25**

For $a = 1$, find the numerical solution for the initial-boundary problem defined by

$$u_{xx} - u_t = xu_x, \qquad 0 \leq x \leq 1 \qquad (1)$$

$$u(0,t) = g_1(t) = 0, \qquad t \geq 0 \qquad (2)$$

$$u(x,0) = f(x) = x, \qquad 0 \leq x \leq 1 \qquad (3)$$

$$u(1,t) = g_2(t) = e^{-t}, \qquad t \geq 0. \qquad (4)$$

<u>Solution</u>: The steady state form of eq. (1) is

$$\frac{d^2u}{dx^2} - x\frac{du}{dx} = 0. \qquad (5)$$

Define

$$\alpha = \lim_{t \to \infty} g_1(t)$$

and

$$\beta = \lim_{t \to \infty} g_2(t).$$

According to (2) and (4), one obtains

$$\alpha = \lim_{t \to \infty} g_1(t) = 0$$

and

$$\beta = \lim_{t \to \infty} e^{-t} = 0 \ .$$

Thus, the boundary conditions for (5) are

$$u(0) = u(1) = 0 \qquad (6)$$

Next set $\Delta x = h = \frac{1}{3}$ so that $0 \leq x \leq 1$ is divided into three equal parts by the points $x_0 = 0$, $x_1 = \frac{1}{3}$, $x_2 = \frac{2}{3}$, $x_3 = 1$. Since the analytical solution of (5)-(6) is $u(x) = 0$, simply set

$$u(x_0, \infty) = u(x, \infty) = u(x_2, \infty) = u(x_3, \infty) = 0 \ . \qquad (7)$$

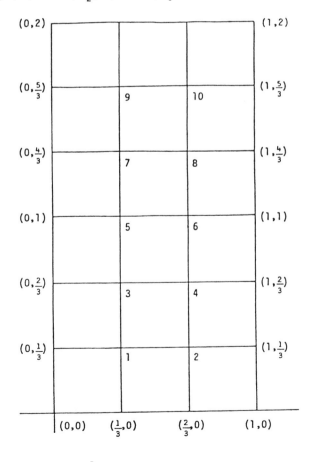

Fig.1

Now let $T = 2$ and $k = \frac{1}{3}$ so that $\bar{R}_h$ and $\bar{S}_h$ are as shown in Fig. 1. On $\bar{S}_h$, one obtains, from

$$u(x_i,t) = u(x_i,\infty), \qquad i = 1,2,\ldots,n-1, \qquad (8)$$

and (2) — (4) as well as (7), the following

$$u(0,2) = 0, \quad u\left(0,\tfrac{5}{3}\right) = 0, \quad u\left(0,\tfrac{4}{3}\right) = 0, \quad u(0,1) = 0,$$

$$u\left(0,\tfrac{2}{3}\right) = 0, \quad u\left(0,\tfrac{1}{3}\right) = 0$$

$$u(0,0) = 0, \quad u\left(\tfrac{1}{3},0\right) = \tfrac{1}{3}, \quad u\left(\tfrac{2}{3},0\right) = \tfrac{2}{3},$$

$$u(1,0) = 1, \qquad (9)$$

$$u\left(1,\tfrac{1}{3}\right) = e^{-\tfrac{1}{3}}, \quad u\left(1,\tfrac{2}{3}\right) = e^{-\tfrac{2}{3}}$$

$$u(1,1) = e^{-1}, \quad u\left(1,\tfrac{4}{3}\right) = e^{-\tfrac{4}{3}}, \quad u\left(1,\tfrac{5}{3}\right) = e^{-\tfrac{5}{3}},$$

$$u(1,2) = e^{-2},$$

$$u\left(\tfrac{2}{3},2\right) = 0, \quad u\left(\tfrac{1}{3},2\right) = 0.$$

Applying the formula:

$$-\frac{2}{h^2} u(x,t) + \frac{1}{h^2} u(x+h,t) - \frac{1}{2k} u(x,t+k)$$

$$+ \frac{1}{h^2} u(x-h,t)$$

to Eq. (1), one obtains

$$-18u(x,t) + 9u\left(x+\tfrac{1}{3},t\right) - \tfrac{3}{2} u\left(x,t+\tfrac{1}{3}\right) + 9u\left(x-\tfrac{1}{3},t\right)$$

$$+ \tfrac{3}{2} u\left(x,t-\tfrac{1}{3}\right) = \tfrac{3x}{2} \left[u\left(x+\tfrac{1}{3},t\right) - u\left(x-\tfrac{1}{3},t\right)\right],$$

$$(10)$$

or equivalently, according to Fig. 2,

$$-2u_0 + \left(1 - \tfrac{x}{6}\right)u_1 - \tfrac{1}{6} u_2 + \left(1 + \tfrac{x}{6}\right)u_3 + \tfrac{1}{6} u_4 = 0. \qquad (11)$$

Application of eq. (11) at each point of $\bar{R}_h$, with the known values inserted, yields the following diagonally dominant linear algebraic system,

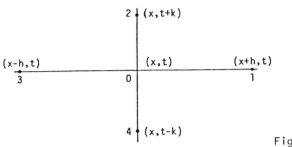

Fig.2

$-2u_1 + \frac{17}{18} u_2 - \frac{1}{6} u_3 = -\frac{1}{18}$

$-2u_2 - \frac{1}{6} u_4 + \frac{10}{9} u_1 = -\frac{8}{9} e^{-1/3} - \frac{1}{9}$

$-2u_3 + \frac{17}{18} u_4 - \frac{1}{6} u_5 + \frac{1}{6} u_1 = 0$

$-2u_4 - \frac{1}{6} u_6 + \frac{10}{9} u_3 + \frac{1}{6} u_2 = -\frac{8}{9} e^{-2/3}$

$-2u_5 + \frac{17}{18} u_6 - \frac{1}{6} u_7 + \frac{1}{6} u_3 = 0$

$-2u_6 - \frac{1}{6} u_8 + \frac{10}{9} u_5 + \frac{1}{6} u_4 = -\frac{8}{9} e^{-1}$

$-2u_7 + \frac{17}{18} u_8 - \frac{1}{6} u_9 + \frac{1}{6} u_5 = 0$

$-2u_8 - \frac{1}{6} u_{10} + \frac{10}{9} u_7 + \frac{1}{6} u_6 = -\frac{8}{9} e^{-4/3}$

$-2u_9 + \frac{17}{18} u_{10} + \frac{1}{6} u_7 = 0$

$-2u_{10} + \frac{10}{9} u_9 + \frac{1}{6} u_8 = -\frac{8}{9} e^{-5/3}$,

whose solution, when found by the generalized Newton's method, agrees with the analytical solution

$$u = xe^{-t}$$

of (1)-(4) to at least two, but usually more, decimal places.

● **PROBLEM 16-26**

Consider the following initial boundary problem where:

$$u_{xx} - u_{tt} = 0 \tag{1}$$

$$u(x,0) = x(1 - x), \quad 0 \leq x \leq 1 \tag{2}$$

$$u(1,t) = 0, \quad t \geq 0 \tag{3}$$

$$u(x,k) = u(x,0) + kf_2(x), \quad (h = \tfrac{1}{4}, \, k = \tfrac{1}{8}) \tag{4}$$

Find its numerical solution if

$$u(x - h, t + k) - 2(1 + \tfrac{h^2}{k^2})u(x, t + k)$$
$$= -u(x - h, t - k) + 2(1 + \tfrac{h^2}{k^2})u(x, t - k)$$
$$- u(x + h, t - k) - 4\tfrac{h^2}{k^2}u(x,t) \tag{5}$$

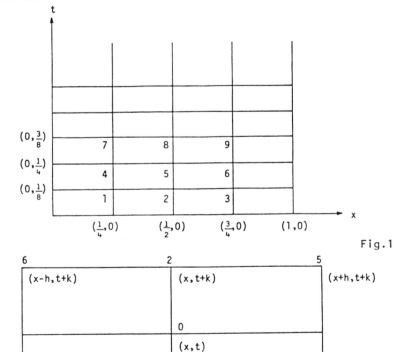

Fig.1

Fig.2

Solution: Consider the initial-boundary problem defined by (1) and (2)-(3). Set $h = \tfrac{1}{4}$, $k = \tfrac{1}{8}$, so that

652

the grid points R_h and S_h are as shown in Figure 1.
From (4), one has

$$u_1 = u(\tfrac{1}{4}, 0) + \tfrac{1}{8}(1) = \tfrac{5}{16}$$

$$u_2 = u(\tfrac{1}{2}, 0) + \tfrac{1}{8}(1) = \tfrac{3}{8}$$

$$u_3 = u(\tfrac{3}{4}, 0) + \tfrac{1}{8}(1) = \tfrac{5}{16}.$$

so $u_1 = \tfrac{5}{16}$, $u_2 = \tfrac{3}{8}$, $u_3 = \tfrac{5}{16}$.

For the given parameter choices, (5) becomes

$$u_6 - 10u_2 + u_5 = -u_7 + 10u_4 - u_8 - 16u_0. \qquad (6)$$

Considering first the points numbered 1, 2, 3 in Figure 1 to be the point (x,y) in Figure 2 yields, in order, by means of (6),

$$u(0,\tfrac{1}{4}) - 10u_4 + u_5 = -u(0,0) + 10u(\tfrac{1}{4},0) - u(\tfrac{1}{2},0)$$
$$- 16u_1$$

$$u_4 - 10u_5 + u_6 = -u(\tfrac{1}{4},0) + 10u(\tfrac{1}{2},0) - u(\tfrac{3}{4},0)$$
$$- 16u_2$$

$$u_5 - 10u_6 + u(1,\tfrac{1}{4}) = -u(\tfrac{1}{2},0) + 10u(\tfrac{3}{4},0) - u(1,0)$$
$$- 16u_3,$$

or, equivalently,

$$-10u_4 + u_5 = -27/8$$
$$u_4 - 10u_5 + u_6 = -31/8$$
$$u_5 - 10u_6 = -27/8$$

the solution of which is

$$u_4 = 43/112, \quad u_5 = 13/28, \quad u_6 = 43/112.$$

One then proceeds to generate the solution on row 3 using (6), the given boundary conditions, and the approximations on rows 1 and 2, and so on.

● **PROBLEM 16-27**

Consider the initial-boundary problem defined by

$$u_{xx} = u_t, \quad a = 1,$$

and

$$u(0,t) = g_1(t) = 0 \qquad (1)$$

$$u(x,0) = f(x) = x, \quad 0 \le x \le 1 \qquad (2)$$

$$u(1,t) = g_2(t) = 1. \qquad (3)$$

Given

$$u(x,t+k) = \lambda u(x+h,t) + (1-2\lambda)u(x,t) + \lambda u(x-h,t), \qquad (4)$$

$$u_2 = \lambda u_1 + (1-2\lambda)u_0 + \lambda u_3, \qquad (5)$$

find the numerical solution for the initial-boundary problem. Assume

$$h = k = \frac{1}{3}.$$

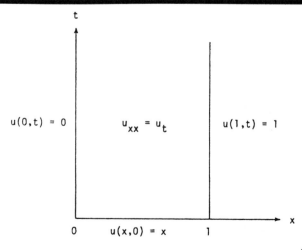

Fig.1

Solution: As shown diagramatically in Figure 1, for

$$h = k = \frac{1}{3},$$

construct R_h and S_h, and number the points of R_h as shown in Figure 2. Note finally that (4) and (5) can be written, respectively, as

$$u(x, t + \tfrac{1}{3}) = 3u(x + \tfrac{1}{3}, t) - 5u(x,t) + 3u(x - \tfrac{1}{3}, t) \quad (4')$$

$$u_2 = 3u_1 - 5u_0 + 3u_3 . \quad (5')$$

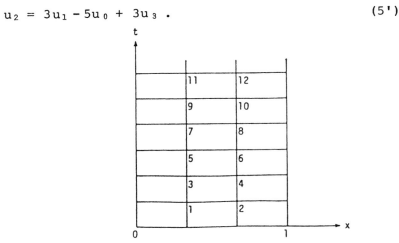

Fig.2

Rounding all numbers to one decimal place, one has, by applying (4'), or (5'), successively at the points numbered 1-11 in Figure 2, that

$$u_1 = 3u(\tfrac{2}{3}, 0) - 5u(\tfrac{1}{3}, 0) + 3u(0,0) = 3(0.7)$$

$$- 5(0.3) + 3 \cdot 0 = 0.6$$

$$u_2 = 3u(1,0) - 5u(\tfrac{2}{3}, 0) + 3u(\tfrac{1}{3}, 0) = 3 \cdot 1 - 5(0.7)$$

$$+ 3(0.3) = 0.4$$

$$u_3 = 3u_2 - 5u_1 + 3u(0, \tfrac{1}{3}) = -1.8$$

$$u_4 = 3u(1, \tfrac{1}{3}) - 5u_2 + 3u_1 = 2.8$$

$$u_5 = 3u_4 - 5u_3 + 3u(0, \tfrac{2}{3}) = 17.4$$

$$u_6 = 3u(1, \tfrac{2}{3}) - 5u_4 + 3u_3 = -16.4$$

$$u_7 = 3u_6 - 5u_5 + 3u(0,1) = -136.2$$

$$u_8 = 3u(1,1) - 5u_6 + 3u_5 = 137.2$$

$$u_9 = 3u_8 - 5u_7 + 3u\left(0, \tfrac{4}{3}\right) = 1092.6$$

$$u_{10} = 3u\left(1, \tfrac{4}{3}\right) - 5u_8 + 3u_7 = -1091.6$$

$$u_{11} = 3u_{10} - 5u_9 + 3u\left(0, \tfrac{5}{3}\right) = -8737.8$$

from which one suspects the development of instability.

• **PROBLEM 16-28**

Consider the general linear boundary value problem given by

$$y'' + P(x)y' + Q(x)y = R(x) \qquad (1)$$

$$y(a) = \alpha, \; y(b) = \beta, \; a < b, \qquad (2)$$

where $P(x)$, $Q(x)$, $R(x)$ are continuous on $a \le x \le b$,

$$Q(x) \le 0. \quad a < x < b \qquad (3)$$

and let (1) be approximated by

$$y_{i-1}[2 - P(x_i)\Delta x] + y_i[-4 + 2(\Delta x)^2 Q(x_i)]$$

$$+ y_{i+1}[2 + P(x_i)\Delta x]$$

$$= 2(\Delta x)^2 R(x_i); \quad i = 1,2,\ldots,n-1. \qquad (4)$$

If $a \le x \le b$ is divided into n equal parts, it follows that (4) yields a tridiagonal linear system. It is not at all obvious, however, that this system has a unique solution. Determine if the system can be made diagonally dominant so that one can have this assurance.

Solution: From (4), one wishes to have

$$|-4 + 2(\Delta x)^2 Q(x_i)| \ge |2 + \Delta x P(x_i)| + |2 - \Delta x P(x_i)|. \qquad (5)$$

For those values of Δx which satisfy both

$$2 - (\Delta x)P(x_i) > 0, \quad 2 + (\Delta x)P(x_i) > 0, \qquad (6)$$

condition (5) reduces to

$$4 - 2(\Delta x)^2 Q(x_i) \geq (2 + (\Delta x)P(x_i)) + (2 - (\Delta x)P(x_i)),$$

or

$$-2(\Delta x)^2 Q(x_i) \geq 0. \qquad (7)$$

Since $Q(x_i) \leq 0$, (7) is valid. Thus, it is sufficient to satisfy (6), or, equivalently,

$$\Delta x |P(x_i)| < 2.$$

If M is any upper bound for $|P(x)|$ on $a \leq x \leq b$, so that

$$|P(x)| \leq M, \quad a \leq x \leq b,$$

then a sufficient condition to assure diagonal dominance is, finally,

$$\Delta x < 2/M.$$

● **PROBLEM 16-29**

Solve the following boundary value problem

$$y'' - y = 1, \quad 0 < x < 3$$

$$y(0) = -1, \quad y(3) = -1.$$

With $\Delta x = \frac{1}{2}$, divide $0 \leq x \leq 3$ into six equal parts so that $x_0 = 0$, $x_1 = 0.5$, $x_2 = 1$, $x_3 = 1.5$, $x_4 = 2$, $x_5 = 2.5$, and $x_6 = 3$.

Solution: Since no approximation for y' must be made, the difference equation can be written as

$$\frac{y_{i-1} - 2y_i + y_{i+1}}{(\Delta x)^2} - y_i = 1, \quad i = 1,2,3,4,5,$$

or, equivalently, as

$$y_{i-1} - \frac{9}{4} y_i + y_{i+1} = \frac{1}{4}, \quad i = 1,2,3,4,5. \tag{1}$$

Using the given boundary values, the tridiagonal system that results for $i = 1,2,3,4,5$ is

$$\begin{aligned}
-\frac{9}{4} y_1 + y_2 &&&&&= \frac{1}{4} \\
y_1 - \frac{9}{4} y_2 + y_3 &&&&&= \frac{1}{4} \\
y_2 - \frac{9}{4} y_3 + y_4 &&&&&= \frac{1}{4} \\
y_3 - \frac{9}{4} y_4 + y_5 &&&&&= \frac{1}{4} \\
y_4 - \frac{9}{4} y_5 &&&&&= \frac{5}{4},
\end{aligned} \tag{2}$$

the exact solution of which is $y_1 = y_2 = y_3 = y_4 = y = -1$. It is not surprising, because of the simplicity of the problem, that the numerical solution happens to coincide with the exact solution, $y = -1$, at the grid points. It is most important to note, however, with regard to the matrix of system (2), that the coefficient of y_i in (1) yields the main diagonal entries, that of y_{i-1} yields the entries just below the main diagonal, and that of y_{i+1} yields the entries just above the main diagonal. These observations also extend to

$$y_{i-1}[2 - P(x_i) \Delta x] + y_i [-4 + 2(\Delta x)^2 Q(x_i)]$$

$$+ y_{i+1}[2 + P(x_i) \Delta x]$$

$$= 2(\Delta x)^2 R(x_i); \quad i = 1,2,\ldots,n-1.$$

• **PROBLEM 16-30**

Find the eigenvalues and eigenfunctions of $y'' + Ky = 0$, for $0 < x < 1$ such that $y(0) = y(1) = 0$.

Solution: The equation

$$y'' + Ky = 0 \tag{a}$$

with the associated boundary conditions is a boundary-value problem (BVP). It is found that the solution to (a) depends vitally on the constant K which is called an eigenvalue or proper value. The problem may thus be viewed as an eigenvalue problem (EVP).

Now K is an unknown constant. Thus one must consider three possibilities for K. One has either $K < 0$, $K = 0$ or $K > 0$.

Case 1: $K < 0$.   Let $K = -K$.

From (a), (since it is a linearly homogeneous equation with constant coefficients, assume a solution of the form $y = e^{mx}$), obtain the characteristic equation

$$m^2 - K = 0 \tag{b}$$

Thus, $m^2 = K$ and $m = \pm\sqrt{K}$; a general solution of (a) is then

$$y = c_1 e^{\sqrt{K}x} + c_2 e^{-\sqrt{K}x} . \tag{c}$$

Applying the boundary conditions $y(0) = y(1) = 0$, one finds

$$c_1 + c_2 = 0 \tag{d}$$

$$c_1 e^{\sqrt{K}} + c_2 e^{-\sqrt{K}} = 0 \tag{e}$$

From (d), $c_2 = -c_1$. Inserting into (e),

$$c_1 e^{\sqrt{K}} - c_1 e^{-\sqrt{K}} = 0$$

or

$$c_1 e^{\sqrt{K}} (1 - e^{-2\sqrt{K}}) = 0.$$

Since

$$(1 - e^{-2\sqrt{K}}) \neq 0$$

(this would be possible only if $\sqrt{K} = 0$) and $e^{\sqrt{K}} \neq 0$, one concludes that $c_1 = 0$; and therefore $c_2 = 0$.

The equation (c) can be written only as

$$y = 0 e^{\sqrt{K}x} + 0 e^{-\sqrt{K}} = 0.$$

Thus, for $K < 0$, $y \equiv 0$ which is the trivial solution yielding no eigenfunctions nor eigenvalues.

Case 2: $K = 0$.

The equation (a) is now

$$y'' = 0 \tag{f}$$

The equation (f) may be integrated twice to give the solution:

$$y = c_1 + c_2 x \tag{g}$$

Applying the boundary conditions

$$c_1 = 0 \tag{h}$$

$$c_1 + c_2 = 0 \tag{i}$$

From (h) and (i) $c_1 = c_2 = 0$ and one gets $y \equiv 0$ as the only solution with $K = 0$.

Case 3: $K > 0$.

So far the trivial solutions have been found in the attempt to find eigenvalues. Now, letting $K > 0$ in the given equation (a), obtain the characteristic equation

$$m^2 + K = 0 \tag{j}$$

From (j) $m = \pm i\sqrt{K}$ and the general solution to (a) is

$$y = c_1 e^{i\sqrt{K}x} + c_2 e^{-i\sqrt{K}x} \tag{k}$$

Using Euler's formula

$$e^{iBx} = \cos Bx + i \sin Bx \tag{ℓ}$$

(k) may be transformed into

$$y = c_1 \cos \sqrt{K}x + c_2 \sin \sqrt{K}x \tag{m}$$

Applying the boundary conditions to (m)

$$y = c_1 = 0 \tag{n}$$

$$y = c_1 \cos \sqrt{K} + c_2 \sin \sqrt{K} = 0 \tag{o}$$

From (n) and (o),

$c_2 \sin \sqrt{K} = 0$ and either $c_2 = 0$ or

$$\sin \sqrt{K} = 0 \qquad (p)$$

If $c_2 = 0$, then (m) reduces to the trivial solution $y \equiv 0$ for all x. On the other hand, from (p), one gets $\sqrt{K} = n\pi$; $n = 1,2,3,\ldots$. Hence, the solution to the EVP is

$$y = y_n = c_n \sin n\pi x, \quad n = 1,2,3,\ldots$$

(which was obtained from (m) by letting $c_1 = 0$)

The eigenvalues are $K = (n\pi)^2$; $n = 1,2,3,\ldots$ and the corresponding eigenfunctions are $y = a_n \sin n\pi x$. Thus the EVP has an infinite number of solutions. Note that if y_n is an eigenfunction, then $z(x) = A(y_n)$ is also an eigenfunction where A is any constant.

● **PROBLEM 16-31**

For which values of λ does the boundary-value problem

$$y'' + \lambda y = 0 \qquad (1)$$

$$y(0) = 0, \quad y(1) = 0$$

have solutions other than $y = 0$?

Solution: The general solution to Eq. (1) is

$y(x) = a \cos(x\sqrt{\lambda}) + b \sin(x\sqrt{\lambda})$,

$y(0) = 0 \Rightarrow a = 0$,

$y(1) = 0 \Rightarrow \sqrt{\lambda} = n\pi, \quad n = 0, \pm 1, \pm 2, \ldots$

Thus

$$\lambda = n^2\pi^2, \quad n = 1,2,3,\ldots$$

(n = 0 yields the trivial solution $y = 0$; n = -k gives the same solution as n = k.) These values are called eigenvalues of the given problem (problem = differential equation + boundary conditions). The solution of the equation when λ is an eigenvalue are called eigenfunctions. In this problem the eigenfunctions $b \sin(n\pi x)$ belong to the eigenvalue $\lambda = n^2\pi^2$.

• **PROBLEM 16-32**

Consider the system

$$y_1' = y_2$$
$$y_2' = -\sin y_1 \tag{1}$$

which has rest points, where $y_1' = y_2' = 0$, on the line $y_2 = 0$ at $y_1 = m\pi$, $m = 0, \pm 1, \pm 2, \ldots$. Near these points the system has the form

$$y_1' = y_2$$
$$y_2' = -(\cos m\pi)(y_1 - m\pi) + O(y_1 - m\pi)^2,$$

so that near $y_1 = 0, \pm 2\pi, \pm 4\pi, \ldots$ the system is similar to that linear system with matrix

$$\begin{pmatrix} 0 & 1 \\ -1 & 0 \end{pmatrix},$$

while the relevant matrix near $y_1 = \pm\pi, \pm 3\pi, \ldots$ is

$$\begin{pmatrix} 0 & 1 \\ 1 & 0 \end{pmatrix};$$

the eigenvalues are $\pm i$ in the first place and ± 1 in the second. Draw solution curves for these conditions.

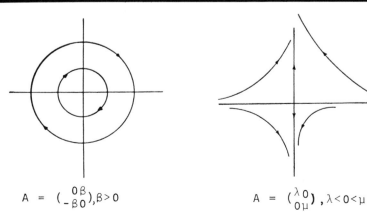

$A = \begin{pmatrix} 0 & \beta \\ -\beta & 0 \end{pmatrix}, \beta > 0$
Fig.1

$A = \begin{pmatrix} \lambda & 0 \\ 0 & \mu \end{pmatrix}, \lambda < 0 < \mu$
Fig.2

Solution: Putting these pictures together gives the partial picture of the solution curves for Eq. (1), as shown in Figures 1 and 2.

• **PROBLEM 16-33**

Consider the equation for a nonlinear spring,

$$y'' + y + y^3 = 0, \quad y(0) = \alpha, \quad y'(0) = 0,$$

and assume that α is small. Letting

$$y = \alpha x(t),$$

one finds

$$x'' + x + \alpha^2 x^3 = 0, \quad x(0) = 1, \quad x'(0) = 0.$$

Apply the expansion

$$x(t) = u_0(t) + \alpha^2 u_1(t) + \ldots,$$

to derive an equation for $y(t)$.

Solution: One gets

$$u_0'' + u_0 = 0, \quad u_0(0) = 1, \quad u_0'(0) = 0,$$

$$u_1'' + u_1 + u_0^3 = 0, \quad u_1(0) = u_1'(0) = 0$$

etcetera. This yields finally

$$y(t) = \alpha\left[\cos t + \alpha^2\left(\frac{\cos 3t}{32} - \frac{\cos t}{32} - \frac{3t \sin t}{8}\right)\right] + O(\alpha^5).$$

• **PROBLEM 16-34**

Consider the equation

$$x'' + x = 0, \quad x(0) = 0, \quad x(T) = x_T,$$

or equivalently,

$$y_1' = y_2$$

> $y_2' = -y_1$
>
> under the conditions $y(0) = 0$, $y(T) = x_T$, where
>
> $$y_1 = x, \quad y_2 = x'.$$
>
> Find the solution to these equations in the general form for linear problems, that is, $y' = Ay$, with
>
> $$A = \begin{pmatrix} 0 & 1 \\ -1 & 0 \end{pmatrix}, \quad M_1 = M_2 = \begin{pmatrix} 1 & 0 \\ 0 & 0 \end{pmatrix},$$
>
> $$k_1 = \begin{pmatrix} 0 \\ 0 \end{pmatrix}, \quad k_2 = \begin{pmatrix} x_T \\ 0 \end{pmatrix}.$$

Solution: The general solution to $y' = Ay$ is

$$y = e^{tA} y(0),$$

in this case

$$x(t) = y_1(t) = a \sin t$$

where a is not determined by the condition at $t = 0$. Thus it is required that

$$x_T = x(T) = a \sin T,$$

so that

$$a = \frac{x_T}{\sin T}$$

Therefore one has the principle of the alternative: (1) if T is not an integer multiple of π, there is a unique solution to the boundary value problem; (2) if T is an integer multiple of π and $x_T \neq 0$, there are no solutions; (3) if T is an integer multiple of π and $x_T = 0$, there are infinitely many solutions, namely $a \sin t$ for all a.

• **PROBLEM 16-35**

Given

$$\frac{d^2y}{dx^2} - (x^2 + 100)^2 y = 0,$$

determine an approximate expression for y over the range 0 < x < 10.

Solution: Let x = 10z, then

$$\frac{d^2y}{dz^2} - 10^6(z^2 + 1)^2 y = 0.$$

Thus h = 1,000 and $\psi = (z^2 + 1)^2$

$$\int^z \psi^{\frac{1}{2}} dz = \frac{z^3}{3} + z$$

and therefore an approximate solution

$$= (z^2 + 1)^{-\frac{1}{2}} + e^{\pm 1000 \left(\frac{z^3}{3} + z\right)}$$

$$= \frac{1}{10}(x^2 + 100)^{-\frac{1}{2}} \cdot e^{\pm \frac{z^3}{3} + 100x}$$

Finally

$$y = (x^2 + 100)^{-\frac{1}{2}}\left[Ae^{\frac{x^3}{3} + 100x} + Be^{-\frac{x^3}{3} - 100x}\right].$$

• **PROBLEM 16-36**

Solve the Dirichlet problem

$$\nabla^2 u = 0 \text{ in } R$$

$$u = x^2 + y^2 \text{ on } \partial R$$

where R is the semicircle $x^2+y^2 < 1$, y > 0 and ∂R is the boundary of R, with h = 1/2.

Solution: The theoretical solution is given by

$$u(x,y) = x^2 - y^2$$

$$+ \sum_{m=0}^{\infty} \frac{-16}{\pi(2m-1)(2m+1)(2m+3)} r^{2m+1} \sin(2m+1)\theta$$

where $x = r\cos\theta$, $y = r\sin\theta$.

The region R is covered by a square network with h = 1/2

(Figure 1). Due to the symmetry about the y-axis, i.e. $u(-x,y) = u(x,y)$, the Dirichlet problem is solved in the first quadrant only.

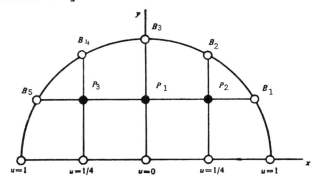

● -internal and ○-boundary pivots Fig.1

The Laplace equation, $\nabla^2 u = 0$, is replaced at the internal boundary pivots by the 5-point difference scheme:

$$h^2 \nabla^2 u_{\ell,m} = 2\left[\frac{1}{(\delta_1+\delta_3)}\left(\frac{u_1}{\delta_1} + \frac{u_3}{\delta_3}\right) - \left(\frac{1}{\delta_2\delta_4} + \frac{1}{\delta_1\delta_3}\right)u_0 \right.$$
$$\left. + \frac{1}{\delta_2+\delta_4}\left(\frac{u_2}{\delta_2} + \frac{u_4}{\delta_4}\right)\right]$$

(where the pivot x_ℓ, y_m is denoted by ℓ,m and $u_{\ell,m}$ is the approximate value of $u(x_\ell,y_m)$).

One gets the following two equations

for $P_1\left(0, \frac{1}{2}\right)$, $u_1 = \frac{1}{2}u_2 + \frac{1}{4}$

for $P_2\left(\frac{1}{2}, \frac{1}{2}\right)$, $u_2 = \frac{\sqrt{3}-1}{2\sqrt{3}} u_1 + \frac{\sqrt{3}+1}{2\sqrt{3}}$

Thus, one finds

$u_1 = 0.631854$, $u_2 = 0.763708$

● **PROBLEM 16-37**

Obtain the numerical solution to the mixed boundary value problem

$$y'' = y - 4xe^x$$
$$y'(0) - y(0) = 1, \quad y'(1) + y(1) = -e$$

with step length $h = 1/4$. Use the Numerov method.

Solution: The analytical solution is given by

$$y(x) = x(1-x)e^x$$

Subdivide the interval [0, 1] into four subintervals; the nodal points are $x_n = nh$, $0 \leq n \leq 4$ and $h = 1/4$.
The Numerov method gives the following system of equations

$$-191y_{n-1} + 394y_n - 191y_{n+1}$$
$$= 4x_{n-1}e^{x_{n-1}} + 40x_n e^{x_n} + 4x_{n+1}e^{x_{n+1}}, \quad 1 \leq n \leq 3$$

The boundary conditions become

$$y'_0 - y_0 = 1, \quad y'_4 + y_4 = -e$$

In order to approximate the boundary conditions, consider the identity

$$y(x_1) = y(x_0) + hy'(x_0) + h^2 PD^2 y(x_0)$$

where the operator P is given by

$$P = (E - 1 - hD)(hD)^{-2}$$
$$= (\Delta - \log(1+\Delta))\left(\Delta - \tfrac{1}{2}\Delta^2 + \tfrac{1}{3}\Delta^3 \ldots\right)^{-2}$$
$$= \tfrac{1}{2}\left(1 + \tfrac{1}{3}\Delta - \tfrac{1}{12}\Delta^2 + \tfrac{2}{45}\Delta^3 - \tfrac{7}{240}\Delta^4 + \ldots\right)$$

Thus, one gets

$$y(x_1) = y(x_0) + h\, y'(x_0) + \tfrac{h^2}{2}\left(1 + \tfrac{1}{3}\Delta - \tfrac{1}{12}\Delta^2 + \tfrac{2}{45}\Delta^3 - \ldots\right) y''(x_0)$$

Similarly, one gets for the second boundary condition

$$y(x_3) = y(x_4) - h\, y'(x_4) + \tfrac{h^2}{2}\left(1 - \tfrac{1}{3}\nabla - \tfrac{1}{12}\nabla^2 - \tfrac{2}{45}\nabla^3 - \ldots\right) y''(x_4)$$

Now, obtain various order approximations to the boundary conditions. The Numerov method has local truncation error of order h^6. Therefore, in order to approximate the boundary conditions to the same order, retain third difference in the previous expressions and get

$$7297\, y_0 - 5646\, y_1 - 39\, y_2 + 8\, y_3$$
$$= -1440 + 4(8\, x_3 e^{x_3} - 39\, x_2 e^{x_3} + 114\, x_1 e^{x_1} + 97\, x_0 e^{x_0})$$

$$8\, y_1 - 39\, y_2 - 5646\, y_3 + 7297\, y_4$$
$$= -1440\, e + 4(97\, x_4 e^{x_4} + 114\, x_3 e^{x_3} - 39\, x_2 e^{x_2} + 8x_1 e^{x_1})$$

The solution of the linear system of equations and the exact values of y(x) at x_i, $0 \le i \le 4$, are given in Table 1.

Table 1

x_n	$y(x_n)$	y_n	$e_n = y(x_n) - y_n$
0	0	0.622525 —03	—0.622525 —03
1	0.240754	0.241416	—0.661249 —03
2	0.412180	0.412886	—0.706125 —03
3	0.396937	0.397682	—0.744520 —03
4	0	0.759196 —03	—0.759196 —03

Solution of $y'' = y - 4x\, e^x$, $y'(0) - y(0) = 1$, $y'(1) + y(1) = -e$ with $h = 1/4$

OTHER METHODS FOR SOLVING INITIAL - VALUE PROBLEMS IN ORDINARY DIFFERENTIAL EQUATIONS

● **PROBLEM** 16-38

(a) Use Picard's method to find a third approximation to the solution of the differential equation $dy/dx = y' = 1 + xy$ with the initial condition being $y(0) = 1$.

(b) To three decimal places, tabulate the solution from $x = 0$ to $x = 0.3$ with $\Delta x = 0.1$.

Solution: (a) The solution of the differential equation $y' = dy/dx = f(x,y)$ can be written in the form

$$y = y_0 + \int_{x_0}^{x} f(x,y)\, dx,$$

where the integration is from (x_0, y_0) to (x,y). Since $y(x)$ is not known, the actual integration can not be performed, but one can get a first approximation for $y(x)$, call it y_1, by substituting y_0 for y in the integrand. That is,

$$y_1 = y_0 + \int_{x_0}^{x} f(x, y_0)\, dx$$

and, similarly,

$$y_2 = y_0 + \int_{x_0}^{x} f(x, y_1)\, dx.$$

This is the Picard (iterative) method and in its general form, for the nth approximation, one gets

$$y_n = y_0 + \int_{x_0}^{x} f(x, y_{n-1}) dx.$$

For the given problem, one gets

$$y_1 = 1 + \int_0^x [1 + x(1)] dx = 1 + x + x^2/2$$

$$y_2 = 1 + \int_0^x [1 + x(1 + x + x^2/2)] dx =$$

$$1 + x + \frac{x^2}{2} + \frac{x^3}{3} + \frac{x^4}{8}$$

$$y_3 = 1 + \int_0^x [1 + x(1 + x + x^2/2 + x^3/3 + x^4/8)] dx$$

$$= 1 + x + x^2/2 + x^3/3 + x^4/8 + x^5/15 + x^6/48$$

(b) The above result should give fairly accurate results for y(x) in the neighborhood of x = 0. If one substitutes x = 0, 0.1, 0.2, 0.3 and computes the resulting values of y(x), the results can be tabulated as follows:

x	0	0.1	0.2	0.3
y	1	1.105	1.223	1.355

● **PROBLEM** 16-39

Apply the Picard method to the solution of

$$dy/dx = y - x$$

with the initial condition that y = 2 at x = 0.

Solution: Write

$$y^{(n+1)} = 2 + \int_0^x (y^{(n)} - x) dx.$$

Taking $y^{(1)} = 2$, one obtains successively

$$y^{(2)} = 2 + \int_0^x (2 - x)\,dx = 2 + 2x - \frac{x^2}{2},$$

$$y^{(3)} = 2 + \int_0^x \left(2 + x - \frac{x^2}{2}\right) dx$$

$$= 2 + 2x + \frac{x^2}{2} - \frac{x^3}{6},$$

$$y^{(4)} = 2 + \int_0^x \left(2 + x + \frac{x^2}{2} - \frac{x^3}{6}\right) dx$$

$$= 2 + 2x + \frac{x^2}{2} + \frac{x^3}{6} - \frac{x^4}{24},$$

and so forth. In this case it is readily shown by induction that

$$y^{(n+1)} = 2 + 2x + \frac{x^2}{2!} + \frac{x^3}{3!} + \ldots + \frac{x^n}{n!} - \frac{x^{n+1}}{(n+1)!},$$

and as n becomes large one is led to a consideration of the infinite series

$$y = 1 + x + \left(1 + \frac{x}{1!} + \frac{x^2}{2!} + \ldots + \frac{x^n}{n!} + \ldots\right),$$

which may be recognized as the expansion of the known exact solution

$$y = 1 + x + e^x,$$

the expansion converging for all values of x.

● **PROBLEM 16-40**

Find the particular solution of $y' = x + y^2$ which passes through the point $(0,1)$.

Solution: It is readily verified that the conditions for a solution to exist hold. The first approximation is the ordinate of the given point, $y_0 = 1$; hence, from

$$y_n = f_n(x) = y_0 + \int_{x_0}^{x} F(x, y_{n-1}) dx, \tag{1}$$

$$y_1 = 1 + \int_0^x (x + 1) dx,$$

or

$$y_1 = 1 + x + \frac{1}{2} x^2.$$

Substituting this second approximation into (1), one gets

$$y_2 = 1 + \int_0^x [x + (1 + x + \frac{1}{2} x^2)^2] dx,$$

or

$$y_2 = 1 + x + \frac{3}{2} x^2 + \frac{2}{3} x^3 + \frac{1}{4} x^4 + \frac{1}{20} x^5.$$

Similarly,

$$y_3 = 1 + x + \frac{3}{2} x^2 + \frac{4}{3} x^3 + \frac{13}{12} x^4 + \frac{49}{60} x^5 + \frac{13}{30} x^6$$
$$+ \frac{233}{1260} x^7 + \frac{29}{480} x^8 + \frac{31}{2160} x^9 + \frac{1}{400} x^{10}$$
$$+ \frac{1}{4400} x^{11}.$$

It is becoming clear that the sequence $y_0, y_1, y_2, \ldots,$ is approaching a function whose Maclaurin expansion starts

$$y = 1 + x + \frac{3}{2} x^2 + \frac{4}{3} x^3 + \ldots .$$

• **PROBLEM 16-41**

Using Picard's Method obtain, with an error not exceeding 10^{-4} when $x < 0.1$, the solution of $dF/dx = 3x^{-\frac{1}{2}} + F - 1$ for which $F = 1$ when $x = 0$.

Solution: Taylor's series cannot be used here, because $\overline{dF/dx}$ is not finite when $x = 0$.

Assuming $F_0 = 1$

then

$$F_1 = 1 + \int_0^x 3x^{-\frac{1}{2}} dx = 1 + 6x^{\frac{1}{2}}$$

$$F_2 = 1 + \int_0^x (3x^{-\frac{1}{2}} + 6x^{\frac{1}{2}}) dx = 1 + 6x^{\frac{1}{2}} + 4x^{3/2}$$

and hence

$$F = 1 + 6x^{\frac{1}{2}} + 4x^{3/2} + (8/5)x^{5/2} + (16/35)x^{7/2}$$
$$+ (32/315)x^{9/2} + \ldots$$

each term being the integral of the preceding one. Although all terms are positive, it is evident that the remainder after the term in $x^{7/2}$ is less than

$$(32/315)x^{9/2}(1 - 2x/11)^{-1}$$

This is less than 5×10^{-5} provided $x < 0.11$, so that the conditions are met by ignoring terms beyond that in $x^{7/2}$.

The solution need not always be expressed in powers of x, fractional or otherwise. Thus if $dF/dx + F = e^{-x}$ and $F = 0$, when $x = 0$, successive approximations to F are

$1 - e^{-x}$

$2(1 - e^{-x}) - x$

$3(1 - e^{-x}) - 2x + (1/2)x^2$

$4(1 - e^{-x}) - 3x + x^2 - (1/6)x^3$

and so on.

The exact solution is of course xe^{-x}, and Taylor's series leads at once to the expansion of this in powers of x.

• PROBLEM 16-42

For the differential equation $y' = 1 + xy$ with initial condition $(0,1)$, determine the first six terms of a Taylor's Series solution and then tabulate the solution for $x = 0, 0.1, 0.2, 0.3$, to three decimal places.

Solution: For solving problems such as the above, it is convenient to put the Taylor Series in the following form:

$$y(x) = y_0 + (y_0'/1!)(x-x_0) + (y_0''/2!)(x-x_0)^2 +$$
$$(y_0'''/3!)(x-x_0)^3 + (y_0^{iv}/4!)(x-x_0)^4 +$$
$$(y_0^V/5!)(x-x_0)^5 + \ldots$$

For the given problem, the initial values are $x_0 = 0$ and $y_0 = 1$. Now, compute the needed derivatives and then their values at $x = x_0 = 0$.

$y' = 1 + xy$

$y'' = xy' + y$

$y''' = xy'' + y' + y' = xy'' + 2y'$

$y^{iv} = xy''' + 3y''$

$y^V = xy^{iv} + 4y'''$

$y_0' = 1 + 0 = 1$

$y_0'' = 0 + y_0 = 1$

$y_0''' = 0 + 2(1) = 2$

$y_0^{iv} = 0 + 3(1) = 3$

$y_0^V = 0 + 4(2) = 8$

Now substituting into the above equation for $y(x)$, one gets

$$y = 1 + x + x^2/2 + x^3/3 + x^4/8 + x^5/15 + \ldots$$

$y(0.1) = 1 + 0.1 + 0.005 + 0.0003 + 0.0000 + 0.0000$

$\qquad = 1.105$

$y(0.2) = 1 + 0.2 + 0.02 + 0.0027 + 0.0002 + 0.0000$

$\qquad = 1.223$

$y(0.3) = 1 + 0.3 + 0.045 + 0.009 + 0.0010 + 0.0002$

$\qquad = 1.355$

The tabulated results are:

x	0	0.1	0.2	0.3
y	1	1.105	1.223	1.355

The following table shows the results obtained by Picard's method, Euler's method, and Taylor Series method, for comparison.

x	0	0.1	0.2	0.3
y (Picard)	1	1.105	1.223	1.355
y (Euler)	1	1.106	1.224	1.357
y (Taylor)	1	1.105	1.223	1.355

It is seen that the Picard and Taylor Series methods agree, but there is a slight difference in the results given by Euler's method.

• **PROBLEM 16-43**

Illustrate the use of Taylor series in differential equations by considering the solution of the differential equation

$$\frac{dy}{dx} = y - x$$

with the initial condition that $y = 2$ at $x = 0$. Choose an interval $h = 0.1$, and then calculate successively the approximate values of y at $x = 0.1, 0.2, 0.3$, and so on.

Solution: By successive differentiation, one obtains

$$y' = y - x, \quad y'' = y' - 1, \quad y''' = y'', \ldots$$

Hence, at $x = 0$, one gets

$$y_0 = 2$$

and

$$y_0' = 2, \quad y_0'' = 1, \quad y_0''' = 1, \ldots,$$

and, with $k = 0$, the following series

$$y_{k+1} = y_k + y_k' \frac{h}{1!} + y_k'' \frac{h^2}{2!} + \ldots \tag{1}$$

becomes

$$y_1 = y_0 + 2h + \frac{h^2}{2} + \frac{h^3}{6} + \ldots$$

$$\cong 2 + 0.2000 + 0.0050 + 0.0002 + \ldots = 2.2052.$$

Next, at $x = 0.1$, one gets

$$y_1 \cong 2.2052$$

and

$$y_1' \cong 2.1052, \qquad y_1'' \cong 1.1052, \qquad y_1''' \cong 1.1052, \ldots,$$

and, with $k = 1$, equation (1) gives

$$y_2 \cong y_1 + 2.1052h + 0.5526h^2 + 0.1842h^3 + \ldots$$

$$\cong 2.2052 + 0.2105 + 0.0055 + 0.0002 + \ldots$$

$$= 2.4214.$$

The procedure may be repeated as often as is required. In this problem the exact solution is readily found to be

$$y = e^x + x + 1,$$

and the above results are found to be accurate to the four decimal places retained.

• **PROBLEM 16-44**

Find the solution, using Taylor series, to the nonlinear differential equation

$$\frac{d^2y}{dx^2} - \frac{dy}{dx} + xy^2 = 0$$

with the initial conditions that $y = 1$ and $y' = -1$ when $x = 0$; take $h = 0.1$.

Solution: The successive derivatives are calculated

$$y'' = y' - xy^2, \qquad y''' = y'' - 2xyy' - y^2,$$

$$y^{iv} = y''' - 2xy'^2 - 2xyy'' - 4yy', \ldots$$

Hence, at $x = 0$, one has $y_0 = 1$, $y_0' = -1$,

and

$$y_0'' = y_0' = -1, \quad y_0''' = y_0'' - y_0^2 = -2, \quad y_0^{iv} = 2, \ldots$$

Then, with $k = 0$,

$$y_{k+1} = y_k + y_k' \frac{h}{1!} + y_k'' \frac{h^2}{2!} + \ldots$$

gives

$$y_1 = 1 - h - \frac{h^2}{2} - \frac{h^3}{3} + \frac{h^4}{12} + \ldots,$$

and, taking $h = 0.1$,

$$y_1 \cong 1 - 0.1 - 0.005 - 0.0003 + \ldots = 0.8947.$$

Now, in order to calculate y_2 it will be necessary next to calculate y_1'', y_1''', and so on. However, the calculation of these values involves knowledge of the value y_1' in addition to the value of y_1, which is now known. The value of y_1' can be calculated by using the series

$$y_1' = -1 - h - h^2 + \frac{h^3}{3} + \ldots,$$

which is obtained by differentiating the series defining $y_1 = y(x_0 + h)$ with respect to h. Hence one obtains

$$y_1' \cong -1 - 0.1 - 0.01 + 0.0003 + \ldots = -1.1097.$$

The values of y_1'', y_1''', and so on, can now be calculated from the forms given, and the calculation of y_2 and y_2' proceeds in the same way.

● **PROBLEM 16-45**

Solve the differential equation

$$y' = t + y, \quad y(0) = 1, \quad t \in [0,1]$$

by Taylor's series method, and determine the number of terms to be included in Taylor's series to obtain an accuracy of 10^{-10}.

Solution: The high order derivatives of $y(t)$ can be calculated by successively differentiating the differential equation

$$y' = t + y$$

$$y'' = 1 + y'$$

$$y''' = y''$$

and

$$y^{(r+1)} = y^{(r)}, \quad r = 2, 3, \ldots$$

Also, one has

$$y'(0) = 1$$

$$y''(0) = 2$$

$$y^{(r)}(0) = 2, \quad r = 3, 4 \ldots$$

Therefore one obtains the truncated Taylor Series

$$y(t) = 1 + t + t^2 + \frac{2t^3}{3!} + \ldots + \frac{2t^p}{p!}.$$

To get results accurate up to 10^{-10} for $t \leq 1$, require by the Taylor Series' Remainder formula that

$$\frac{(t - t_0)^{p+1}}{(p+1)!} \, y^{(p+1)}(\xi_t) \quad \text{for} \quad \xi_t \in [0,1]$$

be less than 5×10^{-11}. Since $y^{(p+1)}(t) = 1 + t + y(t)$, one has

$$\frac{(t)^{p+1}}{(p+1)!} (1 + t + y(t))$$

$$= \frac{(t)^{p+1}}{(p+1)!} (1 + t + 1 + t + t^2 + \frac{2t^3}{3!} + \ldots + \frac{2t^p}{p!} + \ldots)$$

$$= \frac{(t)^{p+1}}{(p+1)!} (2 + 2t + t^2 + \frac{2t^3}{3!} + \ldots + \frac{2t^p}{p!} + \ldots)$$

$$= \frac{(t)^{p+1}}{(p+1)!} 2(1 + t + \frac{t^2}{2!} + \frac{t^3}{3!} + \ldots + \frac{t^p}{p!} + \ldots)$$

677

$$= \frac{(t)^{p+1}}{(p+1)!} 2e$$

which takes on maximum value on [0,1] at t = 1.

Hence, require

$$\frac{2e}{(p+1)!} < 5 \times 10^{-11}$$

which gives $p \cong 15$. Hence it follows that about 15 terms are required to achieve the accuracy in the range $t \leq 1$.

• **PROBLEM 16-46**

For $a = x_0 = 0$ and $h = 0.3$, find y_0, y_1, and y_2 by means of a third order Taylor series approximation for the initial value problem

$$y'' = y^2 - x^2 \qquad (1)$$

$$y(0) = 0, \quad y'(0) = 1. \qquad (2)$$

Solution: Since $a = 0$ and $h = 0.3$, the grid points of interest are $a = x_0 = 0$, $x_1 = 0.3$, $x_2 = 0.6$. From (2)

$$y_0 = 0, \quad y'_0 = 1. \qquad (3)$$

Since third-order Taylor series approximations are sought, the following two equations,

$$y_{k+1} = y_k + hy'_k + \frac{h^2}{2} y''_k + \frac{h^3}{3!} y'''_k + \ldots$$

$$+ \frac{h^n}{n!} y_k^{(n)}; \quad k = 0, 1, \ldots$$

and

$$y'_{k+1} = y'_k + hy''_k + \frac{h^2}{2} y'''_k + \frac{h^3}{3!} y_k^{iv} + \ldots$$

$$+ \frac{h^n}{n!} y_k^{(n+1)}; \quad k = 0, 1, \ldots,$$

take the form of

$$y_{k+1} = y_k + (0.3) y'_k + (0.045) y''_k + (0.0045) y'''_k ;$$

$$k = 0, 1 \qquad (4)$$

$$y'_{k+1} = y'_k + (0.3)y''_k + (0.045)y'''_k + (0.0045)y^{iv}_k;$$

$$k = 0,1. \qquad (5)$$

Also, from (1),

$$y'' = y^2 - x^2 \qquad (6)$$

$$y''' = 2yy' - 2x \qquad (7)$$

$$y^{iv} = 2(y')^2 + 2yy'' - 2. \qquad (8)$$

Now, for $k = 0$, one obtains, from (3) and (6)-(8),

$$y_0 = 0, \; y'_0 = 1, \; y''_0 = y_0^2 - x_0^2 = 0,$$

$$y'''_0 = 2y_0 y'_0 - 2x_0 = 0,$$

$$y^{iv}_0 = 2(y'_0)^2 + 2y_0 y''_0 - 2 = 0,$$

which, upon substitution into (4) and (5) implies

$$y_1 = 0.3, \; y'_1 = 1. \qquad (9)$$

Next, for $k = 1$, one obtains, from (6)-(9),

$$y_1 = 0.3, \; y'_1 = 1, \; y''_1 = y_1^2 - x_1^2 = 0,$$

$$y'''_1 = 2y_1 y'_1 - 2x_1 = 0,$$

$$y^{iv}_1 = 2(y'_1)^2 + 2y_1 y''_1 - 2 = 0.$$

which, upon substitution into (4) and (5), yields

$$y_2 = 0.6, \; y'_2 = 1,$$

and the answer is complete. Note that the numerical solution suggests that the exact solution is $y = x$.

• **PROBLEM** 16-47

The Taylor method has low truncation error and the results are relatively insensitive to the value of the step size. The method is most suitable when the higher order derivatives can be generated recursively.

Consider

$$y' = 1 - y/x$$

for which

$$y'' = (1 - 2y')/x \qquad (1)$$

and higher-order derivatives are calculated from

$$y^{(k)} = -ky^{(k-1)}/x \qquad k = 3,4,5,\ldots \qquad (2)$$

With the initial condition $y(a) = s$, the particular solution of the differential equation is

$$y = (a/x)(s - \tfrac{1}{2}a) + x/2 \qquad (3)$$

(a) Differentiate $y' = 1 - y/x$ successively to derive relationships (1) and (2).

(b) Integrate $y' = 1 - y/x$ analytically to obtain the solution (3). One approach is to let $y = zx$ and replace $y' = dy/dx$ by dz/dx. What is the particular solution for the initial condition $y(0) = 1$? What difficulty could arise in computer coding the problem to include this condition?

Solution: (a) $y' = 1 - y/x$, $y'' = -y'/x + y/x^2 = (1 - 2y')/x$. The higher-order derivatives are obtained in the same way.

(b) Letting $z = y/x$ gives $dz/dx = (1 - 2z)/x$ which integrates to $\ln[x^2(1 - 2z)] = c$. The constant c is found using the condition $z = s/a$ when $x = a$. Replacement of z in terms of y then gives the required answer. $y' = -\infty$ when $x = 0$; the computer would produce a machine overflow.

• **PROBLEM** 16-48

Show that the solution of $y' = y - x$ with the initial condition $y(a) = s$ is

$$y = (s - a - 1)e^{x-a} + x + 1.$$

The substitution $z = y - x$ is helpful. Would the Taylor method be suitable for solving the differential equation?

Solution: Letting $z = y - x$ gives $dz/dx = z - 1$ which integrates to $\ln(z - 1) = x + c$. The constant c is found from the condition $z = s - a$ when $x = a$. Replacement of z in terms of y then gives the required answer. The Taylor method is appropriate since the higher-order derivatives can be written recursively $y^{(k)} = y^{(k-1)}$, $k = 2, 3, 4, \ldots$.

• **PROBLEM 16-49**

Solve the differential system

$$y' = 2y \qquad y(0) = 1$$

by hand using the Taylor series method, of order two:

$$y_{r+1} = y_r + hy_r' + \frac{1}{2}h^2 y_r'',$$

and also the second-order, Runge-Kutta (modified Euler) method, and the predictor-corrector (improved Euler) method.

Find the solutions at $x = 0.1$ and 0.2 with step size $h = 0.1$ and without iterating (s=1) the Euler corrector. Repeat the improved Euler solution by iterating the corrector three times (s=4) at each step. Calculate the exact solution $y = e^{2x}$.

Solution: For the Taylor method, recalling that $y_r' = 2y_r$,

$$y_{r+1} = y_r + hy_r' + h^2 y_r''/2$$

$$= y_r + 2hy_r + 2h^2 y_r$$

$$= y_r[1 + 2h(1 + h)].$$

Proceeding to the second-order Runge-Kutta (modified Euler) method:

$$y_{r+1} = y_r + hk_2$$

$$k_1 = f(x_r, y_r)$$

$$k_2 = f(x_r + \frac{1}{2}h, y_r + \frac{1}{2}hk_1),$$

and the predictor-corrector (improved Euler) method:

$$y_{r+1}^{(0)} = y_r + hf(x_r, y_r)$$

$$y_{r+1}^{(s)} = y_r + \frac{1}{2} h [f(x_r,y_r) + f(x_{r+1},y_{r+1}^{(s-1)})]$$

$$s = 1,2,3,\ldots$$

show that both the modified and improved (non-iterated) Euler methods produce the same second-order recurrence relationship. With $h = 0.1$, $y_{r+1} = 1.22 y_r$, giving $y(0.1) = 1.22000$ and $y(0.2) = 1.48840$. Again using $y_r' = 2y_r$, show that the equations for the predictor-corrector method can be written as

$$y_{r+1}^{(0)} = y_r (1 + 2h)$$

$$y_{r+1}^{(s)} = y_r (1 + h) + hy_{r+1}^{(s-1)} \qquad s = 1,2,3,\ldots$$

On iterating for three cycles at each step, one obtains

x	$y(x)$	$y^{(4)}$	E_r	e_r	(1) pred.	% difference
0.1	1.22140	1.22222	−0.00082	−0.00082	−0.00077	6
0.2	1.49182	1.49382	−0.00200	−0.00118	−0.00094	20

where $y(x)$ is the analytical result and E_r and e_r are, respectively, the global and local truncation errors. The discrepancy between the prediction of the general equation

$$e^*_{r+1} = (y_{r+1}^{(0)} - y^*_{r+1})/29$$

and e_2 can be reduced by reducing the step size.

● **PROBLEM 16-50**

Solve the initial-value problem for a non-linear spring,

$$\ddot{x} + kx + k'x^3 = 0$$

$$x(0) = c$$
$$\dot{x}(0) = 0,$$

using the perturbation series method. For simplicity set spring constants $k = k' = 1$. Assume c, the initial distortion, to be small.

Solution: Perturbation series is a particularly powerful method of approaching a non-linear differential equation because of its wide applicability and its adaptability to numerical or computer work. The method consists of expanding a solution in powers of a small parameter so that for sufficient accuracy in practical work only a few lower order terms need be kept.

Physically what makes this technique appropriate here is that c, the initial distortion be small. Then, let:

$$x = c\nu \quad \text{to give}$$

$$\ddot{\nu} + \nu + c^2\nu^3 = 0 \qquad \begin{array}{l} \nu(0) = 1 \\ \dot{\nu}(0) = 0. \end{array} \qquad (a)$$

(This normalizes the initial condition and displays the expansion parameter.) Now replace c^2 by ε (the usual notation) and solve the equation:

$$\ddot{\nu} + \nu + \varepsilon\nu^3 = 0. \qquad (b)$$

Assume the solution to be

$$\nu = \nu_0 + \varepsilon\nu_1 + \varepsilon^2\nu_2 + \ldots \qquad (c)$$

and substitute into (b):

$$(\ddot{\nu}_0 + \varepsilon\ddot{\nu}_1 + \varepsilon^2\ddot{\nu}_2 + \ldots) + (\nu_0 + \varepsilon\nu_1 + \varepsilon^2\nu_2 \ldots)$$

$$+ \varepsilon(\nu_0 + \varepsilon\nu_1 + \varepsilon^2\nu_2 + \ldots)^3 = 0.$$

Collecting coefficients of powers of ε

$$(\ddot{\nu}_0 + \nu_0) + \varepsilon(\ddot{\nu}_1 + \nu_1 + \nu_0^3) +$$

$$\varepsilon^2(\ddot{\nu}_2 + \nu_2 + 3\nu_0^2\nu_1) + \ldots = 0. \qquad (d)$$

Higher order terms are ignored throughout because of the assumption that $\varepsilon = c^2$ is small.

Now one has a set of simultaneous equations from the coefficients:

$$\ddot{\nu}_0 + \nu_0 = 0 \qquad \nu_0(0) = 1, \qquad \dot{\nu}(0) = 0;$$

$$\ddot{\nu}_1 + \nu_1 + \nu_0^3 = 0 \qquad \nu_1(0) = 0, \qquad \dot{\nu}_1(0) = 0; \qquad (e)$$

683

$$\ddot{v}_2 + v_2 + 3v_0{}^2 v_1 = 0 \qquad v_2(0) = 0, \qquad \dot{v}_2(0) = 0.$$

Notice that one now has a system of equations, linear in the differentiated function, which can be solved by existing techniques. For instance, by inspection $v_0 = \cos t$. This means that for v_1

$$\ddot{v}_1 + v_1 + \cos^3 t = 0. \tag{f}$$

To linearize this equation, substitute

$$\cos^3 t = \frac{3 \cos t}{4} + \frac{\cos 3t}{4}$$

to get:

$$\ddot{v}_1 + v_1 = -\frac{3 \cos t}{4} - \frac{\cos 3t}{4}. \tag{g}$$

By the method of undertermined coefficients the particular solution can be shown to be:

$$v_{1_p} = -\frac{3x \sin t}{8} + \frac{\cos 3t}{32}. \tag{h}$$

● **PROBLEM** 16-51

Find the solution to Bessel's equation of order zero,

$$x \frac{d^2 y}{dx^2} + \frac{dy}{dx} + xy = 0,$$

whose graph has the slope 2 at the point (1,1).

Assume that all the algebraic operations performed on the power series are permissible, at least in some region about the point (1,1).

Solution: Assume the solution can be written in the form

$$y = a_0 + a_1(x - 1) + a_2(x - 1)^2 + \ldots + a_n(x - 1)^n$$
$$+ \ldots .$$

One finds

$$y' = a_1 + 2a_2(x - 1) + \ldots + (n + 1)a_{n+1}(x - 1)^n + \ldots ,$$

684

$$y'' = 1 \cdot 2a_2 + 2 \cdot 3a_3(x-1) + \ldots$$
$$+ (n+1)(n+2)a_{n+2}(x-1)^n + \ldots \quad .$$

Write the differential equation in the form

$$(x-1)y'' + y'' + y' + (x-1)y + y = 0,$$

and substitute the power series expressions for y, y', and y''. Obtain

$$(a_0 + 1^2 a_1 + 1 \cdot 2a_2) + (a_0 + a_1 + 2^2 a_2 + 2 \cdot 3a_3)(x-1)$$

$$+ (a_1 + a_2 + 3^2 a_3 + 3 \cdot 4a_4)(x-1)^2$$

$$+ (a_2 + a_3 + 4^2 a_4 + 4 \cdot 5a_5)(x-1)^3$$

$$+ \ldots\ldots\ldots\ldots\ldots\ldots\ldots\ldots\ldots\ldots\ldots\ldots$$

$$+ (a_{n-1} + a_n + (n+1)^2 a_{n+1}$$

$$+ (n+1)(n+2)a_{n+2})(x-1)^n$$

$$+ \ldots\ldots\ldots\ldots\ldots\ldots\ldots\ldots\ldots\ldots\ldots\ldots$$

$$= 0.$$

The preceding equality yeilds

$$a_0 + 1^2 a_1 + 1 \cdot 2a_2 = 0$$
$$a_0 + a_1 + 2^2 a_2 + 2 \cdot 3a_3 = 0$$
$$a_1 + a_2 + 3^2 a_3 + 3 \cdot 4a_4 = 0$$
$$\ldots\ldots\ldots\ldots\ldots\ldots\ldots\ldots \quad .$$

Obtain from these equations, since $y_0 = f(1) = a_0 = 1$, $y_0' = f'(1) = a_1 = 2$, the values $a_2 = -3/2$, $a_3 = 1/2$, $a_4 = -5/12$, $a_5 = 23/60$, $\ldots$. Hence

$$y = 1 + 2(x-1) - \frac{3}{2}(x-1)^2 + \frac{1}{2}(x-1)^3 - \frac{5}{12}(x-1)^4$$

$$+ \frac{23}{60}(x-1)^5 + \ldots ,$$

is the desired solution.

The alternate solution is also applicable. Write the differential equation in the form $xy'' + y' + xy = 0$. Obtain by successive differentiation

$$xy^{(3)} + 2y'' + xy' + y = 0$$

$$xy^{(4)} + 3y^{(3)} + xy'' + 2y' = 0$$

$$xy^{(5)} + 4y^{(4)} + xy^{(3)} + 3y'' = 0$$

$$\cdots\cdots\cdots\cdots\cdots\cdots\cdots\cdots\cdots\cdots\cdots .$$

Since $x = y = 1$, $y' = 2$, one finds that $y'' = -3$, $y^{(3)} = 3$, $y^{(4)} = -10$, $y^{(5)} = 46, \ldots$. These values yield the same values for $a_0, a_1, a_2, a_3, \ldots$, as before.

• **PROBLEM** 16-52

Find one solution of Bessel's equation of order p using the method of Frobenius.

Solution: Bessel's equation is a second order differential equation with variable coefficients. Thus, it is of the form

$$y'' + p(x)y' + a(x)y = 0. \tag{1}$$

In particular, Bessel's equation of order p is:

$$x^2 y'' + xy' + (x^2 - p^2)y = 0 \tag{2}$$

or, writing (2) to conform with (1),

$$y'' + \frac{1}{x} y' + \frac{x^2 - p^2}{x^2} y = 0, \tag{3}$$

where p is a parameter which denotes the order of the equation. Examining (3) it can be seen that the point $x = 0$ is a singular point, i.e., the coefficients $p(x)$, $a(x)$ are not analytic there. But since $x\,p(x)$ and $x^2\,a(x)$ are analytic, the singularity is weak and hence $x = 0$ is a regular singular point. A solution of (3) around $x = 0$ will be a valid solution in the interval $|x| < R$.

Now, use the method of Frobenius to obtain a solution of (3). Assume that

$$y(x) = x^m (a_0 + a_1 x + \ldots + a_n x^n + \ldots) \qquad (4)$$

$$= x^m \sum_{n=0}^{\infty} a_n x^n$$

$$= \sum_{n=0}^{\infty} a_n x^{m+n} .$$

Then, differentiating (4),

$$y' = m a_0 x^{m-1} + (m + 1) a_1 x^m + \ldots + (n + m) a_n x^{m+n-1} + \ldots$$

$$= \sum_{n=0}^{\infty} (m + n) a_n x^{m+n-1}, \qquad (5)$$

and

$$y'' = m(m - 1) a_0 x^{m-2} + (m + 1)(m) a_1 x^{m-1} + \ldots$$

$$= \sum_{n=0}^{\infty} (m + n)(m + n - 1) a_n x^{m+n-2}. \qquad (6)$$

Substitute these expressions for y, y', y'' into (2). Thus obtain

$$x^2 y'' = \sum_{n=0}^{\infty} (n + m - 1)(n + m) a_n x^{m+n}$$

$$xy' = \sum_{n=0}^{\infty} (n + m) a_n x^{m+n}$$

$$x^2 y = \sum_{n=0}^{\infty} a_n x^{m+n+2} = \sum_{n=0}^{\infty} a_{n-2} x^{n+m}$$

$$-p^2 y = \sum_{n=0}^{\infty} -p^2 a_n x^{n+m},$$

or

$$\sum_{n=0}^{\infty} [(n + m)(n + m - 1) + (n + m) - p^2] a_n x^{n+m}$$

$$+ \sum_{n=0}^{\infty} a_{n-2} x^{n+m} = 0 \qquad (7)$$

where the coefficients of like powers of x have been combined. Equation (7) may be rewritten as

$$[m(m-1) + m - p^2]a_0 x^m + [(m+1)m + (m+1) - p^2]a_1 x^{m+1}$$

$$+ \sum_{n=2}^{\infty} \left[((n+m)(n+m-1) + (n+m) - p^2)a_n + a_{n-2} \right] x^{n+m}$$

$$= 0 \qquad (8)$$

Since, by assumption $a_0 \neq 0$ (if it were, then (4) would take a different form), one finds the indicial equation is

$$m^2 - p^2 = 0. \qquad (9)$$

The roots of (9) are $m_1 = p$ and $m_2 = -p$. Further, one gets, from (8),

$$[(m+1)^2 - p^2]a_1 = 0, \qquad (10)$$

and in general,

$$[(m+n)^2 - p^2]a_n + a_{n-2} = 0, \qquad n \geq 2$$

or,

$$a_n = -\frac{1}{(m+n)^2 - p^2} a_{n-2} \qquad n \geq 2. \qquad (11)$$

Equation (11) is the general recurrence formula. Since $m = \pm p$, (10) implies that,

$$a_1 = 0, \qquad (12)$$

$$a_n = \frac{-1}{n(2p + n)} a_{n-2} \qquad (n \geq 2).$$

Hence,

$$a_1 = a_3 = a_5 = \ldots = 0$$

from (11) and (12) and

$$a_2 = \frac{-1}{2(2p + 2)} a_0 = \frac{-1}{2^2 1!(p + 1)} a_0,$$

$$a_4 = \frac{-1}{2^2 \cdot 2(p + 2)} a_2 = \frac{1}{2^4 2!(p + 2)(p + 1)} a_0,$$

$$a_6 = \frac{-1}{2^2 3(p + 3)} a_4 = \frac{-1}{2^6 3!(p + 3)(p + 2)(p + 1)} a_0$$

and generally,

$$a_{2k} = \frac{(-1)^k}{2^{2k} k!(p + k)(p + k - 1) \ldots (p + 1)} a_0, \quad k \geq 1.$$

Therefore, the solution to (2) is

$$y_1(x) = x^m \sum_{n=0}^{\infty} a_n x^n = x^p \left[a_0 + \sum_{k=1}^{\infty} a_{2k} x^{2k} \right]$$

$$= a_0 x^p \left[1 + \sum_{k=1}^{\infty} \frac{(-1)^k x^{2k}}{2^{2k} k!(p + k)(p + k - 1) \ldots (p + 2)(p + 1)} \right].$$

If

$$a_0 = \frac{1}{2^p \Gamma(p + 1)},$$

the solution may be expressed as

$$y_1(x) = \frac{1}{2^p \Gamma(p + 1)} x^p + \sum_{k=1}^{\infty} \frac{(-1)^k x^{2k + p}}{2^{2k+p} k! \Gamma(p + k + 1)}$$

where the property of the gamma function, namely

$$\Gamma(p + k + 1) = (p + k)(p + k - 1) \cdots$$

$$(p + 2)(p + 1)\Gamma(p + 1)$$

has been used.

Thus, the final solution is

$$y_1(x) = \sum_{k=0}^{\infty} \frac{(-1)^k x^{2k+p}}{2^{2k+p} k! \Gamma(p + k + 1)} J_p(x).$$

● **PROBLEM** 16-53

Derive asymptotic solutions of the equation

$$d^2G/dx^2 + (1 + 1/4x^2)G = 0$$

In this case $p^2 = 1$, $\sigma = 0$ and $b_1 = 0$, $b_2 = 1/4$.

Solution: Assuming that

$$G(x) = e^{ix}(1 + B_1/x + B_2/x^2 + \cdots)$$

it follows that

$$iB_1 = \tfrac{1}{2} b_2 = \tfrac{1}{8}$$

$$iB_2 = \tfrac{1}{4}(b_2 + 2)B_1 = -\frac{1^2 \cdot 3^2}{2! \, 8^2} i$$

$$iB_3 = \tfrac{1}{6}(b_2 + 2 \cdot 3)B_2 = -\frac{1^2 \cdot 3^2 \cdot 5^2}{3! \, 8^3}$$

$$iB_4 = \tfrac{1}{8}(b_2 + 3 \cdot 4)B_3 = \frac{1^2 \cdot 3^2 \cdot 5^2 \cdot 7^2}{4! \, 8^4} i$$

Hence

$$G(x) = e^{ix}(P(x) + iQ(x))$$

where

$$P(x) \cong 1 - \frac{1^2 \cdot 3^2}{2!(8x)^2} + \frac{1^2 \cdot 3^2 \cdot 5^2 \cdot 7^2}{4!(8x)^4} - \cdots$$

$$Q(x) \cong -\frac{1}{8x} + \frac{1^2 \cdot 3^2 \cdot 5^2}{3!(8x)^3} - \cdots$$

A second solution is the conjugate function

$$\bar{G}(x) = e^{-ix}(P(x) - iQ(x))$$

and the general solution in real form is obtained from the combination

$$\tfrac{1}{2} C\{(G + \bar{G}) \cos \beta - i(G - \bar{G}) \sin \beta\}$$

$$\cong C\{P(x) \cos(x - \beta) - Q(x) \sin(x - \beta)\}$$

The equation being considered is the normal form of Bessel's equation of order zero, one solution of which is $x^{1/2} J_0(x)$. Since

$$J_0(x) \cong (2/\pi x)^{1/2} \cos(x - \pi/4),$$

as may be proved from the integral expression

$$J_0(x) = (1/\pi) \int_0^\pi \cos(x \sin \phi) d\phi,$$

$J_0(x)$ corresponds to the values $C = (2/\pi)^{1/2}$, $\beta = \pi/4$.

• **PROBLEM 16-54**

Find the values of y corresponding to the values x = 0.1, 0.2, ..., 0.10 that satisfy the solution of

$$\frac{dy}{dx} = 2 + y - 2x$$

with initial conditions y = 1 when x = 0. Use the formula

$$y_{n+1} = y_{n-3} + \tfrac{4}{3} h(2y'_{n-2} - y'_{n-1} + 2y'_n) \qquad (1)$$

for advancing, and its associated formula

$$y_{n+1} = y_{n-1} + \frac{1}{3} h(y'_{n-1} + 4y'_n + y'_{n+1}); \qquad (2)$$

for correcting to solve this problem.

Table 1

x	y	y'
0	1.0000	3.0000
0.1	1.3052	3.1052
0.2	1.6214	3.2214
0.3	1.9499	3.3499

Solution: From the first formula it is clear that four "starting" values are needed. These are obtained from the Maclaurin expansion of the desired solution. From $y' = 2 + y - 2x$ it follows that

$$y'' = y' - 2$$

$$y^{(3)} = y''$$

$$\cdots\cdots\cdots\cdots$$

$$y^{(n)} = y^{(n-1)}.$$

At $x = 0$ one has

$$y = 1, \quad y' = 3, \quad y'' = y^{(3)} = \ldots = y^{(n)} = 1.$$

Hence

$$y = 1 + 3x + \frac{x^2}{2!} + \frac{x^3}{3!} + \ldots + \frac{x^n}{n!} + \ldots .$$

Use the notation $x_0 = 0$, $x_1 = 0.1$, $x_2 = 0.2$, etc., then $y_1 = 1.3052$, $y_2 = 1.6214$, $y_3 = 1.9499$. Enter these starting values and the corresponding values of y' in Table 1:

It can be seen from one form of the error term,

$$\frac{14}{45} h^5 f^{(5)}(X),$$

that the error inherent in formula (1) will not affect the fourth decimal place of y_{n+1} since in this problem

$h = 0.1$ and $f^{(5)}(X) = 1$. It turns out that formula (2) is not needed to correct, but shall be used as a check on the calculations.

From (1) one gets

$$y_4 = 1.0000 + \frac{4}{30} [2(3.1052) - 3.2214 + 2(3.3499)]$$

$$= 2.2918.$$

Substitute this tentative value of y_4 into the differential equation and obtain $y' = 3.4918$. Now use (2) and find

$$y_4 = 1.6214 + \frac{1}{30} [3.2214 + 4(3.3499) + 3.4918]$$

$$= 2.2918,$$

the same value as before as anticipated. Enter this value in an extension of Table 1 (row 5 of Table 2):

Table 2

x	y	y'
0	1.0000	3.0000
0.1	1.3052	3.1052
0.2	1.6214	3.2214
0.3	1.9499	3.3499
0.4	2.2918	3.4918
0.5	2.6488	3.6488
0.6	3.0221	3.8221
0.7	3.4138	4.0138
0.8	3.8255	4.2255
0.9	4.2596	4.4596
1	4.7182	4.7182

Again,

$$y_5 = 1.3052 + \frac{4}{30} [2(3.2214) - 3.3499 + 2(3.4918)]$$

$$= 2.6487.$$

$$y_5' = 2 + 2.6487 - 1 = 3.6487.$$

Hence

$$y_5 = 1.9499 + \frac{1}{30} [3.3499 + 4(3.4918) + 3.6487] = 2.6488.$$

This time the corrected value is one unit more in the fourth decimal place than the tentative value. Enter these values in the table and continue in the same fashion to find the remaining values of y.

• **PROBLEM 16-55**

Find, correct to five decimal places, the ordinates $y_1, y_2, \ldots, y_5$ corresponding to the abscissas 0.1, 0.2, ..., 0.5 such that each point is on the curve which passes through (0,1) and whose equation is a solution of the differential equation $y' = x + y^2$.

Table 1

x	y	y'	y''	$y^{(3)}$	$y^{(4)}$
0	1	1	3	8	· 34
0.1	1.11649 236	1.34655 519	4.00683 72	12.67362 79	60.449283
0.2	1.27356 426	1.82196 592	5.64078 14	21.00691 47	115.17118
0.3	1.48802 214	2.51420 989	8.48240 00	37.88650 06	240.71111
0.4	1.78936 167	3.60181 519	13.88990 01	75.65425 50	570.91877
0.5	2.23453 307				

x	$y^{(5)}$	$y^{(6)}$	$y^{(7)}$	$y^{(8)}$	$y^{(9)}$	$y^{(10)}$
0	186	1192	8956	77076		
0.1	366.751	2640.54	22250.7	214×10^3	680×10^3	84×10^4
0.2	790.457	6481.68	62107.4	680×10^3	84×10^4	
0.3	1910.111	18163.93	201647.7	2558×10^3	365×10^4	
0.4	5380.677	60836	802649	1212×10^4	205×10^4	386×10^4

Solution: Let $y = f(x)$ be the equation of the required curve, yielding

$y' = x + y$

$y'' = 1 + 2yy'$

$y^{(3)} = 2(yy'' + y'^2)$

$y^{(4)} = 2(yy^{(3)} + 3y'y'')$

$y^{(5)} = 2(yy^{(4)} + 4y'y^{(3)} + 3y''^2)$

$y^{(6)} = 2(yy^{(5)} + 5y'y^{(4)} + 10y''y^{(3)})$ (1)

$y^{(7)} = 2(yy^{(6)} + 6y'y^{(5)} + 15y''y^{(4)} + 10y^{(3)^2})$

$y^{(8)} = 2(yy^{(7)} + 7y'y^{(6)} + 21y''y^{(5)} + 35y^{(3)}y^{(4)})$

$y^{(9)} = 2(yy^{(8)} + 8y'y^{(7)} + 28y''y^{(6)} + 56y^{(3)}y^{(5)}$
$\qquad + 35y^{(4)^2})$

$y^{(10)} = 2(yy^{(9)} + 9y'y^{(8)} + 36y''y^{(7)} + 84y^{(3)}y^{(6)}$
$\qquad + 126y^{(4)}y^{(5)})$

$y^{(11)} = 2(yy^{(10)} + 10y'y^{(9)} + 45y''y^{(8)} + 120y^{(3)}y^{(7)}$
$\qquad + 210y^{(4)}y^{(6)} + 126y^{(5)^2})$

$y^{(12)} = 2(yy^{(11)} + 11y'y^{(10)} + 55y''y^{(9)} + 165y^{(3)}y^{(8)}$
$\qquad + 330y^{(4)}y^{(7)} + 462y^{(5)}y^{(6)}$ ..

Since $x_0 = 0$, $y_0 = 1$, it follows that at this point, $y'' = 3$, $y^{(3)} = 8$, $y^{(4)} = 34$, $y^{(5)} = 186$, $y^{(6)} = 1192$, $y^{(7)} = 8956$, $y^{(8)} = 77076$; therefore, from

$$y_{k+1} = y_k + \frac{y'_k}{1!} h + \frac{y''_k}{2!} h^2 + \ldots + \frac{y^{(n)}_k}{n!} h^n \qquad (2)$$

($h = 0.1$ and $k = 0$),

$$y_1 = 1 + 1(0.1) + \frac{3}{2}(0.1)^2 + \frac{4}{3}(0.1)^3 + \frac{17}{12}(0.1)^4$$
$$+ \frac{31}{20}(0.1)^5 + \frac{149}{90}(0.1)^6 + \frac{2239}{1260}(0.1)^7 + \frac{2141}{1120}(0.1)^8$$

$$= 1.11649236,$$

with an error which can be approximated by use of

$$\frac{f^{(n+1)}(X)}{(n+1)!} (x - x_0)^{n+1}, \qquad (3)$$

of at most one or two units in the last decimal place. Now use the value of y_1 just obtained and $x_1 = 0.1$ to derive by use of (1) the values of y_1', y_1'', ... These values are given in the second row of Table 1. The value of y_2 corresponding to $x = 0.2$ is then calculated from (2) where now $k = 1$; and so on. The necessary entries are shown in Table 1.

Consider the magnitude of the errors in y_1, y_2, ... Since $x_0 = 0$, $y_0 = 1$ were exact values, the only error in y_1 (aside from rounding off errors) is caused by the use of a polynomial of finite degree rather than an infinite power series for its determination. The error is given by (3) where, in this problem, $n = 8$. The maximum value of $f^{(9)}(X)$ (between $x = 0$ and $x = 0.1$) is $f^{(9)}(0.1)$ which is about 3×10^6. The error is then approximately $(3 \times 10^6)(0.1)^9/9!$ which is less than 9×10^{-9} and is therefore certainly less than one unit in the eighth decimal place.

The error in y_2 (again aside from rounding off errors) has a twofold cause: as in the preceding step, y_2 may be in error because a finite series was used instead of an infinite series but unlike the preceding step, y_2 may be in error because the initial value $y_1 = 1.11649236$ is approximate (the initial value $x_1 = 0.1$ is, of course, exact). Now it has been seen that the differentials dy, dy', dy'', ..., are equal to the errors in y, y', y'', ..., respectively, when neglecting infinitesimals of higher order. But from (1),

$$dy' = 1 + 2y \, dy$$

$$dy'' = 2(y\, dy + y'\, dy)$$

$$dy^{(3)} = 2(y\, dy'' + 2y'\, dy' + y''\, dy)$$

$$\cdots\cdots\cdots\cdots\cdots\cdots\cdots\cdots\cdots\cdots\cdots\cdots\cdots$$

$$dy^{(k)} = 2\left[\binom{k-1}{0} y\, dy^{(k-1)} + \binom{k-1}{1} y'\, dy^{(k-2)} + \cdots\right.$$

$$\left. + \binom{k-1}{k-1} y^{(k-1)}\, dy\right]$$

$$\cdots\cdots\cdots\cdots\cdots\cdots\cdots\cdots\cdots\cdots\cdots\cdots\cdots$$

Knowing the error dy_1 in y_1 one can compute the errors in y_1', y_1'', $y_1^{(3)}$, ..., and then, from

$$y_2 = y_1 + \frac{y_1'}{1!} h_2 + \frac{y_1''}{2!} h_2^2 + \cdots + \frac{y_1^{(n)}}{n!} h_2^n,$$

the resulting error in y_2. It develops that this error is not more than one unit in the eighth decimal place of y_2.

Corresponding computations verify that the calculated values of y_3, y_4, and y_5 are certainly correct to five decimal places. Again, because of the accumulation of errors in the progressive tabulations and calculations, it is necessary to start with at least eight decimal places to be assured of five place precision in y_5.

● **PROBLEM 16-56**

Apply the modified midpoint method to the equation $y' = -y, y(0) = 1$ using

$$y_{n+1} = y_{n-1} + 2hf(x_n, y_n)$$

with $h = 0.1$, $y_0 = 1$.

Solution: Algorithm for the modified midpoint method:
Put $y_0 = y(a)$, $y_1 = y_0 + hf(x_0, y_0)$.

$$y_{n+1} = y_{n-1} + 2hf(x_n, y_n), \quad n = 1, 2, \ldots \qquad (1)$$

For certain values of x, one damps the oscillating error component by means of the following formula, where N is even:

$$\hat{y}_N = \frac{1}{2}(y_N + y_{N-1} + hf(x_N, y_N)), \quad n = 1, 2, \ldots$$

The stepwise integration then proceeds by Eq. (1) from the point x_N with

$$\hat{y}_N, \text{ and } \hat{y}_{N+1} := \hat{y}_N + hf(x_N, \hat{y}_N)$$

as new starting values.

The solution corresponding to $y_1 = 0.9$ is indicated by black circles in Fig. 1, while the solution corresponding to $y_1 = 0.85$ is shown with white circles. Note that the perturbation of the initial value gives rise to growing oscillations with a growth of approximately 10 percent per step. This phenomenon is sometimes called weak stability, sometimes weak instability.

In more realistic problems the oscillations become visible much later. For instance,

$$y_1 = e^{-0.1}$$

correct to ten decimals yields the table shown.

(The exact solution is $y(x) = \exp(-x)$.)

x	0	0.1	0.2	... 5.0	5.1	5.2	5.3
y_n	1	0.90484	0.81903	... 0.01803	−0.00775	0.01958	−0.01166
$y_n - y(x_n)$	0.00000	0.00000	0.00030	... 0.01129	−0.01385	0.01406	−0.01665
$y(x_n)$	1.00000	0.90484	0.81873	... 0.00674	0.00610	0.00552	0.00499

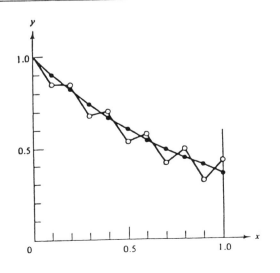

Fig. 1

● **PROBLEM** 16-57

Find the solution of the initial value problem

$$y' = \lambda y, \quad \lambda = \pm 1,$$

$$y(0) = 1, \quad 0 \le t \le 2$$

using the first order method with h = .1.

Table 1

t	$\lambda=1$		$\lambda=-1$	
	First order method	Exact solution	First order method	Exact solution
0	1	1	1	1
0.2	1.21000	1.22140	0.81	0.81873
0.4	1.46410	1.49182	0.6561	0.67032
0.6	1.77156	1.82212	0.53144	0.54881
0.8	2.14359	2.22554	0.43047	0.44933
1.0	2.59374	2.71828	0.34868	0.36788
1.2	3.13843	3.32012	0.28243	0.30119
1.4	3.79750	4.05520	0.22877	0.24660
1.6	4.59497	4.95303	0.18530	0.20190
1.8	5.55992	6 04965	0.15009	0.16530
2.0	6.72750	7.38906	0.12158	0.13534

Solution of $y' = \lambda y$, $y(0) = 1$, $0 \le t \le 2$ with h = 0.1

Solution: Using first order method

$$y_{n+1} = (1 + \lambda h) y_n, \quad n = 0,1,2,\ldots,N-1$$

with h = .1, one gets

$$y_1 = (1 + .1\lambda) y_0 = (1 + .1\lambda)$$
$$y_2 = (1 + .1\lambda) y_1 = (1 + .1\lambda)^2$$
$$y_3 = (1 + .1\lambda) y_2 = (1 + .1\lambda)^3$$

$$\vdots$$

$$y_N = (1 + .1\lambda) y_{N-1} = (1 + .1\lambda)^N$$

where N = 20.

The values of y_n for $\lambda = \pm 1$ are listed in Table 1, together with the true values obtained from $y(t_n) = e^{\lambda nh}$. These values are plotted in Figure 1, and compared with the true values. For $\lambda = 1$, it is observed that the approximate solution increases as fast as the exact solution whereas for $\lambda = -1$, the approximate solution decreases at least as fast as the exact solution.

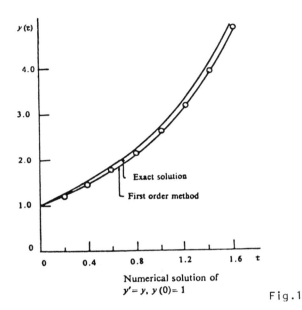

Numerical solution of
$y' = y$, $y(0) = 1$

Fig.1

Numerical solution of
$y' = -y$, $y(0) = 1$

Fig.2

• **PROBLEM 16-58**

Consider the differential equation

$$\frac{dy}{dx} = x^2 - y^2.$$

The integral curves cross $y = \pm x$ horizontally. Find an approximation of y at all points of the line $y = x$ and then at all points of $y = -x$. Hint: for the former case, remove the origin to (h,h) so that

$$x = h + \xi \qquad y = h + \eta.$$

Solution: The equation becomes

$$\frac{d\eta}{d\xi} = (h + \xi)^2 - (h + \eta)^2 = 2h(\xi - \eta) + \xi^2 - \eta^2.$$

Let $\eta = A\xi^n$ be the first approximation at the point.

$$An\xi^{n-1} = 2h\xi - 2hA\xi^n + \xi^2 - A^2\xi^{2n}.$$

The term ξ^2 may be neglected in comparison with ξ.

" ξ^{2n} "   "   "   " ξ^n.

" ξ^n "   "   "   " ξ^{n-1}.

Thus to the first order of small quantities the equation will be satisfied if

$$An\xi^{n-1} = 2h\xi$$

i.e., $n - 1 = 1, An = 2h$

giving as the required approximation at (h,h)

$$\eta = h\xi^2.$$

Hence along $y = x$ the integral curves have a series of minima in the first quadrant and a series of maxima in the third. The maxima and minima become flatter and flatter as they approach the origin at which point this approximation becomes invalid. To determine the shape of the integral curve passing through $(0,0)$, let

700

$y = Ax^n$, then, substituting in the original differential equation

$$A n x^{n-1} = x^2 - 2Ax^{2n}.$$

Equating powers, one gets an approximation if

(i) $n - 1 = 2$ provided x^{2n} is of higher order than x^2.

(ii) $n - 1 = 2n$ " x^2 " " " x^{n-1}.

(iii) $2 = 2n$ " x^{n-1} " " " x^2.

The first gives $n = 3$ and makes $x^{2n} = x^6$, which is of higher order than x^2. The second and third possibilities do not give justifiable approximations.

Thus the equation becomes reduced to

$$An \, x^{n-1} = x^2$$

which with $n = 3$ gives $A = \frac{1}{3}$, and the approximation at the origin, instead of being simply parabolic, is

$$y = \frac{x^3}{3}.$$

In precisely the same way, the approximations at all points of the line $y = -x$ may be found.

• **PROBLEM** 16-59

Let

$$x \frac{dy}{dx} = x - y$$

be the differential equation. Given that when $x = 2$, $y = 2$, find a solution of the equation in the range of $2 \leq x \leq 2.5$, to four places of decimals using the Adams-Bashforth process.

Solution: Let the region be divided into ten equal parts:

$$x_0, \, x_0 + h, \, \ldots \, x_0 + 10h.$$

Then

$$x_0 = 2$$

$$h = 0.05.$$

The values of y corresponding to x equal to 2, 2.05, 2.1, 2.15, 2.2, ... 2.5 are required.

Expanding y in the neighborhood of $x = 2$

$$y = y_0 + (x-2)y_0' + \frac{1}{2!}(x-2)^2 \cdot y_0'' + \frac{1}{3!}(x-2)^3 y_0''' + \ldots$$

where y_0, y_0', etc., are the values of y, $\frac{dy}{dx}$, etc., at $x = 2$.

$$y_0' = [(x-y)/x]_{(x=2,y=2)} = 0$$

$$y'' = (y/x^2) - (y'/x) = (2y-x)/x^2,$$

$$y_0'' = \frac{1}{2}$$

$$y''' = \{x^2(2y'-1) - (2y-x) \cdot 2x\}/x^4$$

$$= 3(x-2y)/x^4, \qquad y_0''' = -\frac{3}{4}$$

$$y^{iv} = -12(x-2y)/x^4, \qquad y_0^{iv} = \frac{3}{2}$$

$$y^v = 60(x-2y)/x^5, \qquad y_0^v = -3\frac{3}{4}$$

$$y^{vi} = -360(x-2y)/x^6 \qquad y_0^{vi} = +11\frac{1}{4}$$

Thus

$$y = y_0 + (x-2) \cdot 0 + \frac{1}{4}(x-2)^2 - \frac{1}{8}(x-2)^3$$

$$+ \frac{1}{16}(x-2)^4 - \frac{1}{32}(x-2)^5 + \ldots \ldots \qquad (a)$$

From series (a) it is required to find y_0, y_1, y_2, y_3, and y_4 and therefore the greatest numerical value of the term

$$-\frac{1}{32}(x-2)^5 \text{ is } -\frac{1}{32}(0 \cdot 2)^5 = -0.00001$$

and of

$$+\frac{1}{64}(x-2)^6 \text{ is } +\frac{1}{64}(0 \cdot 2)^6 = +0.000001.$$

Since the terms in the series are alternately positive and negative, it follows that the values of y from series (a) will be correct to the fifth decimal place.

The values of y calculated in this way are $y_1 = 2.00061$, $y_2 = 2.00248$, $y_3 = 2.00523$, $y_4 = 2.00909$.

From the differential equation

$$h \cdot y^I = q = h\left(1 - \frac{y}{x}\right)$$

$$q_n = 0.05\{1 - (y_n/x_n)\}$$

Using this equation, Table 1 is obtained

Table 1

$y_0 = 2$	$q_0 = 0.00000$					
		$\Delta q_0 = 0.00120$				
$y_1 = 2.00061$	$q_1 = 0.00120$		$\Delta^2 q_0 = -0.00008$			
		$\Delta q_1 = 0.00112$		$\Delta^3 q_0 = +0.00001$		
$y_2 = 2.00238$	$q_2 = 0.00232$		$\Delta^2 q_1 = -0.00007$		$\Delta^4 q_0 = -0.00002$	
		$\Delta q_2 = 0.00105$		$\Delta^3 q_1 = -0.00001$		
$y_3 = 2.00523$	$q_3 = 0.00337$		$\Delta^2 q_2 = -0.00008$			
		$\Delta q_3 = 0.00097$		$\Delta^3 q_2 = +0.00002$		
$y_4 = 2.00909$	$q_4 = 0.00434$		$\Delta^2 q_3 = -0.00006$			
		$\Delta q_4 = 0.00091$				
	$q_5 = 0.00525$					

From Table 1 it is clear that the second differences of q are nearly constant, and therefore the third and fourth differences should be taken as zero to the fifth decimal place.

Now

$$y_{n+1} - y_n = q_n + \frac{1}{2} \Delta q_{n-1} + \frac{5}{12} \Delta^2 q_{n-2} + \frac{3}{8} \Delta^3 q_{n-3}$$

$$+ \frac{251}{720} \Delta^4 q_{n-4} + \ldots$$

$$y_5 = y_4 + q_4 + \frac{1}{2} \Delta q_3 + \frac{5}{12} \Delta^2 q_2 + \frac{3}{8} \Delta^3 q_1$$

$$+ \frac{251}{720} \Delta^4 q_0 + \ldots$$

$$= 2.00909 + 0.00434 + \frac{1}{2}(0.00097) + \frac{5}{12}(-0.00008)$$

= 2.00909
0.00434
0.000485
―――――
2.013915
0.000033
―――――
= 2.01388
―――――

This now enables one to calculate

$$q_5 = 0.05[1 - y_5/x_5]$$
$$= 0.05[1 - 2.01388/2.25]$$
$$= 0.00525$$

$$y_6 = y_5 + q_5 + \frac{1}{2} q\Delta_4 + \frac{5}{12} \Delta^2 q_3 + \ldots$$

$$= 2.01388 + 0.00525 + 0.000455 + (-0.000025)$$

0.00525
0.00043
―――――

704

= 2.01956

Proceeding as above, Table 2 shows the final results.

Table 2

y	Error			
		$q_4 = 0.00434$		
$y_5 = 2.01388$	-0.00001		$\Delta q_4 = 0.00091$	
		$q_5 = 0.00525$		$\Delta^2 q_4 = -0.00006$
			$\Delta q_5 = 0.00085$	
$y_6 = 2.01956$	-0.00001	$q_6 = 0.00610$		$\Delta^2 q_5 = -0.00006$
			$\Delta q_6 = 0.00079$	
$y_7 = 2.02606$	-0.00001	$q_7 = 0.00689$		$\Delta^2 q_6 = -0.00004$
			$\Delta q_7 = 0.00075$	
$y_8 = 2.03332$	-0.00001	$q_8 = 0.00764$		$\Delta^2 q_7 = -0.00005$
			$\Delta q_8 = 0.00070$	
$y_9 = 2.04132$	-0.00001	$q_9 = 0.00834$		
$y_{10} = 2.04999$	-0.00001			

The constancy of the error indicates that the method maintains the accuracy of the value of y at which the Adams-Bashforth method was applied. Greater accuracy may be obtained by taking the first calculations for y_0, y_1, y_2, etc., to a higher degree of accuracy, and if the second differences are not constant to this accuracy, the terms with the higher differences should be taken into account.

The complete solution of the equation

$$x \frac{dy}{dx} = x - y$$

where $x = 2$ when $y = 2$ is

$$y = \frac{x}{2} + \frac{2}{x}.$$

When $x = 2.5$ $y = 2.05$

which compares with $y = 2.04999$.

• **PROBLEM 16-60**

Consider

$$y'' + y/(1 + x^2) = 0$$

where $y = 0$ and $y' = 1$ when $x = 0$; find the value of y at $x = 0.5$.

Solution: Now

$$1 \geq \frac{1}{1+x^2} \geq \frac{4}{5} \qquad 0 \leq x \leq 0.5$$

For $M = 1$ $\qquad y_1'' + y_1 = 0,$

the solution is $\qquad y_1 = \sin x.$

For $m = \frac{4}{5}$ $\qquad y_2'' + \frac{4}{5} y_2 = 0,$

the solution is $\qquad y_2 = 1.12 \sin 0.894 x.$

Thus at $x = 0.5$, y lies between

$$\sin 0.5 = \sin 28° 39' = 0.480$$

and

$$1.12 \sin (0.894 \times 0.5) = 1.12 \sin 25° 37'$$
$$= 0.484.$$

Hence the value lies between

$$y_1(0.5) = 0.480$$

and

$$y_2(0.5) = 0.484.$$

If the average of these values are taken, viz.

$$y(0.5) = 0.482,$$

the error is approximately 0.4 per cent.

CHAPTER 17

PARTIAL DIFFERENTIAL EQUATIONS

SOLUTION OF GENERAL PARTIAL DIFFERENTIAL EQUATIONS

● PROBLEM 17-1

Solve

$$x + yz_x = 0. \tag{a}$$

Solution: The given equation is a homogeneous, partial differential equation of the first order, which one can rewrite as

$$x + y \frac{\partial z}{\partial x} = 0. \tag{b}$$

Since (b) is a first order equation, seek a general solution, $z(x, y)$, which contains one arbitrary function. This is analogous to solving a first order linear differential equation, the general solution of which must contain one arbitrary constant.

Since, in the given equation, y is held constant, rewrite (b) as

$$\frac{\partial z}{\partial x} = \frac{dz}{dx} = -\frac{x}{y}, \tag{c}$$

and integrate with respect to x. Integrating (c), obtain

$$\int \frac{\partial z}{\partial x} dx = \int -\frac{x}{y} dx = -\frac{x^2}{2y} + \phi(y). \tag{d}$$

Therefore, the general solution of the given partial differential equation (a) is

$$z(x, y) = -\frac{x^2}{2y} + \phi(y), \tag{e}$$

where $\phi(y)$ is an arbitrary function of y alone, and is

necessary if (e) is to represent the general solution of (a).

Consider $\phi(y)$ a "constant of integration," in (d). One is carrying out a partial integration with respect to x, thereby treating y as a constant. Since y, like x, is not a constant, but a dependent variable of the function z(x, y), one must add $\phi(y)$ to obtain the general solution.

Demonstrate this by taking the partial derivative of the general solution, (e), with respect to x:

$$\frac{\partial}{\partial x} z(x, y) = \frac{\partial}{\partial x}\left[-\frac{x^2}{2y} + \phi(y)\right] = -\frac{x}{y}, \tag{f}$$

Since (f) and (c) are identical, one can say that (e) is a solution of (a). Furthermore, since z(x, y), as given in (e), contains one arbitrary function, one can say that (e) gives the general solution of (a).

• **PROBLEM 17-2**

Solve the linear, partial differential equation

$$xz_x = x + 2y + 2z. \tag{a}$$

Solution: Write (a) as

$$z_x - \frac{2z}{x} = 1 + \frac{2y}{x}. \tag{b}$$

Introduce an integrating factor, $\phi(x)$, such that it satisfies the following property:

$$\frac{\partial}{\partial x}(\phi z) = \phi z_x - \phi\frac{2z}{x} \tag{c}$$

or $\phi z_x + \phi' z = \phi z_x - \phi\frac{2z}{x}$.

Subtracting ϕz_x and dividing by z one obtains

$$\phi' = -\frac{2\phi}{x}..$$

Separating variables,

$$\frac{d\phi}{\phi} = -\frac{2dx}{x}.$$

Integrating,

$$\ln \phi = -2 \ln x.$$

(Take the constant of integration to be zero since one wants the simplest ϕ which makes (c) true.) Thus,

$$\phi = x^{-2}$$

Hence, multiply (b) by this ϕ and substitute from (c):

$$\frac{\partial}{\partial x}(x^{-2} z) = \frac{1}{x^2} + \frac{2y}{x^3}.$$

Integrating with respect to x,

$$\int \frac{\partial}{\partial x}(x^{-2} z) \, dx = \int \left[\frac{1}{x^2} + \frac{2y}{x^3}\right] dx + \psi(y),$$

where $\psi(y)$ is the arbitrary function of integration.

$$x^{-2} z = -x^{-1} - yx^{-2} + \psi(y),$$

or $\quad z = x^2 \psi(y) - x - y.$

● **PROBLEM 17-3**

Solve the partial differential equation:

$$z_{xy} = x^2 + y^2. \tag{a}$$

Solution: Let

$$u = z_y,$$

then, (a) can be written

$$u_x = x^2 + y^2.$$

Integrating,

$$\int u_x \, dx = \int x^2 \, dx + \int y^2 \, dx + \phi(y),$$

where $\phi(y)$ is the arbitrary function of integration.

$$u = \frac{x^3}{3} + y^2 x + \phi(y),$$

or $\quad z_y = \frac{x^3}{3} + y^2 x + \phi(y).$

Integrating now with respect to y,

$$\int z_y \, dy = \frac{x^3}{3} \int dy + x \int y^2 \, dy + \int \phi \, dy + g(x),$$

where $g(x)$ is an arbitrary function of integration.

$$z = \frac{x^3 y}{3} + \frac{xy^3}{3} + f(y) + g(x). \tag{b}$$

where $f(y) = \int \phi(y) \, dy$.

Equation (b) is the required solution.

● **PROBLEM 17-4**

Find a solution of

$$\frac{\partial^2 u}{\partial x^2} - 5 \frac{\partial^2 u}{\partial x \partial y} + 6 \frac{\partial^2 u}{\partial y^2} = 0 \qquad (a)$$

which contains two arbitrary functions.

Solution: Seek solutions of (a) of the form

$$u = f(y + mx), \qquad (b)$$

where f is an arbitrary function of its argument, and m is a constant. Differentiating (b), obtain

$$\frac{\partial^2 u}{\partial x^2} = m^2 \, f''(y + mx),$$

$$\frac{\partial^2 u}{\partial x \partial y} = m \, f''(y + mx),$$

$$\frac{\partial^2 u}{\partial y^2} = f''(y + mx).$$

Substituting these expressions into the differential equation (a), obtain

$$m^2 \, f''(y + mx) - 5m \, f''(y + mx) + 6 f''(y + mx) = 0,$$

or, $\quad f''(y + mx)[m^2 - 5m + 6] = 0.$

Thus $f(y + mx)$ will be a solution of (a). if m satisfies the quadratic equation

$$m^2 - 5m + 6 = 0. \qquad (c)$$

Solving (c), obtain the distinct roots $m_1 = 2$, $m_2 = 3$, and one concludes that (a) has as solutions:

$$f_1(y + m_1 x) = f_1(y + 2x), \qquad (d)$$

and $\quad f_2(y + m_2 x) = f_2(y + 3x), \qquad (e)$

where f_1 and f_2 are arbitrary functions of their respective arguments. Summing (d) and (e), a solution of the differential equation (a) is obtained,

$$u = f_1(y + 2x) + f_2(y + 3x),$$

which contains two arbitrary functions.

• **PROBLEM** 17-5

Find a solution of

$$\frac{\partial^2 u}{\partial x^2} - 4 \frac{\partial^2 u}{\partial x \partial y} + 4 \frac{\partial^2 u}{\partial y^2} = 0 \qquad (a)$$

which contains two arbitrary functions.

Solution: Since (a) is a second order, homogeneous, partial differential equation with constant coefficients, assume solutions of the form

$$u = f(y + mx)$$

where m is a constant and will be determined. Therefore,

$$u_{xx} = m^2 f''(y + mx);$$

$$u_{yy} = f''(y + mx);$$

$$u_{xy} = m f''(y + mx).$$

Substituting into (a), obtain

$$(m^2 - 4m + 4) f''(y + mx) = 0.$$

Since $f(y + mx) \neq 0$, one requires that

$$m^2 - 4m + 4 = 0. \qquad (b)$$

Equation (b) has a double root $m_1 = m_2 = 2$. Therefore, the following solution of (a) is obtained:

$$u = f_1(y + 2x) + x f_2(y + 2x), \qquad (c)$$

where f_1 and f_2 are arbitrary functions of their arguments. Note that in (c) one must multiply f_2 by x; this always results when a double root $m_1 = m_2$ is obtained.

• **PROBLEM** 17-6

Find the general solution of

$$z_{xx} - 5z_x + 6z = 12x. \qquad (a)$$

Solution: Note first, that in equation (a), all derivatives are taken with respect to x alone. Therefore treat (a) as a second order, linear, ordinary differential equation (O.D.E.) with constant coefficients

$$\frac{d^2 z}{dx^2} - 5 \frac{dz}{dx} + 6z = 12, \qquad (b)$$

and seek a solution.

$$z(x) = c_1 z_1 + c_2 z_2 + z_p, \qquad (c)$$

where z_1 and z_2 are linearly independent solutions of the homogeneous equation,

$$\frac{d^2 z}{dx^2} - 5 \frac{dz}{dx} + 6z = 0, \qquad (d)$$

c_1 and c_2 are arbitrary constants, and z_p is a particular solution of (b). Once (c), the solution of the O.D.E. (b), has been obtained, substitute for the constants, c_1 and c_2, two arbitrary functions of y in order to obtain the general solution of the partial differential equation (P.D.E.) given in (a). Thus, one will have a solution,

$$z(x, y) = \phi_1(y) z_1 + \phi_2(y) z_2 + z_p,$$

where $\phi_1(y)$ and $\phi_2(y)$ are arbitrary functions of y, and z_1, z_2, and z_p are as before.

Begin by solving (d), the homogeneous equation corresponding to the O.D.E. (b). Using its characteristic, quadratic equation,

$$p^2 - 5p + 6 = 0,$$

obtain roots $p_1 = 3$, $p_2 = 2$, and therefore solutions e^{3x}, e^{2x}. The complementary solution, $z_c(x)$, of the O.D.E. is

$$z_c(x) = c_1 e^{3x} + c_2 e^{2x}. \qquad (e)$$

Now seek a particular solution $z_p(x)$ of the O.D.E. (b). Use the method of undetermined coefficients and assume a solution of the form

$$z_p(x) = Ax + B, \qquad (f)$$

where A and B are constants that one must determine. Substituting (f) into the O.D.E. (b), one gets

$$-5A + 6(Ax + B) = 12x,$$

or $\quad (-5A + 6B) + 6Ax = 12x.$

Equating coefficients of like terms, one gets

$$6A = 12,$$

$$-5A + 6B = 0,$$

from which one obtains $A = 2$, $B = \frac{5}{3}$. Substituting these values into (f), one obtains a particular solution of the O.D.E. (b),

$$z_p(x) = 2x + \frac{5}{3}.$$

Summing this with (e), one obtains the general solution of (b):

$$z(x) = c_1 e^{3x} + c_2 e^{2x} + 2x + \frac{5}{3}. \tag{g}$$

Equation (g) is of the form given in (c). If one substitutes $\phi_1(y)$ and $\phi_2(y)$ for the constants c_1 and c_2, the general solution of the given partial differential equation (a) is obtained.

$$z(x, y) = \phi_1(y) e^{2x} + \phi_2(y) e^{3x} + 2x + \frac{5}{3}. \tag{h}$$

It is known that (h) gives the most general solution of the P.D.E. (a) because (a) is second order and (h) contains two arbitrary functions.

● **PROBLEM 17-7**

Solve

$$9(p^2 z + q^2) = 4, \tag{1}$$

where $p = \dfrac{\partial z}{\partial x}$ and $q = \dfrac{\partial z}{\partial y}$.

Solution: Assume $q = ap$, i.e. $\dfrac{\partial z}{\partial y} = a \dfrac{\partial z}{\partial x}$. Then, substituting into (1)

$$9(p^2 z + a^2 p^2) = 4. \tag{2}$$

Solving (2) for p,

$$p = \frac{\pm 2}{3\sqrt{z + a^2}}.$$

From elementary calculus, it is known that, if $z = f(x, y)$, then

$$dz = \frac{\partial z}{\partial x} dx + \frac{\partial z}{\partial y} dy. \tag{3}$$

Now, $p = \dfrac{\partial z}{\partial x} = \dfrac{\pm 2}{3\sqrt{z + a^2}}$, and

$$q = ap = a \frac{\partial z}{\partial x} = \frac{\pm 2a}{3\sqrt{z + a^2}}.$$

Substituting into (3), one obtains,

$$dz = \frac{\pm 2}{3\sqrt{z + a^2}} dx + \frac{\pm 2a}{3\sqrt{z + a^2}} dy,$$

or, $\quad dz = \dfrac{\pm 2}{3\sqrt{z+a^2}}(dx + a\,dy)$. \hfill (4)

Rewriting (4) as a "separable" equation,

$$\left[\pm \frac{3}{2}\sqrt{z+a^2}\right] dz = dx + a\,dy. \tag{5}$$

Integrating (5),

$$\pm (z+a^2)^{3/2} + b = x + ay$$

or, squaring both sides,

$$(x + ay - b)^2 = (z + a^2)^3. \tag{6}$$

Note: (6) is the solution of (1) only with the assumption:

$$\frac{\partial z}{\partial y} = a \frac{\partial z}{\partial x}$$

Other solutions may exist.

FORMULATION OF PARTIAL DIFFERENTIAL EQUATIONS

● **PROBLEM 17-8**

Find the partial differential equation whose solution is the family of surfaces of revolution about the z-axis expressed as

$$z = f(x^2 + y^2) \tag{a}$$

where f is an arbitrary, differentiable function.

Solution: Let $u = x^2 + y^2$; then equation (a) is

$$z = f(u).$$

Compute,

$$\frac{\partial z}{\partial x} = \frac{df}{du}\frac{\partial u}{\partial x} = \frac{df}{du}(2x). \tag{b}$$

One also finds

$$\frac{\partial z}{\partial y} = \frac{df}{du}\frac{\partial u}{\partial y} = \frac{df}{du}(2y). \tag{c}$$

Multiplying (b) by y and (c) by x one gets

$$y\frac{\partial z}{\partial x} = x\frac{\partial z}{\partial y}$$

which is the required, partial differential equation.

● **PROBLEM 17-9**

Find the partial differential equation whose solution is the two parameter family of planes:

$$z = ax + by + ab, \tag{a}$$

where a and b are arbitrary constants.

Solution: Taking the partial derivatives of (a) with respect to x and y one obtains:

$$z_x = a \tag{b}$$

and $z_y = b$.

Substituting (b) and (c) into (a),

$$z = x z_x + y z_y + z_x z_y ,$$

which is the required partial differential equation.

● **PROBLEM 17-10**

Find the second order, partial differential equation whose general solution is expressed in terms of arbitrary functions $\phi(x, y)$ and $\psi(x, y)$:

$$z = \phi(y + ax) + \psi(y - ax) \tag{a}$$

for a fixed constant a.

Solution: Seek an equation relating the partial derivatives z_{xx} and z_{yy}. Differentiating (a) twice with respect to x one gets

$$z_{xx} = \frac{\partial^2 z}{\partial x^2} = \frac{\partial^2 \phi}{\partial x^2} + \frac{\partial^2 \psi}{\partial x^2}, \tag{b}$$

$$z_{yy} = \frac{\partial^2 z}{\partial y^2} = \frac{\partial^2 \phi}{\partial y^2} + \frac{\partial^2 \psi}{\partial y^2}. \tag{c}$$

Since $\phi = \phi(x,y)$ and $\psi = \psi(x,y)$, and both ϕ and ψ are arbitrary, one cannot evaluate (b) and (c). Therefore one needs another approach. Examining (a), the substitutions

$$u = y + ax,$$

$$v = y - ax$$

suggest themselves. Then (a) becomes

$$z = \phi(u) + \psi(v). \tag{d}$$

If one can relate x and y (and their derivatives), to u and v (and their derivatives), one can find z_{xx} and z_{yy} from (d).

Now, $\phi = \phi(u)$, where $u = f(x,y)$. When computing the partial derivative of ϕ with respect to x assume y is held constant. That is, $u = f(x)$. Then, by the chain rule of ordinary differentiation,

$$\frac{d\phi}{dx} = \frac{d\phi}{du}\frac{du}{dx} = a\,\phi'(u).$$

Similarly, $\frac{d\psi}{dx} = \frac{d\psi}{dv}\frac{dv}{dx} = -a\,\psi'(v).$

Thus, $\qquad z_x = a\,[\phi'(u) - \psi'(v)]. \tag{e}$

For purposes of taking the partial derivative with respect to y treat x as constant. Then,

$$z_y = \phi' + \psi'. \tag{f}$$

Now, one wishes to find z_{xx} and z_{yy}. First,

$$z_{xx} = \frac{d}{dx}\left[a\frac{d\phi}{du} - a\frac{d\psi}{dv}\right] = a^2\frac{d^2\phi}{du^2} + a^2\frac{d^2\psi}{dv^2}$$

$$= a^2\,[\phi'' + \psi''], \tag{g}$$

since $\quad \frac{d}{dx}\left[\frac{d\phi}{du}\right] = \frac{d^2\phi}{du^2}\cdot\frac{du}{dx} = a\frac{d^2\phi}{du},$

and $\quad \frac{d}{dx}\left[\frac{d\psi}{dv}\right] = \frac{d^2\psi}{dv^2}\cdot\frac{du}{dx} = -a\frac{d^2\psi}{du}.$

Similarly, $z_{yy} = \frac{d}{dy}[\phi'(u) + \psi'(v)]$

$$= [\phi'' + \psi'']. \tag{h}$$

Comparing (g) and (h),

$$z_{xx} = a^2\,z_{yy}. \tag{i}$$

Equation (i) is a second order, partial differential equation that satisfies (a).

• **PROBLEM 17-11**

Solve the partial differential equation:

$$py + qx = pq, \tag{a}$$

where $p = \frac{\partial z}{\partial x}$ and $q = \frac{\partial z}{\partial y}$.

Solution: One would like to separate (a). Note that

$$(p - x)(q - y) = pq - py - qx + xy.$$

Thus, add xy to both sides of (a) to obtain

$$pq - py - qx + xy = xy,$$

or, $(p - x)(q - y) = xy$.

Separating variables,

$$\frac{p - x}{x} = \frac{y}{q - y}. \tag{b}$$

Now, the left side of (b) is independent of y and the right side independent of x, so, if they are to be equal for any x or y they must equal the same constant value, a.

$$\frac{p - x}{x} = a = \frac{y}{q - y}. \tag{c}$$

Taking each equation in (c) separately,

$$\frac{p - x}{x} = a,$$

$$p = (a + 1)x.$$

Integrating,

$$\int z_x \, dx = (a + 1) \int x \, dx + f(y),$$

where $f(y)$ is an arbitrary function of integration.

$$z = (a + 1) \frac{x^2}{2} + f(y). \tag{d}$$

Next, from (c),

$$\frac{y}{q - y} = a,$$

$$q = \frac{(a + 1)}{a} y.$$

Integrating,

$$\int z_y \, dy = \frac{(a + 1)}{a} \int y \, dy + g(x),$$

where g(x) is the function of integration.

$$z = \frac{(a+1)}{a} \frac{y^2}{2} + g(x).$$

Comparing (d) and (e) the final solution is

$$z = (a+1)\left[\frac{x^2}{2} + \frac{y^2}{2a}\right] + b,$$

where b is an arbitrary constant.

HEAT EQUATIONS

• **PROBLEM** 17-12

Consider the heat equation in rectangular coordinates

$$\frac{\partial u}{\partial t} = h^2 \left[\frac{\partial^2 u}{\partial x^2} + \frac{\partial^2 u}{\partial y^2} + \frac{\partial^2 u}{\partial z^2}\right]. \tag{1}$$

A particular form of (1) is

$$\frac{\partial u}{\partial t} = h^2 \frac{\partial^2 u}{\partial x^2}, \tag{2}$$

where the change in temperature with respect to time is assumed to depend only on the x variable.

Solve (2) subject to the boundary conditions:

As $t \to 0^+$, $u \to f(x)$, $0 < x < c$; (3)

As $x \to 0^+$, $u \to 0$, $0 < t$; (4)

As $x \to c^-$, $u \to 0$, $0 < t$. (5)

Solution: The method of separation of variables will be used to convert the partial differential equation into two ordinary differential equations. Then, the boundary conditions will be used to obtain the formal solution.

Thus, assume that a solution of (2) is a product of a function of t alone by a function of x alone, i.e.,

$$u = f(t) g(x). \tag{6}$$

Substituting for u in (2),

$$f'(t) g(x) = h^2 f(t) g''(x).$$

Dividing through by f(t) g(x),

$$\frac{f'(t)}{f(t)} = h^2 \frac{g''(x)}{g(x)}. \tag{7}$$

Examining (7) one sees that the left hand side is a function of t alone, while the right hand side is a function of x alone. If they are to be equal to each other for arbitrary values of x and t, they must be equal to a constant. Thus,

$$\frac{f'(t)}{f(t)} = k; \quad h^2 \frac{g''(x)}{g(x)} = k. \tag{8}, (9)$$

Rewriting (8) and (9),

$$f'(t) - k f(t) = 0, \tag{10}$$

$$g''(x) - \frac{k}{h^2} g(x) = 0. \tag{11}$$

Equation (10) is a separable first order equation. Its solution is

$$f(t) = c_1 e^{kt}. \tag{12}$$

Let $k = h^2 \beta^2$, a convenient choice for what is, after all, an arbitrary constant. Then, from (11)

$$g''(x) - \beta^2 g(x) = 0 \tag{13}$$

Equation (13) is a second order, linear, homogeneous, differential equation with constant coefficients. Assuming a solution of the form $g = e^{mx}$, obtain the characteristic equation,

$$m^2 - \beta^2 = 0,$$

from which

$$g(x) = c_1 e^{\beta x} + c_2 e^{-\beta x}. \tag{14}$$

By forming appropriate linear combinations of (14) one obtains the general solution

$$g(x) = c_2 \cosh \beta x + c_3 \sinh \beta x. \tag{15}$$

It has already been shown that the solution of (10) is

$$f(t) = c_1 e^{(h^2 \beta^2)t} \tag{16}$$

Combining (15) and (16) one sees that the partial differential equation (2) has the solution:

$$u(x, t) = e^{(h^2 \beta^2)t} [a \cosh \beta x + b \sinh \beta x] \tag{17}$$

where a, β, b are arbitrary constants. If, instead of $k = h^2 \beta^2$, one lets $k = -h^2 \alpha^2$ so that k is a negative constant the solution of (2) is found to be

$$u = e^{(-h^2 \alpha^2)t} [A \cos \alpha x + B \sin \alpha x] \tag{18}$$

where α, A, B are arbitrary constants. Now find which solutions satisfy the boundary conditions (3), (4) and (5). Applying the condition (4) to (17) one finds

$$0 = e^{(h^2\beta^2)t} [a + 0], \quad 0 < t,$$

since $a \cosh \beta(0) = a$ and $b \sinh \beta(0) = 0$. Since $e^{(h^2\beta^2)t} \neq 0$, this implies $a = 0$ and (17) reduces to

$$u = b\, e^{h^2\beta^2 t} \sinh \beta x. \tag{19}$$

Applying condition (5) to (19),

$$b\, e^{h^2\beta^2 t} \sinh \beta c = 0, \quad 0 < t.$$

This forces $\sinh \beta c$ to be equal to zero which implies $\beta c = 0$ or, since $c \neq 0$, $\beta = 0$. Hence $u \equiv 0$ is the only solution satisfying the boundary conditions (4) and (5). But this means that this solution cannot satisfy the final boundary condition (3). Therefore, reject (17) as a possible solution of (2) and turn to (18).

Applying (4) to (18),

$$0 = e^{-h^2\alpha^2 t} [A + 0], \quad 0 < t,$$

since $A \cos \alpha(0) = A$ and $B \sin \alpha(0) = 0$. Thus, (18) reduces to

$$u = B\, e^{-h^2\alpha^2 t} \sin \alpha x. \tag{20}$$

Applying condition (5) to (20),

$$0 = Be^{-h^2\alpha^2 t} \sin \alpha c, \quad 0 < t. \tag{21}$$

If one chooses $B = 0$, then $u \equiv 0$ and one cannot find a non-trivial function to satisfy the third condition. Since $e^{-h^2\alpha^2 t} \neq 0$, this implies

$$\sin \alpha c = 0. \tag{22}$$

The equation (22) holds only when

$$\alpha c = n\pi \quad (n = 0, \pm 1, \pm 2, \ldots).$$

Since c is given, one finds $\alpha = \frac{n\pi}{c}$. Thus the solution (20) becomes

$$u = B \exp\left[-\left[\frac{n\pi h}{c}\right]^2 t\right] \sin \frac{n\pi x}{c}$$

or, since for different n, different arbitrary constants, B, can be used,

$$u_n = B_n \exp\left[-\left[\frac{n\pi h}{c}\right]^2 t\right] \sin \frac{n\pi x}{c}. \tag{23}$$

Note that restriction of n to the positive integers does not change the nature of the solutions.

Now, attempt to satisfy the boundary condition (3) using the solutions (23). That is, at $t \to 0^+$, $u = f(x)$, $0 < x < c$. By the superposition principle, any linear combination, $u_1, \ldots, u_n$, of solutions of the form (23) is also a solution. In particular, assuming convergence, the infinite series,

$$u = \sum_{n=1}^{\infty} u_n,$$

or, $$u(x, t) = \sum_{n=1}^{\infty} B_n \exp\left[-\left[\frac{n\pi h}{c}\right]^2 t\right] \sin \frac{n\pi x}{c} \quad (24)$$

is also a solution of (2). Now (24) satisfies the boundary conditions (4), (5). If it is to satisfy (3) as $t \to 0^+$, one must have

$$\sum_{n=1}^{\infty} B_n \exp\left[-\left[\frac{n\pi h}{c}\right]^2 t\right] \sin \frac{n\pi x}{c} = f(x).$$

As $t \to 0^+$, $\exp\left[-\left[\frac{n\pi h}{c}\right]^2 t\right] = 1$. Hence, it is required that

$$f(x) = \sum_{n=1}^{\infty} B_n \sin \frac{n\pi x}{c}, \quad 0 < x < c. \quad (25)$$

The problem now becomes the determination of the coefficients B_n so that the infinite series $\sum_{n=1}^{\infty} B_n \sin \frac{n\pi x}{c}$ does indeed converge to $f(x)$ for all x in $(0, c)$. Once these B_n are found they may be substituted into (24) which then becomes the final solution of the equation (2) subject to the boundary conditions, (3), (4) and (5). Note that solving for the B_n involves construction of an appropriate Fourier series for given $f(x)$.

• **PROBLEM 17-13**

Find the solution of the one-dimensional heat equation,

$$\frac{\partial u}{\partial t} = h^2 \frac{\partial^2 u}{\partial x^2} \quad 0 < x < 1, \; 0 < t, \quad (1)$$

subject to the boundary conditions:

as $t \to 0^+$, $u = u_0 x$, $\quad 0 < x < 1$ $\quad (2)$

as $x \to 0^+$, $u \to 0$, $\quad 0 < t$ $\quad (3)$

$$\text{as } x \to 1^-, \quad \frac{\partial u}{\partial x} \to 0, \qquad 0 < t. \qquad (4)$$

Solution: The problem stated above arises from the following physical considerations. Suppose one is asked to find the temperature in a flat slab of unit width such that,

(a) its initial temperature varies uniformly from zero at one face to u_0 at the other,

(b) the temperature of the face initially at zero remains at zero for $t > 0$,

(c) the face initially at temperature u_0 is insulated for $t > 0$.

The mathematical form of these conditions would be the above problem.

One may solve (1) using the method of separation of variables. Thus, assume the solution $u(x, t)$ may be written as the product of two functions of single variables i.e.

$$u(x, t) = f(x) \, g(t). \qquad (5)$$

Substituting (5) into (1), one obtains,

$$f(x) \, g'(t) = h^2 \, g(t) \, f''(x).$$

Separating variables,

$$\frac{g'(t)}{g(t)} = \frac{h^2 \, f''(x)}{f(x)}. \qquad (6)$$

If (6) is to hold for arbitrary values of t and x, the two sides must equal a constant k.

Thus, $\quad \dfrac{g'(t)}{g(t)} = k; \quad \dfrac{h^2 \, f''(x)}{f(x)} = k. \qquad (7), (8)$

Solving (7) first, note that it is a separable, first order equation. The solution is

$$g(t) = c e^{kt}, \qquad (9)$$

where c is an arbitrary constant.

Now consider (8). Here one finds it convenient to give a value to the constant k. Thus, let $k = -h^2 \alpha^2$. Then (8) becomes

$$f''(x) + \alpha^2 \, f(x) = 0. \qquad (10)$$

Equation (10) is a second order, linear, homogeneous, differential equation with constant coefficients.

The solution to (10) is

$$f(x) = [A \cos \alpha x + B \sin \alpha x]. \qquad (11)$$

Substituting (9) and (11) into (5),

$$u(x, t) = \exp(-h^2\alpha^2 t) [A \cos \alpha x + B \sin \alpha x], \qquad (12)$$

where one has substituted $-h^2\alpha^2$ for k. Now, apply conditions (2), (3) and (4) to (12). First, consider the homogeneous conditions, i.e., the conditions that require the solution and its rate of change to approach zero for given values of the independent variables. From condition (3),

$$0 = A \exp(-h^2\alpha^2 t), \qquad 0 < t$$

since $B \sin x = 0$ as $x \to 0^+$. A is forced to equal zero because $\exp(-h^2\alpha^2 t) \neq 0$. Hence, (12) reduces to

$$u(x, t) = \exp(-h^2\alpha^2 t) [B \sin \alpha x]. \qquad (13)$$

Taking the partial derivative of (13) with respect to x,

$$\frac{\partial u(x, t)}{\partial x} = \exp(-h^2\alpha^2 t) [\alpha B \cos \alpha x]. \qquad (14)$$

Applying condition (4) to (14),

$$\lim_{x \to 1^-} \exp(-h^2\alpha^2 t) [\alpha B \cos \alpha x] = 0. \qquad (15)$$

Since $\exp(-h^2\alpha^2 t) \neq 0$ and α, $B \neq 0$, this forces

$$\lim_{x \to 1^-} \cos \alpha x = 0. \qquad (16)$$

Solving for α in (16), one finds

$$\alpha = \frac{(2n + 1)\pi}{2} \qquad (n = 0, 1, 2, \ldots). \qquad (17)$$

Substituting (17) into (13),

$$u_n = B_n \exp\left[-\frac{1}{4} h^2 (2n + 1)^2 \pi^2 t\right] \sin\left[(2n + 1)\frac{\pi x}{2}\right]. \qquad (18)$$

Note that expressions of the form (18) satisfy (1), (3) and (4). Also, linear combinations of solutions of the form (18) (different values of n produce different values for u_n) satisfy (1), (3) and (4). In particular, one forms the infinite series

$$u(x, t) = \sum_{n=0}^{\infty} B_n \exp\left[-\frac{1}{4} h^2 (2n + 1)^2 \pi^2 t\right] \sin \frac{(2n + 1)\pi x}{2},$$

in the hope that it will satisfy (2). Thus, one requires

$$u_0 x = \sum_{n=0}^{\infty} B_n \sin\left[\frac{(2n+1)\pi x}{2}\right], \quad 0 < x < 1, \quad (19)$$

since as $t \to 0^+$, $\exp\left[-\frac{1}{4} h^2 (2n+1)^2 \pi^2 t\right] = 1$.

Equation (19) is similar to the Fourier series sine expansion of a function in the interval $0 < x < 2$, i.e.

$$f(x) \sim \sum_{n=0}^{\infty} b_n \sin \frac{n\pi x}{2}, \quad 0 < x < 2, \quad (20)$$

where $f(x) = u_0 x$.

Solving for the b_n in (20) and substituting into (19) one finds

$$u(x, t) = \frac{8u_0}{\pi^2} \sum_{n=0}^{\infty} \frac{(-1)^k}{(2k+1)^2}$$

$$\exp\left[-\frac{1}{4} h^2 (2k+1)^2 \pi^2 t\right] \sin \frac{(2k+1)\pi x}{2}$$

which is the solution to the problem (1) subject to the conditions (2) - (4).

• **PROBLEM 17-14**

Given an exchanger in which the shell-side fluid enters at 400° and leaves at 200° and the tube-side fluid enters at 100° and leaves at 200°.

Let U = overall heat transfer coefficient (Btu/hr - ft^2- °F)

w_h = hot fluid mass flow rate (lb/hr)

w_c = cold fluid mass flow rate (lb/hr)

c_{ph} = specific heat of the hot fluid (Btu/lb - °F)

c_{pc} = specific heat of the cold fluid (Btu/lb - °F)

Assuming that U, w_h, c_{ph}, w_c, and c_{pc} are constant and that heat losses are negligible, determine the mean overall temperature difference from hot to cold fluid (a) for a counterflow apparatus and (b) for a reversed-current exchanger with one well-baffled pass in the shell and two passes in the tube. (See Fig. 1)

Solution: (a) With counterflow, the terminal differences are 400 - 200°, or 200°, at the hot end and 200 - 100° or

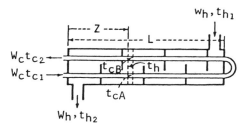

Diagram of 1-2 exchanger, one well-baffled shell pass and two tube passes. Fig. 1

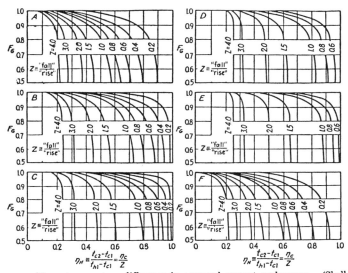

Mean temperature difference in reversed-current exchangers. (Shell-side fluid well mixed at a given cross section.) (A) One shell pass and 2, 4, 6, etc., tube passes. (B) Two shell passes, and 4, 8, 12, etc., tube passes. (C) Three shell passes, and 6, 12, 18, etc., tube passes. (D) Four shell passes, and 8, 16, 24, etc., tube passes. (E) Six shell passes, and 12, 24, 36, etc., tube passes. (F) One shell pass, and 3, 6, 9, etc., tube passes. It is immaterial whether the warmer fluid flows through the shell or the tubes.

Fig. 2

100°, at the cold end. The logarithmic-mean difference for counterflow is given by the equation

$$\Delta t_{o\ell} = \frac{(t_{h1} - t_{c2}) - (t_{h2} - t_{c1})}{\ln[(t_{h1} - t_{c2})/(t_{h2} - t_{c1})]}$$

$$= \frac{(400 - 200) - (200 - 100)}{\ln[(400 - 200)/(200 - 100)]}$$

$$\Delta t_{o\ell} = 144°$$

F_G is a dimensionless correction factor in the relation $F_G = \frac{\Delta t_{om}}{\Delta t_{o\ell}}$, where $\Delta t_{o\ell}$ is the logarithmic-mean temperature

725

difference and Δt_{om} is the true mean Δt_o. Referring to Fig. 2, η_H, the heating effectiveness, is defined as

$$\eta_H = \frac{t_{c2} - t_{c1}}{t_{h1} - t_{c1}},$$

and Z, the hourly heat-capacity ratio is

$$Z = \frac{w_c c_{pc}}{w_h c_{ph}} = \frac{t_{h1} - t_{h2}}{t_{c2} - t_{c1}}$$

Substituting the appropriate values,

$$\eta_H = \frac{(200 - 100)}{(400 - 100)} = \frac{1}{3}$$

$$Z = \frac{(400 - 200)}{(200 - 100)} = 2$$

From curve A of Fig. 2, $F_G = 0.80 = \frac{\Delta t_{om}}{144}$,

$$\Delta t_{om} = 115°.$$

● **PROBLEM 17-15**

Derive the solution for Laplace's eq.:

$$\frac{\partial^2 T}{\partial x^2} + \frac{\partial^2 T}{\partial y^2} = 0 \qquad (1)$$

for the boundary conditions as shown in the figure. Three sides of the plate are maintained at the constant temperature T_1, and the upper side has some temperature distribution impressed upon it. This distribution could be simply a constant temperature or something more complex, such as a sine-wave distribution. Consider both cases.

Isotherms and heat-flow lines in a rectangular plate.

Solution: To solve, the separation-of-variables method is used. The essential point of this method is that the solution to the differential equation is assumed to take a product form

$$T = XY \quad \text{where } X = X(x) \tag{2}$$
$$Y = Y(y)$$

The boundary conditions are then applied to determine the form of the functons X and Y. The basic assumption as given by Eq. (2) can be justified only if it is possible to find a solution of this form which satisfies the boundary conditions.

First consider the boundary conditions with a sine-wave temperature distribution impressed on the upper edge of the plate. Thus

$$\begin{aligned} T &= T_1 & \text{at } y &= 0 \\ T &= T_1 & \text{at } x &= 0 \\ T &= T_1 & \text{at } x &= W \\ T &= T_m \sin\frac{\pi x}{W} + T_1 & \text{at } y &= H \end{aligned} \tag{3}$$

where T_m is the amplitude of the sine function. Substituting Eq. (2) in (1) gives

$$-\frac{1}{X}\frac{d^2X}{dx^2} = \frac{1}{Y}\frac{d^2Y}{dy^2} = \lambda^2 \tag{4}$$

Observe that each side of Eq. (4) is independent of the other since x and y are independent variables. This requires that each side be equal to some constant. Therefore two ordinary differential equations in terms of this constant are obtained.

$$\frac{d^2X}{dx^2} + \lambda^2 X = 0 \tag{5}$$

$$\frac{d^2Y}{dy^2} - \lambda^2 Y = 0 \tag{6}$$

where λ^2 is called the separation constant. Its value must be determined from the boundary conditions. Note that the form of the solution to Eqs. (5) and (6) will depend on the sign of λ^2; a different form would also result if λ^2 were zero. The only way that the correct form can be determined is through an application of the boundary conditions of the problem. First write down all possible solutions and then see which one fits the problem under consideration.

For $\lambda^2 = 0$: $\quad X = C_1 + C_2 x$
$$Y = C_3 + C_4 y$$

$$T = (C_1 + C_2 x)(C_3 + C_4 y)$$

This function cannot fit the sine-function boundary condition, so that the $\lambda^2 = 0$ solution may be excluded.

For $\lambda^2 < 0$: $X = C_5 e^{-\lambda x} + C_6 e^{\lambda x}$

$$Y = C_7 \cos \lambda y + C_8 \sin \lambda y$$

$$T = (C_5 e^{-\lambda x} + C_6 e^{\lambda x})(C_7 \cos \lambda y + C_8 \sin \lambda y)$$

Again, the sine-function boundary condition cannot be satisfied, so that this solution is excluded also.

For $\lambda^2 > 0$: $X = C_9 \cos \lambda x + C_{10} \sin \lambda x$

$$Y = C_{11} e^{-\lambda y} + C_{12} e^{\lambda y}$$

$$T = (C_9 \cos \lambda x + C_{10} \sin \lambda x)(C_{11} e^{-\lambda y} + C_{12} e^{\lambda y})$$

Now, it is possible to satisfy the sine-function boundary condition; so attempt to satisfy the other conditions. The algebra is somewhat easier to handle when the substitution

$$\theta = T - T_1$$

is made. The differential equation and the solution then retain the same form in the new variable θ, and the boundary conditions transform to:

$\theta = 0$ at $y = 0$

$\theta = 0$ at $x = 0$

$\theta = 0$ at $x = W$

$\theta = T_m \sin \frac{\pi x}{W}$ at $y = H$

Applying these conditions,

$0 = (C_9 \cos \lambda x + C_{10} \sin \lambda x)(C_{11} + C_{12})$ (a)

$0 = C_9 (C_{11} e^{-\lambda y} + C_{12} e^{\lambda y})$ (b)

$0 = (C_9 \cos \lambda W + C_{10} \sin \lambda W)(C_{11} e^{-\lambda y} + C_{12} e^{\lambda y})$ (c)

$T_m \sin \frac{\pi x}{W} = (C_9 \cos \lambda x + C_{10} \sin \lambda x)(C_{11} e^{-\lambda H} + C_{12} e^{\lambda H})$ (d)

Accordingly, $C_{11} = -C_{12}$

$C_9 = 0$

and from (c),

$$0 = C_{10} C_{12} \sin \lambda W (e^{\lambda y} - e^{-\lambda y})$$

This requires that

$$\sin \lambda W = 0 \tag{7}$$

Recall that λ was an undetermined separation constant. There are several values which will satisfy Eq. (7), and these may be written

$$\lambda = \frac{n\pi}{W}$$

where n is an integer. By the principle of superposition, the solution to the differential equation may thus be written as a sum of the solutions for each value of n. This is an infinite sum, so that the final solution is the infinite series

$$\theta = T - T_1 = \sum_{n=1}^{\infty} C_n \sin \frac{n\pi x}{W} \sinh \frac{n\pi y}{W} \tag{8}$$

where the constants have been combined and the exponential terms converted to the hyperbolic function. The final boundary condition may now be applied.

$$T_m \sin \frac{\pi x}{W} = \sum_{n=1}^{\infty} C_n \sin \frac{n\pi x}{W} \sinh \frac{n\pi H}{W}$$

which requires that $C_n = 0$ for $n > 1$. The final solution is therefore

$$T = T_m \frac{\sinh (\pi y/W)}{\sinh (\pi H/W)} \sin \left(\frac{\pi x}{W}\right) + T_1$$

The temperature field for this problem is shown in the figure. Note that the heat-flow lines are perpendicular to the isotherms.

Consider now the set of boundary conditions

$T = T_1$ at $y = 0$

$T = T_1$ at $x = 0$

$T = T_1$ at $x = W$

$T = T_2$ at $y = H$

Using the first three boundary conditions, obtain the solution in the form of Eq. (8).

$$T - T_1 = \sum_{n=1}^{\infty} C_n \sin \frac{n\pi x}{W} \sinh \frac{n\pi y}{W}$$

Applying the fourth boundary condition gives

$$T_2 - T_1 = \sum_{n=1}^{\infty} C_n \sin \frac{n\pi x}{W} \sinh \frac{n\pi H}{W} \tag{9}$$

This is a Fourier sine series, and the values of the C_n may be determined by expanding the constant temperature difference $T_2 - T_1$ in a Fourier series over the interval $0 < x < W$. This series is

$$T_2 - T_1 = (T_2 - T_1)\frac{2}{\pi}\sum_{n=1}^{\infty}\frac{(-1)^{n+1}+1}{n}\sin\frac{n\pi x}{W} \qquad (10)$$

Upon comparison of Eq. (9) with Eq. (10), it is found that

$$C_n = \frac{2}{\pi}(T_2 - T_1)\frac{1}{\sinh(n\pi H/W)}\frac{(-1)^{n+1}+1}{n}$$

and the final solution is expressed as

$$\frac{T-T_1}{T_2-T_1} = \frac{2}{\pi}\sum_{n=1}^{\infty}\frac{(-1)^{n+1}+1}{n}\sin\frac{n\pi x}{W}\frac{\sinh(n\pi y/W)}{\sinh(n\pi H/W)}$$

• **PROBLEM** 17-16

a) A rubber sheet ½-in thick is to be cured at 292°F for 50 min. If the sheet is initially at 70°F and heat is applied from both surfaces, find the time required for the temperature at the center of the sheet to reach 290°F. It can be assumed that the surfaces are brought to 292°F as soon as curing is begun and held at that temperature throughout the process. The thermal diffusivity $k/\rho C_p$ of rubber can be taken as 0.0028 ft²/h.

b) Find the solution to the problem in part a) using Fig. 3.

Unsteady-state heat conduction in a rubber sheet.

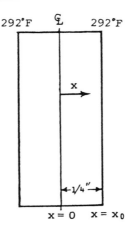

Fig. 1

Solution: a) The differential equation for unsteady heat conduction in one direction is

$$\frac{\partial t}{\partial \theta} = \frac{k}{C_p \rho}\frac{\partial^2 t}{\partial x^2} \qquad (A)$$

The origin is chosen at the center of the sheet, so that x is the distance from the center, as shown in Fig. 1 and x_0 is the half thickness.

In solving partial differential equations it is usually convenient to define the variables so that they range between zero and 1 or between zero and infinity and are dimensionless. Therefore Eq. (A) will be expressed in terms of the following redefined variables:

$$Y = \frac{292 - t}{292 - 70} \qquad n = \frac{x}{x_0} \qquad \tau = \frac{k\theta}{C_p \rho x_0^2} \qquad (1)$$

Equation (A) becomes

$$\frac{\partial Y}{\partial \tau} = \frac{\partial^2 Y}{\partial n^2} \qquad (2)$$

as can be proved by substituting the new variables (1) in Eq. (2) and performing the indicated partial differentiations.

A solution to Eq. (2) is obtained by postulating that such a solution can be written as the product of two terms,

$$Y = TN \qquad (3)$$

where T is a function of only the quantity τ, in which time is the variable, and N is a function of only n, in which distance is the variable. If $Y = TN$ is a solution, it can be differentiated as stated in (2) and yields

$$N \frac{\partial T}{\partial \tau} = T \frac{\partial^2 N}{\partial n^2} \qquad (4)$$

This equation can be rearranged to give only terms which depend on τ on the left side of the equation and only terms which are a function of n on the right side.

$$\frac{1}{T} \frac{\partial T}{\partial \tau} = \frac{1}{N} \frac{\partial^2 N}{\partial n^2} \qquad (5)$$

Since time and distance are independent of each other in this problem, the left side of this equation is independent of distance and the right side is independent of time. Hence each side must be constant. This constant shall be called, for convenience, $-a^2$. Thus one can write two ordinary differential equations,

$$\frac{dT}{d\tau} + a^2 T = 0 \qquad (6)$$

and

$$\frac{d^2 N}{dn^2} + a^2 N = 0 \qquad (7)$$

The solution of Eq. (6) is simply

$$T = C_1 e^{-a^2 \tau} \qquad (8)$$

The student of differential equations will recognize Eq. (7) as a second-order, ordinary differential equation with constant coefficients. The solution is written as

$$N = C_2 \sin an + C_3 \cos an \tag{9}$$

Having obtained expressions for τ and N, now write the solution for the partial differential equation (2), which was originally postulated as Eq. (3).

$$Y = \tau N$$
$$= C_1 e^{-a^2 \tau} (C_2 \sin an + C_3 \cos an) \tag{10}$$

This equation contains four constants, C_1, C_2, C_3, and a; however, C_1 can be combined with C_2 and C_3, so that only three boundary conditions are needed to complete the solution. These are:

1. The system is symmetrical, so no heat is conducted across the center line ($x = 0$). Therefore, at $n = 0$, $\partial Y / \partial n = 0$.

2. At the surface ($x = x_0 = \frac{1}{48}$), the temperature is constant at 292°F at all times. Thus, when $n = 1$, $Y = 0$.

3. Initially ($\theta = 0$), the entire sheet is at 70°F. Therefore, when $\tau = 0$, $Y = 1$.

Boundary condition 1 is used first.

$$\frac{\partial Y}{\partial n} = C_1 e^{-a^2 \tau} (C_2 \, a \cos an - C_3 \, a \sin an)$$

When $n = 0$, $\sin an = 0$; however, $\cos an = 1$. Therefore

$$\frac{\partial Y}{\partial n} = a C_1 C_2 e^{-a^2 \tau}$$

For this to be equal to zero and still preserve a general solution, the term C_2 must equal zero. After this simplification the general solution can be written

$$Y = A e^{-a^2 \tau} \cos an \tag{11}$$

where A includes both C_1 and C_3.

Boundary condition 2 is next employed.

$$0 = A e^{-a^2 \tau} \cos a$$

This condition is realized for many values of a such as $\pi/2$, $3\pi/2$, $5\pi/2$, etc.

The substitution of one of these values for a in Eq. (11) would meet the requirement of the second boundary condition, but it is not possible to represent an arbitrary temperature distribution in the slab by a cosine curve.

However, it is possible to represent an arbitrary function by an infinite series of cosine terms, provided each term is multiplied by a suitable coefficient. Therefore, write the general solution as

$$Y = A_1 e^{-(\pi/2)^2 \tau} \cos\left(\frac{\pi}{2}\right) n + A_2 e^{-(3\pi/2)^2 \tau} \cos\left(\frac{3\pi}{2}\right) n$$

$$+ A_3 e^{-(5\pi/2)^2 \tau} \cos\left(\frac{5\pi}{2}\right) n + \ldots$$

$$+ A_i e^{-[(2i-1)\pi/2]^2 \tau} \cos\left[\frac{(2i-1)\pi}{2}\right] n + \ldots \quad (12)$$

where i is an integer. It can be verified by differentiation and substitution that this sum still satisfies the partial differential equation and the first two boundary conditions.

The final step is to evaluate the constant which serves as a coefficient of each term in the series. This is done using the third boundary condition. Equation (12) becomes

$$1 = A_1 \cos\left(\frac{\pi}{2}\right)n + A_2 \cos\left(\frac{3\pi}{2}\right)n + A_3 \cos\left(\frac{5\pi}{2}\right)n + \ldots$$

$$+ A_i \cos\left(\frac{2i-1}{2}\right)\pi n + \ldots \quad (13)$$

Both sides of this equation are multiplied by $\cos[(2i - 1)/2]\pi n \, dn$ and integrated over the range of 0 to 1. The left-hand side becomes

$$\int_0^1 \cos\left(\frac{2i-1}{2}\right)\pi n \, dn = \frac{2}{2i-1} \frac{1}{\pi} \left[\sin\left(\frac{2i-1}{2}\right)\pi n\right]_0^1$$

$$= -\frac{2}{2i-1} \frac{1}{\pi} (-1)^i$$

The first term on the right-hand side of the equation can be integrated with the aid of a table of integrals to give

$$\int_0^1 A_1 \cos\left(\frac{\pi}{2}\right)n \cos\left(\frac{2i-1}{2}\right)\pi n \, dn$$

$$= A_1 \left[\frac{\sin(i-1)\pi n}{(2i-2)\pi} + \frac{\sin i\pi n}{2i\pi}\right]_0^1$$

$$= A_1 \left[\frac{\sin(i-1)\pi}{(2i-2)\pi} + \frac{\sin i\pi}{2i\pi}\right]$$

$$= 0$$

The integral is zero for all values of the integer i other than 1. The second term on the right-hand side vanishes in the same manner as the first, and so do all succeeding terms except the ith term. This becomes

$$A_i \int_0^1 \cos^2\left(\frac{2i-1}{2}\right)\pi n \, dn$$

$$= A_i \frac{2}{2i-1} \frac{1}{\pi} \left[\frac{\pi}{2} \frac{2i-1}{2} n + \frac{1}{4}\sin(2\pi)\frac{2i-1}{2}n\right]_0^1$$

$$= A_i \frac{2}{2i-1} \frac{1}{\pi}\left(\frac{\pi}{2} \frac{2i-1}{2} + 0 - 0 - 0\right)$$

$$= \frac{A_i}{2}$$

Thus Eq. (13) reduces to

$$-\frac{2}{2i-1} \frac{1}{\pi}(-1)^i = \frac{A_i}{2}$$

from which

$$A_i = \frac{-4(-1)^i}{(2i-1)\pi}$$

$$\left(\text{That is, } A_1 = \frac{4}{\pi}, \; A_2 = -\frac{4}{3\pi}, \; A_3 = \frac{4}{5\pi}, \; \text{etc.}\right)$$

The general solution is therefore

$$Y = \frac{4}{\pi}e^{-(\pi/2)^2\tau}\cos\left(\frac{\pi n}{2}\right) - \frac{4e^{-(3\pi/2)^2\tau}}{3\pi}\cos\left(\frac{3\pi n}{2}\right)$$

$$+ \frac{4e^{-(5\pi/2)^2\tau}}{5\pi}\cos\left(\frac{5\pi n}{2}\right) - \cdots \qquad (14)$$

Equation (14) written in compact form is

$$Y = \sum_{i=1}^{i=\infty} \frac{-2(-1)^i}{[(2i-1)/2]\pi} e^{-[(2i-1)\pi/2]^2\tau}\cos\left(\frac{2i-1}{2}\right)\pi n \qquad (15)$$

Equation (15) is now applied to the specific problem.

$$Y = \frac{292 - 290}{292 - 70} = 0.0090$$

$$n = 0$$

Therefore $0.0090 = \frac{4}{\pi}e^{-(\pi/2)^2\tau} - \frac{4}{3\pi}e^{-(3\pi/2)^2\tau}$

$$+ \frac{4}{5\pi}e^{-(5\pi/2)^2\tau} - \cdots$$

The solution for τ (which contains the time variable θ) must be obtained by trial and error. As a first approximation, only the first term on the right-hand side will be considered. This gives

$$\tau = -\left(\frac{2}{\pi}\right)^2 \ln \frac{(\pi)(0.0090)}{4}$$

$$= 2.01$$

$$\theta = \frac{c_p \rho}{k} x_0^2 \tau$$

$$= \frac{2.01}{(0.0028)(48)^2}$$

$$= 0.313 \text{ h } (18.7 \text{ min})$$

It is necessary to check the relative magnitude of the terms in the series solution to see if all terms other than the first one are negligible.

When $\tau = 2.01$, the series becomes

$$Y = \frac{4}{\pi} e^{-4.97} - \frac{4}{3\pi} e^{-44.8} + \frac{4}{5\pi} e^{-124} - \cdots$$

$$= (1.27)(0.00694) - (0.424)(3.50 \times 10^{-20}) + \cdots$$

Table 1 TEMPERATURE DISTRIBUTION IN RUBBER SHEET

Time elapsed, min	$\frac{4}{\pi} e^{-(\pi/2)^2 \tau}$	$n=0$ $\cos\frac{n\pi}{2}$ $=1$ (center line)	$n=\frac{1}{4}$ $\cos\frac{n\pi}{2}$ $=0.924$	$n=\frac{1}{2}$ $\cos\frac{n\pi}{2}$ $=0.707$	$n=\frac{3}{4}$ $\cos\frac{n\pi}{2}$ $=0.383$	$n=1$ $\cos\frac{n\pi}{2}$ $=0$ (surface)
0	1	70	70	70	70	292
1	9.76×10^{-1}	75	92	139	209	292
5	3.37×10^{-1}	217	223	239	263	292
10	9.02×10^{-2}	272	273	278	284.3	292
20	6.37×10^{-3}	290.6	290.7	291	291.5	292
30	4.52×10^{-4}	291.9	291.9	291.9	292	292
40	3.19×10^{-5}	292	292	292	292	292

The validity of the approximation which employed only the first term of the series is apparent. Although not all series solutions converge so rapidly as in the problem shown here, they frequently do so. This can often be established by a visual inspection of the exponents in the series (as in this case).

The temperature at any point in the rubber sheet at any instant can be determined by substitution of the appropriate values for n and τ in Eq. (15). Table 1 was prepared to show the development of the temperature profiles as heating proceeds. It is apparent that the results are contradictory at the surface at zero time. This is a result of the boundary conditions imposed on the mathematical solution.

The results are illustrated graphically in Fig. 2. At the center the temperature gradient is zero since the

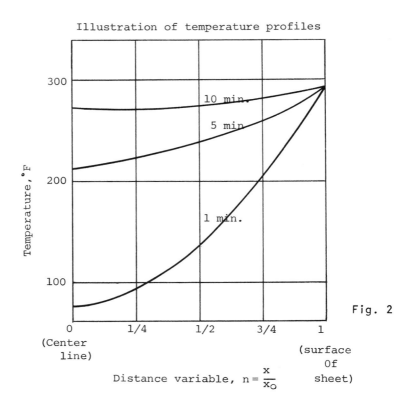

Fig. 2

slab is being heated uniformly from both sides. The same condition would result if a sheet of the half the thickness of the sheet in the problem were heated from one side only, while the other side was covered with a layer of perfect insulation.

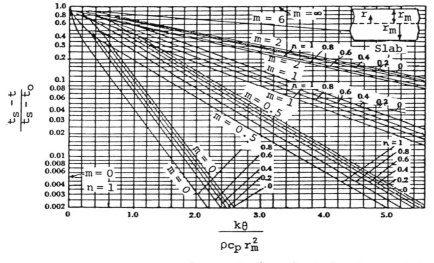

Gurney-Luries chart for large slab.   Fig. 3

b) $\quad \dfrac{\alpha \theta}{x_0^2} = \dfrac{0.0028 \theta}{(1/48)^2} = 6.44 \theta$

$$\dfrac{t_s - t}{t_s - t_0} = \dfrac{292 - 290}{292 - 70} = 0.0090$$

$$n = \dfrac{x}{x_0} = 0$$

$$m = \dfrac{k}{h x_0} = 0$$

The modulus m is zero for this case because the assumption of constant surface temperature implies that surface resistance to heat transfer is negligible (h = ∞). From Fig. 3 is obtained

$$\dfrac{\alpha \theta}{x_0^2} = 2.0$$

Therefore

$$\theta = \dfrac{2.0}{6.44}$$

$$= 0.31 \text{ h } (19 \text{ min})$$

MISCELLANEOUS PROBLEMS AND APPLICATIONS

• **PROBLEM** 17-17

Apply the method of separation of variables to obtain a formal solution, u(x, y), of the problem which consists of the two-dimensional Laplace equation,

$$\dfrac{\partial^2 u}{\partial x^2} + \dfrac{\partial^2 u}{\partial y^2} = 0, \tag{a}$$

and the four boundary conditions:

$$u(0, y) = 0, \qquad 0 \leq y \leq b; \tag{b}$$

$$u(a, y) = 0, \qquad 0 \leq y \leq b; \tag{c}$$

$$u(x, 0) = 0, \qquad 0 \leq x \leq a; \tag{d}$$

$$u(x, b) = f(x), \qquad 0 \leq x \leq a. \tag{e}$$

Solution: The Laplace equation (a), has a wide range of applicability in physics. In the theory of heat conduction, for example, the function u is the steady state temperature, and is equivalent to putting $\dfrac{\partial u}{\partial t} = 0$ in the Diffusion equation. In the theory of gravitation or electricity, u represents the gravitational or electric potential. For this reason equation (a) is often referred to as the potential equation.

To solve the Laplace equation using the method of separation of variables, assume that (a) has product solutions of the form

$$u(x, y) = X(x) Y(y),\qquad(f)$$

where X is a function of x alone and Y is a function of y alone.

Differentiating (f), one finds

$$\frac{\partial^2 u}{\partial x^2} = Y \frac{d^2 X}{dx^2}, \qquad \frac{\partial^2 u}{\partial y^2} = X \frac{d^2 Y}{dy^2}.$$

Substituting these expressions into the differential equation (a), one obtains

$$Y \frac{d^2 X}{dx^2} + X \frac{d^2 Y}{dy^2} = 0.$$

Rearranging terms, one gets

$$\frac{\frac{d^2 X}{dx^2}}{X} = - \frac{\frac{d^2 Y}{dy^2}}{Y}.\qquad(g)$$

The left member of (g) is a function only of x, and is therefore independent of y. Smilarly, the right member of (g) is a function of y alone, and therefore independent of x. From this one may conclude that the two equal expressions in (g) must both be equal to a constant k:

$$\frac{\frac{d^2 X}{dx^2}}{X} = k = - \frac{\frac{d^2 Y}{dy^2}}{Y}.$$

From this one obtains the two ordinary differential equations,

$$\frac{d^2 X}{dx^2} - kX = 0,\qquad(h)$$

and

$$\frac{d^2 Y}{dy^2} + kY = 0.\qquad(i)$$

Thus, one has reduced the given partial differential equation, or P.D.E., (a) to two ordinary differential equations, or O.D.E.'s (h) and (i), both of which contain the unknown constant, k.

In order to determine k, turn your attention to the four boundary conditions contained in the problem. One sees that the first three of these (b), (c), (d) are homogeneous, but the fourth (e) is not. First, try to satisfy homogeneous, but the fourth (e) is not. First, try to satisfy homogeneous boundary conditions. Since $u(x, y) = X(x)Y(y)$, the boundary conditions (b), (c), and (d) are reduced, respectively, to

$$X(0) \, Y(y) = 0, \qquad 0 \leq y \leq b; \qquad (j)$$

$$X(a) \, Y(y) = 0, \qquad 0 \leq y \leq b; \qquad (k)$$

and $\quad X(x) \, Y(0) = 0, \qquad 0 \leq x \leq a. \qquad (\ell)$

If $X(x) = 0$ in $(0, a)$ or if $Y(y) = 0$ in $(0, b)$, the assumed product solution (f) is reduced to the trivial solution. Therefore, one requires that

$$X(x) \neq 0, \qquad (m)$$

and $\quad Y(y) \neq 0. \qquad (n)$

In order to satisfy the boundary conditions (j), (k), and (ℓ), one is required to set

$$X(0) = 0, \qquad (o)$$

$$X(a) = 0, \qquad (p)$$

$$Y(0) = 0. \qquad (q)$$

Only when all five conditions, (m) to (q), are met can one hope to find a nontrivial solution of the given P.D.E. (a).

One has two conditions (o) and (p) on the function $X(x)$. In order for $X(x)$ to be a solution of the given P.D.E. (a), it must fulfill (o) and (p) and satisfy the O.D.E. (h) that one obtained using separation of variables. Therefore, rewriting (h), (o), and (p),

$$\frac{d^2X}{dx^2} - kX = 0, \qquad (0 \leq x \leq a)$$

$$X(0) = 0, \qquad X(a) = 0.$$

If $k = 0$ or if $k > 0$, it is seen that $X(x)$ is reduced to the trivial solution. If $k < 0$, however, one obtains a nontrivial solution of (h):

$$X(x) = c_1 \cos \sqrt{k} \, x + c_2 \sin \sqrt{k} \, x,$$

where c_1 and c_2 are arbitrary constants. From (o) one finds

$$X(0) = c_1 = 0.$$

Setting $c_1 = 0$, and using (p), one finds

$$X(a) = c_2 \sin \sqrt{k} \, a = 0.$$

For a nontrivial solution, one must have $c_2 \neq 0$. Therefore,

$$\sin \sqrt{k} \, a = 0,$$

or $\sqrt{k}\, a = n\pi$.

Solving, the values of k appropriate for the given problem are deterimed:

$$k = -\frac{n^2 \pi^2}{a^2} \qquad (n = 1, 2, 3, \ldots), \qquad (r)$$

and the corresponding nontrivial solution,

$$x_n(x) = c_n \sin \frac{n\pi x}{a} \qquad (n = 1, 2, 3, \ldots), \qquad (s)$$

where each c_n is an arbitrary constant corresponding to a value of n (n = 1, 2, 3, . . .).

One sees that the set of functions $\{x_n\}$ have two very important properties. First, they are solutions of one of the O.D.E.'s obtained by reducing the given P.D.E. Secondly, they obey the given boundary conditions. One can conclude that $\{x_n\}$ is a set of solutions of (a), and that the general solution of (a) is the product, XY, where $X = \{x_n\}$ and $Y = Y(y)$ is yet to be determined.

Substituting the expression for k, i.e. (r), into the ordinary differential equation involving y (i), one gets

$$\frac{d^2 Y}{dy^2} - \frac{n^2 \pi^2}{a^2} Y = 0,$$

where n = 1, 2, 3, ... For each value of n, this O.D.E. has the general solution,

$$Y_n(y) = c_{n,1} \exp\left[\frac{n\pi y}{a}\right] + c_{n,2} \exp\left[-\frac{n\pi y}{a}\right],$$

where $c_{n,1}$ and $c_{n,2}$ are arbitrary constants. In order to satisfy the homogeneous boundary condition obtained in (q), one must have

$$c_{n,1} + c_{n,2} = 0 \qquad (n = 1, 2, 3, \ldots)$$

Using this relation, rewrite the solution using only one artibrary constant:

$$Y_n(y) = c_{n,1}\left[e^{\frac{n\pi y}{a}} - e^{-\frac{n\pi y}{a}}\right].$$

Using the indentity $e^{\theta} - e^{-\theta} = 2 \sinh \theta$, this becomes,

$$Y_n(y) = c'_{n,1} \sinh \frac{n\pi y}{a} \qquad (n = 1, 2, 3, \ldots), \qquad (t)$$

where $c'_{n,1} = 2c_{n,1}$ are arbitrary constants. Since (t) satisfies the O.D.E. given in (i) with the boundary condition (q), one is guaranteed that for each positive integral value of n, one can obtain functions, Y_n, of the form (t) which will serve as the function, Y, in the product solution, (f).

Hence, for each value of n (n = 1, 2, 3, . . .), one can obtain solutions of the given P.D.E. by multiplying (s) and (t):

$$X_n Y_n = \left[c_n \sin \frac{n\pi x}{a} \right] \left[c'_{n,1} \sinh \frac{n\pi y}{a} \right].$$

If $A_n = c_n c'_{n,1}$, one can write these solutions as

$$u_n(x, y) = A_n \sin \frac{n\pi x}{a} \sinh \frac{n\pi y}{a} \quad (n = 1, 2, 3, \ldots). \quad \text{(u)}$$

For each value of n, each one of these solutions (u) satisifes both the partial differential equation (a) and the three homogeneous boundary conditions (b), (c), and (d) for all values of the constant, A_n.

Now, apply the inhomogeneous boundary condition (e). In order to do this, form the infinite series

$$\sum_{n=1}^{\infty} u_n(x, y) = \sum_{n=1}^{\infty} A_n \sin \frac{n\pi x}{a} \sinh \frac{n\pi y}{a}.$$

Assuming this series converges in the region $(0 \leq x \leq a; 0 \leq y \leq b)$, one can obtain a solution of the P.D.E. (a) by taking the sum of this series:

$$u(x, y) = \sum_{n=1}^{\infty} A_n \sin \frac{n\pi x}{a} \sinh \frac{n\pi y}{a}. \quad \text{(v)}$$

Evaluating (v) at the boundary points, one finds that $u(0, y) = 0$, $u(a, y) = 0$, and $u(x, 0) = 0$, all in agreement with the homogeneous boundary conditions, (b), (c), and (d).

Now, apply the inhomogeneous boundary condition (e) to the series solution (v):

$$\sum_{n=1}^{\infty} A_n \sin \frac{n\pi x}{a} \sinh \frac{n\pi b}{a} = f(x), \quad 0 \leq x \leq a.$$

Letting $B_n = A_n \sinh \left[\frac{n\pi b}{a} \right]$ (n = 1, 2, 3, . . .), this becomes:

$$\sum_{n=1}^{\infty} B_n \sin \frac{n\pi x}{a} = f(x), \text{ where } 0 \leq x \leq a. \quad \text{(w)}$$

One needs only to determine the coefficients B_n in (w)

that satisfy (e). Recognizing (w) as a sine series, use the Fourier coefficient formula to determine B_n. For $n = 1, 2, 3, \ldots,$

$$B_n = \frac{2}{a} \int_0^a f(x) \sin \frac{n\pi x}{a} \, dx.$$

Since $B_n = A_n \sinh \frac{n\pi b}{a}$ ($n = 1, 2, 3, \ldots$), one finds that

$$A_n = \frac{B_n}{\sinh \left[\frac{n\pi b}{a}\right]} = \frac{2}{a \sinh \left[\frac{n\pi b}{a}\right]} \int_0^a f(x) \sin \left[\frac{n\pi x}{a}\right] dx$$

$$(n = 1, 2, \ldots). \qquad (x)$$

Thus, in order for the series solution (v) to satisfy the inhomogeneous boundary condition (e), the coefficients A_n in the series must be given by (x).

Therefore, the formal solution of the problem consisting of the partial differential equation (a) and the four boundary conditions (b) through (e) is

$$u(x, y) = \sum_{n=1}^{\infty} A_n \sin \frac{n\pi x}{a} \sinh \frac{n\pi y}{a},$$

where

$$A_n = \frac{2}{a \sinh \frac{n\pi b}{a}} \int_0^a f(x) \sin \frac{n\pi x}{a} \, dx$$

$$(n = 1, 2, 3, \ldots).$$

● **PROBLEM 17-18**

Solve the partial, differential equation

$$\frac{\partial^2 u}{\partial x^2} + \frac{1}{x} \frac{\partial u}{\partial x} = \frac{\partial u}{\partial t}, \qquad (1)$$

subject to the three conditions:

(1) $u(L, t) = 0$, $\qquad t > 0$

(2) $u(x, 0) = f(x)$, $\qquad 0 < x < L$,

where $f(x)$ is a function of x defined on $0 < x < L$, and

(3) $\lim_{t \to \infty} u(x, t) = 0$

for each x, $0 \leq x \leq L$.

Solution: The method of solution is separation of variables. That is, assume that the equation possesses product solutions. Hence if $f = f(x_1, x_2, \ldots, x_n)$ the solutions are in the form $X_1, X_2, \ldots X_n$, where X_i is a function of x_i alone.

The given equation is a function of two variables, $u = u(x, t)$. Assume solutions of the form:

$$u(x, t) = X(x) T(t). \qquad (2)$$

Differentiating (2),

$$\frac{\partial u}{\partial t} = X \frac{dT}{dt} ; \quad \frac{\partial u}{\partial x} = T \frac{dX}{dx} ;$$

$$\frac{\partial^2 u}{\partial t^2} = X \frac{d^2 T}{dt^2}; \quad \frac{\partial^2 u}{\partial x^2} = T \frac{d^2 X}{dx^2}.$$

Upon substitution of these results into (1) one obtains

$$T\frac{d^2 X}{dx^2} + \frac{1}{X} T \frac{dX}{dx} = X \frac{dT}{dt} \qquad (3)$$

Separating variables,

$$\frac{1}{X}\left[\frac{d^2 X}{dx^2} + \frac{1}{x} \frac{dX}{dx}\right] = \frac{1}{T} \frac{dT}{dt} . \qquad (4)$$

Examining (4) one observes that the left hand side of (4) is a function of x alone while the right hand side is a function of t alone. If they are to be equal for arbitrary values of x and t, they must be equal to a constant, k. Thus,

$$\frac{1}{X}\left[\frac{d^2 X}{dx^2} + \frac{1}{x} \frac{dX}{dx}\right] = k. \qquad (5)$$

$$\frac{1}{T}\left[\frac{dT}{dt}\right] = k. \qquad (6)$$

From (5) and (6) one obtains the two ordinary differential equations

$$\frac{d^2 X}{dx^2} + \frac{1}{x} \frac{dX}{dx} - kX = 0. \qquad (7)$$

$$\frac{dT}{dt} - kT = 0. \qquad (8)$$

Thus, using the assumption that the solution is in the form of a product solution one has converted the partial differential equation of two variables into two ordinary differential equations. Solve (8) first. It is a first order, separable equation. Separating variables,

$$\frac{dT}{T} = k \, dt. \qquad (9)$$

The solution to (9) is

$$T = Ce^{kt}, \tag{10}$$

where C is an arbitrary constant.

From the third condition stated at the beginning of the problem,

$$\lim_{t \to \infty} u(x, t) = 0.$$

But, by assumption, $u(x, t) = X(x) T(t)$. Hence,

$$\lim_{t \to \infty} u(x, t) = X(x) \lim_{t \to \infty} T(t) = 0$$

for all x, $0 \leq x \leq L$. Since $X(x)$ is an arbitrary function one cannot assume it is identically zero. Therefore, one must have

$$\lim_{t \to \infty} T(t) = 0.$$

This implies that $k < 0$ in (10). Let $k = -\lambda^2$ where λ is real. Then $k < 0$ always ($\lambda \neq 0$) and (10) takes the form

$$T = Ce^{-\lambda^2 t}. \tag{10a}$$

Now, consider the solution of (7), a second order, ordinary differential equation with variable coefficients. Since $k = -\lambda^2$, rewrite (7) as

$$\frac{d^2 X}{dx^2} + \frac{1}{x} \frac{dX}{dx} + \lambda^2 X = 0$$

or, $$x^2 \frac{d^2 X}{dx^2} + x \frac{dX}{dx} + \lambda^2 x^2 X = 0 \tag{11}$$

Let $\theta = \lambda x$. Then $X = X(\theta)$ and $\frac{dX}{d\theta} = \lambda \frac{dX}{dx}$. Finally,

$$\frac{d^2 X}{d\theta^2} = \frac{d}{d\theta}\left[\lambda \frac{dX}{dx}\right] = \lambda^2 \frac{d^2 X}{dx^2}.$$

Substituting these results into (11),

$$\theta^2 \frac{d^2 X}{d\theta^2} + \theta \frac{dX}{d\theta} + \theta^2 X = 0. \tag{12}$$

Equation (12) is in the form of a Bessel equation, i.e.,

$$x^2 \frac{d^2 y}{dx^2} + x \frac{dy}{dx} + (x - p)^2 y = 0,$$

where p is the order of the equation. Here p is zero, so (12) is a Bessel equation of order zero. The general solution of (12) is

$$X = c_1 J_0(\lambda x) + c_2 Y_0(\lambda x) \tag{13}$$

where J_0 is a Bessel function of the first kind, Y_0 is a Bessel function of the second kind, and c_1 and c_2 are arbitrary constants. Note that λx has been substituted for θ.

Now, see whether the imposition of the three conditions stated at the beginning of the problem creates any new information. Condition (3) states that

$$\lim_{t \to \infty} u(x, t) = 0, \text{ for each } x, \ 0 \le x \le L.$$

When $x = 0$, $\lim_{t \to \infty} u(0, t) = X(0) [\lim_{t \to \infty} T(t)] = 0$

This implies $X(0)$ is bounded. But as $x \to 0$, $J_0(0) = 1$ and $\lim_{x \to 0} Y_0(\lambda x) = -\infty$, i.e. the Bessel function of order zero of the second kind is unbounded. Hence, in order for $X(0)$ to be finite set $c_2 = 0$. Then, (13) is reduced to

$$X = c_1 J_0(\lambda x). \tag{14}$$

Now, apply condition (1) to (14). Requiring that $u(L, t) = 0$, $t > 0$ is equivalent to requiring that

$$X(L) T(t) = 0, \quad t > 0;$$

which implies $X(L) = 0$ or

$$J_0(\lambda L) = 0. \tag{15}$$

As $x \to +\infty$, the function $J_0(\lambda x)$ exhibits dampened oscillatory behavior. Thus, the equation $J_0(x) = 0$ has an infinite number of positive roots x_n ($n = 1, 2, 3, \ldots$). Arranging these roots in a monotonically increasing sequence ($x_{n+1} > x_n$) one obtains a sequence of numbers,

$$\lambda_n = \frac{x_n}{L} \quad (n = 1, 2, 3, \ldots)$$

each of which satisfies (15). Thus, for each positive integer n, there is a function of the form (14) which is the function X in the product solution. Write

$$X_n = c_{1,n} J_0(\lambda_n x) \quad (n = 1, 2, 3, \ldots) \tag{16}$$

to indicate the solutions.

Similarly, for each positive integer n there corresponds solutions of (8) in the form of (10a),

$$T_n = c_{2,n} e^{-\lambda_n^2 t}. \tag{17}$$

The functions in (17) serve as the function T in the product solution (2).

Combining (16) and (17) one finds that for each postive integer n, the product solution (2) takes the form

$$u_n(x, t) = A_n J_n(\lambda_n x) e^{-\lambda_n^2 t} \quad (n = 1, 2, \ldots), \quad (18)$$

where the $A_n = c_{1,n}$, c_{2n} ($n = 1, 2, \ldots$) are arbitrary constants.

Now, come to condition (2), that $u(x, 0) = f(x)$ within the open interval $0 < x < L$. Recall that a common procedure is to express a function as an infinite series, provided that the series converges for all values over which the function is defined. Thus, form an infinite series of the solutions (18):

$$u(x, t) = \sum_{n=1}^{\infty} A_n J_0(\lambda_n x) e^{-\lambda_n^2 t}. \quad (19)$$

In order for (19) to satisfy condition (3),

$$\sum_{n=1}^{\infty} A_n J_0(\lambda_n x) = f(x), \quad 0 < x < L. \quad (20)$$

Thus, one must find the value of the coefficients, A_n, such that the relation (20) is satisfied. The analysis by which the A_n are found is similar to the method whereby the coefficients, c_n are obtained for the trigonometric Fourier series of a function defined over an interval.

First, show that the set $\{J_0(\lambda_n x): n = 1, 2, \ldots\}$ is an orthogonal system with respect to the weight function $r(x) = x$ on the interval $0 \le x \le L$. Therefore,

$$\int_0^L x J_0(\lambda_m x) J_0(\lambda_n x) = 0 \quad m \ne n;$$

$$\int_0^L x [J_0(\lambda_m x)]^2 \, dx = T_n > 0 \quad (m = 1, 2, \ldots).$$

From the orthogonal system one forms the orthonormal system $\{\phi_n\}$ where

$$\phi_n = \frac{J_0(\lambda_n x)}{\sqrt{T_n}}. \quad (21)$$

Equation (21) indicates that every function has been normalized. Recall that if a function, f, is expanded in a series of orthonormal functions,

$$f(x) = \sum_{n=1}^{\infty} c_n \psi_n, \qquad (22)$$

the coefficients, c_n, are given by:

$$c_n = \frac{1}{\sqrt{k}} \int_0^L f(x) \psi_n(x) r(x) \, dx, \qquad (n = 1, 2, \ldots)$$

where $\sqrt{k}$ denotes the value of the integral that normalizes the function. Hence,

$$c_n = \frac{1}{\sqrt{T_n}} \int_0^L x \, f(x) \, J_0(\lambda_n x) \, dx \qquad (n = 1, 2, \ldots)$$

Substituting into (22), one gets

$$\sum_{n=1}^{\infty} \left[\frac{1}{\sqrt{T_n}} \int_0^L x \, f(x) \, J_0(\lambda_n x) \, dx \right] \left[\frac{J_0(\lambda_n x)}{\sqrt{T_n}} \right]$$

and $f(x) = \sum_{n=1}^{\infty} \left[\frac{1}{T_n} \int_0^L x \, f(x) \, J_0(\lambda_n x) \, dx \right] J_0(\lambda_n x). \qquad (23)$

Comparing (20) and (23), the condition (3) will be satisfied if

$$A_n = \frac{1}{T_n} \int_0^L x \, f(x) \, J_0(\lambda_n x) \, dx \quad (n = 1, 2, \ldots)$$

To find the T_n evaluate the integral

$$T_n = \int_0^L x [J_0(\lambda_n x)]^2 \, dx \qquad (n = 1, 2, \ldots).$$

One finds $T_n = \frac{L^2}{2} [J_1(\lambda_n L)]^2$ $(n = 1, 2, \ldots)$, where $J_1(\lambda_n L)$ represents Bessel functions of the first kind of order one. Thus, the values of the coefficients are

$$A_n = \frac{2}{L^2 [J_1(\lambda_n L)]^2} \int_0^L x \, f(x) \, J_0(\lambda_n x) \, dx$$

$$(n = 1, 2, \ldots) \qquad (24)$$

The solution of (1) subject to the conditions (1), (2) and (3) is the series

$$u(x,t) = \sum_{n=1}^{\infty} A_n \, J_0(\lambda_n x) \, e^{-\lambda_n^2 t}$$

where A_n are given by (24).

• **PROBLEM 17-19**

Derive an expression for the Laplacian of a function $u(r, \theta)$ by applying a transformation of coordinate systems to the function, $u(x, y)$, whose Laplacian is given by the formula:

$$\nabla^2 u(x, y) = \frac{\partial^2 u}{\partial x^2} + \frac{\partial^2 u}{\partial y^2}. \qquad (a)$$

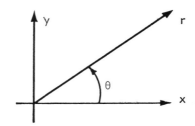

Fig. 1

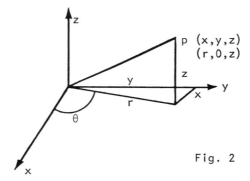

Fig. 2

Solution: A transformation of coordinates $(x, y) \to (r, \theta)$ is obtained using the following equivalent sets of equations:

$$x = r \cos \theta,$$
$$y = r \sin \theta, \qquad (b)$$

or, $r = +\sqrt{x^2 + y^2}$,

$$\theta = \tan^{-1} \frac{y}{x}. \qquad (c)$$

One demonstrates (b) and (c) in Figure 1:

Since x and y may be any real numbers, one gets

$$-\infty < x < \infty,$$
$$-\infty < y < \infty,$$

in R^2. From Figure 1, one sees that θ, too, assumes all positive and negative values, but in cycles of 2π. The variable r, however, must always be positive. The transformation of coordinate systems, $(x, y) \to (r, \theta)$, therefore, confines one to the region,

$$-\infty < \theta < \infty \; ; \qquad r > 0.$$

Now, examine the effect of such a transformation on a given function, $u(x, y)$. Suppose that in planar, Cartesian coordinates, one has a function,

$$u(x, y) = xy^2. \qquad (d)$$

Using the transformation $(x, y) \to (r, \theta)$ given in (b), express this function in polar coordinates, (r, θ):

$$u(r, \theta) = (r \cos \theta)(r \sin \theta)^2$$

$$= r^3 \cos \theta \sin^2 \theta, \qquad (e)$$

and the two forms (d) and (e) are equivalent.

Equation (a) gives the Laplacian $\nabla^2 u(x, y)$, of a function in Cartesian coordinates. One wishes to determine an equivalent form of (a) in polar coordinates, that is, to determine $\nabla^2 u(r, \theta)$.

First, compute the necessary derivatives. First one gets

$$\frac{\partial u(x, y)}{\partial x} = \frac{\partial u}{\partial r} \frac{\partial r}{\partial x} + \frac{\partial u}{\partial \theta} \frac{\partial \theta}{\partial x}, \qquad (f)$$

by the chain rule. Using (c), one gets

$$\frac{\partial r}{\partial x} = \frac{\partial}{\partial x} (\sqrt{x^2 + y^2}) = \frac{x}{\sqrt{x^2 + y^2}}.$$

Using (b), this becomes

$$\frac{\partial r}{\partial x} = \frac{r \cos \theta}{r} = \cos \theta. \qquad (g)$$

Again, from (c), one finds that

$$\frac{\partial \theta}{\partial x} = \frac{\partial}{\partial x} \left[\tan^{-1} \frac{y}{x} \right]$$

$$= \frac{1}{1 + \frac{y^2}{x^2}} \frac{\partial}{\partial x} \left[\frac{y}{x} \right] = \frac{1}{1 + \left[\frac{y}{x} \right]^2} \left[-\frac{y}{x^2} \right]$$

$$= -\frac{y}{x^2 + y^2}.$$

Using (b), this becomes

$$\frac{\partial \theta}{\partial x} = \frac{-r \sin \theta}{r^2} = \frac{-\sin \theta}{r}. \qquad (h)$$

Substituting (g) and (h) into (f):

$$\frac{\partial u}{\partial x} = \frac{\partial u}{\partial r}(\cos\theta) + \frac{\partial u}{\partial \theta}\left[-\frac{\sin\theta}{r}\right]$$

$$= (\cos\theta)\, u_r - \left[\frac{\sin\theta}{r}\right] u_\theta, \qquad (i)$$

where one has substituted the notation $\frac{\partial u(\theta, r)}{\partial r} = u_r$ and $\frac{\partial u(r, \theta)}{\partial \theta} = u_\theta$.

In general, write

$$\frac{\partial}{\partial x} = \cos\theta \frac{\partial}{\partial r} - \frac{\sin\theta}{r}\frac{\partial}{\partial \theta}.$$

Now, to determine $\frac{\partial^2 u}{\partial x^2}$ in terms of r and θ, compute the partial derivative of (i) with respect to x:

$$\frac{\partial^2 u}{\partial x^2} = \frac{\partial}{\partial x}\left[\cos\theta \cdot u_r - \frac{\sin\theta}{r} \cdot u_\theta\right]$$

$$= \left[\cos\theta \frac{\partial}{\partial r} - \frac{\sin\theta}{r}\frac{\partial}{\partial \theta}\right]\left[\cos\theta \cdot u_r - \frac{\sin\theta}{r} \cdot u_\theta\right]$$

$$= \cos\theta \frac{\partial}{\partial r}\left[\cos\theta \cdot u_r - \frac{\sin\theta}{r} \cdot u_\theta\right]$$

$$- \frac{\sin\theta}{r}\frac{\partial}{\partial \theta}\left[\cos\theta \cdot u_r - \frac{\sin\theta}{r} u_\theta\right].$$

Using the product rule for differentiation,

$$\frac{\partial^2 u}{\partial x^2} = \cos\theta\,\{(\cos\theta)u_{rr}\} + \cos\theta$$

$$\left\{\left[\frac{\sin\theta}{r^2}\right] u_\theta - \left[\frac{\sin\theta}{r}\right] u_{\theta r}\right\}$$

$$- \frac{\sin\theta}{r}\{(\cos\theta)u_{r\theta} - (\sin\theta)u_r\}$$

$$- \frac{\sin\theta}{r}\left\{\left[-\frac{\cos\theta}{r}\right] u_\theta - \left[\frac{\sin\theta}{r}\right] u_{\theta\theta}\right\}.$$

Simplifying,

$$\frac{\partial^2 u}{\partial x^2} = \cos^2\theta \cdot u_{rr} + \frac{\sin^2\theta}{r^2} \cdot u_{\theta\theta}$$

$$- \frac{2\cos\theta\sin\theta}{r} u_{r\theta} + \frac{\sin^2\theta}{r} u_r$$

$$+ \frac{2\cos\theta\sin\theta}{r^2} u_\theta. \qquad (j)$$

Performing similar operations on the variable y, one finds that

$$\frac{\partial u}{\partial y} = \frac{\partial u}{\partial r}\frac{\partial r}{\partial y} + \frac{\partial u}{\partial \theta}\frac{\partial \theta}{\partial y} \quad . \tag{k}$$

It is seen from (b) and (c) that

$$\frac{\partial r}{\partial y} = \frac{r \sin \theta}{r} = \sin \theta,$$

and $\frac{\partial \theta}{\partial y} = \frac{x}{x^2 + y^2} = \frac{\cos \theta}{r}$.

Substituting these values into (k), one gets:

$$\frac{\partial u}{\partial y} = \sin \theta \frac{\partial u}{\partial r} + \frac{\cos \theta}{r}\frac{\partial u}{\partial \theta}$$

$$= \sin \theta \, u_r + \frac{\cos \theta}{r} u_\theta, \tag{ℓ}$$

where $u_r = \frac{\partial u}{\partial r}$ and $u_\theta = \frac{\partial u}{\partial \theta}$, as before, or

$$\frac{\partial}{\partial y} = \sin \theta \frac{\partial}{\partial r} + \frac{\cos \theta}{r}\frac{\partial}{\partial \theta} \quad .$$

To obtain $\frac{\partial^2 u}{\partial y^2}$, differentiate ($\ell$) with respect to y:

$$\frac{\partial^2 u}{\partial y^2} = \left[\sin \theta \frac{\partial}{\partial r} + \frac{\cos \theta}{r}\frac{\partial}{\partial \theta} \right]\left[\sin \theta \, u_r + \frac{\cos \theta}{r} u_\theta \right]$$

$$= \sin \theta \frac{\partial}{\partial r}\left[\sin \theta \, u_r + \frac{\cos \theta}{r} u_\theta \right]$$

$$+ \frac{\cos \theta}{r}\frac{\partial}{\partial \theta}\left[\sin \theta \, u_r + \frac{\cos \theta}{r} u_\theta \right] \quad .$$

Using the product rule, this derivative becomes

$$\frac{\partial^2 u}{\partial y^2} = \sin \theta \left\{ \sin \theta \, u_{rr} + \frac{\cos \theta}{r} u_{r\theta} - \frac{\cos \theta}{r^2} u_\theta \right\}$$

$$+ \frac{\cos \theta}{r}\left\{ \sin \theta \, u_{r\theta} + \cos \theta \, u_r + \frac{\cos \theta}{r} u_{\theta\theta} - \frac{\sin \theta}{r} u_\theta \right\} \quad .$$

Simplifying, one finds that

$$\frac{\partial^2 u}{\partial y^2} = \sin^2 \theta \cdot u_{rr} + \frac{\cos^2 \theta}{r^2} \cdot u_{\theta\theta}$$

$$+ \frac{2 \sin \theta \cos \theta}{r} u_{r\theta} + \frac{\cos^2 \theta}{r} u_r$$

$$- \frac{2 \cos \theta \sin \theta}{r^2} u_\theta \quad . \tag{m}$$

Now substitute (j) and (m) into equation (a), thus ob-

751

taining the Laplacian of the transformed function $u(r,\theta)$. To simplify the computations, group the coefficients of like terms as one adds them:

$$\nabla^2 u(r,\theta) = \{(\cos^2\theta + \sin^2\theta)u_{rr}$$

$$+ \left[\frac{\sin^2\theta}{r^2} + \frac{\cos^2\theta}{r^2}\right]u_{\theta\theta}$$

$$+ \left[-\frac{2\cos\theta\sin\theta}{r} + \frac{2\cos\theta\sin\theta}{r}\right]u_{r\theta}$$

$$+ \left[\frac{\sin^2\theta}{r} + \frac{\cos^2\theta}{r}\right]u_r$$

$$+ \left[\frac{2\cos\theta\sin\theta}{r^2} - \frac{2\cos\theta\sin\theta}{r^2}\right]u_\theta \}$$

Therefore, $\nabla^2 u(r,\theta) = u_{rr} + \frac{1}{r^2}u_{\theta\theta} + \frac{1}{r}u_r$

$$= u_{rr} + \frac{1}{r}u_r + \frac{1}{r^2}u_{\theta\theta}. \quad (n)$$

Equation (n) thus expresses the Laplacian of a function u in polar coordinates (r, θ).

This problem could be extended to 3 dimensions by using a transformation,

$$(x, y, z) \rightarrow (r, \theta, z), \quad (o)$$

where (x, y, z) denotes a point P in a rectangular, Cartesian coordinate system for which:

$-\infty < x < \infty$,

$-\infty < y < \infty$,

$-\infty < z < \infty$,

and (r, θ, z) denotes that point P in a circular, cylindrical coordinate system for which,

$-\infty < \theta < \infty$,

$r > 0$,

$-\infty < z < \infty$.

This transofrmation of coordinate systems (o), is demonstrated in Figure 2.

One obtains this transformation by using the equations:

$x = r\cos\theta$,

$y = r\sin\theta$,

$z = z$, $\quad (p)$

and their corresponding inverse forms:

$$r = \sqrt{x^2 + y^2},$$

$$\theta = \tan^{-1} \frac{y}{x}.$$

Note that the transformation equations (p) and (q) are identical to the transformation equations in the original problem (b) and (c) except for the addition of the variable z; the transformation $(x, y) \to (r, \theta)$ is exactly the transformation $(x, y, z) \to (r, \theta, z)$ when z is held constant at $z = 0$.

Adding the term, $\frac{\partial^2 u}{\partial z^2}$, to equation (a), one obtains the Laplacian of a function u in rectangular Cartesian coordinates:

$$u(x, y, z) = \frac{\partial^2 u}{\partial x^2} + \frac{\partial^2 u}{\partial y^2} + \frac{\partial^2 u}{\partial z^2} = u_{xx} + u_{yy} + u_{zz}, \quad (r)$$

where the notation $\frac{\partial^2 u}{\partial v^2} = u_{vv}$ is used for all variables.

If the transformation $(x, y, z) \to (r, \theta, z)$ is applied to (r), one will obtain the Laplacian of the function u in circular, cylindrical coordinates. Noting that z remains unchanged in this transformation (o) and that r and θ are z independent, one has only to add the term $\frac{\partial^2 u}{\partial z^2}$ to the Laplacian in polar coordinates obtained in equation (n):

$$\nabla^2 u(r, \theta, z) = u_{rr} + \frac{1}{r} u_r + \frac{1}{r^2} u_{\theta\theta} + u_{zz}. \quad (s)$$

Equation (s) is thus the formula from which one obtains the Laplacian of a function u in a circular cylindrical coordinate system (r, θ, z).

• **PROBLEM 17-20**

The hyperbolic equation

$$\frac{\partial^2 u}{\partial t^2} - \frac{\partial^2 u}{\partial x^2} = 0, \quad 0 \leq x \leq 1, \quad 0 \leq t \leq T \quad (1)$$

with boundary conditions $u(0,t) = u(1,t) = 0$, $\quad 0 \leq t \leq T \quad (2)$
and initial conditions
$$u(x,0) = \sin\pi x, \quad 0 \leq x \leq 1,$$
$$\frac{\partial u}{\partial t}(x,0) = 0 \quad (3)$$

is given. Solve this problem using the finite-difference method. Set $T = 1$, $M = 10$, $N = 10$.

Solution: Consider the wave equation

$$\frac{\partial^2 u}{\partial t^2} - \lambda^2 \frac{\partial^2 u}{\partial x^2} = 0, \quad \begin{array}{l} 0 \leq x \leq \ell \\ 0 \leq t \leq T \end{array} \quad (4)$$

with the boundary conditions

$$u(0,t) = u(\ell,t) = 0, \quad 0 \leq t \leq T \quad (5)$$

and the initial conditions

$$u(x,0) = f(x), \quad 0 \leq x \leq \ell,$$
$$\frac{\partial u(x,0)}{\partial t} = g(x) \quad (6)$$

The Finite-Difference Algorithm

INPUT:
$\quad\quad 0,\ell$ - endpoints $(0 \leq x \leq \ell)$
$\quad\quad 0,T$ - time $(0 \leq t \leq T)$
$\quad\quad \lambda$ - constant in (4)
$\quad\quad M,N$ - integers

OUTPUT:
$\quad\quad \omega_{m,n}$ - approximations to $u(x_m,t_n)$
$\quad\quad$ for $m = 0,1,\ldots,M$, and $n = 0,1,\ldots,N$

Step 1: Set
$$s = \frac{T}{N}$$
$$h = \frac{\ell}{M}$$
$$\alpha = \frac{s\lambda}{h}$$

Step 2: For $n = 1,\ldots,N$, set
$$\omega_{M,n} = \omega_{0,n} = 0$$

Step 3: Set
$$\omega_{0,0} = f(0)$$
$$\omega_{M,0} = f(\ell)$$

Step 4: For $m = 1,\ldots,M-1$ set
$$\omega_{m,0} = f(mh)$$
$$\omega_{m,1} = (1-\alpha^2) f(mh) + \frac{\alpha^2}{2}\left[f\big((m+1)h\big) + f\big((m-1)h\big)\right] + sg(mh)$$

754

Step 5: For n = 1,...,N-1
for m = 1,...,M-1

set
$$\omega_{m,n+1} = 2(1-\alpha^2)\omega_{m,n} + \alpha^2(\omega_{m+1,n} + \omega_{m-1,n}) - \omega_{m,n-1}$$

Step 6: For n = 0,...,N

set
t = ns

for m = 0,...,M

set x = mh

OUTPUT: $(x, t, \omega_{m,n})$

Step 7: STOP.

In our case, we have

$$s = .1 \quad h = .1 \quad \alpha = \lambda = 1$$

Values of $\omega_{m,n}$

n=	0	1	2	3 ...
m = 0	0	0	0	0
↓ 1	0.309016994	0.293892626		
2	0.587785252	0.559016995		
3	0.809016994	0.769420885		
4	0.951056517	0.904508497		
5	1	0.951056517		
6	0.951056516			
7	0.809016994			
8	0.587785252			
9	0.309016994			
10	0	0	0	0 ...

Note that the exact solution of (1) is

$$u(x,t) = \sin(\pi x)\cos(\pi t) \tag{7}$$

● **PROBLEM 17-21**

Develop and solve the problem of the vibrating string by considering an elastic, flexible string stretched tightly and clamped at the two ends. Use separation of variables in determining the solution.

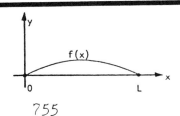

Solution: Consider a perfectly elastic flexible string of length L stretched tightly and clamped at the two ends. For convenience, fix the ends on the x axis at x = 0 and x = L. Assume that for each x in the interval 0 < x < L, the string is displaced into the xy plane and that for each such x, the displacement from the x axis is given by f(x), a known function of x. This is illustrated in Figure 1.

Suppose that at t = 0 the string is released from the initial position defined by f(x), with an initial velocity given at each point in the interval 0 ≤ x ≤ L by g(x), where g(x) is a known function of x. The string vibrates, and its displacement in the y direction at any point x at any time t will be a function of both x and t. Determine this displacement as a function of x and t; denote it by y(x, t).

This assumption, that y = y(x, t), is valid provided that the system, as given in Figure 1, fulfills certain conditions.

Concerning the vibrations, assume that all motion is confined to the xy plane and that each point on the string moves on a straight line perpendicular to the x axis as the string vibrates. Also, assume that the displacement y at each point of the string is small compared with the length L. (This is equivalent to assuming that the length of any segment of the string is considered constant.) In addition, it is required that at times during motion, the angle between the string and the x axis at each point is sufficiently small.

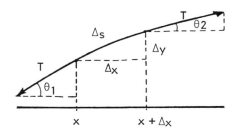

Fig. 2

As for the string itself, assume that its linear density ρ, and its tension T, are both constant at all times. It is required that the tension T of the string be great enough so that the string acts as though it were perfectly flexible.

These assumptions are illustrated in Figure 2.

Let Δs represent an element of arc of the string. Since the tension is assumed constant, the net upward vertical force acting on Δs is given by

$$T \sin \theta_2 - T \sin \theta_1. \tag{a}$$

Since it was required that the angle between the string and the x axis be small, assume

$$\sin \theta \cong \tan \theta.$$

The the force given in (a) becomes

$$T \tan \theta_2 - T \tan \theta_1. \qquad (b)$$

Since, in Figure 2, the slope is given by

$$\frac{\partial y}{\partial x} = \tan \theta,$$

rewrite (b) as

$$T \left.\frac{\partial y}{\partial x}\right|_{x+\Delta x} - T \left.\frac{\partial y}{\partial x}\right|_{x} \qquad (c)$$

Newton's Law requires that this net force (c) be equal to the mass of the string times the acceleration. For a segment of string of arc length Δs, the mass is given by $\rho \Delta s$, where ρ is the linear density, which has been assumed constant. The acceleration of the arc element Δs is given by $\left[\frac{\partial^2 y}{\partial t^2} + \varepsilon\right]$, where the term $(\rho \Delta s)\varepsilon$ can be interpreted as the force difference at the points (x, y) and $(x + \Delta x, y + \Delta y)$, such that

$$\lim_{\Delta s \to 0} \varepsilon = 0. \qquad (d)$$

For the problem, Newton's Law assumes the form

$$\text{Force} = \text{mass} \times \text{acceleration}$$

$$= (\rho \Delta s) \left[\frac{\partial^2 y}{\partial t^2} + \varepsilon\right]$$

Substituting this expression into (c), one gets, for the net force acting on an element of arc Δs,

$$T \left[\left.\frac{\partial y}{\partial x}\right|_{x+\Delta x} - \left.\frac{\partial y}{\partial x}\right|_{x}\right]$$

$$= (\rho \Delta s) \left[\frac{\partial^2 y}{\partial t^2} + \varepsilon\right] \qquad (e)$$

By assuming small vibrations, the condition $\Delta x \cong \Delta s$ has been imposed.

Making this substitution, (e) becomes

$$T \left[\left.\frac{\partial y}{\partial x}\right|_{x+\Delta x} - \left.\frac{\partial y}{\partial x}\right|_{x}\right]$$

$$= (\rho \Delta x) \left[\frac{\partial^2 y}{\partial t^2} + \varepsilon\right].$$

Dividing by $\rho \Delta x$ yields

$$\frac{T}{\rho} \frac{\left[\frac{\partial y}{\partial x}\Big|_{x+\Delta x} - \frac{\partial y}{\partial x}\Big|_{x} \right]}{\Delta x} = \left[\frac{\partial^2 y}{\partial t^2} + \varepsilon \right].$$

Taking the limit as $\Delta x \to 0$ (in which case $\varepsilon \to 0$, by (d)) gives

$$\frac{T}{\rho} \frac{\partial}{\partial x} \left[\frac{\partial y}{\partial x} \right] = \frac{\partial^2 y}{\partial t^2},$$

or $\quad \dfrac{\partial^2 y}{\partial t^2} = a^2 \dfrac{\partial^2 y}{\partial x^2},$ \hfill (f)

where $a^2 = \dfrac{T}{\rho}$.

This is the one dimensional wave equation, a second order partial differential equation that can be solved by using separation of variables. Before determining the solution, however, establish the boundary and initial conditions.

Since one imposed the condition that the ends of the string are fixed at $x = 0$ and $x = L$ for all time t, the displacement $y(x, t)$ must satisfy the two boundary conditions

$$y(0, t) = 0, \qquad 0 \leq t < \infty;$$
$$y(L, t) = 0, \qquad 0 \leq t < \infty. \qquad (g)$$

To determine the initial conditions that $y(x, t)$ must satisfy, recall that at $t = 0$ the string was released from the initial position defined by $f(x)$ ($0 \leq x \leq L$), with initial velocity $g(x)$ ($0 \leq x \leq L$). Thus, $y(x, t)$ must fulfill the two initial conditions

$$y(x, 0) = f(x), \qquad 0 \leq x \leq L;$$
$$\frac{\partial y(x, 0)}{\partial t} = g(x), \qquad 0 \leq x \leq L. \qquad (h)$$

The problem may now be stated: seek a function $y(x, t)$ which simultaneously satisfies the partial differential equation (f), the boundary conditions (g), and the initial conditions (h).

In order to apply the method of separation of variables, assume that the P.D.E. (f) has product solutions of the form XT, where X is a function of x alone, and T is a function of t alone:

$$y(x, t) = X(x)T(t). \qquad (i)$$

Differentiating and substituting (i) into the differential equation (f) yields

$$a^2 T \frac{d^2 X}{dx^2} = X \frac{d^2 T}{dt^2},$$

from which is obtained

$$a^2 \frac{\frac{d^2X}{dx^2}}{X} = k, \tag{j}$$

and

$$\frac{\frac{d^2T}{dt^2}}{T} = k, \tag{k}$$

where k is a constant that will have to be determined.

Rewriting (j) and (k), obtain, respectively, the two ordinary differential equations,

$$\frac{d^2X}{dx^2} - \frac{k}{a^2} X = 0, \tag{ℓ}$$

and

$$\frac{d^2T}{dt^2} - kT = 0. \tag{m}$$

Now call upon the boundary conditions (g). Since $y(x, t) = X(x)T(t)$, seemingly $y(0, t) = X(0)T(t)$ and $y(L, t) = X(L)T(t)$. Thus the boundary conditions become

$$X(0)T(t) = 0, \qquad 0 \leq t < \infty;$$

$$X(L)T(t) = 0, \qquad 0 \leq t < \infty.$$

In order for this system to have a nontrivial solution $T(t)$, one must require that

$$X(0) = 0 \text{ and } X(L) = 0. \tag{n}$$

The boundary conditions thus impose a third set of restrictions on the function $X(x)$, for it is already required that $X(x)$ must satisfy the given P.D.E. (j) as well as the O.D.E., (ℓ). Furthermore, it is seen that the solution of the O.D.E. (ℓ) depends upon the value of the unknown constant k.

If $k = 0$, the general solution of (ℓ) is

$$X(x) = c_1 + c_2 x, \tag{o}$$

where c_1 and c_2 are arbitrary constants. Applying the boundary conditions (n) to the solution (o) gives

$$X(0) = 0 = c_1,$$

$$X(L) = c_1 + c_2 L = 0 = c_2 L.$$

Since this requires that both c_1 and c_2 be zero, dismiss the case $k = 0$, for it only yields the trivial solution.

If $k > 0$, the general solution of the O.D.E. (ℓ) is

$$X(x) = c_1 \exp\left[\frac{\sqrt{k}x}{a}\right] + c_2 \exp\left[-\frac{\sqrt{k}x}{a}\right]. \qquad (p)$$

Applying the boundary conditions to (p):

$$c_1 + c_2 = 0,$$

$$c_1 \exp\left[\frac{\sqrt{k}L}{a}\right] + c_2 \exp\left[-\frac{\sqrt{k}L}{a}\right] = 0.$$

Using Cramer's Rule, solve this system for c_1 and c_2:

$$c_1 = \frac{\begin{vmatrix} 0 & 1 \\ 0 & \exp\left[-\frac{\sqrt{k}L}{a}\right] \end{vmatrix}}{\begin{vmatrix} 1 & 1 \\ \exp\left[\frac{\sqrt{k}L}{a}\right] & \exp\left[-\frac{\sqrt{k}L}{a}\right] \end{vmatrix}} = 0,$$

$$c = \frac{\begin{vmatrix} 1 & 0 \\ \exp\left[\frac{\sqrt{k}L}{a}\right] & 0 \end{vmatrix}}{\begin{vmatrix} 1 & 1 \\ \exp\left[\frac{\sqrt{k}L}{a}\right] & \exp\left[-\frac{\sqrt{k}L}{a}\right] \end{vmatrix}} = 0.$$

Since $c_1 = c_2 = 0$, one has only the trivial solution for $k > 0$.

For $k < 0$, equation (ℓ) has the general solution

$$X(x) = c_1 \sin\frac{\sqrt{-k}}{a}x + c_2 \cos\frac{\sqrt{-k}}{a}x. \qquad (q)$$

In order for (q) to satisfy the boundary conditions (n), one must have

$$X(0) = c_2 = 0; \quad \text{i.e.,} \quad c_2 = 0$$

and $X(L) = c_1 \sin\frac{\sqrt{-k}L}{a} + c_2 \cos\frac{\sqrt{-k}L}{a} = 0.$

Therefore, $c_1 \sin\frac{\sqrt{-k}L}{a} = 0.$ \hfill (r)

Since the solution (q) is to be nontrivial, it is required that $c_1 \neq 0$. In order to satisfy equation (r), it is concluded that

$$\sin\frac{\sqrt{-k}L}{a} = 0.$$

This implies

$$\frac{\sqrt{-kL}}{a} = n\pi, \qquad (n = 1, 2, 3, \ldots).$$

Therefore, determine k by restricting it to those values for which

$$k = -\frac{n^2 \pi^2 a^2}{L^2}, \qquad (n = 1, 2, 3, \ldots). \qquad (s)$$

One says that (s) yields the characteristic values k_n ($n = 1, 2, 3, \ldots$) of the system that are being considered, and one notes that (s) can generate only negative numbers, which is in agreement with the assumption that $k < 0$. Substituting (s) into (q), obtain the corresponding non-trivial solutions

$$X_n(x) = c_n \sin \frac{n\pi x}{L}, \qquad (n = 1, 2, 3, \ldots), \qquad (t)$$

where c_n ($n = 1, 2, 3, \ldots$) are arbitrary constants. Refer to (t) as the characteristic functions of the system, since the function $X(x)$ in the assumed solution (i) must be of the form given in (t). More generally, it can be concluded that for each value of n ($n = 1, 2, 3, \ldots$), one obtains functions X_n of the form (t) which serve as the function X in the product solution of the system, which is given by (i).

Now attention is turned to the differential equation that T, the other member of the product solution, must satisfy. Substituting k from (s) into (m), obtain the O.D.E.

$$\frac{d^2 T}{dt^2} + \frac{n^2 \pi^2 a^2}{L^2} T = 0,$$

where $n = 1, 2, 3, \ldots$. Since the coefficient of T in this equation is always positive, obtain a solution valid for each $n = 1, 2, 3, \ldots$

$$T_n = c_{n,1} \sin \frac{n\pi at}{L} + c_{n,2} \cos \frac{n\pi at}{L}, \qquad (u)$$

where $c_{n,1}$ and $c_{n,2}$ are arbitrary constants. As in the case of X_n given in (t), the function T_n given in (u) yields, for each value of n, ($n = 1, 2, 3, \ldots$) an appropriate function to serve in the product solution (i).

Therefore, for each positive integral value of n, obtain corresponding solutions as the product

$$X_n T_n = \left[c_n \sin \frac{n\pi x}{L} \right] \left[c_{n,1} \sin \frac{n\pi at}{L} + c_{n,2} \cos \frac{n\pi at}{L} \right].$$

If one sets $a_n = c_n c_{n,1}$ and $b_n = c_n c_{n,2}$ ($n = 1, 2, 3,$

. . .) one may write the product solutions as

$$Y_n(x, t) = \left[\sin \frac{n\pi x}{L}\right]\left[a_n \sin \frac{n\pi at}{L} + b_n \cos \frac{n\pi at}{L}\right],$$

$$(n = 1, 2, 3, \ldots), \qquad (v)$$

and it is guaranteed that each of these solutions (v) satisfies both the partial differential equation (f) and the two boundary conditions (g) for all values of the constants a_n and b_n.

Now try to satisfy the two initial conditions (h). In general no single one of the solutions (v) will satisfy these conditions. For example, if the first initial condition (h) is applied to a solution of the form (v) one must have

$$b_n \sin \frac{n\pi x}{L} = f(x), \quad 0 \leq x \leq L$$

where n is some positive integer; and this is impossible unless f happens to be a sine function of the form $A \sin\left[\frac{n\pi x}{L}\right]$ for some positive integer n.

Approach this problem from a theoretical standpoint. It is known from the Principle of Superposition, that every finite linear combination of the solutions of (f) is also a solution of (f); furthermore, assuming appropriate convergence, an infinite series of solutions of (f) is also a solution of (f). This suggests that one should form either a finite linear combination or an infinite series of the solutions (v) and attempt to apply the initial conditions (h) to the "more general" solutions thus obtained. In general no finite linear combination will satisfy these conditions, and one must resort to an infinite series.

Therefore, form an infinite series

$$\sum_{n=1}^{\infty} Y_n(x, t) = \sum_{n=1}^{\infty} \left[\sin \frac{n\pi x}{L}\right]\left[a_n \sin \frac{n\pi at}{L} + b_n \cos \frac{n\pi at}{L}\right]$$

of the solutions (v). Assuming appropriate convergence, the sum of this series is also a solution of the differential equation (f). Denoting this sum by $y(x, t)$, write

$$y(x, t) = \sum_{n=1}^{\infty} \left[\sin \frac{n\pi x}{L}\right]\left[a_n \sin \frac{n\pi at}{L} + b_n \cos \frac{n\pi at}{L}\right], \qquad (w)$$

and note that $y(0, t) = 0$ and $y(L, t) = 0$. Thus, assuming appropriate convergence, the function y given by (w) satisfies both the differential equation (f) and the two boundary conditions (g).

Now apply the initial conditions (h) to the series solution (w). The first condition $y(x, 0) = f(x)$, $0 \leq x \leq L$ reduces to

$$\sum_{n=1}^{\infty} b_n \sin \frac{n\pi x}{L} = f(x), \quad 0 \leq x \leq L. \tag{x}$$

Thus to satisfy the first initial condition (h), determine the coefficients b_n so that (X) is satisfied.

This is recognized as a problem in Fourier sine series. Using the coefficient formula yields

$$b_n = \frac{2}{L} \int_0^L f(x) \sin \frac{n\pi x}{L} dx \quad (n = 1, 2, 3, \ldots). \tag{y}$$

Thus in order for the series solution (w) to satisfy the initial condition $y(x, 0) = f(x)$, $0 \leq x \leq L$, the coefficients b_n in the series must be given by formula (y).

The only condition which remains to be satisfied is the second initial condition (h), which is

$$\frac{\partial y(x, 0)}{\partial t} = g(x), \quad 0 \leq x \leq L.$$

From (w) it is found that

$$\frac{\partial y}{\partial t}(x, t) = \sum_{n=1}^{\infty} \left[\frac{n\pi a}{L}\right] \left[\sin \frac{n\pi x}{L}\right] \left[a_n \cos \frac{n\pi at}{L} - b_n \sin \frac{n\pi at}{L}\right].$$

The second initial condition reduces this to

$$\sum_{n=1}^{\infty} \frac{a_n n\pi a}{L} \sin \frac{n\pi x}{L} = g(x), \quad 0 \leq x \leq L.$$

Letting $A_n = \frac{a_n n\pi a}{L}$ $(n = 1, 2, 3, \ldots)$, this takes the form

$$\sum_{n=1}^{\infty} A_n \sin \frac{n\pi x}{L} = g(x), \quad 0 \leq x \leq L. \tag{z}$$

Thus to satisfy the second initial condition (h), determine the coefficinets A_n so that (z) is satisfied.

This is another problem in Fourier sine series; use the coefficient formula to obtain

$$A_n = \frac{2}{L} \int_0^L g(x) \sin \frac{n\pi x}{L} dx \quad (n = 1, 2, 3, \ldots).$$

Since $A_n = \dfrac{a_n n\pi a}{L}$ ($n = 1, 2, 3, \ldots$), it is found that

$$a_n = \dfrac{L}{n\pi a} A_n = \dfrac{2}{n\pi a} \int_0^L g(x) \sin \dfrac{n\pi x}{L}\, dx.$$

$$(n = 1, 2, 3, \ldots). \qquad (aa)$$

Thus in order for the series solution (w) to satisfy the second initial condition (h), the coefficients a_n in the series must be given by (aa).

Therefore, the formal solution of the problem consisting of of the partial differential equation (j), the two boundary conditions (g), and the two initial conditions (h) is

$$y(x, t) = \sum_{n=1}^{\infty} \left[\sin \dfrac{n\pi x}{L} \right]\left[a_n \sin \dfrac{n\pi a t}{L} + b_n \cos \dfrac{n\pi a t}{L} \right],$$

where $a_n = \dfrac{2}{n\pi a} \int_0^L g(x) \sin \left[\dfrac{n\pi x}{L} \right] dx \quad (n = 1, 2, 3, \ldots),$

and $b_n = \dfrac{2}{L} \int_0^L f(x) \sin \left[\dfrac{n\pi x}{L} \right] dx \quad (n = 1, 2, 3, \ldots).$

In summary, the principal steps in the solution of this problem were, first, to assume the product solution XT given by (i). This led to the ordinary differential equation (j) for the function X and the ordinary differential equation (k) for the function T. Then the boundary conditions (g) were considered and were found to reduce to the boundary conditions (n) on the function X. Thus the function X(x) had to be a nontrivial solution of the problem consisting of (j) and (n) in order to be part of the product solution (i) of the partial differential equation (f). Solving this problem, one obtained for solutions, the functions $X_n(x)$ given by (t). Then one returned to the differential equation (k) for the function T(t) and obtained the solutions $T_n(t)$ given by (u). Thus for each positive integral value of n, the product solutions $X_n T_n$ denoted by y_n and given by (v) were found. Each of these solutions y_n satisfied both the partial differential equation (f) and the boundary conditions (g), but no one of them satisfied the initial conditions (h). In order to satisfy these initial conditions, an infinite series of the solutions Y_n was formed. Thus one obtained the formal solution y given by (w), in which the coefficients a_n and b_n were arbitrary. The initial conditions to this series solution were applied and thereby the coefficients a_n and b_n were determined. One thus obtained the formal solution Y given by (w) in which the coefficients a_n and b_n are given by (aa) and (y), respectively. This solution is a formal one, for in the process of obtaining it, assumptions of convergence were made which one did not justify.

CHAPTER 18

MONTE CARLO METHODS

MONTE CARLO RANDOM NUMBER DISTRIBUTION

● **PROBLEM** 18-1

Obtain random numbers with discrete distribution with the supposition that a random variable X takes on the values 0, 1, and 2 with probabilities 0.18, 0.44, and 0.38, respectively.

Solution: Since the probabilities are given to two decimals, it is sufficient, in hand computation, to use two-digit rectangularly distributed random numbers R.

Rule:

R	X
$0 \leq R < 0.18$	0
$0.18 \leq R < 0.62$	1
$0.62 \leq R < 1.00$	2

Using a random digit table to get values for 100R, the following table is generated:

	55	69	30	29	45	81	72	34	35	88
X	1	2	1	1	1	2	2	1	1	2

	81	35	07	63	02
X	2	1	0	2	0

(Notice that the above rule is formulated in such a way that no "round-off correction" need be used.)

● **PROBLEM 18-2**

Illustrate by graphical means a general method for generating random numbers with a continuous (cumulative) distribution, function F(x). Proof that X has the desired distribution, and is a ramdom number with (cumulative) distribution F(x).

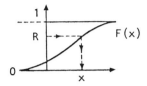

Fig. 1

Solution: Take a rectangularly distributed random number R in the interval [0, 1], and solve the equation F(X) = R. Then X has the desired distribution, and is called a random number with (cumulative) distribution F(X); see Fig. 1.

Proof. Since F(x) is a nondecreasing continuous function then

$$P[X \leq x] = P[F(X) \leq F(x)] = P[R \leq F(\cdot)],$$

but the last expression is equal to F(X), since by the definition of a rectangular distribution one has P[R ≤ r] = r, where r is an arbitrary number between 0 and 1. Thus P[X ≤ x] = F(x), which (by definition) means that X has the distribution function F(x). In practice, one often tabulates X for a sequence of values of R and then interpolates linearly between them.

Note: One can equally well solve the equation

$$F(X) = 1 - R.$$

● **PROBLEM 18-3**

Show how one can generate a series of exponentially distributed random numbers with the distribution function

$$F(x) = 1 - e^{-\lambda x}, \quad \left(\text{mean} = \frac{1}{\lambda}\right).$$

Solution: Take a rectangularly distributed random number R in [0, 1], and form

$$X = -\lambda^{-1} \ln R. \tag{1}$$

Then, it can be seen from the above problem that X can be obtained by solving the equation

$$1 - e^{-\lambda X} = 1 - R.$$

Hence

$$e^{-\lambda X} = R, \quad -\lambda X = \ln R,$$

and one gets the rule of Eq. (1).

Thus, it follows that a series of exponentially distributed random numbers antithetic to the above series can be obtained from

$$X' = -\lambda^{-1} \ln (1 - R).$$

SAMPLING SIMULATION

• **PROBLEM** 18-4

In a classroom of 100 men, assume that heights are distributed as in Table 1(a).

Draw a "Monte Carlo" random sample of n = 5 men, and calculate $\bar{X}$.

TABLE 1

Height X (Inches)	(a) Frequency	(b) Serial Number (Address)
60	1	0
63	6	1–6
66	24	7–30
69	38	31–68
72	24	69–92
75	6	93–98
78	1	99

Solution: The most practical method begins by assigning each man a serial number (address). This could be done alphabetically, or according to where each person sits in class, or in any other arbitrary way, since the numbers are to be drawn at random in any case. Therefore number the students from shortest to tallest, so that this enumerating can easily be shown in Table 1(b).

Now consult the random digits in Table 2. The first random pair is 39, which is the address of an individual 69 inches tall. Continuing to draw the remaining observations in the same way, the following sample is obtained:

Random Pair (Address)	Height
39	$X_1 = 69$
65	$X_2 = 69$
76	$X_3 = 72$
45	$X_4 = 69$
45	$X_5 = 69$

$$\bar{X} = 69.6$$

Note that the random address 45 is repeated, and so this student is included both times in the sample. This is because simple random sampling involves drawing with replacement. Thus, this student is sampled and then "replaced" in the population, and by the luck of the draw sampled again.

TABLE 2

Random Digits
(Blocked merely for convenience)

```
39 65 76 45 45   19 90 69 64 61   20 26 36 31 62   58 24 97 14 97   95 06 70 99 00
73 71 23 70 90   65 97 60 12 11   31 56 34 19 19   47 83 75 51 33   30 62 38 20 46
72 20 47 33 84   51 67 47 97 19   98 40 07 17 66   23 05 09 51 80   59 78 11 52 49
75 17 25 69 17   17 95 21 78 58   24 33 45 77 48   69 81 84 09 29   93 22 70 45 80
37 48 79 88 74   63 52 06 34 30   01 31 60 10 27   35 07 79 71 53   28 99 52 01 41

02 89 08 16 94   85 53 83 29 95   56 27 09 24 43   21 78 55 09 82   72 61 88 73 61
87 18 15 70 07   37 79 49 12 38   48 13 93 55 96   41 92 45 71 51   09 18 25 58 94
98 83 71 70 15   89 09 39 59 24   00 06 41 41 20   14 36 59 25 47   54 45 17 24 89
10 08 58 07 04   76 62 16 48 68   58 76 17 14 86   59 53 11 52 21   66 04 18 72 87
47 90 56 37 31   71 82 13 50 41   27 55 10 24 92   28 04 67 53 44   95 23 00 84 47

93 05 31 03 07   34 18 04 52 35   74 13 39 35 22   68 95 23 92 35   36 63 70 35 33
21 89 11 47 99   11 20 99 45 18   76 51 94 84 86   13 79 93 37 55   98 16 04 41 67
95 18 94 06 97   27 37 83 28 71   79 57 95 13 91   09 61 87 25 21   56 20 11 32 44
97 08 31 55 73   10 65 81 92 59   77 31 61 95 46   20 44 90 32 64   26 99 76 75 63
69 26 88 86 13   59 71 74 17 32   48 38 75 93 29   73 37 32 04 05   60 82 29 20 25

41 47 10 25 03   87 63 93 95 17   81 83 83 04 49   77 45 85 50 51   79 88 01 97 30
91 94 14 63 62   08 61 74 51 69   92 79 43 89 79   29 18 94 51 23   14 85 11 47 23
80 06 54 18 47   08 52 85 08 40   48 40 35 94 22   72 65 71 08 86   50 03 42 99 36
67 72 77 63 99   89 85 84 46 06   64 71 06 21 66   89 37 20 70 01   61 65 70 22 12
59 40 24 13 75   42 29 72 23 19   06 94 76 10 08   81 30 15 39 14   81 83 17 16 33

63 62 06 34 41   79 53 36 02 95   94 61 09 43 62   20 21 14 68 86   94 95 48 46 45
78 47 23 53 90   79 93 96 38 63   34 85 52 05 09   85 43 01 72 73   14 93 87 81 40
87 68 62 15 43   97 48 72 66 48   53 16 71 13 81   59 97 50 99 52   24 62 20 42 31
47 60 92 10 77   26 97 05 73 51   88 46 38 03 58   72 68 49 29 31   75 70 16 08 24
56 88 87 59 41   06 87 37 78 48   65 88 69 58 39   88 02 84 27 83   85 81 56 39 38

22 17 68 65 84   87 02 22 57 51   68 69 80 95 44   11 29 01 95 80   49 34 35 86 47
19 36 27 59 46   39 77 32 77 09   79 57 92 36 59   89 74 39 82 15   08 58 94 34 74
16 77 23 02 77   28 06 24 25 93   22 45 44 84 11   87 80 61 65 31   09 71 91 74 25
78 43 76 71 61   97 67 63 99 61   80 45 67 93 82   59 73 19 85 23   53 33 65 97 21
03 28 28 26 08   69 30 16 09 05   53 58 47 70 93   66 56 45 65 79   45 56 20 19 47

04 31 17 21 56   33 73 99 19 87   26 72 39 27 67   53 77 57 68 93   60 61 97 22 61
61 06 98 03 91   87 14 77 43 96   43 00 65 98 50   45 60 33 01 07   98 99 46 50 47
23 68 35 26 00   99 53 93 61 28   52 70 05 48 34   56 65 05 61 86   90 92 10 70 80
15 39 25 70 99   93 86 52 77 65   15 33 59 05 28   22 87 26 07 47   86 96 98 29 06
58 71 96 30 24   18 46 23 34 27   85 13 99 24 44   49 18 09 79 49   74 16 32 23 02

93 22 53 64 39   07 10 63 76 35   87 03 04 79 88   08 13 13 85 51   55 34 57 72 69
78 76 58 54 74   92 38 70 96 92   52 06 79 79 45   82 63 18 27 44   69 66 92 19 09
61 81 31 96 82   00 57 25 60 59   46 72 60 18 77   55 66 12 62 11   08 99 55 64 57
42 88 07 10 05   24 98 65 63 21   47 21 61 88 32   27 80 30 21 60   10 92 35 36 12
77 94 30 05 39   28 10 99 00 27   12 73 73 99 12   49 99 57 94 82   96 88 57 17 91
```

● **PROBLEM 18-5**

In a class of 80 students, each student was asked how many novels he had read in the past three months (X). The distribution represented graphically is shown in Figure 1.

(a) Draw a Monte Carlo random sample of 8 students,

(b) derive the mean and standard deviation of the sampling distribution of $\bar{X}$. Graph it.

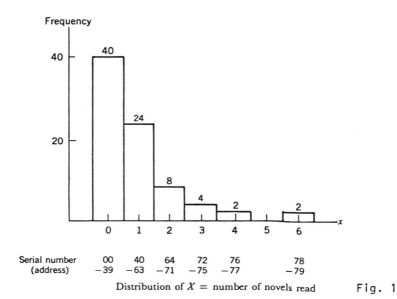

Distribution of X = number of novels read   Fig. 1

Solution: (a) It takes two digits to number everybody from 00 to 79, as shown below the graph. Draw the first observation by consulting the first pair of digits in Table I; they turn out to be 39, which is the address of a student who read 0 books. This is recorded, and continue to sample until the address 90 is drawn, which corresponds to no actual student. Then, simply ignore such an address, and go on to the next.

Thus obtain the following random sample:

Random Pair (Address)	X_i
39	0
65	2
76	4
45	1
45	1
19	0
ignore 90	
69	2
64	2
$\bar{X} = 12/8 = 1.50$	

(b) To theoretically obtain the sampling distribution of $\bar{X}$, begin by calculating the moments of the population, by first converting the Figure 1 to tabular form:

x	f	$p(x) = \dfrac{f}{N}$	$xp(x)$	$x^2 p(x)$
0	40	.50	0	0
1	24	.30	.30	.30
2	8	.10	.20	.40
3	4	.05	.15	.45
4	2	.025	.10	.40
5	0	0·	0	0
6	2	.025	.15	.90
	$N = 80$	$1.00 \checkmark$	$\mu = .90$	$E(X^2) = 2.45$ $-\mu^2 = -.81$ $\sigma^2 = 1.64$

Now one can find the sampling moments of $\bar{X}$:

$$E(\bar{X}) = \mu = .90$$

$$\sigma_{\bar{X}} = \frac{\sigma}{\sqrt{n}} = \frac{\sqrt{1.64}}{\sqrt{8}} = .45$$

Using these moments and the central limit theorem, one graphs the approximate normal distribution of $\bar{X}$ in Figure 2. The Monte Carlo distribution of $\bar{X}$ in part (a) will be approximately like this.

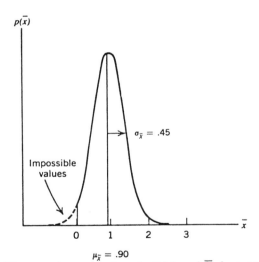

$\mu_{\bar{x}} = .90$
The approximate sampling distribution of $\bar{X}$, for a sample of $n = 8$ drawn from the population in Figure 1

Fig. 2

● **PROBLEM 18-6**

In a classroom of 100 men, assume that heights are distributed as in Table 1(a).

Given that n = 5, derive the mean and standard deviation of the sampling distribution of $\bar{X}$. Graph it.

TABLE 1

Height X (Inches)	(a) Frequency	(b) Serial Number (Address)
60	1	0
63	6	1-6
66	24	7-30
69	38	31-68
72	24	69-92
75	6	93-98
78	1	99

Solution: First calculate the moments of the population:

x	$p(x)$	$xp(x)$	$(x-\mu)^2 p(x)$
60	.01	.60	.81
63	.06	3.78	2.16
66	.24	15.84	2.16
69	.38	26.22	0
72	.24	17.28	2.16
75	.06	4.50	2.16
78	.01	.78	.81

$$\mu = 69.00 \qquad \sigma^2 = 10.26$$

Now one can find the sampling moments of $\bar{X}$:

$$E(\bar{X}) = \mu = 69$$

$$\sigma_{\bar{X}} = \frac{\sigma}{\sqrt{n}} = \frac{\sqrt{10.26}}{\sqrt{5}} = 1.43$$

Using these moments and the central limit theorem, the normal distribution of $\bar{X}$ in the accompanying figure is graphed. This graph demonstrates the Monte Carlo distribution of $\bar{X}$ for a sample of n = 5 observations drawn from the population in Table 1.

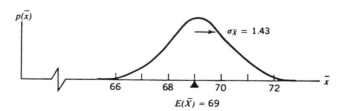

QUEUING MODELS AND GAME THEORY

● **PROBLEM** 18-7

In a single station service facility, the mean arrival rate (λ) is one customer every 4 minutes and the mean service time (μ) is 2½ minutes. The calculations for arrival and service times in minutes and, on the basis of an hour, are as follows:

$\lambda = \frac{1}{4} = 0.25$ arrivals per minute or 15 arrivals per hour

$\mu = \frac{1}{2.5} = 0.4$ service time per minute or 24 service times per hour.

What is the average number of cutomers in the system, the average queue length, the average time a customer spends in the system, and the average time a customer waits in line to be served?

Solution: The average number of customers in the system is:

$$E(n) = \frac{\lambda}{\mu - \lambda} = \frac{0.25}{0.4 - 0.25} = \frac{0.25}{0.15} = 1.66 \text{ customers}$$

$$\frac{15}{24 - 15} = \frac{15}{9} = 1.66 \text{ customers}$$

The average number of customers waiting to be served or average queue length is:

$$E(w) = \frac{\lambda^2}{\mu(\mu - \lambda)} = \frac{(0.25)^2}{0.4(0.4 - 0.25)} = \frac{0.0625}{0.06}$$

$$= 1.04 \text{ customers}$$

$$\frac{(15)^2}{24(24 - 15)} = \frac{225}{216} = 1.04 \text{ customers}$$

Calculation for the average number being served:

1.66 (average number in system) - 1.04 (average queue length)

= 0.62 (average number being served)

The average time a customer spends in the system is:

$$E(v) = \frac{1}{\mu - \lambda} = \frac{1}{0.4 - 0.25} = \frac{1}{0.15} = 6.66 \text{ minutes}$$

$$\frac{1}{24 - 15} = \frac{1}{9} = 0.111 \text{ hour}$$

The average time a customer waits before being served is:

$$E(y) = \frac{\lambda}{\mu(\mu - \lambda)} = \frac{0.25}{0.4(0.4 - 0.25)} = 4.16 \text{ minutes}$$

$$\frac{15}{24(24 - 15)} = \frac{15}{216} = 0.07 \text{ hour}$$

• **PROBLEM 18-8**

Arrivals at a telephone booth are considered to be Poisson, with an average time of 10 minutes between one arrival and the next. The length of a phone call is assumed to be distributed exponentially, with mean 3 minutes.

(a) What is the probability that a person arriving at the booth will have to wait?

(b) What is the average length of the queues that form from time to time?

(c) The telephone company will install a second booth when convinced that an arrival would expect to have to wait at least three minutes for the phone. By how much must the flow of arrivals be increased in order to justify a second booth?

Solution:

Here $\lambda = 0.1$ arrival per minute

$\mu = 0.33$ service per minute

(a) P {an arrival has to wait} $= 1 - P_0$

$$= \lambda/\mu, \quad \text{or}$$

by $P_n = \left(1 - \frac{\lambda}{\mu}\right)\left(\frac{\lambda}{\mu}\right)^n$

$$= 0.1/0.33 = 0.3$$

(b) $E(m|m > 0) = \dfrac{\mu}{\mu - \lambda}$

$$= \frac{0.33}{0.23} = 1.43 \text{ persons}$$

(c) $E(\text{waiting}) = \dfrac{\lambda}{\mu(\mu - \lambda)}$

$$= \frac{\lambda}{0.33(0.33 - \lambda)} \quad \text{if } \mu \text{ is fixed at } 0.33$$

Seek the new value λ' for which $E(w) = 3$ minutes. Solving the equation

$$3 = \frac{\lambda'}{0.33(0.33 - \lambda')}$$

one obtains an answer of $\lambda' = 0.16$ arrival per minute. So one must increase the flow of arrivals from 6 per hour, the present figure, to 10 per hour.

• **PROBLEM 18-9**

An insurance company has three claims adjusters in its branch office. People with claims against the company are found to arrive in a Poisson fashion, at an average rate of 20 per 8-hour day. The amount of time that an adjuster spends with a claimant is found to have an exponential distribution, with mean service time 40 minutes. Claimants are processed in the order of their appearance.

(a) How many hours a week can an adjuster expect to spend with claimants?

(b) How much time, on the average, does a claimant spend in the branch office?

Solution:

(a) Here $\lambda = \frac{5}{2}$ arrivals per hour

$\mu = \frac{3}{2}$ services per hour for each adjuster

$$P_0 = \frac{1}{1 + \frac{5}{3} + \frac{1}{2}\left(\frac{5}{3}\right)^2 + \frac{1}{6}\left(\frac{5}{3}\right)^3 \frac{\frac{9}{2}}{\frac{4}{2}}}$$

$$= \frac{24}{139}$$

The expected number of idle adjusters, at any specified instant, is

$$3P_0 + 2P_1 + 1P_2 = 3\left(\frac{24}{139}\right) + 2\left(\frac{40}{139}\right) + 1\left(\frac{100}{3 \times 139}\right) = \frac{4}{3} \text{ adjusters}$$

Then the probability that any one adjuster will be idle at any specified time is 4/9; and the expected weekly time an adjuster spends with claimants is $(5/9)40 = 22.2$ hours.

(b) The average time an arrival spends in the system is found from the equation

$$E(v) = \frac{\mu(\lambda/\mu)^k}{(k-1)!(k\mu-\lambda)^2} P_0 + \frac{1}{\mu}$$

to be 49.0 minutes.

● **PROBLEM** 18-10

Find the probability that there are no customers in the system, given that:

(i) number of channels in parallel = 3

(ii) mean arrival rate = 24 per hour

(iii) mean service rate of each channel = 10 per hour

Solution: The traffic intensity, $\rho = \frac{\lambda}{c\mu} = \frac{24}{3 \times 10} = 0.8$
The number of channels, $c = 3$

Hence, substituting these values into the equation for P_0,

$$P_0 = \frac{c!(1-\rho)}{(\rho c)^c + c!(1-\rho)\left\{\sum_{x=0}^{c-1} \frac{1}{n!}(\rho c)^n\right\}}$$

$$P = \frac{3 \times 2 \times 1 \times (1 - 0.8)}{(0.8 \times 3)^3 + 3 \times 2 \times 1 \times (1 - 0.8)\{1 + (0.8 \times 3) + (0.8 \times 3)^3/2\}}$$

$$= \frac{6(0.2)}{(2.4)^3 + 6(0.2)(1 + 2.4 + (2.4)^2/2)}$$

$$= \frac{1.2}{13.82 + 1.2(6.3)}$$

$$= 0.056$$

Thus, with a traffic intensity of 0.8, the system will be completely idle for about 6% of the time.

● **PROBLEM** 18-11

Assume that a telephone switchboard, with $\lambda = 10$ calls per minute, has a large number of channels. The calls are of random and independent length, but average 4 minutes each. Approximate the probability that, after the switchboard has been in service for a long time, there will be no busy channels at some specified time t_0.

Solution: Use the formula

$$\int_0^\infty [1 - F(s)]ds = E(Y).$$ Thus, for a large value t_0,

$$E[X(t_0)] = \lambda \int_0^{t_0} [1 - F(s)] \, ds$$

$$\doteq \lambda E(Y)$$

$$= 10(4) = 40.$$

Since $X(t)$ has a Poisson distribution,

$$P[X(t_0) = 0] = \exp\{-\lambda \int_0^{t_0} [1 - F(s)]ds\}$$

$$\doteq \exp(-40).$$

● **PROBLEM 18-12**

You are involved in a game of chess with Bobby Fischer. There are three possible outcomes: you win, event A, you draw, event B, or you lose, event C. Setting $u(A) = 1$ and $u(C) = -1$, what should $u(B)$ be? Because of the abilities of your opponent, much satisfaction would be gained from a draw, and so clearly $u(B)$ should be positive. More precisely, suppose you feel equally disposed to a draw and a lottery in which you have a probability of $\frac{19}{20}$ of being accorded a victory over Fischer and a probability of $\frac{1}{20}$ of being accorded a loss.

Solution: The desirability of B lies somewhere between the desirability of C and of A. It is reasonable to assume that there exists a particular r, $0 < r < 1$, such that you are indifferent to the events B and the lottery with outcomes A and C, A occurring with probability r and C with probability $1 - r$. Denote this particular lottery by the symbol $rA + (1 - r)C$.

Since you are indifferent to these two events, the utility of B, $u(B)$, should equal the utility of the lottery, denoted by $u(rA + (1 - r)C)$. Analogous to the definition of expected value in probability theory, the desirability of the lottery $rA + (1 - r)C$ can be given by $ru(A) + (1 - r)u(C)$, since this can be considered to be the "expected utility value" of the lottery. Thus, to determine $u(B)$, all you need determine is $u(A)$, $u(C)$, and the above, r, and then set

$$u(B) = ru(A) + (1 - r)u(C).$$

Since $r = \frac{19}{20}$ in this problem,

$$u(B) = u\left[\frac{19}{20}A + \frac{1}{20}C\right]$$

$$= \frac{19}{20}u(A) + \frac{1}{20}u(C)$$

$$= \frac{19}{20} - \frac{1}{20} = \frac{9}{10}$$

● **PROBLEM 18-13**

Show how a game with payoff matrix below can be converted to a linear programming problem.

	B's strategies	
A's strategies	2	4
	6	1

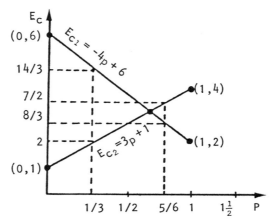

Solution: Since A's maximin strategy is to play row 1 while B's minimax strategy is to play column 2, we see that maximin ≠ minimax, i.e., no saddle point exists. However, if we let A choose row 1 with probability p and row 2 with probability $1-p$ then we can compute the expected value of the game to A. Now A has an expected value E_{C_1} against the column player playing column 1 of

$$E_{C_1} = 2p + 6(1-p) = -4p + 6$$

Similarly, $E_{C_2} = 4p + (1)(1-p) = 3p + 1$. Graph E_{C_i} versus p: (See fig.)

For any choice of p, $0 \leq p \leq 1$, $E_{C_1}, E_{C_2} > 0$. Let M denote the minimum expected value to the row player corresponding to his choosing row 1 with probability p and row 2 with probability $1-p$. For example, if $p = 1/3$, $M = E_{C_2}$ since $E_{C_2} = 2$ is smaller than $E_{C_1} = 14/3$. Conversely, if $p = 5/6$, $M = E_{C_1}$ since $E_{C_1} = 10/3$ is smaller than $E_{C_2} = 7/2$. Define s and t as follows:

$$s = p/M, \quad t = \frac{1-p}{M}. \tag{1}$$

Then, $s + t = \frac{p}{M} + \frac{1-p}{M} = 1/M$. Thus, the maximization of

M is equivalent to the minimization of s + t where s and t are both non-negative. Further restrictions on s and t are obtained by noting that for any p, E_{C_1} and $E_{C_2} \geq M$, i.e.,

$$2p + 6(1 - p) \geq M \tag{2}$$

$$4p + (1)(1 - p) \geq M.$$

But, from (1), p = sM and 1 - p = tM. Hence the inequalities (2) become

$$2sM + 6tM \geq M$$

$$4sM + tM \geq M.$$

Dividing through by M,

$$2s + 6t \geq 1$$

$$4s + t \geq 1.$$

Thus, A's problem is:

Minimize s + t

subject to

$$2s + 6t \geq 1$$
$$4s + t \geq 1 \tag{3}$$
$$s \geq 0, \ t \geq 0.$$

By analogous logic, B's problem is:

Maximize x + y

subject to

$$2x + 4y \leq 1$$
$$6x + y \leq 1 \tag{4}$$
$$x \geq 0, \ y \geq 0$$

Note that (4) is the dual of (3). From the theory of duality one knows that if they exist, the optimal values of (3) and (4) are the same. From game theory one knows that the value of a game computed from either player's point of view must coincide.

Finally, George Dantzig, (the inventor of the simplex method), has shown that any game be converted to a linear program and, conversely, any linear program can be converted to a game.

• **PROBLEM 18-14**

Consider the following game. Two players A and B must each select a number ouf of 1, 2, or 3. If both have chosen the same number, A will pay B the amount of the chosen number. Otherwise A receives the amount of his own number from B. The payoff table for this game is shown in Table 1. What are the best strategies for A and B? Use Linear Programming to solve.

TABLE 1

A	1	B 2	3
1	-1 B	1	1
2	2	-2 B	2 A
3	3 A	3 A	-3 B

Solution: The linear program for A's best random strategy, x_1, x_2, x_3, (fractions of time during which row 1, 2 and 3 will be chosen so as to maximize A's gains) is

$$\max v$$

subject to

$$-x_1 + 2x_2 + 3x_3 \geq v$$
$$x_1 - 2x_2 + 3x_3 \geq v$$
$$x_1 + 2x_2 - 3x_3 \geq v$$
$$x_1 + x_2 + x_3 = 1$$

and

$$x_1, x_2, x_3 \geq 0$$

The optimal v could be negative because the payoff table does contain negative payoffs. However, if the number 3 is added to all α_{ij}'s, then all will be nonnegative. Then define $v' = v + 3$ or $v = v' - 3$. The problem can be rewritten with all non-negative variables as

$$\max v' - 3$$

subject to

$$2x_1 + 5x_2 + 6x_3 \geq v'$$
$$4x_1 + x_2 + 6x_3 \geq v'$$
$$4x_1 + 5x_2 \geq v'$$
$$x_1 + x_2 + x_3 = 1$$

780

$$x_1, x_2, x_3, v' \geq 0$$

It is best with hand calculation to elminate the equality constraint by substituting

$$x_3 = 1 - x_1 - x_2$$

and replacing the nonengativity condition $x_3 \geq 0$ by

$$x_1 + x_2 \leq 1$$

The resulting problem is

$$\max v' - 3$$

subject to

$$4x_1 + x_2 + v' \leq 6$$
$$2x_1 + 5x_2 + v' \leq 6$$
$$4x_1 + 5x_2 - v' \geq 0$$
$$x_1 + x_2 \leq 1$$

and

$$x_1, x_2, v' \geq 0$$

Slack variables are then used to convert the four inequality constraints to equalities. The problem becomes

$$\max v' - 3$$

subject to

$$4x_1 + x_2 + v' + s_1 = 6$$
$$2x_1 + 5x_2 + v' + s_2 = 6$$
$$4x_1 + 5x_2 - v' - s_3 = 0$$
$$x_1 + x_2 + s_4 = 1$$

The initial BFS will have basic variables s_1, s_2, s_3, s_4 and zero variables x_1, x_2, v'. This corner point is degenerate because the third constraint passes through the origin. However, the degeneracy gives no trouble. The first tableau of the simplex method is

Tableau A   Pivot

	Const.	x_1	x_2	v'	Ratio	
E	−3	0	0	1		
s_1	6	−4	−1	−1	−6	
s_2	6	−2	−5	−1	−6	
s_3	0	4	5	−1	−0	Pivot
s_4	1	−1	−1	0	∞	

The next tableau shows no gain in objective because of the degneracy:

Tableau B   Pivot

	Const.	x_1	x_2	s_3	Ratio	
E	-3	$+4$	5	-1		
s_1	6	-8	-6	$+1$	-1	
s_2	6	-6	-10	$+1$	$-\dfrac{6}{10}$	Pivot
v'	0	$+4$	$+5$	-1	0	
s_4	1	-1	-1	0	-1	

The next tableau is

Tableau C   Pivot

	Const.	x_1	s_2	s_3	Ratio	
E	0	1	$-\dfrac{1}{2}$	$-\dfrac{1}{2}$		
s_1	$\dfrac{24}{10}$	$-\dfrac{44}{10}$	$+\dfrac{6}{10}$	$+\dfrac{4}{10}$	$-\dfrac{24}{44}$	Pivot
x_2	$\dfrac{6}{10}$	$-\dfrac{6}{10}$	$-\dfrac{1}{10}$	$+\dfrac{1}{10}$	-1	
v'	3	1	$-\dfrac{1}{2}$	$-\dfrac{1}{2}$	$+3$	
s_4	$+\dfrac{4}{10}$	$-\dfrac{4}{10}$	$+\dfrac{1}{10}$	$-\dfrac{1}{10}$	-1	

The next tableau gives the optimal solution:

Tableau D

	Const.	s_1	s_2	s_3	Ratio
E	$\dfrac{6}{11}$	$-\dfrac{10}{44}$	$-\dfrac{4}{11}$	$-\dfrac{9}{22}$	
x_1	$\dfrac{6}{11}$	$-\dfrac{10}{44}$	$\dfrac{6}{44}$	$\dfrac{1}{11}$	
x_2	$\dfrac{3}{11}$	$+\dfrac{6}{44}$	$-\dfrac{2}{11}$	$\dfrac{1}{22}$	
v'	$\dfrac{39}{11}$	$-\dfrac{10}{44}$	$-\dfrac{4}{11}$	$-\dfrac{9}{22}$	
s_4	$\dfrac{2}{11}$	$+\dfrac{4}{44}$	$+\dfrac{1}{22}$	$-\dfrac{7}{110}$	

The result is

$$x_1 = \frac{6}{11}$$

$$x_2 = \frac{3}{11}$$

$$x_3 = \frac{2}{11}$$

and

$$v = \frac{6}{11}$$

The first three constraints hold as equalities in the optimal solution. Therefore it is known that B uses all three of his courses of action. The best random strategy for B can be found from these results by solving three simultaneous equations in three unknowns. The ith equation represents the expected loss to B if A uses A_i. All these must equal the expected gain to A that is known. The equations are

$$-y_1 + y_2 + y_3 = \frac{6}{11}$$

$$2y_1 - 2y_2 + 2y_3 = \frac{6}{11}$$

$$3y_1 + 3y_2 - 3y_3 = \frac{6}{11}$$

The solution is

$$y_1 = \frac{5}{22}$$

$$y_2 = \frac{4}{11}$$

$$y_3 = \frac{9}{22}$$

This completes the solution to the game problem.

MONTE CARLO APPLICATIONS TO INTEGRAL PROBLEMS

• **PROBLEM** 18-15

Compute the integral by employing group sampling,

$$J = \int_0^1 e^x dx.$$

Solution: Divide the interval (0, 1) into two equal parts and take 4 points in (0, ½) and 6 points in (½, 1). One gets the value:

$$J_5 = \frac{1}{8} \sum_{i=1}^{4} e^{\xi_i^{(1)}} + \frac{1}{12} \sum_{i=1}^{6} e^{\xi_i^{(2)}}.$$

The variance of J_5 is:

$$DJ_5 = \frac{1}{16} D_1 + \frac{1}{24} D_2,$$

where

$$D_1 = 2\int_0^{\frac{1}{2}} e^{2x}dx - \left[2\int_0^{\frac{1}{2}} e^x dx\right]^2$$

$$= e - 1 - 4(\sqrt{e} - 1)^2 = 0.03492;$$

$$D_2 = 2\int_{\frac{1}{2}}^{1} e^{2x}dx - \left[2\int_{\frac{1}{2}}^{1} e^x dx\right]^2$$

$$= e^2 - e - 4(e - \sqrt{e})^2 = 0.09493.$$

Accordingly, $DJ_5 = 0.006138$, and

$$\delta_p = 0.675\sqrt{0.00614} = 0.053.$$

The values of $\xi^{(1)}$ and $\xi^{(2)}$ may be computed from the values of γ:

$$\xi^{(1)} = 0.5\gamma, \quad \xi^{(2)} = 0.5(1 + \gamma).$$

• **PROBLEM** 18-16

Compute the integral by extraction of the regular part method

$$J = \int_0^1 e^x dx.$$

Solution: Since $e^x = 1 + x + \ldots$, then one takes $g(x) = 1 + x$ for extraction of the regular part. According to

$$\theta_3 = \frac{b-a}{N}\sum_{i=1}^{N}[f(\xi_i) - g(\xi_i)] + I,$$

where $\xi_1, \xi_2, \ldots$ are values of a random variable ξ, uniformly distributed over the range (a, b), one gets the value:

$$J_3 = \frac{1}{N}\sum_{i=1}^{N}(e^{\gamma_i} - \gamma_i) + \frac{1}{2},$$

where $\gamma_1, \gamma_2, \ldots$ are the values of a random variable, uniformly distributed over the range $(0, 1)$. The variance of the averaged quantity is:

$$D\zeta^{(3)} = \frac{1}{2}(e - 1)(5 - e) - \frac{23}{12} = 0.0437.$$

• **PROBLEM** 18-17

Consider: $J = \int_0^1 \sin \pi x \, dx = \frac{2}{\pi}$.

Solve by using (1) and (2) as an approximate estimate of the integral J:

$$\theta_1 = \frac{b-a}{N} \sum_{i=1}^{N} f(\xi_i) \tag{1}$$

$$\theta_7 = \frac{1}{4N} \sum_{i=1}^{N} \left[f\left(\frac{Y_i}{2}\right) + f\left(1 - \frac{Y_i}{2}\right) \right.$$

$$\left. + f\left(\frac{1}{2} + \frac{Y_i}{2}\right) + f\left(\frac{1}{2} - \frac{Y_i}{2}\right) \right]. \tag{2}$$

Which one of the approximate estimates is more effective?

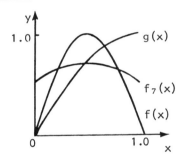

Fig. 1

Solution: Since in this problem, (Fig. 1):

$$f_7 = \frac{\sqrt{2}}{2} \cos \pi \left(\frac{1}{4} - \frac{x}{2}\right),$$

then the variance is:

$$D\zeta^{(7)} = \frac{1}{2} \int_0^1 \cos^2 \pi \left(\frac{1}{4} - \frac{x}{2}\right) dx - \left(\frac{2}{\pi}\right)^2$$

$$= \frac{1}{4} + \frac{1}{2\pi} - \frac{4}{\pi^2} = 0.003871.$$

At the same time,

$$D\zeta^{(1)} = \int_0^1 \sin^2 \pi x \, dx - \left(\frac{2}{\pi}\right)^2 = \frac{1}{2} - \frac{4}{\pi^2} = 0.09472,$$

so that $D\zeta^{(1)} : D\zeta^{(7)} = 24.5$. It follows from this that, in the problem which is being considered, the esimate (2) is 6.1 times as effective as the estimate (1).

CHAPTER 19

APPLICATIONS OF FORTRAN LANGUAGE IN NUMERICAL ANALYSIS

COMPUTER MANIPULATIONS WITH MATRICES

● **PROBLEM** 19-1

Write a FORTRAN program to load and add two given m×n matrices.

Solution: An m×n matrix is an ordered array of real numbers containing m rows and n columns. Let $A = [a_{i,j}]$, $B = [b_{i,j}]$ $(i = 1,\ldots,m; j = 1,\ldots,n)$ be the two matrices to be added. Their sum $C = [c_{i,j}] = A + B = [a_{i,j} + b_{i,j}]$.
To load the matrices, a dimension statement reserving two sets of mn places each, is required. Also, one more set of mn places must be dimensioned for the finally obtained array. Then, two nested DO-loops must be constructed to read in the values of $a_{i,j}$ and $b_{i,j}$.

After that, the two arrays are added using two nested DO-loops again. Finally, the resulting array is printed out by the similar method.

Note, that the format statements 20 and 30 are left out because they depend on the actual data used.

```
          DIMENSION A(10,15), B(10,15), C(10,15)
   C      THE ARRAY C(10,15) WILL CONTAIN THE SUM
   C      OF ARRAYS A(10,15) AND B(10,15).
   C      LOAD THE ARRAYS
          DO 50 I = 1,10
          DO 50 J = 1,15
          READ (5,20) A(I,J), B(I,J)
   50     CONTINUE
   C      ADD THE ARRAYS
          DO 60 I = 1,10
          DO 60 J = 1,15
          C(I,J) = A(I,J) + B(I,J)
   60     CONTINUE
```

```
C         OUTPUT THE RESULT
          DO 70 I = 1,10
          DO 70 J = 1,15
          WRITE (6,30) C(I,J)
70        CONTINUE
          STOP
          END
```

• **PROBLEM** 19-2

Given two Matrices A and B, each with N columns and M rows. Write FORTRAN statements which would form the sum, C = A+B.

Solution: The matrix A can be represented by a single subscripted variable where I runs from 1 to M and J runs from 1 to N. The same is true for C. Then the required FORTRAN statements are

```
          DO 20 I = 1,M
          DO 20 J = 1,N
20        C(I,J)=A(I,J)+B(I,J)
```

A total of N x M additions are required to obtain C.

• **PROBLEM** 19-3

For the following two matrices A and B,

$$A = \begin{bmatrix} 1 & 1 & 1 \\ 2 & 2 & 1 \\ 3 & 2 & 1 \end{bmatrix} \quad B = \begin{bmatrix} 0 & -1 & 1 \\ -1 & 2 & -1 \\ 2 & -1 & 0 \end{bmatrix}.$$

Write a FORTRAN program to perform the following operations:

(a) Find the sum $[A] + [B]$
(b) Find the difference $[A] - [B]$
(c) Find the product $[A][B]$

Solution:
```
C         MATRIX ADDITION, SUBTRACTION, AND MULTIPLI-
C         CATION
          DIMENSION A(3,3), B(3,3), C(3,3)
C         DEFINE MATRIX ELEMENTS
          A(1, 1) = 1
          A(1, 2) = 1
          A(1, 3) = 1
          A(2, 1) = 2
          A(2, 2) = 2
          A(2, 3) = 1
          A(3, 1) = 3
          A(3, 2) = 2
          A(3, 3) = 1
          B(1, 1) = 0
          B(1, 2) = -1
```

```
              B(1, 3) = 1
              B(2, 1) = -1
              B(2, 2) = 2
              B(2, 3) = -1
              B(3, 1) = 2
              B(3, 2) = -1
              B(3, 3) = 0
       C      CALULATE AND PRINT SUM
              DO 10 IROW = 1, 3
              DO 10 JCOL = 1, 3
           10 C(IROW, JCOL) = A(IROW, JCOL) + B(IROW,JCOL)
              WRITE (3, 14)
           14 FORMAT (' SUM')
              WRITE (3, 15) ((C(I,J), J = 1, 3), I = 1, 3)
           15 FORMAT (3(3F6. 1, /))
       C      CALCULATE AND PRINT OUT DIFFERENCE
              DO  20 IROW = 1, 3
              DO  20 JCOL = 1, 3
           20 C(IROW, JCOL) = A(IROW, JCOL) - B(IROW,JCOL)
              WRITE (3, 25)
           25 FORMAT (' DIFFERENCE')
              WRITE (3, 15) ((C(I, J), J = 1, 3) I = 1, 3)
       C      CALCULATE AND PRINT OUT PRODUCT A * B
              DO 50 IROW = 1, 3
              DO 50 JCOL = 1, 3
              SUM = 0
              DO 40 K = 1, 3
           40 SUM = SUM + A(IROW, K) * B(K, JCOL)
           50 C( ROW, JCOL) = SUM
              WRITE (3, 55)
           55 FORMAT (' PRODUCT A * B')
              WRITE (3, 15) ((C(I, J), J = 1, 3), I = 1, 3)
              CALL EXIT
              END
```

The sum, difference, and product matrices provided by the computer are as follows:

```
       SUM
         1.0    0.0    2.0
         1.0    4.0    0.0
         5.0    1.0    1.0

       DIFFERENCE
         1.0    2.0    0.0
         3.0    0.0    2.0
         1.0    3.0    1.0

       PRODUCT A * B
         1.0    0.0    0.0
         0.0    1.0    0.0
         0.0    0.0    1.0
```

An interesting point is that the product of [A] and [B] gives the unit matrix [I] . If this happened with two real numbers, one would say that they are reciprocals of each other; in matrix terminology, [A] and [B] are inverses of each other if their product is the identity matrix [I] .

● **PROBLEM** 19-4

Show that the inverse of

$$\begin{pmatrix} 2 & -1 \\ -1 & 1 \end{pmatrix}$$

is

$$\begin{pmatrix} 1 & 1 \\ 1 & 2 \end{pmatrix}$$

and write a FORTRAN subroutine for calculating the inverse of the matrix.

Solution: Form the product

$$\begin{pmatrix} 2 & -1 \\ -1 & 1 \end{pmatrix} \begin{pmatrix} 1 & 1 \\ 1 & 2 \end{pmatrix} = \begin{pmatrix} 1 & 0 \\ 0 & 1 \end{pmatrix}$$

Since the product is the identity matrix, the second matrix is indeed the inverse of the first.

It can be shown that any square matrix A has a unique inverse if and only if its determinant is different from zero. It can also be shown that A commutes with its inverse; that is

$$AA^{-1} = A^{-1}A = I$$

The inverse is not defined for nonsquare matrices.
A formula for the inverse of a matrix A can be found as follows. Consider the set of linear equations

$$y_1 = a_{11} x_1 + a_{12} x_2 + \ldots + a_{1n} x_n$$
$$y_2 = a_{21} x_1 + a_{22} x_2 + \ldots + a_{2n} x_n$$
$$\vdots \qquad\qquad\qquad\qquad\qquad\qquad (1)$$
$$y_n = a_{n1} x_1 + a_{n2} x_2 + \ldots + a_{nn} x_n$$

connecting one set of variables $x_1, x_2, \ldots, x_n$ with another set $y_1, y_2, \ldots, y_n$. In matrix form write this set of equations as

$$y = Ax$$

where

$$y = \begin{pmatrix} y_1 \\ y_2 \\ \vdots \\ y_n \end{pmatrix} \qquad x = \begin{pmatrix} x_1 \\ x_2 \\ \vdots \\ x_n \end{pmatrix}$$

If one multiplies this set of equations by A^{-1}, one obtains

$$A^{-1}y = A^{-1}Ax$$

or

$$A^{-1}y = Ix = x$$

Hence the elements of A^{-1} are just the coefficients of the y's if one solves the set of equations (1) for the x's in terms of the y's. Start with the n by 2n array of the form

$$\begin{array}{cccccccccc}
a_{11} & a_{12} & a_{13} & \cdots & a_{1n} & 1 & 0 & 0 & \cdots & 0 \\
a_{21} & a_{22} & a_{23} & \cdots & a_{2n} & 0 & 1 & 0 & \cdots & 0 \\
a_{31} & a_{32} & a_{33} & \cdots & a_{3n} & 0 & 0 & 1 & \cdots & 0 \\
\vdots & & & & & & & & & \\
a_{n1} & a_{n2} & a_{n3} & \cdots & a_{nn} & 0 & 0 & 0 & \cdots & 1
\end{array}$$

Then proceeding gives an array of the form

$$\begin{array}{cccccccc}
1 & 0 & 0 & \cdots & 0 & a_{1n+1} & a_{1n+2} & a_{1n+3} & \cdots & a_{12n} \\
0 & 1 & 0 & \cdots & 0 & a_{2n+1} & a_{2n+2} & a_{2n+3} & \cdots & a_{22n} \\
0 & 0 & 1 & \cdots & 0 & a_{3n+1} & a_{3n+2} & a_{3n+3} & \cdots & a_{32n} \\
\vdots & & & & & & & & & \\
0 & 0 & 0 & \cdots & 1 & a_{nn+1} & a_{nn+2} & a_{nn+3} & \cdots & a_{n2n}
\end{array}$$

The solution, instead of being the single column a_{1n+1} to a_{nn+1}, is the entire right-hand side of the above array. The subroutine can be written as

```
      SUBROUTINE MATINV(AA,N,AINV)
      DIMENSION AA(20,20), AINV(20,20),A(20,40),ID(20)
      NN=N+1
      N2=2*N
      DO 100 I=1,N
      ID(I)=I
      DO 100 J=1,N
  100 A(I,J)= AA(I,J)
      DO 200 I=1,N
      DO 200 J=NN,N2
  200 A(I,J)=0.
      DO 300 I=1,N
  300 A(I,N+I)=1.
      K=1
    1 CALL EXCH3(A,N,N2,K,ID)
    2 IF(A(K,K))3,999,3
    3 KK=K+1
```

```
      DO 4 J=KK,N2
      A(K,J)=A(K,J)/A(K,K)
      DO 4 I=1,N
      IF(K-I)41,4,41
   41 W=A(I,K)*A(K,J)
      A(I,J)=A(I,J)-W
      IF(ABS(A(I,J)-.0001*ABS(W))42,4,4
   42 A(I,J)=0.
    4 CONTINUE
      K=KK
      IF(K-N)1,2,5
    5 DO 10 J=1,N
      DO 10 J=1,N
      IF(ID(J)-I)10,8,10
    8 DO 10 K=1,N
      AINV(I,K)=A(J,N+K)
   10 CONTINUE
      RETURN
  999 PRINT 1000
      RETURN
 1000 FORMAT(19H MATRIX IS SINGULAR)
      END
```

• **PROBLEM** 19-5

Consider the Matrix

$$B = \begin{bmatrix} 5 & 2.6 & 3.1538 & 2.9512 & 3.0165 \\ 2 & 3.5 & 2.8571 & 3.5 & 2.9836 \\ 1 & 1 & 1 & 1 & 1 \end{bmatrix}$$

Let $\lambda_1 = 3$ and $u = \{i_1, i_2, i_3\}$.
Write a FORTRAN program to carry out inverse iteration by choosing $u = B^m z$ at the mth step.

Solution: Here is the FORTRAN subroutine for carrying out the inverse iteration. At the mth step, assume $u = B^m z$, that is, calculate the Rayleigh quotient at each step.

```
      SUBROUTINE INVITR ( B,N, EGUESS, VGUESS, W,D, IPIVOT,
     *                         EVALUE, VECTOR, IFLAG)
   C     CALLS FACTOR , SUBST.
      INTEGER IFLAG, IPIVOT(N),   I,ITER,ITERMX,J
      REAL B(N,N), D(N), EGUESS,EVALUE,VECTOR(N),VGUESS(N),
     *         W(N,N),EPSLON, EVNEW,EVOLD,SQNORM
C******* INPUT *******
   C  B  THE MATRIX OF ORDER N WHOSE EIGENVALUE/VECTOR IS
         SOUGHT.
   C  N  ORDER OF THE MATRIX B.
   C  EGUESS A FIRST GUESS FOR THE EIGENVALUE.
   C  VGUESS N-VECTOR CONTAINING A FIRST GUESS FOR THE
   C  EIGENVECTOR.
```

```
C******* W O R K   A R E A *******
C   W    MATRIX OF ORDER N
C   D    VECTOR OF LENGHT N
C   IPIVOT INTEGER VECTOR OF LENGHT N
C******* O U T P U T*******
C   EVALUE COMPUTED APPROXIMATION TO EIGENVALUE
C   VECTOR COMPUTED APPROXIMATION TO EIGENVECTOR
C   IFLAG AN INTEGER,
C         = 1 OR -1 (AS SET IN FACTOR), INDICATES THAT ALL
C                                       IS WELL,
C         = 0, INDICATES THAT SOMETHING WENT WRONG.  SEE
C             PRINTED ERROR MESSAGE.
C******* M E T H O D *******
C     INVERSE ITERATION, AS DESCRIBED IN THE TEXT, IS USED.
C******
C   THE FOLLOWING TERMINATION PARAMETERS ARE SET
C   HERE, A TOLERANCE E P S L O N  ON THE DIFFERENCE BETWEEN
C
C   SUCCESSIVE EIGENVALUE ITERATES, AND AN UPPER BOUND I T E R M X
C   ON THE NUMBER OF ITERTION STEPS.
      DATA EPSLON, ITERMX/.000001,20/
C                         PUT B - (EGUESS)*IDENTITY INTO W
      DO 10 J=1,N
         DO 9 I=1,N
    9       W(I,J) = B(I,J)
   10    W(J,J) = W(J,J) - EGUESS
      CALL FACTOR ( W, N, D, IPIVOT, IFLAG)
      IF (IFLAG .EQ. 0) THEN
          PRINT 610
  610     FORMAT(' EIGENVALUE GUESS TOO CLOSE.'
     *           'NO EIGENVECTOR CALCULATED.')
                                               RETURN
      END IF
C                         ITERATION STARTS HERE
      PRINT 619
  619 FORMAT(' ITER EIGENVALUE    EIGENVECTOR COMPONENTS'/)
      EVOLD = 0.
      DO 50 ITER=1,ITERMX
C                         NORMALIZE CURRENT VECTOR GUESS
         SQNORM = 0.
         DO 20 I=1,N
   20       SQNORM = VGUESS(I)**2 + SQNORM
         SQNORM = SQRT(SQNORM)
         DO 21 I=1,N
   21       VGUESS(I) = VGUESS(I)/SQNORM
C                         GET NEXT VECTOR GUESS
         CALL SUBST ( W, IPIVOT, VGUESS, N, VECTOR)
C                         CALCULATE RAYLEIGH QUOTIENT
         EVNEW = 0.
         DO 30 I=1,N
   30       EVNEW = VGUESS(I)*VECTOR(I) + EVNEW
         EVALUE = EGUESS + 1./EVNEW
         PRINT 630, ITER,EVALUE,VECTOR
  630    FORMAT(I3,E15.7,2x,3E14.7/(20X,3E14.7))
C           STOP ITERATION IF CURRENT GUESS IS CLOSE TO
C           PREVIOUS GUESS FOR EIGENVALUE
         IF ( ABS(EVNEW-EVOLD) .LE. EPSLON*ABS(EVNEW) )
     *                                         RETURN
         EVOLD = EVNEW
```

```
              DO 50 I=1,N
    50           VGUESS(I) = VECTOR(I)
C
              IFLAG = 0
              PRINT 660,EPSLON,ITERMX
    660       FORMAT(' NO CONVERGENCE TO WITHIN',E10.4,' AFTER
              I3,' STEPS.')
                                              RETURN

              END
```

• **PROBLEM 19-6**

Write a FORTRAN program for the calculation of the inverse of an NxN Matrix A. Display the input and output for the data N=3

$$A = \begin{bmatrix} 2 & 3 & -1 \\ 4 & 4 & -3 \\ -2 & 3 & -1 \end{bmatrix}$$

Solution:
```
C     PROGRAM FOR CALCULATING THE INVERSE OF A GIVEN
C     MATRIX
C     CALLS F A C T O R, S U B S T.
          PARAMETER NMAX=30, NMAXSQ=NMAX*NMAX
          INTEGER I,IBEG, IFLAG,IPIVOT(NMAX),J,N,NSQ
          REAL A(NMAXSQ),AINV(NMAXSQ),B(NMAX)
        1 READ 501, N
      501 FORMAT(I2)
          IF (N .LT. 1 .OR. N .GT. NMAX)    STOP
C                  READ IN MATRIX ROW BY ROW
          NSQ =N*N
          DO 10 I=1,N
       10    READ 510, (A(J),J=I,NSQ,N)
      510 FORMAT(5E15.7)
C
          CALL FACTOR (A, N, B, IPIVOT, IFLAG )
          If (IFLAG .EQ. 0) THEN
              PRINT 611
      611     FORMAT('1MATRIX IS SINGULAR')
                                              GO TO 1
          END IF
          DO 21 I=1,N
       21   B(I) = 0.
          IBEG =1
          DO 30 J=1,N
            B(J) = 1.
            CALL SUBST ( A, IPIVOT, B,N, AINV(IBEG) )
            B(J) = 0.
       30   IBEG =IBEG + N
          PRINT 630
      630 FORMAT('1THE COMPUTED INVERSE IS '//)
          DO 31 I=1,N
       31    PRINT 631, I, (AINV(J),J=I,NSQ,N)
      631 FORMAT('ØROW',12,8E15.7/(7X,8E15.7))
                                              GO TO 1
          END
```

SAMPLE INPUT
   3
   2.   3.   -1.
   4.   4.   -3.
  -2.   3.   -1.
RESULTING OUTPUT
   THE COMPUTED INVERSE IS

   ROW 1   0.2500000E 00   0.0   -0.2499999E 00
   ROW 2   0.5000000E 00   -0.1999998E 00   0.9999996E -01
   ROW 3   0.1000000E 01   -0.6000000E 00 -0.2000000E 00

• **PROBLEM 19-7**

Evaluate $D_4 = \begin{bmatrix} 2 & 4 & .6 & 8 \\ 3 & 1 & 2 & 1 \\ 1 & 2 & -2 & 2 \\ 2 & 3 & 4 & 1 \end{bmatrix}$ by pivotal condensation

(see flow chart). Display the input and output of a program that could be written according to the flow chart.

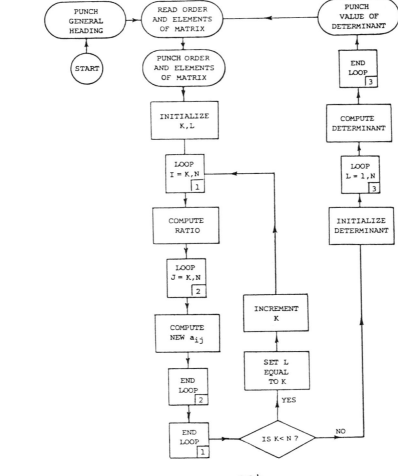

Solution:

$$D_4 = \begin{bmatrix} 2 & 4 & 6 & 8 \\ 3 & 1 & 2 & 1 \\ 1 & 2 & -2 & 2 \\ 2 & 3 & 4 & 1 \end{bmatrix} = 2 \begin{bmatrix} 1 & 2 & 3 & 4 \\ 3 & 1 & 2 & 1 \\ 1 & 2 & -2 & 2 \\ 2 & 3 & 4 & 1 \end{bmatrix}$$

$$= 2 \begin{bmatrix} 1 - 3(\tfrac{2}{1}) & 2 - 3(\tfrac{3}{1}) & 1 - 3(\tfrac{4}{1}) \\ 2 - 1(\tfrac{2}{1}) & -2 - 1(\tfrac{3}{1}) & 2 - 1(\tfrac{4}{1}) \\ 3 - 2(\tfrac{2}{1}) & 4 - 2(\tfrac{3}{1}) & 1 - 2(\tfrac{4}{1}) \end{bmatrix} = 2 \begin{bmatrix} -5 & -7 & -11 \\ 0 & -5 & -2 \\ -1 & -2 & -7 \end{bmatrix}$$

$$= 2(-5) \begin{bmatrix} 1 & \tfrac{7}{5} & \tfrac{11}{5} \\ 0 & -5 & -2 \\ -1 & -2 & -7 \end{bmatrix} = 2(-5) \begin{bmatrix} -5 - 0(\tfrac{7}{5}) & -2 - 0(\tfrac{11}{5}) \\ -2 + 1(\tfrac{7}{5}) & -7 + 1(\tfrac{11}{5}) \end{bmatrix}$$

$$= -10 \begin{bmatrix} -5 & -2 \\ -\tfrac{3}{5} & -\tfrac{24}{5} \end{bmatrix} = \frac{-10}{-5} \begin{bmatrix} -5 & -2 \\ 3 & 24 \end{bmatrix} = 2(-120 + 6) = -228$$

Program input and output:

MATRIX OF ORDER 4

```
2.0000000E+00    4.0000000E+00    6.0000000E+00    8.0000000E+00
3.0000000E+00    1.0000000E+00    2.0000000E+00    1.0000000E+00
1.0000000E+00    2.0000000E+00   -2.0000000E+00    2.0000000E+00
2.0000000E+00    3.0000000E+00    4.0000000E+00    1.0000000E+00
```

THE DETERMINANT OF THIS MATRIX IS -2.2800000E+02

● **PROBLEM 19-8**

Find the rank of

$$\begin{bmatrix} 1 & -1 & -1 & -2 \\ 2 & 1 & -2 & 2 \\ 4 & 3 & -4 & 6 \end{bmatrix}.$$

Write the FORTRAN subroutine to find the rank, K, of a matrix having N rows and M columns, where neither N nor M exceed 20.

Solution: Use the following theorem: The rank of a matrix is unchanged if any multiple of the elements of one row (or column) is added to the corresponding elements of another row (or column). One may proceed as follows:

$$\text{rank} \begin{bmatrix} 1 & -1 & -1 & -2 \\ 2 & 1 & -2 & 2 \\ 4 & 3 & -4 & 6 \end{bmatrix} = \text{rank} \begin{bmatrix} 1 & -1 & -1 & -2 \\ 0 & 3 & 0 & 6 \\ 4 & 3 & -4 & 6 \end{bmatrix} \text{(Twice first row subtracted from second)}$$

$$= \text{rank} \begin{bmatrix} 1 & -1 & -1 & -2 \\ 0 & 3 & 0 & 6 \\ 0 & 7 & 0 & 14 \end{bmatrix} \text{(four times first row subtracted from third)}$$

$$= \text{rank} \begin{bmatrix} 1 & -1 & -1 & -2 \\ 0 & 3 & 0 & 6 \\ 0 & 0 & 0 & 0 \end{bmatrix} \text{(7/3 times second row subtracted from third)}$$

It is obvious in this last matrix all third-order determinants are zero, but at least one second-order determinant,

$$\begin{bmatrix} 1 & -1 \\ 0 & 3 \end{bmatrix}$$

is not zero. Hence the rank of the original matrix is 2.

The FORTRAN subroutine below finds the rank, K, of a matrix having N rows and M columns, where neither N nor M exceed 20.

```
      SUBROUTINE MARANK(AA,N,M,K)
      DIMENSION AA(20,20),A(20,20)
      DO 100 I=1,N
      DO 100 J=1,M
  100 A(I,J)=AA(I,J)
      K=1
    1 CALL EXCH2(A,N,M,K)
      IF(A(K,K))2,10,2
    2 IF(K-N)3,11,11
    3 IF(K-M)40,11,11
   40 KK =K +1
      DO 4 J=KK,M
      A(K,J)=A(K,J)/A(K,K)
      DO 4 I=KK,N
      W=A(I,K)*A(K,J)
      A(I,J)=A(I,J)-W
      IF(ABS(A(I,J))-.0001*ABS(W))42,4,4
   42 A(I,J)=0
    4 CONTINUE
      K=KK
      GO TO 1
   10 K=K-1
   11 RETURN
      END
```

• **PROBLEM 19-9**

Write a FORTRAN program to implement Gauss's elimination method for solving the system of equations.

Solution: Consider the system of n linear equations with n unknowns:

$$a_{11}x_1 + a_{12}x_2 + \ldots + a_{1n}x_n = b_1$$
$$a_{21}x_1 + a_{22}x_2 + \ldots + a_{2n}x_n = b_2 \quad (1)$$
$$\vdots$$
$$a_{n1}x_1 + a_{n2}x_2 + \ldots + a_{nn}x_n = b_n$$

Gauss's elimination method is used to find a solution of (1) i.e., a set $\{x_1,\ldots,x_n\}$ such that when it is substituted into (1), all the equations are satisfied. The method is as follows:
1) Divide the first equation by a_{11} to obtain

$$x_1 + \frac{a_{12}}{a_{11}} x_2 + \ldots + \frac{a_{1n}}{a_{11}} x_n = b_1/a_{11}. \quad (2)$$

2) Now subtract a_{21} times the first equations from row 2, a_{31} times the first equation from row 3,..., a_{n1} times the first equation from row n to obtain

$$x_1 + \frac{a_{12}}{a_{11}} x_2 + \ldots + \frac{a_{1n}}{a_{11}} x_n = b_1/a_{11}$$
$$w_{22}x + \ldots + w_{2n}x_n = c_2 \quad (3)$$
$$\vdots$$
$$w_{n2}x_2 + \ldots + w_{nn}x_n = c_n .$$

where the result of adding $-(a_{j1}(a_{1j}/a_{11})x_j)$ from $a_{ij}x_i$ has been written as $w_{ij}x_j$ and the result of adding $-(a_{i1}b_1/a_{11})$ to b_i has been written as c_i (i = 2,...,n). The method is applied again to eliminate x_2 from the third, forth,...,n-th equations. Repeated application yields the tridiagonal system:

$$x_1 + v_{12}x_2 + \ldots + v_{1n}x_n = d_1$$
$$x_2 + v_{23}x_3 + \ldots + v_{2n}x_n = d_2$$
$$\vdots \quad (4)$$
$$x_{n-1} + v_{n-1,n}x_n = d_{n-1}$$
$$x_n = d_n$$

Now the system can be solved by back-substituting. The X,Y and Z's need not be presented to solve such systems on a computer. The coefficients can be loaded into an array A(N,N+1) and solved in a similar manner.

```
            DIMENSION A(N,N+1)
C           AN EXTRA COLUMN IS NEEDED FIRST TO STORE
C           THE CONTINUOUSLY CHANGING VALUES (b_1...b_n) OF THE
C           EQUATIONS, AND THEN THE ROOTS OF THE SYSTEM.
            READ (5,10)N
            DO 40 I = 1,N
            DO 40 J = 1,N+1
            READ (5,20) A(I,J)
    40      CONTINUE

            M = N - 1
            L = N + 1
            DO 60 K = 1,M
            K1 = K + 1
            DO 60 J = K1,L
            DO 60 I = K1,N
            A(I,J) = A(I,J) - A(I,K)/A(K,K)*A(K,J)
    60      CONTINUE
            I = N
            J = I + 1
            A(I,N + 1) = A(I,N + 1)/A(I,I)
   100      I = I - 1
   110      J = J - 1
            B(I,N + 1) = B(I,N + 1) + A(I,J)*A(J,N + 1)
            IF (J.EQ.1) GO TO 120
            GO TO 110
   120      A(I,N + 1) = 1/A(I,I)*(A(I,N + 1) - B(I,N + 1))
   130      IF (I.EQ.0) GO TO 140
            GO TO 100
   140      WRITE (3,30) (A(I,N + 1), I = 1,N)
            STOP
            END
```

Note: The formats 10,20,30 are not included because they depend on actual data.

• **PROBLEM** 19-10

Develop a FORTRAN program which solves a system of linear simultaneous equations according to the Gauss-Seidel method.

Solution: To apply the Gauss-Seidel method, the system of linear equations AX =C is rewritten in the form:

$$x_1 = \frac{1}{a_{11}} (c_1 - a_{12}x_2 - a_{13}x_3 - \ldots - a_{1n}x_n) \quad (1)$$

$$x_2 = \frac{1}{a_{22}} (c_2 - a_{21}x_1 - a_{23}x_3 - \ldots - a_{2n}x_n) \quad (2)$$

$$\vdots$$

$$x_n = \frac{1}{a_{nn}} (c_n - a_{n1}x_1 - a_{n2}x_2 - \ldots - a_{n,n-1}x_{n-1}) \quad (n)$$

Next, a set of starting values $x_1^0, x_2^0, \ldots, x_n^0$ is chosen. While the program developed will take 0 as starting values, $x_i^0 = c_i/a_{ii}$ are also frequently used as starting values.

Substituting $x_2 = x_3 = \ldots = x_n = 0$ into (1) yields an approximation for x_1. This approximation is used in (2) where $x_3 = x_4 = \ldots = x_n = 0$. Now one gets approximation for x_1 and x_2. Continuing, observe that x_n is found by substituting the approximation for $x_1, x_2, \ldots, x_{n-1}$ into (n). This completes the first iteration. For the k-th iteration:

$$x_1^{k+1} = \frac{1}{a_{11}} (c_1 - a_{12}x_2^k - a_{13}x_3^k - \ldots - a_{1n}x_n^k)$$

$$x_2^{k+1} = \frac{1}{a_{22}} (c_2 - a_{21}x_1^{k+1} - a_{13}x_3^k - \ldots - a_{2n}x_n^k)$$

$$\vdots$$

$$x_n^{k+1} = \frac{1}{a_{nn}} (c_n - a_{n1}x_1^{k+1} - a_{n2}x_2^{k+1} - \ldots - a_{n,n-1}x_{n-1}^{k+1}).$$

A Sufficient condition for convergence (and the one which will be assumed to hold) is that

$$\sum_{j \neq i} |a_{ij}| < |a_{ii}| \quad \text{for } i = 1, 2, \ldots, n$$

as an illustrative example, let

$$x_1 = \frac{1}{10} (9 - 2x_2 - x_3)$$
$$x_2 = \frac{1}{20} (-44 - 2x_1 + 2x_3)$$
$$x_3 = \frac{1}{10} (22 + 2x_1 - 3x_2).$$

These equations correspond to the determinant system

$$\begin{bmatrix} 10 & 2 & 1 \\ 2 & 20 & -2 \\ -2 & 3 & 10 \end{bmatrix} \begin{bmatrix} x_1 \\ x_2 \\ x_3 \end{bmatrix} = \begin{bmatrix} 9 \\ -44 \\ 22 \end{bmatrix}$$

For $x_i^{(0)} = 0$ one gets

$$x_1^{(1)} = 9/10$$

$$x_2^{(1)} = \frac{1}{20}(-44 - 2x_1^{(1)} + 3x_3^{(0)}) = + \frac{1}{20}(-44 - \frac{18}{10}) = -2.29$$

$$x_3^{(1)} = \frac{1}{10}(22 + 2x_1^{(1)} - 3x_2^{(1)}) = \frac{1}{10}(22 + \frac{18}{10} - 3(-2.29))$$

= 3.067

etc.
The program itself is given below.

```
C         GAUSS-SEIDEL ITERATION OF SIMULTANEOUS EQUATIONS
          DIMENSION A(30,30), X(30), Y(30)
   1      READ 999, N, ITLAST, ((A(I,J),J=1,N), Y(I),I =1,N)
          PUNCH 996, N, ((A(I,J),J =1,N), I = 1,N)
          PUNCH 995 (Y(I),I =1,N)
          DO 10 I = 1,N
  10      X(I) = 0.
          IT = 1.
  20      PUNCH 994, IT
          DO 60 I = 1,N
          P = Y(I)
          DO 50 J = 1,N
          IF (I - J) 40,50,40
  40      P = P - A(I,J)*X(J)
  50      CONTINUE
          X(I) = P/A(I,I)
  60      PUNCH 998, I, X(I)
          IT = IT + 1
          IF (IT - ITLAST) 20,20,1
 994      FORMAT(/24X, 9HITERATION I2)
 995      FORMAT(/22X,15HCONSTANT VECTOR/(3E18.7)//22X,
   1      15HSOLUTION VECTOR)
 996      FORMAT(//21X, 15HMATRIX OF ORDER 12 //(3E18.7))
 998      FORMAT(20X, I2, E16.7)
 999      FORMAT(2I5/(8F10.0))
          END
```

● **PROBLEM** 19-11

Write a suitable FORTRAN program for the following:
Assume that a matrix A has one eigenvalue λ, which is larger than the others in absolute value, and y is any nonzero column vector conformable with A. Let $y_1, y_2,$ etc. be defined by

$$y_1 = Ay$$
$$y_2 = Ay_1$$
$$\vdots$$
$$y_n = Ay_{n-1}$$

The vectors y_i defined in this manner can lead to the value of λ_1 and to eigen value corresponding to λ_1.

Solution: A suitable program is

```
   1      DIMENSION A(10,10),Y(10),YN(10)
   2      PRINT, "INPUT N, TEN OR LESS"
```

```
3           INPUT ,N
4           PRINT,     "INPUT A(1,1)A(1,2),,,A(N,N)"
5           INPUT,  ((A(I,J),J =1,N),I=1,N)
6           PRINT, "INPUT Y(1),Y(2),,,Y(N)"
7           INPUT,  (Y(I),I=1,N)
8        1  DO 2 I=1,N
9           YN(I)=0.
10          DO 2 J=1,N
11       2  YN(I)=YN(I)+A(I,J)*Y(J)
12          PRINT,(YN(I),I=1,N)
13          INPUT, Q
14          DO 3 I=1,N
15       3  Y(I)=YN(I)
16          GO TO 1
17          END
```

In this program, the statements at lines 2 through 7 allow the user to input an initial matrix A and vector y of order up to 10. The statements at lines 8 through 12 compute and print the vector $y_1 = Ay$. At line 13, the user is allowed to specify whether another step of the process is required. If the typed entry is the letter S, the program will terminate. If the entry is any number whatsoever, the program will cause y_1 to replace y, and will repeat lines 8 through 12, thereby computing and printing $y_2 = Ay_1$, and so on.

• **PROBLEM** 19-12

Consider the matrix $B = \begin{bmatrix} 1 & 2 & 0 \\ 2 & 1 & 0 \\ 0 & 0 & -1 \end{bmatrix}$. Use the FORTRAN routine to find an eigenvalue and eigenvector of B.

HINT: INVITR with $z = [1, 1, 1]^T$ and $p = 3.0165$, which is the best guess for $\lambda_1 = 3$ from the sequence of ratios 5, 2.6, 3.1538 ....., 2.9512 ....., 3.0165.

Solution:

```
ITER  EIGENVALUE              EIGENVECTOR COMPONENTS

 1    0.2991801 + 01   -0.3499093+ 02   -0.3499093 + 02   -0.1437446 + 00
 2    0.3000000 + 01    0.4285478+ 02    0.4285478 + 02    0.7232219 - 03
 3    0.3000000 + 01   -0.4285496+ 02   -0.4285496 + 02   -0.2971047 - 05
 4    0.3000000 + 01    0.4285496+ 02    0.4285496 + 02    0.1220522 - 07
              EIGENVALUE = 0.3000000 + 01
              EIGENVECTOR =
                 0.4285496 + 02   0.4285496 + 02   0.1220522 - 07
```

The output shows very rapid convergence of the eigenvector (a gain of about two decimal places per iteration step), and an even more rapid convergence of the eigenvalue, because B is symmetric and a Rayleigh quotient was computed.

As an illustration of the fact that, in contrast to the power method itself, inverse iteration may be used for any eigenvalue, also start with z = $1,1,1^T$ AND P = 0, hoping to catch thereby an absolute smallest eigenvalue of B.

ITER	EIGENVALUE	EIGENVECTOR COMPONENTS		
1	−0.9000000 + 01	0.1924501 + 00	0.1924501 + 00	−0.5773503 + 00
2	−0.1320000 + 01	0.1005038 + 00	0.1005038 + 00	0.9045340 + 00
3	−0.1033195 + 01	0.3658808 − 01	0.3658809 − 01	−0.9878783 + 00
4	−0.1003661 + 01	0.1232878 − 01	0.1232877 − 01	0.9986311 + 00
5	−0.1000406 + 01	0.4114594 − 02	0.4114604 − 02	−0.9998476 + 00
6	−0.1000045 + 01	0.1371724 − 02	0.1371714 − 02	0.9999831 + 00
7	−0.1000005 + 01	0.4572417 − 03	0.4572513 − 03	−0.9999981 + 00
8	−0.1000001 + 01	0.1524206 − 03	0.1524109 − 03	0.9999998 + 00
9	−0.1000000 + 01	0.5080043 − 04	0.5081010 − 04	−0.1000000 + 01

EIGENVALUE = −0.1000000 + 01
EIGENVECTOR =
0.5080043 − 04   0.5081010 − 04   −0.1000000 + 01

The convergence is much slower since 0 is not particularly close to the eigenvalue −1, but we have convergence after nine iterations, with the computed eigenvector of the form $[0,0,1]^T$ (rather than of the more general form $[a,-a,b]^T$ possible for the eigenvalue −1 of B).

• **PROBLEM 19-13**

Write a program that will input sets of values x_1, x_2, x_3 and print y_1, y_2, y_3 according to the equations

$$y_1 = a_{11}x_1 + a_{12}x_2 + a_{13}x_3$$

$$y_2 = a_{21}x_1 + a_{22}x_2 + a_{23}x_3$$

$$y_3 = a_{31}x_1 + a_{32}x_2 + a_{33}x_3$$

for given values of the a_{ij}.

Solution: A suitable program is

```
      DIMENSION X(3),Y(3),A(3,3)
      DO 1 I=1,3
    1 READ 101,A(I,1),A(I,2),A(I,3)
    2 READ 101,X(1),X(2),X(3)
      IF(X(1)-1.E30)3,4,4
    3 DO 5 I=1,3
      Y(I)=0.
      DO 5 J=1,3
    5 Y(I)=Y(I)+A(I,J)*X(J)
      PRINT 101,Y(1),Y(2),Y(3)
      GO TO 2
    4 STOP
  101 FORMAT(3E12.4)
      END
```

● **PROBLEM 19-14**

(a) Determine the lowest eigenvalue k of the following problem:

$$y'' + ky = 0, \quad y(0) = 0, \quad y(1) = 0.$$

(b) Construct a flow chart for the problem in (a) and write a corresponding FORTRAN program. Display the output.

Flowchart

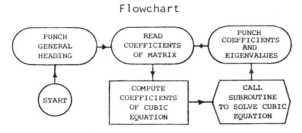

Solution: By equation

$$y_{i+1} - 2y_i + y_{i-1} + \tfrac{h}{2} f_i (y_{i+1} - y_{i-1}) + h^2 g_i y_i = h^2 F_i,$$

or

$$\left(1 - \tfrac{h}{2} f_i\right) y_{i-1} - (2 - h^2 g_i) y_i + \left(1 + \tfrac{h}{2} f_i\right) y_{i+1} = h^2 F_i$$

with $f(x) = 0, g(x) = k$, and $F(x) = 0$,

$$y_{i-1} - (2 - kh^2) y_i + y_{i+1} = 0, \quad y_0 = y_n = 0.$$

With $h = \tfrac{1}{2}$:

$$0 - (2 - k/4) y_1 + 0 = 0,$$

$$y_1 \neq 0 \quad \therefore \quad 2 - \tfrac{k}{4} = 0, \quad k^{(1)} = 8.$$

With $h = \tfrac{1}{3}$ and assuming y to be symmetrical about $x = \tfrac{1}{2}$ so that $y_2 = y_1$:

$$0 - \left(2 - \tfrac{k}{9}\right) y_1 + y_1 = -\left(1 - \tfrac{k}{9}\right) y_1 = 0,$$

$$y_1 \neq 0 \quad \therefore \quad 1 - \tfrac{k}{9} = 0, \quad k^{(2)} = 9.$$

With $h = \tfrac{1}{4}$ and $y_1 = y_3$:

$$0 - \left(2 - \tfrac{k}{16}\right) y_1 + y_2 = 0,$$

$$y_1 - \left(2 - \tfrac{k}{16}\right) y_2 + y_1 = 0.$$

For y_1 and y_2 to be different from zero, the determinant of this set of equations must be equal to zero:

$$\begin{vmatrix} (2 - k/16) & -1 \\ -2 & (2 - k/16) \end{vmatrix} = \left(2 - \frac{k}{16}\right)^2 - 2 = 0,$$

$$2 - \frac{k}{16} = \pm \sqrt{2}, \quad k^{(3)} = 16(2 \pm \sqrt{2}) = \begin{cases} 9.38, \\ 54.62. \end{cases}$$

[The value k = 54.62 is the first approximation to the third eigenvalue of problem (a), and corresponds to deflections $y_1 = y_3$ of sign opposite to the sign of y_2]. The h^2-extrapolations of

$$Q_{12} = \frac{(h_1/h_2)^n Q_2 - Q_1}{(h_1/h_2)^n - 1}$$

may be used to estimate the true value of k (= π^2 = 9.87):

$$h_1 = \frac{1}{2}, \quad h_2 = \frac{1}{3}, \quad \frac{h_1}{h_2} = \frac{3}{2}, \quad k_{1,2} = \frac{\left(\frac{3}{2}\right)^2 9 - 8}{\left(\frac{3}{2}\right)^2 - 1} = 9.8;$$

$$h_1 = \frac{1}{3}, \quad h_2 = \frac{1}{4}, \quad \frac{h_1}{h_2} = \frac{4}{3}, \quad k_{2,3} = \frac{\left(\frac{4}{3}\right)^2 9.38 - 9}{\left(\frac{4}{3}\right)^2 - 1} = 9.87.$$

```
C      PROGRAM
C      EIGENVALUES OF A SYMMETRIC MATRIX OF THIRD ORDER
C
       DIMENSION C(4), EIGEN(3)
       PUNCH 997
     1 READ 999, A11, A12, A13, A22, A23, A33
       C(1) = 1.
       C(2) = -(A11 + A22 + A33)
       C(3) = A11*A22 + A11*A33 + A22*A33 - A12*A12 - A13*A13 - A23*A23
       C(4) = -A11*A22*A33 - 2.*A12*A13*A23 + A11*A23*A23 + A22*A13*A13
     1       + A33*A12*A12
       CALL CUBIC (C, EIGEN, XI)
       PUNCH 998, A11, A12, A13, A22, A23, A33, EIGEN
       GO TO 1
C
   997 FORMAT ( // 24X, 24HRESULTS FROM PROGRAM)
   998 FORMAT ( // 33X, 6HMATRIX // 9X, 3E17.7 / 26X, 2E17.7 / 43X,
     1         E17.7 // 31X, 11 HEIGENVALUES / (26X, E17.7))
   999 FORMAT (8E10.0)
       END
```

RESULTS FROM PROGRAM

MATRIX

```
          1.0000000E+00    0.0000000E-99    0.0000000E-99
                           1.0000000E+00    0.0000000E-99
                                            1.0000000E+00
```

```
                EIGENVALUES
              1.0000000E+00
              1.0000000E+00
              1.0000000E+00


                  MATRIX

  1.0000000E+01  2.0000000E+00  1.0000000E+00
                 1.0000000E+01  2.0000000E+00
                                1.0000000E+01

                EIGENVALUES
              1.3372281E+01
              9.0000001E+00
              7.6277189E+00
```

EVALUATING FUNCTIONS • PROBLEM 19-15

> Write two FORTRAN programs which will compute n! in two different ways:
> (a) First according to the standard formula for a positive interger $n \geq 1$,
>
> $$n! = n(n-1) \cdots (3)(2)(1)$$
>
> (b) then use Stirling's approximation formula
>
> $$n! \cong (2\pi)^{1/2} \, n^{n+1/2} \, e^{-n}$$

<u>Solution</u>: The answers will be formulated as two function routines. Denote these two functions by FACT(N) and FACTST(N). Then by applying the above formulas directly one gets the programs below:

```
(a)            INTERGER FUNCTION FACT(N)
               IPROD = 1
               IF (N.LE.1) GO TO 200
               DO 100 I = 2,N
               IPROD = IPROD*I
       100     CONTINUE
               FACT = IPROD
               GO TO 300
       200     FACT = 1
       300     RETURN
               END

(b)            FUNCTION FACTST(N)
               Z1 = EXP(-FLOAT(N))
               Z2 = (FLOAT(N))**(FLOAT(N) + 0.5)
       C       FLOAT CONVERTS INTEGERS TO REALS
               Z3 = SQRT(2*3.14159)
               FACTST =Z1*Z2*Z3
               RETURN
               END
```

• **PROBLEM** 19-16

Write a FORTRAN program to evaluate the polynominal expression

$$a_1 x^{n-1} + a_2 x^{n-2} + \ldots + a_n$$

for given values of n, $a_1, a_2, \ldots, a_n$ using various values of x which are read in. Let the program terminate when a zero value of x is read in.

Solution: The most efficient polynomial evaluation procedure is based on the nesting

$$(\ldots((a_1 x + a_2) x + a_3) x + \ldots + a_{n-1}) x + a_n.$$

Assume $n \leq 25$. The data deck consist of a value for n followed by the n coefficients a_i on one or more data cards. These are followed by successive data cards, each with a value for x; the final card contains the zero value.

This nesting procedure is known as Horner's method for evaluating polynomials. The program segment is presented below.

```
C          POLYNOMIAL EVALUATION
           DIMENSION A(25)
           READ, N,(A(J), J =1,N)
16         READ, X
           IF (X.EQ.0.) STOP
           POLY = A(1)
           DO 12 I = 2,N
12         POLY = POLY*X + A(1)
           GO TO 16
           STOP
           END
```

• **PROBLEM** 19-17

Develop a FORTRAN program to use Lagrangian interpolation to evaluate $f(x) = x^3$ at x=3, given the table below:

i	1	2	3	4
x_i	1	2	4	7
f_i	1	8	64	343

$f_i \equiv f(x_i)$

Solution: The Lagrange polynomials of degree m are defined as

$$P_j(x) = A_j \prod_{\substack{k=1 \\ k \neq j}}^{m+1} (x - x_k) \quad \text{where} \quad A_j = \prod_{\substack{k=1 \\ k \neq j}}^{m+1} \frac{1}{(x_j - x_k)}$$

4

806

Here $m = 3$ and one approximates $f(x)$ by $\sum_{i=1}^{ } f(x_i) P_i(x)$. The evaluations of $P_j(x)$ are shown below, for $x = 3$.

$$P_1(x) \equiv P(x,1) = \frac{(x-2)(x-4)(x-7)}{(x_1-x_2)(x_1-x_3)(x_1-x_4)} = \frac{(x-2)(x-4)(x-7)}{-18}$$

$$P_2(x) \equiv P(x,2) = \frac{(x-1)(x-4)(x-7)}{(x_2-x_1)(x_2-x_3)(x_2-x_4)} = \frac{(x-1)(x-4)(x-7)}{10}$$

$$P_3(x) \equiv P(x,3) = \frac{(x-1)(x-2)(x-7)}{(x_3-x_1)(x_3-x_2)(x_3-x_4)} = \frac{(x-1)(x-2)(x-7)}{18}$$

$$P_4(x) \equiv P(x,4) = \frac{(x-1)(x-2)(x-4)}{(x_4-x_1)(x_4-x_2)(x_4-x_3)} = \frac{(x-1)(x-2)(x-4)}{90}$$

Thus $f(3) \cong \sum_{i=1}^{4} f(x_i) P(3,I)$
$= 1(-2/9) + 8(4/5) + 64(4/9) + 343(-2/90)$
$= 26.9999980$ which is very close to $3^3 = 27$.

The program and sample output are given below:

```
C         PROGRAM FOR LAGRANGIAN INTERPOLATION
C         UNEVENLY SPACED PIVOTAL POINTS
          DIMENSION X(50), P(50), F(50)
          PUNCH 994
          READ 999, N, (X(I), F(I), I = 1,N)
     1    READ 998, X0
          DO 10 J = 1,N
          P(J) = 1
          DO 10 I = 1,N
          IF (I - J) 9, 10 9
     9    P(J) = P(J) * (X0 -X(I))/(X(J) -X(I))
     10   CONTINUE
          FO = 0
          DO 20 I = 1,N
     20   FO = FO + P(I) * F(I)
          PUNCH 997
          PUNCH 996, (I,X(I),F(I), P(I), I =1,N)
          PUNCH 995, X0, FO
          GO TO 1
     994  FORMAT (//, 19X, 24H RESULTS FROM PROGRAM)
     995  FORMAT (19X, 6H AT X = F5.2, 21H THE VALUE OF F(X)
         1 IS F12.7)
     996  FORMAT (I10, 3F14.7)
     997  FORMAT (//, 9X 1H1, 8X, 4HX(I), 10X, 4HF(I), 9X
     998  FORMAT (8F10.0)
     999  FORMAT (15/8F10.0)
          END
```

[SAMPLE] RESULTS FROM PROGRAM

	X(I)	F(I)	P(X,I)
1	1.0000000	1.0000000	-0.2222222
2	2.0000000	8.0000000	0.8000000
3	4.0000000	64.0000000	0.4444444
4	7.0000000	343.0000000	-0.0222222

AT X = 3.00 THE VALUE OF F(X) IS 26.9999980

• **PROBLEM** 19-18

Write a FORTRAN program that solves the following problem: Let $f(x) = (1 + x^2)^{-1}$. Use piecewise-cubic Hermite interpolation to approximate $f(x)$ in $[-5,5]$. That is, interpolate $f(x)$ at the points
$$x_i = \frac{(i-1)10}{N} - 5, \quad i=1,2,\ldots,N+1$$
for $N=2,4,\ldots,16$. Estimate the maximum interpolation error in $[-5,5]$.

Solution: The following FORTRAN program solves this problem:

```
C         MAX INTERPOLATION ERROR PROGRAM
          INTERGER I,J,K,N
          REAL C(4,17), ERRMAX,H,X(17),Y
C         PIECEWISE CUBIC HERMITE INTERPOLATION AT EQUALLY
C         SPACED POINTS
C         TO THE FUNCTION
              F(Y) = 1./(1. + Y*Y)
C
          PRINT 600
    600   FORMAT('1   N',5X,'MAXIMUM ERROR')
          DO 40 N=2,16,2
              H = 10./FLOAT (N)
              DO 10 I=1,N+1
                  X(I) = FLOAT (I - 1)*H - 5.
                  C(1,I) = F(X(I))
C                 C(2,I) = F'(X(I))
     10           C(2,I) = -2.*X(I)*C(1,I)**2
              CALL CALCCF ( X, C, N )
C                     ESTIMATE MAXIMUM INTERPOLATION ERROR
C                        ON (-5,5).
              ERRMAX = 0.
              DO 30 I=1,101
                  Y = .1*I - 5.
                  ERRMAX = MAX(ERRMAX, ABS(F(Y)-PCUBIC(Y,X,C,N)))
     30       CONTINUE
     40       PRINT 640, N, ERRMAX
    640   FORMAT(I5,E18.7)
          STOP
          END
COMPUTER OUTPUT
```

N	MAXIMUM ERROR
2	4.9188219E - 01
4	2.1947326E - 01
6	9.1281965E - 02
8	3.5128250E - 02
10	1.2705882E - 02
12	4.0849234E - 03
14	1.6011164E - 03
16	1.6953134E - 03

● **PROBLEM** 19-19

Use a Taylor series expansion to compute e^x to 8 significant digits. Also write a FORTRAN program to approximate the exponential function.

Solution: In order to determine the error involved in approximating a function by its Taylor expansion, it is necessary to establish the maximum value of the remainder term. For the exponential function

$$e^x = 1 + x + \frac{x^2}{2!} + \frac{x^3}{3!} + \ldots + \frac{x^{n-1}}{(n-1)!} + \frac{x^n}{n!} e^{\theta x}$$

where $0 < \theta < 1$, this remainder is $x^n e^{\theta x}/n!$. If one is to establish an upper limit on the value of this quanity without further knowledge of the value of θ, one must choose that value of θ which will maximize the expression. For this expression the value of $\theta = 1$ is the proper choice. If one uses n terms of the Taylor expansion to approximate e^x, the error will certainly not be greater than $x^n e^x/n!$. If one is interested in writing a computer routine which will compute e^x to eight significant figures, then, one should insist on a relative error of less than $1/(2 \times 10^8)$, where $\frac{\Delta Q}{|Q|}$ = relative error.

In the case of e^x, one requires

$$\frac{\frac{x^n}{n!} e^x}{e^x} < \frac{1}{2 \times 10^8}$$

or

$$\frac{x^n}{n!} < \frac{1}{2 \times 10^8}$$

It is clear that the value of n required depends on how big a value of x one desires to be able to handle. For example, if $x = 1$ is the largest value required, one must have

$$\frac{1}{n!} < \frac{1}{2 \times 10^8}$$

or $n! > 2 \times 10^8$. This is satisfied by $n = 12$. If one desired to have an expansion suitable for values of x up to $x = 10$, one would need

$$\frac{10^n}{n!} < \frac{1}{2 \times 10^8}$$

which will require $n = 41$, or 41 terms. This would be a much longer execution. To avoid having to compute 41 terms, simply apply the laws of exponents. For example,

$$e^{5.632} = e^5 \cdot e^{.632}$$

The factor e^5 can be computed by multiplying e by itself five times and the factor $e^{.632}$ can be computed to sufficient accuracy with 12 terms of Maclaurin expansion. To the accuracy required,

$$e^x = 1 + x + \frac{x^2}{2!} + \frac{x^3}{3!} + \frac{x^4}{4!} + \frac{x^5}{5!} + \frac{x^6}{6!} + \frac{x^7}{7!} + \frac{x^8}{8!} + \frac{x^9}{9!} + \frac{x^{10}}{10!} + \frac{x^{11}}{11!}$$

or, in a form requiring a smaller number of multiplications to compute,

$$e^x = 1 + x\left(1 + x\left(\frac{1}{2!} + x\left(\frac{1}{3!} + x\left(\frac{1}{4!} + x\left(\frac{1}{5!} + x\left(\frac{1}{6!} + x\left(\frac{1}{7!} + x\left(\frac{1}{8!} + x\left(\frac{1}{9!} + x\left(\frac{1}{10!} + \frac{x}{11!}\right)\right)\right)\right)\right)\right)\right)\right)\right)\right)$$

In FORTRAN, the coefficients for this expression can be determined by the statements:

```
      DIMENSION A(11)
      A(1) = 1
      DO 2 J =2,11
      FJ = J
    2 A(J)=A(J-1)/FJ
```

With these coefficients, e^x could be calculated by the loop

```
      Y=X*A(11)+A(10)
      DO 4 I=1,9
      J = 10-I
    4 Y=Y*X+A(J)
      Y=Y*X+1
```

or by the single lengthy statement

```
      Y =(((((((((X*A(11)+A(10))*X+A(9))* X+A(8))*X+A(7))*X
     +A(6))*X+A(5))*X+A(4))*X+A(3))*X+A(2))*X
     +A(1))*X+1
```

● **PROBLEM** 19-20

Write a program which uses fixed-point iteration to find the smallest positive zero of the function

$$f(x) = e^{-x} - \sin x.$$

OUTPUT

XN	F(XN)	ERROR.
6.00000000E − 01	−1.58308373E − 02	
5.84169163E − 01	6.06240576E − 03	2.70997483E − 02
5.90231568E − 01	−2.35449276E − 03	1.02712326E − 02
5.87877076E − 01	9.09583240E − 04	4.00507667E − 03
5.88786659E − 01	−3.52118178E − 04	1.54484349E − 03
5.88434541E − 01	1.36203144E − 04	5.98398213E − 04
5.88570744E − 01	−5.27011849E − 05	2.31413378E − 04
5.88518043E − 01	2.03892661E − 05	8.95489706E − 05
5.88538432E − 01	−7.88865463E − 06	3.46438992E − 05
5.88530543E − 01	3.05208415E − 06	1.34039851E − 05
5.88533595E − 01	−1.18084550E − 06	5.18591321E − 06
5.88532415E − 01	4.56865632E − 07	2.00642389E − 06
5.88532871E − 01	−1.76760146E − 07	7.76278869E − 07
5.88532695E − 01	6.83880224E − 08	3.00340402E − 07
5.88532763E − 01	−2.64591478E − 08	1.16200876E − 07
5.88532737E − 01	1.02369739E − 08	4.49578182E − 08
5.88532747E − 01	−3.96065403E − 09	1.73940600E − 08
5.88532743E − 01	1.53236357E − 09	6.72970888E − 09

Solution: The first step is to select an iteration function and an initial value which will lead to a convergent iteration. Rewrite $f(x) = 0$ in the form

$$x = x + e^{-x} - \sin x = :g(x)$$

Now since $f(0.5) = 0.127 \ldots$ and $f(0.7) = -0.147 \ldots$ the smallest positive zero lies in the interval $I = [0.5, 0.7]$. To verify that $g(x)$ is a convergent iteration function it should be noted that with

$$g'(x) = 1 - e^{-x} - \cos x$$

$g'(0.5) = -0.48 \ldots$, $g'(0.7) = -0.26 \ldots$ and since $g'(x)$ is a monotonic function on I, then $|g'(x)| < 1$ for $x \in I$. It can similarly be verified that $0.5 < g(x) < 0.7$ for all $x \in I$. Hence fixed-point iteration will converge if x_0 is chosen in I.

The program below was run on a CDC 6500. Note that successful termination of this program requires that both of the following error tests be satisfied

$$|x_n - x_{n-1}| < \text{XTOL } |x_n|$$
$$|f(x_n)| < \text{FTOL}$$

Where XTOL is the error tolerance for x and FTOL is the variation of $f(x_n)$ from the ideal $f(x) = 0$, one is willing to to tolerate. The program also terminates if the convergence tests are not satisfied within 20 iterations.

811

```fortran
C     PROGRAM FOR FINDING SMALLEST POSITIVE ZERO OF
C     FUNCTION
      INTERGER J
      REAL ERROR,FTOL,XNEW,XOLD,XTOL,Y
C     THIS PROGRAM SOLVES THE EQUATION
C         EXP(-X) =SIN(X)
C     BY FIXED POINT ITERATION, USING THE ITERATION
C     FUNCTION
          G(X) = EXP(-X) - SIN(X) + X
          DATA XTOL, FTOL/ 1.   8, 1.E-8/
          PRINT 600
  600     FORMAT (9X,'XNEW',12X,'F(XNEW)',10X,'ERROR')
          XOLD = .6
          Y = G(XOLD) - XOLD
          PRINT 601, XOLD,Y

  601     FORMAT(3X,3E16.8)
          DO 10 J=1,20
              XNEW = G(XOLD)
              Y = G(XNEW) - XNEW
              ERROR = ABS(XNEW - XOLD)/ABS(XNEW)
              PRINT 601, XNEW,Y,ERROR
              IF (ERROR .LT. XTOL.OR.ABS(Y).LT.FTOL) STOP
              XOLD = XNEW
   10     CONTINUE
          PRINT 610
  610     FORMAT('FAILED TO CONVERGE IN 20 ITERATIONS')
          STOP
          END
```

● **PROBLEM** 19-21

Consider the following problem. Find the particular value of x which causes the function y= xcos(x) to be maximized within the interval bounded by x=0 on the left and x=π on the right. It is required that the maximing value of x be known quite accurately. It is also required that the search scheme be relatively efficient in the sense that the function y= xcos(x) should be evaluated as few times as possible. Write a FORTRAN program to search for a maximum of the function y = xcos(x) and apply the program for the case XL = 0, XR=3.141593, and EPSLN = 0.0001.

Use the following elimination scheme, which is a highly efficient computational procedure for all functions which have only one "peak" within the search interval.

    XL   =Left end of the search interval
    XR   =right end of the search interval
    XL1   =left-hand interior search point
    XR1   =right-hand interior search point
    EPSLN =distance beween XL1 and XR1.

<u>Solution</u>: There are certain mathematical calculations which are needed over and over again when writing FORTRAN programs. Since this arises so often, it is neccessary to outline the following for reference.

Function	Application	Description
EXP	Y=EXP(X)	Raise e to the x power: $y=e^x$
ALOG	Y=ALOG(X)	Compute the natural logarithm of x; $y=\log_e x$, $x>0$
ALOG10	Y=ALOG10(X)	Compute the common logarithm of x; $y=\log_{10} x$, $x>0$
SIN	Y=SIN(X)	Compute the sine of x; y=sinx, where x is in radians
COS	Y=COS(X)	Compute the cosine of x; y=cos x, x in radians
TAN	Y=TAN(X)	Compute the tangent of x; y=tan x, x in radians
SQRT	Y=SQRT(X)	Compute the square root of x; $y=\sqrt{x}$, $x>0$
ABS	Y=ABS(X)	Calculate the absolute value of x; $y=|x|$
INT	I=INT(X)	Convert x to the largest interger which does not exceed $|x|$, preserving the sign of x.
IFIX	I=IFIX(X)	Convert x to an interger quantity (as explained above)
FLOAT	X=FLOAT(I)	Convert i to a real quantity
AMAX1	Y1=AMAX1(U,V,W,X,Y,..)	Determine the largest of 2 or more real constants
MAX0	I1=MAX0(I,J,K,L,M,..)	Determine the largest of 2 or more interger constants
AMIN1	Y1=AMIN1(U,V,W,X,Y,..)	Determine the smallest of 2 or more real constants
MIN0	I1=MIN0(I,J,K,L,M,..)	Determine the smallest of 2 or more interger constants

Fig. 1

Evaluate the function y= x cos(x) at XL1 and XR1, and call these values YL1 and YR1 repectively. Suppose YL1 turns out to be greater than YR1. Then it is known that the maximum that is sought will lie somewhere between XL and XR1. Hence retain only that portion of the search interval which ranges from x = XL to x =XR1(actually one now refers to the old point XR1 as XR), and generate two new search points XR1 and XL1. These points will be located at the center of the new search interval, a distance EPSLN apart, as shown in Fig 2.

On the other hand, suppose now that in the original search interval the value of YR1 turned out to be greater than YL1. This would indicate that the new search interval should lie between XL1 and XR. Hence rename the point which was originally called XL1 to be XL, and generate two new search points, XL1 and XR1, at the center of the new search interval as shown in fig. 3.

Fig. 2 Fig. 3

Continue to generate a new pair of search points at the center of each new interval, compare the respective values of y, and eliminate a portion of the search interval until the search interval becomes smaller than 3.*EPSLN. Once this happens one cannot distinguish the interior points from the boundaries. Hence the search is ended.

Each time a comparison is made between YL1 and YR1 one eliminates that portion of the search interval which does not contain the larger value of y. If both interior values of y should happen to be identical (which can happen, though it is unusual), then the search procedure stops, and the maximum is assumed to ccur at the center of the two search points.

Once the search has ended, either because the search has become sufficiently small or because the two interior points yield identical values of y, the approximate value of YMAX can be calculated as YMAX=.5*(YL1+YR1) which accurs at approximately XMAX=.5*(XL1+XR1).

The Proceedure can easily be programmed by adhering to the following outline:

1. Define the function y= xcos(x).

2. Read the initial values of XL and XR, and a value for EPSLN.

3. Calculate a pair of interior points.

4. Write out the values of x at the ends of the interval and at the interior points, and write out the interior values for y.

5. Compare YL1 with YR1:
   (a) If YL1 is greater than YR1, let XR1 be called XR, thus defining a new search interval, and proceed to Step 6.
   (b) If YR1 is greater than YL1, let XL1 be called XL, thus defining a new search interval, and proceed below.

6. Test to see if the search should be terminated.
   (a) If the computation is to continue, go to Step 3 above
   (b) It XR-XL is smaller than 3. * EPSLN, calculate
   $$YMAX=.5*(YL1+YR1)$$
   $$XMAX=.5*(XL1+XR1)$$
   then write out the final results and stop.
   (c) If an excessive number of search points have been generated without narrowing down the interval sufficiently, write an appropriate message and stop.

A flowchart of the procedure is given in fig. 4.

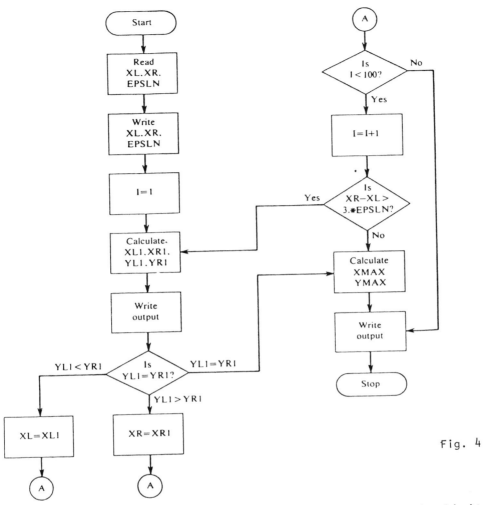

Fig. 4

The actual FORTRAN program appears in fig.5. For simplicity, a statement function has been used to evaluate the formula $y = x \cos(x)$.

```
              1234567
              C SEARCH FOR A MAXIMUM OF THE FUNCTION Y=X*COS(X)
0001              100 FORMAT (3E12.5)
0002              200 FORMAT (57H1 SEARCH FOR A MAXIMUM OF Y=X*COS(X) WITHIN
                 1THE INTERVAL,F8.6,7H < X < ,F8.6//,11H EPSILON =,F8.6////)
0003              250 FORMAT (1H0,16X,4HYL1=,F8.6,3X,4HYR1=,F8.6/,3X,3HXL=,
                 1F8.6,3X,4HXL1=, F8.6,3X,4HXR1=,F8.6,3X,3HXR=,F8.6/)
0004              300 FORMAT (26H0   THE MAXIMUM VALUE IS Y=,F8.6,18H   OCURR
                 1ING AT X=,F8.6/) .
0005              350 FORMAT (54H0   THE SOLUTION HAS NOT CONVERGED AFTER 100
                 1ITERATIONS/,24H0   TERMINATE COMPUTATION)
              C DEFINE THE FUNCTION Y(X)
0006                  Y(X)=X*COS(X)
              C READ AND WRITE OUTPUT
0007                  READ (5,100)   XL,XR,EPSLN
0008                  WRITE (6,200)  XL,XR,EPSLN
0009                  I=1
              C CALCULATE INTERIOR POINTS
```

```
0010              1 XL1=XL+.5*(XR-XL-EPSLN)
0011                XR1=XL1+EPSLN
0012                YL1=Y(XL1)
0013                YR1=Y(XR1)
              C WRITE OUTPUT
0014                WRITE (6,250)  YL1,YR1,XL,XL1,XR1,XR
                    IF (YL1-YR1)  2,5,3
              C YR1 GREATER THAN YL1
0016              2 XL=XL1
0017                GO TO 4
              C YL1 GREATER THAN YR1
0018              3 XR=XR1
              C TEST FOR END OF SEARCH
0019              4 IF (I.GE.100) GO TO 6
0020                I=I+1
0021                IF (XR-XL.GT.3.*EPSLN) GO TO 1
0022              5 XMAX=.5*(XL1+XR1)
0023                YMAX=.5*(YL1+YR1)
              C WRITE FINAL SOLUTION
0024                WRITE (6,300) YMAX,XMAX
0025                GO TO 7
              C WRITE OUTPUT - COMPUTATION TERMINATED BECAUSE OF MAXIMUM
              C     ITERATION COUNT
0026              6 WRITE (6,350)
0027              7 STOP
0028                END
```

Fig. 5

Figure 6 shows the output generated by the program for the case XL = 0, XR=3.141593, and EPSLN=0.0001. One sees that the maximum value of y is approximately 0.5611, occuring at x=0.8606. Notice that this result has been obtained to a high degree of accuracy using only 11 pairs of search points.

SEARCH FOR A MAXIMUM OF Y=X*COS(X) WITHIN THE INTERVAL 0.0 < x < 3.141593
EPSILON=0.000100

	YL1=0.000078	YR1= -.000077	
XL=0.0	XL1=1.570746	XR1=1.570846	XR=3.141593
	YL1=0.555356	YR1=0.555372	
XL=0.0	XL1=0.785373	XR1=0.785473	XR=1.570846
	YL1=0.545438	YR1=0.545412	
XL=0.785373	XL1=0.981715	XR1=0.981815	XR=1.178158
	YL1=0.560534	YR1=0.560529	
XL=0.785373	XL1=0.883544	XR1=0.883644	XR=0.981815
	YL1=0.560405	YR1=0.560410	
XL=0.785373	XL1=0.834458	XR1=0.834558	XR=0.883644

	YL1=0.561095	YR1=0.561095	
XL=0.834458			XR=0.883644
	XL1=0.859001	XR1=0.859101	
	YL1=0.560972	YR=0.560969	
XL=0.859001			XR=0.883644
	XL1=0.871273	XR1=0.871372	
	YL1=0.561072	YR1=0.561071	
XL=0.859001			XR=0.871372
	XL1=0.865137	XR1=0.865237	
	YL1=0.561093	YR1=0.561093	
XL=0.859001			XR=0.865237
	XL1=0.862069	XR1=0.862169	
	YL1=0.561096	YR1=0.561096	
XL=0.859001			XR=0.862169
	XL1=0.860535	XR1=0.860635	

THE MAXIMUM VALUE IS Y=0.561096, OCCURRING AT X=0.860585

FIG. 6

● **PROBLEM** 19-22

Consider the problem regarding a game of chance (shooting craps).
There are two ways a player can win in craps. He can throw the dice once and obtain a score of either 7 or 11; or he can obtain a 4,5,6,8,9, or 10 on the first throw, and then come up with the same score on a subsequent throw before obtaining a score of 7. Conversely, there are two ways a player can lose. Either he can throw the dice once and obtain a score of 2,3, or 12, or he can obtain a 4,5,6,8,9, or 10 on the first throw, and then obtain a score of 7 on a subsequent throw before coming up with the same score as he had on the first throw.

Write a FORTRAN program that will read the necessary input, generate a specified number of plays, count the total number of wins, and finally, calculate the odds of winning. Then, apply the program to show a portion of the output that will be obtained for the input values N = the number of plays = 1000, KRAND = the starting value for FUNCTION XRAND = 123456789, and output indicator IOUT = 1, signifying the choice of the option to write out the results of each play.

Hint: The odds of winning are calculated by dividing the number of wins by the total number of plays. An output indicator should be included in the main program so that one will have the option of either writing out the results of each play, or simply writing out the final calculated odds of winning.

Solution: In order to computerize the craps game, 3 FUNCTION subprograms will be required. The first of these will generate a sequence of uniformly distributed random numbers be-

tween zero and one. By uniformly distributed one means that
any number between zero and one is just as likely to appear
as any other number. The numbers which one generates will
not really be random because one will always generate the
same sequence of numbers, given the same starting value,
whenever one runs the program. However, the sequence which
is generated will appear to be random and will have many of
the statistical characteristics of numbers which are truly
random.

The actual subprogram, however, is quite simple. For a computer with a 32-bit word, such as an IBM 360 or 370 series
computer, one has

```
          FUNCTION XRAND(KX)
          IF(KX.GT.0) IX = KX
          IY = 65539*IX
          IF (IY.LT.0)IY = IY + 2147483647 + 1
          XRAND = .4656613E-9*FLOAT(IY)
          IX = IY
          RETURN
          END
```

The first time the subprogram is referenced, the argument KX
should be assigned a positive, odd, integer value having not
more than 9 digits. This is a starting value which is required by the subprogram. Whenever the subprogram is referenced thereafter, a value of zero should be assigned to KX.

This subprogram is valid only for a computer with a 32-bit
word. In general, if b is the number of bits per word on a
particular computer, the number 65539 should be replaced by
the integer quantity calculated from the formula $2^{b-2} + 3$;
2147483647 is replaced by the integer quantity calculated by
the formula $2^{b-1}-1$, and .4656613E - 9 is replaced by the
floating point equivalent of $2^{-(b-1)}$, expressed in terms
of the largest permissible number of significant figures.

The second FUNCTION subprogram will simulate one throw of a
pair of dice. To do this, the subprogram will reference
XRAND twice, obtaining two random numbers between zero and
one. If one of these numbers falls between 0 and $\frac{1}{6}$, assign
a score of 1 to one of the dice; if the random number falls
between $\frac{1}{6}$ and $\frac{1}{3}$, assign a score of 2 to the die; if the random number is between $\frac{1}{3}$ and $\frac{1}{2}$, assign a score of 3, and so
on. Each die will thus be assigned to a score of 1,2,3,4,5
or 6, and the combined score of the two dice will represent
one throw. Call this subprogram KTOSS.

The third FUNCTION subprogram, called KSCORE, will simulate
one play. That is, one or more throws will be simulated, by
referring to KTOSS, until either a win or a loss is obtained.
If the result is a win, a value of 1 is assigned to KSCORE
and then transmitted back to the calling program. If the
result of the play is a loss, then KSCORE is assigned a value
of zero.

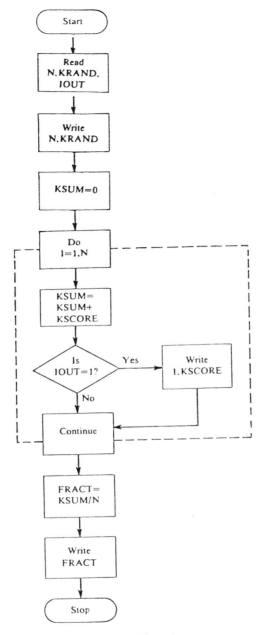

Fig. 1

Rather than write one outline of the entire program, it is helpful to write a separate outline for the main program and each subprogram. The outline of the main program can be written as follows.

1. Read the number of plays (N), the initial value for KX required by the FUNCTION subprogram XRAND(KRAND), and the output indicator (IOUT).
2. Write out N and KRAND to identify the problem.
3. Set up the random number generator by transmitting the

initial value for KX. (This number will be called KRAND in all programs and subprograms except subprogram XRAND, where it is called KX.)

4. Initialize the number of wins (KSUM) to zero, and do the following N times: generate a random play by referring to subprogram KSCORE, and then add the value obtained to the win counter (KSUM = KSUM + KSCORE(0)). Write each value for KSCORE if IOUT = 1.

5. Calculate the final percentage of wins (FRACT = FLOAT(KSUM)/FLOAT(N)) and write out.

6. STOP.

A flowchart of the procedure is shown in Figure 1.

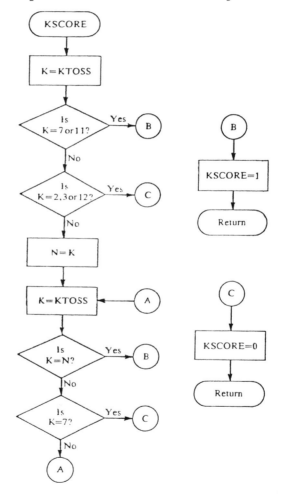

Fig. 2

Subprogram KSCORE will have the following structure:

1. Generate the outcome of one random toss by referring to subprogram KTOSS. If the outcome is 7 or 11, set KSCORE

equal to 1 (signifying a win) and return. If the outcome
is a 2,3, or 12, set KSCORE equal to zero (indicating a
loss) and return.

2. If the outcome of the first toss is 4,5,6,8,9 or 10, save
the outcome of the first toss and simulate subsequent
tosses until either
   (a) A 7 is obtained, which signifies a loss. Set KSCORE
       equal to zero and return.
   (b) A toss has the same outcome as the original toss, and
       a 7 has not yet been obtained. This signifies a win.
       Hence set KSCORE equal to one and return.

Figure 2 shows a flowchart of this subprogram.

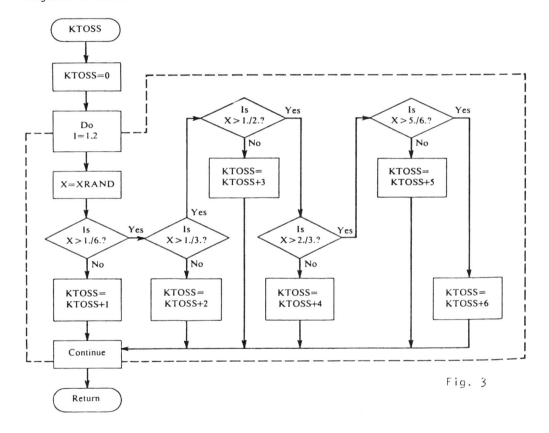

Fig. 3

Subprogram KTOSS can be outlined as follows:

1. Set KTOSS equal to zero, and then do the following 2 times.
2. Generate a random number between zero and one by referring
   to subprogram XRAND.
   (a) If the value obtained for XRAND lies between zero and
       $\frac{1}{6}$, add 1 to KTOSS.
   (b) IF XRAND lies between 1/6 and 1/3, add 2 to KTOSS.
   (c) IF XRAND lies between 1/3 and 1/2 add 3 to KTOSS.
   (d) IF XRAND lies between 1/2 and 2/3, add 4 to KTOSS.
   (e) IF XRAND lies between 2/3 and 5/6, add 5 to KTOSS.
   (f) If XRAND exceeds $\frac{5}{6}$, add 6 to KTOSS.

3. Each random number will thus simulate the throwing of one of the dice. When this has been carried out 2 times, KTOSS will represent the outcome of having thrown a pair of dice; hence control is returned to subprogram KSCORE.

A flowchart of this program is shown in fig. 3.

THE FORTRAN PROGRAM

The actual program is shown in Figs. 4-7.

```
            C COMPUTER SIMULATION OF A GAME OF .CRAPS
            C
            C N IS THE NUMBER OF PLAYS TO BE SIMULATED
            C KRAND IS A POSITIVE ODD NUMBER WHICH STARTS THE
            C RANDOM NUMBER GENERATOR
            C IOUT INDICATES THE LEVEL OF OUTPUT DESIRED
            C      IOUT = 0: ONLY THE FINAL PERCENTAGE OF WINS
            C      WILL BE WRITTEN OUT
            C      IOUT = 1: OUTPUT THE RESULT OF EACH PLAY AS
            C      WELL AS THE FINAL PERCENTAGE OF WINS
            C
0001          100 FORMAT (3.I10)
0002          200 FORMAT (41H1 COMPUTER SIMULATION OF A GAME OF
                 1CRAPS/,25H0 SIMULATION BASED UPON,I5,13H RANDOM
                 1PLAYS/,9H0 KRAND = I10//)
0003          300 FORMAT (10X,2HI=,I5,10X,7HKSCORE=,I1)
0004          400 FORMAT (32H0 THE PERCENTAGE OF WINS AFTER,
                 3I5, 10H PLAYS IS ,F8.6)
            C READ AND WRITE OUTPUT
0005           10 READ (5,100) N,KRAND,IOUT
0006              WRITE(6,200) N,KRAND
            C SET UP THE RANDOM NUMBER GENERATOR
0007              X=XRAND(KRAND)
            C GENERATE N RANDOM PLAYS
0008              KSUM=0
0009              DO 1 I=1,N
0010              K=KSCORE(0)
0011              KSUM=KSUM+K
0012              IF (IOUT.EQ.1)WRITE (6,300) I,K
0013            1 CONTINUE
            C CALCULATE FINAL PERCENTAGE OF WINS
0014              XNUM=FLOAT(KSUM)
0015              XDENOM=FLOAT(N)
0016              FRACT=XNUM/XDENOM
0017              WRITE(6,400) N,FRACT
0018              GO TO 10
0019              END
```

FIG. 4

```
0001          FUNCTION KSCORE(KRAND)
         C
         C THIS FUNCTION GENERATES THE OUTCOME OF ONE PLAY OF
         C CRAPS
```

```
        C
        C KSCORE=0 INDICATES A LOSS
        C KSCORE=1 INDICATES A WIN
        C
0002          K=KTOSS(KRAND)
0003          IF((K.EQ.7).OR.(K.EQ.11)) GO TO 2
0004          IF((K.EQ.2).OR.(K.EQ.3).OR.(K.EQ.12)) GO TO 3
        C CONDITIONAL SITUATION - MORE TOSSES REQUIRED
0005          N=K
0006        1 K=KTOSS(KRAND)
0007          IF (K.EQ.N) GO TO 2
0008          IF (K.EQ.7) GO TO 3
0009          GO TO 1
        C WIN
0010        2 KSCORE=1
0011          RETURN
        C LOSE
0012        3 KSCORE=0
0013          RETURN
0014          END
```

FIG. 5

```
0001          FUNCTION KTOSS(KRAND)
        C
        C THIS FUNCTION GENERATES THE OUTCOME OF A RANDOM
        C TOSS OF TWO DICE
        C
0002          KTOSS=0
0003          DO 6 I=1,2
0004          X=XRAND(KRAND)
0005          IF (X.GT.0.1666667) GO TO 1
0006          KTOSS=KTOSS+1
0007          GO TO 6
0008        1 IF (X.GT.0.3333333) GO TO 2
0009          KTOSS=KTOSS+2
0010          GO TO 6
0011        2 IF (X.GT.0.5) GO TO 3
0012          KTOSS=KTOSS+3
0013          GO TO 6
0014        3 IF (X.GT.0.6666667) GO TO 4
0015          KTOSS=KTOSS+4
0016          GO TO 6
0017        4 IF (X.GT.0.8333333) GO TO 5
0018          KTOSS=KTOSS+5
0019          GO TO 6
0020        5 KTOSS=KTOSS+6
0021        6 CONTINUE
0022          RETURN
0023          END
```

FIG. 6

```
0001          FUNCTION XRAND(KX)
        C
        C THIS FUNCTION GENERATES A UNIFORMLY DISTRIBUTED
        C RANDOM NUMBER BETWEEN ZERO AND ONE
```

```
              C
              C  KX CAN BE ANY POSITIVE ODD INTERGER, NOT EXCEEDING
              C  NINE DIGITS, THE FIRST TIME THE FUNCTION IS CALLED
              C  THEREAFTER KX SHOULD BE SET TO ZERO
              C
              C  THIS RANDOM NUMBER GENERATOR IS VALID ONLY FOR A
              C  32-BIT WORD COMPUTER
              C
0002                IF (KX.GT.O)  IX=KX
0003                IY=65539*IX
0004                IF (IY.LT.0)  IY=IY+2147483647+1
0005                XRAND=.4656613E-9*FLOAT(IY)
0006                IX=IY
0007                RETURN
0008                END
```

FIG. 7

Figure 8 shows a portion of the output obtained for the input values N=1000, KRAND=123456789 and IOUT=1. Note that the results of 1000 plays show that the odds of winning are about 0.499; in other words, 499 plays out of 1000 will be wins. It can be shown mathematically that the odds are exactly 0.493. The exact answer would be approached more closely if the number of plays were increased.

COMPUTER SIMULATION OF A GAME OF CRAPS

SIMULATION BASED UPON 1000 RANDOM PLAYS

KRAND=123456789

I=	1	KSCORE=0
I=	2	KSCORE=0
I=	3	KSCORE=1
I=	4	KSCORE=0
I=	5	KSCORE=0
I=	6	KSCORE=1
I=	7	KSCORE=1
I=	8	KSCORE=0
I=	9	KSCORE=0
I=	10	KSCORE=0
I=	11	KSCORE=1
I=	12	KSCORE=0
I=	13	KSCORE=0
I=	14	KSCORE=1
I=	15	KSCORE=1
I=	16	KSCORE=0
I=	17	KSCORE=0
I=	18	KSCORE=1
I=	19	KSCORE=0
I=	20	KSCORE=0
I=	21	KSCORE=1
I=	22	KSCORE=1
I=	23	KSCORE=0
I=	24	KSCORE=0

I=	25	KSCORE=1
I=	26	KSCORE=1
I=	27	KSCORE=0
I=	28	KSCORE=0
I=	29	KSCORE=0
I=	30	KSCORE=0
I=	31	KSCORE=0
I=	32	KSCORE=1
⋮	⋮	⋮
I=	991	KSCORE=0
I=	992	KSCORE=1
I=	993	KSCORE=0
I=	994	KSCORE=1
I=	995	KSCORE=1
I=	996	KSCORE=0
I=	997	KSCORE=0
I=	998	KSCORE=1
I=	999	KSCORE=0
I=	1000	KSCORE=0

The Percentage of Wins After 1000 Plays is 0.499000

IHC900I EXECUTION TERMINATING DUE TO ERROR COUNT FOR ERROR NUMBER 217

IHC217I FIOCS-END OF DATA SET ON UNIT   5

TRACEBACK ROUTINE CALLED FROM ISN REG.14   REG.15   REG.0   REG.1

    I BCOM   00049BOC   0004A2E8 000003E8 000003E8

    MAIN   00009984   01049900 FFFFFFB5 00068FF8

FIG. 8

ROOTS OF EQUATIONS

• **PROBLEM** 19-23

Write a FORTRAN program that uses Newton's method to find the real positive root of the polynomial equation

$$x^5 - 3.7x^4 + 7.4x^3 - 10.8x^2 + 10.8x - 6.8 = 0$$

Take into consideration the situation in which the method fails to converge for certain initial guesses.

Hint: It is easily verified that the root lies between 1 and 2. Choose $x_0 = 1.5$.

Solution: FORTRAN program for finding real positive root of polynomial equation:

```
C     NEWTON'S METHOD FOR FINDING A REAL ZERO OF A CERTAIN POLY-
C     NOMIAL
```

```
C       THE COEFFICIENTS ARE SUPPLIED IN A DATA STATEMENT.
C       A FIRST GUESS X FOR THE ZERO IS READ IN.
        PARAMETER N=6
        INTEGER J,K
        REAL A(N), B,C,DELTAX,X
        DATA A/-6.8, 10.8, -10.8, 7.4, -3.7, 1./
    1   READ 500, X
  500   FORMAT (E16.8)
        PRINT 601
  601   FORMAT ('1 NEWTONS METHOD FOR FINDING A REAL ZERO
       *         OF A POLYNOMIAL' //4X, 'I', 10X, 'X', 14X,
       *         'AP(0)', 12X, 'APP(1)'/)
        DO 10 J=1,20
           B = A(N)
           C = B
           DO 5 K=N, 3, -1
             B = A(K-1) + X*B
             C = B + X*C
    5      CONTINUE
           B = A(1) + X*B
           PRINT 605,J,X,B,C
  605      FORMAT (I5,3(1PE17.7))
           DELTAX = B/C
           IF (ABS(DELTAX) .LT. 1.E-7 .OR. ABS(B) .LT.
       *       1.E-7) STOP
           X = X - DELTAX
   10   CONTINUE
        PRINT 610
  610   FORMAT ('FAILED TO CONVERGE IN 20 ITERATIONS')
                                                  GO TO 1
        END
```

COMPUTER RESULTS

I	X	AP	APP
1	1.5000000E 00	−1.0625001E − 00	3.7124998E 00
2	1.7861953E 00	7.2393334E − 01	9.6004875E 00
3	1.7107894E 00	8.0013633E − 02	7.5470622E 00
4	1.7001875E 00	1.3663173E − 03	7.2905675E 00
5	1.7000000E 00	4.7683716E − 07	7.2861013E 00
6	1.7000000E 00	−1.1920929E − 07	7.2860994E 00
7	1.7000000E 00	−5.9604645E − 08	7.2860998E 00

• **PROBLEM** 19-24

Figure 1 shows the parabola $y = x^2$ and the hyperbola $y = (x + 1)/x$. Find their intersections using the Modified Newton-Raphson Method and write a FORTRAN program. Let $x_0 = y_0 = 3$

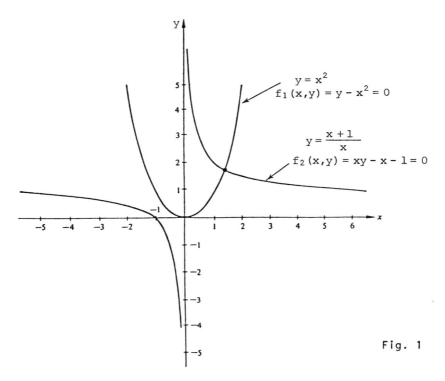

Fig. 1

Solution: Since both of these equations give y explicitly in terms of x, one way would be by setting

$$x^2 = \frac{x+1}{x}$$

$$x^3 = x + 1$$

$$x^3 - x - 1 = 0$$

and solving the resulting cubic equation to find its real root. The problem can also be approached by solving two simultaneous nonlinear equations

$$y = x^2 \qquad\qquad y = \frac{x+1}{x}$$

$$xy = x + 1$$

$$f_1(x,y) = y - x^2 = 0 \qquad f_2(x,y) = xy - x - 1 = 0$$

$$\frac{\partial f_1}{\partial x} = -2x \qquad\qquad \frac{\partial f_2}{\partial x} = y - 1$$

$$\frac{\partial f_1}{\partial y} = 1 \qquad\qquad \frac{\partial f_2}{\partial y} = x$$

Now note that at the solution (Figure 1) the curve defined by $f_1(x,y) = 0$ is steeper, and therefore should be used for finding x, while f_2 should be used for finding y. Therefore prepare the iteration

$$x_{i+1} = x_i - \frac{f_1(x_i,y_i)}{\partial f_1/\partial x} \qquad y_{i+1} = y_i - \frac{f_2(x_{i+1},y_i)}{\partial f_2/\partial y}.$$

As an initial guess, pick $x_0 = y_0 = 3$ as being a reasonable set of values, and write the following program

```
      X = 3
      Y = 3
      WRITE (3, 5)
    5 FORMAT (14X, 'X', 14X, 'Y')
      WRITE (3, 10) X, Y
   10 FORMAT (5X, 2F15.8)
      DO 100 I = 1, 15
      X = X - (Y - X**2) / (-2. * X)
      Y = Y - (X * Y - X - 1.) / X
      WRITE (3, 20) I, X, Y
   20 FORMAT (I5, 2F15.8)
  100 CONTINUE
      CALL EXIT
      END
```

As the following results show, 13 iterations produce an answer accurate to eight decimal places.

	X	Y
	3.00000000	3.00000000
1	2.00000000	1.50000000
2	1.37500000	1.72727272
3	1.31559917	1.76010993
4	1.32673810	1.75372825
5	1.32428632	1.75512370
6	1.32481092	1.75482469
7	1.32469796	1.75488905
8	1.32472225	1.75487521
9	1.32471703	1.75487819
10	1.32471815	1.75487755
11	1.32471791	1.75487769
12	1.32471796	1.75487766
13	1.32471795	1.75487766
14	1.32471795	1.75487766
15	1.32471795	1.75487766

If f_2 had been chosen to find the next x, and f_1 to find the next y, only two statements would have been changed:

```
      X = X - (X * Y - X - 1.) / (Y - 1.)
      Y = Y - (Y - X**2)
```

It would have been found that the program diverges quite badly:

	X	Y
	3.00000000	3.00000000
1	0.50000000	0.25000000
2	-1.33333333	1.77777778
3	1.28571428	1.65306121
4	1.53125001	2.34472660
5	0.74364558	0.55300875
6	-2.23718028	5.00497560
7	0.24968940	0.06234480
8	-1.06649011	1.13740116
9	7.27795873	52.96868340
10	0.01924235	0.00037026

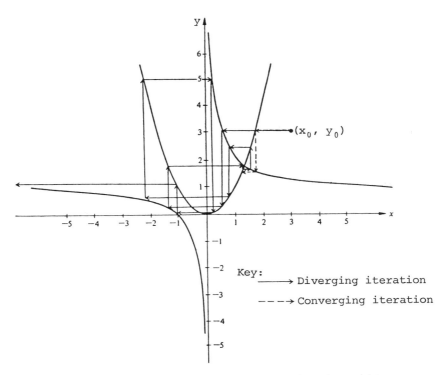

Paths taken by the converging and diverging iterations.

Fig. 2

Figure 2 shows the paths taken by the converging and diverging iterations; the diverging iteration starts off badly because of the wrong slopes, and soon starts to hit the other hyperbola defined by f_2. The converging iteration, on the other hand, goes directly to the solution.

● **PROBLEM 19-25**

As shown in the figure, the function

$$f(x) = x^3 - 11x^2 + 39x - 45 = 0$$

has a double root at $x = 3$, and a single root at $x = 5$. Given only the information that the roots are between -10 and $+10$, write a program which will approximately find the location of the roots.

Hint: First decide on the step size. Assume one agrees on using 200 steps of size 0.1; this will start from $x = -10$ to $x = +10$.

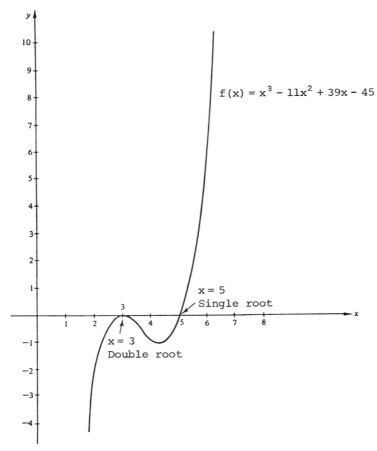

Graph of the function $f(x) = x^3 - 11x^2 + 39x - 45$.

<u>Solution</u>: At each step calculate $f(x)$, and also the derivative

$$f'(x) = 3x^2 - 22x + 39.$$

The criteria for indicating a possible root shall be either:
(1) $f(x)$ either becomes zero or (more likely) changes sign between two successive steps, indicating a crossover; or
(2) $f'(x)$ either becomes zero or changes sign between two successive steps, and $f(x)$ is very small. A zero derivative merely indicates a maximum or minimum in the curve, and is not indicative of a root unless this minimum or maximum occurs on the x axis or very near it.

All the calculations in the following program are contained inside a DO loop which counts off exactly 200 repetitions. During each repetition calculate the function $f(x_n)$, which is called FNEW in the program, and the derivative $f'(x_n)$, which is called DNEW. The previous values of $f(x_{n-1})$ and $f'(x_{n-1})$, calculated during the previous repetition of the loop, are called FOLD and DOLD.

First in the loop calculate the product FOLD · FNEW. If the

product is negative a root has just been passed and so print "root is near"; if the product is zero then either FNEW is zero (indicating a root has accidentally been stumbled upon), or FOLD is zero (which is ignored since it would have been detected in the previous repetition), or else there is an underflow and the product is zero although neither number is zero; this may indicate a close root.

If, on the other hand, the product FNEW · FOLD is positive, then check the product DOLD · DNEW. If the product is positive then no change of derivative exists, and go on. If on the other hand the product is negative or zero there might be a maximum or minimum in the function, and so check whether FNEW, the value of the function $f(x_n)$ is less than an arbitrary constant Z in magnitude; that is, whether the maximum or minimum is close to the x axis. If so, then there may be a possible root.

Finally, at the end of each loop set

$$FOLD = FNEW$$
$$DOLD = DNEW$$

to have the old values of the function and derivatives during the next repetition of the loop.

Now write a program for the function

$$f(x) = x^3 - 11x^2 + 39x - 45$$
$$f'(x) = 3x^2 - 22x + 39$$

```
         F(X) = X**3 - 11. * X**2 + 39. * X - 45.
         D(X) = 3. * X **2 - 22. * X + 39.
         H = 0.1
         XG = -10.
         Z = 0.01
         FOLD = F(XG)
         DOLD = D(XG)
         DO 1000 N = 1, 200
         X = XG + N * H
         FNEW = F(X)
         DNEW = D(X)
         IF (FNEW * FOLD) 10, 30, 70
   10    WRITE (3, 20) X
   20    FORMAT ('ROOT IS NEAR', F15.8)
         GO TO 900
   30    IF (FOLD) 40, 900, 40
   40    IF (FNEW) 10, 50, 10
   50    WRITE (3, 60) X
   60    FORMAT  ('ROOT IS AT', F15.8)
         GO TO 900
   70    IF (DNEW * DOLD) 80, 80, 900
   80    IF (ABS(FNEW) - Z) 90, 90, 900
   90    WRITE (3, 100) X, FNEW
  100    FORMAT ('POSSIBLE ROOTS AT', F15.8, 'FNEW IS', E15.8)
  900    FOLD = FNEW
 1000    DOLD = DNEW
         CALL EXIT
         END
```

When the program is run, one gets the strange result

$$\begin{array}{lll} \text{ROOT IS NEAR} & & 2.99999952 \\ \text{ROOT IS NEAR} & & 3.09999943 \\ \text{ROOT IS NEAR} & & 5.09999943 \end{array}$$

Roundoff is causing strange results indeed. First of all, note that X is calculated by the statement

$$X = XG + N * H \quad [\text{which equals } -10.0 + n(0.1)]$$

and should therefore go up only in steps of 0.1 whereas the x values printed out are 2.99999952 and so on; obviously the result of roundoff error.

Another strange result is the apparent fact that all three roots seem to be separate, whereas it is known that there are two identical roots at x = 3, and there is no root x = 3.09999943 at all.

The reason is that the double root at x = 3 is being picked up by the program twice, again because of roundoff error. At x = 2.99999952 a slight roundoff is making f(x) positive, whereas it should be just slightly negative. As a result, the program decides it has just stepped over a root since f(x) has changed sign, and prints a message that a "root is near 2.99999952". The next step, at x = 3.09999943, again yeilds a negative f(x), and so the program decides it has again stepped over a root and prints another message. Thus the slight error in calculating f(x) at x = 2.99999952 causes two sign changes in f(x), leading the program into printing two messages.

There is also an error in the third root, which is at x = 5.0 instead of x = 5.09999943; the reason here is the large step size. The program cannot sense a root unless it either happens to hit it very closely, or else it steps over it. Because of the large steps, it reaches almost 5.1 before it can print out that a root has been found.

• **PROBLEM 19-26**

Using the Newton-Raphson Method, solve the simultaneous equations

$$y = \cos x$$
$$x = \sin y.$$

Also write a FORTRAN program which will solve the simultaneous equations. Let $x_0 = y_0 = 1$.

Solution: When the equations are put into standard form, one gets

$$f_1(x,y) = \cos x - y = 0$$
$$f_2(x,y) = x - \sin y = 0$$

Figure 1 shows the curves specified by the two equations; the solution is at the intersection of the two curves.

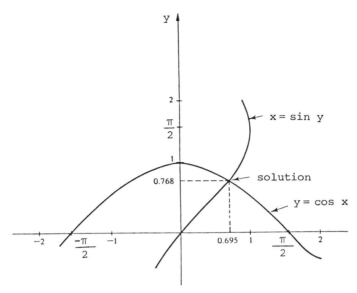

Fig. 1

Solution of the simultaneous equation $y = \cos x$ and $x = \sin y$.

Taking the partial derivatives of each function, one gets

$$f_1(x,y) = \cos x - y = 0 \qquad f_2(x,y) = x - \sin y = 0$$

$$\frac{\partial f_1}{\partial x} = -\sin x \qquad \qquad \frac{\partial f_2}{\partial x} = 1$$

$$\frac{\partial f_1}{\partial y} = -1 \qquad \qquad \frac{\partial f_2}{\partial y} = -\cos y,$$

so that the Jacobian is

$$J = \begin{vmatrix} \dfrac{\partial f_1}{\partial x} & \dfrac{\partial f_1}{\partial y} \\ \dfrac{\partial f_2}{\partial x} & \dfrac{\partial f_2}{\partial y} \end{vmatrix} = \begin{vmatrix} -\sin x & -1 \\ 1 & -\cos y \end{vmatrix} = \sin x \cos y + 1.$$

In order for the Jacobian to be zero, one must have

$$\sin x \cos y + 1 = 0$$

$$\sin x \cos y = -1,$$

so that either $\sin x$ or $\cos y$ must be negative, with the other being positive; this does not happen in the first quadrant, and so, as long as the values of x_i and y_i are either in the first quadrant or on the x and y axes, solve for h and k:

$$h = \frac{\begin{vmatrix} -f_1(x,y) & \dfrac{\partial f_1}{\partial y} \\ -f_2(x,y) & \dfrac{\partial f_2}{\partial y} \end{vmatrix}}{J} = \frac{\begin{vmatrix} -(\cos x - y) & -1 \\ -(x - \sin y) & -\cos y \end{vmatrix}}{J}$$

$$= \frac{-(\cos x - y)(-\cos y) + (x - \sin y)(-1)}{\sin x \cos y + 1}$$

$$k = \frac{\begin{vmatrix} \dfrac{\partial f_1}{\partial x} & -f_1(x,y) \\ \dfrac{\partial f_2}{\partial x} & -f_2(x,y) \end{vmatrix}}{J} = \frac{\begin{vmatrix} -\sin x & -(\cos x - y) \\ 1 & -(x - \sin y) \end{vmatrix}}{J}$$

$$= \frac{-(x - \sin y)(-\sin x) + (\cos x - y)(1)}{\sin x \cos y + 1}$$

Now write the following program, starting with the initial guess that $x_0 = y_0 = 1$. Note that a test is included to stop the program if the Jacobian JACOB becomes zero.

```
          REAL JACOB, K
          X = 1.0
          Y = 1.0
          WRITE (3, 5)
    5     FORMAT (14X, 'X', 14X, 'Y')
          WRITE (3, 10) X, Y
   10     FORMAT (5X, 2F15.8)
          DO 100 I = 1, 5
          JACOB = SIN (X) * COS (Y) + 1.
          IF (JACOB) 15, 200, 15
   15     H = (-(COS (X) - Y) * (-COS (Y)) + (X - SIN (Y)) * (-1.))/ JACOB
          K = (-(X - SIN (Y)) * (-SIN (X)) + (COS (X) - Y) * 1.)/ JACOB
          X = X + H
          Y = Y + K
          WRITE (3, 20) I, X, Y
   20     FORMAT (I5, 2F15.8)
  100     CONTINUE
  200     CALL EXIT
          END
```

At each step of the program, the new value of the Jacobian is calculated, tested for zero, and then used to calculate the new values of h and k; these values are then added to the old values of x and y, respectively, to give new values which are printed.

The program gives the following results:

	X	Y
	1.00000000	1.00000000
1	0.72027285	0.77568458
2	0.69495215	0.76832706
3	0.69481970	0.76816915
4	0.69481969	0.76816915
5	0.69481969	0.76816915

● **PROBLEM** 19-27

Show by the damped Newton's Method that the system $f(\xi) = 0$ with

$$f_1(x) = x_1 + 3 \ln|x_1| - x_2^2 \qquad f_2(x) = 2x_1^2 - x_1 x_2 - 5x_1 + 1$$

has several solutions. Hint: For that reason, the initial guess has to be picked carefully to ensure convergence to a particular solution, or, to ensure convergence at all.

Solution: The Newton equations are

$$\begin{bmatrix} 1 + 3/x_1 & -2x_2 \\ 4x_1 - x_2 - 5 & -x_1 \end{bmatrix} h = -f(x)$$

Starting with the initial guess $x^{(0)} = \begin{bmatrix} 2 & 2 \end{bmatrix}^T$, obtain the following sequence of iterates.

m	$x^{(m)}$		$\|f(x^{(m)})\|_2$	$\|h\|_2$
0	2.	2.	0.500 + 1	0.238 + 2
1	−18.1588	−10.5794	0.572 + 3	0.112 + 2
2	−8.3710	−5.2287	0.142 + 3	0.543 + 1
3	−3.5525	−2.7191	0.351 + 2	0.266 + 1
4	−1.2015	−1.4728	0.860 + 1	0.198 + 1
5	−0.0004	0.0945	0.234 + 2	0.187 + 4
6	0.0451	−1866.2415	0.348 + 7	0.933 + 3
7	0.0233	−933.1179	0.871 + 6	0.467 + 3
8	0.0108	−466.5520	0.218 + 6	0.233 + 3
		etc.		

Clearly, the iteration is not settling down at all. But, now employ the damped Newton's Method, starting with the same first guess.

m	$x^{(m)}$		$\|f(x^{(m)})\|_2$	$\|h\|_2$	i
0	2.	2.	0.500 + 1	0.238 + 2	4
1	0.7400698	1.2137849	0.299 + 1	0.160 + 1	1
2	0.5310238	0.4415855	0.205 + 1	0.217 + 1	2
3	0.5178341	−0.1001096	0.178 + 1	0.372 + 1	3
4	0.5584838	−0.5637875	0.173 + 1	0.832 + 1	6
5	0.5847026	−0.6910621	0.172 + 1	0.967 + 1	6
6	0.6215780	−0.8376443	0.171 + 1	0.937 + 1	6
7	0.6657612	−0.9772562	0.171 + 1	0.684 + 1	5
8	0.7448782	−1.1760004	0.169 + 1	0.328 + 1	3
9	0.9489394	−1.5313175	0.163 + 1	0.676	0
10	1.5501608	−1.8410875	0.105 + 1	0.315	0
11	1.3892191	−1.5703845	0.132	0.473 − 1	0
12	1.3735386	−1.5257440	0.249 − 2	0.781 − 3	0
13	1.3734783	−1.5249650	0.608 − 6	0.156 − 6	0
14	1.3734783	−1.5249648	0.843 − 7		

Also listed here for each iteration, is the interger i determined by

$$i = \min\{j: 0 \leq j, ||f(x^{(m)} + h/2^j)||_2 < ||f(x^{(m)})||_2\}$$

of the damped Newton's Method for a system. Initially, the proposed steps h are rather large and are damped by as much as $1/2^6 = 1/64$. Correspondingly, the size $||f(x^{(m)})||_2$ of the residual error barely decreases from step to step. But, eventually, the full Newton step is taken and the iteration converges quadratically, as it should.

The calculations were run in single precision on a UNIVAC 1110. The error $||f(x^{(14)})||_2$ is therefore at noise level.

● **PROBLEM** 19-28

Write a FORTRAN program segment to find a root of the continuous function f(x) via the bisection method.

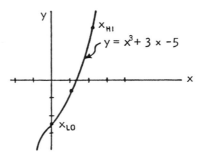

Solution: The bisection method starts with two values of X, called HI and LO, for which the values of f(x) have opposite signs. The object is to find X when f(x) = 0. This is done by halving the interval between HI and LO, alternately from the negative side (LO) and from the positive side (HI). As one continues the process, X can be approached until an approximation with any desired degree of accuracy is reached. Note that rounding errors can alter the accuracy of the approximation. The program logic is given below, using the function X^3 + 3X - 5 (Figure) as a sample f(x).

It has been chosen that

HI = 2.0, so f(HI) = $(2)^3$ + 3(2) - 5 = 9, and
LO = 0.0, so f(LO) = $(0)^3$ + 3(0) - 5 = -5.

```
      REAL F, X, HI, LO, ERROR
C     CHOOSE SOME INITIAL HI AND LO VALUES,
C     SUCH AS HI = 2.0 and LO = 0.0.
C     ALSO CHOOSE ERROR TO BE, SAY, = .002.
      DO 20 J = 1, 1000
      IF ((HI - LO) .LE. ERROR) GO TO 30
      X = (LO + HI)/2.0
```

```
      F = (X ** 3) + (3.0 * X) - 5.0
      IF (F.GT.0.0) GO TO 15
      LO = X
      GO TO 20
  15  HI = X
  20  CONTINUE
  30  WRITE (5,*) X
```

Note: This WRITE statement includes an asterisk, a feature provided on some machines. It simply indicates that only variables are to be outputted, without any literals. Its purpose is to save time by eliminating a FORMAT statement.

NUMERICAL INTEGRATION METHODS IN FORTRAN PROGRAMS

● **PROBLEM** 19-29

Write a trapezoidal integration routine in FORTRAN to evaluate

$$\int_0^{.4} \sin^2 x \; dx + \int_{.1}^{.3} \cos^2 x \; dx.$$

Solution: Let the problem be structured as follows: there shall be a main routine which calls a trapezoid function routine, which in turn references two function routines defining the functions $\sin^2 x$ and $\cos^2 x$. First note that according to the trapezoid rule

$$\int_a^b f(x)\,dx \approx \frac{h}{2}\Big[f(a) + 2f(a+h) + 2f(a+2h) + \ldots + 2f(a+(n-1)h) + f(b)\Big],$$

where h = (b-a)/n is the number of trapezoids. Thus one gets

```
      FUNCTION TRAP (A,B,N,FX)
      H = (B - A)/N
      SUM = 0
      K = N - 1
      DO 6 I = 1,K
  6   SUM = SUM + FX(A + I*H)
      TRAP = (FX(A) + FX(B) + 2.*SUM)*(H/2.)
      RETURN
      END
```

The main calling program and function routines are given below.

```
      EXTERNAL FX1, FX2
      APPROX = TRAP(0.,.4,5,FX1) + TRAP(.1,.3,5,FX2)
      PRINT, APPROX
      STOP
      END
      FUNCTION FX1(X)
      FX1 = (SIN X)**2
      RETURN
      END
      FUNCTION FX2(X)
      FX2 = (COS X)**2
      RETURN
      END
```

Notice that the main program uses an EXTERNAL statement. This declaration must be used in every calling program which passes the name of a subprogram or built-in function to another subprogram. Also remember that the main program is entered first. After the END statement in the main program, all subroutines may be entered.

In this program, it is up to the user to define the accuracy of the integration. In other words, the user chooses the value of N, which determines the number of iterations to be done.

● **PROBLEM** 19-30

Develop a FORTRAN subprogram to evaluate the integration of f(x)dx between the limits of A and B using GAUSSIAN quadrature, which is expressible as

$$\int_A^B f(x)\,dx = \frac{B-A}{2} \sum_{i=1}^{N} w_i f\left[\frac{(B-A)t_i + (B+A)}{2}\right]$$

where $w_1, w_2, \ldots, w_N$ are the weighting coefficients and $t_1, t_2, \ldots, t_N$ are the roots of the Legendre polynomial $P_N(t) = 0$

Solution: One needs the result from elementary numerical analysis that the weights w_K can be expressed in the form

$$w_K = \int_{-1}^{1} L_K(x)\,dx = \frac{1}{P_{n+1}(x_K)} \int_{-1}^{1} \frac{P_{n+1}(x)\,dx}{x - x_K}$$

where $L_K(x)$ is a Lagrange polynomial and $P_n(x)$ is a Legendre polynomial.

The program will let the value of N range from 3 to 6. The computation starts with N = 3. The program will compare the result based on N = 3 with that based on N = 4. The results must satisfy the criterion

$$\varepsilon \geq \frac{A_{n+1} - A_n}{A_n}$$

where A_{n+1} is the answer based on N + 1 points, and A_n is the answer based on N points, and ε is taken to be 10^{-4}. If the result fails to pass the above test, the value of N will be increased by one. The maximum value of N is set to be 6.

```
      SUBROUTINE GAUSS (A, B, X, F, KOUNT)
C     INTEGRATION OF F(X). DX BY GAUSSIAN QUADRATURE
C     BETWEEN LIMITS OF A AND B
C     NOMENCLATURE
C     F = F(X) UNDER INTEGRAL SIGN
C     KOUNT = INTEGER USED TO CONTROL METHOD
C     OF EXECUTION
C     ANS = ANSWER TO INTEGRATION
      DIMENSION W(4,6), T(4,6)
      IF (KOUNT) 8, 10, 8
    8 GO TO 30
   10 ANS = 1.0
      IPOINT = 3
      EPS = 10.E - 05
C     STORE WEIGHT COEFF AND LEGENDRE ROOTS
      DO 2 I = 1,4
      IP2 = I + 2
    2 READ 1, (W(I,K), K = 1,IP2), (T(I,K), K = I,IP2)
    1 FORMAT (5F15.10)
C     CHANGE INTEGRATION LIMITS TO (-1 TO +1).
C     EVALUATE COEFF OF NEW FUNCTION
      C = (B - A)/2.
   18 IPOINT = IPOINT + 1
      TEMP = ANS
      ANS = 0
      IPM2 = IPOINT - 2
      KOUNT = 1
C     EVALUATE NEW VARIABLES WHICH ARE EXPRESSED
C     IN TERMS OF LEGENDRE ROOT.
   20 X = C*T(IPM2,KOUNT) + (B + A)/2.
      RETURN
C     CARRY OUT INTEGRATION BY CALCULATING
C     ANS = C*(W1*F(X1) + W2*F(X2) + ...).
   30 ANS = ANS + C*W(IPM2,KOUNT)* F
      KOUNT = KOUNT + 1
      IF (KOUNT - IPOINT) 20,20,40
   40 IF (IPOINT - 3) 18, 18, 50
C     NEXT DETERMINE WHETHER THE DEVIATION
C     OF ANSWER IS WITHIN THE LIMIT
C     IF NOT, TAKE 1 MORE INTEGRATION POINT
   50 DELT = ABSF(ANS - TEMP)
      RATIO = DELT/TEMP
    7 IF (RATIO - EPS) 70,70,80
C     PUNCH OUT ANSWER IF DEVIATION WITHIN LIMIT
   70 PUNCH 72, IPOINT, ANS
   72 FORMAT (///5X, 16H BY CONVERGENCE, , I2, 24H
     1 POINT GAUSS. QUADRATURE, 15H GIVES ANSWER = ,
     2 E14.8//)
      KOUNT = 7
      RETURN
   80 IF (IPOINT - 6) 18, 100, 100
  100 PUNCH 102, IPOINT, ANS
C     ANSWER PUNCHED OUT AFTER 6 POINT
```

```
C         GAUSSIAN INTEGRATION IS STILL NOT
C         WITHIN THE LIMIT.
  102     FORMAT (///5X, 22H BY LIMITS OF PROGRAM, , I2,
         1 14H POINT GAUSS., 23H QUADRATURE GIVES ANS. =,
         2 E14.8//)
          KOUNT = 7
          RETURN
C         DATA FOR WEIGHTING COEFFICIENTS
C         AND LEGENDRE ROOTS
          END
```

• **PROBLEM** 19-31

Using power series, find an approximate solution to the definite integral

$$\int_0^1 \sin x \, dx$$

Write a FORTRAN program segment.

Solution: First, one must remember that

$$\sin x = x - \frac{x^3}{3!} + \frac{x^5}{5!} - \frac{x^7}{7!} \ldots$$

Then, write the integral like this:

$$\int_0^1 \sin x \, dx \approx \int_0^1 (x - \frac{x^3}{3!} + \frac{x^5}{5!} - \frac{x^7}{7!}) \, dx,$$

assuming that it is necessary to compute only 4 terms, the generalized expansion would be

$$\int_0^1 (x - \frac{x^3}{3!} + \frac{x^5}{5!} - \frac{x^7}{7!} \ldots - \frac{x^{2n+1}}{(2n+1)!} + \ldots) \, dx.$$

When integration is performed, one gets the generalized primitives as

$$\left[\frac{x^2}{2!} - \frac{x^4}{4(3!)} + \frac{x^6}{6(5!)} - \frac{x^8}{8(7!)} \ldots \frac{x^{2n+2}}{(2n+2)((2n+1)!)} \right]_0^1$$

One can make use of the FACT(N) function which finds N! to calculate the integral. Notice that the accuracy parameter ERROR has not been defined. You may choose that value and insert it at the start of the program. Since the lower boundary of this integral is zero, you need only compute the terms for x = 1.

```
          SINDX = (X**2)/2.0
          I = 1
  20      IF (ABS(TERM).LT.ERROR) GO TO 50
          N = FLOAT (I)
```

```
          SINDX = SINDX + TERM
          TERM = -1.0*(X**(2*N + 2)/(2*N+2)*(FACT(2*N+1)))
C         IF I IS ODD, TERM IS NEGATIVE.  IF I IS
C         EVEN, TERM IS POSITIVE.
          IF (2*(I/2).EQ.I) ABS(TERM) = TERM
          I = I + 1
          GO TO 20
   50     WRITE (S,*) SINDX
```

• **PROBLEM 19-32**

Evaluate the integral by the Trapezoidal method

$$\int_0^1 (6 - 6x^5)\,dx = \left[6x - x^6\right]_0^1 = (6 - 1) = 5,$$

which is shown in Figure 1. Write a FORTRAN program to evaluate the integral of the function $f(x) = 6 - 6x^5$ using the Trapezoidal method.

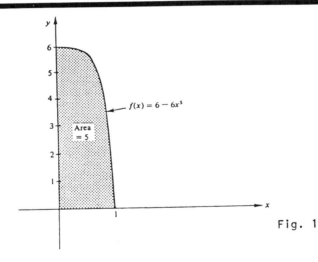

Fig. 1

Solution: The program will use 1, 2, 4, 8, 16, ..., 16348 trapezoids so that one may compare results for different numbers of trapezoids. For each trapezoid, the program calculates XL and XR, the left and right boundaries repectively, and the area A of that trapezoid. This area is then added into the total AREA at the end of the inner DO loop, and the total is printed after all the trapezoids are counted.

Use the following program

```
          F(X) = 6. - 6. * X ** 5
          WRITE (3, 10)
   10     FORMAT ('TRAPEZOIDS', 6X, 'AREA', 12X, 'ERROR')
          DO 50 I = 1, 15
          NODIV = 2 ** (I - 1)
          AREA = 0.
          WIDTH = 1./NODIV
```

```
          DO 40 J = 1, NODIV
          XL = (J - 1) * WIDTH
          XR = J * WIDTH
          A = (WIDTH/2.) * (F(XL) + F(XR))
   40     AREA = AREA + A
          ERROR = 5. - AREA
   50     WRITE (3, 60) NODIV, AREA, ERROR
   60     FORMAT (I8, F16.8, E19.8)
          CALL EXIT
          END
```
and get the following results:

TRAPEZOIDS	AREA	ERROR
1	3.00000048	0.20000004E 01
2	4.40625096	0.59375012E 00
4	4.84570408	0.15429690E 00
8	4.96106053	0.38940437E-01
16	4.99023820	0.97627658E-02
32	4.99755193	0.24490361E-02
64	4.99937249	0.62847149E-03
128	4.99981881	0.18215182E-03
256	4.99990082	0.10013581E-03
512	4.99985410	0.14686587E-03
1024	4.99971677	0.28419500E-03
2048	4.99947263	0.52833568E-03
4096	4.99893094	0.10700228E-02
8192	4.99783326	0.21677021E-02
16384	4.99563695	0.43640146E-02

Looking over these results note that the error seems to be decreasing until 256 trapezoids are reached, and then again increases. To see the effect of roundoff error on these results, repeat the program using double-precision arithmetic. (These results were obtained on an IBM 1130 computer in so-called "extended precision" which, on that machine, is somewhat less than double-precision.)

TRAPEZOIDS	AREA	ERROR
1	3.00000000	0.20000000E 01
2	4.40625000	0.59375000E 00
4	4.84570312	0.15429687E 00
8	4.96105957	0.38940429E-01
16	4.99024200	0.97579956E-02
32	4.99755904	0.24409629E-02
64	4.99938962	0.61037764E-03
128	4.99984729	0.15271455E-03
256	4.99996161	0.38392841E-04
512	4.99998995	0.10050833E-04
1024	4.99999659	0.34086406E-05
2048	4.99999732	0.26822090E-05
4096	4.99999565	0.43511390E-05
8192	4.99999164	0.83558261E-05
16384	4.99998331	0.16693025E-04

This time a minimum error is reached at about 2048 trapezoids, and this minimum is significantly lower than the error using standard single-precision arithmetic.

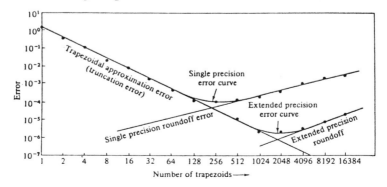

Error curve.   Fig. 2

To understand the meaning of roundoff error in this calculation plot the error against the number of trapezoids as in Figure 2, for both standard and extended precision. In each case the error for both precisions is the same for small numbers of trapezoids, and would in fact decrease linearly (on the logarithmic plot of Figure 2) with additional trapezoids were it not for roundoff error. The standard precision roundoff error is larger, and therefore starts increasing the error near 256 trapezoids, whereas the extended precision calculations can use more trapezoids before roundoff error becomes a problem; hence the minimum is lower and farther to the right.

● **PROBLEM** 19-33

Write a program that uses the Corrected Trapezoid Rule to find the value of

$$\int_0^1 e^{-x^2} dx$$

correct to six digits after the decimal point. Compute the approximation for N = 10, 11, ..., 14, 15.

Solution: FORTRAN PROGRAM

```
C     CORRECTED TRAPEZOID RULE
      INTEGER I,N
      REAL A,B,CORTRP,H,TRAP
      F(X) = EXP (-X*X)
      FPRIME (X) = -2.*X*F(X)
      DATA A,B /0., 1. /
      PRINT 600
  600 FORMAT (9X,'N',7X,'TRAPEZOID SUM',7X,
     *' CORR. TRAP.SUM')
      DO 10 N = 10,15
         H = (B - A)/FLOAT(N)
         TRAP = (F(A) + F(B))/2.
         DO 1 I=1,N-1
    1       TRAP = TRAP + F(A + FLOAT(I)*H)
         TRAP = H*TRAP
         CORTRP = TRAP + H*H*(FPRIME(A) - FPRIME(B))/12.
   10    PRINT 610, N,TRAP,CORTRP
  610 FORMAT(I10,2E20.7)
      STOP
      END
```

Single precision output

N	TRAPEZOID SUM	CORR.TRAP.SUM
10	0.7462108E 00	0.7468239E 00
11	0.7463173E 00	0.7468240E 00
12	0.7463983E 00	0.7468240E 00
13	0.7464612E 00	0.7468240E 00
14	0.7465112E 00	0.7468240E 00
15	0.7465516E 00	0.7468241E 00

Double precision output

N	TRAPEZOID SUM	CORR.TRAP.SUM
10	7.4621080E-01	7.4682393E-01
11	7.4631727E-01	7.4682399E-01
12	7.4639825E-01	7.4682403E-01
13	7.4646126E-01	7.4682406E-01
14	7.4651126E-01	7.4682408E-01
15	7.4655159E-01	7.4682409E-01

● **PROBLEM 19-34**

Consider the integral

$$\int_{-1}^{+1} \frac{1}{\sqrt{2\pi}} e^{-\frac{1}{2}x^2} dx = \frac{1}{\sqrt{2\pi}} \int_{-1}^{+1} e^{-\frac{1}{2}x^2} dx,$$

which cannot be solved by ordinary calculus. Since the integrand is an exponential, use a Maclaurin series for e^a and solve. Prepare programs using Integration of Power Series, Integrations by the Trapezoidal Rule, Romberg Integration, Integration by Simpson's Rule and Integration by Gauss quadrature.

Solution: Since

$$e^a = 1 + a + \frac{a^2}{2!} + \frac{a^3}{3!} + \frac{a^4}{4!} + \frac{a^5}{5!} + \frac{a^6}{6!} + \frac{a^7}{7!} + \ldots,$$

let $a = -\frac{1}{2}x^2$ and write the series as

$$e^{-\frac{1}{2}x^2} = 1 - \frac{x^2}{2} + \frac{x^4}{2^2 \cdot 2!} - \frac{x^6}{2^3 \cdot 3!} + \frac{x^8}{2^4 \cdot 4!} - \frac{x^{10}}{2^5 \cdot 5!} + \frac{x^{12}}{2^6 \cdot 6!}$$

$$- \frac{x^{14}}{2^7 \cdot 7!} + \ldots$$

$$= 1 - \frac{x^2}{2} + \frac{x^4}{8} - \frac{x^6}{48} + \frac{x^8}{384} - \frac{x^{10}}{3840} + \frac{x^{12}}{46080} - \frac{x^{14}}{645120} + \ldots$$

The integral is now

$$\frac{1}{\sqrt{2\pi}} \int_{-1}^{+1} e^{-\frac{1}{2}x^2} dx = \frac{1}{\sqrt{2\pi}} \int_{-1}^{+1} \left\{ 1 - \frac{x^2}{2} + \frac{x^4}{8} - \frac{x^6}{48} + \frac{x^8}{384} - \frac{x^{10}}{3840} \right.$$

$$\left. + \frac{x^{12}}{46080} - \frac{x^{14}}{645120} + \ldots \right\} dx$$

Since the series converges uniformly to the exponential in the range $-1 \leq x \leq +1$ integrate the series term by term, to get

$$\frac{1}{\sqrt{2\pi}} \int_{-1}^{+1} e^{-\frac{1}{2}x^2} dx$$

$$= \frac{1}{\sqrt{2\pi}} \left[\int_{-1}^{+1} 1\, dx - \int_{-1}^{+1} \frac{x^2}{2} dx + \int_{-1}^{+1} \frac{x^4}{8} dx - \int_{-1}^{+1} \frac{x^6}{48} dx \right.$$

$$+ \int_{-1}^{+1} \frac{x^8}{384} dx - \int_{-1}^{+1} \frac{x^{10}}{3840} dx + \int_{-1}^{+1} \frac{x^{12}}{46080} dx$$

$$\left. - \int_{-1}^{+1} \frac{x^{14}}{645120} dx + \ldots \right].$$

Each of these integrals is easy to integrate, except that there is an infinite number of them. If one starts to integrate the first few terms, one gets

$$\frac{1}{\sqrt{2\pi}} \int_{-1}^{+1} e^{-\frac{1}{2}x^2} dx$$

$$= \frac{1}{\sqrt{2\pi}} \left\{ \left[x \right]_{-1}^{+1} - \left[\frac{x^3}{6} \right]_{-1}^{+1} + \left[\frac{x^5}{40} \right]_{-1}^{+1} - \left[\frac{x^7}{336} \right]_{-1}^{+1} + \left[\frac{x^9}{3456} \right]_{-1}^{+1} \right.$$

$$\left. - \left[\frac{x^{11}}{42240} \right]_{-1}^{+1} + \left[\frac{x^{13}}{599040} \right]_{-1}^{+1} - \left[\frac{x^{15}}{9676800} \right]_{-1}^{+1} + \ldots \right\}.$$

Finally, substituting the limits into each bracket one gets

$$\frac{1}{\sqrt{2\pi}} \int_{-1}^{+1} e^{-\frac{1}{2}x^2} dx = \frac{2}{\sqrt{2\pi}} \left[1 - \frac{1}{6} + \frac{1}{40} - \frac{1}{336} + \frac{1}{3456} - \frac{1}{42240} \right.$$

$$\left. + \frac{1}{599040} - \frac{1}{9676800} + \ldots \right].$$

With some analysis, it can be seen that this series can be written in the compact form

$$\frac{1}{\sqrt{2\pi}} \int_{-1}^{+1} e^{-\frac{1}{2}x^2} dx = \frac{2}{\sqrt{2\pi}} \sum_{n=0}^{\infty} \frac{(-1)^n}{2^n \cdot n! \cdot (2n+1)}.$$

Since this is an alternating series which obviously converges, the computer can find the sum.

To permit us to compare the five methods of integration, the following program sums the above series for the first eight terms (accurate to at least 10^{-7} since it is an alternating series and the first term neglected is smaller than 10^{-7}), and also performs the integration by each of the other four methods. Since definite knowledge of the series truncation error is known, integration of the series is possibly the safest method to use in this case.

```
      DIMENSION T(13, 13)
      EFUNC (XI) = EXP(-XI * XI / 2.) / SQRT (2. * PI)
      PI = 3.1415926
C     AREA BY INTEGRATION OF THE SERIES
      WRITE (3, 10)
   10 FORMAT ('INTEGRATION OF THE POWER SERIES', /)
      AREA = 2.*(1. -1./6. + 1./40. - 1./336. + 1./3456.
     1   -1./42240. + 1./599040.-1./9676800.)/SQRT(2.*PI)
      WRITE (3, 20) AREA
   20 FORMAT (20X, F10.7)
C AREA BY THE TRAPEZOIDAL RULE
      WRITE (3, 30)
   30 FORMAT ('0INTEGRATION BY TRAPEZOIDAL RULE', /)
      DO 50 I = 1, 13
      NODIV = 2 ** I / 2
      AREA = 0.
      WIDTH = 2. / NODIV
      DO 40 J = 1, NODIV
      XL = -1. + (J - 1) * WIDTH
      XR = -1. + J * WIDTH
      A = (WIDTH / 2.) * (EFUNC (XL) + EFUNC (XR))
   40 AREA = AREA + A
      T(I, 1) = AREA
   50 WRITE (3, 60) NODIV, AREA
   60 FORMAT ('WITH', I6, 'TRAPEZOIDS, THE AREA IS', F11.7)
C     CALCULATE ROMBERG INTEGRATION TERMS THROUGH T-SUB-16,
C     USING THE TRAPEZOIDAL AREAS STORED IN THE ARRAY T
      WRITE (3, 61)
   61 FORMAT ('0ROMBERG INTEGRATION', /)
      DO 62 J = 2, 5
      DO 62 I = J, 5

   62 T(I, J) = (4.**(J-1) * T(I,J-1) - T(I-1,J-1))/
     1          (4.**(J-1) - 1)
      DO 65 I = 1, 5
   65 WRITE (3, 67) (T(I, J), J = 1, I)
   67 FORMAT (5F12.7)
C AREA BY SIMPSON'S RULE
      WRITE (3, 70)
   70 FORMAT ('0INTEGRATION BY SIMPSON''S RULE', /)
      DO 90 I = 1, 12
      NODIV = 2** I
      NOSIM = NODIV / 2
      AREA = 0.
      WIDTH = 2. / NODIV
      DO 80 J = 1, NOSIM
      XL = -1. + (J-1) * 2. * WIDTH
      XR = -1. + J * 2. * WIDTH
      XM = (XL + XR) / 2.
      A = (WIDTH/3.) * (EFUNC (XL) + 4. * EFUNC (XM)
     1    + EFUNC (XR))
   80 AREA = AREA + A
   90 WRITE (3, 100) NODIV, AREA
  100 FORMAT ('WITH', I6, 'DIVISIONS, THE AREA IS', F11.7)
C AREA BY GAUSS QUADRATURE
```

```
      WRITE (3, 110)
110   FORMAT ('0INTEGRATION BY GAUSS QUADRATURE', /)
      I = 2
      AREA = EFUNC (-0.57735027) + EFUNC (+0.57735027)
      WRITE (3, 120) I, AREA
120   FORMAT ('WITH', I3, 'POINTS, THE AREA IS', F11.7)
      I = 3
      AREA = 0.555555 * EFUNC (-.77459667)
     1      + 0.8888889 * EFUNC (0.0)
     2      + 0.5555555 * EFUNC (+.77459667)
      WRITE (3, 120) I, AREA
      I = 4
      AREA = 0.34785485 * EFUNC (-.86113631)
     1      + 0.65214515 * EFUNC (-.33998104)
     2      + 0.65214515 * EFUNC (+.33998104)
     3      + 0.34785485 * EFUNC (+.86113631)
      WRITE (3, 120) I, AREA
      I = 5
      AREA = 0.23692689 * EFUNC (-.90617985)
     1      + 0.47862867 * EFUNC (-.53846931)
     2      + 0.56888889 * EFUNC (0.0)
     3      + 0.47862867 * EFUNC (+.53846931)
     4      + 0.23692689 * EFUNC (+.90617985)
      WRITE (3, 120) I, AREA
      I = 6
      AREA = 0.17132449 * EFUNC (-.93246951)
     1      + 0.36076157 * EFUNC (-.66120939)
     2      + 0.46791393 * EFUNC (-.23861919)
     3      + 0.46791393 * EFUNC (+.23861919)
     4      + 0.36076157 * EFUNC (+.66120939)
     5      + 0.17132449 * EFUNC (+.93246951)
      WRITE (3, 120) I, AREA
      CALL EXIT
      END
```

Using extended precision, the program gives the following results:

```
INTEGRATION OF THE POWER SERIES

              0.6826894

INTEGRATION BY TRAPEZOIDAL RULE

WITH     1 TRAPEZOIDS, THE AREA IS   0.4839414
WITH     2 TRAPEZOIDS, THE AREA IS   0.6409130
WITH     4 TRAPEZOIDS, THE AREA IS   0.6725218
WITH     8 TRAPEZOIDS, THE AREA IS   0.6801636
WITH    16 TRAPEZOIDS, THE AREA IS   0.6820590
WITH    32 TRAPEZOIDS, THE AREA IS   0.6825319
WITH    64 TRAPEZOIDS, THE AREA IS   0.6826501
WITH   128 TRAPEZOIDS, THE AREA IS   0.6826796
WITH   256 TRAPEZOIDS, THE AREA IS   0.6826870
WITH   512 TRAPEZOIDS, THE AREA IS   0.6826888
WITH  1024 TRAPEZOIDS, THE AREA IS   0.6826892
WITH  2048 TRAPEZOIDS, THE AREA IS   0.6826891
WITH  4096 TRAPEZOIDS, THE AREA IS   0.6826889


ROMBERG INTEGRATION

0.4839414
0.6409130   0.6932368
0.6725218   0.6830581   0.6823795
0.6801636   0.6827109   0.6826878   0.6826927
0.6820590   0.6826908   0.6826894   0.6826894   0.6826894
```

```
INTEGRATION BY SIMPSON'S RULE

WITH     2 DIVISIONS, THE AREA IS    0.6932368
WITH     4 DIVISIONS, THE AREA IS    0.6830581
WITH     8 DIVISIONS, THE AREA IS    0.6827109
WITH    16 DIVISIONS, THE AREA IS    0.6826908
WITH    32 DIVISIONS, THE AREA IS    0.6826895
WITH    64 DIVISIONS, THE AREA IS    0.6826895
WITH   128 DIVISIONS, THE AREA IS    0.6826894
WITH   256 DIVISIONS, THE AREA IS    0.6826894
WITH   512 DIVISIONS, THE AREA IS    0.6826894
WITH  1024 DIVISIONS, THE AREA IS    0.6826894
WITH  2048 DIVISIONS, THE AREA IS    0.6826893
WITH  4096 DIVISIONS, THE AREA IS    0.6826892

INTEGRATION BY GAUSS QUADRATURE

WITH  2 POINTS, THE AREA IS    0.6753947
WITH  3 POINTS, THE AREA IS    0.6829970
WITH  4 POINTS, THE AREA IS    0.6826798
WITH  5 POINTS, THE AREA IS    0.6826897
WITH  6 POINTS, THE AREA IS    0.6826894
```

A comparison of the results is interesting. Of all the methods the fastest (in computer time) and most reliable is the summation of the series, since it is an alternating series and one knows what its maximum error will be; for least roundoff error, the series should probably be summed backward, however. The trapezoidal rule gives a very close answer with 1024 trapezoids, but roundoff error prevents getting any closer. Romberg integration provides the right answer to seven decimal places on the last line, which represents $T_{16}^{(2)}$ through $T_{16}^{(4)}$. Here, only 16 trapezoids followed by repeated application of the Romberg method is sufficient to give the right answer, with very little work.

Simpson's method provides 7-place accuracy if the right number of divisions is picked, but gives a slight error if too many are picked. Gauss quadrature gives the exact answer with only 6 points and probably represents the least amount of actual computation; the trouble with it is that the error analysis is difficult, and it is difficult to recognize the right answer when one gets it.

SOLUTIONS OF DIFFERENTIAL EQUATIONS

● **PROBLEM** 19-35

Consider the following differential equation with the initial condition $y(0) = 1$:

$$\frac{dy}{dx} = y^2 - x^2 \qquad (1)$$

Develop a FORTRAN program to get a solution for $y = y(x)$ in the interval $0 \le x \le .5$ applying the increment method. Let $dx = \Delta x = 0.05$, and output a table containing the following quantities:

$$x, \ y, \ y^2, \ x^2, \ y^2 - x^2, \ dy.$$

Solution: First develop the problem from a mathematical viewpoint. Before one begins solving the equation one must know that a solution exists. Furthermore, this solution should be unique. The existence and uniqueness conditions for the solution of a given differential equation are contained in the following theorems:

1) Existence:
    Let $y' = \phi(x,y)$. $0 \leq x \leq 1$, $-\infty < y < \infty$, $y(0) = c$

If ϕ is continuous and bounded, then $y = y(x)$ is a solution of $\frac{dy}{dx} = \phi(x,y)$.

2) Uniqueness:
    If there exists a constant A (Lipschitz number) such that

$$|\phi(x,y_2) - \phi(x,y_1)| \leq A |Y_2 - Y_1|$$

(where (x,y_1), (x,y_2) are in the domain), then $y = y(x)$ is a unique solution to

$$y' = \phi(x,y).$$

The given differential equation is

$$\phi(x,y) = \frac{dy}{dx} = y^2 - x^2 \quad 0 \leq x \leq 0.5, \; y(0) = 1$$

$\phi(x,y)$ is a difference of continuous functions; hence it is continuous. Clearly $|y^2 - x^2|$ is bounded (i.e., less than some constant K) in $[0,0.5]$. Thus a solution exists.

Let $A = .3$. Then $|\phi(x,y_2) - \phi(x,y_1)| \leq .3(Y_2 - Y_1)$ for all (x,y_1), (x,y_2) in the domain. Thus the solution is unique. Rewrite equation (1) in the differential form

$$dy = (y^2 - x^2)dx \tag{2}$$

The increment method entails the use of differentials to find approximate values of y. Thus, let $\Delta x = dx$, $\Delta y = dy$. Since the initial condition is $y(0) = 1$, substitute into equation (2) and obtain

$$\Delta y = (1^2 - 0^2)(0.05)$$

$$\Delta y = 0.05$$

thus, at $x = 0.05$, $\dot{y} = y_0 + \Delta y$ becomes $1 + 0.05 = 1.05$. This means that now $y(1.00)$ becomes 1.05. The next value of Δy becomes, therefore

$$\Delta y = ((1.05)^2 - (0.05)^2)(0.05)$$

$$\Delta y = 0.055$$

As can be seen, the method readily lends itself to computer implementation. One can use a FORTRAN main program to print the values in tabular form. A subroutine, INCR, can do each calculation and pass the values back to the main program for output.

```
            Y = 1.000
            DX = 0.05
      C     WRITE TABLE HEADINGS
            WRITE (5,100)
        100 FORMAT (2X, 'X', 5X, 'Y', 6X, 'Y**2', 3X,
           1'X**2', 3X, 'Y**2-X**2', 3X, 'DY')
      C     DO FOR X FROM 1 TO 10 BY 0.05
            DO 10 I = 1,10
            X = (FLOAT (I) - 1.0)*0.05
            CALL INCR (DX, DY, X, Y)
            XX = X**2
            YY = Y**2
            DYDX = YY - XX
            WRITE (5,101) X, Y, YY, XX, DYDX, DY
        101 FORMAT (1X, F4.2,3(2X,F5.3), 4X, F5.3, 5X, F5.3)

         10 CONTINUE

      C     END DO - FOR
            STOP
            END
      C     SUBROUTINE INCR TO PERFORM CALCULATIONS
            SUBROUTINE INCR (DX, DY, X, Y)
            Y = Y + DY
            DY = ((Y**2) - (X**2))*DX
            RETURN
            END
```

● **PROBLEM** 19-36

Use the Euler method to solve the equation of motion of a damped harmonic oscillator. Let the ititial conditions be that at t = 0, x(0) = 10 cm and dx/dt = v = 0.

Write program in BASIC for the solution of this problem

Solution: The equation of motion is given by

$$\frac{md^2x}{dt^2} = -\frac{cdx}{dt} - kx$$

where $\frac{-cdx}{dt}$ is the damping force and $-kx$ is the restoring force. Since Euler's method is based on first order differential equations, rewrite the equation of motion as two first order equations:

$$\frac{dx}{dt} = v \qquad (1)$$

$$\frac{dv}{dt} = -\frac{c}{m}v - \frac{k}{m}x \qquad (2)$$

According to Euler's method, the solution is given by

$$dx \approx \Delta x = x_{new} - x_{old}; \quad dx = vdt$$

$$x_{new} = x_{old} + v_{old} \Delta t \qquad (1')$$

$$v_{new} = v_{old} + \left(\frac{-c}{m} v_{old} - \frac{k}{m} x_{old}\right) \Delta t \qquad (2')$$

Here one has used the general result that if $dy/dx = f(x,y)$, then

$$y_{i+1} = y_i + y'_i \Delta x, \text{ where } y'_i = f(x_i, y_i).$$

For the sake of concreteness, use the following data:

$$m = k = 10, \; c = 2, \; x_0 = 10, \; v_0 = 0, \; D \equiv \Delta t = 0.1$$

and follow the motion for 2 seconds.

```
100  REM DAMPED HARMONIC OSCILLATOR
101  REM EULER'S METHOD
105  PRINT "TIME", "VELOCITY", "POSITION"
106  PRINT
110  READ M, K, C, X0, V0, D
120  LET X1 = X0
121  LET V1 = V0
130  FOR T = 0 TO 2 STEP D
140  PRINT T, V1, X1
150  LET X2 = X1 + V1*D
151  LET V2 = V1 + (-C*V1/M - K*X1/M)*D
160  LET X1 = X2
161  LET V1 = V2
162  NEXT T
800  DATA 10, 10, 2, 10, 0, 0.1
999  END
```

Note that the exact solution of $t = 2$ gives $x = -2.5807$, compared with the Euler method result of -2.96507. More sophisticated numerical methods (such as the improved Euler method or Runge-Kutta type methods) are necessary for better accuracy.

Finally, v_{new} can also be solved with the quadratic equation. If one rearranges the equation of motion, one can obtain

$$m \frac{d^2 x}{dt^2} + c \frac{dx}{dt} + kx = 0,$$

which is the quadratic form. This equation has a solution of the form $x = e^{\phi t}$. If one substitutes x into the first equation and divides by $e^{\phi t}$, one gets the simpler form $m\phi^2 + c\phi + k = 0$. By using the discriminant $c^2 - 4mk$, one can determine what type of motion is present:

If $c^2 - 4mk$ is	the motion is
positive	over-damped
zero	critically-damped
negative	oscillatory-damped

● **PROBLEM** 19-37

Solve the heat flow equation $K \frac{\partial T}{\partial t} = \frac{\partial^2 T}{\partial x^2}$, where K is the thermal diffusivity of the homogeneous medium. Apply the algorithm to the case of a copper bar of thermal diffusivity of $1.14 \text{cm}^2/\text{sec}$ with 1 end maintained at $100°C$ while the other end is held at $0°C$. Let the bar be 8cm long. Write a program in BASIC that solves the heat transfer problem.

Solution: The solution can be derived from a central finite difference approximation for the spatial derivative and a forward difference approximation for the time derivative. Thus make use of

$$\frac{\partial^2 T}{\partial x^2} = \frac{T_{i-1,j} - 2T_{i,j} + T_{i+1,j}}{\Delta x^2}$$

$$\frac{\partial T}{\partial t} = \frac{T_{i,j+1} - T_{i,j}}{\Delta t}$$

(These results are derive from Taylor's formula.) So, the heat flow equation becomes

$$\frac{T_{i-1,j} - 2T_{i,j} + T_{i+1,j}}{\Delta x^2} = K \left(\frac{T_{i,j+1} - T_{i,j}}{\Delta t} \right)$$

which can be solved for $T_{i,j+1}$ to give

$$T_{i,j+1} = \alpha (T_{i-1,j} + T_{i+1,j}) + (1 - 2\alpha) T_{i,j}$$

where $\alpha = \Delta t / (K \Delta x^2)$.

This last equation is self-starting in that if the boundary conditions on the space-time grid are known, that is, if the values of the function are known for all positions at $t = 0$, and for all times at $x = 0$ and $x = L$, then the remainder of the solution can be produced from the algorithm.

An illustrative program which takes $\Delta x = 1 \text{cm}$ and $\Delta t = 0.4$ sec is given below. Note that the dummy variable V_i is used to handle intermediate results for T. This process allows one to avoid storing values of T in a large two-dimensional array.

```
100   REM PARABOLIC PARTIAL DIFFERENTIAL EQUATION
101   REM HEAT TRANSFER PROBLEM
102   PRINT
110   READ F0, D, X1, K, T2
120   LET F(1) = F0
121   LET F(9) = 0
130   LET A = D*K/X1↑2
140   FOR I = 2 TO 8
150   LET F[I] = 0
151   NEXT I
160   LET C = 5
```

```
170  FOR T = 0 TO T2 STEP D
180  IF C <5 THEN 220
190  PRINT "AT"; T; "SEC, THE TEMP DISTRIBUTION IS"
191  PRINT
200  FOR I = 1 TO 9
201  PRINT F [I],
202  NEXT I
203  PRINT
210  LET C = 0
220  FOR I = 2 TO 8
230  LET V [I] = A*(F(I - 1) + F(I + 1)) + (1 - 2*A)*F(I)
231  NEXT I
240  FOR I = 2 TO 8
250  LET F [I] = V [I]
251  NEXT I
260  LET C = C + 1
261  NEXT T
800  DATA 100, 0.2, 1, 1.14, 5
999  END
```

• **PROBLEM** 19-38

Construct a flowchart and write a FORTRAN program to compute the solution of the parabolic equation

$$k \frac{\partial^2 f}{\partial x^2} - \frac{\partial f}{\partial t} = 0$$

by the Bender-Schmidt recurrence equation

$$f_{i,j+1} = \frac{1}{2}(f_{i+1,j} + f_{i-1,j}),$$

with boundary conditions

$$f_1 = A, \quad f_{n+1} = B,$$

and the initial condition

$$f_{i,0} = g_{0i} \quad (i = 2, n).$$

The correspondence between variables is

$$f_{i,j+1} \rightarrow F1(I)$$
$$f_{i,j} \rightarrow F0(I)$$
$$g_{0i} \rightarrow G0(1)$$

The process continues until the number of steps NSTEP becomes greater than NLAST.

Solution: Note since only two values of i appear at each step of the integration process, the introduction of the variables F0(I), F1(I) allows them to be simply-subscripted rather than doubly-subscripted as $f_{i,j}$.

FLOW CHART

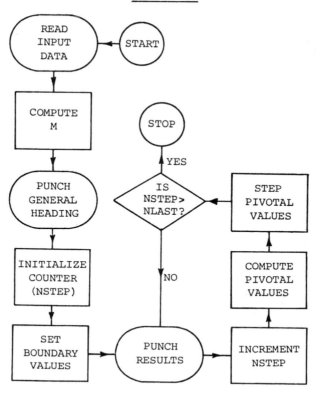

Two variables with A-type format are used to introduce
parentheses in the output of the program. The variable
LEFT contains the two-character field, F(, while the
variable RIGHT contains the one-character ,). Thus the
instruction

$$\text{PUNCH } n_1, \text{ LEFT, I, RIGHT}$$

causes the output of F(I).

Notice that it is possible to use either fixed-point
variables (like LEFT) or floating-point variables (like RIGHT)
with the A-type format. It is sometimes necessary to distinguish between these two types of variables when using a computer with variable word length, since the amount of storage
assigned to a fixed-point variable may differ from that assigned to a floating-point variable.

```
C       PROGRAM
C       SOLUTION OF PARABOLIC DIFFERENTIAL EQUATION
C
        DIMENSION F0(25), F1(25)
        READ 999, N, NLAST, LEFT, RIGHT, A, B, (F0(I),
       1I = 2,N)
        M = N + 1
        PUNCH 997, (LEFT, I, RIGHT, I = 1,M)
        NSTEP = 0
        F0(1) = A
        F0(M) = B
```

```
    1   PUNCH 998, NSTEP, (F0(I), I = 1,M)
        NSTEP = NSTEP + 1
        DO 10 I = 2,N
   10   F1(I) = (F0(I + 1) + F0(I - 1))/2.
        DO 20 I = 2,N
   20   F0(I) = F1(I)
        IF(NSTEP - NLAST) 1,1,30
   30   STOP
C
  997   FORMAT ( // 24X, 24HRESULTS FROM PROGRAM //
       $          11X, 1HN 5X, 6(A2, I1, A1, 6X))
  998   FORMAT (7X, I5, 6F10.4)
  999   FORMAT (2I5, A2, A1/(8E10.0))
        END
```

RESULTS FROM PROGRAM

N	F(1)	F(2)	F(3)	F(4)	F(5)	F(6)
0	-10.0000	-10.0000	-10.0000	-10.0000	-10.0000	10.0000
1	-10.0000	-10.0000	-10.0000	-10.0000	0.0000	10.0000
2	-10.0000	-10.0000	-10.0000	-5.0000	0.0000	10.0000
3	-10.0000	-10.0000	-7.5000	-5.0000	2.5000	10.0000
4	-10.0000	-8.7500	-7.5000	-2.5000	2.5000	10.0000
5	-10.0000	-8.7500	-5.6250	-2.5000	3.7500	10.0000
6	-10.0000	-7.8125	-5.6250	-.9375	3.7500	10.0000
7	-10.0000	-7.8125	-4.3750	-.9375	4.5312	10.0000
8	-10.0000	-7.1875	-4.3750	.0781	4.5312	10.0000
9	-10.0000	-7.1875	-3.5546	.0781	5.0390	10.0000
10	-10.0000	-6.7773	-3.5546	.7421	5.0390	10.0000
11	-10.0000	-6.7773	-3.0175	.7421	5.3710	10.0000
12	-10.0000	-6.5087	-3.0175	1.1767	5.3710	10.0000
13	-10.0000	-6.5087	-2.6660	1.1767	5.5883	10.0000
14	-10.0000	-6.3330	-2.6660	1.4611	5.5883	10.0000
15	-10.0000	-6.3330	-2.4359	1.4611	5.7305	10.0000
16	-10.0000	-6.2179	-2.4359	1.6473	5.7305	10.0000
17	-10.0000	-6.2179	-2.2853	1.6473	5.8236	10.0000
18	-10.0000	-6.1426	-2.2853	1.7691	5.8236	10.0000
19	-10.0000	-6.1426	-2.1867	1.7691	5.8845	10.0000
20	-10.0000	-6.0933	-2.1867	1.8489	5.8845	10.0000
21	-10.0000	-6.0933	-2.1222	1.8489	5.9244	10.0000
22	-10.0000	-6.0611	-2.1222	1.9011	5.9244	10.0000
23	-10.0000	-6.0611	-2.0799	1.9011	5.9505	10.0000
24	-10.0000	-6.0399	-2.0799	1.9352	5.9505	10.0000
25	-10.0000	-6.0399	-2.0523	1.9352	5.9676	10.0000
26	-10.0000	-6.0261	-2.0523	1.9576	5.9676	10.0000
27	-10.0000	-6.0261	-2.0342	1.9576	5.9788	10.0000
28	-10.0000	-6.0171	-2.0342	1.9722	5.9788	10.0000
29	-10.0000	-6.0171	-2.0224	1.9722	5.9861	10.0000
30	-10.0000	-6.0112	-2.0224	1.9818	5.9861	10.0000
31	-10.0000	-6.0112	-2.0146	1.9818	5.9909	10.0000
32	-10.0000	-6.0073	-2.0146	1.9881	5.9909	10.0000
33	-10.0000	-6.0073	-2.0096	1.9881	5.9940	10.0000
34	-10.0000	-6.0048	-2.0096	1.9922	5.9940	10.0000
35	-10.0000	-6.0048	-2.0062	1.9922	5.9961	10.0000
36	-10.0000	-6.0031	-2.0062	1.9949	5.9961	10.0000
37	-10.0000	-6.0031	-2.0041	1.9949	5.9974	10.0000
38	-10.0000	-6.0020	-2.0041	1.9966	5.9974	10.0000
39	-10.0000	-6.0020	-2.0026	1.9966	5.9983	10.0000
40	-10.0000	-6.0013	-2.0026	1.9978	5.9983	10.0000
41	-10.0000	-6.0013	-2.0017	1.9978	5.9989	10.0000
42	-10.0000	-6.0008	-2.0017	1.9985	5.9989	10.0000
43	-10.0000	-6.0008	-2.0011	1.9985	5.9992	10.0000
44	-10.0000	-6.0005	-2.0011	1.9990	5.9992	10.0000
45	-10.0000	-6.0005	-2.0007	1.9990	5.9995	10.0000

• **PROBLEM** 19-39

Construct a flowchart and write a FORTRAN program to compute the solution of the wave equation

$$a^2 \frac{\partial^2 f}{\partial x^2} - \frac{\partial^2 f}{\partial t^2} = 0, \qquad (a)$$

using such spacings h in the x-direction and τ in the t-direction that $a^2\tau^2/h^2 = 1$. The recurrence equation for (a) becomes:

$$f_{i,j+1} = -f_{i,j-1} + f_{i+1,j} + f_{i-1,j}.$$

The boundary conditions are

$$f_{1,j} = f_{1,0}; \quad f_{n,j} = f_{n,0}.$$

The initial conditions give

$$f_{i,0} = g_{0i} \quad (i = 2, n-1),$$

and the starting formula

$$f_{i,1} = \frac{1}{2}(f_{i+1,0} + f_{i-1,0}).$$

FLOW CHART

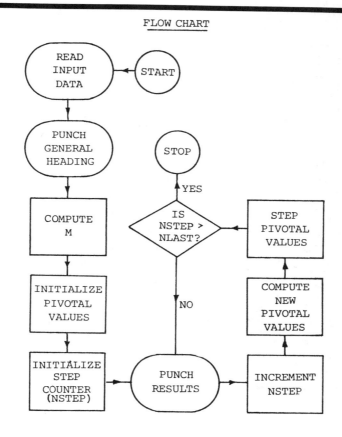

Solution:

The variables $f_{i,j}$ are singly-subscripted, not doubly-subscripted, because only three values of the j subscript need be considered at one time; the correspondence between $f_{i,j}$ and F is given by

$$f_{i,j+1} \rightarrow F2(I)$$

$$f_{i,j} \rightarrow F1(I)$$

$$f_{i,j-1} \rightarrow F0(I)$$

```
C       PROGRAM
C       SOLUTION OF A HYPERBOLIC PARTIAL DIFFERENTIAL EQUATION
C
        DIMENSION F0(25), F1(25), F2(25)
        READ 999, N, NLAST, LEFT, RIGHT, (F0(I), I = 1,N)
        PUNCH 997, (LEFT, I, RIGHT, I = 1,N)
        M = N - 1
        F1(1) = F0(1)
        DO 1 I = 2,M
    1   F1(I) = (F0(I + 1) + F0(I - 1))/2.
        F1(N) = F0(N)
        NSTEP = 0
    2   PUNCH 998, NSTEP, (F0(I), I = 1,N)
        NSTEP = NSTEP + 1
        DO 3 I = 2,M
    3   F2(I) = -F0(I) + F1(I + 1) + F1(I - 1)
        DO 4 I = 2,M
        F0(I) = F1(I)
    4   F1(I) = F2(I)
        IF (NSTEP - NLAST) 2,2,5
    5   STOP
C
  997   FORMAT ( // 24X, 24HRESULTS FROM PROGRAM //
       $           11X, 1HN 5X, 6(A2, I1, A1, 6X))
  998   FORMAT (7X, I5, 6F10.4)
  999   FORMAT 2I5, A2, A1/8E10.0))
        END
```

RESULTS FROM PROGRAM

N	F(1)	F(2)	F(3)	F(4)	F(5)
0	0.0000	3.0000	4.0000	3.0000	0.0000
1	0.0000	2.0000	3.0000	2.0000	0.0000
2	0.0000	0.0000	0.0000	0.0000	0.0000
3	0.0000	-2.0000	-3.0000	-2.0000	0.0000
4	0.0000	-3.0000	-4.0000	-3.0000	0.0000
5	0.0000	-2.0000	-3.0000	-2.0000	0.0000
6	0.0000	0.0000	0.0000	0.0000	0.0000
7	0.0000	2.0000	3.0000	2.0000	0.0000
8	0.0000	3.0000	4.0000	3.0000	0.0000
9	0.0000	2.0000	3.0000	2.0000	0.0000
10	0.0000	0.0000	0.0000	0.0000	0.0000
11	0.0000	-2.0000	-3.0000	-2.0000	0.0000
12	0.0000	-3.0000	-4.0000	-3.0000	0.0000

(ADDITIONAL) RESULTS FROM PROGRAM

N	F(1)	F(2)	F(3)	F(4)	F(5)	F(6)
0	0.0000	15.0000	30.0000	45.0000	60.0000	30.0000
1	0.0000	15.0000	30.0000	45.0000	37.5000	30.0000
2	0.0000	15.0000	30.0000	22.5000	15.0000	30.0000
3	0.0000	15.0000	7.5000	0.0000	15.0000	30.0000
4	0.0000	-7.5000	-15.0000	0.0000	15.0000	30.0000
5	0.0000	-30.0000	-15.0000	0.0000	15.0000	30.0000
6	0.0000	-7.5000	-15.0000	0.0000	15.0000	30.0000
7	0.0000	15.0000	7.5000	0.0000	15.0000	30.0000
8	0.0000	15.0000	30.0000	22.5000	15.0000	30.0000
9	0.0000	15.0000	30.0000	45.0000	37.5000	30.0000
10	0.0000	15.0000	30.0000	45.0000	60.0000	30.0000
11	0.0000	15.0000	30.0000	45.0000	37.5000	30.0000
12	0.0000	15.0000	30.0000	22.5000	15.0000	30.0000
13	0.0000	15.0000	7.5000	0.0000	15.0000	30.0000
14	0.0000	-7.5000	-15.0000	0.0000	15.0000	30.0000

15	0.0000	-30.0000	-15.0000	0.0000	15.0000	30.0000
16	0.0000	-7.5000	-15.0000	0.0000	15.0000	30.0000
17	0.0000	15.0000	7.5000	0.0000	15.0000	30.0000
18	0.0000	15.0000	30.0000	22.5000	15.0000	30.0000
19	0.0000	15.0000	30.0000	45.0000	37.5000	30.0000
20	0.0000	15.0000	30.0000	45.0000	60.0000	30.0000
21	0.0000	15.0000	30.0000	45.0000	37.5000	30.0000
22	0.0000	15.0000	30.0000	22.5000	15.0000	30.0000
23	0.0000	15.0000	7.5000	0.0000	15.0000	30.0000
24	0.0000	-7.5000	-15.0000	0.0000	15.0000	30.0000
25	0.0000	-30.0000	-15.0000	0.0000	15.0000	30.0000
26	0.0000	-7.5000	-15.0000	0.0000	15.0000	30.0000
27	0.0000	15.0000	7.5000	0.0000	15.0000	30.0000
28	0.0000	15.0000	30.0000	22.5000	15.0000	30.0000
29	0.0000	15.0000	30.0000	45.0000	37.5000	30.0000
30	0.0000	15.0000	30.0000	45.0000	60.0000	30.0000

APPENDIX

SERIES

SERIES OF CONSTANTS

ARITHMETIC SERIES

$$a + (a+d) + (a+2d) + \cdots + \{a+(n-1)d\} = \tfrac{1}{2}n\{2a+(n-1)d\}$$
$$= \tfrac{1}{2}n(a+l)$$

where $l = a + (n-1)d$ is the last term.

Some special cases are

$$1 + 2 + 3 + \cdots + n = \tfrac{1}{2}n(n+1)$$

$$1 + 3 + 5 + \cdots + (2n-1) = n^2$$

GEOMETRIC SERIES

$$a + ar + ar^2 + ar^3 + \cdots + ar^{n-1} = \frac{a(1-r^n)}{1-r} = \frac{a-rl}{1-r}$$

where $l = ar^{n-1}$ is the last term and $r \neq 1$.

If $-1 < r < 1$, then

$$a + ar + ar^2 + ar^3 + \cdots = \frac{a}{1-r}$$

ARITHMETIC-GEOMETRIC SERIES

$$a + (a+d)r + (a+2d)r^2 + \cdots + \{a+(n-1)d\}r^{n-1} = \frac{a(1-r^n)}{1-r} + \frac{rd\{1-nr^{n-1}+(n-1)r^n\}}{(1-r)^2}$$

where $r \neq 1$.

If $-1 < r < 1$, then

$$a + (a+d)r + (a+2d)r^2 + \cdots = \frac{a}{1-r} + \frac{rd}{(1-r)^2}$$

SUMS OF POWERS OF POSITIVE INTEGERS

$$1^p + 2^p + 3^p + \cdots + n^p = \frac{n^{p+1}}{p+1} + \tfrac{1}{2}n^p + \frac{B_1 p n^{p-1}}{2!} - \frac{B_2 p(p-1)(p-2)n^{p-3}}{4!} + \cdots$$

where the series terminates at n^2 or n according as p is odd or even, and B_k are the Bernoulli numbers.

Some special cases are

$$1 + 2 + 3 + \cdots + n = \frac{n(n+1)}{2}$$

$$1^2 + 2^2 + 3^2 + \cdots + n^2 = \frac{n(n+1)(2n+1)}{6}$$

$$1^3 + 2^3 + 3^3 + \cdots + n^3 = \frac{n^2(n+1)^2}{4} = (1+2+3+\cdots+n)^2$$

$$1^4 + 2^4 + 3^4 + \cdots + n^4 = \frac{n(n+1)(2n+1)(3n^2+3n-1)}{30}$$

If $S_k = 1^k + 2^k + 3^k + \cdots + n^k$ where k and n are positive integers, then

$$\binom{k+1}{1}S_1 + \binom{k+1}{2}S_2 + \cdots + \binom{k+1}{k}S_k = (n+1)^{k+1} - (n+1)$$

SERIES INVOLVING RECIPROCALS OF POWERS OF POSITIVE INTEGERS

$$1 - \frac{1}{2} + \frac{1}{3} - \frac{1}{4} + \frac{1}{5} - \cdots = \ln 2$$

$$1 - \frac{1}{3} + \frac{1}{5} - \frac{1}{7} + \frac{1}{9} - \cdots = \frac{\pi}{4}$$

$$1 - \frac{1}{4} + \frac{1}{7} - \frac{1}{10} + \frac{1}{13} - \cdots = \frac{\pi\sqrt{3}}{9} + \frac{1}{3}\ln 2$$

$$1 - \frac{1}{5} + \frac{1}{9} - \frac{1}{13} + \frac{1}{17} - \cdots = \frac{\pi\sqrt{2}}{8} + \frac{\sqrt{2}\ln(1+\sqrt{2})}{4}$$

$$\frac{1}{2} - \frac{1}{5} + \frac{1}{8} - \frac{1}{11} + \frac{1}{14} - \cdots = \frac{\pi\sqrt{3}}{9} - \frac{1}{3}\ln 2$$

$$\frac{1}{1^2} + \frac{1}{2^2} + \frac{1}{3^2} + \frac{1}{4^2} + \cdots = \frac{\pi^2}{6}$$

$$\frac{1}{1^4} + \frac{1}{2^4} + \frac{1}{3^4} + \frac{1}{4^4} + \cdots = \frac{\pi^4}{90}$$

$$\frac{1}{1^6} + \frac{1}{2^6} + \frac{1}{3^6} + \frac{1}{4^6} + \cdots = \frac{\pi^6}{945}$$

$$\frac{1}{1^2} - \frac{1}{2^2} + \frac{1}{3^2} - \frac{1}{4^2} + \cdots = \frac{\pi^2}{12}$$

$$\frac{1}{1^4} - \frac{1}{2^4} + \frac{1}{3^4} - \frac{1}{4^4} + \cdots = \frac{7\pi^4}{720}$$

$$\frac{1}{1^6} - \frac{1}{2^6} + \frac{1}{3^6} - \frac{1}{4^6} + \cdots = \frac{31\pi^6}{30{,}240}$$

$$\frac{1}{1^2} + \frac{1}{3^2} + \frac{1}{5^2} + \frac{1}{7^2} + \cdots = \frac{\pi^2}{8}$$

$$\frac{1}{1^4} + \frac{1}{3^4} + \frac{1}{5^4} + \frac{1}{7^4} + \cdots = \frac{\pi^4}{96}$$

$$\frac{1}{1^6} + \frac{1}{3^6} + \frac{1}{5^6} + \frac{1}{7^6} + \cdots = \frac{\pi^6}{960}$$

$$\frac{1}{1^3} - \frac{1}{3^3} + \frac{1}{5^3} - \frac{1}{7^3} + \cdots = \frac{\pi^3}{32}$$

$$\frac{1}{1^3} + \frac{1}{3^3} - \frac{1}{5^3} - \frac{1}{7^3} + \cdots = \frac{3\pi^3\sqrt{2}}{128}$$

$$\frac{1}{1\cdot 3} + \frac{1}{3\cdot 5} + \frac{1}{5\cdot 7} + \frac{1}{7\cdot 9} + \cdots = \frac{1}{2}$$

$$\frac{1}{1\cdot 3} + \frac{1}{2\cdot 4} + \frac{1}{3\cdot 5} + \frac{1}{4\cdot 6} + \cdots = \frac{3}{4}$$

$$\frac{1}{1^2\cdot 3^2} + \frac{1}{3^2\cdot 5^2} + \frac{1}{5^2\cdot 7^2} + \frac{1}{7^2\cdot 9^2} + \cdots = \frac{\pi^2 - 8}{16}$$

$$\frac{1}{1^2\cdot 2^2\cdot 3^2} + \frac{1}{2^2\cdot 3^2\cdot 4^2} + \frac{1}{3^2\cdot 4^2\cdot 5^2} + \cdots = \frac{4\pi^2 - 39}{16}$$

$$\frac{1}{a} - \frac{1}{a+d} + \frac{1}{a+2d} - \frac{1}{a+3d} + \cdots = \int_0^1 \frac{u^{a-1}\,du}{1+u^d}$$

$$\frac{1}{1^{2p}} + \frac{1}{2^{2p}} + \frac{1}{3^{2p}} + \frac{1}{4^{2p}} + \cdots = \frac{2^{2p-1}\pi^{2p}B_p}{(2p)!}$$

$$\frac{1}{1^{2p}} + \frac{1}{3^{2p}} + \frac{1}{5^{2p}} + \frac{1}{7^{2p}} + \cdots = \frac{(2^{2p}-1)\pi^{2p}B_p}{2(2p)!}$$

$$\frac{1}{1^{2p}} - \frac{1}{2^{2p}} + \frac{1}{3^{2p}} - \frac{1}{4^{2p}} + \cdots = \frac{(2^{2p-1}-1)\pi^{2p}B_p}{(2p)!}$$

$$\frac{1}{1^{2p+1}} - \frac{1}{3^{2p+1}} + \frac{1}{5^{2p+1}} - \frac{1}{7^{2p+1}} + \cdots = \frac{\pi^{2p+1}E_p}{2^{2p+2}(2p)!}$$

MISCELLANEOUS SERIES

$$\tfrac{1}{2} + \cos\alpha + \cos 2\alpha + \cdots + \cos n\alpha = \frac{\sin(n+\tfrac{1}{2})\alpha}{2\sin(\alpha/2)}$$

$$\sin\alpha + \sin 2\alpha + \sin 3\alpha + \cdots + \sin n\alpha = \frac{\sin[\tfrac{1}{2}(n+1)]\alpha \sin \tfrac{1}{2}n\alpha}{\sin(\alpha/2)}$$

$$1 + r\cos\alpha + r^2\cos 2\alpha + r^3\cos 3\alpha + \cdots = \frac{1 - r\cos\alpha}{1 - 2r\cos\alpha + r^2}, \quad |r|<1$$

$$r\sin\alpha + r^2\sin 2\alpha + r^3\sin 3\alpha + \cdots = \frac{r\sin\alpha}{1 - 2r\cos\alpha + r^2}, \quad |r|<1$$

$$1 + r \cos \alpha + r^2 \cos 2\alpha + \cdots + r^n \cos n\alpha$$
$$= \frac{r^{n+2} \cos n\alpha - r^{n+1} \cos (n+1)\alpha - r \cos \alpha + 1}{1 - 2r \cos \alpha + r^2}$$

$$r \sin \alpha + r^2 \sin 2\alpha + \cdots + r^n \sin n\alpha = \frac{r \sin \alpha - r^{n+1} \sin (n+1)\alpha + r^{n+2} \sin n\alpha}{1 - 2r \cos \alpha + r^2}$$

THE EULER-MACLAURIN SUMMATION FORMULA

$$\sum_{k=1}^{n-1} F(k) = \int_0^n F(k)\, dk - \frac{1}{2}\{F(0) + F(n)\}$$
$$+ \frac{1}{12}\{F'(n) - F'(0)\} - \frac{1}{720}\{F'''(n) - F'''(0)\}$$
$$+ \frac{1}{30{,}240}\{F^{(v)}(n) - F^{(v)}(0)\} - \frac{1}{1{,}209{,}600}\{F^{(vii)}(n) - F^{(vii)}(0)\}$$
$$+ \cdots (-1)^{p-1}\frac{B_p}{(2p)!}\{F^{(2p-1)}(n) - F^{(2p-1)}(0)\} + \cdots$$

THE POISSON SUMMATION FORMULA

$$\sum_{k=-\infty}^{\infty} F(k) = \sum_{m=-\infty}^{\infty} \left\{ \int_{-\infty}^{\infty} e^{2\pi i m x} F(x)\, dx \right\}$$

TAYLOR SERIES

TAYLOR SERIES FOR FUNCTIONS OF ONE VARIABLE

$$f(x) = f(a) + f'(a)(x-a) + \frac{f''(a)(x-a)^2}{2!} +$$
$$\cdots + \frac{f^{(n-1)}(a)(x-a)^{n-1}}{(n-1)!} + R_n$$

where R_n, the remainder after n terms, is given by either of the following forms:

Lagrange's form $\quad R_n = \dfrac{f^{(n)}(\xi)(x-a)^n}{n!}$

Cauchy's form $\quad R_n = \dfrac{f^{(n)}(\xi)(x-\xi)^{n-1}(x-a)}{(n-1)!}$

The value ξ, which may be different in the two forms, lies between a and x. The result holds if $f(x)$ has continuous derivatives of order n at least.

If $\lim_{n \to \infty} R_n = 0$, the infinite series obtained is called the *Taylor series* for

$f(x)$ about $x = a$. If $a = 0$ the series is often called a *Maclaurin series*. These series, often called power series, generally converge for all values of x in some interval called the *interval of convergence* and diverge for all x outside this interval.

BINOMIAL SERIES

$$(a+x)^n = a^n + na^{n-1}x + \frac{n(n-1)}{2!}a^{n-2}x^2 + \frac{n(n-1)(n-2)}{3!}a^{n-3}x^3 + \cdots$$

$$= a^n + \binom{n}{1}a^{n-1}x + \binom{n}{2}a^{n-2}x^2 + \binom{n}{3}a^{n-3}x^3 + \cdots$$

Special cases are

$$(a+x)^2 = a^2 + 2ax + x^2$$

$$(a+x)^3 = a^3 + 3a^2x + 3ax^2 + x^3$$

$$(a+x)^4 = a^4 + 4a^3x + 6a^2x^2 + 4ax^3 + x^4$$

$$(1+x)^{-1} = 1 - x + x^2 - x^3 + x^4 - \cdots \qquad -1 < x < 1$$

$$(1+x)^{-2} = 1 - 2x + 3x^2 - 4x^3 + 5x^4 - \cdots \qquad -1 < x < 1$$

$$(1+x)^{-3} = 1 - 3x + 6x^2 - 10x^3 + 15x^4 - \cdots \qquad -1 < x < 1$$

$$(1+x)^{-1/2} = 1 - \frac{1}{2}x + \frac{1 \cdot 3}{2 \cdot 4}x^2 - \frac{1 \cdot 3 \cdot 5}{2 \cdot 4 \cdot 6}x^3 + \cdots \qquad -1 < x \leq 1$$

$$(1+x)^{1/2} = 1 + \frac{1}{2}x - \frac{1}{2 \cdot 4}x^2 + \frac{1 \cdot 3}{2 \cdot 4 \cdot 6}x^3 - \cdots \qquad -1 < x \leq 1$$

$$(1+x)^{-1/3} = 1 - \frac{1}{3}x + \frac{1 \cdot 4}{3 \cdot 6}x^2 - \frac{1 \cdot 4 \cdot 7}{3 \cdot 6 \cdot 9}x^3 + \cdots \qquad -1 < x \leq 1$$

$$(1+x)^{1/3} = 1 + \frac{1}{3}x - \frac{2}{3 \cdot 6}x^2 + \frac{2 \cdot 5}{3 \cdot 6 \cdot 9}x^3 - \cdots \qquad -1 < x \leq 1$$

SERIES FOR EXPONENTIAL AND LOGARITHMIC FUNCTIONS

$$e^x = 1 + x + \frac{x^2}{2!} + \frac{x^3}{3!} + \cdots \qquad -\infty < x < \infty$$

$$a^x = e^{x \ln a} = 1 + x \ln a + \frac{(x \ln a)^2}{2!} + \frac{(x \ln a)^3}{3!} + \cdots \qquad -\infty < x < \infty$$

$$\ln(1+x) = x - \frac{x^2}{2} + \frac{x^3}{3} - \frac{x^4}{4} + \cdots \qquad -1 < x \leq 1$$

$$\frac{1}{2}\ln\left(\frac{1+x}{1-x}\right) = x + \frac{x^3}{3} + \frac{x^5}{5} + \frac{x^7}{7} + \cdots \qquad -1 < x < 1$$

$$\ln x = 2\left\{\left(\frac{x-1}{x+1}\right) + \frac{1}{3}\left(\frac{x-1}{x+1}\right)^3 + \frac{1}{5}\left(\frac{x-1}{x+1}\right)^5 + \cdots\right\} \qquad x > 0$$

$$\ln x = \left(\frac{x-1}{x}\right) + \frac{1}{2}\left(\frac{x-1}{x}\right)^2 + \frac{1}{3}\left(\frac{x-1}{x}\right)^3 + \cdots \qquad x \geq \tfrac{1}{2}$$

SERIES FOR TRIGONOMETRIC FUNCTIONS

$$\sin x = x - \frac{x^3}{3!} + \frac{x^5}{5!} - \frac{x^7}{7!} + \cdots \qquad -\infty < x < \infty$$

$$\cos x = 1 - \frac{x^2}{2!} + \frac{x^4}{4!} - \frac{x^6}{6!} + \cdots \qquad -\infty < x < \infty$$

$$\tan x = x + \frac{x^3}{3} + \frac{2x^5}{15} + \frac{17x^7}{315} + \cdots + \frac{2^{2n}(2^{2n}-1)B_n x^{2n-1}}{(2n)!} + \cdots \qquad |x| < \frac{\pi}{2}$$

$$\cot x = \frac{1}{x} - \frac{x}{3} - \frac{x^3}{45} - \frac{2x^5}{945} - \cdots - \frac{2^{2n}B_n x^{2n-1}}{(2n)!} - \cdots \qquad 0 < |x| < \pi$$

$$\sec x = 1 + \frac{x^2}{2} + \frac{5x^4}{24} + \frac{61x^6}{720} + \cdots + \frac{E_n x^{2n}}{(2n)!} + \cdots \qquad |x| < \frac{\pi}{2}$$

$$\csc x = \frac{1}{x} + \frac{x}{6} + \frac{7x^3}{360} + \frac{31x^5}{15{,}120} + \cdots + \frac{2(2^{2r-1}-1)B_n x^{2n-1}}{(2n)!} + \cdots \quad 0 < |x| < \pi$$

$$\sin^{-1} x = x + \frac{1}{2}\frac{x^3}{3} + \frac{1\cdot 3}{2\cdot 4}\frac{x^5}{5} + \frac{1\cdot 3\cdot 5}{2\cdot 4\cdot 6}\frac{x^7}{7} + \cdots \qquad |x| < 1$$

$$\cos^{-1} x = \frac{\pi}{2} - \sin^{-1} x = \frac{\pi}{2} - \left(x + \frac{1}{2}\frac{x^3}{3} + \frac{1\cdot 3}{2\cdot 4}\frac{x^5}{5} + \cdots\right) \qquad |x| < 1$$

$$\tan^{-1} x = \begin{cases} x - \dfrac{x^3}{3} + \dfrac{x^5}{5} - \dfrac{x^7}{7} + \cdots & |x| < 1 \\ \pm\dfrac{\pi}{2} - \dfrac{1}{x} + \dfrac{1}{3x^3} - \dfrac{1}{5x^5} + \cdots & [+\text{ if } x \geq 1, \; -\text{ if } x \leq -1] \end{cases}$$

$$\cot^{-1} x = \frac{\pi}{2} - \tan^{-1} x$$

$$= \begin{cases} \dfrac{\pi}{2} - \left(x - \dfrac{x^3}{3} + \dfrac{x^5}{5} - \cdots\right) & |x| < 1 \\ p\pi + \dfrac{1}{x} - \dfrac{1}{3x^3} + \dfrac{1}{5x^5} - \cdots & [p = 0 \text{ if } x > 1,\; p = 1 \text{ if } x < -1] \end{cases}$$

$$\sec^{-1} x = \cos^{-1}(1/x) = \frac{\pi}{2} - \left(\frac{1}{x} + \frac{1}{2\cdot 3x^3} + \frac{1\cdot 3}{2\cdot 4\cdot 5x^5} + \cdots\right) \qquad |x| > 1$$

$$\csc^{-1} x = \sin^{-1}(1/x) = \frac{1}{x} + \frac{1}{2\cdot 3x^3} + \frac{1\cdot 3}{2\cdot 4\cdot 5x^5} + \cdots \qquad |x| > 1$$

SERIES FOR HYPERBOLIC FUNCTIONS

$$\sinh x = x + \frac{x^3}{3!} + \frac{x^5}{5!} + \frac{x^7}{7!} + \cdots \qquad -\infty < x < \infty$$

$$\cosh x = 1 + \frac{x^2}{2!} + \frac{x^4}{4!} + \frac{x^6}{6!} + \cdots \qquad -\infty < x < \infty$$

$$\tanh x = x - \frac{x^3}{3} + \frac{2x^5}{15} - \frac{17x^7}{315} + \cdots \frac{(-1)^{n-1}2^{2n}(2^{2n}-1)B_n x^{2n-1}}{(2n)!} + \cdots \qquad |x| < \frac{\pi}{2}$$

$$\coth x = \frac{1}{x} + \frac{x}{3} - \frac{x^3}{45} + \frac{2x^5}{945} + \cdots \frac{(-1)^{n-1}2^{2n}B_n x^{2n-1}}{(2n)!} + \cdots \qquad 0 < |x| < \pi$$

$$\operatorname{sech} x = 1 - \frac{x^2}{2} + \frac{5x^4}{24} - \frac{61x^6}{720} + \cdots \frac{(-1)^n E_n x^{2n}}{(2n)!} + \cdots \qquad |x| < \frac{\pi}{2}$$

$$\operatorname{csch} x = \frac{1}{x} - \frac{x}{6} + \frac{7x^3}{360} - \frac{31x^5}{15,120} + \cdots \frac{(-1)^n 2(2^{2n-1}-1)B_n x^{2n-1}}{(2n)!} + \cdots \qquad 0 < |x| < \pi$$

$$\sinh^{-1} x = \begin{cases} x - \frac{x^3}{2\cdot 3} + \frac{1\cdot 3 x^5}{2\cdot 4\cdot 6} - \frac{1\cdot 3\cdot 5 x^7}{2\cdot 4\cdot 6\cdot 7} + \cdots & |x| < 1 \\ \pm \left(\ln|2x| + \frac{1}{2\cdot 2x^2} - \frac{1\cdot 3}{2\cdot 4\cdot 4x^4} + \frac{1\cdot 3\cdot 5}{2\cdot 4\cdot 6\cdot 6x^6} - \cdots \right) & \begin{bmatrix} + \text{ if } x \geq 1 \\ - \text{ if } x \leq -1 \end{bmatrix} \end{cases}$$

$$\cosh^{-1} x = \pm \left\{ \ln(2x) - \left(\frac{1}{2\cdot 2x^2} + \frac{1\cdot 3}{2\cdot 4\cdot 4x^4} + \frac{1\cdot 3\cdot 5}{2\cdot 4\cdot 6\cdot 6x^6} + \cdots \right) \right\}$$

$$\begin{bmatrix} + \text{ if } \cosh^{-1} x > 0,\ x \geq 1 \\ - \text{ if } \cosh^{-1} x < 0,\ x \geq 1 \end{bmatrix}$$

$$\tanh^{-1} x = x + \frac{x^3}{3} + \frac{x^5}{5} + \frac{x^7}{7} + \cdots \qquad |x| < 1$$

$$\coth^{-1} x = \frac{1}{x} + \frac{1}{3x^3} + \frac{1}{5x^5} + \frac{1}{7x^7} + \cdots \qquad |x| > 1$$

MISCELLANEOUS SERIES

$$e^{\sin x} = 1 + x + \frac{x^2}{2} - \frac{x^4}{8} - \frac{x^5}{15} + \cdots \qquad -\infty < x < \infty$$

$$e^{\cos x} = e\left(1 - \frac{x^2}{2} + \frac{x^4}{6} - \frac{31x^6}{720} + \cdots \right) \qquad -\infty < x < \infty$$

$$e^{\tan x} = 1 + x + \frac{x^2}{2} + \frac{x^3}{2} + \frac{3x^4}{8} + \cdots \qquad |x| < \frac{\pi}{2}$$

$$e^x \sin x = x + x^2 + \frac{2x^3}{3} - \frac{x^5}{30} - \frac{x^6}{90} + \cdots + \frac{2^{n/2} \sin(n\pi/4) x^n}{n!} + \cdots \qquad -\infty < x < \infty$$

$$e^x \cos x = 1 + x - \frac{x^3}{3} - \frac{x^4}{6} + \cdots + \frac{2^{n/2} \cos(n\pi/4) x^n}{n!} + \cdots \qquad -\infty < x < \infty$$

$$\ln|\sin x| = \ln|x| - \frac{x^2}{6} - \frac{x^4}{180} - \frac{x^6}{2835} - \cdots - \frac{2^{2n-1} B_n x^{2n}}{n(2n)!} + \cdots \qquad 0 < |x| < \pi$$

$$\ln|\cos x| = -\frac{x^2}{2} - \frac{x^4}{12} - \frac{x^6}{45} - \frac{17x^8}{2520} - \cdots - \frac{2^{2n-1}(2^{2n}-1)B_n x^{2n}}{n(2n)!} + \cdots \qquad |x| < \frac{\pi}{2}$$

$$\ln |\tan x| = \ln |x| + \frac{x^2}{3} + \frac{7x^4}{90} + \frac{62x^6}{2835} + \cdots + \frac{2^{2n}(2^{2n-1}-1)B_n x^{2n}}{n(2n)!} + \cdots \quad 0 < |x| < \frac{\pi}{2}$$

$$\frac{\ln(1+x)}{1+x} = x - (1+\tfrac{1}{2})x^2 + (1+\tfrac{1}{2}+\tfrac{1}{3})x^3 - \cdots \qquad |x| < 1$$

REVERSION OF POWER SERIES

If

$$y = c_1 x + c_2 x^2 + c_3 x^3 + c_4 x^4 + c_5 x^5 + c_6 x^6 + \cdots$$

then

$$x = C_1 y + C_2 y^2 + C_3 y^3 + C_4 y^4 + C_5 y^5 + C_6 y^6 + \cdots$$

where

$$c_1 C_1 = 1$$

$$c_1^3 C_2 = -c_2$$

$$c_1^5 C_3 = 2c_2^2 - c_1 c_3$$

$$c_1^7 C_4 = 5c_1 c_2 c_3 - 5c_2^3 - c_1^2 c_4$$

$$c_1^9 C_5 = 6c_1^2 c_2 c_4 + 3c_1^2 c_3^2 - c_1^3 c_5 + 14c_2^4 - 21c_1 c_2^2 c_3$$

$$c_1^{11} C_6 = 7c_1^3 c_2 c_5 + 84c_1 c_2^3 c_3 + 7c_1^3 c_3 c_4 - 28c_1^2 c_2 c_3^2 - c_1^4 c_6 - 28c_1^2 c_2^2 c_4 - 42c_2^5$$

TAYLOR SERIES FOR FUNCTIONS OF TWO VARIABLES

$$\begin{aligned} f(x,y) = {} & f(a,b) + (x-a)f_x(a,b) + (y-b)f_y(a,b) \\ & + \frac{1}{2!}\{(x-a)^2 f_{xx}(a,b) + 2(x-a)(y-b)f_{xy}(a,b) + (y-b)^2 f_{yy}(a,b)\} + \cdots \end{aligned}$$

where $f_x(a,b)$, $f_y(a,b)$, ... denote partial derivatives with respect to $x, y, \ldots$ evaluated at $x = a$, $y = b$.

REPRESENTATION OF NUMBERS

Any positive real number x can be uniquely represented in the scale of some integer $b>1$ as

$$x=(A_m \ldots A_1A_0 \cdot a_{-1}a_{-2} \ldots)_{(b)},$$

where every A_i and a_{-j} is one of the integers 0, 1, ..., $b-1$, not all A_i, a_{-j} are zero, and $A_m>0$ if $x\geq 1$. There is a one-to-one correspondence between the number and the sequence

$$x=A_mb^m+ \ldots +A_1b+A_0+\sum_1^\infty a_{-j}b^{-j}$$

where the infinite series converges. The integer b is called the base or radix of the scale.

The sequence for x in the scale of b may terminate, i.e., $a_{-n-1}=a_{-n-2}= \ldots =0$ for some $n\geq 1$ so that

$$x=(A_m \ldots A_1A_0 \cdot a_{-1}a_{-2} \ldots a_{-n})_{(b)};$$

then x is said to be a finite b-adic number.

A sequence which does not terminate may have the property that the infinite sequence a_{-1}, a_{-2}, ... becomes periodic from a certain digit $a_{-n}(n\geq 1)$ on; according as $n=1$ or $n>1$ the sequence is then said to be pure or mixed recurring.

A sequence which neither terminates nor recurs represents an irrational number.

NAMES OF SCALES

Base	Scale	Base	Scale
2	Binary	8	Octal
3	Ternary	9	Nonary
4	Quaternary	10	Decimal
5	Quinary	11	Undenary
6	Senary	12	Duodenary
7	Septenary	16	Hexadecimal

GENERAL CONVERSION METHODS

Any number can be converted from the scale of b to the scale of some integer $\bar{b} \neq b$, $\bar{b} > 1$, by using arithmetic operations in either the b-scale or the $\bar{b}$-scale. Accordingly, there are four methods of conversion, depending on whether the number to be converted is an integer or a proper fraction.

INTEGERS $X = (A_m \ldots A_1 A_0)_{(b)}$

(I) b-scale arithmetic. Convert $\bar{b}$ to the b-scale and define

$$X/\bar{b} = X_1 + \overline{A}_0'/\bar{b},$$
$$X_1/\bar{b} = X_2 + \overline{A}_1'/\bar{b},$$
$$\vdots$$
$$X_{\bar{m}}/\bar{b} = 0 + \overline{A}_{\bar{m}}'/\bar{b},$$

where $\overline{A}_0'$, $\overline{A}_1'$, ..., $\overline{A}_{\bar{m}}'$ are the remainders and X_1, X_2, ..., $X_{\bar{m}}$ the quotients (in the b-scale) where X, X_1, ..., $X_{\bar{m}-1}$, respectively are divided by $\bar{b}$ in the b-scale. Then convert the remainders to the $\bar{b}$-scale,

$$(\overline{A}_0')_{(\bar{b})} = \overline{A}_0, \ (\overline{A}_1')_{(\bar{b})} = \overline{A}_1, \ \ldots, \ (\overline{A}_{\bar{m}}')_{(\bar{b})} = \overline{A}_{\bar{m}}$$

and obtain

$$X = (\overline{A}_{\bar{m}} \ldots \overline{A}_1 \overline{A}_0)_{(\bar{b})}.$$

(II) $\bar{b}$-scale arithmetic. Convert b and A_0, A_1, ..., A_m to the $\bar{b}$-scale and define, using arithmetic operations in the $\bar{b}$-scale,

$$X_{m-1} = A_m b + A_{m-1},$$
$$X_{m-2} = X_{m-1} b + A_{m-2},$$
$$X_1 = X_2 b + A_1,$$

868

then
$$X = X_1 b + A_0.$$

PROPER FRACTIONS $x = (0.a_{-1} a_{-2} \cdots)_{(b)}$

To convert a proper fraction x, given to n digits in the b-scale, to the scale of $\bar{b} \neq b$ such that inverse conversion from the $\bar{b}$-scale may yield the same n rounded digits in the b-scale, the representation of x in the $\bar{b}$-scale must be obtained to $\bar{n}$ rounded digits where n satisfies $\bar{b}^{\bar{n}} > b^n$.

(III) b-scale arithmetic. Convert $\bar{b}$ to the b-scale and define

$$x\bar{b} = x_1 + \bar{a}'_{-1}$$
$$x_1 \bar{b} = x_2 + \bar{a}'_{-2}$$
$$x_{\bar{n}-1} \bar{b} = x_{\bar{n}} + \bar{a}'_{-\bar{n}}$$

where $\bar{a}'_{-1}, \bar{a}'_{-2}, \ldots, \bar{a}'_{-\bar{n}}$ are the integral parts and $x_1, x_2, \ldots, x_{\bar{n}}$ the fractional parts (in the b-scale) of the products $x\bar{b}, x_1 \bar{b}, \ldots, x_{\bar{n}-1}\bar{b}$, respectively. Then convert the integral parts to the $\bar{b}$-scale,

$$(\bar{a}'_{-1})_{(\bar{b})} = \bar{a}_{-1}, \ (\bar{a}'_{-2})_{(\bar{b})} = \bar{a}_{-2}, \ \ldots, \ (\bar{a}'_{-\bar{n}})_{(\bar{b})} = \bar{a}_{-\bar{n}},$$

and obtain

$$x = (0.\bar{a}_{-1} \bar{a}_{-2} \cdots \bar{a}_{-\bar{n}})_{(\bar{b})}.$$

(IV) $\bar{b}$-scale arithmetic. Convert b and $a_{-1}, a_{-2}, \ldots, a_{-n}$ to the $\bar{b}$-scale and define, using arithmetic operations in the $\bar{b}$-scale,

$$x_{-n+1} = a_{-n}/b + a_{-n+1},$$
$$x_{-n+2} = x_{-n+1}/b + a_{-n+2},$$
$$x_{-1} = x_{-2}/b + a_{-1};$$

then
$$x = x_{-1}/b.$$

$2^{\pm n}$ IN DECIMAL

2^n	n	2^{-n}
1	0	1.0
2	1	0.5
4	2	0.25
8	3	0.125
16	4	0.0625
32	5	0.03125
64	6	0.01562 5
128	7	0.00781 25
256	8	0.00390 625
512	9	0.00195 3125
1024	10	0.00097 65625
2048	11	0.00048 82812 5
4096	12	0.00024 41406 25
8192	13	0.00012 20703 125
16384	14	0.00006 10351 5625
32768	15	0.00003 05175 78125
65536	16	0.00001 52587 89062 5
1 31072	17	0.00000 76293 94531 25
2 62144	18	0.00000 38146 97265 625
5 24288	19	0.00000 19073 48632 8125
10 48576	20	0.00000 09536 74316 40625
20 97152	21	0.00000 04768 37158 20312 5
41 94304	22	0.00000 02384 18579 10156 25
83 88608	23	0.00000 01192 09289 55078 125
167 77216	24	0.00000 00596 04644 77539 0625
335 54432	25	0.00000 00298 02322 38769 53125
671 08864	26	0.00000 00149 01161 19384 76562 5
1342 17728	27	0.00000 00074 50580 59692 38281 25
2684 35456	28	0.00000 00037 25290 29846 19140 625
5368 70912	29	0.00000 00018 62645 14923 09570 3125
10737 41824	30	0.00000 00009 31322 57461 54785 15625
21474 83648	31	0.00000 00004 65661 28730 77392 57812 5
42949 67296	32	0.00000 00002 32830 64365 38696 28906 25
85899 34592	33	0.00000 00001 16415 32182 69348 14453 125
1 71798 69184	34	0.00000 00000 58207 66091 34674 07226 5625
3 43597 38368	35	0.00000 00000 29103 83045 67337 03613 28125
6 87194 76736	36	0.00000 00000 14551 91522 83668 51806 64062 5
13 74389 53472	37	0.00000 00000 07275 95761 41834 25903 32031 25
27 48779 06944	38	0.00000 00000 03637 97880 70917 12951 66015 625
54 97558 13888	39	0.00000 00000 01818 98940 35458 56475 83007 8125
109 95116 27776	40	0.00000 00000 00909 49470 17729 28237 91503 90625
219 90232 55552	41	0.00000 00000 00454 74735 08864 64118 95751 95312 5
439 80465 11104	42	0.00000 00000 00227 37367 54432 32059 47875 97656 25
879 60930 22208	43	0.00000 00000 00113 68683 77216 16029 73937 98828 125
1759 21860 44416	44	0.00000 00000 00056 84341 88608 08014 86968 99414 0625
3518 43720 88832	45	0.00000 00000 00028 42170 94304 04007 43484 49707 03125
7036 87441 77664	46	0.00000 00000 00014 21085 47152 02003 71742 24853 51562 5
14073 74883 55328	47	0.00000 00000 00007 10542 73576 01001 85871 12426 75781 25
28147 49767 10656	48	0.00000 00000 00003 55271 36788 00500 92935 56213 37890 625
56294 99534 21312	49	0.00000 00000 00001 77635 68394 00250 46467 78106 68945 3125
112589 99068 42624	50	0.00000 00000 00000 88817 84197 00125 23233 89053 34472 65625

2^x IN DECIMAL

x	2^x	x	2^x	x	2^x
0.001	1.00069 33874 62581	0.01	1.00695 55500 56719	0.1	1.07177 34625 36293
0.002	1.00138 72557 11335	0.02	1.01395 94797 90029	0.2	1.14869 83549 97035
0.003	1.00208 16050 79633	0.03	1.02101 21257 07193	0.3	1.23114 44133 44916
0.004	1.00277 64359 01078	0.04	1.02811 38266 56067	0.4	1.31950 79107 72894
0.005	1.00347 17485 09503	0.05	1.03526 49238 41377	0.5	1.41421 35623 73095
0.006	1.00416 75432 38973	0.06	1.04246 57608 41121	0.6	1.51571 65665 10398
0.007	1.00486 38204 23785	0.07	1.04971 66836 23067	0.7	1.62450 47927 12471
0.008	1.00556 05803 98468	0.08	1.05701 80405 61380	0.8	1.74110 11265 92248
0.009	1.00625 78234 97782	0.09	1.06437 01824 53360	0.9	1.86606 59830 73615

$10^{\pm n}$ IN OCTAL

10^n	n	10^{-n}	n	10^n	10^{-n}
1	0	1.000 000 000 000 000 000 00	10	112 402 762 000	0.000 000 000 000 006 676 337 66
12	1	0.063 146 314 631 463 146 31	11	1 351 035 564 000	0.000 000 000 000 000 537 657 77
144	2	0.005 075 341 217 270 243 66	12	16 432 451 210 000	0.000 000 000 000 000 043 136 32
1 750	3	0.000 406 111 564 570 651 77	13	221 411 634 520 000	0.000 000 000 000 000 003 411 35
23 420	4	0.000 032 155 613 530 704 15	14	2 657 142 036 440 000	0.000 000 000 000 000 000 264 11
303 240	5	0.000 002 476 132 610 706 64	15	34 327 724 461 500 000	0.000 000 000 000 000 000 022 01
3 641 100	6	0.000 000 206 157 364 055 37	16	434 157 115 760 200 000	0.000 000 000 000 000 000 001 63
46 113 200	7	0.000 000 015 327 745 152 75	17	5 432 127 413 542 400 000	0.000 000 000 000 000 000 000 14
575 360 400	8	0.000 000 001 257 143 561 06	18	67 405 553 164 731 000 000	0.000 000 000 000 000 000 000 01
7 346 545 000	9	0.000 000 000 104 560 276 41			

$n \log_{10} 2$, $n \log_2 10$ IN DECIMAL

n	$n \log_{10} 2$	$n \log_2 10$	n	$n \log_{10} 2$	$n \log_2 10$
1	0.30102 99957	3.32192 80949	6	1.80617 99740	19.93156 85693
2	0.60205 99913	6.64385 61898	7	2.10720 99696	23.25349 66642
3	0.90308 99870	9.96578 42847	8	2.40823 99653	26.57542 47591
4	1.20411 99827	13.28771 23795	9	2.70926 99610	29.89735 28540
5	1.50514 99783	16.60964 04744	10	3.01029 99556	33.21928 09489

ADDITION AND MULTIPLICATION TABLES

Addition   Multiplication
Binary Scale

$$0 + 0 = 0$$
$$0 + 1 = 1 + 0 = 1$$
$$1 + 1 = 10$$

$$0 \times 0 = 0$$
$$0 \times 1 = 1 \times 0 = 0$$
$$1 \times 1 = 1$$

Octal Scale

0	01	02	03	04	05	06	07
1	02	03	04	05	06	07	10
2	03	04	05	06	07	10	11
3	04	05	06	07	10	11	12
4	05	06	07	10	11	12	13
5	06	07	10	11	12	13	14
6	07	10	11	12	13	14	15
7	10	11	12	13	14	15	16

1	02	03	04	05	06	07
2	04	06	10	12	14	16
3	06	11	14	17	22	25
4	10	14	20	24	30	34
5	12	17	24	31	36	43
6	14	22	30	36	44	52
7	16	25	34	43	52	61

MATHEMATICAL CONSTANTS IN OCTAL SCALE

$\pi = (3.11037\ 552421)_{(8)}$ $e = (2.55760\ 521305)_{(8)}$

$\pi^{-1} = (0.24276\ 301556)_{(8)}$ $e^{-1} = (0.27426\ 530661)_{(8)}$

$\sqrt{\pi} = (1.61337\ 611067)_{(8)}$ $\sqrt{e} = (1.51411\ 230704)_{(8)}$

$\ln \pi = (1.11206\ 404435)_{(8)}$ $\log_{10} e = (0.33626\ 754251)_{(8)}$

$\log_2 \pi = (1.51544\ 163223)_{(8)}$ $\log_2 e = (1.34252\ 166245)_{(8)}$

$\sqrt{10} = (3.12305\ 407267)_{(8)}$ $\log_2 10 = (3.24464\ 741136)_{(8)}$

$\gamma = (0.44742\ 147707)_{(8)}$ $\sqrt{2} = (1.32404\ 746320)_{(8)}$

$\ln \gamma = -(0.43127\ 233602)_{(8)}$ $\ln 2 = (0.54271\ 027760)_{(8)}$

$\log_2 \gamma = -(0.62573\ 030645)_{(8)}$ $\ln 10 = (2.23273\ 067355)_{(8)}$

INDEX

Numbers on this page refer to PROBLEM NUMBERS, not page numbers

Absolute convergence criterion, 6-7, 12-18
Absolute error, 1-7, 2-2, 2-4, 2-6
   function, 2-20, 2-21
Adams-Bashforth process, 16-59
Adam's method, 16-20
Aitken process (see Aitken's method)
Aitken's method, 5-11, 5-20, 5-24
Algorithm:
   linear shooting, 16-21
   nonlinear shooting, 16-22
Amplitude, 16-5
Annual Payment, 5-14
Annuity, 5-14
Approximate value, 2-10
Approximating polynomial, 2-14
Approximation, methods of, 2-13 to 2-23
Approximation of functions:
   nonpolynomial, 2-16, 2-17
   polynomial, 2-14
Arbitrary constant, 17-11
Arbitrary functions, 17-4 to 17-6, 17-10
Arithmetic IF statement, 1-10
Array, 19-1
Associative property:
   addition, 11-2, 11-3, 11-5
   multiplication, 11-2, 11-3
Asymptotic solution, 16-53
Augmented matrix, 6-5, 6-24, 6-25
Auxiliary equation, 16-5, 16-8

Backward difference:
   representation, 4-16, 4-18
   table, 5-28
Backward substitution, 6-24
Banded coefficient matrix, 6-10
Basic equation, 6-11
Basis, 12-15, 12-24 to 12-26
Bender-Schmidt recurrence equation, 19-38
Bernoulli numbers, 3-33
Bessel's equation, 16-51 to 16-53, 17-18
Bessel's interpolation formula, 5-4, 5-7, 5-8
Binary number system, 1-12 to 1-14, 1-16, 1-18
Binding power rule, 1-19
Bisection, 2-20, 7-35
   methods, 19-27
Bit, 1-12
Boundary conditions, 15-10, 15-12, 16-6, 16-22, 16-25, 16-27. 16-30, 16-37, 17-12, 17-13, 17-15 to 17-17,

Numbers on this page refer to **PROBLEM NUMBERS**, not page numbers

17-20, 17-21
Boundary-value problem, 15-8, 15-9, 15-12, 16-6, 16-21 to 16-24, 16-29 to 16-31, 16-34
  mixed, 16-37
Bounds, 2-18

Calculated value, 2-4
Cartesian coordinates, 17-19
Central difference, 4-17, 5-6
  representation, 4-16, 4-17
Central limit theorem, 18-5, 18-6
Chain rule, 17-10, 17-19
Characteristic:
  determinant, 6-5
  equation, 12-2 to 12-5, 12-7 to 12-9, 12-12, 12-14, 12-15, 12-20, 12-23, 12-24, 12-26, 16-9, 16-30, 17-12
  function, 17-21
  polynomial, 6-15, 12-2, 12-3, 12-12, 12-26
  values, 6-9, 6-12, 6-15, 12-10 to 12-12, 12-16, 12-19
  vector, 12-10, 12-11, 12-16, 12-25, 12-26
Chebyshev:
  series, 2-23
  set, 8-12
Cholesky factorization, 11-24
Chopping, 2-8
  methods, 2-9
Circular cylindrical coordinate system, 17-19
Circular frequency, 16-7
Closure:
  under addition, 11-2
  under multiplication, 11-2
Coefficient matrix, 6-10, 6-23, 8-8
Coefficient of damping, 16-9
Cofactor matrix, 6-23
Cold fluid, 17-4
Column, 6-16, 6-17, 6-20
  scaling, 2-11
  shifting, 6-10
Commutative property:

  addition, 11-2, 11-3, 11-5
  multiplication, 11-2, 11-3
Computer operations, 2-6
Computer precision, 1-7
Continuous function, 2-16, 11-19
Control unit, 1-1
Convergence, 3-1 to 3-4, 3-8, 3-16, 3-22, 3-23, 3-29, 7-1, 7-2, 7-12, 7-21, 7-32, 12-18, 12-21, 14-6 to 14-10, 14-12 to 14-16, 14-19, 14-21, 14-24, 19-10, 19-12, 19-23, 19-28
  interval of, 3-16, 3-22, 3-23, 3-29
  in the mean, 14-6, 14-12
  pointwise, 14-6
  rate, 12-18
  test for, 3-1 to 3-4, 3-8, 3-16, 3-22, 3-23, 3-29, 6-7
  uniform, 3-29, 14-6, 14-8, 14-9, 14-14, 14-15, 14-21
Convergence criterion, 10-15, 12-18
  absolute, 6-7, 12-18
Convergent iteration, 19-20
Convergent series, 3-1 to 3-4, 3-8, 3-16, 3-22, 3-23, 3-29
Convolution theorem, 14-27
Coplanar vectors, 11-1
Correction formula, 7-14
Correction method of Newton, 7-15
Correct significant digits, 2-1
Cramer's rule, 11-7, 17-21
Critically damped motion, 16-9
Critical points, 2-19, 2-20, 7-38
Crout's method, 8-7, 11-9 to 11-11
Cubic interpolation polynomials, 5-20
Cubic Lagrange polynomial, 5-12
Cubic polynomial, 5-15
Cubic splines method, 5-12
Curve fitting, 8-1 to 8-24

Numbers on this page refer to **PROBLEM NUMBERS**, not page numbers

Decimal:
    floating arithmetic, 6-3
    numbers, 2-9
Degenerate set, 4-15
Degree to radian, 2-2
Dependent variables, 6-14
Derived matrix, 11-10
Descartes' rule, 7-24, 7-37
Determinant, 4-15, 6-21, 6-23, 6-26
    second-order, 6-21
    third-order, 6-22
Diagonal elements, 6-7
Diagonalizable, 12-25
Difference operators, 4-19 to 4-24
Difference representation:
    backward, 4-16
    central, 4-16, 4-17
    higher order backward, 4-16
    higher order forward, 4-16
Difference table, 5-28
    backward, 5-28
Differential correction, method of, 8-20
Differential equation, 2-7, 16-1 to 16-60
Diffusion constant, 16-6
Digital computer, 1-1, 5-15
    input to a, 1-1
    output from a, 1-1
Digit chopping, 2-6
Direct step-by-step method, 15-12
Dirichlet problem, 16-36
Displacement, 16-5
Distribution:
    discrete, 18-1
    exponential, 18-9
    Monte Carlo, 18-6
    normal, 18-5, 18-6
    Poisson, 18-8, 18-9, 18-11, 18-13
Distributive law, 11-2, 11-3, 11-5
Divergence, 3-1 to 3-4, 3-8, 3-16, 3-22, 3-23, 3-29, 7-1
    test for, 3-2 to 3-4, 3-16, 3-22, 3-23, 3-29

Divergent series, 3-1 to 3-4, 3-8, 3-29
Divided difference(s), 4-4, 4-6, 4-10, 4-15, 5-11, 5-20, 5-25
    table, 4-6, 4-7, 4-10, 4-13
DO-loop, 19-1, 19-32
Dominance ratio, 12-16, 12-18
Dominant:
    characteristic value, 12-16
    characteristic vector, 12-16
    diagonal, 6-11
    eigenvalue, 12-4, 12-16, 12-18
Double root, 17-5
Duality, 18-13

Economized approximations, 2-3
Eigenspace, 12-1, 12-14, 12-15, 12-24
Eigenvalue, 12-1 to 12-9, 12-13 to 12-16, 12-18, 12-20 to 12-24, 16-30, 16-31, 19-11, 19-12, 19-14
Eigenvector, 12-1, 12-3 to 12-9, 12-14 to 12-18, 12-23, 12-24, 12-26, 19-12
Elimination, 6-28
Elliptic function, 5-20
Empirical function, 9-2
Error, 2-2, 2-3, 5-22
    magnitude of, 2-3
    maximum absolute, 2-21
    maximum relative, 2-21
    minimax absolute, 2-18, 2-20, 2-22
    minimax relative, 2-18, 2-19
    percentage, 2-4
Euclidean norm, 11-6
Euler corrector, 16-49
Euler differential equation, 10-26
Euler's:
    constant, 2-14
    identity, 16-7
    method, 2-7, 16-11, 16-12, 16-42, 16-49, 19-36
    polynomial of the first order, 4-22

Numbers on this page refer to **PROBLEM NUMBERS**, not page numbers

summation formula, 2-14
Even function, 2-20, 2-22, 2-23
Event, 18-12
Everett's formula, 5-8
Exponential distribution, 18-9
Extrapolation, 5-12, 5-28,
   5-29, 6-27
   linear, 5-28
   Richardson, 10-15, 16-24
Extrema, 5-22
Extreme points, 2-19

Factorials, 4-2
Finite difference(s), 4-1 to
   4-8
   method, 9-1, 17-20
   solution, 15-8
First order differential
   equation, 16-2
First order method, 16-57
Fixed-point:
   arithmetic, 7-4
   iteration, 19-20
Floating-point:
   arithmetic, 2-10
   representations, 2-6
Flowchart, 1-3 to 1-5
   general description, 1-4
Force constant, 16-9
Format, 19-1 to 19-39
FORTRAN:
   arithmetic operations, 1-20
   assignment statement, 1-2
   binding power of operations
   in, 1-19
   coding, 1-11, 1-19, 1-20,
   1-21, 1-22
   expressions, 1-21
   IF statement, 1-10
   program, 19-1 to 19-39
   READ statement, 1-20
   statements, 1-19 to 1-22
   subroutine, 19-5, 19-8,
   19-29
   variable names, 1-9, 1-20
   WRITE statement, 1-20
Forward difference, 4-3, 4-5,
   4-9, 4-12

operator, 4-24
representation, 4-16
table, 4-3, 5-1, 5-3, 5-15,
   5-23
Fourier:
   coefficient formula, 17-17
   coefficients, 14-1, 14-6,
   14-8 to 14-16, 14-19, 14-23,
   14-24, 14-28, 14-30, 14-31
   sine series, 17-13, 17-21
   spectrum, 14-1
Fourier's law, 15-6
Fourier transform, 14-1 to
   14-31
   attenuation property, 14-3
   derivative property, 14-3
   elastic curve of beam using,
   14-11
   shifting property, 14-3,
   14-26
Free damped motion, 16-9
Frequency, 16-5
Frobenius' method, 16-52

Game, 18-12 to 18-14
Gauss' formula, 5-26
Gaussian:
   elimination method, 2-19,
   2-20, 5-12, 6-1 to 6-3, 6-5,
   6-24, 19-9
   quadrature, 19-30, 19-34
   scheme, 6-4
Gauss-Jordan elimination:
   with maximization, 6-10
   without maximization, 6-10
Gauss' quadrature formulas,
   10-24
Gauss-Seidel iteration (see
   Gauss-Seidel's method)
Gauss-Seidel method, 6-7 to
   6-9, 6-11, 6-13, 19-10
Gauss-Seidel process (see
   Gauss-Seidel's method)
Gauss' three point formula,
   10-25
General solution, 17-6
Gerschgorin circles, 6-11

Numbers on this page refer to **PROBLEM NUMBERS**, not page numbers

Givens-Householder method, 6-15
Global minimum, 7-38
Good approximation, 2-16, 2-17
Gregory-Newton formula, 9-13
   backward, 5-28
Group sampling, 8-15

Half-open Lagrange integration formula, 15-15
Halley's formula (see Halley's method)
Halley's method, 7-33
Harmonic analysis, 8-6
Hessenberg matrix, 12-12
Heat:
   capacity ratio, 17-14
   conduction, 17-16
   equations, 17-12 to 17-16
   flow lines, 17-15
   loss, 17-14
   transfer, 17-16
Heating effectiveness, 17-14
Hermitian forms, 6-9
Hessenberg matrix, 12-12
Hexadecimal number, 1-13 to 1-15
Higher order backward difference representation, 4-16
Higher order forward difference representation, 4-16
Homogeneous:
   equation, 6-26, 16-1, 16-2, 16-8
   partial differential equation (P.D.E.), 17-1
   solution, 16-8
Hooke's law, 16-5, 16-7
Horner's method, 19-16
Horner's rule, 7-34
Hot fluid, 17-14
Householder vectors, 6-15
Hyperbolic problem, 17-20

IBM 360/67, 6-10
Identity matrix, 19-3, 19-4

Increment method, 19-35
Independent variable, 6-14
Infinite series, 3-1 to 3-4, 17-12, 17-13, 17-17, 17-21
Inherent error, 9-14, 9-19
Initial approximation, 7-11, 7-16
Initial boundary problem, 16-25 to 16-28
Initial conditions, 17-20, 17-21
Initial guess, 6-7
Initial value problem, 2-7, 16-8, 16-12, 16-16 to 16-22, 16-46, 16-50, 16-57
Initial vector, 12-4
Integral equations, 14-25
Integral mean value theorem, 1-6
Integral test, 3-2
Interest rate, 5-14
Interior point, 19-21
Interpolating polynomial, 5-5, 5-10, 5-12
Interpolation, 5-12, 5-15, 7-7 to 7-9, 7-19
   error, 19-18
   iterated, 5-18
   iterative inverse, 5-23
   linear, 5-7, 5-16 to 5-20
   polynomial, 5-20, 9-8
Interval approximation, 2-15
Interval halving, 7-35
Interval of differencing, 5-4
Inverse interpolation:
   by divided differences, 5-25
   by reversal of series, 5-27
   by successive approximation, 5-26
Inverse matrix, 6-13, 6-23 to 6-25, 6-28, 19-3 to 19-6
Iteration, 7-1, 7-10 to 7-13, 7-29, 12-4, 12-18, 12-22
Iterative correction process, 11-18
Iterative method, 7-1, 7-2, 11-25 to 11-29, 12-4
   Jacobi, 11-29
   Seidal, 11-28
Iterative process, 5-7, 6-7

Numbers on this page refer to PROBLEM NUMBERS, not page numbers

Jacobian, 19-26
    determinants, 6-14
    matrix, 7-11
Jacobi and Gauss-Seidel
    methods, 6-8, 6-11, 6-13
Jacobi's iteration (see Jacobi's method)
Jacobi's method, 6-7, 6-8, 6-11, 6-13
Jacobi's process (see Jacobi's method)
Jacobi rotations, 6-12

K-digit arithmetic, 2-6
Kutta's formula, 16-15, 16-16

Lagrange:
    coefficient polynomial, 9-10
    coefficients, 5-12, 9-4
    differentiation formula, 9-4
    polynomial, 19-17, 19-30
    quadratics, 5-12
Lagrange's interpolation
    formula, 5-9, 5-10, 5-29
Lagrangian:
    formula, 9-9
    interpolation, 5-11, 19-17
Laplace's equation, 16-36, 17-15, 17-19
    two-dimensional, 15-10
Laplacian, 17-19
Latent roots, 6-27
Latent vectors, 6-27
Laurent series, 3-34 to 3-37
Law of diffusion, 16-6
Laws of exponents, 19-19
Leading zeroes, 2-9
Least-squares:
    approximation, 8-1 to 8-24
    polynomial, 5-12
Limit, 11-20
    test, 3-4
Linear:
    equations, 6-7, 6-24
    extrapolation, 5-28
    interpolation, 5-7, 5-16 to 5-20
    shooting algorithm, 16-21
    systems, 11-1 to 11-29
Linearly dependent columns, 11-1
Linear programming, 13-1 to 13-18, 18-13
    big-M technique, 13-12
    curve fitting using, 13-17
    dual simplex method, 13-13, 13-15
    formulation, 13-1 to 13-3, 13-6
    graphical solution, 13-4, 13-7, 13-11
    simplex algorithm, 13-5, 13-7 to 13-10, 13-13 to 13-15
    transportation problem, 13-18
Line printer, 1-1
Lipschitz number, 19-35
Local truncation error, 16-37
Logarithm, 2-13

Maclaurin series expansion, 3-5, 3-10, 3-20 to 3-24, 16-40, 16-54, 19-34
Maehly's method, 2-23
Magic square, 11-5
Magnetic tape, 1-1
Matrix, 6-1, 6-16, 6-18
    banded coefficient, 6-10
    coefficients, 5-12
    derived, 11-10
    factorization, 11-21 to 11-24
    Hessenberg, 12-12
    identity, 19-3, 19-4
    inverse, 6-13, 6-25, 6-28
    Jacobian, 7-11
    mathematics, 6-16 to 6-29
    method, 16-23
    multiplication techniques, 6-24
    positive definite, 6-9
    rank, 6-21, 6-22, 12-9, 19-8
Maximum error:
    absolute, 2-21
    relative, 2-21

Numbers on this page refer to **PROBLEM NUMBERS**, not page numbers

Mean, 2-2, 18-6
Memory, 1-1
Method of:
   averages, 8-10
   determinants, 11-14
   false position, 7-22, 7-26, 7-28
   partial fractions, 16-4
   undetermined coefficients, 17-6
Midpoint method, 16-56
Minimax:
   absolute error, 2-18, 2-20, 2-22
   polynomial approximation, 2-18 to 2-20
   rational approximation, 2-22
   relative error, 2-18, 2-19
Mixed boundary value problem, 16-37
Mixed mode, 1-22
Modulus, 17-16
Moment, 18-5, 18-6
Monte Carlo:
   distribution, 18-6
   random sample, 18-4, 18-5
Muller's method, 7-39

Natural:
   cubic splines method, 5-12
   frequency, 16-7
   logarithm, 2-5
Negative peaks, 2-3
Neville polynomials, 5-12
Neville's method, 5-12
Neville's process of iteration, 5-11, 5-21
Newton:
   correction method, 7-15
   equation, 19-28
Newton-Raphson iterative formula (see Newton-Raphson method)
Newton-Raphson method, 7-17, 7-20 to 7-25, 19-24, 19-26
Newton's:
   backward formula, 4-9
   divided difference formula, 5-5
   formula (see Newton's method)
   forward difference formula, 5-1, 5-2, 5-15
   law, 17-21
   law of cooling, 15-7
   method, 6-15, 7-10 to 7-13, 7-15, 16-22, 16-25, 19-23, 19-28
   second law, 16-3, 16-4, 16-7, 16-9
   series, 4-13
Node, 15-5 to 15-8
Nonhomogeneous set, 6-5
Nonlinear:
   equations, 2-22
   shooting algorithm, 16-22
Non-polynomial approximation, 2-16
Norm, 6-6, 11-15 to 11-17
Normal equations, 5-12, 8-5, 8-7, 8-12, 8-15 to 8-17, 8-20, 8-24
   weighted, 8-5, 8-16
Normal probability distribution, 8-13, 18-5, 18-6
Number:
   Bernoulli, 3-33
   Lipschitz, 19-35
   random, 18-1 to 18-3
   Stirling, 4-5
Number sequence, 1-3
   sum of, 1-4
Number systems:
   binary, 1-12, 1-13, 1-14, 1-16, 1-18
   decimal, 1-12, 1-13, 1-14, 1-17
   hexadecimal, 1-13 to 1-15
   octal, 1-13, 1-14, 1-16
Numerical differentiation, 9-1 to 9-20
Numerical integration, 10-1 to 10-26
Numerov method, 16-37

Octal number, 1-13, 1-14, 1-16

Numbers on this page refer to PROBLEM NUMBERS, not page numbers

Odd-degree coefficients, 2-22
One-dimensional heat equation, 17-13
One-dimensional wave equation, 17-21
Operations:
    addition, 2-6
    subtraction, 2-6
    multiplication, 2-6
    division, 2-6
Operator series, 4-24
Ordinary differential equation (O.D.E.), 17-6
    power series solution, 3-28
Orthogonality conditions, 6-27
Orthogonality property, 8-16
Orthogonalization method, 8-15
Orthogonal set, 14-8
Orthogonal system, 8-16, 17-18
Orthonormal system, 17-18
Oscillatory damped motion, 16-9
Over damped motion, 16-9

Paper tape, 1-1
Parabola, 2-15, 6-29
Parseval theorem, 14-4
Partial differential equation (P.D.E.), 17-1 to 17-21
Partial fractions, 16-4
Partial pivoting, 2-12, 6-6
Particular solution, 16-8, 16-40, 16-50, 17-6
Pathological function, 14-6, 14-8
Payoff matrix, 18-13
Percentage error, 2-4
Period, 16-5
    interval, 2-16
Periodic function, 2-16
Perturbation series method, 16-50
Picard's method, 16-20, 16-38 to 16-42
Piecewise continuous function, 14-17, 14-19, 14-29
Piecewise-cubic Hermite interpolation, 19-18
Piecewise polynomial, 5-12

Pivotal condensation, 19-7
Pivot element, 6-6, 6-10
Pivoting strategy, 6-25
Plateau problem, 10-5
Poisson distribution, 18-8, 18-9, 18-11
Polar coordinates, 17-19
Pole, 3-36, 3-37
Polynomial, 4-2, 4-3
    approximation, 2-13, 2-14, 2-21
    cubic, 5-15
    cubic interpolation, 5-20
    cubic lagrange, 5-12
    extrapolation, 5-28, 5-29
    interpolating, 5-5, 5-10, 5-12
    interpolation, 5-1, 5-2, 5-4, 5-5, 5-9, 5-28
    least-squares, 5-12
Positive definite Hermitian, 6-9
Positive peaks, 2-3
Positive root, 5-19
Potential:
    equation, 17-17
    function, 11-4
Power method, 12-16 to 12-19
Power series:
    expansion, 2-16, 2-21, 3-25 to 3-33, 4-5, 5-27
    error in, 2-21, 3-25, 3-27
    solution to differential equation, 3-28
Principal part, 3-36, 3-37
Principle of superposition, 17-15, 17-21
Probability, 18-1, 18-8 to 18-13
Product matrix, 6-20
Punched cards, 1-1

QL algorithm, 12-18, 12-22
QR algorithm, 12-18, 12-21
Quadratic:
    approximations, 2-3
    equation, 2-19, 5-24
    formula, 2-19, 16-5, 16-8
    interpolation polynomials, 5-20

Numbers on this page refer to PROBLEM NUMBERS, not page numbers

polynomial, 5-12, 5-22

Random:
  number, 18-1 to 18-3
  variable, 18-1, 18-16
Rank of a matrix, 6-21, 6-22, 12-9, 19-8
Rate of convergence, 7-5, 7-12, 7-28
Rate of interest, 5-22
Rational approximation, 2-23
Rational fraction, 4-15
Ratio test, 3-3, 3-16, 3-22, 3-23, 3-29
Rayleigh quotient, 19-5
READ statement, 1-4
Real root, 4-8
Reciprocal difference, 4-15
  table, 4-14
Rectangular distribution, 18-2
Rectangularly distributed random number, 18-1 to 18-3
Recurrence method, 6-15
Regula falci, method of (see method of false position)
Regular part method, extraction of, 18-16
Relative convergence criterion, 6-7
Relative error, 1-7, 2-1, 2-2, 2-4, 2-6, 2-8, 2-10
  function, 2-19, 2-21
  maximum, 1-7
  root-mean-square, 1-7
Relaxation method, 15-10
Remez method, 2-19, 2-20, 2-22
Residual, 6-28, 8-5, 8-6, 8-9, 8-20
  vector, 11-18
  weighted, 8-5
Richardson extrapolation, 10-15, 16-24
Romberg algorithm, 10-15
Root, 2-19
  test, 3-3
Rounding, 2-8
  decimal numbers, 2-20, 2-21
Round-off:

error, 2-10, 2-12, 6-25, 9-14, 9-15
  method, 2-9
Row, 6-16, 6-17, 6-20
  operations, 6-1, 6-24
  scaling, 2-11, 2-12
Rule of false position (see method of false position)
Runge-Kutta formula (see Runge-Kutta method)
Runge-Kutta method, 16-12 to 16-15, 16-17, 16-18 to 16-22, 16-30, 16-49, 19-36
Runge's formula (see Runge-Kutta method)

Saddle point, 18-13
Sampling:
  distribution, 18-5, 18-6
  moment, 18-6
Scaling, 2-11, 6-6
  column, 2-11
Secant method, 7-31
Second-order determinant, 6-21
Semi-magic square, 11-5
Separable equation, 17-7
Separation constant, 17-15
Separation of variables method, 17-12, 17-13, 17-15, 17-17, 17-18, 17-21
Series, 3-1 to 3-37, 14-5
  Chebyshev, 2-23
  Fourier, 14-6 to 14-16
  infinite, 3-1 to 3-4, 17-12, 17-13, 17-17, 17-21
  Laurent, 3-34 to 3-37
  Maclaurin, 3-5, 3-10, 3-20 to 3-24, 16-40, 16-54
  operator, 4-24
  power, 2-21
Set:
  Chebyshev, 8-12
  degenerate, 4-15
  nonhomogeneous, 6-5
Shell-side fluid, 17-14
Sheppard's zigzag rule, 5-5
Shifting operator, 4-24
Significant figures, 2-1 to 2-3,

Numbers on this page refer to **PROBLEM NUMBERS**, not page numbers

2-8
Simplex method, 18-13, 18-14
Simpson's rule, 10-6 to 10-14, 10-16, 10-17, 10-19, 10-21 to 10-23, 11-8
Simultaneous equations, 6-5
Sine curve, 6-29
Sine-function boundary condition, 17-15
Sine-wave distribution, 17-15
Slack variable, 18-14
Smoothing, 8-18, 8-19
Solution's error (see round-off error)
Spectral radii, 6-13
Spectrum:
 discrete, 14-1
 continuous, 14-1
Standard-error function, 2-19, 2-20
Starting value, 15-14 to 15-16
START statement, 1-4
Steepest descent algorithm, 7-38
Stirling numbers, 4-5
Stirling's approximation formula, 19-15
Stirling's formula, 5-4, 5-13, 9-11, 9-13, 10-21
Stokes' law, 7-2
Stopping criterion, 16-22
String:
 linear density, 17-21
 mass, 17-21
 tension, 17-21
Sums and series, 4-19 to 4-21
Superposition principle, 17-12, 17-15, 17-21
Surplus points, 4-15
Symmetric:
 approximation interval, 2-20
 elements, 6-15
Synthetic division, 4-1
System of inconsistent linear equations, 2-17

Table of reciprocal differences, 4-14

Taylor polynomial, 9-7, 16-24
Taylor's:
 formula, 9-4, 9-5, 19-37
 method, 9-8
 series, integration using, 10-2
Taylor-series expansion, 1-5, 2-3, 2-7, 3-5 to 3-19, 4-24, 15-15, 16-41 to 16-49, 19-19
Tchebycheff set (see Chebyshev set)
t-distribution, 8-13
Tetrad, 8-11
Theorem:
 central limit, 18-5, 18-6
 convolution, 14-27
 Parseval, 14-4
 Wierstrauss approximation, 9-4
Theory:
 game, 18-13
 of continued fractions, 4-14
Thermal diffusivity, 17-16
Third order determinant, 6-22
Three-digit:
 arithmetic, 2-8
 chopping, 2-8
 rounding, 2-8
Transformation equation, 5-2, 5-3
Transposed matrix, 5-12, 12-26
Trapezoidal rule, 10-10, 10-11, 10-14, 10-18, 10-20, 19-32, 19-34
Trapezoids, 19-29, 19-32, 19-33
Triangulation, 11-12
Tridiagonal matrix, 6-10
Trigonometric:
 Fourier series, 17-18
 table, 5-16
Trigonometry, 6-12
True value, 2-4
Truncating series, 2-21
Truncation error, 2-7, 2-12, 3-6, 3-7, 3-10, 3-11 to 3-13, 3-21, 3-24 to 3-27, 9-4 to 9-10, 9-19, 9-20, 11-8, 15-5, 15-16
 local, 16-37

Numbers on this page refer to **PROBLEM NUMBERS**, not page numbers

Tube-side fluid, 17-14
Two-dimensional Laplace equation, 17-17
Two-dimensional Newton-Raphson method, 7-18
Two parameter family of planes, 17-9
Two-stage Gauss elimination, 5-12

Undetermined coefficients, method of, 17-6
Unit circle, 6-9

Variable:
  dependent, 6-14
  independent, 6-14

random, 18-1, 18-16
slack, 18-14
Vector:
  coplanar, 11-1
  Householder, 6-15
  space, 11-1 to 11-6, 12-2
  space axiom's, 11-2
Vibrating string problem, 17-21

Wave equation, 17-20
Weddle's rule, 10-6, 10-8, 10-9, 10-14, 10-17, 10-21, 10-22
Wegstein's method, 7-32
Weierstrass' approximation theorem, 9-4
Weierstrass M-test, 3-29, 14-9, 14-10, 14-22
Weighted fit, 8-5
Weighted normal equations, 8-5

THE PROBLEM SOLVERS

 Research and Education Association has published Problem Solvers in:

- ADVANCED CALCULUS
- ALGEBRA & TRIGONOMETRY
- AUTOMATIC CONTROL SYSTEMS / ROBOTICS
- BIOLOGY
- BUSINESS, MANAGEMENT, & FINANCE
- CALCULUS
- CHEMISTRY
- COMPUTER SCIENCE
- DIFFERENTIAL EQUATIONS
- ECONOMICS
- ELECTRICAL MACHINES
- ELECTRIC CIRCUITS
- ELECTROMAGNETICS
- ELECTRONIC COMMUNICATIONS
- ELECTRONICS
- FINITE and DISCRETE MATH
- FLUID MECHANICS/DYNAMICS
- GENETICS
- GEOMETRY: PLANE • SOLID • ANALYTIC
- HEAT TRANSFER
- LINEAR ALGEBRA
- MECHANICS: STATICS • DYNAMICS
- NUMERICAL ANALYSIS
- OPERATIONS RESEARCH
- OPTICS
- ORGANIC CHEMISTRY
- PHYSICAL CHEMISTRY
- PHYSICS
- PRE-CALCULUS
- PSYCHOLOGY
- STATISTICS
- STRENGTH OF MATERIALS & MECHANICS OF SOLIDS
- TECHNICAL DESIGN GRAPHICS
- THERMODYNAMICS
- TRANSPORT PHENOMENA: MOMENTUM • ENERGY • MASS
- VECTOR ANALYSIS

HANDBOOK OF MATHEMATICAL, SCIENTIFIC, AND ENGINEERING FORMULAS, TABLES, FUNCTIONS, GRAPHS, TRANSFORMS

If you would like more information about any of these books, complete the coupon below and return it to us.

RESEARCH and EDUCATION ASSOCIATION
505 Eighth Avenue • New York, N.Y. 10018
Phone: (212)695-9487

Please send me more information about your Problem Solver Books.

Name ....

Address ....

City .... State ....